Ecology of Himalayan Treeline Ecotone

S P Singh • Zafar Ahmad Reshi •
Rajesh Joshi
Editors

Ecology of Himalayan Treeline Ecotone

Springer

Editors
S P Singh
Central Himalayan Environment Association
Dehradun, Uttarakhand, India

Zafar Ahmad Reshi
Department of Botany
University of Kashmir
Srinagar, Jammu and Kashmir, India

Rajesh Joshi
G B Pant National Institute of Himalayan Environment
Sikkim Regional Centre, Pangthang
Gangtok, Sikkim, India

ISBN 978-981-19-4475-8 ISBN 978-981-19-4476-5 (eBook)
https://doi.org/10.1007/978-981-19-4476-5

This Springer imprint is published by the registered company Springer Nature Singapore Pte Ltd.
The registered company address is: 152 Beach Road, #21-01/04 Gateway East, Singapore 189721, Singapore

Dedicated to
Late Dr. Ranbeer Singh Rawal
A renowned biodiversity expert, lead investigator of Indian Himalayan Timberline Project, and Former Director G B Pant National Institute of Himalayan Environment

Foreword

सत्यमेव जयते

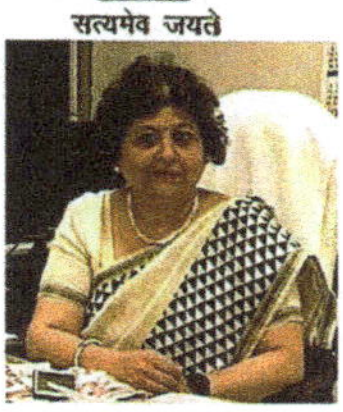

The Ministry of Environment, Forest and Climate Change launched the National Mission on Himalayan Studies to support the sustenance and enhancement of the ecological, natural, cultural, and socio-economic capital assets and values.

One among these is a 5-year long multi-location, multi-institutional, coordinated project taken up under the Mission on Indian Himalayan Timberline. The Himalayan Timberline is the highest and one of the most fragile ecosystems in the Northern Hemisphere. The project was initiated to study the impact of climate change or integrated biotic zones adjacent to snow covered areas, particularly timberline, alpine meadows, and nival zones in the Indian Himalaya.

The results of the study will be useful to draw inferences, devise strategic conservation frameworks, and implement remedial measures at the target sites across the IHR States/UTs. After thorough investigative analysis of vulnerable biological and physical features of the Himalayas Timberline, the project brings forth a set of key learnings, lessons, and recommendations in the form of the present edited book entitled *Ecology of Himalayan Treeline Ecotone*, which sheds light on multiple dynamics of Himalayan Timberline and suggests the way forward.

I hope that the multi-faceted interventions and suggestive future imperatives will help ensure all-inclusive protection and development of Himalayan timberline, interdependent resources and lives. It is also anticipated that the research outcomes embodied in the book will lay the foundation for future research and motivate young researchers to take up further studies in the Indian Himalaya.

Indira Paryavaran Bhawan
New Delhi, India

Manju Pandey

List of Reviewers

List of domain experts who served as reviewers for the book:

1. A. Bhattacharya
2. C. P. Singh
3. Hitendra Padalia
4. P. S. Ranhotra
5. Rajesh Joshi
6. S. P. Singh
7. Subrat Sharma
8. Zafar A. Reshi
9. S. P. S. Kushwaha

Preface

Treeline ecotones show remarkable variation in vegetation structure over small elevation changes and bringing out such changes in vegetation and other allied biotic and abiotic attributes in relation to climate change is the main theme of the book. It is timely in view of lack of any such detailed work on the Himalayan treeline ecotone and also because of the current surge in the study of such sensitive transition zones in the context of climate change. In particular, the treelines in the Himalayas, being the highest in the Northern Hemisphere (up to 4900 m), are among the least investigated systems. Treeline advance to higher elevation results in substantial physiognomic, structural, and functional changes in mountain landscapes and ecosystems. Equally interesting is to investigate the treelines which are stationary despite the decades of rapid warming in the Himalayas. In the Himalayas, such treelines prevail! Thus, sensitivity and response of the Himalayan treelines to climate change is being increasingly discussed. Considering that the knowledge on the treelines in the Himalayas is very limited, this book for the first time summarizes evidence-based multidisciplinary aspects of treeline in the Himalayas, through a team science approach engaging 12 experts to carry out a multisite, and multidisciplinary research on the Indian Himalayan Treelines under the aegis of Indian Himalayan Timberline Project (IHTP). It may lead to the establishment of an important entity of conservation and management.

The outcomes of the team research on the Himalayan timberline are published in an edited book form. While each chapter of the book reflects outcomes of a complete study on specific aspect, the entire book provides a comprehensive account of ecological dimensions at the Himalayan timberline under changing climate. Towards making the scope of this book wider and bring a more cohesive picture of timberline in the Himalayas, representative chapters have also been included from Bhutan, Nepal, China, and Pakistan as well. With this inclusion, the book substantially (i) covers geographical extent and diversity therein of the Himalayan timberline, (ii) improves global understanding on this *hitherto* less explored ecotone, and (iii) increases availability of evidence to decipher changes in diverse ecological dimensions of timberline/treeline under warming climate.

The focus of the book is on extraordinarily large spatial scale and various dynamics of the Himalayan treeline under warming climate. It covers various aspects of the ecology of treeline ecotone; length and elevation range of treelines, treeline species pool, species diversity, tree morphology, climatology and temperature lapse rates, tree phenology, water relations, stress physiology, tree ring width chronology, and climate relationships and dimensions of human sustenance. The volume addresses a major gap in the field of the Himalayan ecosystems and climate change.

This book is of interest and useful to scholars, academicians, researchers, foresters having interest on treeline ecology, and valuable source of reference to the relevant researchers and policy planners. It constitutes bedrock for any further investigation on the ecology and dynamics of the treeline ecotone. It further provides a rationale for importance of pursuing team science approach for the macroecological investigations at appropriate spatial scales. It should serve ecologists, general scientists, students of forestry, botany and climate change, policy makers, forest and wildlife managers.

Varanasi, UP, India

Prof. J. S. Singh, (Emeritus) FNA

Acknowledgments

The editors would like to thank all the lead investigators and researchers of Indian Himalayan Timberline Project (IHTP) for their dedicated efforts in producing the quality knowledge on timberline in the Himalayas and successful completion of the project. We also extend our gratitude to all the investigators and researchers of Himalayan Alpine Dynamics research Initiative (HIMADRI) under PRACRITI-II and SHRESTI program of Indian Space Research Organization (ISRO) and authors and co-authors of different chapters; without their support the idea of producing comprehensive and holistic knowledge on the Himalayan timberline in the form of an edited book would not have been possible. We would like to acknowledge our sincere gratitude to the reviewers and domain expert who helped in enhancing the quality of the content of the book by their critical reviews and assessments. Our sincere thanks to the Director GBPNIHE, heads of different Institutions, Forest departments of Government of Jammu and Kashmir, Himachal Pradesh, Uttarakhand, and Sikkim in India for their kind support in implementation of the project. Additionally, we are immensely grateful to the Ministry of Environment, Forest and Climate Change (MoEF&CC), Government of India, for financially supporting the IHTP under the National Mission on Himalayan Studies (NMHS).

Contents

Editors and Contributors

About the Editors

S P Singh is a renowned Mountain Ecologist who pioneered ecological studies on Himalayan vegetation with major emphasis on forest ecosystem processes and on the application of ecology to the solution of environmental problems. He obtained his PhD in Botany from Agra University (UP). More recently, he has been involved with research on climate change impacts on Himalayan biodiversity. He was conferred as Fellow of Indian National Science Academy (FNA); Fellow of National Academy of Science, India; Courtesy Professor, Department of Botany and Plant Pathology, Oregon State University, Corvallis, USA; Dr Birbal Sahni Medal of 2003, and Honoured with Uttarakhand Ratna Award by Govt. of Uttarakhand for his outstanding contribution in the field of education and research. He served as Chair of Excellence in Biodiversity and Ecology at Forest Research Institute, Dehradun, Uttarakhand; Advisor, Planning Commission, Uttarakhand; Vice Chancellor, HNB Garhwal University, Srinagar-Garhwal, Uttarakhand; and Professor and Head, Department of Botany, Kumaun University, Nainital, Uttarakhand. During his 37 years of teaching and research, he supervised 35 PhD students and published more than 180 research papers.

Zafar Ahmad Reshi is a well-known ecologist who holds a PhD degree in Botany and is currently serving as a Professor of Botany at the University of Kashmir, Jammu & Kashmir, India. He has over three decades of experience working on northwest Himalayan biodiversity and has contributed significantly to the understanding of the causes and consequences of plant invasions in Kashmir Himalaya, species distributions under changing climate and biotic interactions in community assembly. He has held various key positions at the University of Kashmir such as Dean of Research, Registrar, Head of Department, and Honorary Director, Centre of Plant Taxonomy and Biodiversity. He is a Fellow of the National Institute of Ecology and an expert member on various committees of the Department of Biotechnology, Govt. of India, University Grants Commission (UGC), New Delhi. During his 32 years of teaching and research, he has published more than 210 publications in journals of international repute and guided 20 students for their PhD programme and 19 students for the MPhil programme. He has successfully completed 16 research projects funded by national and international agencies and has been associated with 12 consultancy projects as a Functional area expert/ Coordinator.

Rajesh Joshi holds a PhD in applied Mathematics from GBPUAT Pantnagar, Uttarakhand, India. Being a mathematical modeller blended with ecological modelling, his research work is focused on the environment and climate including climate change impacts and vulnerability assessment on critical ecosystems, forest hydrology, spring shed management, dendrochronology, etc., in the Indian Himalayan Region (IHR). Presently he is serving as Regional Head of G B Pant National Institute of Himalayan Environment, Sikkim Regional Centre (SRC) Sikkim. He has over 15 years of teaching and research experience in IHR and published over 50 publications including peer-reviewed research papers in journals of national and international repute. He has also contributed to drafting key policy documents; e.g. Climate change and India: a sectoral and regional analysis for the 2030s, Progression of Development in Himalaya, Spring revival with Sustainable Land Management (SLM) practices in the mid-hills Northwestern Himalayas, etc., and also served as a reviewer for international/national journals. He is the recipient of the Young Scientist Award conferred by the Uttarakhand Council for Science & Technology. He has successfully completed 10 research projects and presently coordinating 5 R&D projects as a principal investigator/nodal person.

Contributors

Bhupendra S. Adhikari Department of Habitat Ecology, Wildlife Institute of India, Dehradun, Uttarakhand, India

Ritesh Agrawal EPSA, Space Applications Centre (SAC), ISRO, Ahmedabad, Gujarat, India

Rameez Ahmad Centre for Biodiversity and Taxonomy, University of Kashmir, Srinagar, Jammu and Kashmir, India
Centre for Biodiversity & Taxonomy, Department of Botany, University of Kashmir, Srinagar, Jammu and Kashmir, India

Sonam Bahuguna Forestry and Climate Change Division, Uttarakhand Space Application Centre, Dehradun, Uttarakhand, India

Rajesh Bajpai Environment, Agricultural & Education Society, Bareilly, Uttar Pradesh, India
Lichenology Laboratory, CSIR-National Botanical Research Institute, Lucknow, Uttar Pradesh, India

Manisha Bhandari Mahila Degree College, Haldwani, Uttarakhand, India

A. Bhattacharyya Birbal Sahni Institute of Palaeosciences, Lucknow, Uttar Pradesh, India

Dinesh Raj Bhuju Nepal Academy of Science and Technology, Lalitpur, Nepal

B. S. Bisht G. B. Pant National Institute of Himalayan Environment, Sikkim Regional Centre, Pangthang, Gangtok, Sikkim, India

Kapil Bisht G.B. Pant National Institute of Himalayan Environment, Almora, Uttarakhand, India

M. Bobrowski CEN Center for Earth System Research and Sustainability, Institute of Geography, Universität Hamburg, Hamburg, Germany

J. Böhner CEN Center for Earth System Research and Sustainability, Institute of Geography, Universität Hamburg, Hamburg, Germany

B. Bürzle CEN Center for Earth System Research and Sustainability, Institute of Geography, Universität Hamburg, Hamburg, Germany

Vaishali Chandola High Altitude Plant Physiology Research Center, Hemwati Nandan Bahuguna Garhwal University, Srinagar, Garhwal, Uttarakhand, India

Naveen Chandra Forestry and Climate Change Division, Uttarakhand Space Application Centre, Dehradun, Uttarakhand, India

Sudeep Chandra High Altitude Plant Physiology Research Center, Hemwati Nandan Bahuguna Garhwal University, Srinagar, Garhwal, Uttarakhand, India

R. P. Chaudhary RECAST Research Centre for Applied Science and Technology, Tribhuvan University, Kirtipur, Nepal

Parveen K. Chhetri Department of Earth Science and Geography, California State University, Carson, CA, USA

Bency D. Chinthala Birbal Sahni Institute of Palaeosciences, Lucknow, Uttar Pradesh, India
Institute of Geography, University of Erlangen-Nürnberg, Erlangen, Germany

Firdous A. Dar Centre for Biodiversity & Taxonomy, Department of Botany, University of Kashmir, Srinagar, Jammu and Kashmir, India

Yub Raj Dhakal Siddhartha Environmental Services, Kathmandu, Nepal

Sanjay Kumar Dwivedi Forestry and Climate Change Division, Uttarakhand Space Application Centre, Dehradun, Uttarakhand, India

Ze-Xin Fan CAS Key Lab of Tropical Forest Ecology, Xishuangbanna Tropical Botanical Garden, Chinese Academy of Sciences, Mengla, Yunnan, People's Republic of China

Narayan Prasad Gaire CAS Key Lab of Tropical Forest Ecology, Xishuangbanna Tropical Botanical Garden, Chinese Academy of Sciences, Mengla, Yunnan, People's Republic of China
Department of Environmental Science, Patan Multiple Campus, Tribhuvan University, Lalitpur, Nepal

Bandan Gajmer Sikkim State Council of Science and Technology, Vigyan Bhawan, Gangtok, Sikkim, India

Aadil Gulzar Department of Environmental Science, University of Kashmir, Srinagar, Jammu and Kashmir, India

Surabhi Gumber Department of Forestry & Environmental Science, Kumaun University, Nainital, Uttarakhand, India

Priyadarshini Gurung Sikkim State Council of Science and Technology, Vigyan Bhawan, Gangtok, Sikkim, India

Maroof Hamid Centre for Biodiversity & Taxonomy, Department of Botany, University of Kashmir, Srinagar, Jammu and Kashmir, India

Rajesh Joshi G. B. Pant National Institute of Himalayan Environment, Sikkim Regional Centre, Pangthang, Gangtok, Sikkim, India

Vinod Joshi G.B. Pant National Institute of Himalayan Environment, Almora, Uttarakhand, India

Yogesh Joshi Department of Botany, University of Rajasthan, Jaipur, Rajasthan, India

B. S. Kholia Botanical Survey of India, Northern Circle, Dehradun, Uttarakhand, India

Anzar A. Khuroo Centre for Biodiversity & Taxonomy, Department of Botany, University of Kashmir, Srinagar, Jammu and Kashmir, India

Amit Kumar CSIR-Institute of Himalayan Bioresource Technology (IHBT), Palampur, Himachal Pradesh, India

Devendra Kumar G.B. Pant National Institute of Himalayan Environment, Sikkim Regional Centre, Gangtok, Sikkim, India

Rahul Kumar Department of Habitat Ecology, Wildlife Institute of India, Dehradun, Uttarakhand, India

Avantika Latwal G.B. Pant National Institute of Himalayan Environment (NIHE), Ladakh Regional Centre, Leh, Ladakh, India

A. H. Malik Centre for Biodiversity and Taxonomy, University of Kashmir, Srinagar, Jammu and Kashmir, India
Centre for Biodiversity & Taxonomy, Department of Botany, University of Kashmir, Srinagar, Jammu and Kashmir, India

Rayees A. Malik Centre for Ecological Sciences, Indian Institute of Science (IISc), Bangalore, India

Jincy Rachel Mathew EPSA, Space Applications Centre (SAC), ISRO, Ahmedabad, Gujarat, India
Department of Environmental Sciences, Gujarat University, Ahmedabad, Gujarat, India

Chetna Mehra G. B. Pant National Institute of Himalayan Environment (NIHE), Almora, Uttarakhand, India

Amit Mittal Department of Allied Sciences, Graphic Era Hill University, Bhimtal Campus, Nainital, Uttarakhand, India

Jakesh Mohapatra Earth, Ocean, Atmosphere, Planetary Sciences and Applications Area (EPSA), Space Applications Centre (SAC), Indian Space Research Organisation (ISRO), Ahmedabad, Gujarat, India
Department of Botany, Bioinformatics, and Climate Change Impacts Management, University School of Sciences, Gujarat University, Ahmedabad, Gujarat, India
North Eastern Space Applications Centre, Department of Space, Government of India, Umiam, Meghalaya, India

M. Müller Department of Geosciences, Chair of Soil Science and Geomorphology, University of Tübingen, Tübingen, Germany

Subzar Ahmad Nanda Department of Botany, University of Kashmir, Srinagar, Jammu and Kashmir, India

M. C. Nautiyal High Altitude Plant Physiology Research Center, Hemwati Nandan Bahuguna Garhwal University, Srinagar, Garhwal, Uttarakhand, India

G. C. S. Negi G.B. Pant National Institute of Himalayan Environment, Almora, Uttarakhand, India

Aseesh Pandey G.B. Pant National Institute of Himalayan Environment, Sikkim Regional Centre, Gangtok, Sikkim, India

Pushpendra Pandey Department of Geography, Lucknow University, Lucknow, Uttar Pradesh, India

Himanshu A. Pandya Department of Botany, Bioinformatics, and Climate Change Impacts Management, University School of Sciences, Gujarat University, Ahmedabad, Gujarat, India

Mohit Phulara G. B. Pant National Institute of Himalayan Environment, Almora, Uttarakhand, India

Parminder S. Ranhotra Birbal Sahni Institute of Palaeosciences, Lucknow, Uttar Pradesh, India

Ranbeer S. Rawal G.B. Pant National Institute of Himalayan Environment, Almora, Uttarakhand, India

Renu Rawal G.B. Pant National Institute of Himalayan Environment, Almora, Uttarakhand, India

Shaik Rehana International Institute of Information Technology, Hyderabad, Telangana, India

Zafar A. Reshi Department of Botany, University of Kashmir, Srinagar, Jammu and Kashmir, India

Priyanka Sah G.B. Pant National Institute of Himalayan Environment (NIHE), Ladakh Regional Centre, Leh, Ladakh, India

Kumar Sambhav G. B. Pant National Institute of Himalayan Environment (NIHE), Almora, Uttarakhand, India

U. Schickhoff CEN Center for Earth System Research and Sustainability, Institute of Geography, Universität Hamburg, Hamburg, Germany

T. Scholten Department of Geosciences, Chair of Soil Science and Geomorphology, University of Tübingen, Tübingen, Germany

N. Schwab CEN Center for Earth System Research and Sustainability, Institute of Geography, Universität Hamburg, Hamburg, Germany

Santosh K. Shah Birbal Sahni Institute of Paleosciences, Lucknow, Uttar Pradesh, India

Shruti Shah Department of Forestry & Environmental Science, D.S.B. Campus, Kumaun University, Nainital, Uttarakhand, India

Bimal Sharma CAS Key Lab of Tropical Forest Ecology, Xishuangbanna Tropical Botanical Garden, Chinese Academy of Sciences, Mengla, Yunnan, People's Republic of China

Narpati Sharma Sikkim State Council of Science and Technology, Vigyan Bhawan, Gangtok, Sikkim, India

Subrat Sharma G.B. Pant National Institute of Himalayan Environment (NIHE), Ladakh Regional Centre, Leh, Ladakh, India

Mayank Shekhar Birbal Sahni Institute of Palaeosciences, Lucknow, Uttar Pradesh, India

Peili Shi Institute of Geographic Sciences and Natural Resources Research, Chinese Academy of Sciences, Beijing, People's Republic of China

Dhiren G. Shrestha Sikkim State Council of Science and Technology, Vigyan Bhawan, Gangtok, Sikkim, India

Ankit Singh High Altitude Plant Physiology Research Center, Hemwati Nandan Bahuguna Garhwal University, Srinagar Garhwal, Uttarakhand, India

Ayushi Singh Birbal Sahni Institute of Palaeosciences, Lucknow, Uttar Pradesh, India

C. P. Singh Earth, Ocean, Atmosphere, Planetary Sciences and Applications Area (EPSA), Space Applications Centre (SAC), Indian Space Research Organisation (ISRO), Ahmedabad, Gujarat, India
AED/BPSG/EPSA, Space Applications Centre, ISRO, Ahmedabad, Gujarat, India

Gajendra Singh Forestry and Climate Change Division, Uttarakhand Space Application Centre, Dehradun, Uttarakhand, India

Nandan Singh Department of Forestry & Environmental Science, D.S.B. Campus, Kumaun University, Nainital, Uttarakhand, India

Pradeep Singh G.B. Pant National Institute of Himalayan Environment, Almora, Uttarakhand, India

Ripu Daman Singh Department of Forestry & Environmental Science, Kumaun University, Nainital, Uttarakhand, India

S. P. Singh Central Himalayan Environment Association (CHEA), Dehradun, Uttarakhand, India

Hitesh Solanki Department of Environmental Sciences, Gujarat University, Ahmedabad, Gujarat, India

Raman Sukumar Centre for Ecological Sciences, Indian Institute of Science (IISc), Bangalore, India

Ninchhen Dolma Tamang G. B. Pant National Institute of Himalayan Environment, Sikkim Regional Centre, Pangthang, Gangtok, Sikkim, India

Krishna Kumar Tamta Central Himalayan Environment Association (CHEA), Nainital, Uttarakhand, India

Ashish Tewari Department of Forestry & Environmental Science, D.S.B. Campus, Kumaun University, Nainital, Uttarakhand, India

Nidhi Tomar Birbal Sahni Institute of Palaeosciences, Lucknow, Uttar Pradesh, India

Manish Tripathi Soban Singh Jeena Campus, Kumaun University, Almora, Uttarakhand, India

Shashi Upadhyay G.B. Pant National Institute of Himalayan Environment, Almora, Uttarakhand, India

Dalip Kumar Upreti Lichenology Laboratory, CSIR-National Botanical Research Institute, Lucknow, Uttar Pradesh, India

Anirudh Verma CSIR-Institute of Himalayan Bioresource Technology (IHBT), Palampur, Himachal Pradesh, India

Jinniu Wang Chengdu Institute of Biology, Chinese Academy of Sciences, Chengdu, People's Republic of China

Sajad A. Wani Centre for Biodiversity & Taxonomy, Department of Botany, University of Kashmir, Srinagar, Jammu and Kashmir, India

J. Weidinger CEN Center for Earth System Research and Sustainability, Institute of Geography, Universität Hamburg, Hamburg, Germany

Arti Yadav Department of Geography, Lucknow University, Lucknow, Uttar Pradesh, India

Treeline Research in the Himalaya: Current Understanding and Future Imperatives

1

S. P. Singh, Zafar A. Reshi, and Rajesh Joshi

Abstract

Treeline ecotone is of particular ecological significance because of its distinct physiognomy, species composition, rapid changes in the community organization over a small spatial scale and characteristic treeline species. The Himalayan treelines are under-researched and under-represented in global analyses, in spite of being the highest in the Northern Hemisphere and distinctively diverse. This book is an attempt to address this scientific knowledge gap. Encompassing 24 chapters, the book sheds light on various basic aspects of treeline ecotones, as well as novel features, such as treeline dimension at a regional level, temperature lapse rate along the elevation transects, water relations and phenology of treeline species, tree responses to climate change using dendrochronological analysis, vegetation changes in relation to early snow melt, and elevation. The book is largely based on the primary data generated under the multi-locational pan-Indian Himalaya project on treeline ecotone ecology, funded by the Ministry of Environment, Forest and Climate Change (MoEF & CC) under National Mission on Himalayan Studies (NMHS), and contributions from other scientists who have worked on the treeline ecotone in the Himalayan region. The book is important as it brings out the significance of this transition zone in terms of its species richness, community structure, and sensitivity to climate change and other

S. P. Singh
Central Himalayan Environment Association (CHEA), Dehradun, India

Z. A. Reshi (✉)
Department of Botany, University of Kashmir, Srinagar, Jammu and Kashmir, India
e-mail: zreshi@uok.edu.in

R. Joshi
G. B. Pant National Institute of Himalayan Environment, Sikkim Regional Centre, Pangthang, Gangtok, Sikkim, India

S. P. Singh et al. (eds.), *Ecology of Himalayan Treeline Ecotone*,
https://doi.org/10.1007/978-981-19-4476-5_1

anthropogenic stresses, as a centre of speciation, and also highlights the interplay between geomorphological, physical, climatic and biological components in structural and functional organization of the ecotone. An attempt has also been made to collate all available information on the Himalayan treeline ecotone for its more holistic understanding, particularly under the changing climate. A brief synthesis is included at the end of the book, which presents a way forward for future research and also flags policy interventions that are required for better management of this transition zone that is critical to the ecology of the Himalaya and well-being of people whose lives and livelihoods are intricately linked to the goods and services provided by this young Himalayan mountain system.

Keywords

Dendrochronology · Natural snow melt · Phenology · Temperature lapse rate · Treeline shift · Vegetation dynamics · Water relations

1.1 Introduction

Treeline ecotone is a transition zone between contiguous subalpine forest and open alpine tundra (Elliott 2017; Batllori et al. 2009; Montesano et al. 2020). This ecotone is relatively narrow due to the rapid changes in environmental factors such as temperature and is characterized by a multitude of interactions between different co-occurring species, vegetation and environmental factors (Malanson et al. 2001, 2007). Feedback and interactions between ecological patterns and processes result in a variety of distinctive alpine treeline patterns with considerable spatial complexity (Zeng and Malanson 2006). Notwithstanding the importance of ecotones, in general, and treeline ecotone, in particular, there exists inconsistency and lack of uniformity in defining the spatial dimensions of this transition zone in the published literature. The definition followed more or less consistently in this book refers to the treeline ecotone as a transition zone between a closed-canopy forest below and the alpine tundra above (Fig. 1.1). Notwithstanding ambiguities associated with the use of different terms, such as 'treeline ecotone', 'timberline ecotone', 'alpine treeline ecotone' and 'alpine timberline ecotone', this transition zone has recently attracted a lot of attention world over in view of its sensitivity to global climate change.

Globally, the initiation in treeline ecotone research (until the end of the eighteenth century) is linked to the explorations of high mountain ranges in different parts of the world (Holtmeier and Broll 2020). Somehow, such early explorations in the Himalaya did not pay attention to treeline ecotones though describing alpine meadows and livestock grazing in them was a common practice. In a way, in the Himalaya the research on treelines is very recent, largely a feature of the twenty-first century that has spread now in several Himalayan countries, like China and Nepal (Gaire et al. 2020; Zhang et al. 2021). In the Indian Himalayan region (IHR), it received focused attention only recently with the initiative of the Indian Himalayan Timberline Research Project (IHTRP), supported by the Ministry of Environment,

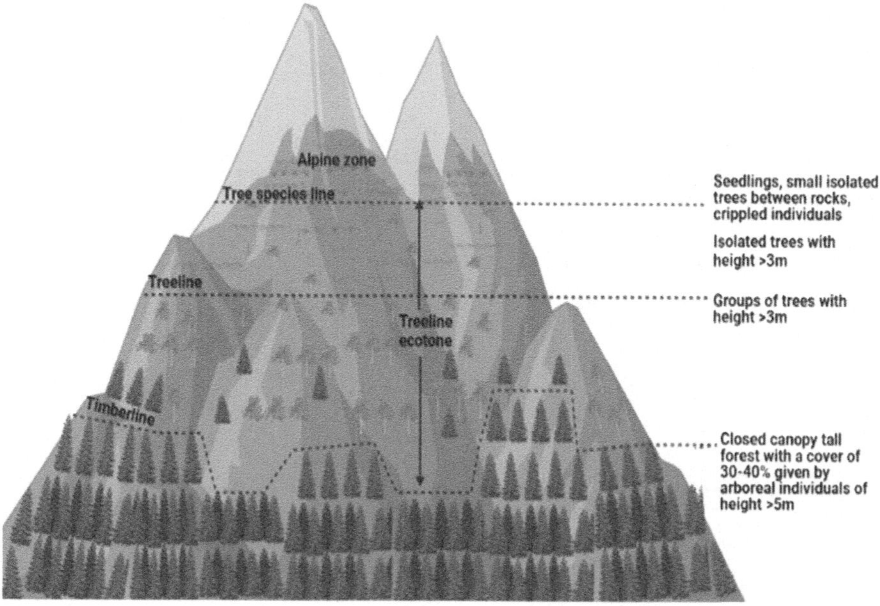

Fig. 1.1 A diagrammatic representation of treeline ecotone (*Source*: Körner and Paulsen 2004; Batllori et al. 2009)

Forest and Climate Change (MoEF & CC), Government of India, under the National Mission on Himalayan Studies (NMHS).

The changing climate gave a new impetus to research on treeline ecotones all over the world. But in the Himalaya, particularly in the IHR, studies on the treeline ecotone have largely been scattered and incoherent notwithstanding the fact that the Himalaya, owing to its complexity, dynamic nature and sensitivity to changing environment (Aryal et al. 2014; Telwala et al. 2013) is likely to undergo significant changes in structural and functional attributes of its ecosystems, including the treeline ecotone, with consequences for invaluable goods and services that these ecosystems provide to the dependent human population. Debilitating changes in the treeline ecotone would be also due to change in pastoralism, wood logging, fuel-wood gathering, grazing and exploitation of economically important species (Schickhoff 2005; Miehe et al. 2015), which would also significantly impact the Himalayan treelines that are mostly anthropogenic in nature (Schickhoff et al. 2016a) (Plate 1.1).

Treeline ecotones range from abrupt treelines to extended zones wherein the trees become increasingly small, stunted and/or dispersed with increasing elevation (Bader et al. 2021). These variable spatial treeline patterns are a consequence of underlying ecological processes, and many attempts have been made to distinguish and describe these treeline patterns. Holtmeier (2009) distinguished four types of treeline form, namely abrupt forest limit, transition zone (ecotone), 'true krummholz belt' and gradual transition from high-stemmed forest to crippled trees of the same

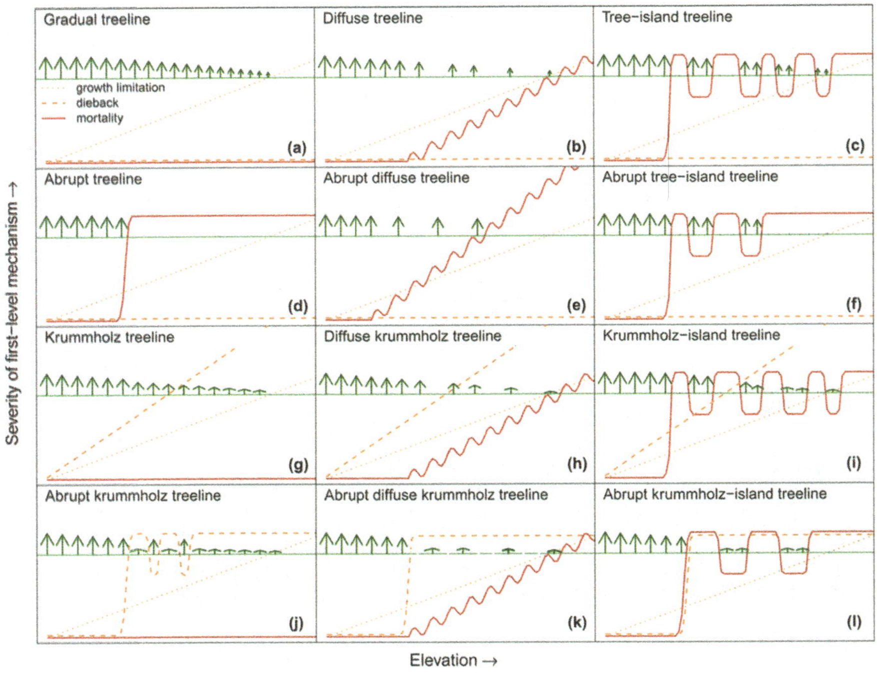

Plate 1.1 Diagrammatic representation of treeline forms (Reproduced from Bader et al. 2021)

species (Table 1.1). 'True krummholz' was used for genetic krummholz, i.e. species that can grow only as shrubs even under conditions favourable for tree growth, but now krummholz usually refers to environmentally induced stunted and crippled trees (Bader et al. 2021).

Subsequently, four treeline forms – diffuse, abrupt, island and krummholz – were recognized by Harsch and Bader (2011), with growth limitation being dominant only in the most common diffuse treeline, and dieback and seedling mortality in other treeline forms. Recently, Bader et al. (2021) proposed a framework of 12 treeline forms at the hillslope scale based on the spatial pattern of tree cover as seen from above and changes in tree stature and physiognomy (Table 1.1; Plate 1.1).

Globally, treeline ecotone has attracted considerable attention and several hypotheses related to it have been proposed. For example, Stevens and Fox (1991) proposed size-related, growth-related and stature-related hypotheses, all of which considered tree size, growth or stature as the limiting factors of alpine treeline formation. They emphasized the disadvantages and resource limitations associated with individual tree growth, rather than the pattern–process interaction at larger scales. Körner (1999) focused more on environmental constraints and impacts on tree growth and advanced several hypotheses, namely stress hypothesis, carbon balance hypothesis, growth limitation hypothesis and reproduction hypothesis. In addition to focussing on individual trees on a local level, these hypotheses are

Table 1.1 Types of treeline form

Author	Treeline form	Characteristics
Holtmeier (2009)	Abrupt forest	Closed-canopy forest limit bordering alpine vegetation
	Transition zone (ecotone)	Gradual decrease in the forest cover from the timberline to a tree species line
	True krummholz belt	True 'krummholz' is applied to the bush-like growth forms of species that are genetically determined
	Gradual transition from high-stemmed forest to crippled trees	Gradual transition from high-stemmed forest to crippled trees of the same species bordering alpine vegetation
Harsch and Bader (2011)	Diffuse	Characterized by a gradual decrease in height of single-stemmed trees along the treeline ecotone
	Abrupt	Characterized by a continuous forest >3 m tall directly bordering low alpine vegetation; tree height, as well as density, thus changes rapidly; trees may be present above the continuous forest but their presence is infrequent
	Island	Characterized by clumped patches or linear strips ('fingers') of krummholz or trees above the continuous forest limit
	Krummholz	Krummholz, characterized by severely stunted or deformed multi-stemmed trees; the krummholz growth form can occur in clumped patches above the upright forest, in which case we class the treeline as an island treeline, or it can occur as a dispersed or contiguous band above the upright forest, in which case we class the treeline as a 'krummholz treeline'; the characteristics of krummholz treelines also apply to krummholz-island treelines
Bader et al. (2021)	Gradual treeline	Decline in maximum woody plant height across the treeline ecotone is gradual without much change in density
	Diffuse treeline	Gradual height decline and diffuse density change
	Tree-island treeline	Gradual height decline and trees clumping into islands
	Abrupt treeline	Decline in maximum woody plant height across the treeline ecotone is abrupt
	Abrupt diffuse treeline	Abrupt height and diffuse density change
	Abrupt tree-island treeline	Islands form at elevations where growth is not yet much limited and dieback does not play a large role
	Krummholz treeline	Characterized by severely and progressively stunted or deformed multi-stemmed trees

(continued)

Table 1.1 (continued)

Author	Treeline form	Characteristics
	Diffuse krummholz treeline	Diffuse krummholz treelines are a rather common treeline form, as single trees above the closed forest are prone to wind damage and to winter desiccation
	Krummholz-island treeline	Severely stunted trees clumping into islands
	Abrupt krummholz treeline	When tall trees change directly to low-stature krummholz
	Abrupt diffuse krummholz treeline	When tall trees change to krummholz with diffuse density change
	Abrupt krummholz-island treeline	When tall trees change to krummholz and stunted or deformed trees are clumped into islands

concerned more with the ultimate causes of the formation of alpine treelines than on the evolution of alpine treeline ecotones. More precisely, they give importance to how spatial patterns and processes interact to create the complex spatial patterns with or without exogenous factors.

Changes in ecotones often result from a complex interplay between different species and between species and the elements of the physical environment like geomorphology, soil, climate, hydrology or CO_2 (Cairns and Malanson 1998; Körner 1998; Malanson 1999; Walsh et al. 2003). A key aspect in ecotone research is to integrate temporal and spatial aspects of landscapes for a more holistic understanding of landscape changes, which is particularly important in the context of changing climate.

1.2 Treeline Ecotone Research in the Himalaya

Although several facets of the Himalayan treeline ecotone have been investigated, yet the studies are scattered, fragmentary and mostly not at appropriate spatiotemporal scales, and hence a meaningful understanding of patterns and processes in the Himalayan treeline ecotone is still lacking. The Himalaya has not only the highest treeline (of *Juniperus tibetica* at 4900 m) in the Northern Hemisphere, but also the treelines vary widely in elevation (Table 1.2), generally from about 3000 m to about 4500 m (Singh et al. 2019a; b, 2021). Because of anthropogenic factors, natural treelines in the Himalaya are rare notwithstanding the remoteness of treeline areas (Schickhoff 2005, 2011). The treeline elevation in the Himalaya increases from northwest (NW) to southeast (SE) along the Himalayan Arc (Singh 2018) despite an eastward increase in the mesic condition, which is known to supress treelines (Körner 2012). It seems to be largely because of a decrease in latitude from NW to SE (range being about 10° N latitude). Treelines are higher in inner dry valleys than in outer mountain ranges, receiving the thrust of monsoon rainfall.

Table 1.2 Elevation range of treeline ecotone in the Himalaya

The Himalayan region	Elevational extent of treeline ecotone (m amsl)	Ecotone elevation range (m amsl)	Location	Reference
India Himalaya	3200–3700	3228–3900	Daksum-Sinthan Top (Kashmir)	Nanda et al. (2018)
	3000–4000		Uttarakhand and Himachal Pradesh	Rai et al. (2013)
	2800–4000		Bhaderwah (Jammu and Kashmir)	Singh et al. (2019a, b)
	3200–3300		Tungnath (Uttarakhand)	Adhikari and Kumar (2020)
	3300–3900		Nanda Devi National Park (Uttarakhand)	Bharti et al. (2012)
	3600–3900		Garhwal (Uttarakhand)	Adhikari and Rawat (2004)
	3500–4500		Khangchendzonga National Park (Sikkim)	Pandey et al. (2018)
Nepal Himalaya	3870–4180	3781–4118	Barun, Manang and Dhorpatan (Nepal)	Chhetri et al. (2020)
	3700–4200		Manang and Rasuwa district (Central Nepal)	Shrestha et al. (2015)
	3700–4000		Rolwaling Valley (East Central Nepal)	Schwab et al. (2018)
	3856–4095		Barun Valley (Eastern Nepal)	Chhetri and Cairns (2015)
China Himalaya	3300–3400	2766–3283	Qilian Mountains (North-western China)	He et al. (2013)
	2500–2750		Tianchi Nature Reserve (North-western China)	Wang et al. (2006)
	2500–3700		Balang Mountain, Wolong Natural Reserve (Sichuan Province)	Shi et al. (2006)

m amsl: metres above mean sea level

1.3 Treeline Forms

Treeline forms reported in the Himalaya are listed in Table 1.3 (Plates 1.2 to 1.5). Although diffuse-type treeline form is common, other types, such as island (finger) type, or abrupt and krummholz types, have also been reported. However, the treeline forms need to be characterized in terms of characteristics proposed by Bader et al. (2021) that are presented in Table 1.1 and Plate 1.1.

1.4 Climate Change in the Region

The Himalayan mountain system is undergoing rapid and elevation-dependent warming (EDW), particularly since the 1980s (Dimri et al. 2022; Ren et al. 2017; Sun et al. 2020). The mean annual temperature has increased by approximately 0.1 °C/decade over the 1901–2014 period (Ren et al. 2017). Compared to 1961–1990, the mean annual temperature is predicted to increase by about 1–2 °C during 2021–2050. Even without climate change, the Himalaya was warm due to the mass elevation effect (MEE) in the mountains of middle latitudes (Holtmeier 2009). MEE often overlaps with the positive effect of continental climate characters in the middle portion of mountains of large sizes, such as the Himalaya (Holtmeier and Broll 2020). In such mountain areas, high thermal amplitude, high solar radiations, less cloudiness and less precipitation combine to contribute to high temperatures (Holtmeier and Broll 2020). Because of warm and relatively moist conditions in outer Himalayan ranges, forests even at 2700–3000 m are closer in their functional attributes to those of tropical forests than temperate forests of the world (Zobel and Singh 1995).

Table 1.3 Treeline forms in the Himalaya

The Himalayan region	Treeline form/ Type	Location	References
India Himalaya	Diffuse	Daksum-Sinthan Top (Kashmir)	Nanda et al. (2018)
	Finger-like	Sonamarg	Singh et al. (2018)
	Abrupt; finger-like	Chaudas Valley (Uttarakhand)	Singh et al. (2018)
	Diffuse	Khangchendzonga National Park (Sikkim)	Pandey et al. (2018)
Nepal Himalaya	Diffuse	Barun, Manang and Dhorpatan (Nepal)	Chhetri et al. (2020)
	Abrupt	Manang district (Central Nepal)	Shrestha et al. (2007)
	Krummholz	Rolwaling Valley (East Central Nepal)	Schwab et al. (2016)
	Abrupt	Barun Valley (Eastern Nepal)	Chhetri and Cairns (2015)
China Himalaya	Diffuse	Changbai Mountains (Northeast China)	Zong et al. (2014)

Plate 1.2 Treeline in Kashmir (Photo credit@ Rajesh Joshi)

Plate 1.3 Treeline along Daksum-Sinthan Top transect, Jammu and Kashmir (Photo credit@ Rajesh Josh)

Plate 1.4 Treeline along Yuksam-Dzongri transect, Sikkim (Photo credit@ Rajesh Joshi)

Plate 1.5 Treeline along Chopta-Tungnath transect, Uttarakhand (Photo credit@ B S Adhikari)

Based on simulations of Global Circulation Models (GCM) and Regional Climate Models (RCM), the following trends can be summarized for the Himalayan region (Rangwala et al. 2019).

- Data from 1901 to 2014 indicate that temperature has increased throughout the Hindu Kush Himalaya (HKH), but is more prominent in Tibetan Plateau (Ren et al. 2017).
- Significantly greater than the global change warming rates have been reported in Hindu Kush Himalaya (HKH), particularly Tibetan Plateau, since the second half of the twentieth century (e.g. Liu and Chen 2000, Ohmura 2012; Yan and Liu 2014). The climate model experiment indicates that the mean annual temperature in the Himalaya is predicted to increase in the range of 1–4 °C by mid-twenty-first century and 2–6 °C by the late twenty-first century. The projection includes a greater increase for minimum temperatures than maximum temperatures (Rangwala et al. 2013; Wu et al. 2017).
- The projected increase in daily minimum temperatures is higher during winter and spring than during summer and autumn (Rangwala et al. 2013).
- The phenomenon of elevation-dependent warming (EDW) is being observed in several mountain areas. The amplified warming at higher elevation has been observed and indicated by modelling (Liu and Chen 2000; Qin et al. 2009a; b; Rangwala et al. 2009; Ohmura 2012). EDW is partly the product of strong positive feedbacks from the depletion of snow and ice cover (Pepin and Lundquist 2008), and the greening of landscapes, all leading to decreased albedo and increased absorption of solar radiation. The warming rate increases with elevation between 3000 and 4800 m, then stabilizes above that (Qin et al. 2009a; b) because of snow albedo feedback (SAF). The strength of snow/ice albedo feedback (SAF) depends on the difference between the albedo of snow and that of the land surface beneath the snow. SAF in mountains is also affected by aerosols.
- Generally, minimum temperature increases at a faster rate than maximum temperature, but between 1971 and 2007 maximum temperature increased more than minimum temperature (Kothawale et al. 2010). It is related to the snow albedo feedback effect apart from changes in the monsoon circulation (Yan et al. 2011).
- Snow cover season has become shorter, and precipitation is more as rain than before (Ramanathan and Carmichael 2008).
- According to Bhutiyani et al. (2010), Black Carbon could account for half of the total warming during the last several decades.
- Increasing water vapour partially accounts for winter warming rates.
- An increase in mean precipitation, inter-annual variability and extremes are predicted.
- A decrease in the number of wet days, i.e. intense and episodic monsoon season, is being seen.
- A 22% increase in monsoon precipitation over SE Himalaya and much of the Tibet Plateau is predicted by the end of the twenty-first century under Representative Concentration Pathways (RCP) 8.5.

- Mid-latitude extratropical storms (westerlies) during winter and spring are a primary source of moisture to the NW and northern parts of the HKH.
- Due to the heavy summer monsoon rainfall (~100 mm/day), several catastrophic flooding events have occurred during the last decade, e.g. devastating floods in Northern Pakistan, the Indus Basin, Uttarakhand flood in 2013.

1.5 Temperature Lapse Rate (TLR)

Temperature **lapse rate** (TLR) refers to the rate of temperature change observed while moving upward through the Earth's atmosphere. It is regarded as one of the important features of local and regional climates. Several studies carried out world over (Laughlin 1982; Thyer 1985; Barry 2002; Pepin 2001; Rolland 2003; Tang and Fang 2006; Marshall et al. 2007; Blandford et al. 2008; Gardner et al. 2009; Gouvas et al. 2011) have revealed a relationship between TLR and macro-topography, with the magnitude of this relationship varying with energy balance regimes, such as surface condition, elevation, air moisture content, wind speed, cloudiness, radiative conditions and distance from the sea. Surface air temperature lapse rate is of much significance in determining temperature range in many fields, such as glaciology, hydrology, ecology, forestry and agriculture (Diaz and Bradley 1997; Barry 2002; Pepin 2001; Rolland 2003; Tang and Fang 2006; Mokhov and Akperov 2006; Marshall et al. 2007; Blandford et al. 2008; Minder et al. 2010). Many studies have been carried out on the annual cycle of surface air TLR and its controlling factors in many mountain regions, except the Himalaya where the meteorological data including location-specific precise temperature data in high ranges of the Himalaya are scarce, making it difficult to estimate temperature lapse rate (TLR). In the absence of empirically estimated TLR, a constant value of environmental lapse rate (from −0.65 °C/100 m to −0.70 °C/100 m) is often applied to determine the temperature at an ungauged site (Joshi et al. 2018). Apart from the lack of realization of the scientific importance of such data, there are several logistic problems in setting up a system of temperature measurement in the difficult terrain of the Himalaya. This gap in data was addressed to an extent in the Indian Himalayan Timberline Research Project (IHTRP). Temperature data collected from three elevational transects representing the principal segments of the Himalaya, namely north-western (Kashmir), central (Uttarakhand) and eastern (Sikkim), indicate marked variabilities in TLR values across regions and seasons with implications for species distribution patterns along the elevation gradient, including treeline ecotone. Across the three elevation transects mentioned above, TLR ranged from −0.50 °C/100 m in Sikkim to 0.66 °C/100 m in Kashmir (Table 1.4). It brings out that TLR decreases from arid to humid regions. TLR values in Bhutan and Arunachal Pradesh in the Eastern Himalaya are similar (0.46–0.51 °C/100 m) to that of Sikkim (Joshi et al. 2018; Joshi 2021). The TLR decreased from west to east and is dissimilar for different precipitation regimes in the Himalaya.

Higher growing season temperatures (GSTs) were observed (8.4 ± 1.8 °C, 10.3 ± 1.4 °C and 7.5 ± 2.7 °C) for treelines in the Himalaya than the defined GST (6.6 ±

Table 1.4 A summary of characters of the Himalayan treelines (largely from the chapters of this book, where references are indicated)

Treeline elevation: range and variation	Up to 4900 m (in Tibet), but the range is wide, 1500–2000 m, generally from 3000 to 4500 m elevation; elevation more on the south-facing slope than north-facing slope, more in dry inner valleys than in outer ranges and more towards the east
Treeline length: an estimate based on remote sensing techniques	Cumulative length in several thousands of kilometres, generally 6–8 km per km crow fly distance
Timberline upward shift during last four decades or so	5–11% across the Indian states, the rest being stationary
Number of tree species across all treelines in the Himalaya	58 species; but Khuroo et al. (2022) reported 69 species
Change in structural and functional features of tree species with an increase in elevation	Shorter leaves have higher density on twigs; increase in stem diameter to height ratio of trees
Peak species richness along an elevation gradient	Varies across growth forms and taxonomical groups; generally tree species richness peaking at a lower elevation, and herbs at a higher elevation
Trends in species richness from west to east	Tree species richness increases considerably from west to east, while that of herbs increases from east to west
Temperature lapse rate (TLR°C/100 m) with elevation	Increases with aridity 0.51 °C in Sikkim to 0.66 °C in Kashmir; marked seasonal variation; winter TLR being about 50% lower than pre-monsoon summer TLR
Tree water relations: predawn tree water potential and leaf conductance	Trees not stressed, water potential always more than −1 MPa; pre-monsoon $\Psi_{predawn}$ is the lowest, indicating the snow melt water has only a moderate positive effect; leaf water conductance is higher in treelines than in forests
Dendrochronology	Supports the observation based on remote sensing that treelines are not moving upslope in a majority of instances; and shows stress as the adverse impact of pre-monsoon droughts
Early snow melt effect on vegetation	Tends to increase species richness and growth in herbaceous vegetation, but long-term prediction may be different
Phenological characters of trees	Three leaf forms: broadleaved deciduous, broadleaved evergreen, and conifer multi-year; leaf expansion period extends to monsoon
Growing period length and mean temperature	~200 days; growing period mean temperature generally 1.5–3.5 °C higher than climatic treeline

0.8 °C) based on a global survey of air temperatures at treelines suggesting that treelines in the Himalaya are relatively warmer. The observed shallow TLRs signify that EDW is amplified with elevation in the Himalayan region under the influence of climate change. The lower values of TLR may have several possible impacts on the dynamics of treeline ecotone in the Himalaya, such as a change in snow and moisture regime, increased evapotranspiration and water stress, change in albedo and surface energy balance acting in concert to modify distribution patterns, range shift and growing season of alpine vegetation.

The seasonal changes in TLR in the Himalaya are profound, and their impact on species distribution and adaptation needs to be analysed. TLR in December is recorded at about half of that during pre-monsoon (March–May) (Joshi et al. 2018). Not much is known about the cause of so low TLR in December and its effect on the distribution of different species. Overall, TLR is relatively low in monsoonal regions of the Himalaya. The phenomenon of elevation-dependent warming (EDW) due to global climate change may contribute to decreased TLR. Pollution and consequent reduction in solar radiation in lower areas during winter have also emerged as a modifier of TLR. In addition, decreased albedo because of snow depletion in the treeline and alpine areas in the warming climate is expected to further increase the warming rate in those areas.

In brief, more sites representing different climatic conditions are required to be investigated on a long-term basis to understand the TLR patterns and their influence on high-elevation vegetation and ecosystem services. Not much is known about the situation in inner valleys. A study in Ladakh has shown that the cold desert TLR can be relatively very high, 0.86 °C/100 m (Thayyen and Dimri). In short, the relationships among TLR, climate change and treeline dynamics have not attracted the interest of researchers to the extent these deserve.

1.6 Treeline Dimensions and Dynamics

1.6.1 Treeline Dimensions

Treeline research is mostly site-based and a researcher selects suitable sites for examining treeline shifts or changes in species composition over a period. Somehow, dimensions and spread of treelines/timberlines in this region have drawn little attention. However, the importance of large-scale changes due to climate change, such as the retreat of glaciers, colonization of sub-nival areas released by retreating glaciers and the resultant change in albedo and soil carbon is being increasingly realized. Changes in treeline and adjoining vegetation in large mountain regions like the Himalaya are expected to have implications for regional carbon balance and plant community dynamics. Keeping this in view, the timberlines in the IHTRP were mapped in five Indian Himalayan States having treelines and alpine meadows (Sharma 2021). Traversing high ranges in a zigzag way, the treelines/timberlines in the Himalaya are several thousand kilometres long. Roughly, their length is 6–8 km per km of crow flight distance, which would convert into close to 15–20 thousand kilometre length in the Himalayan region. The project findings (Sharma 2021) based on remote sensing techniques show that only a small fraction mostly below 10% of the length of the timberline has shown upslope shift, giving the impression that the Himalayan timberlines are largely static despite the decades of warming. A global survey of the literature shows that the upward shifts in treelines due to climate change has been found in about 50% of instances, which is much higher than the fraction of timberline length that was found to move upslope in the Himalaya (Sharma 2021; Table 1.4). The portions of timberline showing upslope shift differed from the stationary portions in terms of change in temperature and precipitation during the study time interval (Sharma 2021). To find out how climate change drives elevational treeline dynamics, Lu et al. (2021) analysed data of 143 treeline sites of the Northern Hemisphere and found that in 88.8% of sites treelines ascended during 1901–2018 but at half the rate (0.354 m/year) expected from climatic warming alone. They found that the precipitation effect was relatively higher than the temperature effect.

The absence of treeline shift despite temperature rise indicates that there are several factors, other than temperature change, that influence change in treelines. The treeline elevation in Himalaya varies considerably, the difference between the lowest and highest treelines being ~1800 m. In Uttarakhand, treelines occur between 2800 and 4200 m, but above 50% are concentrated between 3200 and 3600 m elevation (Sah and Sharma 2018). Some watersheds have more timberlines than others. The Ganga and Yamuna watersheds alone account for about one-third of the treeline length in Uttarakhand (Sah and Sharma 2018).

In present approach treeline is being considered as an ecosystem type with its presence throughout the Himalayan Arc, providing a vegetal cover for wild mammals and several plant forms, and microbes. A system that runs parallel to permanent snowlines provides long corridors for lateral (east to west) management of species adversely affected by climate change and other anthropogenic factors. Model-based research on niche characterization of treeline species and likely shift along the Himalayan Arc has been conducted on the Himalayan birch (*Betula utilis*).

Looking at treelines at a regional or other macro-scale may help us develop better management of these precious and fragile systems. Keeping in view this approach, the number of tree species in treelines and in the Himalaya has been estimated. Khuroo et al. (2022, this volume) have extended the idea of species richness assessment by considering all growth forms, such as trees, shrubs and herbs.

Equally important, however, are micro-level processes that lead to the establishment of species in the face of climate change. The highly stressful environment leads to facilitation involving physiognomically different taxa. For example, the establishment of *Pinus wallichiana* seedlings is facilitated by a juniper mat (Reshi et al. 2021), which might also need specific microbes, particularly mycorrhiza.

While treeline position and dynamics are an outcome of myriad interactions between a multitude of factors, such as topography, ecology of treeline species, site characteristics, and current natural and anthropogenic influences (Gaire et al. 2014), climate has an overriding influence because of which treelines are sensitive to changing climatic conditions, particularly temperature and precipitation, and hence treeline shifts are expected in response to these changes. Palaeoecological studies have revealed that the position of the Himalayan treeline ecotones has shifted in response to Holocene climate change with treelines advancing several hundred metres higher than today under warm-humid conditions and reaching uppermost limits in the early Holocene (Schickhoff et al. 2016b). In the recent past also, altitudinal shifts in the treeline position have been reported (Schickhoff 2005; Panigrahy et al. 2010; Negi 2012; Rawat 2012; Singh et al. 2012; Telwala et al. 2013; Aryal et al. 2014; Gaire et al. 2014; Schickhoff et al. 2015). But Singh et al. (2018) reported a static treeline of the Himalayan silver fir at Tungnath for last several decades. Such inconsistent response has also been shown by high-altitude plant species in the Himalayan region to recent climate change under various ecological contexts (Gaire et al. 2014; Schickhoff et al. 2015). For example, Gaire et al. (2014) reported a shift to higher elevations in *Abies spectabilis* in central Nepal with a rate of 2.61 m/year since the 1850s, with episodic recruitment being higher during 1950s and 1980s. On the contrary, the response of *Abies spectabilis* to recent climate change in Barun Valley in Eastern Nepal has been almost negligible (Chhetri and Cairns 2015). The Himalayan birch (*Betula utilis*) in central Nepal also has not shown any elevation change in recruitment pattern (Shrestha et al. 2007; Gaire et al. 2014) in recent decades.

Treeline responses to climate change have made them an effective indicator akin to glacier melt. There is general agreement that treelines are formed at certain soil or/and atmosphere temperatures of the growing season (Holtmeier and Broll 2005, 2020; Körner and Paulsen 2004). However, considerable importance in it is given to the upright tree form, as the status of the tree crown is coupled with air temperature (Millar et al. 2020). A climatic treeline is formed where atmospheric conditions enforce the limits of tree growth of trees with ≥3 m height and ≥94 days growing season (Körner and Paulsen 2004). Recently, research on *Pinus longaeva* emphasized that temperature is the key controlling factor in its growth at cooler sites while precipitation seems to be an overriding factor on warmer sites (Tran et al. 2017; Bunn et al. 2018). So, moisture sensitivity could be a new controlling factor for tree growth, particularly under climate change (Millar et al. 2020). In several treelines, tree densification is being observed because of climate change. While the relationship between treeline position and temperature is now understood to a great extent, the debate has not completely faded away as the assessment of possible underlying causes at finer levels is still a challenge (Holtmeier and Broll 2020).

1.7 Treeline Phenology and Water Relations

Phenology has been a focal research area in recent years in view of climate change (Dutta et al. 2020; Kumar et al. 2022). Advancement in the timing of phenophases, such as leaf formation and flowering, is expected to be a common feature. In treelines of European and American mountains where historical data of treeline position are available, analysis of changes due to climatic warming is quite rewarding. In the Himalaya, such comparisons are difficult to make because of the lack of past data.

Another area of interest of treelines has been how morphological characters of leaf, stem and tree forms change with increasing elevation leading to treelines. The Himalayan tree species have all those physiognomic forms that occur in lower-elevation forests: deciduous broadleaf evergreen with one leaf life span and multi-year evergreens. In several instances, all trees occur within a short treeline strip. They differ significantly in the period of stable leaf population. They also differ significantly in nutrient resorption from senescing leaves. One of the common strategies the species of the different growth forms employ is to differ in the timing of phenophases despite the restricted growth period. For example, at the Tungnath treeline site, vegetative bud-break was recorded as early as in early April in *A. spectabilis* and *B. utilis*, and in mid-May in the remaining species. Interestingly, the two early bud-breaking species differed most in leaf longevity, the birch being deciduous and fir a multi-year evergreen. In the warmer spring of 2017 (mean temperature, 8.33°C), bud-break and flowering occurred earlier by about 2–3 weeks in all the species than in 2020, which was cooler by 1.8°C (mean temperature, 6.59°C).

The study showed that in a harsh treeline environment characterized by heat deficiency, functional traits vary widely despite a low species diversity although the length of phenophases remains within a limit. The maximum value is 166 times the minimum value for leaf area, 141 times for the leaf mass and 76 times for stable leaf population and leaf life span (derived from the data of Pradeep Mehta and GCS Negi, Final Technical Report, IHTRP, 2021, unpubl.). Evidently, the Himalayan treeline communities consist of highly niche-differentiated species, warranting a complex management scheme, and deeper analysis of the adaptational significance of morphological traits. Climate warming has triggered snowmelt, which results in more water availability to trees. However, in dry continental climates availability of snow meltwater stops soon, thus affecting growth adversely (Öberg and Kullman 2012). The future scenarios appear to be very uncertain, warranting long-term monitoring.

Research on tree water relations has attracted little attention worldwide because trees in treelines are not considered to be water-stressed (Körner 2012), though several studies on tree-ring chronology emphasize the role of drought in treeline dynamics (Singh 2018). The findings of IHTRP on tree water relations show that in all treeline species across all seasons and years, predawn water potential remained above −1 MPa (Tewari et al. 2018; also Table 1.4; Tewari 2021). Predawn tree water potential lower than −1 MPa is quite common among the forest tree species of the Central Himalaya (Singh et al. 2006). In contrast, studies based on the relationship between tree-ring growth and climate change over hundreds of years have highlighted the frequent suppression of tree-ring growth during pre-monsoon drought (March to June beginning). Pre-monsoon soil drying has been observed during the last several decades of climatic warming (Li et al. 2016; Sigdel et al. 2018; Tewari 2021). Global warming may further intensify pre-monsoon drought and hence deplete the tree-ring width. However, the effect of climate change on tree growth is not straightforward. For example, tree growth during early summer or pre-monsoon is usually adversely affected because cold soil restricts water availability to trees, particularly in early day hours, resulting in a fall in the tree water potential until the soil gets warmer in the day (Li et al. 2016). Low soil temperature limits water uptake by increasing water viscosity and decreasing root permeability, which influences the soil–plant–atmosphere continuum (Goldstein et al. 1985). Climate warming may increase early-season soil availability and hence tree growth.

The IHTRP study (Singh et al. 2021) showed that some treeline species and their seedlings often attain unusually high leaf water conductance (exceeding 1000 mmol/m^2/s). How common is this very high leaf conductance in treeline species, and which factors enable species to attain it, needs more data from a range of sites. Another Himalayan treeline feature about water stress is that trees in treelines respond differently from trees of lower forests during the monsoon. The treeline species undergo a marked daily decline (by 0.8–1.2 MPa; Tewari 2021) in tree water potential during monsoon when predawn water potential is high, while forest trees show little daily change in water potential during this period (0.2–0.5 MPa). How tree water relations vary spatially from east to west along the Himalayan Arc and from south to north with a steep decline in precipitation is of considerable scientific interest. The impact of climatic warming may be manifested through changes in water stress to an extent. We argue that tree water relations in treelines need to be given importance for understanding the impact of climate change on treelines across different precipitation regimes.

1.8 Species Diversity in the Himalayan Treeline Ecotone

Despite being a biodiversity hotspot, detailed studies on species diversity of plants (Bargali et al. 2022), animals (Kashyap et al. 2022) and microbes (Bhattacharya et al. 2022) at appropriate spatiotemporal scales in the Himalaya are lacking. However, some sporadic studies on various facets of diversity and distribution have been carried out. For example, phytosociological studies in Western Himalaya and Karakoram have been carried out by Hartmann (1968; 1999), Schickhoff (1993; 1996), Eberhardt (2004) and Khan (2012). Based on floristic inventory and numerous phytosociological relevés in the Langtang-Helambu region, Miehe (1990) described in detail the upper elevational plant communities. Likewise, Kikuchi and Ohba (1988) also carried out a phytosociological study of major alpine plant communities near Panch Pokhari Lake in the south-eastern part of Rolwaling Himalalaya (Nepal). Other standard works are of Schweinfurth (1957), Champion and Seth (1968), Stainton (1972), Dobremez (1976), Miehe (1982; 1990), Puri et al. (1983; 1989), Schickhoff (1993; 1996), Dickoré and Nüsser (2000), Eberhardt et al. (2007) and Miehe et al. (2015). The most comprehensive account of subalpine and alpine vegetation types with detailed floristic and ecological information has been provided by Miehe (1990). In addition, Miehe and Miehe (2000) also presented information on treeline floristic, structural and spatial patterns, in addition to human impact (Miehe et al. 2015). While summarizing the available information about the geographical and ecological aspects of the Himalayan treelines, Schickhoff (2005) observed that treeline ecological patterns and processes, including regeneration, carbon balance, frost, drought, snow cover, wind, soil physical and chemical conditions, have received very little attention.

Some empirical studies on patterns of plant species distribution along elevational gradients in the Himalaya are also available. Such studies on ferns (Bhattarai et al. 2004), trees (Bhattarai and Vetaas 2006) and vascular plants (Panthi et al. 2007) have been undertaken in Nepal Himalaya. In addition, studies on ecotonal species bordering subalpine forest (Shrestha and Vetaas 2009), woody plant species under different land-use types and slopes in trans-Himalayan valleys in central Nepal (Paudel and Vetaas 2014), the richness of different plant life-forms along an elevational gradient in Eastern Nepal (Bhattarai and Vetaas 2003), and variation in forest biodiversity in central Nepal (Christensen and Heilmann-Clausen 2009) are also on record. Some studies related to floristic characteristics of communities and plant diversity have been carried out in the high-elevation region of Indian Western Himalaya (Rawal et al. 1991; Rawal and Pangtey 1994; Rawal and Dhar 1997; Joshi and Samant 2004; Oommen and Shanker 2005; Behera and Kushwaha 2006; Rana et al. 2021; Rawat et al. 2021). More recently, Rawal et al. (2018) reported heterogeneity in patterns across altitudinal transects (2000–3500 m) and life-forms, thereby suggesting the stronger influence of micro-level factors in Western Himalaya. The increasing focus on climate change that is taking place in the Himalaya has resulted in more interest in scientific exploration in the Himalayan treelines (Schickhoff et al. 2015, 2016a) with a focus on treeline response to climate change. Nevertheless, we still lack a detailed characterization of treeline vegetation–environment relationships based on correlations of vegetation and site-ecological data (Bürzle et al. 2018).

A critical appraisal of published literature has revealed that vascular plant species diversity has received more attention compared to lower groups of plants, animals and microbes. Even the studies on vascular plants do not uniformly cover the entire expanse of the Himalaya. This has been brought out in one of the chapters in this book on 'Patterns of Plant Species Richness Across the Himalayan Treeline Ecotone'. Based on a detailed analysis of literature published over the last 30 years, a total of 593 vascular plant species belonging to 232 genera in 70 families were reported in the scientific literature published on the Himalayan treeline ecotone with a higher number of plant species in Western Himalaya as compared to that of Eastern Himalaya. Thus, it becomes apparent that studies detailing the species richness and diversity of different taxa, particularly lower plants, animals and microbes, in the treeline ecotone are urgently required. In particular, the knowledge of vegetation–environment relationships in the Himalayan treeline ecotones is still very deficient.

1.9 Litterfall and Litter Decomposition Treelines

Globally not much is known about litterfall and litter decomposition for mountain treelines though they directly affect carbon values. A study on *Polylepis*, one of the most endangered endemic forest ecosystems of Ecuadorian Central Andes (Pinos et al. 2017), shows that the leaf litterfall was 3.77 Mg/ha/year, which was 51.1% of the leaf biomass present in the canopy. The litter decomposition rate was 0.38/year and annual weight loss was 30% (Pinos et al. 2017). The forest located at 3735–3930 m elevation was evergreen with little seasonal variations in litterfall.

The annual precipitation of the sites was 875 mm, and the annual mean temperature was 5.44 °C. In the Himalaya (Pindar valley), Garkoti and Singh (1995) have reported the leaf litterfall of 1.67 Mg/ha at a site close to timberline. Here the litter weight loss was about 42.3%/year at treeline for *Rhododendron campanulatum*. Seeing that leaf litter weight loss in *Quercus leucotrichophora*, *Quercus semecarpifolia* and *Rhododendron arboreum* at mid-elevation sites is 71%, 66% and 60%, respectively, the value for *R. campanulatum* in treeline is relatively high.

Since it largely depends on climatic factors like temperature and precipitation that are changing under the impact of global climate change, litter decomposition needs to be thoroughly investigated for changes that are already occurring or predicted to occur. Implications of litter decomposition for the carbon cycle are likely to be significant as treelines are subject to accelerated climate change.

1.10 Dendrochronology and Climate Change

Dendrochronology is being increasingly used in recent years to understand the relationship between climate change and tree growth. The historical knowledge could be applied to find out future trends about climatic warming with evaluation. Tree-ring counting has helped considerably to reconstruct the population structure and changes in treeline species during different phases of climate. It is used even to find out treeline elevation shift, densification and impact of anthropogenic activities. Dendrochronology can shed light on glaciation and deglaciation, the tree rings being narrower when glaciers swell and wider when glaciers melt and retreat. In conjunction with data on soil moisture and Normalized Difference Vegetation Index (NDVI), tree-ring analysis can help in analysing the relationship between climate and ecological processes. Apart from heat deficiency, treeline formation is driven by physical factors like snowfall and wind, which could stop trees to move upslope by damaging tree canopy, branches and stems. Tree ting width changes could be used to predict future changes.

1.11 Future Imperatives

Treeline ecotone in the Himalaya needs more detailed long-term studies over relevant spatiotemporal scales to identify the processes that structure this ecotone and control its functioning, particularly under the changing climate. Unfortunately, climate change studies have largely focused on glaciers and the hydrological changes in the Himalaya that are likely to result from changes to the cryosphere. Somehow, the treeline ecotones and treelines that are adjacent to the snowfields in the Himalaya have not received desired attention and prominence in the climate change discourse and research programmes.

- Treeline ecotones along with closed-canopy forests below and alpine tundra above make up a highly complex continuum with variously interconnected components, which includes some of the most attractive but threatened mammals like the snow leopard as well as rare lower plant groups like lichens growing on a leaf blade. The 'treeline ecotone' is a dynamic system and understanding its intricacies and complex interactions will require taxonomists of various disciplines, ecologists, researchers and practitioners of deep understanding who can work in remote areas who can address various issues ranging from tourism to glacier retreat and the thawing of permafrost.

- There is a need to recognize 'treeline as a separate and principal entity of conservation and management' because of the following reasons: (1) treelines are present and apparent throughout the Himalaya, with an extraordinary length and variability in geographical position, (2) treelines vary a lot along their length in their response to climate change, and (3) treelines have high corridor and habitat values for all forms of organisms.
- An independent 'Treeline Ecotone Conservation and Management Authority' (TECMA) may be established. This institutional framework would provide for focused attention on this so far neglected but ecologically novel and dynamic system. Such an authority can build upon the knowledge products developed under the IHTRP and establish long-term ecological research (LTER) sites in the entire Indian Himalayan region, such as in Kashmir, Himachal Pradesh, Uttarakhand, Sikkim and Arunachal Pradesh, for a better understanding of the structure and functioning of this dynamic system over both space and time, including TLR, phenological patterns, water relations, and ecosystem goods and services provided by this system.
- While developing the TECMA, it is important to involve all stakeholders including researchers and local communities, nomads and army personnel as partners and collaborators, given their presence in these high-elevation systems and traditional knowledge to develop highly informative citizen science.
- The treelines being biotic systems are a storehouse of varied information that can help us understand changes and responses of treelines to climate change and other anthropogenic stresses. Tree population dynamics and tree-ring growth could reveal considerable information about the past pattern of climate change, and other proxies could be used to reconstruct past climatic history for better comparisons and modelling.
- Treeline ecotone should be expanded to include adjoining biotic and abiotic zones for the study of their interaction and impact on the treeline ecotone species assembly. The management of treelines should be considered along with connected biotic zones namely alpine meadows and nival/sub-nival belts, and abiotic components such as permanent snow/glaciers and permafrost, which provide water to biotic systems.
- There is a need to develop training material on treeline management and conservation for stakeholders, such as forest department personnel, biodiversity management committees, nongovernmental organizations (NGOs), and army and paramilitary personnel.
- Develop short-term courses on treeline ecology and climate change for school and college-level students.

- Studies on treeline-linked socioeconomic activities need to be encouraged. The cultural linkages with treeline areas also include religious tourism besides pastoralism. All these anthropogenic activities need to be made sustainable keeping in view the issues related to livelihoods, religion and environmental sustainability. Treeline-based eco-tourism should be promoted with adequate safeguards against any environmental degradation of this fragile system.
- What would be the 'asylum value of the treeline ecotone' for threatened species? Do we need to develop a science of 'assisted migration' to save critically threatened species? How do we create 'topographical refugia' for species threatened by climate change? These are questions that beg answers and understanding, and require some novel ideas that need consideration in any future long-term study on treeline ecotones.

Acknowledgements The IHTRP project was financially supported by the Ministry of Environment, Forest and Climate Change (MoEF & CC), Government of India, under its National Mission on Himalayan Studies (NMHS). The authors would like to thank MoEF & CC and NMHS Project Management Unit for their support. Further, all the relevant departments of the Government of Jammu and Kashmir, Uttarakhand, and Sikkim and stakeholders are thanked for their cooperation received, which helped in the successful implementation of the project.

References

Adhikari BS, Kumar R (2020) Effect of snowmelt regime on phenology of herbaceous species at and around treeline in Western Himalaya, India. Not Sci Biol 12(4):901–919

Adhikari BS, Rawat GS (2004) Assessment of Garhwal Himalayan forests with special reference to climate change. In: Proc of the workshop on vulnerability assessment and adaptation due to Climate Change on Indian Agriculture, Forestry and Natural Ecosystems. Indian Inst Sci, Bangalore, MoEF, Govt of India, New Delhi

Aryal A, Brunton D, Raubenheimer D (2014) Impact of climate change on human-wildlife-ecosystem interactions in the Trans-Himalaya region of Nepal. Theor Appl Climatol 115(3–4):517–529

Bader MY, Llambí LD, Case BS, Buckley HL, Toivonen JM, Camarero JJ, Cairns DM, Brown CD, Wiegand T, Resler LM (2021) A global framework for linking alpine-treeline ecotone patterns to underlying processes. Ecography 44(2):265–292

Bargali H, Kumar A, Singh P (2022) Plant studies in Uttarakhand, Western Himalaya: a comprehensive review. Trees Forests People 8:100203

Barry RG (2002) Past and potential future changes in mountain environments: a review. In: Mountain environments in changing climates, vol 1, pp 27–50

Batllori E, Bianco-Moreno JM, Ninot JM, Gutierrez E, Carrillo E (2009) Vegetation patterns at the alpine treeline ecotone: the influence of tree cover on abrupt change in species composition of alpine communities. J Veg Sci 20(5):814–825

Behera MD, Kushwaha SP (2006) An analysis of altitudinal behaviour of tree species in Subansiri district, Eastern Himalaya. In: Hawksworth DL, Bull AT (eds) Plant conservation and biodiversity. Springer, Dordrecht, pp 277–291

Bharti RR, Adhikari BS, Rawat GS (2012) Assessing vegetation changes in timberline ecotone of Nanda Devi National Park, Uttarakhand. Int J Appl Earth Obs Geoinf 18:472–479

Bhattacharya P, Tiwari P, Rai ID, Talukdar G, Rawat GS (2022) Edaphic factors override temperature in shaping soil bacterial diversity across an elevation-vegetation gradient in Himalaya. Appl Soil Ecol 170:104306

Bhattarai KR, Vetaas OR (2003) Variation in plant species richness of different life forms along a subtropical elevation gradient in the Himalayas, east Nepal. Glob Ecol Biogeogr 12(4):327–340
Bhattarai KR, Vetaas OR (2006) Can Rapoport's rule explain tree species richness along the Himalayan elevation gradient, Nepal? Divers Distrib 12(4):373–378
Bhattarai KR, Vetaas OR, Grytnes JA (2004) Fern species richness along a central Himalayan elevational gradient Nepal. J Biogeogr 31(3):389–400
Bhutiyani MR, Kale VS, Pawar NJ (2010) Climate change and the precipitation variations in the northwestern Himalaya. Int J Climatol 30(4):535–548
Blandford TR, Humes KS, Harshburger BJ, Moore BC, Walden VP, Ye H (2008) Seasonal and synoptic variations in near-surface air temperature lapse rates in a mountainous basin. J Appl Meteorol Climatol 47(1):249–261
Bunn AG, Salzer MW, Anchukaitis KJ, Bruening JM, Hughes MK (2018) Spatiotemporal variability in the climate growth response of high elevation bristlecone pine in the White Mountains of California. Geophys Res Lett 45(24):13–312
Bürzle B, Schickhoff U, Schwab N, Wernicke LM, Müller YK, Böhner J, Chaudhary RP, Scholten T, Oldeland J (2018) Seedling recruitment and facilitation dependence on safe site characteristics in a Himalayan treeline ecotone. Plant Ecol 219(2):115–132
Cairns DM, Malanson GP (1998) Environmental variables influencing the carbon balance at the alpine treeline: a modelling approach. J Veg Sci 9(5):679–692
Champion HG, Seth SK (1968) A revised survey of the forest types of India. Manager of Publications, p 404
Chhetri PK, Cairns DM (2015) Contemporary and historic population structure of Abies spectabilis at treeline in Barun valley, eastern Nepal Himalaya. J Mt Sci 12(3):558–570
Chhetri PK, Bista R, Shrestha KB (2020) How does the stand structure of treeline-forming species shape the treeline ecotone in different regions of the Nepal Himalayas? J Mt Sci. 17(10): 2354–2368
Christensen M, Heilmann-Clausen J (2009) Forest biodiversity gradients and the human impact in Annapurna Conservation Area, Nepal. Biodivers Conserv 18(8):2205–2221
Diaz HF, Bradley RS (1997) Temperature variations during the last century at high elevation sites. In: Climatic change at high elevation sites. Springer, Dordrecht, pp 21–47
Dickoré WB, Nüsser M (2000) Flora of Nanga Parbat (NW Himalaya, Pakistan): an annotated inventory of vascular plants with remarks on vegetation dynamics. Englera 1:3–253
Dimri AP, Palazzi E, Daloz AS (2022) Elevation dependent precipitation and temperature changes over Indian Himalayan region. Clim Dyn 29:1–21
Dobremez JF (1976) Le Népal. Ècologie et biogéographie. CNRS, Paris, FR
Dutta P, Chakraborti S, Chaudhuri KM, Mondal S (2020) Physiological responses and resilience of plants to climate change. In: Rakshit A, Singh HB, Singh AK, Singh US, Fraceto L (eds) New frontiers in stress management for durable agriculture. Springer, pp 3–20
Eberhardt E (2004) Plant life of the Karakorum. The vegetation of the upper Hunza catchment (Northern Areas, Pakistan). Diversity, syntaxonomy, distribution. Dissertationes Botanicae 387:1–223
Eberhardt E, Dickoré WB, Miehe G (2007) Vegetation map of the Batura Valley (Hunza Karakorum, North Pakistan) (Die Vegetation des Batura-Tals (Hunza-Karakorum, Nord-Pakistan). Erdkunde 1:93–112
Elliott GP (2017) Treeline ecotones. In: Castree N, Goodchild MF, Kobayashi A, Liu W, Marston RA (eds) International Encyclopaedia of geography. John Wiley & Sons, Ltd, pp 1–10
Gaire NP, Koirala M, Bhuju DR, Borgaonkar HP (2014) Treeline dynamics with climate change at the central Nepal Himalaya. Clim of the Past 10(4):1277–1290
Gaire NP, Fan ZX, Shah SK, Thapa UK, Rokaya MB (2020) Tree-ring record of winter temperature from Humla, Karnali, in central Himalaya: a 229 years-long perspective for recent warming trend. Geogr Ann Ser B 102(3):297–316

Gardner AS, Sharp MJ, Koerner RM, Labine C, Boon S, Marshall SJ, Burgess DO, Lewis D (2009) Near-surface temperature lapse rates over Arctic glaciers and their implications for temperature downscaling. J Clim 22(16):4281–4298

Garkoti SC, Singh SP (1995) Variation in net primary productivity and biomass of forests in the high mountains of Central Himalaya. J Veg Sci 6(1):23–28

Goldstein GH, Brubaker LB, Hinckley TM (1985) Water relations of white spruce (*Picea glauca* (Moench) Voss) at tree line in north central Alaska. Can J For Res 15(6):1080–1087

Gouvas MA, Sakellariou NK, Kambezidis HD (2011) Estimation of the monthly and annual mean maximum and mean minimum air temperature values in Greece. Meteorog Atmos Phys 110 (3):143–149

Harsch MA, Bader MY (2011) Treeline form–a potential key to understanding treeline dynamics. Glob Ecol Biogeogr 20(4):582–596

Hartmann H (1968) Über die Vegetation des Karakorum. 1. Teil: Gesteinsfluren, subalpine Strauchbestände und Steppengesellschaften im Zentral-Karakorum. Vegetatio 15:297–387

Hartmann H (1999) Studien zur Flora und Vegetation im östlichen Transhimalaya von Ladakh (Indien). Candollea 54:171–230

He Z, Zhao W, Zhang L, Liu H (2013) Response of tree recruitment to climatic variability in the alpine treeline ecotone of the Qilian Mountains, northwestern China. For Sci 59(1):118–126

Holtmeier FK (ed) (2009) Mountain timberlines. Springer Netherlands, Dordrecht

Holtmeier FK, Broll G (2005) Sensitivity and response of northern hemisphere altitudinal and polar treelines to environmental change at landscape and local scales. Glob Ecol Biogeogr 14(5): 395–410

Holtmeier FK, Broll G (2020) Treeline research—From the roots of the past to present time. A review. Forests 11(1):38

Joshi R (2021) Temperature lapse rate (TLR) and pattern of precipitation along altitudinal gradients in different precipitation regimes across the IHR. Final Technical Report. Indian Himalayan Timberline Research Project (IHTRP). National Mission of Himalayan Studies, Ministry of Environment, Forests and Climate Change, Govt. of India, unpublished

Joshi HC, Samant SS (2004) Assessment of forest vegetation and conservation priorities of communities in part of Nanda Devi Biosphere Reserve, West Himalaya. Part I. Int J Sustain Dev World Ecol 11(3):326–336

Joshi R, Sambhav K, Singh SP (2018) Near surface temperature lapse rate for treeline environment in western Himalaya and possible impacts on ecotone vegetation. Trop Ecol 59(2):197–209

Kashyap P, Afzal S, Rizvi AN, Ahmad W, Uniyal VP, Banerjee D (2022) Nematode community structure along elevation gradient in high altitude vegetation cover of Gangotri National Park (Uttarakhand), India. Sci Rep 12(1):1–3

Khan SM (2012) Plant communities and vegetation ecosystem services in the Naran Valley, Western Himalaya. PhD Thesis. University of Leicester, Leicester

Khuroo AA, Dar FA, Hamid M, Ahmad R, Wani SA, Gulzar A, Malik HA, Singh CP (2022) Patterns of plant species richness across the Himalayan treeline ecotone. In: Singh SPS, Reshi ZA, Joshi R (eds) Ecology of Himalayan timberline ecotone. Springer Nature Singapore Pvt. Ltd.

Kikuchi T, Ohba H (1988) Preliminary study of alpine vegetation of the Himalayas, with special reference to the small-scale distribution patterns of plant communities. The Himalayan Plants 1:47–70

Kothawale DR, Munot AA, Kumar KK (2010) Surface air temperature variability over India during 1901–2007, and its association with ENSO. Clim Res 42(2):89–104

Körner C (1998) A re-assessment of high elevation treeline positions and their explanation. Oecologia 115(4):445–459

Körner C (1999) Alpine plants: stressed or adapted? In: Physiological Plant Ecology: 39th Symposium of the British Ecological Society (pp. 297–311). Cambridge University Press, Cambridge, UK

Körner C (2012) Alpine treelines: functional ecology of the global high elevation tree limits. Springer, Cham

Körner C, Paulsen J (2004) A world-wide study of high altitude treeline temperatures. J Biogeogr 31(5):713–732

Kumar M, Phukon SN, Paygude AC, Tyagi K, Singh H (2022) Mapping phenological functional types (PhFT) in the Indian Eastern Himalayas using machine learning algorithm in google earth engine. Comput Geosci 158:104982

Laughlin GP (1982) Minimum temperature and lapse rate in complex terrain: influencing factors and prediction. Arch Meteorol Geophys Bioclimatol B 30(1):141–152

Li K, Liu Y, Yang Y, Li Z, Liu B, Xue L, Yu W (2016) Possible role of pre-monsoon sea surface warming in driving the summer monsoon onset over the Bay of Bengal. Clim Dyn 47(3): 753–763

Liu X, Chen B (2000) Climatic warming in the Tibetan Plateau during recent decades. Int J Climatol 20(14):1729–1742

Lu X, Liang E, Wang Y, Babst F, Camarero JJ (2021) Mountain treelines climb slowly despite rapid climate warming. Glob Ecol Biogeogr 30(1):305–315

Malanson GE (1999) Considering complexity. Ann Assoc Am Geogr 89(4):746–753

Malanson GP, Xiao N, Alftine KJ (2001) A simulation test of the resource-averaging hypothesis of ecotone formation. J Veg Sci 12(6):743–748

Malanson GP, Butler DR, Fagre DB, Walsh SJ, Tomback DF, Daniels LD, Resler LM, Smith WK, Weiss DJ, Peterson DL, Bunn AG (2007) Alpine treeline of western North America: linking organism-to-landscape dynamics. Phys Geogr 28(5):378–396

Marshall SJ, Sharp MJ, Burgess DO, Anslow FS (2007) Near-surface-temperature lapse rates on the Prince of Wales Icefield, Ellesmere Island, Canada: implications for regional downscaling of temperature. Int J Climatol J R Meteorol Soc 27(3):385–398

Miehe G (1982) Vegetationsgeographische Untersuchungen im Dhaulagiri- und Annapurna-Himalaya. Dissertationes Botanicae 66:1–224

Miehe G (1990) Langtang Himal – Flora und Vegetation als Klimazeiger und -zeugen im Himalaya. A prodromus of the vegetation ecology of the Himalayas. Mit einer kommentierten Flechtenliste von Josef Poelt. Dissertationes Botanicae 158:1–529

Miehe G, Miehe S (2000) Comparative High Mountain Research on the Treeline Ecotone under Human Impact: Carl Troll's "Asymmetrical Zonation of the Humid Vegetation Types of the World" of 1948 Reconsidered (Studien zu natürlichen und anthropogen geprägten Strukturen an der oberen Waldgrenze—eine kritische Würdigung von Carl Trolls Konzept der "asymmetrischen Verteilung der immerfeuchten Vegetationsytpen der Erde". Erdkunde 1:34–50

Miehe G, Miehe S, Böhner J, Bäumler R, Ghimire SK, Battarai K, Chaudhary RP, Subedi M, Jha PK, Pendry C (2015) Vegetation ecology. In: Miehe G, Pendry C, Chaudhary RP (eds) Nepal. An introduction to the natural history, ecology and human environment of the Himalayas. A companion to the Flora of Nepal. Royal Botanic Garden, Edinburgh, pp 144–168

Millar CI, Delany DL, Westfall RD (2020) From treeline to species line: thermal patterns and growth relationships across the krummholz zone of whitebark pine, Sierra Nevada, California, USA. Arct Antarct Alp Res 52(1):390–407

Mokhov II, Akperov MG (2006) Tropospheric lapse rate and its relation to surface temperature from reanalysis data. Izvestiya Atmos Oceanic Phys 42(4):430–438

Montesano PM, Neigh CS, Macander M, Feng M, Noojipady P (2020) The bioclimatic extent and pattern of the cold edge of the boreal forest: the circumpolar taiga-tundra ecotone. Environ Res Lett 15(10):105019

Nanda SA, Reshi ZA, Ul-haq M, Lone A, Mir SA (2018) Taxonomic and functional plant diversity patterns along an elevational gradient through treeline ecotone in Kashmir. Trop Ecol 59(2): 211–224

Negi PS (2012) Climate change, alpine treeline dynamics and associated terminology: focus on northwestern Indian Himalaya. Trop Ecol 53(3):371

Öberg L, Kullman L (2012) Contrasting short-term performance of mountain birch (*Betula pubescens* ssp. czerepanovii) treeline along a latitudinal continentality-maritimity gradient in the southern Swedish Scandes. Fennia 190(1):19–40

Ohmura A (2012) Enhanced temperature variability in high-altitude climate change. Theor Appl Climatol 110(4):499–508

Oommen MA, Shanker K (2005) Elevational species richness patterns emerge from multiple local mechanisms in Himalayan woody plants. Ecology 86(11):3039–3047

Pandey A, Badola HK, Rai S, Singh SP (2018) Timberline structure and woody taxa regeneration towards treeline along latitudinal gradients in Khangchendzonga National Park, Eastern Himalaya. PLoS One 13(11):e0207762

Panigrahy S, Anitha D, Kimothi MM, Singh SP (2010) Timberline change detection using topographic map and satellite imagery. Trop Ecol 51(1):87–91

Panthi MP, Chaudhary RP, Vetaas OR (2007) Plant species richness and composition in a trans-Himalayan inner valley of Manang district, central Nepal. Himalayan J Sci 4(6):57–64

Paudel S, Vetaas OR (2014) Effects of topography and land use on woody plant species composition and beta diversity in an arid Trans-Himalayan landscape, Nepal. J Mt Sci 11(5):1112–1122

Pepin N (2001) Lapse rate changes in northern England. Theor Appl Climatol 68(1):1–6

Pepin NC, Lundquist JD (2008) Temperature trends at high elevations: patterns across the globe. Geophys Res Lett 35(14):14701

Pinos J, Studholme A, Carabajo A, Gracia C (2017) Leaf Litterfall and decomposition of *Polylepis reticulata* in the Treeline of the Ecuadorian Andes. Mt Res Dev 37(1):87–96

Puri GS, Meher-Homji VM, Gupta RK, Puri S (1983) Forest ecology. Volume I: Phytogeography and forest conservation, 2nd edn. Oxford & IBH Publishing, New Delhi, IN

Puri GS, Gupta RK, Meher-Homji VM, Puri S (1989) Forest ecology. Volume II: Plant form, diversity, communities and succession, 2nd edn. Oxford & IBH Publishing, New Delhi, IN

Qin J, Yang K, Liang S, Guo X (2009a) The altitudinal dependence of recent rapid warming over the Tibetan Plateau. Clim Chang 97(1):321–327

Qin J, Yang K, Liang S, Guo X, Xiaofeng QJ, Yang K, Guo X, Yang K, Guo X, Liang S (2009b) The altitudinal dependence of recent rapid warming over the Tibetan Plateau. Clim Chang 97(1):321–327

Rai ID, Bharti RR, Adhikari BS, Rawat GS (2013) Structure and functioning of timberline vegetation in the Western Himalaya: a case study. In: Ning W, Rawat GS, Joshi S, Ismail M, Sharma E (eds) High-altitude rangelands and their interfaces in the Hindu Kush Himalayas. International Centre for Integrated Mountain Development (ICIMOD)

Ramanathan V, Carmichael G (2008) Global and regional climate changes due to black carbon. Nat Geosci 1(4):221–227

Rana SK, Rawal RS, Dangwal B, Bhatt ID, Price TD (2021) 200 years of research on Himalayan biodiversity: trends, gaps, and policy implications. Front Ecol Evol 516. https://doi.org/10.3389/fevo.2020.603422

Rangwala I, Miller JR, Xu M (2009) Warming in the Tibetan Plateau: possible influences of the changes in surface water vapor. Geophys Res Lett 36(6):6703

Rangwala I, Sinsky E, Miller JR (2013) Amplified warming projections for high altitude regions of the northern hemisphere mid-latitudes from CMIP5 models. Environ Res Lett 8(2):024040

Rangwala I, Palazzi E, Miller JR (2019) Projected climate change in the Himalayas during the twenty-first century. In: Himalayan Weather and Climate and their Impact on the Environment 2020. Springer, Cham, pp 51–71

Rawal RS, Dhar U (1997) Sensitivity of timberline flora in Kumaun Himalaya, India: conservation implications. Arct Alp Res 29(1):112–121

Rawal RS, Pangtey YP (1994) Distribution and structural-functional attributes of trees in the high altitude zone of Central Himalaya, India. Vegetatio 112(1):29–34

Rawal RS, Bankoti NS, Samant SS, Pangtey YPS (1991) Phenology of tree layer species from the timberline around Kumaun in central Himalaya, India. Vegetatio 93:109–118

Rawal RS, Rawal R, Rawat B, Negi VS, Pathak R (2018) Plant species diversity and rarity patterns along altitude range covering treeline ecotone in Uttarakhand: conservation implications. Trop Ecol 59(2):225–239

Rawat DS (2012) Monitoring ecosystem boundaries in the Himalaya through an 'eye in the sky'. Curr Sci 102(10):1352–1354

Rawat B, Gaira KS, Gairola S, Tewari LM, Rawal RS (2021) Spatial prediction of plant species richness and density in high-altitude forests of Indian west Himalaya. Trees For People 6: 100132

Ren S, Yi S, Peichl M, Wang X (2017) Diverse responses of vegetation phenology to climate change in different grasslands in Inner Mongolia during 2000–2016. Remote Sens 10(1):17

Reshi ZA, Rawal RS, Kumar D (2021) Plant diversity, community structure, tree diameter changes and natural recruitment pattern along the three principal sites in the IHR. Final Technical Report. Indian Himalayan Timberline Research Project (IHTRP). National Mission of Himalayan Studies, Ministry of Environment, Forests and Climate Change, Govt. of India, unpublished

Rolland C (2003) Spatial and seasonal variations of air temperature lapse rates in Alpine regions. J Clim 16(7):1032–1046

Sah P, Sharma S (2018) Topographical characterisation of high altitude timberline in the Indian Central Himalayan region. Trop Ecol 59(2):187–196

Schickhoff U (1993) Das Kaghan-Tal im Westhimalaya (Pakistan): Studien zur landschaftsökologischen Differenzierung und zum Landschaftswandel mit vegetationskundlichem Ansatz. Bonner Geographische Abhandlungen 87

Schickhoff U (1996) Contributions to the synecology and syntaxonomy of West Himalayan coniferous forest communities. Phytocoenologia 26:537–581

Schickhoff U (2005) The upper timberline in the Himalayas, Hindu Kush and Karakorum: a review of geographical and ecological aspects. In: Mountain ecosystems. Springer, Cham, pp 275–354

Schickhoff U (2011) Dynamics of mountain ecosystems. In: Millington A, Blumler M, Schickhoff U (eds) Handbook of biogeography. SAGE, London, pp 313–337

Schickhoff U, Bobrowski M, Böhner J, Bürzle B, Chaudhary RP, Gerlitz L, Heyken H, Lange J, Müller M, Scholten T, Schwab N, Wedegärtner R (2015) Do Himalayan treelines respond to recent climate change? An evaluation of sensitivity indicators. Earth Syst Dyn 6(1):245–265

Schickhoff U, Bobrowski M, Böhner J, Bürzle B, Chaudhary RP, Gerlitz L, Lange J, Müller M, Scholten T, Schwab N (2016a) Climate change and treeline dynamics in the Himalaya. In: Climate change, glacier response, and vegetation dynamics in the Himalaya. Springer, Cham, pp 271–306

Schickhoff U, Singh RB, Mal S (2016b) Climate change and dynamics of glaciers and vegetation in the Himalaya: an overview. In: Climate change, glacier response, and vegetation dynamics in the Himalaya. Springer, Cham, pp 1–26

Schwab N, Schickhoff U, Müller M, Gerlitz L, Bürzle B, Böhner J, Chaudhary RP, Scholten T (2016) Treeline responsiveness to climate warming: insights from a krummholz treeline in Rolwaling Himal, Nepal. In: Climate change, glacier response, and vegetation dynamics in the Himalaya. Springer, Cham, pp 307–345

Schwab N, Kaczka RJ, Janecka K, Böhner J, Chaudhary RP, Scholten T, Schickhoff U (2018) Climate change-induced shift of tree growth sensitivity at a central Himalayan treeline ecotone. Forests 9(5):267

Schweinfurth U (1957) Die horizontale und vertikale Verbreitung der Vegetation im Himalaya. Bonner Geographische Abhandlungen 20

Sharma S (2021) Mapping treeline using remote sensing method across the Indian Himalayan Arc. Final Technical Report. Indian Himalayan Timberline Research Project (IHTRP). National Mission of Himalayan Studies, Ministry of Environment, Forests and Climate Change, Govt. of India, unpublished

Shi P, Körner C, Hoch G (2006) End of season carbon supply status of woody species near the treeline in western China. Basic Appl Ecol 7(4):370–377

Shrestha KB, Vetaas OR (2009) The forest ecotone effect on species richness in an arid Trans-Himalayan landscape of Nepal. Folia Geobot 44(3):247–262

Shrestha BB, Ghimire B, Lekhak HD, Jha PK (2007) Regeneration of treeline birch (*Betula utilis* D. Don) forest in a trans-Himalayan dry valley in central Nepal. Mt Res Dev 27(3):259–267

Shrestha KB, Hofgaard A, Vandvik V (2015) Recent treeline dynamics are similar between dry and mesic areas of Nepal, central Himalaya. J Plant Ecol 8(4):347–358

Sigdel SR, Wang Y, Camarero JJ, Zhu H, Liang E, Peñuelas J (2018) Moisture-mediated responsiveness of treeline shifts to global warming in the Himalayas. Glob Chang Biol 24(11):5549–5559

Singh SP (2018) Research on Indian Himalayan treeline ecotone: an overview. Trop Ecol 59(2): 163–176

Singh SP, Zobel DB, Garkoti SC, Tewari A, Negi CM (2006) Patterns in water relations of central Himalayan trees. Trop Ecol 47(2):159–182

Singh CP, Panigrahy S, Thapliyal A, Kimothi MM, Soni P, Parihar JS (2012) Monitoring the alpine treeline shift in parts of the Indian Himalayas using remote sensing. Curr Sci 102(4):559–562

Singh U, PhularaM DB, Ranhotra PS, Shekhar M, Bhattacharyya A, Dhyani R, Joshi R, Pal AK (2018) Static tree line of Himalayan silver fir since last several decades at Tungnath, western Himalaya. Trop Ecol 59(2):351–363

Singh D, Sharma A, Sharma N (2019a) Composition, richness and floristic diversity along an elevational gradient in a semi-disturbed treeline ecotone, Bhaderwah, Jammu and Kashmir. J Appl Nat Sci 11(1):23–34

Singh SP, Sharma S, Dhyani PP (2019b) Himalayan arc and treeline: distribution, climate change responses and ecosystem properties. Biodivers Conserv 28(8):1997–2016

Singh SP, Bhattacharyya A, Mittal A, Pandey A, Tewari A, Latwal A, David B, Adhikari BS, Kumar D, Negi GC, Mir IA, Tamta KK, Sambhav K, Shekhar M, Phulara M, Manzoor M, Singh N, Tewari P, Ranhotra PS, Singh P, Dhaila P, Sah P, Kumar R, Joshi R, Rawal RS, Rawal R, Singh RD, Shah S, Sharma S, Nanda SA, Gumber S, Singh U, Reshi Z (2021) Indian Himalayan timberline ecotone in response to climate change - initial findings. Curr Sci 120(5): 859–871

Stainton JDA (1972) Forests of Nepal. London: Murray xvi. Coloured illustrations, maps, dot maps Geog, 6

Stevens GC, Fox JF (1991) The causes of treeline. Annu Rev Ecol Syst 22(1):177–191

Sun YL, Shan M, Pei XR, Zhang XK, Yang YL (2020) Assessment of the impacts of climate change and human activities on vegetation cover change in the Haihe River basin, China. Phys Chem Earth A/B/C 115:102834

Tang Z, Fang J (2006) Temperature variation along the northern and southern slopes of Mt. Taibai, China. Agric For Meteorol 139(3–4):200–207

Telwala Y, Brook BW, Manish K, Pandit MK (2013) Climate-induced elevational range shifts and increase in plant species richness in a Himalayan biodiversity epicentre. PLoS One 8(2):e57103

Tewari A (2021) Water relations of the Indian Himalayan treeline species. Final Technical Report. Indian Himalayan Timberline Research Project (IHTRP). National Mission of Himalayan Studies, Ministry of Environment, Forests and Climate Change, Govt. of India, unpublished

Tewari A, Shah S, Singh N, Mittal A (2018) Treeline species in Western Himalaya are not water stressed: a comparison with low elevation species. Trop Ecol 59(2):313–325

Thyer N (1985) Looking at western Nepal's climate. Bull Am Meteorol Soc 66(6):645–650

Tran TJ, Bruening JM, Bunn AG, Salzer MW, Weiss SB (2017) Cluster analysis and topoclimate modeling to examine bristlecone pine tree-ring growth signals in the Great Basin, USA. Environ Res Lett 12(1):014007

Walsh SJ, Butler DR, Malanson GP, Crews-Meyer KA, Messina JP, Xiao N (2003) Mapping, modelling, and visualization of the influences of geomorphic processes on the alpine treeline ecotone, Glacier National Park, MT, USA. Geomorphology 53(1–2):129–145

Wang T, Zhang QB, Ma K (2006) Treeline dynamics in relation to climatic variability in the central Tianshan Mountains, northwestern China. Glob Ecol Biogeogr 15(4):406–415

Wu J, Miao C, Zhang X, Yang T, Duan Q (2017) Detecting the quantitative hydrological response to changes in climate and human activities. Sci Total Environ 586:328–337
Yan L, Liu X (2014) Has climatic warming over the Tibetan Plateau paused or continued in recent years. Earth Ocean Atmos Sci 1(1):13–28
Yan L, Liu X, Yang P, Yin ZY, North GR (2011) Study of the impact of summer monsoon circulation on spatial distribution of aerosols in East Asia based on numerical simulations. J Appl Meteorol Climatol 50(11):2270–2282
Zeng Y, Malanson GP (2006) Endogenous fractal dynamics at alpine treeline ecotones. Geogr Anal 38(3):271–287
Zhang Y, Miao N, Liu S (2021) Has tree density increased at alpine treelines on the eastern Tibetan Plateau? Environ Res Comm 3(12):121005
Zobel DB, Singh SP (1995) Tree water relations along the vegetational gradient in the Himalayas. Curr Sci 68(7):742–745
Zong S, Wu Z, Xu J, Li M, Gao X, He H, Du H, Wang L (2014) Current and potential tree locations in tree line ecotone of Changbai Mountains, Northeast China: the controlling effects of topography. PLoS One 9(8):e106114

Capturing Himalayan Timberline Dimension and Ecological Attributes in Warming Climate Through Team Science

2

S. P. Singh, Ripu Daman Singh, Surabhi Gumber, and Rajesh Joshi

Abstract

We took up a team science approach to carry out a multisite and multidisciplinary study on the 'Indian Himalayan Timberline Ecotones'. Timberlines in the Himalayas reach highest elevation (4900 m) in the Northern Hemisphere, but vary widely in elevation, being 300–600 m lower in moist outer ranges than in dry inner ranges. The zigzag, twisting and curvy timberlines are together more than ~20,000 km long and have a total of 58 tree species. In moist regions, the annual temperature lapse rate is rather low (e.g. 0.53 °C/100 m in Uttarakhand) and varies seasonally (e.g. in December, 0.24 °C/100 m). Along 83–95% of their length, the timberlines are stationary in spite of decades of rapid warming. An analysis of tree ring growth and climate relationship indicates that the intensified pre-monsoon drought could suppress the upward shift of treeline in a warming world. Paradoxically, tree water relations indicate the lack of water stress in treelines. The other changes observed were: upward shift of *Rhododendron campanulatum* and increase in plant species richness because of an early snow-melt. At the end, this chapter discusses the learning from our team science approach.

S. P. Singh (✉)
Central Himalayan Environment Association (CHEA), Dehradun, Uttarakhand, India

R. D. Singh · S. Gumber
Department of Forestry & Environmental Science, Kumaun University, Nainital, Uttarakhand, India

R. Joshi
G. B. Pant National Institute of Himalayan Environment, Sikkim Regional Centre, Pangthang, Gangtok, Sikkim, India

S. P. Singh et al. (eds.), *Ecology of Himalayan Treeline Ecotone*,
https://doi.org/10.1007/978-981-19-4476-5_2

Keywords

Snowmelt · Species richness · Team science · Temperature lapse rate · Timberline elevation and length · Tree water relations

2.1 Introduction

A timberline in mountains represents a major 'physiognomic discontinuum' along an elevation gradient, in which, because of heat deficiency, forests give way to communities of low stature, consisting of herbs and scrubs. Regardless of growing season length, the treeline is generally formed at an elevation where growing season mean air temperature is 6.7 °C, generally between 5 °C and 7 °C (Körner and Paulsen 2004). Here, reference may be made to two terms that we have used interchangeably in this article: 'timberline', which is defined as the upper limit of forests with 30% and more crown cover, and 'treeline', which could be considered as an imaginary line connecting the uppermost tree patches (trees are generally considered >2 m tall woody plants with a single stem).

Though the Himalayas are known to have the highest timberline in the Northern Hemisphere (at 4900 m, formed by *Juniperus tibetica*; Miehe et al. 2015), these biotic boundaries and ecotones have remained less investigated, partly because of their remoteness. With regard to the Indian Himalayan region, the terms timberline and treeline have remained missing until recently, when studies on climate change impact on vegetation, particularly on treeline dynamics, began to draw global attention. To develop a meaningful understanding of the Himalayan treelines, we need a team research approach that considers various relevant research components and several representative Himalayan sites. Worldwide, science and social science studies have become increasingly collaborative and team-oriented in approach; that is why about 90% of science and engineering publications have two or more authors (generally six to ten authors, but also in hundreds), and solo publications are becoming rare (Stokols et al. 2008). Multifaceted and complex problems, and complex systems, call for a strong team science effort. Moreover, complex problems involving large areas and varied factors require team science efforts and high technologies. However, in India progress on team-based research has been sluggish. Indian scientists collaborate less frequently nationally as well as internationally. For example, the number of papers based on collaboration between Indian and US scientists is not only very small, but has also declined in the recent years. Neither universities and institutes, nor granting agencies have taken interest in strengthening collaboration and networking, only exception being the interest taken by Department of Science and Technology (DST), Government of India, in recent years (Singh et al. 2018). Overall, researchers in India have remained stuck to their spheres of disciplines and homesteads.

In recent years, the Himalayas have attracted global interest because of the rapid warming (Wester et al. 2019) and resultant shrinkage of the glaciers. Having more snow than any other region outside the two poles, the Himalayas are important source of water for about 1.7 billion people (Bajracharya et al. 2015) living in ten river basins originating from the region (Immerzeel et al. 2010; Bolch et al. 2012). The snowmelt discharges are also connected with several biotic zones namely timberlines, treelines, alpine meadows and subnival and nival species groups occurring above treeline and below permanent snowline. These biotic systems are experiencing marked alterations in their expanse and species composition with long-term consequences for treeline ecology, biodiversity and carbon cycles (Anderson et al. 2020). They are integral parts of larger hydrological systems that snow and ice melts and monsoon rain generate.

Historically, treeline research in different parts of mountains has largely been influenced by plant physiology, and a search for a universal value of air and soil temperatures that limit tree growth has continued to be of general interest for a long time (Körner 2012; Müller et al. 2015). The other question that has been in focus is related to plant carbon reserve and carbon sink as a limiting factor for tree growth at treeline. Recent observations indicate that the postulated 6.4 °C ± 0.7 °C for growing season mean soil temperature at treeline may not hold good universally, as several soil factors, such as decreasing soil nutrient availability with elevation, can affect tree growth in mountain treelines (Müller et al. 2015). Several other factors that are unrelated to temperature but influence treeline formations are: soil moisture, soil erosion, snow avalanches, the height of nearest mountain top, interference and facilitation by other plant species and anthropogenic pressure (Singh 2018). Patterns in treeline elevation and species composition in relation to wider environmental, phytogeographical and historical factors received less importance until recently (Schickhoff 2005; Holtmeier 2009). The global climate change generated interest in treeline dynamics, leading to many studies on change in treeline elevation due to climate warming (Schickhoff et al. 2015; Bhuju et al. 2016; Tewari et al. 2018; Tiwari and Jha 2018). Generally, these studies were based on two to three points along a timberline.

This study is based on a team science approach involving investigations on geographical spread and dimensions of timberline/treelines at Indian Himalayan scale using remote sensing technology. It sheds light on patterns of plant species richness, temperature lapse rate (TLR) through treeline ecotones, tree water relations, tree phenology, tree ring chronology and plant growth in relation to snowmelt under the influence of climate change. The main objectives of this chapter are: (1) to construct a cohesive ecological and phytogeographical picture of Indian Himalayan timberline/treeline ecotone, using relevant findings of our team research, and (2) to share experiences and learning from the application of team research approach in a difficult terrain of a considerable remoteness.

Here, we have emphasized: (1) that the timberline ecotone in the Himalayas is a huge and diverse system with a large structural variability, (2) that the treeline dynamics under the influence of climate change is rather complex because of several factors, particularly related to elevation-dependent warming (EDW) and anthropogenic activities and (3) that there are several new questions that emerge from the team research approach. While shedding light on these issues, we have drawn attention to the fact that the treeline is not only a huge ecological ecosystem, but also it is accompanied by equally important and extensive nival and other biotic components, which are under flux because of climatic changes in a highly sensitive geological settings of the Himalayas.

2.2 Material and Methods

Extending from Afghanistan in the northwest (ca. 26° N and 70° E) to Yunnan in the southeast (ca. 26° N and 100° E), the Himalayas are vast, massive and highly heterogeneous. The region has more snow and ice than any parts of the planet, outside the two poles. That is why it is referred to as the third pole and water tower of Asia. While annual precipitation above 3000 mm is common in the outer ranges receiving direct thrust of monsoon air masses, several areas in the north of the main Himalayan ranges represent some of the largest rain shadows with annual precipitation as low as 300 mm or less. In general, moisture decreases from east to west and from south to north, i.e. from low to high elevations (Singh et al. 2017).

Generally, the mean annual temperature declines from 22 °C to 24 °C in foothills to 18–20 °C at 1000 m, 10–15 °C at 2000 m, 7–10 °C at 3000 m and less than 7 °C in alpine zone. However, temperature and elevation relationship is not straightforward. For example, the elevated heating surface of the large Tibetan Plateau raises temperature by its mass elevation effect (Zhang and Yao 2016). This is one of reasons for the occurrence of the highest treeline in Tibet. The common treeline genera are: *Abies*, *Betula*, *Juniperus* and *Rhododendron* (Singh 2018). The timberline ecotone is followed by alpine meadows and subnival zone of small-statured plants (Fig. 2.1), which occupy (>4150 m elevation) 5–15 times more area than permanent snow and ice (Anderson et al. 2020). So the biotic layers, though not given as much importance as snow and ice when climate change impact is discussed, are important in the Himalayas in the context of carbon balance in the changing climate.

Sheep and goats have been grazing in alpine areas of Kashmir, Himachal Pradesh and Uttarakhand, since time immemorial, but are rare in Sikkim and Arunachal Pradesh (Singh and Thadani 2015). However, even in western Himalayan states and Nepal, now livestock density is on decline, giving an opportunity for trees to move upslope (Suwal et al. 2016). In all these regions, glaciers are generally shrinking (Singh et al. 2011; Yao et al. 2012; Chudley et al. 2017).

This study sites were located in Kashmir Valley, Uttarakhand and Sikkim, which broadly cover the entire range of variation along the Indian Himalayan Arc. While Kashmir is relatively dry (600 mm annual precipitation) and non-monsoonal, with monsoon months (June–September) accounting for only 28.9% annual precipitation, Sikkim is wet and monsoonal, and Uttarakhand is moderately moist, with average annual precipitation of about 1500 mm (Fig. 2.2).

To conduct the study, first we formed a multidisciplinary team of investigators, with expertise in various relevant areas: vegetation analysis, application of remote sensing techniques to treeline distribution, tree ring width chronology, tree water relations, plant phenology, climatology, ecology and community livelihoods. The

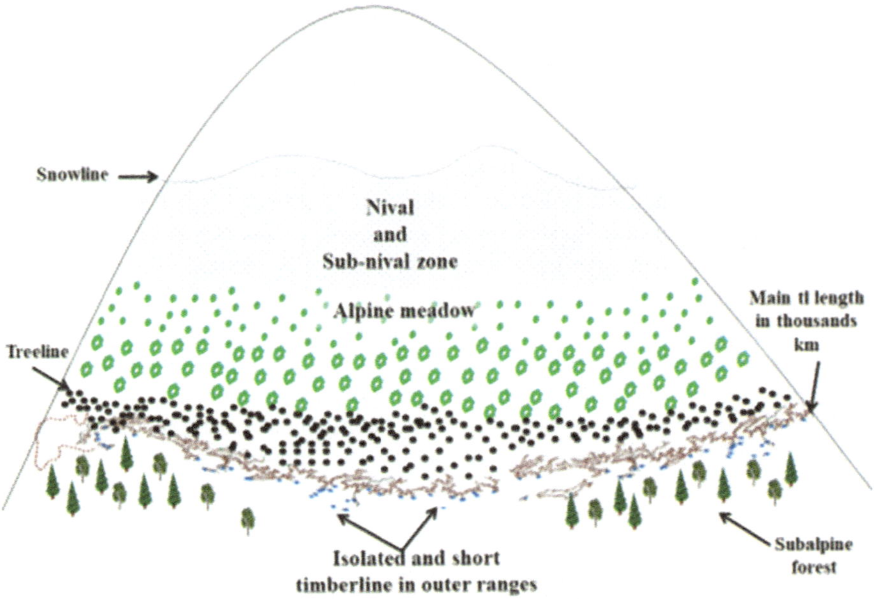

Fig. 2.1 A representation of section of the series of biotic zones below permanent snowline: nival and subnival belt, alpine meadows, treeline, timberline (tl) and forests

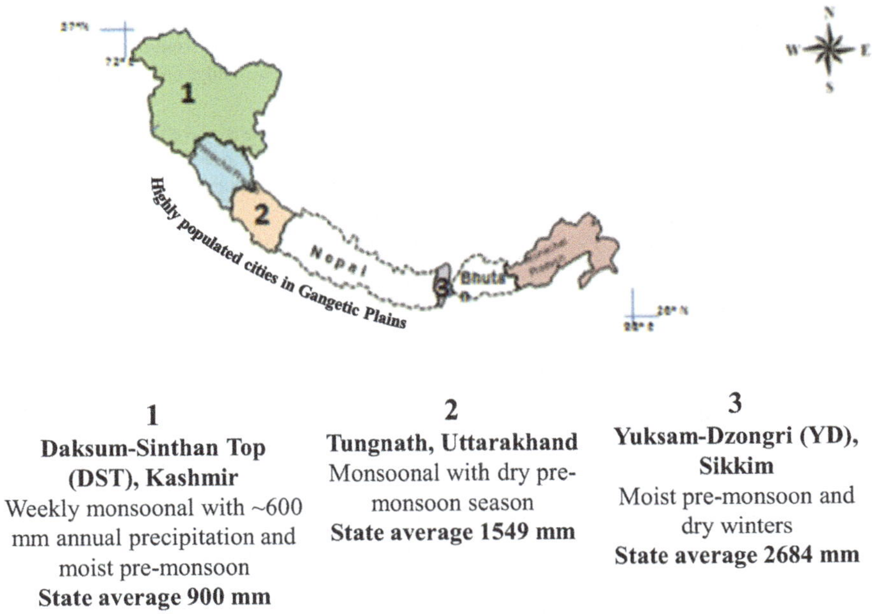

Fig. 2.2 The three sites for detailed 'Indian Himalayan Timberline Ecotones' study, which cover much of the range of variation across the Himalayan Arc, and represent three precipitation regimes (Courtesy: Subrat Sharma, GBPNIHESD, Almora)

team members represented following organizations: Kashmir University, Srinagar; Kumaun University, Nainital; Wildlife Institute of India, Dehradun; GB Pant National Institute of Himalayan Environment and Sustainable Development, Almora; Birbal Sahni Institute of Palaeobotany, Lucknow; and Central Himalayan Environment Association (CHEA), Nainital. Several brain storming workshops along with field visits were conducted to work out methods, time schedule and study sites for detailed investigation. Field visits in groups helped finalizing sampling design, number of replicates, timing and sample points. Several exercises were carried out confirming interdependence in studies. For example, to find out the impact of pre-monsoon drought on treeline ecosystems we measured and analysed data on soil and tree water status, tree ring width changes, phenological phases and species regeneration. Subsequently, a manual on timberline study methods was developed (Singh and Rawal 2017). The coordinator continuously monitored and influenced the progress on each research component during the entire course of the study. The methods used in various components of the study are listed in Table 2.1.

Table 2.1 A list of study components along with purpose with regard to Indian Himalayan Timberline Research Project (IHTRP)

Study components	Purpose and remarks
Mapping the timberline in the Indian Himalayan region (IHR) using satellite imageries, and comparing with past records	To compare changes in elevation of timberline at the regional level, and over time, and provide region-level dimensions of timberline
Temperature lapse rate (TLR) based on observed temperature data along an elevation transect	To develop the first observed data-based TLR; examining how it differs from the West to East along the Arc, seasonally, and because of the climate change
Phenology of trees, namely *Abies spectabilis*, *Quercus semecarpifolia* and *Rhododendron campanulatum*	To find out the timing of various phenophases such as leaf expansion, leaf nutrient resorption and upslope advancement of krummholz species
Tree and soil water relations	To find out the severity of water stress in treeline, and species response to various seasons in comparison to lower elevation trees
Tree ring width chronology; changes in soil water status in time due to climate change	To find out longer climate and tree growth relationships and to predict future changes
Patterns of community and ecosystem changes along altitudinal gradient centred around treeline	To establish the patterns in species richness in relation to elevation in different taxonomical groups and growth forms; how non-tree species populations are distributed in relation to the physiognomic discontinuum of treeline?
Species richness and plant density in relation to snow	To find out how early snowmelt is likely to affect species growth and species richness in meadows

Note: The study sites were Kashmir, Uttarakhand and Sikkim

2.3 Results and Discussion

2.3.1 Length of Timberlines

The Himalayan timberline is long, high and yet depressed. It moves back and forth and sideways, varying in elevation and across the breadth of the Himalayas from south to north. In the Indian Himalayan state of Uttarakhand, the timberline length, as an example, is estimated to be 2839 km within the horizontal distance of 455.1 km (Table 2.2), giving an average length of 6.24 km/km horizontal distance (Latwal et al. 2018). Similar patterns are being observed in other Indian Himalayan regions (Indian Himalayan Timberline Research Project [IHTRP] Annual Report 2019–2020, Dr. Subrat Sharma and his associates, Govind Ballabh Pant National Institute of Himalayan Environment and Sustainable Development [GBPNIHESD]). At this scale the total timberline length of the Himalayan Arc from Pakistan to Yunnan (72° E to 97°30′ E) is likely to be approximately 24,000 km. In addition to this long principal timberline, there also occur timberline strips of a few kilometres length around mountain summits generally in the outer ranges (Table 2.2). For example, in Uttarakhand, the combined length of timberline strips of this type is <50 km, compared to 2839 km length of the principal timberline (Latwal et al. 2018).

2.3.2 Timberline Elevation and Shift

The timberline elevations are lower in outer ranges than in inner areas around the main Himalayan ranges, their mean values being 3687.88 m and 4272 m, respectively (Fig. 2.3). In the main Himalayan ranges, timberline elevations are higher because of drier conditions and greater mass elevation effect. The mountains with large masses are able to separate their atmosphere from free atmosphere more than mountains with small masses. According to Zhang and Yao (2016), timberlines in the Himalayas could not have been higher than 3500–3700 m elevation without the

Table 2.2 The extent of timberline length where no elevation change occurred from 1976 to 2015 in Uttarakhand

	Uttarakhand
Total timberline (continuous long timberline) length (km) in 1976	2839.0
Percentage of above timberline length remaining stationary	89%[a]
The upper extent of upward shift (m) in above timberline	189 m
Island type/short strip timberlines (no.)	32
Remaining stationary (no.)	22 (70.6%)
Upward shift	4 partially and 6 fully

Note: Data from IHTRP- Annual Report 2019–2020, Dr. Subrat Sharma and his associates from GBPNIHESD

[a]Values similar to this have been recorded for several other regions of the Himalayas (Dr. Subrat Sharma and his associates from GBPNIHESD)

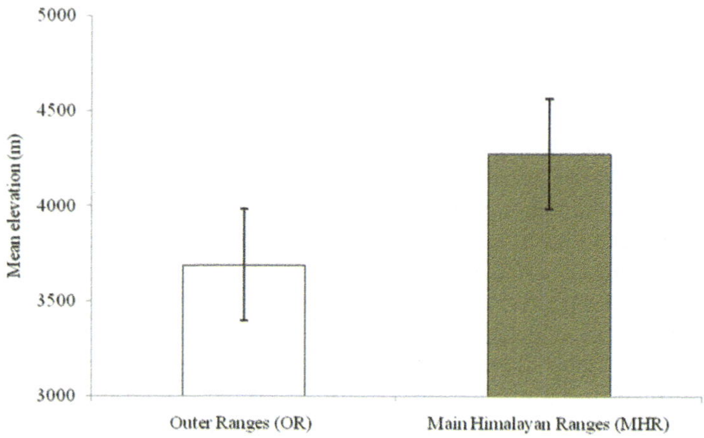

Fig. 2.3 Mean timberline elevation in Hindu Kush Himalaya in Outer Ranges ($n = 73$), and Main Himalayan Ranges ($n = 80$) (in each case, n is the number of sample site)

mass elevation effect. The mass elevation effect is the highest (above 11 °C) with middle part of the Tibetan Plateau and declines towards peripheral areas (Han et al. 2018).

The Himalayan timberlines have remained stationary in most parts (89%) for last four decades (Latwal et al. 2018; Dr. Subrat Sharma and and his associates) (Table 2.2) in spite of decades of global warming. Several factors unrelated to direct temperature impact are known to check the upward shift of treelines, of which soil nutrients, soil moisture (Müller et al. 2015), intensified pre-monsoon drought, nature of tree species and biotic interferences (Holtmeier 2009; Gaire et al. 2014; Singh et al. 2018) are quite common. Increase in growth and density of existing plants are shown to restrict the upward shift of treeline species.

The spread of krummholz may facilitate the establishment of *Betula utilis*, and thus promote the upward shift of treeline (Prof. Zafar Reshi and his associates). However, when densified, a species may also restrict tree establishment and advancement (Liang et al. 2016). Such a tree species-to-species interaction helps to explain why treelines have generally not advanced in response to climatic warming in the Himalayas. So the decline in grazing pressure may not necessarily lead to the upslope progress, as the let-up in grazing pressure may help some other species to spread and thicken and form a barrier to treeline advancement. Furthermore, it could reduce the cover of meadow particularly around mountain summits, and may deplete meadow species richness.

2.3.3 Arrested Timberlines and Associated Plant Communities

Treelines in the Himalayas are generally depressed, largely because of historical anthropogenic activities, such as pastoralism, cutting of trees and burning (Schickhoff 2005). In other words, the existing treeline temperatures are warmer

Table 2.3 Growing season mean temperature of three treeline sites (located in Kashmir Valley, Uttarakhand and Sikkim; they are the main study sites of National Mission on Himalayan Studies [NMHS] project; Singh 2018), and estimates of treeline depression (in m) compared to climatic treelines (elevation at which growing season mean temperature is 6.7 °C; Körner 2012)

Study site: elevation (m) and latitude/longitude	Growing season (May–September) mean temperature (GST) (°C)	Difference between TT and GST of the site (°C)	Treeline depression (m)
Kashmir: 3900 m elevation; 33°32′–33°41′ N/75°25′–75°31′ E	8.6	1.9	292.3
Uttarakhand: 3300 m elevation; 30°47′–30°51′ N/79°15′–79°22′ E	10.3	3.6	679.2
Sikkim: 4000 m elevation; 27°30′ N/88°30′ E	7.5	0.8	158.8

Note: The mean temperature lapse rate (TLR, per 100 m elevation) is 0.53 °C for Sikkim, 0.51 °C for Uttarakhand and 0.65 °C for Kashmir (Joshi et al. 2018), referred to as climatic treeline temperatures. Difference between climatic treeline elevation and existing treeline elevation was calculated as: existing growing season temperature = climatic treeline temperature (6.7 °C)/ temperature lapse rate (TLR) × 100 m. This difference is referred to as treeline depression

than that ascribed to climatic treelines. Considering that globally mean growing season temperature at climatic treeline is 6.7 ± 0.85 °C (Körner 2012), we can calculate how much depressed treelines are in comparison to climatic treeline by using mean annual temperature lapse rate. Accordingly, the three treeline sites being studied in detail (located in Kashmir, Uttarakhand and Sikkim) by us (Singh 2018) are lower than climatic treelines by 292 m, 680 m and 159 m, respectively (Table 2.3). Treeline sites of Uttarakhand and Kashmir are affected by heavy summer time grazing, while in Sikkim pastoralism at present is low. Possibly, because of this, Sikkim treeline is close to that of climatic treeline. Currently, grazing pressure is on decline in the Himalayas as a whole because of declining livestock, and this is considered one of the reasons of upslope movement of treeline apart from climatic warming (Gaire et al. 2014). However, increase in tourism can affect treelines in opposite direction.

The general validity of certain soil and air threshold temperature for all treeline location without a clearly broader error (6.4 ± 0.7 °C) is questionable (Müller et al. 2015). In the Himalayas the timberline/treeline turns and twists to track a threshold temperature, but the range of temperature along their length is very wide, particularly on a higher side (up to 12 °C mean growing season temperature), because of extraordinarily wide elevation range (~2000 m). The lack of upward timberline shifts in spite of the decades of warming shows that the timberlines now are warmer than in the past. It may lead to densification of trees and enhanced tree growth as long as snowmelt water is available, but subsequently as snowmelt water depletes, the sites would become too dry for tree survival. Such 'arrested treelines' need to be monitored and the tipping point of treeline degradation identified. In a warmer timberline, litter decomposition and nutrient cycling are likely to be higher and vulnerable to change.

Though considerable amounts of data have been generated on the effect of climate change and snow and ice in the Himalayas (Gautam et al. 2009; Gardner et al. 2013; Bajracharya et al. 2015) and its consequences on snowmelt hydrology and water availability (Immerzeel et al. 2012), how they are being impacted by changes in various vegetation zones in the Himalayas is hardly investigated (Gaire et al. 2014; Anderson et al. 2020). Generally, when the Himalayan timberlines are not moving up they are arrested, but marked changes could occur in timberline ecotones and species and communities that make them (Singh et al. 2018), alpine meadows (Adhikari et al. 2018) and subnival zone, which, according to an estimate, cover 5–15 times area of permanent glaciers and snow cover (Anderson et al. 2020). The upward advancement of *Rhododendron campanulatum* into meadows above treeline has been found at the rate of 3.4 m/year during last two to three decades at Tungnath in Uttarakhand (Singh et al. 2018). The upslope advancement of *R. campanulatum* has also been reported from Nepal (Prabinarana et al. 2017). The advancement of woody plants into vegetation of small-statured plants may lead to changes in the carbon storage in biomass and soil pools and lead to a rapid vegetation dynamic.

Changes in the plant growth form can result in marked changes in snow and ice cover and storage, and hydrological parameters (Anderson et al. 2020). The development of plant cover on bare soil surfaces in the subnival zone results in reduced albedo and more warming, outstripping the cooling effect caused by increase in evapo-transpiration (Shen et al. 2015). The expansion of subnival zone due to the decrease in temperature-limited areas in the warming climate (Gonzalez et al. 2010) needs to be studied for its impact on snow- and ice-driven high-altitude hydrology. It may be pointed out that the area above 4150 m, which was used to estimate changes in subnival plant cover in the Himalayas by Anderson et al. (2020), is being affected by the spread of krummholz and shrubs in several areas, particularly below 5000 m. The occurrence of the treeline above 4000 m is quite common in the eastern part of the Himalayan Arc. There, under-canopy woody layer (several rhododendrons, and junipers) is getting denser and spreading, thus restricting canopy tree species like fir to be sparser (<100 trees/ha). These structural factors are known to influence forest hydrology (Mestre et al. 2017). In brief, the climate-change-driven impacts on the high Himalayas would not limit to cryosphere, and to understand species and growth form dynamics and the issues of water management, studies are needed to consider large-scale changes occurring in vegetation layers right from subnival communities to alpine meadows adjacent to treelines and timberline, and their dynamic hydrological connections.

2.3.4 Relationship Between Global Warming, Air Pollution and Temperature Lapse Rate

The temperature lapse rate (TLR) estimated from observed data at Tungnath site (Uttarakhand) was relatively low (mean annual rate 0.53 °C/100 m; Joshi et al. 2018). This low TLR could be partly because of the elevation-dependent warming

(EDW), which emphasizes that the higher and cooler parts of an elevation transect is warming at a higher rate than warmer lower areas (Palazzi et al. 2019). Warming is observed to be high in elevations where snow mass loss and albedo reduction are high, suggesting the snow-albedo feedback as the main cause of elevation-dependent warming (EDW) (Minder et al. 2018; Palazzi et al. 2019). At the Uttarakhand site, the TLR was lowest in December (0.24 °C/100 m) when air pollution in lower elevation areas has been high for last several years, resulting in solar radiation dimming and cooler temperatures. This in combination with EDW seems to change the seasonal pattern, as the decline in temperature from autumn months to December (the first winter month) is less in higher elevation areas than in lower elevation areas. The solar dimming effect of aerosol particles is particularly high in low-elevation areas near the big cities of Gangetic plains (Wester et al. 2019). The pollutants can spread to snow-clad mountain areas resulting in an albedo reduction. In Tungnath, because of delay in snowfall, the dark cover of senescent and decayed plants was made visible, resulting in a reduced albedo (Joshi et al. 2018) than following winter months when snowfall occurred. There is a need to observe these temperature change patterns at more representative sites and over a longer period to understand the TLR effect in relation to elevation-dependent warming. It may be pointed out that plant cover changes considerably affect regional carbon balance, species composition and community development. The EDW-linked snow depletion partly explains the greening of subnival zones in the Himalayas, as observed recently (Anderson et al. 2020). The reduced TLR and change in its timing are expected to cause not only more climate change, but also profound changes in vegetation cover and species composition.

2.3.5 Change in Growth Periodicity

In monsoon climate, grasses and other herbs begin growth with the onset of monsoon rain, generally in June (Zobel and Singh 1997). However, in treeline communities and alpine meadows, most herb species start growing well before the arrival of monsoon, as now melt water supply begins in April and May. Because of the synchronization of plant growth with snowmelt, these herbs have a long growth period, generally 150–200 days. The advancement in snowmelt is expected to further prolong growth period of such species. However, we do not know the fate of such early growing species when snow depletion renders meadows dry. The site might be taken over by late growth initiating species, which wait for monsoon to arrive to start growth (23 species at Tungnath) (Adhikari et al. 2018; Dr. B. S. Adhikari and his associates). The shift from the species that depend on snowmelt for growth initiation to those that depended on monsoon rain is expected to change the community structure and species diversity in meadows. Fast-growing species of lower elevations may arrive and outcompete some of the species.

2.3.6 Treeline Water Relations

Global warming, if not accompanied by increased precipitation, can adversely affect plant growth, plant cover and growth form (Hatfield and Prueger 2015). In monsoonal regions of the Himalayas, pre-monsoon (March, April and May) is the driest and warmest time of the year (Singh et al. 2017), yet trees and shrubs put forth new shoots and expand leaves during that period so that canopy is fully developed to take the advantage of monsoon (June–September), for fixing carbon and increasing carbon stock (Singh and Singh 1992; Zobel and Singh 1997). Studies on tree ring growth generally indicate that in certain cases growth has been constantly high during last two to three decades of climate warming. However, growth suppression and the lack of upward treeline shifts due to water stress have also been recorded (Gaire et al. 2014).

Tree water relation studies at Tungnath, a site located in outer ranges receiving over 2000 mm annual rainfall, indicate that water stress at treeline is not severe (pre-dawn tree water potential mostly being above −1 MPa; Tewari et al. 2018). But this may not apply to dry inner ranges where pre-dawn water potential is frequently below −1 MPa. Tree water relations at the Tungnath treelines differ in some important aspects from those of low-elevation forest trees of the region (Fig. 2.4). First, at treelines, species undergo a high daily change (measured as difference between pre-dawn and mid-day water potential) in tree water potential during monsoon months (1.1 MPa, compared to 0.16 MPa in low-elevation forests). This conspicuous difference is difficult to explain, and needs to be further investigated. The high daily change in tree water potential in monsoon at treeline corresponds with high leaf conductance values. Second, species exhibit very high leaf water conductance values (>1000 m mol m^2/s) in treelines in dry inner areas. What enables a species to realize much higher leaf conductance at a drier site than a moist site? What enables them to keep stomata open at a dry high-elevation site? What is the role of low humidity combined with low atmospheric pressure in it? We need to do more research to understand these patterns. To predict forest hydrological changes in a warming climate, detailed studies on tree ring growth and tree water relations encompassing various Himalayan habitats might be required. Our present study on tree water relations and tree ring width chronology has enabled us to raise more questions. More studies are required to understand basic processes and controlling environmental factors related to plant water status in treeline areas. Increase in dry season stomatal conductance with elevation (1700–2700 m) has been reported for tropical forests in dry season (Mujawamariya et al. 2018). In China, Wang et al. (2016) found that among trees, shrubs and leaves, trees have the highest stomatal density but the lowest stomata length, and the latter increased with elevation. Actual measured photosynthetic rates are usually as high as or even higher than those at lower elevations (Körner and Diemer 1987; Shi et al. 2006).

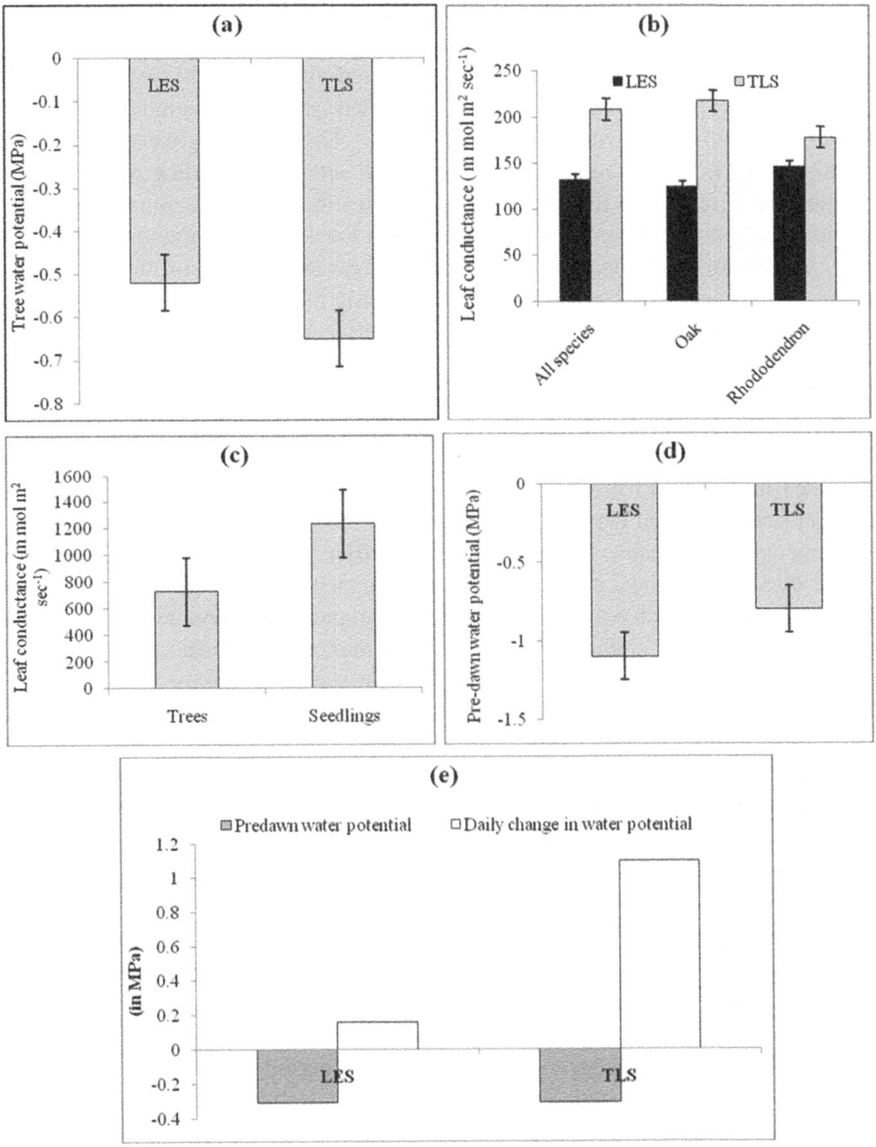

Fig. 2.4 (**a**) A comparison of mean daily change in tree water potential in low-elevation species (LES; <2400 m) and treeline species (TLS; >3000 m). (**b**) A comparison of tree leaf conductance between LES and TLS. For LES, the mean values are based on 11 species measured in 6 seasons (fall, from mid-September to mid-November; winter, from mid-November to mid-January; spring, from mid-January to mid-February; early summer, from March to April; summer, from May to mid-June; and rainy, from June to mid-September) and for TLS the mean is of the values of 5 species measured in 4 seasons (pre-monsoon, from March to June; monsoon, from July to September; autumn, from October to November; and winter, from December to February). (**c**) Leaf conductance values of *Betula utilis* in TLS at Chitkul (dry inner valleys): a comparison between trees and seedlings. (**d**) A comparison of pre-dawn water potential in pre-monsoon season

2.3.7 Learning by Doing Team Research

Our study on the Himalayan timberlines involved about over 30 researchers, dispersed across six institutions, which included national institutes and state universities and non-government organizations. The team originated from team members, not from any authority or institutional setup. The leading researchers of the timberline team had diverse expertise and institutional background, involving disciplines as varied as tree ring chronology and local level livelihoods (Table 2.4). The coordinator of the team conceptualized a research project on the Himalayan timberline keeping in view the fact that even basic information on the subject was not available. Then, he shared the first project draft with identified experts in relevant fields. Needless to say, he made some efforts to make them interested to work together on a common framework. The first draft of research proposal received the inputs of members, resulting in a polished draft of research proposal for the National Mission on Himalayan Studies (NMHS) of the Ministry of Environment, Forest and Climate Change (MoEF & CC). The mission's research part was managed by a small team of scientists from Govind Ballabh Pant National Institute of Himalayan Environment and Sustainable Development (GBPNIHESD), Kosi-Katarmal, Almora, whose director and nodal person had requisite capacity and sensitivity to appreciate the significance of such a team approach to investigate an important research area of the Himalayas. That perhaps played a role in making the MoEF & CC Ministry convinced in providing a large grant to the Himalayan timberline research. Even before the project was approved, the researchers had begun to exchange their ideas, identifying study sites and developing field sampling designs. The research team assembled at Tungnath, Uttarakhand, one of the study sites, soon after the grant was approved, to fine-tune the methodological details keeping in view the difficulties that the remote and high Himalayan timberlines generally pose. For example, to reach a timberline suitable for research in Sikkim one required to walk a full day. Timberline was easily accessible in Kashmir, but there the insurgency was a serious obstacle. In Uttarakhand, getting to timberline was not a problem, but tourism was a serious threat to timberline environment and data collection, and the maintenance of permanent plots. Since the Himalayan sites are generally affected by human disturbance, one cannot be sure that devices installed to measure temperature and other parameters would be safe over a sufficiently long period for data collection. The solidarity of the team individuals encouraged us to install equipment. That some of the leading researchers belonged to study regions, helped a lot in terms of site selection, conducting research and understanding the meaning of data generated. The team members used to meet periodically, share data, brainstorm, improve methods and add new studies and sites, and plan strategies to further improve the

Fig. 2.4 (continued) between LES and TLS. (**e**) Monsoon season (July–September) pattern in pre-dawn water potential and daily change in tree water potential of LES and TLS (Source: Singh et al. 2006; Tewari et al. 2018; and Dr. Ashish Tewari and his associates)

Table 2.4 Team composition and study designs

Science team composition and study designs	Features/remarks
Size and composition of research team: One coordinator, 11 principal investigators (PIs)/ researchers and 17 research scholars from various disciplines (students with master degree)	A team of >10 researchers is considered a large team; generally, research teams consist of <10 researchers (including both PIs and students) but in exceptional cases the number could be in hundreds and thousands; attempts were made to include other Himalayan countries, but it could not materialize
Age structure and gender ratio of the team: 60+ = 2, 45–60 = 7, 30–45 = 3, <30 =17 (research scholars); of the 29 members, 6 were female	The team was uneven-aged; women were represented in research scholar category
Extent of inter-disciplinarity, and trans-disciplinarity in the team: PIs had research expertise in general ecology, biodiversity and plant taxonomy, vegetation science, remote sensing, tree water relations, sustainable development, phenology, meteorology and physics, paleoecology and dendrochronology, and local livelihoods	Researchers from different disciplines co-developed research questions and designed sampling to answer them
Geographical distribution of scientists was wide, and included scientists from institutions located in western, central and eastern parts of the Himalayas, plains and foothills of mountains within India	Efforts were made to include scientists of other Himalayan countries, but could not materialize
Proximity of team members	Not co-located
Study sites: 3 timberline (>3000 m) sites within 75°26′6″ E–88°08′58.69″ E longitude and 27°29′04.79″ N–33°36′43″ N latitude; in addition, secondary data were collected from the entire West-to-East Arc (~3000 km length) of the Himalayas (extending from Afghanistan in the northwest [ca. 70°E] to Yunnan in the southeast [ca. 100°E]; the latitudinal range of the arc is of about 10°, from 36° N [Afghanistan] to 26° N [Yunnan])	It helped to identify various drivers of change of biodiversity, tree species composition and timberline elevations across the Himalayan Arc, which is very heterogeneous
Task interdependence	High among biodiversity experts, phenology and water relations; in others, it was moderate or low
Goal alignment	Strong

project. The research findings were shared with decision makers including the Chief Minister of Uttarakhand, various scientists and officials, both at national and state levels. In this regard the meetings at Dehradun, Indian National Science Academy (INSA), Delhi, Gangtok and Srinagar were notable. A short film was made and shared widely. Research training was another area that was used frequently to develop human resource.

The group realized soon that historically timberline studies generally had a strong physiological leaning, and ecological and geographical dimensions were peripheral. So, while working out the details of methods, the group had to work together to apply ecological concepts, and develop terminologies and sampling designs.

The capacity of the coordinator to make the team members interested in timberline research contributed significantly to the progress of the research. However, the team science did not develop in a formal way, as there was no separate provision for it under the NMHS programme, and the members lacked academic familiarity with the nuances of mechanisms required for its structuring and making it operative. That is why weaknesses pertaining to the lack of trans-disciplinarity and synthesis of data were left unattended. The team could not do enough to strengthen collaboration among young researchers; most of joint efforts were limited to individual level. Academic discussions and exchange of ideas could not be mainstreamed. A serious drawback at this system level is that there is no well thought-out scheme to build on the platform this project has created. The average age of investigation was a bit higher; perhaps a younger team could have sustained research enthusiasm longer.

2.4 Conclusion

The study carried out by the research team has contributed substantially to the basic understanding of the Himalayan timberline ecotone from the standpoints of macroecology. It emphasizes that the presence of the Himalayan timberline ecotones should be considered at a regional scale; their role in the functioning of the high Himalayan systems is far more than generally perceived. The lack of upward shift of timberline ecotone in spite of decades of warming shows that there are several more factors that influence timberline dynamics in the Himalayas. It also sheds light on why the Himalayan treelines are not only very high but also vary widely in elevation. Evidently, treeline species can survive a wide range of temperature and other environmental conditions. A given treeline species could be a part of treeline with temperatures ascribed to it as well as warmer temperatures. Temperature lapse rate (TLR), which is linked to treeline elevation and climate change pattern, deserves far more attention than generally given; knowledge about tree water relations of treeline ecotone remains scanty in spite of our present efforts. It has raised several difficult questions, rather than giving solutions. Several research approaches in the Himalayas have largely focussed on glaciers, which would be of only a limited use; it should expand to include hydrological implications of melt water flow, and community and ecosystem level changes that occur in and around treeline and timberline areas under the influence of climate change. Our findings shed light on changes in phenology of treeline species and those occurring in alpine meadows due to warming and early snowmelt, but the data collected are too scarce to bring about a major advancement in our understanding.

2.4.1 Way Forward: Research Questions and Future Strategies

The present NMHS-sponsored research enterprise enabled us to establish a research base, which could be used to build a long-term and elaborated research programme on the Himalayan treelines. In this regard, following observations may be of help:

1. With regard to research and management, treelines should not be considered in isolation of connected biotic zones, namely alpine meadow and nival species, and abiotic systems such as permanent snow area that provide water to biotic systems should also be included. Though physiognomically these biotic components are divided functionally, they can be considered as interconnected components of a continuum consisting of forests, timberline, treeline and short-statured plants and barren snow and ice area, all under state of flux due to climate change. In this continuum, wild and domestic animals should also be included.
2. The question how climate change is affecting elevation-dependent warming and temperature lapse rate (TLR) calls for long-term data from several representative sites.
3. How seasonal changes in TLR are likely to affect biotic components, and how TLR is influenced by widespread pollution in the big cities of adjacent plains are the related important questions for both basic understanding and management. A macro-ecological approach that gives a coherent picture of various interconnected systems might serve the purpose of managing the high-landscape Himalayas.
4. Among the seasons, the research focus should be on the pre-monsoon period (April–June), which is of critical importance because while water stress during this period is the highest, most species shed leaves as well as begin growth. This is also important culturally as it is the time of peak tourism and water demand, and forest fires. There are evidences to suggest that tree ring width is severely affected by pre-monsoon droughts.

Treeline and mountain summits are likely to experience changes in species composition, species accumulation, migration and extinction as climate change impact increases. We need to mark and maintain permanent plots to monitor species flux on a long-term basis. While doing so, sites should be chosen to appropriately represent outer ranges as well as inner and main ranges, which differ in treeline elevations, climatic conditions, and treeline dimensions and scales. To capture the Himalaya-level variability, research networking among the Himalayan countries is necessary. Efforts would be required to enhance interdependence and alignment in research purposes and approaches. Team research should be expanded to include researches for several Himalayan countries and measures should be taken to hone up team science culture.

Acknowledgements This study was carried out as part of the Indian Himalayan Timberline Research Project (IHTRP). The financial support received under the National Mission on Himalayan Studies (NMHS), Ministry of Environment, Forest & Climate Change, Government of India, is duly acknowledged. We thank the Coordinator, IHTRP project, and Director, GBPNIHE, Almora, for providing the necessary facilities to conduct this study. We give great importance to the research contributions of all the Principal Investigators, and the research scholars of IHTRP who worked in difficult terrains of the treeline areas of the Himalayas. We extend our special thanks to Indian National Science Academy (INSA) and Central Himalayan Environment Association (CHEA) for providing facilities for this study.

References

Adhikari BS, Kumar R, Singh SP (2018) Early snowmelt impact on herb species composition, diversity and phenology in a western Himalayan treeline ecotone. Trop Ecol 59(2):365–382

Anderson K, Fawcett D, Cugulliere A, Benford S, Jones D, Leng R (2020) Vegetation expansion in the subnival Hindu Kush Himalaya. Glob Chang Biol 26(3):1608–1625

Bajracharya SR, Maharjan SB, Shrestha F, Guo W, Liu S, Immerzeel W, Shrestha B (2015) The glaciers of the Hindu Kush Himalayas: current status and observed changes from the 1980s to 2010. Int J Water Resour Dev 31(2):161–173. https://doi.org/10.1080/07900627.2015.1005731

Bhuju DR, Shah SK, Gaire NP (2016) Environmental reconstruction and impact of climate change on vegetation at treelines of Nepal Himalaya. Annu Rep Pro Nat Foundation Jpn 24:169–180

Bolch T, Kutusov S, Li X (2012) Updated GLIMS Glacier Database for the Tian Shan. National Snow and Ice Data Center, Boulder

Chudley TR, Miles ES, Willis IC (2017) Glacier characteristics and retreat between 1991 and 2014 in the Ladakh Range, Jammu and Kashmir. Remote Sens Lett 8(6):518–527. https://doi.org/10.1080/2150704X.2017.1295480

Gaire NP, Koirala M, Bhuju DR, Borgaonkar HP (2014) Treeline dynamics with climate change at the central Nepal Himalaya. Clim Past 10:1277–1290. https://doi.org/10.5194/cp-10-1277-2014

Gardner AS, Moholdt G, Cogley JG, Wouters B, Arendt AA, Wahr J, Berthier E, Hock R, Pfeffer WT, Kaser G, Ligtenberg SRM, Bolch T, Sharp MJ, Hagen JO, Van den Broeke MR, Paul F (2013) A reconciled estimate of glacier contributions to sea level rise: 2003 to 2009. Science 340:852–857

Gautam R, Hsu NC, Lau KM, Tsay SC, Kafatos M (2009) Enhanced pre-monsoon warming over the Himalayan-Gangetic region from 1979 to 2007. Geophys Res Lett 36(7). https://doi.org/10.1029/2009GL037641

Gonzalez P, Neilson RP, Lenihan JM, Drapek RJ (2010) Global patterns in the vulnerability of ecosystems to vegetation shifts due to climate change. Glob Ecol Biogeogr 19(6):755–768. https://doi.org/10.1111/j.1466-8238.2010.00558.x

Han F, Zhang BP, Zhao F, Wan L, Tan J, Liang T (2018) Characterizing the mass elevation effect across the Tibetan Plateau. J Mt Sci 15:2651–2665. https://doi.org/10.1007/s11629-018-4978-2

Hatfield J, Prueger JH (2015) Temperature extremes: Effect on plant growth and development. Weather Clim Extremes 10(PA):WACED140046. https://doi.org/10.1016/j.wace.2015.08.001

Holtmeier FK (2009) Mountain timberlines. Ecology, patchiness, and dynamics. In: Advances in global change research. Springer, Dordrecht, p 36

Immerzeel WW, Van Beek LPH, Bierkens MFP (2010) Climate change will affect the Asian water towers. Science 328(5984):1382–1385. https://doi.org/10.1126/science.1183188

Immerzeel WW, Van Beek LPH, Konz M, Shrestha AB, Bierkens MFP (2012) Hydrological response to climate change in a glacierized catchment in the Himalayas. Clim Chang 110:721–736. https://doi.org/10.1007/s10584-011-0143-4

Joshi R, Sambhav K, Singh SP (2018) Near surface Temperature Lapse Rate for Treeline environment in Western Himalaya and possible impacts on ecotone vegetation. Trop Ecol 59(2):197–209

Körner C (2012) Treelines will be understood once the functional difference between a tree and a shrub is. Ambio 41:197–206

Körner C, Diemer M (1987) In situ photosynthetic responses to light, temperature and carbon dioxide in herbaceous plants from low and high altitude. Funct Ecol 1:179–194. https://doi.org/10.2307/2389420

Körner C, Paulsen J (2004) A world-wide study of high altitude treeline temperatures. J Biogeogr 31:713–732

Latwal A, Sah P, Sharma S (2018) A cartographic representation of a timberline, treeline and woody vegetation around a central Himalayan summit using remote sensing method. Trop Ecol 59(2): 177–186

Liang E, Wang Y, Piao S, Lu X, Camarero JJ, Zhu H, Zhu L, Ellison AM, Ciais P, Peñuelas J (2016) Species interactions slow warming-induced upward shifts of treelines on the Tibetan Plateau. Proc Natl Acad Sci U S A 113:4380–4385

Mestre L, Toro-Manríquez M, Soler R, Huertas-Herrera A, Martínez-Pastur G, Lencinas MV (2017) The influence of canopy-layer composition on understory plant diversity in southern temperate forests. For Ecosyst. 4:6. https://doi.org/10.1186/s40663-017-0093-z

Miehe G, Pendry CA, Chaudhary RP (eds) (2015) Nepal: An Introduction to the Natural History, Ecology and Human Environment of the Himalayas. Edinburgh: Royal Botanic Garden, Edinburgh

Minder JR, Letcher TW, Liu C (2018) The character and causes of elevation-dependent warming in High-resolution simulations of rocky mountain climate change. J Clim 31:2093–2113. https://doi.org/10.1175/JCLI-D-17-0321.1

Mujawamariya M, Manishimwe A, Nitrugulirwa B, Zibera E, Ganszky D, Bahati EN, Nyirambangutse B, Nsabimana D, Wallin G, Uddling J (2018) Climate sensitivity of tropical trees along elevation gradient in Rwanda. Forests 9:647. https://doi.org/10.3390/f9100647

Müller M, Schickhoff U, Scholten T, Drollinger S, Böhner J, Chaudhary RP (2015) How do soil properties affect alpine treelines? General principles in a global perspective and novel findings from RolwalingHimal, Nepal. Prog Phys Geogr 40:135–160. https://doi.org/10.1177/0309133315615802

Palazzi E, Mortarini L, Terzago S, Hardenberg JV (2019) Elevation-dependent warming in global climate model simulations at high spatial resolution. Clim Dyn 52:2685–2702. https://doi.org/10.1007/s00382-018-4287-z

Prabinarana A, Bhuju D, Koirala M, Boonchird C (2017) Dendroecological studies of *Rhododendron campanulatum* D. Don along the elevational gradient of Manaslu Conservation area, Nepal Himalaya. Pak J Bot 49(5):1749–1755

Schickhoff U (2005) The upper timberline in the Himalayas, Hindu Kush and Karakorum: A review of geographical and ecological aspects. In: Broll G, Keplin B (eds) Mountain ecosystems: studies in treeline ecology. Springer, Berlin, pp 275–354

Schickhoff U, Bobrowski M, Böhner J, Bürzle B, Chaudhary RP, Gerlitz L, Heyken H, Lange J, Müller M, Scholten T, Schwab N, Wedegärtner R (2015) Do Himalayan treelines respond to recent climate change? An evaluation of sensitivity indicators. Earth Syst Dynam 6:245–265

Shen M, Piao S, Jeong S-J et al (2015) Evaporative cooling over the Tibetan Plateau induced by vegetation growth. Proc Natl Acad Sci U S A. 112:9299–9304. https://doi.org/10.1073/pnas.1504418112

Shi P, Körner C, Hoch G (2006) End of season carbon supply status of woody species near the treeline in western China. Basic Appl Ecol 7:370–377. https://doi.org/10.1016/j.baae.2005.06.005

Singh SP (2018) Research on Indian Himalayan Treeline Ecotone: an overview. Trop Ecol 59(2): 163–176

Singh SP, Rawal RS (2017) Manuals of field methods. Central Himalayan Environment Association, Nainital

Singh SP, Singh JS (1992) Forests of the Himalayas: structure, function and impact of man. Gyanodaya Prakashan, Nainital, p 294

Singh SP, Thadani R (2015) Complexities and controversies in Himalayan research: a call for collaboration and rigor for better data. Mt Res Dev 35(4):401–409. https://doi.org/10.1659/MRD-JOURNAL-D-15-00045

Singh SP, Zobel DB, Garkoti S, Tewari A, Negi CMS (2006) Patterns in water relations of central Himalayan trees. Trop Ecol 47(2):159–182

Singh CP, Panigrahy S, Parihar JS (2011) Alpine vegetation ecotone dynamics in Gangotri catchment using remote sensing techniques. International Archives of the Photogrammetry, Remote Sensing and spatial information Sciences, Volume XXXVIII-8/W20, workshop of ISPRS Bhopal, India

Singh SP, Singh RP, Gumber S, Bhatt S (2017) Two principal precipitation regimes in Himalayas and their influence on tree distribution. Trop Ecol 58(4):679–691

Singh SP, Sanwal M, Dimri VP, Dubey SK (2018) The climate change programme of the Department of Science and Technology. Curr Sci 115(1):22–24

Stokols D, Misra S, Moser RP, Hall KL, Taylor BK (2008) The ecology of team science: understanding contextual influences on transdisciplinary collaboration. Am J Prev Med 35(2): 96–115. https://doi.org/10.1016/j.amepre.2008.05.003

Suwal MK, Shrestha KB, Guragain L, Shakya R, Shrestha K, Bhuju DR, Vetaas OR (2016) Land-use change under a warming climate facilitated upslope expansion of Himalayan silver fir (*Abiesspectabilis* (D. Don) Spach). Plant Ecol 217:993–1002. https://doi.org/10.1007/s11258-016-0624-7

Tewari A, Shah S, Singh N, Mittal A (2018) Treeline species in Western Himalaya are not water stressed: a comparison with low elevation species. Trop Ecol 59(2):313–325

Tiwari A, Jha PK (2018) An Overview of Treeline response to environmental changes in Nepal Himalaya. Trop Ecol 59(2):273–285

Wang JS, Feng JG, Chen BX, Shi PL, Zhang JL, Fang JP, Wang ZK, Yao SC, Ding LB (2016) Controls of seed quantity and quality on seedling recruitment of Smith fir along altitudinal gradient in southwestern Tibetan Plateau. J Mt Sci 13(5):81–821

Wester P, Mishra A, Mukherji A, Shrestha AB (eds) (2019) The Hindu Kush Himalaya assessment—mountains, climate change, sustainability and people. Springer, Cham

Yao T, Thompson L, Yang W, Yu W, Gao Y, Guo X, Yang X, Duan K, Zhao H, Xu B, Pu J, Lu A, Xiang Y, Kattel DB, Joswiak D (2012) Different glacier status with atmospheric circulations in Tibetan Plateau and surroundings. Nat Clim Chang 2:663–667. https://doi.org/10.1038/nclimate1580

Zhang B, Yao Y (2016) Implications of mass elevation effect for the altitudinal patterns of global ecology. J Geogr Sci 26(7):871–877. https://doi.org/10.1007/s11442-016-1303-2

Zobel DB, Singh SP (1997) Himalayan forests and ecological generalizations. Bio Sci 47(11): 735–745. https://doi.org/10.2307/1313096

Temperature Lapse Rate in Climatically Different Himalayan Treeline Environments: Regional Analysis of Patterns, Seasonality, and Variability

3

Rajesh Joshi, Ninchhen Dolma Tamang, Kumar Sambhav, Chetna Mehra, B. S. Bisht, and S. P. Singh

Abstract

This chapter presents a regional analysis of spatial and temporal variation in temperature lapse rate (TLR) across the Himalayan arc based on a primary study conducted in three transects and reviews of earlier studies. The annual mean TLR for tree line ranged from −0.50 to −0.66 °C $(100\ m)^{-1}$ and varied across the seasons with highest being in pre-monsoon (0.52–0.66 °C/100 m) and lowest during winters (0.31–0.46 °C/100 m). Mean annual TLR was higher for North-West aspect (−0.46 °C/100 m) than the South-East (SE) aspect (−0.36 °C/100 m). TLR for treeline in Himalaya was distinctly lower in comparison to previous records (−0.65 to −0.7 °C/100 m) and varied for different climate regimes. It was 21–29% shallower for Eastern Himalaya, 20–38% for Central Himalaya, and 1–21% for Western Himalaya. The mean annual TLR increased from east to west along the Himalayan arc (east to central by 4.1%, central to west by 14%, and from the Western Himalaya to the Pir-Panjal and Karakoram ranges by 30%). The shallow TLR reflected that higher regions in Himalaya were warmer than perceived resulting in elevation-dependent warming (EDW) under the influence of climate change. The higher TLR in dry or warmer and lower in humid or cold atmospheric conditions suggests different controlling factors determining TLRs in different seasons. This chapter also discusses possible implications of EDW on vegetation in Himalayan treeline ecotone.

R. Joshi (✉) · N. D. Tamang · B. S. Bisht
G. B. Pant National Institute of Himalayan Environment, Sikkim Regional Centre, Pangthang, Gangtok, Sikkim, India

K. Sambhav · C. Mehra
G. B. Pant National Institute of Himalayan Environment (NIHE), Almora, Uttarakhand, India

S. P. Singh
Central Himalayan Environment Association (CHEA), Dehradun, Uttarakhand, India

S. P. Singh et al. (eds.), *Ecology of Himalayan Treeline Ecotone*,
https://doi.org/10.1007/978-981-19-4476-5_3

Keywords

Climate change · Ecotone · Elevation dependent warming · Himalayan region · Temperature lapse rate · Treeline

3.1 Introduction

The mountains are characterized by rugged terrain and steep topography formed through complex processes and structural deformations. Such topographic constraints combined with other factors, such as inaccessibility, harsh climate, and safety, restricts installation of meteorological stations for ground-based climatological observations. In such circumstances, extrapolation of climate variables along elevation gradient from the data recorded at base stations is often required. Such extrapolation is mostly based on empirical relationship between variables recorded at some high elevated stations and base stations located at low elevation. The gradient of air temperature with altitude, commonly known as the environmental temperature lapse rate (ELR), is a constant value at which temperature changes vertically in the troposphere. The ELRs are required for computation of air temperatures at various elevations within a given geographical area and simulation of various modelling processes in various physical and ecological studies. It is a controlling factor of many environmental and ecological processes, such as change in hydrological and moisture regimes, climate-vegetation interactions, glacier mass balance, and melt runoff. Considering that warming has an unequivocal impact on ecological balance of mountain ecosystems, largely influenced by local scale variations in climate parameters, the altitudinal dependence of temperature in mountainous region is of prime importance to study impacts of climate change on various ecosystems. In a complex Himalayan terrain, climate at regional and local scale is affected by topography, latitude, movement of air, and vegetation patterns (Barry 1992; Pepin 2001). Because of these factors, the climate, particularly temperature and precipitation, varies remarkably over short geographic distance. Hence, the effect of topographical factors leads to change in temperature along altitudinal gradients.

The ground-based meteorological observations in Himalaya are very scanty and location-specific long-term precise records are limited, particularly for treeline environments and other high-altitude ecosystems in Himalaya (Friedland et al. 2003; Shrestha et al. 1999). Analysis of temperature lapse rate (TLR) based on observed data for Himalaya are rather scarce (Joshi et al. 2018). In comparison, several studies have been carried out for other mountain regions across the world (e.g., Washington Cascades Mountains, Northern Italy and Austrian Alps, Mt. Taibai, Colardo Rocky Mountains, and in the Northern Hemisphere) on variability of TLR and its controlling factors (Diaz and Bradley 1997; Pepin and Losleben 2002; Minder et al. 2010; Rolland 2003; Tang and Fang 2006). In the absence of the empirically determined lapse rate, a constant value of environmental lapse rate (i.e., −0.65 °C or −0.70 °C/100 m) is often used to estimate air temperature at ungauged sites when low precision suffices (Rolland 2003). Studies across the globe have confirmed that the TLR and precipitation gradient are highly variable in space and time in mountain regions as it varies with latitude, topographic slope, region, and seasons (Dodson and Marks 1997; Rolland 2003; Kattel et al. 2013; Immerzeel et al. 2014; Joshi et al. 2018). Its magnitude varies across location as a function of energy balance (Rolland 2003; Kattel et al. 2013; Tang and Fang 2006). Therefore, simply adopting the general temperature-elevation (T-E) relationship and rough approximations of constant TLR value may not effectively explain T-E trends, both at spatial and temporal scales. Hence, using a constant value of TLR in mountains may lead to erroneous results about patterns of elevation-dependent warming rate and its impacts on different ecosystems in Himalayan region.

Treeline ecotone is an extremely climate-sensitive transition zone. Change in temperature and other abiotic factors (e.g., radiation, moisture, wind, slope exposure, topography) enhance sensitivity of plant species across this ecotone (Körner 1998; Körner and Paulsen 2004; Liu et al. 2009). Hence, an understanding of gradients of climatic parameters is essential to assess the impacts of climate change on climate-vegetation interactions across this region. However, due to insufficient ground-based meteorological observation, studies on temperature gradient for such climate sensitive ecosystems are lacking in Himalaya. Considering this knowledge gap, maiden attempts were made to estimate TLRs for climatically different treeline environments in Eastern, Central, and Western Himalaya (Joshi et al. 2018; IHTP Report 2020).

Moreover, factors controlling TLR in Himalaya are largely unknown which lead to uncertainty over the warming rates of the mountainous region *vis-a-vis* the rest of the land surface (Beniston 1997; Rangwala and Miller 2012). In recent years, studies from Himalayan region have reported varying rates of TLRs in different seasons (Kattel et al. 2013; Bandyopadhyay et al. 2014; Immerzeel et al. 2014; Kattel et al. 2015; Heynen et al. 2016; Kattel et al. 2017a, b; Thayyen and Dimri 2018; Joshi et al. 2018). These studies have suggested that the magnitude of TLR in Himalayan region is shallower than the commonly used values and varies in different locations as a function of energy balance. It implies that conditions in high Himalayas are likely to be warmer than generally held out. The warmer temperatures as such would promote plant growth and may have several implications on the dynamics of treeline ecotone in Himalaya. The change in snow and moisture regimes, increased evapotranspiration and water stress, change in albedo and surface energy balance, and modified distribution patterns, and shift in range and growing season of alpine vegetation are some examples.

The studies carried out so far in the Himalayan region are site specific; while a regional scale analysis of spatio-temporal variation in TLR is essential to understand patterns of TLR in different climate regimes, their controlling factors, and possible impacts on the vegetation distributed along treeline ecotone in the Himalayan region. Considering the above, the main goals of this study are to:

1. Analyze patterns of TLR for climatically different treeline environments in the Himalaya and its variation across treeline ecotone along the elevation gradient. It is hypothesized that TLR below and above the treeline would differ in magnitude.
2. Analyze variation in seasonal and annual temperature lapse rate (TLR) from the East to West across the Himalayan Arc. We expect that regions differing in moisture and humidity would differ in TLR.
3. Describe the factors controlling temperature lapse rate. It is assumed that temperature distribution along elevation is governed by seasonal variation in climate.
4. Discuss possible implications of shallow TLRs on treeline ecotone vegetation.

This chapter presents an analysis of spatial and temporal variability in TLR in the Himalayan arc based on both primary studies carried out at three transects-representing different climatic regimes of treeline environment in Himalaya and secondary information available from published literature.

3.2 Material and Methods

3.2.1 Study Area

The study was carried out in-and-around the treeline ecotone at three sites located in Jammu & Kashmir, Uttarakhand, and Sikkim states of Indian Himalayan Region and representative of different climate regimes. These sites included Daksum-Sinthan (D-S) transect (30°30.76′–30°27.59′ N; 79°05.54′–79°16.55′ E)—a non-monsoonal site with low annual precipitation in Jammu & Kashmir, North-Western Himalaya), Chopta-Tungnath (C-T) transect (30°30.76′- 30°27.59′ N; 79°05.54′–79°16.55′ E)—a typical monsoon rainfall dominated site with dry pre-monsoon season (in Uttarakhand, Western Himalaya), and Yuksam-Dzongri (Y-D) transect (30° 30.76′–30°27.59′ N; 79°05.54′–79°16.55′ E)—characterized by extended monsoon season, moist pre-monsoon, and relatively dry winter season (Sikkim, Eastern Himalaya). The treeline elevation in these transects ranged between 3400 and 4000 m asl and tended to increase from the west to east, largely because of decreasing latitude. These three study sites were part of a major multi-disciplinary study of the Himalayan treeline under the National Mission on Himalayan Studies (NMHS) of Ministry of Enviornment, Forest & Climate Change, Government of India.

Further, the entire Himalayan arc, including the Pir-Panjal and Karakoram ranges, was considered in this study for analyzing the patterns of spatial and temporal variability in the TLR (Fig. 3.1). This region, a massive storehouse of freshwater and home to 10 major river basins, is experiencing widespread changes driven by climate and anthropogenic factors (Sharma et al. 2019). Studies have shown that temperature in the Himalaya has increased rapidly, whereas the direction of change in precipitation was uncertain in the last century. Higher rates of warming (0.9–1.6 °C) were in this region during the last century (1901–2010) as compared to the global average warming rates (Bhutiyani et al. 2007; Joshi and Kumar 2013; Wester et al.

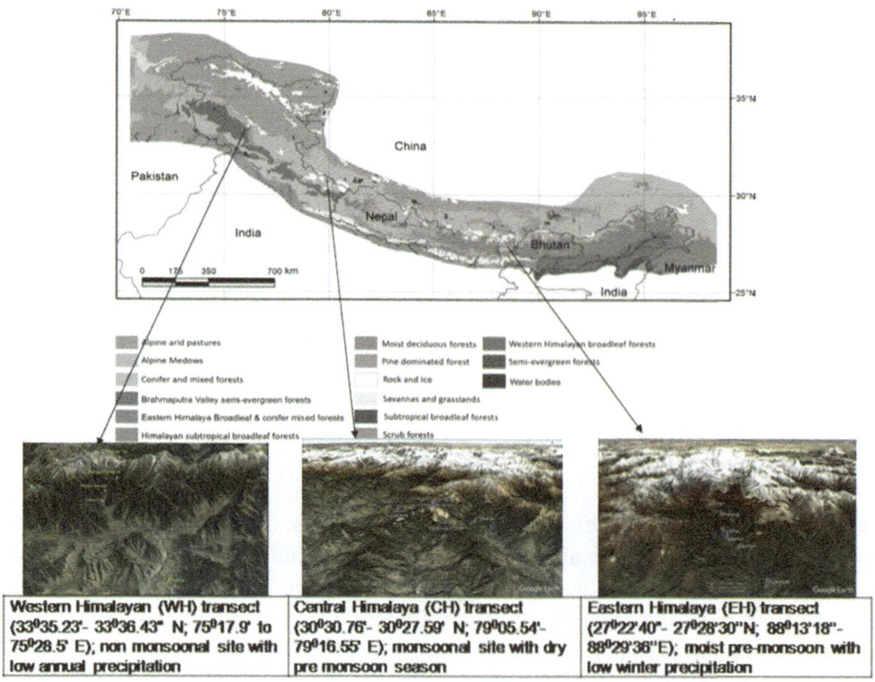

Fig. 3.1 Generalized map of the study region (*Modified after Rawat, 2017*)

2019). For analyzing the spatio-temporal variability, the entire region was divided in Eastern Himalaya (Arunachal, Bhutan, Sikkim), Central Himalaya (Nepal part of Himalaya), Western Himalaya (Uttarakhand, Himachal Pradesh, and parts of J & K), and Pir-Panjal Himalaya and Karakoram Hindu Kush region. The eastern part of Himalaya received high amount of annual rainfall with high annual variability (800–4000 mm; Dorji et al. 2016) and the pre-monsoon season was relatively moist. The climate of Central Himalayan region was dominated by south-easterly (moisture-laden air mass) monsoon circulation with predominant easterly winds in the summer and westerly during October to May (Immerzeel et al. 2014). The Western Himalaya was mostly dominated by Indian Summer Monsoon (ISM) in summer and Indian Western Monsoon (IWM) embedded with the western disturbances during the winter.

3.2.2 Methods and Approaches

Here, we focused on spatio-temporal variation in temperature lapse rate (TLR) in the Himalayan region based on our observed data collected for the different parts of the region. Studies on variation in TLR for the treeline region across the Himalayan arc did not receive much attention, mainly due to lack of observed data and

Table 3.1 Details of the meteorological stations along with altitude (in masl) located along the three study transect in Indian Himalayan region

D-S Transect (Jammu & Kashmir		C-T Transect (Uttarakhand)		Y-D Transect (Sikkim)	
Station	Altitude	Station	Altitude	Station	Altitude
Kokarnag	1900	Ukhimath	1500	Yuksam	1700
Daksum	2400	Siroli	1600	Sachen	2225
Arkhani	2563	Taala	1820	Bakhim	2649
Bampathri	2973	Mandal	2100	Tsokha	3002
Sinthan	3900	Dugalbhitta	2500	Phedang	3690
		Kanchulakharak	2675	Dzongri	4000
		Chopta	2870		
		Saukharak	3100		
		Tungnath	3360		
		Chandrashila	3680		

non-availability of safe sites for installation of sensors. The first study on TLR of treeline areas in Indian Himalaya based on observed data was of Joshi et al. (2018) which was first of its kind in explaining the variation in TLR across treeline in the Himalayan region. It was conducted along Chopta-Tungnath transect in Uttarakhand (Western Himalaya) under the Indian Himalayan Timberline Research project. To understand variations in TLR along treeline environment, the study was further extended to two other transects in Sikkim (Eastern Himalaya) and Jammu & Kashmir (Western Himalaya), and TLRs were estimated using standard methodology followed by Joshi et al. (2018). Five portable ONSET HOBO Pro-V2 temperature loggers, covered with radiation shield, were installed along D-S transect, 10 along C-T transect, and 6 along Y-D transect (Table 3.1). The number of stations across the three sites differed depending on the availability of suitable and safe sites.

Initially, the Pearson's correlation coefficient was calculated between observed monthly mean temperature and elevation of the station and significant correlations were considered for further analysis. Following the methods of Joshi et al. (2018), TLRs (^{0}C/100 m) were estimated by developing a regression equation using all point level observations of mean temperature and elevation. Statistical analyses were performed to test the significance of results obtained. Independent sample t-test was used to test the significance of difference in TLR among all the sites and one-way analysis of variance (ANOVA) was used to compare the monthly and seasonal TLRs between different transects. Both the tests were applied at significance alpha level $\alpha = 0.05$ (or CI = 95%) and *p*-values ($p < 0.05$) using STATISTICA 8.0. To study the reason of seasonal variation in TLRs, a statistical relationship with two other climatic parameters (i.e., rainfall and vapor pressure was analyzed. Further, to investigate the moisture presence in the atmosphere, the saturation vapor pressure was calculated following Kattel et al. (2013, 2015) and using the equation based on the Clausius–Claperon relationship (Tetens 1930).

To analyze the spatio-temporal variability in TLRs across the Himalaya, an extensive review of literature was conducted which showed that only 18 studies have been carried out on TLR across the Himalaya, including Pir-Panjal and Karakoram ranges, out of which six (06) were at spatial level, whereas twelve (12) were site specific and along the elevation gradient (Table 3.4). In Eastern Himalayan region, four (4) studies were from Arunachal Pradesh and Sikkim in India and southern slope of Bhutan. Three (3) studies in Central Himalaya were reported from the southern slope of Central Himalaya (where seasonal rainfall vary considerably from highest in monsoon season to the lowest in winters; Kattel et al. 2013) and few glaciarized catchments (e.g., Lirung glacier and Langtang catchment) in Nepal which received relatively less amount of annual rainfall (approx. 300 mm; Heynen et al. 2016). From Western Himalaya, seven (7) studies on TLR have been reported. High variability in annual average precipitation was observed across these sites, which differed in the magnitude of rainfall (1400–3242 mm). However, in some cases, where western disturbances influenced precipitation pattern, the rainfall was low and varied between 115 and 500 mm annually. In western part of the Himalaya, ground-based long-term observations are deficient. Hence, most of the studies were based on short-term data (1–4 years) observed at a specific site. From Pir-Panjal Himalaya and Karakoram Hindu Kush region, only four (4) studies have been reported on estimation and analysis of TLR.

Based on the available data from the studies carried out so far across the Himalayan region (Table 3.4), this chapter presents an analysis of TLRs for different climate regimes across the Himalayan arc. Considering the fact that temperature in mountains varies topographically in different months and seasons (Barry 1992), TLRs were calculated separately for all the months well as for each of the four seasons; viz. pre-monsoon (MAM: March, April and May), monsoon (JJAS: June, July, August and September), post-monsoon (ON: October and November), and winter (DJF: December, January and February). The annual mean TLR was taken as the average value of TLR from January to December months.

3.3 Results

3.3.1 Patterns of TLR for Different Treeline Environments in Himalaya

The correlation between monthly temperature and elevation and the monthly TLRs across various sites (Table 3.2) showed that temperature decreases significantly with increase in elevation. Further, the strong and significant correlation values indicated that the strong linear relationship existed between altitude and the temperature for all the sites which can be used to estimate the TLR. The annual mean TLR for Y-D transect in Sikkim (−0.50 °C/100 m) was lower than that of C-T transect in Uttarakhand (−0.52 °C/100 m) and D-S transect in Jammu & Kashmir (−0.66 °C/100 m) and increased with decreasing moisture, being markedly higher at D-S transect of Western Himalaya. A distinctly lower annual mean TLR than the commonly used values of 0.65 to 0.7 °C/100 m (Barry and Chorley 1987) existed for treeline regions in Eastern and Central Himalaya. Results showed that the annual

Table 3.2 Monthly TLR (^{0}C/100 m) values and correlation coefficient (*r*) between mean temperature and elevation for three treeline transects across the IHR

Month	D-S Transect, Jammu & Kashmir		C-T Transect, Uttarakhand		Y-D Transect, Sikkim	
	TLR	*r*	TLR	*r*	TLR	*r*
JANUARY	−0.3	−0.86**	−0.42	−0.94*	−0.54	−0.87**
FEBRUARY	−0.38	−0.89**	−0.52	−0.96*	−0.53	−0.91*
MARCH	−0.54	−0.94*	−0.63	−0.95*	−0.52	−0.92*
APRIL	−0.64	−0.93*	−0.67	−0.98*	−0.53	−0.93*
MAY	−0.63	−0.95*	−0.68	−0.99*	−0.52	−0.94*
JUNE	−0.62	−0.97*	−0.63	−0.99*	−0.5	−0.96*
JULY	−0.65	−0.98*	−0.52	−0.99*	−0.51	−0.95*
AUGUST	−0.9	−0.91*	−0.49	−0.98*	−0.5	−0.94*
SEPTEMBER	−0.8	−0.92*	−0.54	−0.98*	−0.49	−0.95*
OCTOBER	−0.78	−0.86**	−0.51	−0.97*	−0.52	−0.93*
NOVEMBER	−0.34	−0.85**	−0.38	−0.94*	−0.53	−0.92*
DECEMBER	−0.26	−0.79**	−0.24	−0.83**	−0.32	−0.81**

*Correlation values significant at $p < 0.01$; **correlation values significant at $p < 0.05$

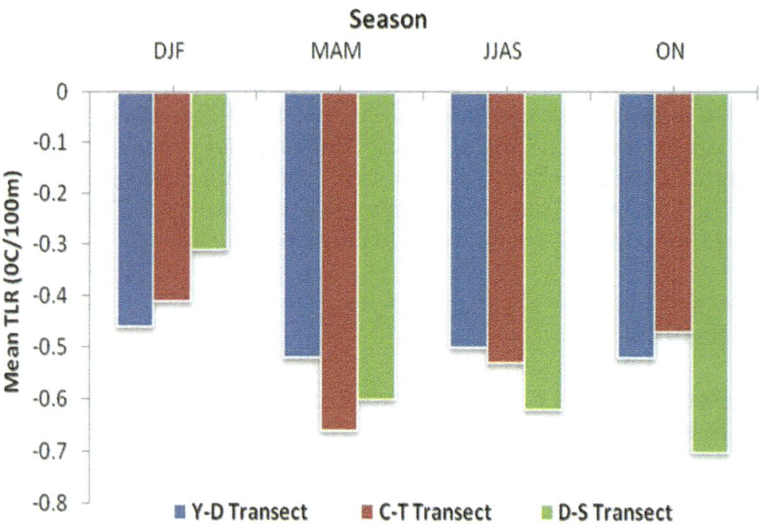

Fig. 3.2 Variation in mean TLRs in four season across three study sites

TLR for treeline environment varied spatially and decreased from West (dry site) to East (moist site) in Himalaya, and was dissimilar for different precipitation regimes in the Himalaya.

Among the three study transects, the highest mean TLR values were observed during pre-monsoon season at Y-D transect (−0.66 °C/100 m) and C-T transect (−0.52 °C/100 m) (Fig. 3.2). During the pre-monsoon season, higher temperature and minimal presence of clouds led to more incoming solar radiation at land surfaces as compared to outgoing radiation. This resulted in daytime surface temperature rise and large sensible heat flux that presumably enhanced strong dry convection in the daytime. The pre-monsoon season also had the highest daytime saturation vapor pressure (e_s) lapse rate; as a result, the TLR reached at maximum value in that season. The lowest mean TLR values were observed during winter season (DJF) on all the transects (Y-D: −0.46 °C/100 m; C-T: −0.41 °C/100 m; D-S: −0.31 °C/100 m) which implied that snow albedo or reflectance played a more important role as controlling factors on TLR. The results of one-way analysis of variance ($F = 3.2175$; $p = 0.03$) confirmed that the TLR significantly varied across the region (sites) and among the seasons. These results authenticated that the seasons had strong influence on TLR. These results were also in agreement with shallow TLRs estimated by Tang and Fang (2006), Kattel et al. (2013), Immerzeel et al. (2014), Kattel et al. (2015) and Joshi et al. (2018) for other parts of Himalaya.

The results from C-T transect in Western Himalaya also showed that patterns of temperature gradient varied along different aspects. Along the North-West (NW) aspect, the mean TLR varied from −0.23 (±0.55)°C/100 m in December to −0.69 (±0.20)°C/100 m in May. Along South East (SE) aspect, the lowest TLR for mean temperature was observed in December (−0.18 ± 0.37 °C/100 m) and highest in May (−0.66± 0.20 °C/100 m). Study showed that TLR for maximum temperatures were significantly higher than that for minimum temperatures both for NW aspect ($t = -6.553$; $p = 0.00004$) and SE aspects ($t = -2.725$; $p = 0.019$). Among the two aspects, TLR values were significantly higher for NW aspect than for SE aspect for mean temperatures ($t = 2.06$; $p = 0.035$) and minimum temperatures (t=2.1483; P=0.04), but not in case of maximum temperatures ($t = 1.44$; $p = 0.16$). The two aspects also differed significantly ($P \leq 0.05$) in seasonal distribution of TLR (Table 3.3). A higher value of mean TLR was observed for NW aspect (−0.46 °C/100 m) than the SE aspect (−0.36 °C/100 m). The TLR varied significantly across all the seasons for the mean ($F = 3.2175$; $p = 0.03$), maximum ($F = 3.675$; $p = 0.019$), and minimum ($F = 9.895$; $p = 0.002$) temperatures on both the aspects indicating that the seasonal variations have strong influence in determining the TLR in the Himalaya. The higher values of TLRs observed for NW aspect in comparison to SE aspect could be related to more rainfall

Table 3.3 TLR (in °C/100 m) for mean, minimum and maximum temperatures in relation to aspects and seasons

	North-West aspect			South-East aspect		
Season	Mean TLR	Maximum TLR	Minimum TLR	Mean TLR	Maximum TLR	Minimum TLR
Winter (DJF)	−0.42	−0.71	−0.35	−0.39	−0.36	−0.35
Pre-monsoon (MAM)	−0.64	−0.85	−0.49	−0.60	−0.81	−0.51
Monsoon (JJAS)	−0.57	−0.67	−0.55	−0.55	−0.63	−0.57
Post-monsoon (ON)	−0.44	−0.41	−0.38	−0.47	−0.42	−0.48

on SE aspect (2175 mm rainfall for monsoon season) than NW aspect (2070 mm average rainfall for monsoon season).

Seasonal changes indicated that the decline in temperature from autumn months to December was less in higher ranges than lower ranges. For CT transect, the effect of darkened vegetation and reduced albedo in meadows in December is also reflected in lower TLR across meadow sites (TLR was −0.18 °C/100 m above 3200 m) than across forest sites (TLR was −0.6 °C/100 m below treeline). In brief, TLR was lowest due to EDW and high albedo in December. It was moderately low during wet monsoon months when the humidity was uniformly high.

3.3.2 Seasonal and Annual Variation in TLRs for Different Climatic Regimes Across the Himalayan Arc

3.3.2.1 TLRs for Eastern Himalayan Region

Summary of the results of various studies from Eastern Himalayan region showed that the TLR attained maximum value during post-monsoon season (ranging −0.5 to −0.52 °C/100 m) followed by pre-monsoon (−0.49 to −0.53 °C/100 m), monsoon season (−0.38 to −0.5 °C/100 m), and winter season (−0.43 to −0.54 °C/100 m). The range of mean annual lapse rate for this region (−0.46 to −0.51 °C/100 m) was relatively lower than that of the values estimated for other part of the Himalaya (Table 3.4). For Bhutan part of Himalaya, lowest TLRs occurred during monsoon season, while higher TLRs were found during winter season (Dorji et al. 2016; Kattel et al. 2017c). Whereas in Arunachal Pradesh and Sikkim, the shallow TLRs were during winter season and steep during the post-monsoon season (Bandyopadhyay et al. 2014; IHTP Annual Report 2020). The annual mean TLR for treeline in Eastern Himalaya (Sikkim) was higher (−0.5 °C/100 m) than that in Arunachal Pradesh and Bhutan. These were less variant during post-monsoon season (−0.52 °C/100 m), pre-monsoon (−0.51 °C/100 m), and monsoon season (−0.5 °C/100 m) but declined to −0.46 °C/100 m during winter season (IHTP Annual Report 2020). The TLR for Eastern Himalaya mostly remained invariant during summer season (i.e., both in pre-monsoon and monsoon) but decline to minimum during winters. The climate of Eastern Himalaya is relatively humid and noted for high precipitation and extended monsoon season (April–September) as compared with other surrounding mountainous regions (Dorji et al. 2016). Higher evapotranspiration due to increase in relativity humidity has been reported to cause temperature reduction (Song et al. 2017) which consequently results in decreased mean TLR in Eastern Himalaya (Karlsson 2000).

3.3.2.2 TLRs for Central Himalayan Region

The seasonal variation in TLR for central part of the Himalaya showed a different pattern. Studies from Langtang valley in Nepal revealed that mean annual TLR varied between 0.4 and 0.54 °C/100 m and the steepest TLR was recorded during summer months or pre-monsoon season (−0.64 °C/100 m) in the valley (Immerzeel et al. 2014; Heynen et al. 2016). The mean TLR for the winter season declined up to

Table 3.4 Seasonal and spatial variation in TLR across the Himalayan Arc

Region	Study area/Site	Temperature lapse rate (°C/100 m)					Reference
		Pre-monsoon	Monsoon	Post-monsoon	Winter	Annual	
Eastern Himalaya	Arunachal Pradesh	−0.49	−0.48	−0.5	−0.43	−0.46	Bandyopadhyay et al. (2014)
	Bhutan	−0.5	−0.38	0.51	−0.54	−0.47	Kattel et al. (2017a, b)
	Yuksam-Dzongri transect, Sikkim	−0.51	−0.5	−0.52	−0.46	−0.50	IHTP (2020)
	Bhutan	−0.53	−0.45	−0.52	−0.53	−0.51	Dorji et al. (2016)
Central Himalaya	Upper Langtang Catchment, Nepal	−0.64	−0.46	−0.49	−0.58	−0.54	Immerzeel et al. (2014)
	Langtang river catchment, Nepal	−0.62	−0.46	−0.52	−0.20	−0.40	Heynen et al. (2016)
	Central Nepal	−0.58	−0.49	−0.52	−0.46	−0.52	Kattel et al. (2013)
	Lirung Glacier Upper Langtang catchment, Nepal	−0.50	−0.66	−0.78	-NA-	-NA-	Steiner et al. (2016)
Western Himalaya	Chopta Tungnath, Uttarakhand	−0.66	−0.53	−0.47	−0.41	−0.53	Joshi et al. (2018)
	Dingad catchment, upper Ganga Basin	−0.62	−0.56	−0.58	−0.53	−0.57	Thayyen and Dimri (2016)
	Daksum Sinthan transect, J & K	−0.60	−0.62	−0.70	−0.31	−0.66	IHTP Report (2020)
	Chandra Basin, Western Himalaya	−0.69	−0.59	−0.46	−0.61	−0.58	Pratap et al. (2013)
	Upper Beas and Sutlej Basin	−0.57	−0.47	−0.54	−0.46	−0.51	Thayyen and Dimri (2016)
Pir-Panjal and Karakoram Himalaya	Kashmir Himalaya	−0.71	−0.72	−0.72	−0.7	−0.71	Romshoo et al. 2018
	Ganglass catchment, Ladakh	−0.97	−0.93	−0.79	−0.69	−0.86	Thayyen et al. (2018)
	Parts of Hindu Kush and Karakoram region	−0.66	−0.66	−0.59	−0.63	−0.66	Ashraf (2020)
	Upper Indus Basin	-NA-	-NA-	-NA-	-NA-	−0.68	Hayat et al. (2019)

−0.20 °C/100 m. The Langtang valley is a glaciated valley with upper region covered under snow, hence low TLR during winters could be possibly due to high snow abledo at upper most valley and surging of cool air in lower elevations of the valley (Immerzeel et al. 2014). Across the southern slope of the Central Himalaya, mean TLR of −0.52 °C/100 m was highest during the pre-monsoon (−0.58 °C/100 m) (Kattel et al. 2013) and the second highest LR (−0.52 °C/100 m) was recorded for post-monsoon season. The shallowest LR occurred in winter season (−0.46 °C/100 m). This pattern was different from the findings from the other mountain regions and indicated different controlling factors operating to govern TLR in the individual seasons. The seasonal variations of TLR in Central Himalaya exhibited bi-modal patterns with two peaks; first during the pre-monsoon and second during the post-monsoon season. Further, two minima of TLR values occurred during the winter and the monsoon season, respectively (Kattel et al. 2013, 2015).

3.3.2.3 TLRs for Western Himalayan Region

Relatively high annual mean TLR (−0.53 °C/100 m) was recorded for treeline in Western Himalaya in comparison to Eastern and Central Himalaya. The highest mean TLR occurred during pre-monsoon (−0.66 °C/100 m) and the lowest (0.41 °C/100 m) during the winter season (Joshi et al. 2018). Notably, the lapse rate of temperature varied across seasons and aspects suggesting that the micro-climatic and physiographic features influenced TLR at local scale. As like Central Himalaya, the annual cycle of TLR in Western Himalaya also revealed a bi-modal pattern (Fig. 3.3) with two maxima in the pre-monsoon and post-monsoon season, whereas two minima in winter and monsoon season, respectively (Joshi et al. 2018). The patterns of TLRs observed for C-T transect in Western Himalaya were similar to the Central Himalaya; in both the cases low TLR rose sharply during pre-monsoon and declined during winter season and it nearly remained moderately close to the mean

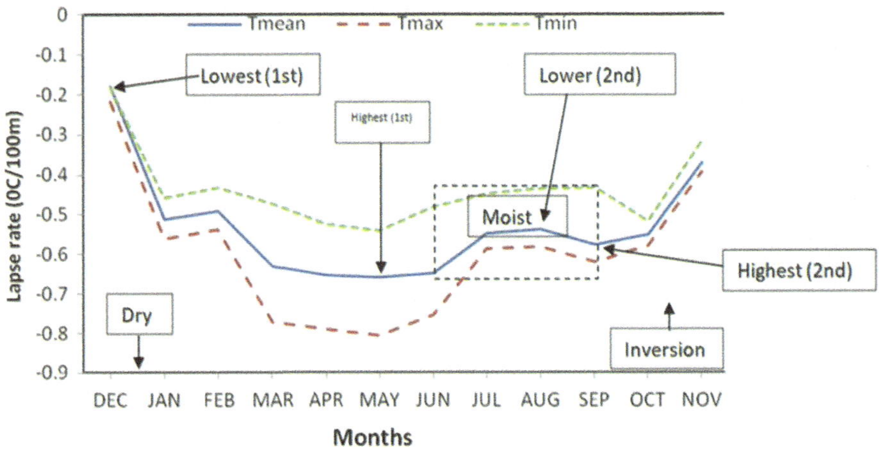

Fig. 3.3 Annual cycle of maximum, minimum, and mean TLR along Chopta-Tungnath transect, Western Himalaya

during monsoon season. The high TLR during pre-monsoon was mainly due to minimal cloud, maximum incoming solar radiation, and daytime temperature resulting into high saturation vapor pressure. Weather conditions during post-monsoon season were similar to pre-monsoon conditions except that the thermal forcing effect was relatively small due to the turbulence between the two climatic phases, that is, retreat of monsoon and the onset of winter westerlies.

Seasonal variations in LRs for Beas, Sutlej and upper Ganga basins ranged from 0.81 to 0.2 °C/100 m (Thayyen and Dimri 2018). For the Daksum-Sinthan transect, a less rainfall-dominated site in Jammu & Kashmir in Western Himalaya, the monthly TLR varied widely from −0.26 °C to −0.90 °C. It decreased from −0.78 °C in October to −0.26 °C in December, where after it increased (−0.62 °C to −0.64 °C) during the pre-monsoon months of April–June (IHTP 2020). The low TLR values during the winter months and high during pre-monsoon (summer) months have emerged as the characteristics of the sites dominated by the monsoon in this region.

3.3.2.4 TLRs for Cold-Arid Regions in Pir-Panjal Range and Karakoram Himalaya

TLR for cold-arid regions varied considerably across various regions of the Himalaya (Table 3.4), and because of elevation-dependent warming may change temporarily. The studies from Pir-Panjal ranges in Kashmir (Thayyen and Dimri 2018; Thayyen 2020; Romshoo et al. 2018) showed relatively higher annual TLR varying between 0.71 and 0.86 °C $(100\ \text{m})^{-1}$ with consistently higher LR during summer months, a common characteristics of the arid conditions. In cold-arid region in Kashmir, the seasonal TLR values were low during winter season and high during summer season. TLR was high from March to August (0.95 °C to 1.11 °C/100 m) and then dipped to 0.67 °C/100 m in December. The mean annual TLRs for some parts of Karakoram region were shallow (ranging between 0.66 °C and 0.68 °C/100 m; Hayat et al. 2019; Ashraf 2020) than the cold-arid region of Kashmir but higher than the values reported from the Himalayan region.

3.4 Discussion

Among the three treeline transects, the least values of mean TLR were recorded for the moist site with extended monsoon (Y-D transect; Sikkim) and highest for a non-monsoonal site with low annual precipitation (D-S transect; Jammu & Kashmir). For CH transect, the effect of darkened vegetation and reduced albedo in meadow in December is also reflected in lower TLR across alpine meadow sites (TLR was −0.18 °C/100 m above treeline) than across forest sites (TLR was −0.6 °C/100 m below treeline). Relatively shallow TLRs were found along these sites. The lower values of TLR for treeline in comparison to those of the past might be due to enhanced elevation-dependent (EDW) warming in higher ranges of Himalaya. Throughout the world, more warming is observed in higher elevation than lower elevation areas and minimum temperatures show a stronger tendency towards EDW than maximum temperatures (Pepin et al. 2015). Such a differential warming is expected to decrease the TLR because of EDW under the influence of climate change.

Most of the studies showed that winter has the lowest mean TLR indicating more winter warming with elevation. In about 50% cases (studies) pre-monsoon has the highest TLR, though some seasonal values were almost invariant (Table 3.4). Because of the absence of clouds and higher temperature during pre-monsoon, land surfaces receive more incoming solar radiation compared with outgoing radiation. This phenomenon results in a rise in daytime surface temperature and large sensible heat flux, which can enhance strong dry convection in the daytime (Blandford et al. 2008). The pre-monsoon season also has the highest daytime saturation vapor pressure (e_s) lapse rate (Kattel et al. 2013); as a result, the TLR is highest and stable during this season. During monsoon season, the difference between maximum and minimum temperature decreases to the presence of moisture.

Various studies from the Himalayan region have indicated that TLR varies significantly for different seasons and across the geographical region (Table 3.4). The mean annual TLR increases significantly from the East to West along the Himalayan arc ($R^2 = 0.86$, $p < 0.05$); the lowest being 0.48 °C $(100\ \text{m})^{-1}$ for Eastern Himalaya (EH), the most wet region, and highest value being 0.74 °C $(100\ \text{m})^{-1}$ for the relatively dry and arid region of Pir-Panjal and Karakoram ranges (Fig. 3.4). The average annual TLR increased from the Eastern Himalaya to the Central Himalaya by 4.1%, from the Central Himalaya to the Western Himalaya by 14%, and from the Western Himalaya to the Pir-Panjal and Karakoram ranges by nearly 30%. The analysis further showed that on a regional scale the TLR depicts similar pattern in seasonal variation along the Himalayan arc with highest being in pre-monsoon season and lowest during winter season (Fig. 3.5). It indicated that winter warming and elevation-dependent warming were high in elevated regions than the low elevation areas. Lapse rate was lower in cold and moist months when moist adiabatic lapse rate prevails and higher in the dry months when dry adiabatic

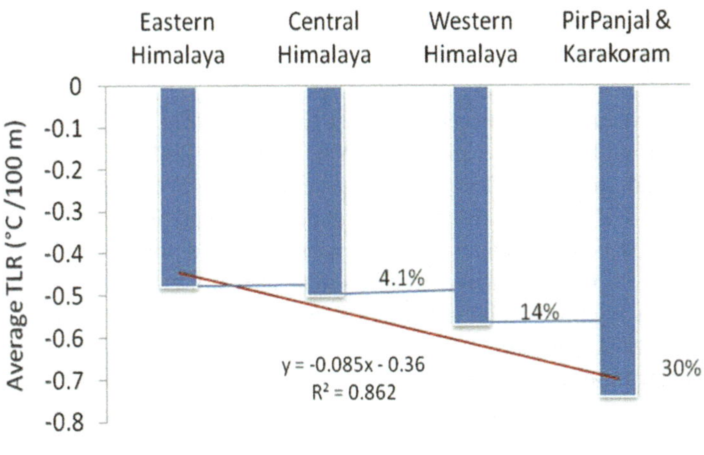

Fig. 3.4 Percentage change in mean TLR from the East to the West along the Himalayan arc

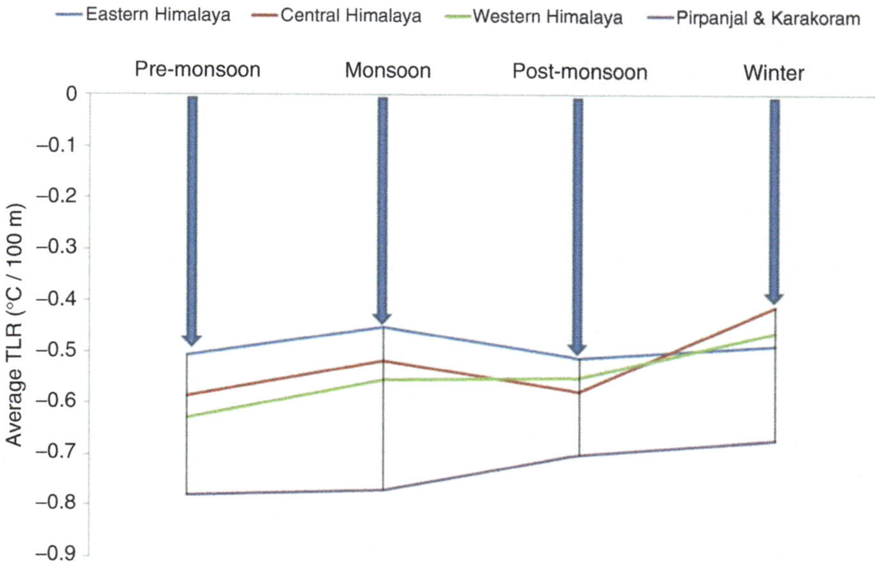

Fig. 3.5 Seasonal pattern of mean TLR from the East to the West along the Himalayan arc

lapse rate dominates. The higher TLR in dry or warmer and lower in humid or cold atmospheric conditions suggested that the TLR does not remain constant for different seasons and varied along different regions and seemingly controlled by diverse factors (e.g., topography and microclimate) in different seasons. Therefore, assumption of persistent values of TLR for all seasons and months is imprecise and may lead to erroneous results when used to study elevational warming and its impacts on different ecosystems in Himalayan region. In view of observed variability in TLRs, the present study advocates for use of seasonally varying lapse rates for analyzing impact of climate change on different ecosystems along elevation gradient across the Himalaya.

The mean annual TLR for Eastern, Central, and Western Himalaya was often lower than the commonly used TLR value of $0.65°(100\ \text{m})^{-1}$ and varied for different climatic regimes across the Himalaya and other mountains ranges in Hindu-Kush Himalaya. It was 21–29% shallower for Eastern Himalaya, 20–38% lower for Central Himalaya, and 1–21% lower for Western Himalaya in comparison to the commonly used mean TLR value. Whereas the mean annual TLRs for the cold-arid region of Pir-Panjal and upper Indus Basin in Karakoram ranges were 9–30% and 2–4%, respectively, these were higher than the commonly used constant value of $0.65°(100\ \text{m})^{-1}$. The higher summer TLR seems to be a common feature of cold-arid climate (Thayyan 2020). Though the lack of water and snow accumulation over a long period within a year are main causes of failure of trees to grow in most parts of Ladakh, the high TLR might also contribute to it.

The cold-arid regime is mostly characterized by very low precipitation. Low values of TLR during winter months and consistently higher LR during summer months clearly reflected the characteristics of the arid conditions and indicated the role of moisture influx and low temperature combination controlling the LR of cold-arid regime in mountains.

The pattern of TLR for the Himalayan region is distinct from other mountain regions (e.g., South entral Idaho, Washington Cascades, Colorado Rocky Mountains, and semiarid southeastern Arizona in the USA; Northern Italy, Greece, and the Austrian Alps in Europe; and northern and southern slope of Mt. Taibai in China). The annual mean TLR for European Alps ranges between −0.54 and −0.64 °C $(100\ \mathrm{m})^{-1}$ manifested by seasonal dependency with shallow gradients in winter and autumn and steeper in spring and summer (Dougue'droit and De Saintignon 1970; Douguédroit and de Saintignon 1981; De Saintignon 1986; Cortemiglia 1989; Rolland 2003; Kirchner et al. 2013; Elisa et al. 2015). As in the Himalayas, the lapse rate values for these mountains were lower in winter (−0.4 °C/100 m) as compared to summer (−0.78 °C/100 m). Inconsistent pattern in TLR across different mountain ranges of the world is mainly due to the relatively short instrumental records and micro-scale climatic variability. Majority of the study regions, other than Himalaya, are located in mid latitude mountain systems, where there is no distinct summer wet season or where winter is the wettest season. However, the precipitation pattern of Himalayan region, particularly in Central and Western Himalaya is governed by two atmospheric circulation patterns: one driven by the Indian summer monsoon or south-easterlies and the other by the westerlies in winter (Singh et al. 2017). Contributions of these two characteristics possibly result in a distinct bimodal pattern of TLR in Central and Western Himalayan region (Joshi et al. 2018). Over the last few decades, change in TLR has been reported across few mountain regions as an effect of climate change (Liu and Chen 2000; Vuille and Bradley 2000). A decline in temperature in the range between 0.41 and 0.59 °C/100 m has been reported for some alpine regions in the Europe, (Rolland 2003) during last 2–3 decades. Most of these studies have found the lapse rates steepening at the higher elevations (>3000 m). Hence, due to change in lapse rates and the climate, high elevation areas may experience higher warming as compared to the lowlands regions (Diaz and Bradley 1997; Pepin 2000; Pepin and Lundquist 2008).

3.4.1 Factors Governing Variation in Temperature Lapse Rate

The presence and absence of moisture play a key role in regulating distribution of temperature and precipitation along elevation, and thus control the climate of mountain slopes (Dimri and Niyogi 2013). The higher relative humidity at the high elevation areas in comparison to the low altitude areas during monsoon months leads to sustained lower LRs in mountains. Further, low relative humidity and corresponding higher Lifting Condensation Level (LCL) during pre-monsoon months result in higher LRs (Thayyen and Dimri 2016). The Eastern, Central, and part of the Western Himalaya experience moisture influx through Indian summer monsoon (ISM) during June to September (Kumar et al. 1999, 2006) and Western and North-Western Himalaya by Indian winter monsoon (IWM) during November to March (Dimri 2013a, 2013b). As these two systems negotiate the Himalayan region from opposite directions, the topography regulates their flow and produces seasonal moisture surplus and deficient zones across the Himalayan arc, forming distinct climatic and hydrological zones. Therefore, the amount of water vapor in the atmosphere and its seasonal variation along an elevation plays a key role in controlling the seasonal variations in TLR along the mountain slopes (Joshi et al. 2018).

Apart from this, other factors such the saturation vapor pressure (e_s) and cloud cover also play key role in controlling TLR of a region. Studies have shown that a strong positive relationship exists between monthly rainfall and cloud cover (Kattel et al. 2013). The highest cloud cover and rainfall occur during the monsoon summer, and the lowest cloud cover and rainfall occur between fall and spring. The lowest mean saturation vapor pressure (e_s) lapse rate (−3.5 hPa/km, $r = -0.96$, and $p < 0.0001$) also exist in January and highest (−9.7 hPa/km, $r = -0.95$, and $p < 0.0001$) in May, and the value remains stable temporarily as well as elevationally, resulting in low TLR for a fairly long time (July to September). The weak negative relationship between mean TLR and rainfall ($r = -0.17$, $r^2 = 0.03\%$) suggests that the variations in lapse rate are not consistently associated with rainfall. On the other hand, a significant positive relationship ($r = 0.63$, $p < 0.05$) between mean TLR and e_s lapse rate indicates that the variation in TLR is consistently associated with the e_s in all seasons. This signifies that variation in TLR is governed by the atmospheric vapor pressure.

Surface temperature oscillations also depend on cloudiness. The largest cloud cover occurs in the monsoon or summer months and is lower in the non-monsoon months. Significant cloud cover leads to a reduction in insolation during the day and a confinement of outgoing long-wave radiation during the night (Bhutiyani et al. 2007) thereby increases minimum surface temperatures. Therefore, cloudy skies and moist conditions may reduce the TLR for maximum temperature and enhance the TLR for minimum temperature. Lapse rates are also lower for minimum temperatures (at night) than for maximum temperatures (during day). The difference between lapse rate for maximum and minimum temperature is due to inversions for the minimum temperature, rather than stronger sunshine warming for maximum temperature. The lapse rate is more stable for maximum temperature than that of minimum values, because minimum temperature is more susceptible to cold air drainage effects that would tend to create local temperature inversions (Dodson and Marks 1997). The lower values of TLR in winter and higher TLR in summer are due to maximum dry convection in summer and minimum in winter. The inversions of temperature appeared to be responsible for low temperatures during winter near valley areas. This phenomenon involves strongly reduced gradients in December and January, especially for minimum night temperatures (Rolland 2003). Higher LR values exists for dry slopes (receiving more sunshine and solar irradiance) than moist and humid slopes (receiving less sunshine and solar irradiance) in the Himalaya (Tang and Fang 2006; Immerzeel et al. 2014; Kattel et al. 2015; Joshi et al. 2018). Hence, topographic features (slope and aspect) and microclimatic conditions also play significant role in distribution of temperature and its lapse rate along mountain slopes.

3.4.2 Possible Implications of Low TLRs and EDW on Treeline Ecotone Vegetation

The observed shallower TLR, an indication of EDW, may have several implications on treeline ecotone vegetation in Himalaya under changing climate scenario. Low TLRs particularly during winter and monsoon months are likely to contribute to high treeline in the Himalaya. Heat deficiency is considered as the main cause of treeline formation (Körner 2012). Winters in Himalaya are mild (Sakai and Malla 1981) partly because day lengths remain long (>10 h), and days are sunny (Zobel and Singh 1997). As for growth of trees in high elevations, monsoon (warm and moist) is the key period, when net primary productivity is higher. Hence, elevation-dependent warming is likely to be a major contributor to elevational rise of treeline. The treeline vegetation is largely constrained by the environmental factors including temperature, radiation, wind, and water stress (Körner and Larcher 1988). Plants in treeline ecotone are influenced by climatic changes by way of morphological and physiological adaptations (such as stunted growth forms and small leaves, low thermal requirements for basic life functions). Location of the Himalayan treeline is usually governed by growing season air temperature combined with topographic factors (Körner 1998). Therefore, the shallow TLRs may have several impacts on treeline ecotone vegetation, such as change in treeline dynamics, loss of potential habitats of many threatened plant species, tree species distribution, and upward migration of species to suitable habitat in response to EDW (McArthur 1972; Peters and Darling 1985). As a consequence of EDW, the upwards migration of the treeline would significantly alter the surface characteristics and local climate of alpine regions by modifying the albedo and surface energy balance (Fitzharris et al. 1996) resulting in possibility of new assemblages of plant and animal species in these regions. Furthermore, EDW may also modify the frequency of fire outbreaks as forests tend to transpire most of the available soil moisture under dry and warm conditions. Hence, it may convert fire sensitive areas into regions of sustained fire hazard.

Under warm and dry conditions, treeline vegetation is likely to be affected by increased evapotranspiration and water stress (Kitayama 1996). Because of EDW, vegetation communities that grow in snow beds could be subjected to summer desiccation (Ozenda and Borel 1991). An upward shifting of the treelines by nearly 200 m has been reported in the European Alps as a effect of EDW (Grabherr et al. 1994; Keller et al. 2000) and increase in the plant cover of dwarf shrubs in alpine areas (Cannone et al. 2007) since the early twentieth century. Further, upward migration of species at a rate of 25–45 m per decade (Walther et al. 2005) and increase in species richness (Pauli et al. 2007) at summits have also been reported during the last century. Therefore, because of more warming at higher elevation, the endemic species occupying the highest elevation zones and having narrow range of extents are likely to face extinction.

3.5 Conclusion

The mean annual TLRs for treeline environments in Eastern and Central Himalaya are shallow than the commonly used values of 0.65–0.7 °C $(100\ m)^{-1}$ and dissimilar for different climate regimes. They vary spatially and increase with decreasing moisture being markedly higher for cold and arid region in the Himalaya. Lapse rate are lower in cold and moist months and higher in the dry months suggesting that seasons have strong influence in determining the TLR which is determined by different controlling factors in different seasons. TLR also varies from one aspect to another; being higher for drier aspect and lower for humid and wet aspect. This partially explains the aspect-related difference in treeline elevation in Himalayas. The micro-scale climatology (e.g., rainfall, moisture level, and cloudiness) of a region and its topographical features (e.g., aspect, surface conditions, and elevation) play a key role in determining the variability in TLRs. The annual mean TLR varies across the Himalayan arc and increases significantly from the East to West by 4.1 to 30%; being lowest for Eastern Himalaya followed by for Central Himalaya and Western Himalaya. The range of variation in mean annual TLR for Pir-Panjal and Karakoram mountains is much higher (−0.66 to −0.86 °C/100 m). Along the Himalayan arc, the seasonal variation in TLR is of higher values in pre-monsoon season and lowest during winter season.

Systematic comparison of earlier studies revealed that the lapse rates steepen towards the higher elevations (>3000 m) than the lower elevation areas. Hence, high elevation areas may significantly get warm in comparison to the lowlands due to enhanced elevation-dependent warming (EDW) contributed by snow albedo, surface-based feedbacks, water vapor, radiative flux changes, temperature change, and aerosols. The observed shallower TLR in Himalaya combined with the EDW may affect the ecotone vegetation of treeline because of climate change. This would promote plant growth in treeline ecotone, provided enhances evapotranspiration does not become limiting factor and suitable conditions are available. Low TLRs particularly during winter and monsoon months are likely to have contributed to high treeline in Himalayas. Hence, EDW may further contribute to elevational rise of treeline.

The inconsistencies in lapse rate, both seasonally and annually, across all the mountain regions were largely due to insufficient climatic stations and shortened temporal coverage which is insufficient for a complete cover of the altitude or spatial ranges. Variation in TLR seems to be strongly depending on the studied area, and no evident agreement appears among mountain ranges and different sites across the world. Therefore, it is necessary to have long-term observations across different mountain ranges to achieve a complete re-examination of temperature variation with altitude to explain such inconsistent results. Information about changing climate and vegetation patterns of climatically sensitive alpine ecosystems is very crucial for a comprehensive understanding of their current and future states. Therefore, long-term climate observations are essential to study the change in TLRs *vis-a-vis* change in climate and its impacts on treeline environments in Himalayan region. Further, there is a need to examine whether the shallow TLRs in Himalaya are due to EDW (being more in higher elevations) under the influence of global climate change. The comprehensive study of TLR is further required to increase understanding of Himalayan climate, their controlling mechanisms and impacts on critical ecosystems through improved and long-term observations.

Acknowledgements This study was part of Indian Himalayan Timberline Research (IHTR) project supported by Ministry of Environment, Forest and Climate Change, Government of India under National Mission on Himalayan Studies (NMHS). We thank the Director, GBPNIHE for providing necessary facilities for conducting the research. We highly acknowledged contribution made by Mr. B.S. Bisht for installation and maintenance of sensors in harsh conditions and regular collection of the data. We also thank Project Management Unit of IHTR, UKFD, Government of Uttarakhand, FEWMD, Government of Sikkim, and Director, HAPPRC, HNBGU, Uttarakhand for their support in implementation of the study.

References

Ashraf A (2020) Altitudinal runoff assessment under variable lapse rates of temperature in the HinduKush, Karakoram and Himalaya ranges of Pakistan. Int J Adv Geosci 8(1):10–17

Bandyopadhyay A, Bhadra A, Maza M, Shelina RK (2014) Monthly variations of air temperature lapse rates in Arunachal Himalaya. J Indian Water Resour Soc 34(3):16–25

Barry RG (1992) Mountain weather and climate. Routledge, London, p 402

Barry RG, Chorley RJ (1987) Atmosphere, Weather and Climate, 5th edn. London, Routledge, p 448

Beniston M (1997) Variations of snow depth and duration in the Swiss Alps over the last 50 years: links to changes in large-scale Climatic forcings. Clim Chang 36:281–300

Bhutiyani MR, Kale VS, Pawar NJ (2007) Long term trends in maximum, minimum and mean annual temperatures across the northwestern Himalaya during the twentieth century. Clim Chang 85:159–177

Blandford TR, Humes KS, Harshburger BJ, Moore BC, Walden VP, Ye H (2008) Seasonal and synoptic variations in near-surface air temperature lapse rates in a Mountainous basin. J Appl Met Climatol 47:249–261

Cannone N, Sgorbati S, Guglielmin M (2007) Unexpected impacts of climate change on alpine vegetation. Front Ecol Environ 5(7):360–364

Cortemiglia GC (1989) Validita` dell'elaborazione statistica nel calcolo del gradiente termico verticale in valle Scrivia. *Julia Dertona, 2d serie,* 68: 63–84

De Saintignon MF (1986) De'croissance de la tempe'rature avec l'altitude dans le Jura franc¸ais en fonction du site des postes de mesure: Comparison avec les Alpes et les Vosges. Proc Int Symp on Topoclimatology and its Applications, Lie'ge, Mont-Rigi, Belgium, IGU, pp. 1–11

Diaz HF, Bradley RS (1997) Temperature variations during the last century at high elevation. Clim Change 36:253–279

Dimri AP (2013a) Interannual variability of Indian Winter Monsoon over the Western Himalaya. Glob Planet Chang 106:39–50. https://doi.org/10.1016/j.gloplacha.2013.03.002

Dimri AP (2013b) Intraseasonal oscillation associated with Indian Winter Monsoon. J Geophys Res Atmos 118:1189–1198. https://doi.org/10.1002/jgrd.50144

Dimri AP, Niyogi D (2013) Regional climate model application at subgrid scale on Indian Winter Monsoon over the western Himalayas. Int J Climatol 33:2185–2205. https://doi.org/10.1002/joc.3584

Dodson R, Marks D (1997) Daily air temperature interpolated at high spatial resolution over a large mountainous region. Clim Res 8(1):1–20

Dorji U, Olesen JE, Bøcher PK, Seidenkrantz MS (2016) Spatial variation of temperature and precipitation in bhutan and links to vegetation and land cover. Mt Res Dev 36(1):66–79

Dougue'droit A, De Saintignon MF (1970) Me'thode d'e'tudede la de'croissance des tempe'ratures en montagne de latitude moyenne: Exemple des Alpes franc¸aises du Sud. Rev Geogr Alp 58: 453–472

Douguédroit A, de Saintignon MF (1981) Décroissance des températures mensuelles et annuelles avec l'altitude dans les Alpes du Sud et en Provence (séries 1959–1978). Eaux et Climat: mélanges géographiques offerts en hommage a` Charles-Pierre Péguy. ER 30:179–194

Elisa G, Colombo N, Acquaotta F, Paro L, Fratianni S (2015) Climate variations in a high altitude Alpine basin and their effects on a glacial environment (Italian Western Alps). Atmósfera 28(2): 117–128

Fitzharris BB, Allison I, Braithwaite RJ, Brown J, Foehn P et al (1996) The cryosphere: changes and their impacts. In: Second Assessment Report of the Intergovernmental Panel on Climate Change (IPCC). Cambridge University Press, Cambridge, pp 241–265

Friedland AJ, Boyce RL, Vostral CB, Herrick GT (2003) Winter and early spring microclimate within a mid-elevation conifer forest canopy. Agric For Meteorol 115:195–200

Grabherr G, Gottfried M, Pauli H (1994) Climate effects on mountain plants. Nature 369:448

Hayat H, Akbar TA, Tahir AA, Hassan QK, Dewan A, Irshad M (2019) Simulating current and future river-flows in the Karakoram and Himalayan Regions of Pakistan using snowmelt-runoff model and RCP scenarios. Water 11:761. https://doi.org/10.3390/w11040761

Heynen M, Miles E, Ragettli S, Buri P, Immerzeel WW, Pellicciotti F (2016) Air temperature variability in a high-elevation Himalayan catchment. Ann Glaciol 57:212–222. https://doi.org/10.3189/2016AoG71A076

IHTP (2020) Annual report of the project "Timberline and altitudinal gradient ecology of Himalayas, and human use sustenance in a warming climate", submitted to MoEF&CC. Government of India

Immerzeel WW, Petersen L, Ragettli S, Pellicciotti F (2014) The importance of observed gradients of air temperature and precipitation for modeling runoff from a glacierized watershed in the Nepalese Himalayas. Water Resour Res 50:2212–2226

Joshi R, Kumar K (2013) Analysis of long term climate variability and changes in North-Western states of Indian Himalayan Region (IHR). In: Sundaresan J, Gupta P, Santosh KM, Boojh R (eds) Climate change and himalaya-natural hazards and mountain resources. Scientific Publisher, New Delhi, pp 130–148

Joshi R, Sambhav K, Singh SP (2018) Near surface temperature lapse rate for treeline environment in western Himalaya and possible impacts on ecotone vegetation. Trop Ecol 59(2):197–209

Karlsson IM (2000) Nocturnal air temperature variations between forest and open areas. J Appl Meteorol 39:851–862

Kattel DB, Yao T, Yang K, Tian L, Yang G, Joswiak D (2013) Temperature lapse rate in complex mountain terrain on the southern slope of the central Himalayas. The Appl Climatol 113(3–4): 671–682

Kattel DB, Yao T, Yang W, Yang W, Tiana L (2015) Comparison of temperature lapse rates from the northern to the southern slopes of the Himalayas. Int J Climatol 35:4431–4443

Kattel DB, Yao T, Panday PK (2017a) Near-surface air temperature lapse rate in a humid mountainous terrain on the southern slopes of the eastern Himalayas. The Appl Climatol. https://doi.org/10.1007/s00704-017-2153-2

Kattel DB, Tandong Y, Pandey PK (2017b) Near-surface air temperature lapse rate in a humid mountainous terrain on the southern slopes of the eastern Himalayas. Theor Appl Climatol. https://doi.org/10.1007/s00704-017-2153-2

Kattel DB, Yao T, Panday PK (2017c) Lapse rate in a humid mountainous terrain on the southern slopes of the eastern Himalayas. Theor Appl Climatol. https://doi.org/10.1007/s00704-017-2153-2

Keller F, Kienast F, Beniston M (2000) Evidence of the response of vegetation to environmental change at high elevation sites in the Swiss Alps. Reg Environ Chang 2:70–77

Kirchner M, Faus-Kessler T, Jakobi G, Leuchner M, Ries L, Scheele HE, Suppan P (2013) Altitudinal temperature lapse rates in an Alpine valley: trends and the influence of season and weather patterns. Int J Climatol 33(3):539–555

Kitayama K (1996) Climate of the Summit Region of Mount Kinabalu (Borneo) in 1992, an El Niño Year. Mt Res Dev 16:65–75

Körner C (1998) A reassessment of high elevation treeline positions and their explanation. Oecologia 115:445–459

Körner C (2012) Treelines will be understood once the functional difference between a tree and a shrub is. Ambio 41:197–206

Körner C, Larcher W (1988) Plant life in cold climates. Society for experimental biology. Cambridge University Press, Cambridge, pp 25–57

Körner C, Paulsen J (2004) A world-wide study of high altitude treeline temperatures. J Biogeogr 31:713–732

Kumar KK, Rajagopalan B, Cane MA (1999) On the weakening relationship between the indian monsoon and ENSO. Science 284:2156–2159. https://doi.org/10.1126/science.284.5423.2156

Kumar KK, Rajagopalan B, Hoerling M, Bates G, Cane MA (2006) Unraveling mystery of Indian monsoon failure during El Ni-o. Science 314:115–119. https://doi.org/10.1126/science.1131152

Liu XD, Chen BD (2000) Climatic warming in the Tibetan Plateau during recent decades. Int J Climatol 20(14):1729–1742

Liu X, Cheng Z, Yan L, Yin Z (2009) Elevation dependency of recent and future minimum surface air temperature trends in the Tibetan Plateau and its surroundings. Glob Planet Chang 68:164–174

McArthur RH (1972) Geographical ecology. Harper and Row, New York

Minder JR, Mote PW, Lundquist JD (2010) Surface temperature lapse rates over complex terrain: lessons from the Cascade Mountains. J Geophys Res 115:D14122

Ozenda P, Borel JL (1991) *Les Conséquences Ecologiques Possibles des Changements Climatiques dans l'Arc Alpin*, Rapport Futuralp No. 1, International Centre for Alpine Environment (ICALP), Le Bourget-du-lac, France

Pauli H, Gottfried M, Reiter K, Klettner C, Grabherr G (2007) Signals of range expansions and contractions of vascular plants in the high Alps: observations (1994–2004) at the GLORIA master site Schrankogel, Tyrol, Austria. Glob Chang Biol 13:147–156

Pepin NC (2000) Twentieth-century change in the climate record for the Front Range, Colorado, USA. Arct Antarct Alp Res 32:135–146

Pepin N (2001) Lapse rate changes in Northern England. Theor Appl Climatol 68:1–16

Pepin N, Losleben M (2002) Climate change in the Colorado Rocky Mountains: Free air versus surface temperature trends. Int J Climatol 22(3):311–329

Pepin NC, Lundquist JD (2008) Temperature trends at high elevations: patterns across the globe. Geophys Res Lett 35:L14701. https://doi.org/10.1029/2008GL034026

Pepin N, Bradley RS, Diaz HF, Baraer M, Caceres EB et al (2015) Elevation-dependent warming in mountain regions of the world. Nat Clim Chang 5:424–430

Peters RL, Darling JD (1985) The Greenhouse Effect and Nature Reserves: Global warming would diminish biological diversity by causing extinctions among reserve species. Bioscience 35(11): 707–717

Pratap B, Dobhal DP, Bhambri R, Mehta M (2013) Near-surface temperature lapse rate in Dokriani Glacier catchment, Garhwal Himalaya, India. Himal Geol 34(2):183–186

Rangwala I, Miller JR (2012) Climate change in mountains: a review of elevation- dependent warming and its possible causes. Clim Chang 114(3–4):527–547

Rolland C (2003) Spatial and seasonal variations of air temperature lapse rates in Alpine regions. J Clim 16(7):1032–1046

Romshoo SA, Rafiq M, Rashid I (2018) Spatio-temporal variation of land surface temperature and temperature lapse rate over mountainous Kashmir Himalaya. J Mt Sci 15(3):563–576

Sakai A, Malla SB (1981) Winter hardiness of tree species at high altitudes in the East Himalaya, Nepal. Ecology 62:1288–1298

Sharma E et al (2019) Introduction to the Hindu Kush Himalaya Assessment. In: Wester P, Mishra A, Mukherji A, Shrestha A (eds) The Hindu Kush Himalaya assessment. Springer, Cham

Shrestha AB, Wake CP, Mayewski PA, Dibb JE (1999) Maximum temperature trends in the Himalaya and its vicinity: an analysis based on temperature records from Nepal for the period 1971-94. J Clim 12:2775–2786

Singh SP, Singh RD, Gumber S, Bhatt S (2017) Two principal precipitation regimes in Himalayas and their influence on tree distribution. Trop Ecol 58(4):679–691

Song L, Zhuang Q, Yin Y, Zhu X, Wu S (2017) Spatio-temporal dynamics of evapotranspiration on the Tibetan Plateau from 2000 to 2010. Environ Res Lett 12:014011. https://doi.org/10.1088/1748-9326/aa527d

Tang Z, Fang J (2006) Temperature variation along the northern and southern slopes of Mt. Taibai, China. Agric For Meteorol 139:200–207

Tetens O (1930) Uber einige meteorologische begrie. J Geophys 6:297–309

Thayyen RJ (2020) Hydrology of cold arid Himalaya. In: Dimri AP, Bookhagen B, Stoffel M, Yasunari T (eds) Himalayan weather and climate and their impact on the environment. Springer, Cham. https://doi.org/10.1007/978-3-030-29684-1

Thayyen RJ, Dimri AP (2016) Modeling Slope Environmental Lapse Rate (SELR) of temperature in the monsoon glacio-hydrological regime of the Himalaya. Cryosphere Discuss. https://doi.org/10.5194/tc-2016-152

Thayyen RJ, Dimri AP (2018) Slope environmental lapse rate (SELR) of temperature in the monsoon regime of the western Himalaya. Front Environ Sci 6:42. https://doi.org/10.3389/fenvs.2018.00042

Vuille M, Bradley RS (2000) Mean annual temperature trends and their vertical structure in the tropical Andes. Geophys Res Lett 27:3885–3888

Walther G R, Beißner S, Pott R (2005) Climate change and high mountain vegetation shifts. In: Mountain ecosystems. Springer, Berlin, pp 77–96

Wester P, Mishra A, Mukherji A, Shrestha AB (eds) (2019) The Hindu Kush Himalaya assessment-mountains, climate change, sustainability and people. Springer, Cham

Zobel DB, Singh SP (1997) Forests of the Himalaya: Their contribution to ecological generalizations. Bioscience 47:735–745

Relationship Between Timberline Elevation and Climate in Sikkim Himalaya

4

Avantika Latwal, Priyanka Sah, Subrat Sharma, and Shaik Rehana

Abstract

This study was undertaken to describe characteristics of climate (air temperature and precipitation) at timberline in relation to topography, and also bring out the changes in climate that have taken place over the past 37 years (1977–2015) in the Sikkim Himalaya. The rate of increase in mean air temperature was more at timberline elevations of the outer Himalayan ranges than the inner Himalayan ranges (0.21 °C decade^{-1} for island type timberline [ITL] vs 0.23 °C decade^{-1} for continuous type timberline [CTL]), and same was true for rainfall (more at elevations of the outer Himalayan timberline [ITL, 192 mm decade^{-1}] than the elevations of the inner Himalayan timberline [CTL, 96 mm decade^{-1}]). Elevational bands of similar elevation (2800–3500 m asl) in a timberline zone that were away from permanent snowline were 3.3–4.5 °C warmer than the same elevation close to permanent snowline (inner ranges), which received less rainfall (438–650 mm) than the outer ranges. Annual mean air temperature was higher (~1 °C) in the locations where timberline moved upwards in 37 years. It was observed that the outer Himalayan ranges became warmer and wetter than the inner Himalayan timberline in the studied period. The warming rate was more at island type timberline (ITL) sites than continuous type timberline (CTL) sites indicating that inner regions in Sikkim are able to hold their own climatic features compared to CTL areas that are more vulnerable or responsive to climate change.

A. Latwal (✉) · P. Sah · S. Sharma
G.B. Pant National Institute of Himalayan Environment, Ladakh Regional Centre, Leh, Ladakh (UT), India

S. Rehana
International Institute of Information Technology, Hyderabad, Telangana, India
e-mail: rehana.s@iiit.ac.in

S. P. Singh et al. (eds.), *Ecology of Himalayan Treeline Ecotone*,
https://doi.org/10.1007/978-981-19-4476-5_4

Keywords

APHRODITE · Climate change · Elevation · Temperature · Timberline · Precipitation

4.1 Introduction

The impacts of global climate change are evident through increase in air temperature and changing patterns of precipitation and melting of the glaciers (Bhutiyani et al. 2010). In recent decades the mountain ecosystems in the Himalaya have warmed more rapidly than many other parts of the Earth (Bhutiyani et al. 2010; IPCC 2014). Therefore, an increasing number of studies have focused on timberline as a field laboratory to analyse impacts of global warming (Qi et al. 2015; Walck et al. 2011; Wielgolaski et al. 2017). Upward movement of treeline is expected due to global warming (Grace et al. 2002; MacDonald et al. 1998; Schickhoff et al. 2015) across altitudinal and latitudinal gradients of the Himalayan Arc. Long-term studies have been shown to be useful in assessing the responses of high-elevation ecosystems to climate change (Rogora et al. 2018).

Climatic conditions in mountains are related to the elevation and geographical locations, particularly where altitude causes steep environmental gradients, as in the Himalaya (Singh et al. 2018). The decrease in temperature with increasing elevation is a primary driver of species biodiversity and the formation of treelines (Mayor et al. 2017). The position of the timberline or treeline depends on multiple factors, and temperature is commonly held to be the main abiotic factor that constrains the growth and regeneration of tree species (Harsch et al. 2009; Holtmeier and Broll 2005; Jump et al. 2007; Körner 1998, 2003, 2012).

Worldwide, high-altitude climatic treeline is associated with the seasonal mean ground temperature of 6.7 ± 0.8 °C during the growing period (Körner and Paulsen 2004) but varies considerably between treelines of Temperate and Mediterranean zones (7–8 °C) and rest of the world (equatorial = 5–6 °C; subarctic and boreal zones = 6–7 °C). Across most of the mountainous regions of the world, air temperature is expected to rise significantly over coming decades (IPCC 2013). The Himalayas are warming relatively more than the global average, particularly during dry winters (Shrestha et al. 2012). In the high mountain ranges and cold deserts of the Trans-Himalayan region, warming episodes are even more intense (Sharma and Tsering 2009; Xu et al. 2009). Temperature and precipitation are the essential environmental parameters that govern timberline dynamics (Singh et al. 2018). A high-elevation treeline is expected to be particularly sensitive to changes in climate as it receives low precipitation (Aryal et al. 2012), hence it provides a unique opportunity to study the impact of different climatic parameters (temperature, precipitation) on vegetation. In the climate change scenarios, the rate of change in temperature with the rise in elevation (temperature lapse rate [TLR]) is also an indication of elevation-dependent warming, which might have several implications on vegetation in the treeline ecotone in the Himalaya (Joshi et al. 2018). In addition, the response of treeline ecotone could vary significantly in relation to climate change, which depends upon warmer temperatures of pre-monsoon months as well as increased evaporation from soil and vegetation which is associated with warming. Thus, climate warming may cause the upslope movement of treeline species, whereas grazing pressure may depress treelines (Singh et al. 2019).

Details of spatial distribution of the two main types of timberlines (parallel to permanent snowline or island-type present in isolated mountain summits away from the snowy ranges) and their temporal changes (1977–2015) in Sikkim Himalaya have been presented in Sah et al. (2022). The main objectives of the study are: (i) to find out variations in the air temperature and precipitation at timberline, and their relationship in different spatial locations of timberline in the Sikkim Himalaya, and (ii) to analyse influence of changing climate (air temperature, precipitation) on timberline altitude between the years 1977 and 2015 (upward/downward movement or stationary during the studied time period).

4.2 Study Area: Climate and Altitude

This study was carried out in Sikkim, which lies between 27°04′46″ to 28°07′48″ N latitudes and 88°00′58″ to 88°55′25″ E longitudes in the north-eastern part of the Indian Himalayan Region (IHR). The geographical area of Sikkim is 7096 km^2, and elevation range is from 194 to 7802 m amsl (for details see Sah et al. 2022). Sikkim represents part of the Himalayan biodiversity hotspot encompassing vegetation from subtropical to alpine climatic conditions. While at some places, temperature drops below 0 °C during winter and occasionally exceeds 28 °C in summer, generally the mean daily minimum temperature ranges from 4.2 °C to 16.6 °C and mean daily maximum temperature from 12.3 °C to 22.2 °C. The rainfall is well distributed from May to early October, and July is the wettest month in most of the places and mean monthly rainfall varies between 19.6 mm and 688.2 mm (based on data for 1971–2000; Fig. 4.1) (www1 n.d.).

4.3 Methods

4.3.1 Spatial Attributes of Timberline Altitude

High elevation limit of forests represents one of the most conspicuous vegetation boundaries in mountains. Timberlines delineated by Sah et al. (2022) for the years 1977 and 2015 were used for this study. Spatially arranged distinct timberlines were classified as (1) '*continuous*' type timberline (parallel to permanent snowline having alpine region in between these two lines; hereafter CTL), and (2) '*island*' type timberline, hereafter ITL, that is found in high-elevation summits away from the snowy ranges that arise due to geological processes where island type shape of alpine zone occurs around a summit, which is surrounded by timberline towards

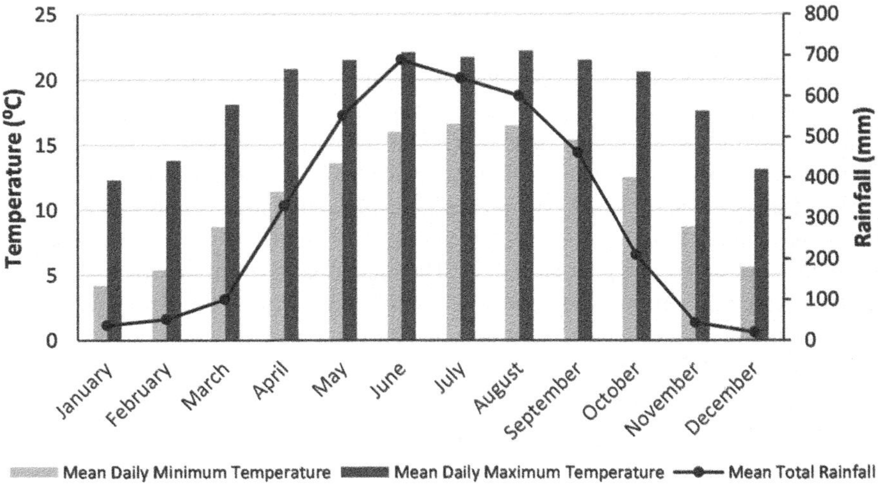

Fig. 4.1 Mean monthly temperature and precipitation at Gangtok India Meteorological Department (IMD) station from 1971 to 2000 (*Source*: imdsikkim.gov.in)

lower elevations. So, this type of timberline is the upper edge of a forest and around the summit, and they are small and isolated like an island (Sah and Sharma 2018).

Temporal change analysis (1977–2015) was done for these two types of timberlines by Sah et al. (2022), and changes were marked as 'shift' (upward/downward) and 'no change' (stationary) in timberline position with respect to the base year (1977). The continuous line data (vector format on Geographic Information System (GIS) platform) were used to create point data (30 m separation between two points in a line) for corresponding years (1977–2015) using Pixel to ASCII Converter feature of Earth Resources Data Analysis System (ERDAS) Imagine 16. Thirty-metre spatial resolution of points was in tune with the resolution of Advanced Spaceborne Thermal Emission and Reflection Radiometer (ASTER) Digital Elevation Model (DEM) (ASTER DEM, 30 m), which was used to extract altitudinal information. Interaction of line points and DEM provided the maximum altitude of timberline occurrence at each location. For each point of timberline, spatial attributes (latitude, longitude, elevation) were also extracted.

4.3.2 Climate and Timberline Homogeneity

The Himalaya is generally data-deficient and it is more true for meteorological data, particularly in the higher altitudes. Gridded/modelled datasets were explored for this study as many global and regional climatic datasets are available; for example, APHRODITE (Asian Precipitation – Highly Resolved Observational Data Integration Towards Evaluation) daily gridded precipitation is the only long-term (1951 onwards) continental-scale product that contains a dense network of daily rain-gauge data for Asia including the Himalaya, South and Southeast Asia and mountainous areas in the Middle East (Yatagai et al. 2012). Number of valid stations ranges between 5000 and 12,000, representing 2.3–4.5 times the data available through the Global Telecommunication System network. These were used for most of the daily grid precipitation products. Climatologically, daily mean precipitation and temperature data are available for the moonsonal Asian region at 0.05° × 0.05° resolution (www2 n.d.). The product contributes to studies related to diagnosis of climate change, evaluation of Asian water resources, statistical downscaling, forecast improvements, and verification of numerical model simulation and satellite precipitation estimates. APHRODITE precipitation data can be used as a benchmark for various estimations of gridded precipitation. Due to its long-term availability (1951 and 1961 for precipitation and temperature, respectively), APHRODITE was preferred over high-resolution (e.g. 1 × 1 km of Worldclim) and lower-resolution data of IMD (gridded data of 1°).

Air temperature (daily mean) and daily precipitation (rainfall) at timberline altitude in Sikkim State were extracted from gridded data (resolution of 0.25°) of APHRODITE for different years (1977 and 2015) using MATLAB R2019a. This dataset is available at http://www.chikyu.ac.jp/. Annual mean temperature and total annual rainfall were derived from the daily dataset. To extract various points of annual mean temperature and annual precipitation at timberline altitude, bilinear interpolation technique was used to harmonize the different resolutions of two datasets, which uses four near-neighbour grids and estimates the distance average with the closer grid being given higher weights. Bilinear interpolation algorithm is popular due to its computational efficiency and quality (Dilip et al. 2014). It is particularly useful for downscaling meteorological input data that are already gridded (Schulla and Jasper 2007). Temperature and precipitation (rainfall) were taken for both the years (1977 and 2015).

APHRODITE data were not available from 2015 onwards. Since November 2016, an automatic meteorological station is operational in a watershed where ground-based studies on timberline are being conducted. We tried to compare daily data for Sikkim timberline from these two different sources: (1) interpolated point data from APHRODITE to nearest location of Automatic Weather Station (AWS) close to timberline for year 2015, and (2) AWS data for 2017 (available from November 2016). Strong correlation was observed for air temperature (coefficient $r = 0.85$, $p < 0.05$) between the same dates of the year (Fig. 4.2a). Daily precipitation also showed some correlation between similar dates ($r = 0.46$, $p < 0.05$; Fig. 4.2b). Thus, interpolated gridded data of APHRODITE were used to generate climatic variables of timberline points for analysis purpose.

To account for orographic effects, comprehensive interpolation methodology was used (www2 n.d.). It is believed to be one of the most realistic datasets for Asia because it uses the largest number of gauge observations (Ménégoz et al. 2013). For two spatially distinct timberlines, total data points (at 30 m interval) generated in

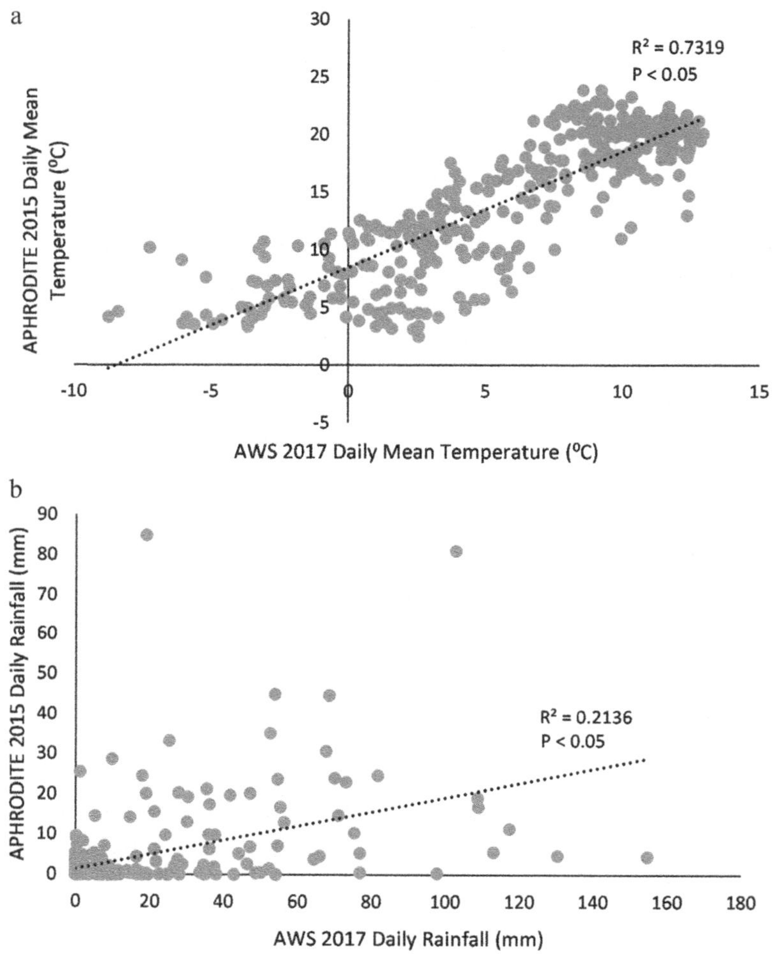

Fig. 4.2 Relationship between (**a**) daily mean temperature (°C), and (**b**) daily rainfall (mm) for the period APHRODITE 2015 and AWS 2017

different years were: (1) ITL = 598 and CTL = 21,903 for year 2015, and (2) ITL = 660 and CTL = 21,042 for year 1977. For each data point, latitude, longitude, altitude, daily air temperature, and precipitation were recorded. This dataset was further categorized on spatial arrangements of timberline (ITL and CTL) and elevation class (100 m interval). All the values in an elevation class were averaged to represent that class. Comparisons were made for stationary and shifted timberline in each category. Statistical analysis (average, standard deviation) was performed using Microsoft Excel 2016.

4.3.3 Timberline Elevation in Relation to Temperature and Precipitation

Relationship between air temperature and precipitation at timberline (year 2015) analysed for the pooled data (all elevation classes) was stronger ($R^2 = 0.99$, $p < 0.0001$; Fig. 4.3a) in ITL elevation than the CTL elevation ($R^2 = 0.79$, $p < 0.0001$; Fig. 4.3b). This was because of specific location of the timberline types, with ITL being in outer ranges and exposed to monsoon thrust while the CTL locations were in the dry inner Himalayan region.

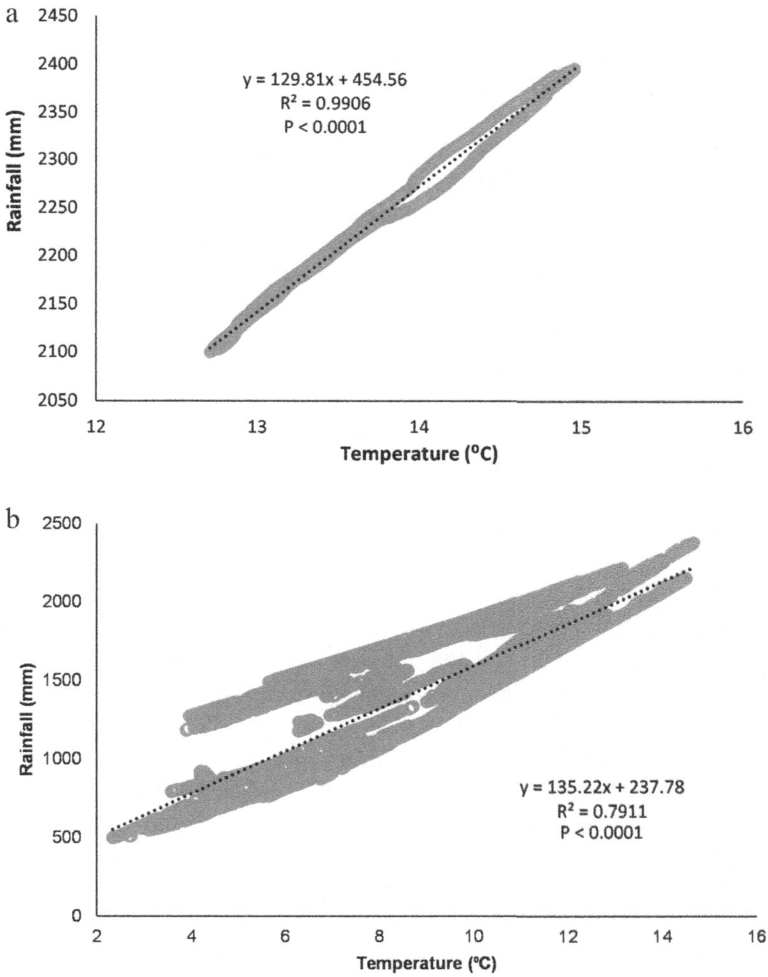

Fig. 4.3 Relationship between temperature and precipitation (year 2015) at timberline altitude of two timberline types/forms: (**a**) ITL and (**b**) CTL

4.4 Results

We analysed two spatially distinct timberline types (ITL and CTL) in relation to air temperature and precipitation. In both the timberline types, we observed similar relationship with temperature (decreasing with altitude) but with different magnitudes (timberline temperature in 2015: ITL, $p = 0.04$, and CTL, $p = 0.001$; timberline temperature in 1977: ITL, $p < 0.0001$, and CTL, $p = 0.001$). Timberline precipitation followed similar pattern (as temperature) in magnitude for the year 2015 (ITL, $p = 0.3$, and CTL, $p = 0.001$), and for the year 1977 (ITL, $p < 0.0001$, and CTL, $p < 0.001$).

4.4.1 Annual Mean Temperature and Total Rainfall at Timberline – Present (Year 2015)

For the year 2015, annual mean temperature and total annual rainfall in different altitudinal zones in two timberline types (island type and continuous type; ITL/CTL) are presented in Fig. 4.4a and b. Timberline altitude of CTL varied considerably between ~2600 and ~4400 m amsl near snowy ranges, while far away from permanent snow, elevation of ITL did occur from 2800 m to 3300 m in Sikkim state (Sah et al. 2022). ITL elevations (away from the permanent snow) were warmer than the CTL elevations (close to permanent snowline) as reflected by average annual mean temperature (ITL = average 13.8 ± 0.7 °C, range 12.7–14.9 °C among the different elevation zones), while this value was 8.6 ± 3.0 °C (average) for CTL (range 2.3–14.7 °C; Fig. 4.4a). This indicates that ITL sites (mostly in the outer Himalaya) are warmer than the CTL sites (close to permanent snow in the inner Himalayan region). This has direct implications to changing environment due to yearly temperature change, temperature lapse rate and anthropogenic activities in the outer Himalayan region.

The outer ranges (ITL) received more rainfall (Fig. 4.4b) than the inner Himalayan timberline elevation (CTL). Between the elevation ranges of ITL (outer ranges), the annual rainfall (range 2099–2396 mm) varied by 300 mm across the ITL sites with an average of 2253 ± 90 mm. In comparison, the annual rainfall observed varied much more (500–2378 mm), with an average of 1397 ± 451 mm, across the CTL sites. The annual rainfall decreased much more with elevation in inner areas where CTL occurred than in outer ranges where ITL occurred (Fig. 4.4b). It appears that in ITL sites, rainfall was almost invariant across the elevation range. Thus, at the highest elevation, annual rainfall in CTL was only about one-fourth of that in ITL.

It appeared that at a given elevation, climate may vary considerably with location, which in the present case is due to distance from snowy ranges (i.e. permanent snowline). Thus, location of timberline altitude (reflected by the outer Himalayan locations, ITL; and the inner Himalayan locations, CTL) also mattered. CTL was cooler and drier than ITL, and the difference magnified with elevation. A given

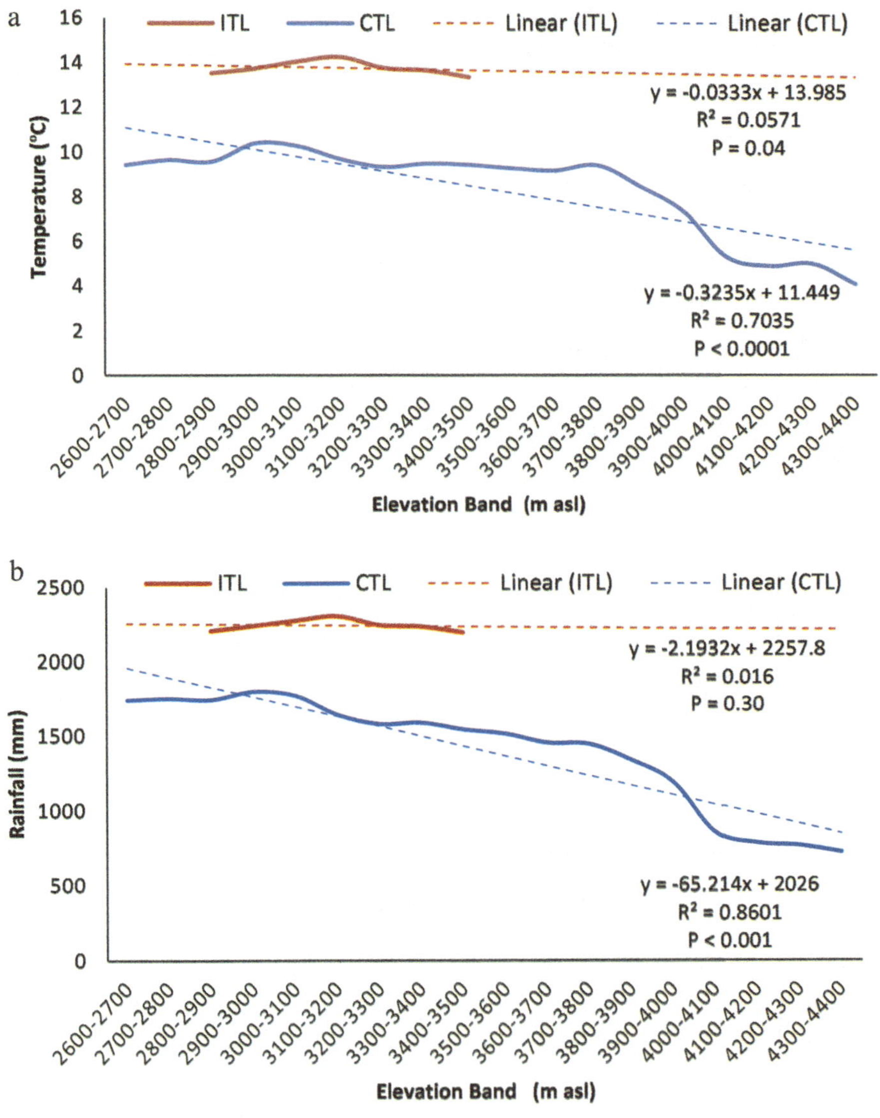

Fig. 4.4 Annual mean temperature (**a**) and annual precipitation (**b**) in relation to timberline elevation (2015)

elevation band was 3.3–4.5 °C warmer in ITL than CTL, and also received 438–650 mm annual rainfall than CTL. Difference in total annual rainfall within an altitudinal range increased with increasing altitude of the inner Himalayan ranges. Thus, high-altitude timberline elevations away from permanent snowline received more rainfall than same elevation of inner ranges.

4.4.2 Annual Mean Temperature and Total Rainfall at Timberline Altitude – Past (Year 1977) and Rate of Change from Year 2015

For the year 1977, annual mean temperature and total annual rainfall in different altitudinal zones in two spatially distinct timberlines (ITL/CTL) are presented in Fig. 4.5a and b. Similar to present time (year 2015), ITL elevations were warmer

a

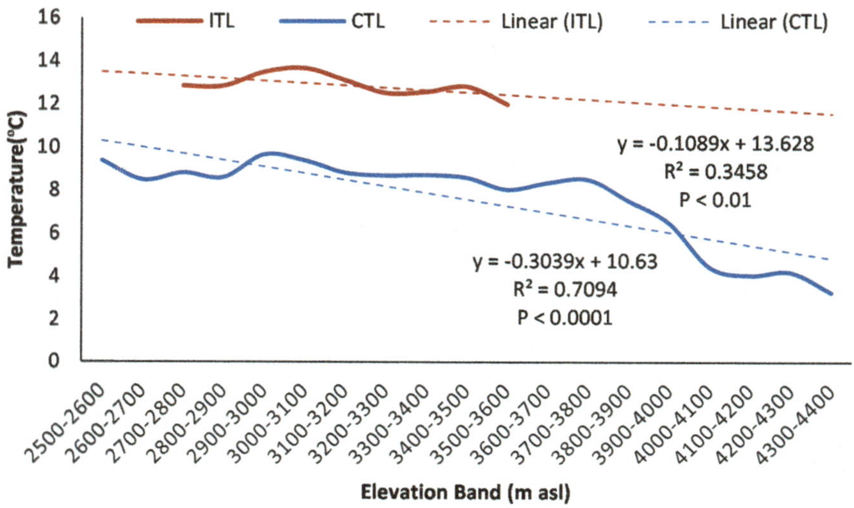

b

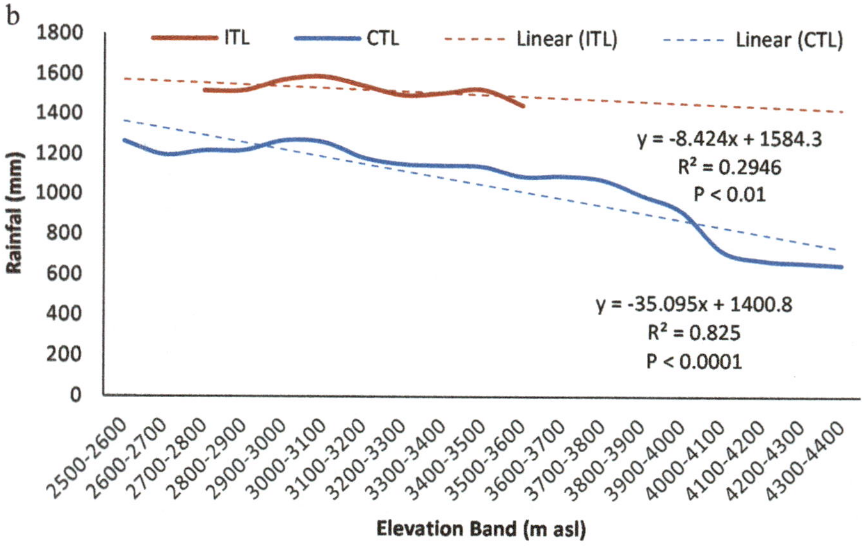

Fig. 4.5 Trends in annual mean temperature (**a**) and total annual rainfall (**b**) in 1977 along elevational gradient where timberline is present in Sikkim (values averaged for each 100 m elevation band)

(range of annual mean temperature, 11.8–14.2 °C; average 13.1 ± 0.8 °C) than the CTL elevations where the annual mean temperature ranged between 3.3 °C and 9.7 °C (average 7.7 ± 3 °C; Fig. 4.5a). It was observed that annual mean temperature at timberline elevation increased at different rates in different spatial arrangements of timberline. For example, at ITL elevation the rate of increase was 0.21 °C decade^{-1} while at CTL elevation the rate was slightly higher, 0.23 °C decade^{-1}. This observation indicated that the warming of inner Himalayan ranges was faster than the outer Himalayan ranges.

Across elevation bands, total annual rainfall varied from 1434 mm to 1636 mm (average 1543 ± 63 mm) for ITL and from 503 mm to 1696 mm (average 1041 ± 296 mm) for CTL (Fig. 4.5b). From 1977 to 2015, generally, annual rainfall increased, but the extent was more for ITL (192 mm decade^{-1}) than for CTL (96 mm decade^{-1}).

4.4.3 Changes in Timberline Altitude Between 1977 and 2015 and Corresponding Changes in Climate at Different Timberline Locations

Shift in timberline has been attributed to the climate change and anthropogenic activities at high elevation. Timberline dynamics between studied years (measured i.e. upward/downward movement or no changes, i.e. stationary) have been described for Sikkim State by Sah et al. (2022). We analysed climatic differences in timberline segments of 1977, which either changed (timberline segment moved upward/downward) or remained stationary (no changes in elevation of timberline segments since 1977).

4.4.3.1 Segments Having No Change in the Timberline Altitude Between the Years

At elevations where timberline has remained stationary (no change) in different elevation bands of CTL, less than 1 °C (0.80–0.98 °C) increase in annual mean temperature (average 0.89 °C) was observed in a time period of 37 years. Temperature difference, with respect to year 1977, slightly increased towards high elevation (Fig. 4.6), and the dips in temperature (Fig. 4.6) may be due to presence of more timberlines on cooler aspects of those elevation bands.

In a time period of 37 years, increase in rainfall was observed in various elevation bands of stationary timberline (CTL). This increment ranged between 108 mm and 545 mm (average 374 mm across elevation bands). The amount of increase in rainfall over time decreased towards higher elevations (Fig. 4.6), and at about 4000 m timberline altitude rainfall increased by less than 175 mm.

4.4.3.2 Upward Shift of CTL

At the places where CTL timberline moved up (since 1977), increase in mean annual temperature was 0.93 °C (average of elevation bands; range 0.88–0.99 °C), which was slightly higher than the stationary CTL (0.89 °C). The average annual mean

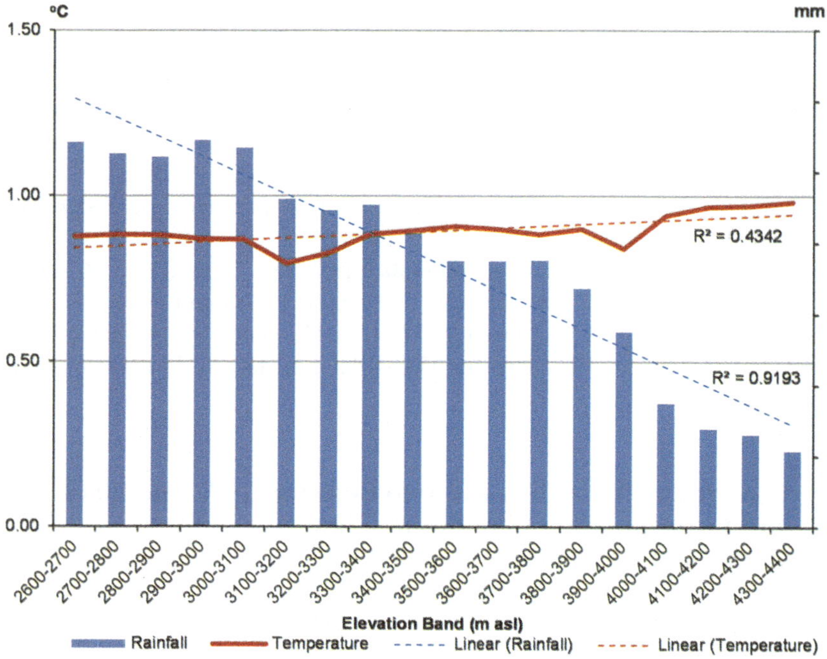

Fig. 4.6 Difference in annual mean temperature and total annual rainfall at the altitude of stationary timberline of CTL between 1977 and 2015

temperature at elevation of upwardly moved timberline (present position) was 8.44 °C while this value was 8.35 °C for stationary CTL in 2015. In brief, the upward movement of timberline was recorded at relatively warmer sites, which also warmed up move.

Total annual rainfall ranged from 597 mm to 1790 mm at upward locations of CTL (moved upwards) in 2015, compared to 651 mm and 1203 mm at stationary CTL. Lower elevations (<3900 m) of CTL altitudes received more rainfall in 2015 than the same timberline altitude in 1977 (Fig. 4.7a). In contrast, temperature difference (present and past) was more (close to 1 °C rise) towards higher timberline elevation (0.96 °C or more, Fig. 4.7a) than lower timberline elevation. A strong inverse relationship (correlation coefficient of −0.918, $p < 0.0001$) was observed between temperature and rainfall of various elevation bands, which indicated that timberline elevation experiencing higher temperature difference had lower rainfall difference in last 37 years.

4.4.3.3 Upward Shift of ITL

In 2015, average annual mean temperature at the elevation to which ITL had moved up was 13.8 °C across the elevation bands (range 13.2–14.3 °C), which was 0.83 °C higher than the temperature of year 1977 (average 12.9 °C). Total annual rainfall of ITL elevation in 2015 varied from 2182 mm to 2306 mm (average 2239 ± 63 mm),

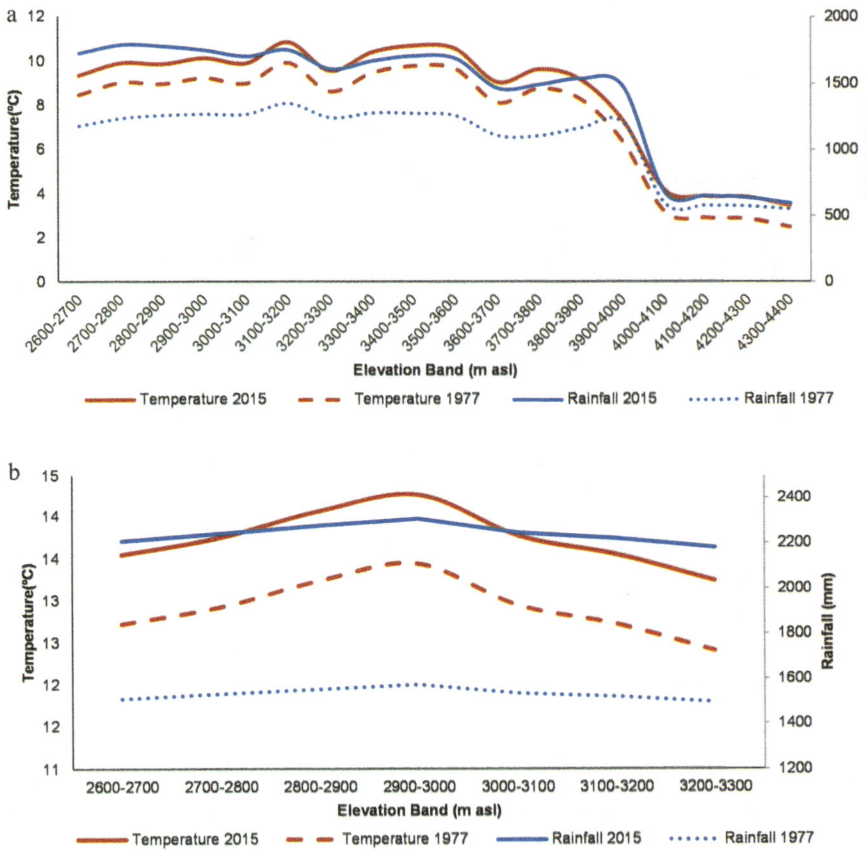

Fig. 4.7 Difference in annual mean temperature and total annual rainfall between 1977 and 2015 at timberline locations that had shifted to upper elevation: (**a**) CTL and (**b**) ITL

which was 708 mm more than the average total rainfall received at same timberline altitude in year 1977 (Fig. 4.7b).

We also made a comparison with CTL timberline that had moved upwards in the same altitude range (2800–3500 m; Fig. 4.8). Greater temperature difference was observed (average 3.54 °C, range 2.54–4.21 °C) in upward-moved ITL than the upward-moved CTL in the same elevation band (Fig. 4.8). Same was true for difference in rainfall (average 529 mm, range 427–636 mm). In the outer Himalayan ranges, same timberline elevations became warmer and wetter in 37 years than the inner Himalayan timberline elevation.

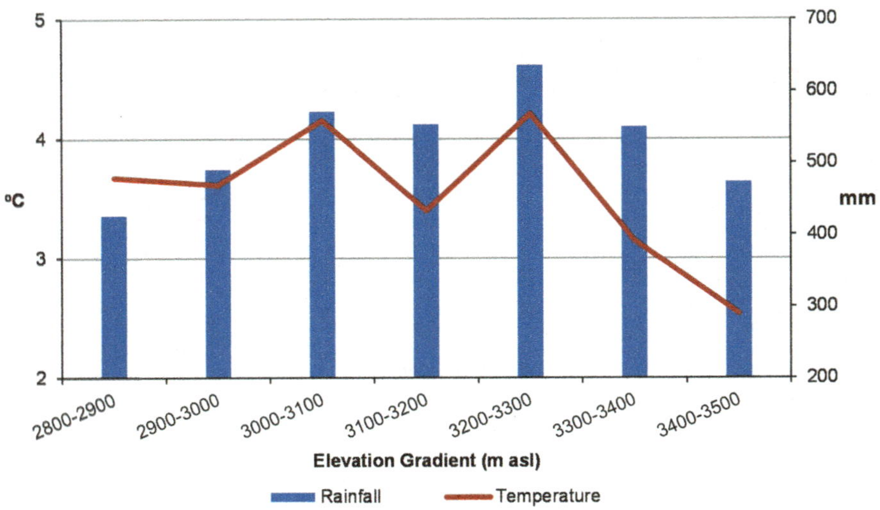

Fig. 4.8 Difference in climatic features (annual mean temperature and total rainfall) at elevations where ITL and CTL had moved upwards

4.5 Discussion

The position of the treeline is influenced by complex interactions with past and current climates and various disturbances (Hofgaard 1997; Körner 1999; Oksanen et al. 1995). The rapid warming of 3–4 °C at the Late Glacial/Holocene transition (11,600 cal. Before Present (BP)) caused significant altitudinal displacements of alpine species that were additionally affected by the rapid upward movement of trees and shrubs (Berthel et al. 2012). In northern Romania, timberline had exceeded an elevation of 1540 m asl by 10,200 cal. Year BP, when summers were warmer than today. The treeline remained below 1810 m asl at this time and reached its maximum elevation after 8500 cal. Year BP, when winter temperatures became milder in the Ronda mountains (Feurdean et al. 2016).

Temperature and precipitation are important climatic parameters that govern timberline/treeline dynamics in climate-sensitive transition zones like timberline areas that are expected to respond to climate warming by advancing beyond their current position (Harsch et al. 2009). Various estimates for the Himalayan region and nearby ranges indicate that warming is occurring in these mountains. In Nepal, mean annual maximum temperature increase (1977–1994) in different ranges varies from >0.06 °C/year (northern and central/western mountain ranges of Middle Mountains) to ≥0.12 °C/year in some pockets; however, rate of increase in annual regional mean for the Himalaya is 0.09 ($p > 0.01$; Shrestha et al. 1999). Over the entire Hindu Kush Himalayan (HKH) region, long-term change (1901–2014) in annual mean surface-air-temperature has increased at a rate of ~0.1 °C per decade but a faster rate of increase has been recorded after 1951 (~0.2 °C per decade; Sabin et al. 2020). Ren et al. (2017) reported rate of mean temperature, mean maximum temperature and mean minimum temperature for same period in whole HKH region as 0.104 °C per decade, 0.077 °C per decade and 0.176 °C per decade, respectively. Our study indicated that since 1977 the average mean temperature at ITL and CTL had increased at a rate of 0.21 °C/decade and 0.23 °C/decade, respectively. This rate is in tune with the studies quoted above but lower than the earlier observations on minimum mean temperature (0.3 °C/decade between 1977 and 2013) calculated from IMD gridded data (1° spatial resolution) for Sikkim State (Singh et al. 2018), and close to Kumar's (2012) station-based observations (station record at Gangtok at an altitude of 1765 m amsl), which showed rate of increase for mean annual minimum temperature to be 0.21 °C/decade (period 1957–2005). Thus, modelled results from APHRODITE are comparable with such analysis.

In a mountain ecosystem, rainfall is also an important factor that governs the growth and phenology of treeline species. It is evident from treeline distribution in Tibetan plateau where at least 500 mm of annual precipitation is required for tree growth (Hou 1982a, 1982b). In central and north-western Tibetan plateau, annual precipitation amounts to about 50–300 mm (Liao 1990; Wang et al. 2011; Zhang et al. 2002; Zheng and Li 1990); therefore, the highest treeline in the Northern Hemisphere occurs in the south-eastern region of the Tibetan plateau (Yao and Zhang 2015). This study indicated that the total annual precipitation at the Himalayan timberline sites (ITL and CTL) was above 500 mm and was increasing at a rate of 192 mm/decade (ITL) and 96 mm/decade (CTL). Therefore, an increase in rainfall between 1977 and 2015 might have contributed to the upward shift in timberline of Sikkim Himalaya. In other words, increase in dry condition due to warming limited the treeline advancement. The increase in rainfall compensated the adverse effect of warming on soil moisture.

Precipitation trend in whole HKH region showed a slight decrease in 114 years but it increased significantly (@5.28% per decade) from 1961 and more rapidly since the mid-1980s (Ren et al. 2017). Pre-monsoon warming accompanied by rain is likely to increase seedling recruitments and upslope rise of treelines in the Himalaya (Singh et al. 2019). Our study indicated that the total annual precipitation at ITL and CTL increased, between 1977 and 2015, at a rate of 192 mm/decade and 96 mm/decade, respectively. This rate was higher than the rate observed between 1957 and 2005 (49.6 mm/decade annual mean precipitation) at lower altitude of 1765 m amsl (IMD station of Gangtok) by Kumar (2012). In contrast to these two studies (present, and Kumar 2012), Singh et al. (2018) observed decreasing trend in annual cumulative precipitation at a rate of 206.5 mm/decade (period 1977–2013; IMD gridded data of cumulative precipitation, 0.25°). Thus, the Himalayan studies require more robust long-term data to study the response and position of timberline to climate change. Our results showed increasingly warming habitats across the high-elevation timberline ecosystems in the Sikkim Himalaya during the last few decades. The European Alps are also experiencing a temperature increase that is stronger than the global mean of about +0.7 °C for the last century (Hansen et al. 2006).

Two spatially distinct timberlines in Sikkim Himalaya (ITL and CTL) occurred over a wide elevation range. While temperature decline with elevation is a universal feature, elevation and moisture relationship varies across the mountain areas of the world. In Sikkim it declined with elevation, and the extent of decline is significantly more in inner ranges (ITL sites).

CTL occurred over a wide elevation range (2600–4400 m asl) whereas ITL elevation range was not as wide as CTL range (2800–3400 m asl). Within the elevation range where timberline occurred, across CTL sites both temperature and rainfall decreased sharply with elevation, while across ITL sites changes were moderate. Consequently, at the highest CTL elevation, rainfall was only one-fourth of that at ITL. In brief, CTL sites were cooler and drier than ITL sites and the difference between them increased with elevation. These contrasts in temperature and moisture between timberlines of outer and inner ranges of the Himalaya have been brought out for the first time.

We also observed greater temperature and precipitation difference in the ITLs that had moved upwards than the CTLs that had also shifted upwards in the same elevation band. Also, the CTL that had shifted upwards had slightly higher mean annual temperature than stationary CTL. It was also observed that the outer Himalayan ranges became warmer and wetter than the inner Himalayan timberline elevation in the studied time period (37 years). Studying *Betula* timberline in central Scandinavian mountains, Odland (2015) observed that latitude explained approximately 71% of the timberline variation but mountain height also contributed significantly with an explanatory rate of nearly 20%, which means if a mountain height is low, an increase in temperature may result in no or small timberline uplift. Yao and Zhang (2015) also demonstrated the effect of mountain height (mass elevation effect) on both timberline elevation and air temperatures for the Tibetan Plateau. The study showed that the mass elevation effect of the central high mountain areas pushed the 10 °C isotherm upwards in the warmest month up to elevations of 4600–4700 m, which shifted the treeline altitude to 500–1000 m higher than along the eastern edge where the mountains reached only 1000 m asl. Therefore, the mass elevation effect contributes to the highest treeline in the Northern Hemisphere, which is present on the south-eastern Tibetan Plateau.

4.6 Conclusion

In this study, variation in climatic parameters (air temperature and precipitation) at timberline elevation was analysed in two spatially distinct timberlines of Sikkim state, and their relationship with timberline altitude was established. The warming rate was more at ITL sites than CTL sites indicating that inner regions in Sikkim are able to hold their own climatic features compared to CTL areas that are more vulnerable or responsive to climate change. During the approximately four decades of climate warming, rainfall has increased in both timberline areas, but more in ITL sites (192 mm/decade, 31.5%) than CTL sites (96 mm/decade, 25.4%).

Generally, upward shift of timberline was limited despite four decades of global warming. It was to an extent related to the warming rate because at the sites where temperature rise was high (average annual), timberline elevation did increase between 1977 and 2015. For example, above 3800 m asl the average temperature difference was 1.33 °C for CTLs that had shifted to upward elevation, whereas below this altitude the observed difference in temperature was 0.84 °C between 1977 and 2015. These findings suggested that the timberline ecotone is sensitive to climate change, and climatic parameters (temperature and precipitation) are natural drivers of changes in the timberline altitude triggered by climate change.

Acknowledgements Authors are thankful to Prof. S. P. Singh for encouragement and guidance to conduct this research; Director, G. B. Pant 'National Institute of Himalayan Environment' (NIHE), Kosi, for providing necessary facility; and Dr. Rajesh Joshi, Sikkim Regional Centre, for providing AWS data. Dr. Rehana Shaik is thankful to IIIT, Hyderabad, for computational and climate data support. Financial grant for this study was supported by National Mission on Himalayan Studies, Ministry of Environment, Forest and Climate Change, Government of India.

References

Aryal A, Hipkins J, Ji W et al (2012) Distribution and diet of brown bear in Annapurna Conservation Area, Nepal. Ursus 23(2):231–236

Berthel N, Schwörer C, Tinner W (2012) Impact of Holocene climate changes on alpine and treeline vegetation at Sanetsch Pass, Bernese Alps, Switzerland. Rev Palaeobot Palynol 174:91–100

Bhutiyani MR, Kale VS, Pawar NJ (2010) Climate change and the precipitation variations in the northwestern Himalaya: 1866–2006. Int J Climatol 30(4):535–548

Dilip PA, Rameshbabu K, Ashok KP et al (2014) Bilinear interpolation image scaling processor for VLSI architecure. Int J Reconfigurable Embedded Syst 3(3). https://doi.org/10.11591/ijres.v3.i3.pp104-113

Feurdean A, Gałka M, Tanţău I et al (2016) Tree and timberline shifts in the northern Romanian Carpathians during the Holocene and the responses to environmental changes. Quat Sci Rev 134:100–113

Grace J, Berninger F, Nagy L (2002) Impacts of climate change on the tree line. Ann Bot 90(4): 537–544

Hansen J, Sato M, Ruedy R et al (2006) Global temperature change. Proc Natl Acad Sci 103(39): 14288–14293

Harsch MA, Hulme PE, McGlone MS et al (2009) Are treelines advancing? A global meta-analysis of treeline response to climate warming. Ecol Lett 12(10):1040–1049

Hofgaard A (1997) Inter-relationships between treeline position, species diversity, land use and climate change in the central Scandes mountains of Norway. Glob Ecol Biogeogr Lett 6:419–429

Holtmeier FK, Broll G (2005) Sensitivity and response of northern hemisphere altitudinal and polar treelines to environmental change at landscape and local scales. Glob Ecol Biogeogr 14(5): 395–410

Hou XY (1982a) China Vegetation Geography and Dominant Plant Composition. Science Press, Beijing

Hou XY (1982b) Vegetational geography and chemical components of dominant plants in China (in Chinese). Science Press, Beijing

IPCC (2013) Climate change 2013: the physical science basis. In: Stocker TF, Qin D, Plattner GK, Tignor M, Allen SK, Boschung J, Nauels A, Xia Y, Bex V, Midgley PM (eds) Contribution of working group I to the fifth assessment report of the intergovernmental panel on climate change. Cambridge University Press, Cambridge

IPCC (2014) Climate Change 2014: Impacts, adaptation, and vulnerability. In: Field CB, Barros VR, Dokken DJ, Mach KJ, Mastrandrea MD, Bilir TE, Chatterjee M, Ebi KL, Estrada YO,

Genova RC, Girma B, Kissel ES, Levy AN, MacCracken S, Mastrandrea PR, White LL (eds) Part A: Global and sectoral aspects. Contribution of Working Group II to the Fifth Assessment Report of the Intergovernmental Panel on Climate Change. Cambridge University Press, Cambridge

Joshi R, Sambhav K, Singh SP (2018) Near surface temperature lapse rate for treeline environment in western Himalaya and possible impacts on ecotone vegetation. Trop Ecol 59(2):197–209

Jump AS, Hunt JM, Peñuelas J (2007) Climate relationships of growth and establishment across the altitudinal range of Fagus sylvatica in the Montseny Mountains, northeast Spain. Ecoscience 14(4):507–518

Körner C (1998) A re-assessment of high elevation treeline positions and their explanation. Oecologia 115:445–459

Körner C (1999) Alpine plant life. Functional plant ecology of high mountain ecosystems. Springer, Berlin

Körner C (2003) Alpine plant life. Springer, Berlin

Körner C (2012) Alpine treelines: Functional ecology of the global high elevation tree limits. Springer, Berlin

Körner C, Paulsen J (2004) A world-wide study of high altitude treeline temperatures. J Biogeogr 31(5):713–732

Kumar P (2012) Assessment of impact of climate change on Rhododendrons in Sikkim Himalayas using Maxent modelling: limitations and challenges. Biodivers Conserv 21(5):1251–1266

Liao K (1990) The Atlas of the Tibetan Plateau. Science Press, Beijing

MacDonald GM, Szeicz JM, Claricoates J et al (1998) Response of the central Canadian treeline to recent climatic changes. Ann Assoc Am Geogr 88(2):183–208

Mayor JR, Sanders NJ, Classen AT et al (2017) Elevation alters ecosystem properties across temperate treelines globally. Nature 542(7639):91–95

Ménégoz M, Gallée H, Jacobi HW (2013) Precipitation and snow cover in the Himalaya: from reanalysis to regional climate simulations. Hydrol Earth Syst Sci. 17:3921–3936. https://doi.org/10.5194/hess-17-3921-2013

Odland A (2015) Effect of latitude and mountain height on the timberline (Betula pubescens ssp. czerpanovii) elevation along the central Scandinavian mountain range. Fennia 193(2):260–270

Oksanen L, Moen J, Helle T (1995) Timberline patterns in northernmost Fennoscandia. Relative importance of climate and grazing. Acta Bot Fenn 153:93–105

Qi Z, Liu H, Wu X et al (2015) Climate-driven speedup of alpine treeline forest growth in the Tianshan Mountains, Northwestern China. Glob Chang Biol 21(2):816–826

Ren Y-Y, Ren GY, Sun XB et al (2017) Observed changes in surface air temperature and precipitation in the Hindu Kush Himalayan region over the last 100-plus years. Adv Clim Chang Res 8(3):148–156

Rogora M, Frate L, Carranza ML et al (2018) Assessment of climate change effects on mountain ecosystems through a cross-site analysis in the Alps and Apennines. Sci Total Environ 624: 1429–1442

Sabin TP, Krishnan R, Vellore R et al (2020) Climate change over the himalayas. In: Krishnan R, Sanjay J, Gnanaseelan C, Mujumdar M, Kulkarni A, Chakraborty S (eds) Assessment of climate change over the Indian region. Springer, Singapore, pp 207–222. https://doi.org/10.1007/978-981-15-4327-2_11

Sah P, Sharma S (2018) Topographical characterisation of high altitude timberline in the Indian Central Himalayan Region. Trop Ecol 59(2):187–196

Sah P, Latwal A, Sharma S (2022) Challenges of timberline mapping in the Himalaya: case of Sikkim and changes between 1977 and 2015. In: Singh SP, Reshi ZA, Joshi RJ (eds) Ecology of Himalayan treeline ecotone. Springer Nature

Schickhoff U, Bobrowski M, Böhner J et al (2015) Do Himalayan treelines respond to recent climate change? An evaluation of sensitivity indicators. Earth Syst Dynam 6:245–265

Schulla J, Jasper K (2007) Model description waSiM-ETH. Institute for Atmospheric and Climate Science. Swiss Federal Institute of Technology, Zürich

Sharma E, Tsering K (2009) Climate change in the Himalayas: the vulnerability of biodiversity. Sustain Mt Dev 55:10–12
Shrestha AB, Wake CP, Mayewski PA et al (1999) Maximum temperature trends in the Himalaya and its vicinity: an analysis based on temperature records from Nepal for the period 1971–94. J Clim 12:2775–2787
Shrestha UB, Gautam S, Bawa KS (2012) Widespread climate change in the Himalayas and associated changes in local ecosystems. PLoS One 7(5):e36741
Singh CP, Mohapatra J, Pandya HA (2018) Evaluating changes in treeline position and land surface phenology in Sikkim Himalaya. Geocarto Int 35(5):453–469. https://doi.org/10.1080/10106049.2018.1524513
Singh SP, Sharma S, Dhyani PP (2019) Himalayan arc and treeline: distribution, climate change 252 responses and ecosystem properties. Biodivers Conserv 28(8–9):1997–2016
Walck JL, Hidayati SN, Dixon KW et al (2011) Climate change and plant regeneration from seed. Glob Chang Biol 17(6):2145–2161
Wang C, Zhou S, Tang X et al (2011) Temporal and spatial distribution of heavy precipitation over Tibetan Plateau in recent 48 years. Sci Geogr Sin 31(4):470–477
Wielgolaski FE, Hofgaard A, Holtmeier FK (2017) Sensitivity to environmental change of the treeline ecotone and its associated biodiversity in European mountains. Clim Res 73(1–2): 151–166
www1 (n.d.) India Meteorological Department, Gangtok, Sikkim. http://imdsikkim.gov.in/monthlymetparam.html
www2 (n.d.) Maeda, Mio, Yasutomi, Natsuko, Yatagai, Akiyo & National Center for Atmospheric Research Staff (Eds). "The Climate Data Guide: APHRODITE: Asian Precipitation - Highly-Resolved Observational Data Integration Towards Evaluation of Water Resources." Retrieved from. https://climatedataguide.ucar.edu/climate-data/aphrodite-asian-precipitation-highly-resolved-observational-data-integration-towards
Xu J, Grumbine RE, Shrestha A et al (2009) The melting Himalayas: cascading effects of climate change on water, biodiversity, and livelihoods. Conserv Biol 23(3):520–530
Yao Y, Zhang B (2015) The mass elevation effect of the Tibetan Plateau and its implications for Alpine treelines. Int J Climatol 35(8):1833–1846
Yatagai A, Kamiguchi K, Arakawa O et al (2012) APHRODITE: Constructing a Long-Term Daily Gridded Precipitation Dataset for Asia Based on a Dense Network of Rain Gauges. Bull Am Meteorol Soc 93:1401–1415. https://doi.org/10.1175/BAMS-D-11-00122.1
Zhang BP, Chen XD, Li BL et al (2002) Biodiversity and conservation in the Tibetan Plateau. J Geogr Sci 12(2):135–143
Zheng D, Li BY (1990) Evolution and differentiation of the natural environment of the Qinghai-Tibet Plateau. Geogr Res 9(2):1–10

The Treeline Ecotone in Rolwaling Himal, Nepal: Pattern-Process Relationships and Treeline Shift Potential

5

U. Schickhoff, M. Bobrowski, J. Böhner, B. Bürzle, R. P. Chaudhary, M. Müller, T. Scholten, N. Schwab, and J. Weidinger

Abstract

Using a landscape approach, the sensitivity, and response of a treeline ecotone to climate warming were addressed in a multiyear study in Rolwaling Himal, Nepal. The Rolwaling treeline ecotone still has a vegetation zonation unaffected by human land use, providing ideal conditions for capturing a climate signal. Along the elevational gradient, upper subalpine forests of tall, upright-growing *Abies spectabilis* and *Betula utilis* give way at the treeline to *Rhododendron campanulatum* krummholz at 3900–4000 m. Warming trends across Nepal have increased to more than 0.2 °C per decade. The treeline growing season mean soil temperature at 10 cm soil depth is 7.5 ± 0.6 °C, considerably higher than the limit of the life form tree. Thus, the treeline position in Rolwaling is lagging behind climatic changes. Lag factors, non-thermal drivers and feedback processes are largely related to the dense krummholz zone, which act as an effective barrier for upslope migration of other tree species. Feedback processes in the self-sustaining *Rhododendron campanulatum* krummholz create site conditions such as nutrient deficiency, reduced light availability and lower soil temperatures that severely restrict the competitiveness of other tree species, further constrained by allelopathic effects of *Rhododendron*. Tree growth-climate relationships show that

U. Schickhoff (✉) · M. Bobrowski · J. Böhner · B. Bürzle · N. Schwab · J. Weidinger
CEN Center for Earth System Research and Sustainability, Institute of Geography, Universität Hamburg, Hamburg, Germany
e-mail: udo.schickhoff@uni-hamburg.de

R. P. Chaudhary
RECAST Research Centre for Applied Science and Technology, Tribhuvan University, Kirtipur, Nepal

M. Müller · T. Scholten
Department of Geosciences, Chair of Soil Science and Geomorphology, University of Tübingen, Tübingen, Germany

S. P. Singh et al. (eds.), *Ecology of Himalayan Treeline Ecotone*,
https://doi.org/10.1007/978-981-19-4476-5_5

climate warming-induced moisture deficits during pre-monsoon seasons have adversely affected radial tree growth of *Abies spectabilis* and *Betula utilis* over recent decades, also playing a role in retarded treeline dynamics. Regardless of stand densification and high levels of recruitment, even above the krummholz zone, a treeline shift to higher elevation is only to be expected in the medium to long term (decades to centuries).

Keywords

Alpine treeline · Climate change · Feedback processes · Himalayas · Krummholz · Recruitment · Soil temperature · Spatial pattern · Tree growth-climate relationships · Treeline dynamics · Vegetation-environment relationships

5.1 Introduction

Current climate warming has become the most important driver of modern treeline research. The recent decade has seen further intensification of the interest in treelines and their dynamics that had experienced a major upturn since the 1990s (Holtmeier and Broll 2020). Recent reviews and global-scale studies corroborated that thermal limitations determine treeline elevation at large scales (Körner 2012, 2021a; Paulsen and Körner 2014; Holtmeier and Broll 2017; Wieser et al. 2019), while newer palaeoecological studies confirmed previous evidence that treeline elevations have responded to warming and cooling episodes throughout the Holocene (Schwörer et al. 2014; Vincze et al. 2017; Li et al. 2019).

The majority of recent studies address the key question of treeline response to climate warming and the potential upslope advance of the climatic treeline. Most of these studies are landscape- and local-scale studies focussing on complex spatial and temporal patterns of treeline response as influenced by the interaction of broad-scale controls (regional temperature) and fine-scale modulators (the whole of abiotic and biotic local site conditions and their interactions including human impact and treeline-landscape history). Landscape- and local-scale studies are strongly advocated in this context by Wieser et al. (2014), Holtmeier and Broll (2007, 2010a, 2017, 2020), Feuillet et al. (2020) and Bader et al. (2021). Studies at this scale have been conducted in an ongoing research project in the Rolwaling Himal, Nepal (e.g. Schickhoff et al. 2015, 2016a; Müller et al. 2016a; Schwab et al. 2016, 2017, 2020, 2022; Bürzle et al. 2017, 2018), the results of which are summarized in this paper.

There is still inconsistency in the currently observed treeline responses with treelines advancing upslope and others showing insignificant responses (Harsch et al. 2009), even though the percentage of ascending treelines, including krummholz treelines, strongly increased in the most recent meta-analyses (Hansson et al. 2020; Lu et al. 2021). The debate is ongoing to what extent treeline form is an indicator of the responsiveness of treelines. Some studies, including our own, are in line with the respective model of Harsch and Bader (2011), while other studies provide no evidence supporting it (cf. Fajardo and McIntire 2012; Compostella and Caccianiga 2017; Treml and Veblen 2017). It is evident, however, that spatial patterns contain much information about treeline-forming processes and about the processes that control treeline dynamics (Bader et al. 2021). In the Himalayas, diffuse treelines are largely limited to less disturbed or near-natural sites in southern aspects which have vanished almost completely. For this mountain system, the occurrence of abrupt and island treelines under natural conditions can be virtually excluded. The far majority of less disturbed or near-natural Himalayan treelines, mainly confined to north-facing slopes, has to be categorized as krummholz treelines (Schickhoff et al. 2015). Since krummholz treelines often show lower responsiveness to climate warming, substantial treeline shifts at these near-natural Himalayan treelines occur most likely in the long term only.

Many recent studies use dendroecological methods in order to detect shifts in tree growth sensitivity at treelines, mostly showing positive effects of climate warming, but also the complexities of growth limitation (cf. Schwab et al. 2018; Gaire et al. 2020; Zhang et al. 2021). In this respect, the promising dendroecological potential of shrubs in treeline ecotones is increasingly exploited (Lu et al. 2015; Weijers et al. 2018; Panthi et al. 2021). A range of other recent studies focus on the role of land use legacies in driving treeline dynamics, aiming at disentangling effects of human impact from effects of multiple ecological and biophysical factors (Treml et al. 2016; Cudlín et al. 2017).

While the findings of many newer studies suggest growth enhancement and treeline shifts to be a response to climate warming (e.g. Camarero et al. 2015; Kullman 2019), an increasing number of studies indicate that climate warming inputs are modulated at local scales by non-thermal factors. Drought stress has emerged as one of the major constraints limiting warming-induced treeline shifts (Camarero et al. 2021a; Elliott et al. 2021). Another research focus is directed at the complex patterns of recruitment and seedling establishment in recent treeline studies (Lett and Dorrepaal 2018; Johnson and Yeakley 2019). Other recent studies highlighting non-thermal factors focus on topographic, geomorphological and edaphic factors (Dawes et al. 2017; Holtmeier and Broll 2018), snow precipitation (Hagedorn et al. 2014), interspecific trait and response variation (Trant and Hermanutz 2014) and biotic interactions (Liang et al. 2016; Camarero et al. 2017).

The results of these studies provide explanations for lagged responses of treelines, and elucidate that treeline dynamics in many treeline environments are to some extent decoupled from overall climate warming inputs (Malanson et al. 2019). To what extent treeline trees that were left behind by the advancing isotherm lost their temperature sensitivity is subject to a current controversial discussion (cf. Camarero et al. 2021b; Körner 2021b). Whether or not a thermal separation of the current treeline from the climatic treeline can be confirmed for near-natural Himalayan treelines will be discussed in this paper. At the Rolwaling treeline, rather high treeline soil temperatures as well as spatial patterns of tree demography suggest such a separation. In Rolwaling, we assume the krummholz zone to play a crucial role for the retarded treeline shift to higher elevations (see below), similar to the buffer effect of other krummholz zones to treeline dynamics (Millar et al. 2020). The role of soil nutrients and soil moisture as driving factors of a lagged treeline response in Rolwaling will also be discussed in this paper. Recent modelling studies support the concept that treeline elevations are not necessarily in equilibrium with the current climate (e.g. Bobrowski et al. 2017, 2018; Bobrowski and Schickhoff 2017).

After treeline research in the Himalayan region had long been neglected (Schickhoff 2005; Singh 2018; Singh et al. 2019), the question of treeline dynamics has generated considerable interest in recent years, fuelled by large warming trends in the past 30–40 years (Gerlitz et al. 2014; Schickhoff et al. 2016b, 2022; Hasson et al. 2016; Krishnan et al. 2019; Schickhoff and Mal 2020; Karki et al. 2020a). According to the complex three-dimensional geoecological differentiation with a high diversity of hygric, thermal, edaphic and biotic conditions, Himalayan treeline ecotones show considerable differences in altitudinal position, as well as in physiognomy and species composition (Schickhoff et al. 2015). In the Himalayan mountain system, we consider the vast majority of treelines to be anthropogenic, and a relatively low percentage to be orographically/edaphically and climatically determined (Schickhoff et al. 2015, 2016a). If anthropogenic treelines are moving upslope, recent land abandonment or declining human impacts are the dominant drivers, whereas climate change plays a subordinate role. Among treeline types, climatic treelines show comparatively high susceptibility and are more likely to reflect climate tracking since increases in temperature sums and growing season length will affect growth patterns, regeneration and treeline position, at least in the long term. By contrast, orographic and edaphic treelines are largely resistant to the effects of climate warming. Since treeline ecotones in the Himalayas are strongly modified by human impact, it is often a challenge to detect a clear climate change signal, and to exclude land use change as a driver of treeline dynamics. Thus, the few remaining near-natural treeline sites, as in the Rolwaling Himal (Fig. 5.1), provide prime study areas for respective research approaches (Schickhoff et al. 2016a). A substantial medium-term treeline response to climate warming is to be expected from the tiny fraction of climatic treelines, and from those anthropogenic treelines which are no longer exposed to significant human disturbances. However, the vast majority of anthropogenic treelines will be subjected to continued intensive land use in the foreseeable future. Thus, the proportional distribution of treeline types in the Himalayas suggests rather low responsiveness to climate warming, at least in terms of treeline shifts.

Substantial treeline advances or shrub encroachments of alpine meadows in recent decades, reported in some studies (e.g. Baker and Moseley 2007; Brandt et al. 2013; Singh et al. 2018), have to be mainly attributed to effects of land use change. Assertions about exceptional short-term climate warming-induced treeline advances, derived from remote-sensing studies (e.g. Mohapatra et al. 2019; Singh et al. 2020), must be viewed with extreme caution, in particular if they are not backed up by field data (Schickhoff et al. 2022). While remote sensing and modelling studies indicate an expansion of treeline forests (Bobrowski et al. 2017; Lamsal et al. 2017; Chhetri et al. 2018), studies based on field data show stand densification and increased number of seedlings in treeline ecotones, but insignificant treeline shifts in recent decades (Chhetri and Cairns 2018; Chhetri et al. 2020; Gaire et al.

Fig. 5.1 The treeline ecotone in the Rolwaling Himal is unaffected by land use, providing ideal conditions for capturing a climate signal when analysing treeline dynamics (Photo © Udo Schickhoff, April 13, 2013)

2017; Sigdel et al. 2020). Gaire et al. (2014) and Suwal et al. (2016) claim higher rates of upslope shifts, based on uppermost seedling occurrences. However, seedling occurrence is not equivalent to effective regeneration or treeline advance. A major deficit in Himalayan treeline research is that the role of *R. campanulatum* krummholz in treeline dynamics has not been properly addressed so far (cf. Mainali et al. 2020; Sharma et al. 2020).

An increasing number of dendroecological studies in Himalayan treeline environments detect a recent sensitivity of tree growth to pre-monsoon temperature and humidity conditions, also in line with our results (Schwab et al. 2018, 2020; see below), pointing to an increasing influence of warming-induced higher evapotranspiration rates and soil moisture deficits during dry spring months on tree growth and on treeline dynamics (Bräuning et al. 2016; Panthi et al. 2017; Sigdel et al. 2018; Tiwari and Jha 2018; Gaire et al. 2020).

In spite of recent advances in global treeline research, large knowledge gaps still prevail. Meaningful predictions of treeline dynamics under climate change conditions are still not possible. To attribute climate warming-induced non-linear treeline responses to specific local-scale modulators and their interactions remains a highly challenging task. In order to analyse the role of local-scale processes in mediating global or regional climate impacts, and to improve the model-based prediction ability regarding changing spatial treeline patterns, complex research approaches in various treeline environments are still highly needed.

5.2 Methodical Approach

Coordinated research activities in the Rolwaling Himal, Nepal, focussed on the sensitivity and response of a near-natural treeline ecotone to climate warming, and integrated field sampling/mapping, experimental treatments, and modelling studies from the perspective of spatial pattern and process. We used a landscape approach derived from the pattern-process paradigm in landscape ecology and hierarchy theory in ecology. We assumed that processes and patterns at treeline develop at multiple spatial and temporal scales, and regarded treeline ecotones as complex patterns in three dimensions, changing through time by pattern-process feedback governed by the interaction of broad-scale, landscape-scale, and local-scale controls. We hypothesized (1) that the warming trend in recent decades coincides with generally enhanced annual growth increment of trees, enhanced height growth of saplings and explicit increases in tree establishment and stand densities; (2) that annual growth increment of treeline trees reflects interannual and multidecadal climate variability, and that tree establishment and increment patterns primarily respond to increased growing season temperatures; and (3) that interacting climatic, physical and ecological factors cause a spatially non-uniform response pattern because multiple environmental factors (topographic and topoclimatic settings, edaphic conditions such as soil temperatures, soil moisture, soil carbon, soil nitrogen and plant interactions of tree seedlings and ground vegetation) control the treeline response.

The Rolwaling valley (27°52′ N; 86°25′ E) is located in Dolakha District, east-central Nepal, adjacent to the border to Tibet Autonomous Region (Fig. 5.2). It is embedded in the Gaurishankar Conservation Area (2035 km^2), which has been a protected area since 2010. The climate is continental with dry and cold winter conditions and a monsoonal summer season, and distinct spatial variability of temperature and precipitation (Böhner et al. 2015; Karki et al. 2016; Weidinger et al. 2018). The monsoon season between June and September accounts for approximately 80% of the total annual precipitation of 1200–1300 mm at treeline

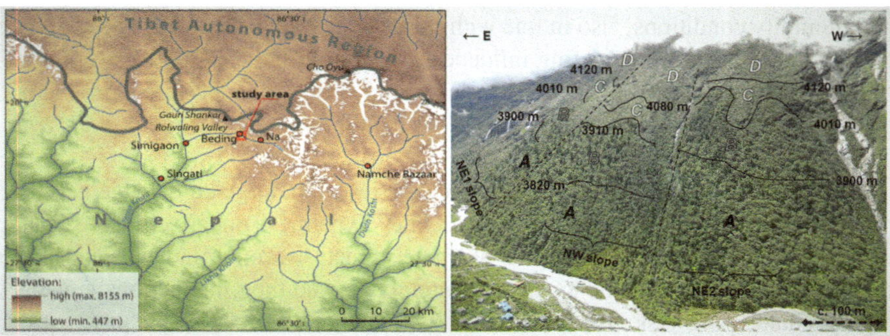

Fig. 5.2 Location of the study area; the slopes were stratified by elevation and aspect, with zones A and B representing the lower and upper sections of the upper subalpine forest, zone C the krummholz belt and zone D the alpine dwarf shrub heath (modified from Schwab et al. 2017)

Fig. 5.3 The central section of the treeline ecotone at north-facing slopes in the Rolwaling Himal; uppermost subalpine forests of *Abies spectabilis* and *Betula utilis* (in April still without leaves) give way to a krummholz zone of *Rhododendron campanulatum* (Photo © Udo Schickhoff, April 13, 2013)

elevation (Gerlitz et al. 2016; Karki et al. 2017a; b). It is preceded by the pre-monsoon season, a dry period with high solar insolation, maximum temperatures, little precipitation and low available soil water capacity (Müller et al. 2016a). Warming trends in Rolwaling correspond to the general Himalayan trends (Schickhoff et al. 2016b, 2022; Hasson et al. 2016; Krishnan et al. 2019). Soils of the treeline ecotone are classified as podzols (Müller et al. 2017). Due to the remote location without connection to the road network (3 days walking distance), the small human population, and the fact that plants and animals in Rolwaling are protected to a certain extent by the recurring Buddhist theme of a sacred hidden valley (Sacherer 1979; Baumgartner 2015; Sherry et al. 2018), the Rolwaling treeline is in a near-natural state and represents a climatic treeline. The study slopes show no signs of fire or grazing by cattle, and wood cutting is negligible. The Rolwaling River separates the uninhabited north-facing study slope from the very sparsely populated south-facing slope where human impact is likewise low. Thus, in view of the fact that most Himalayan treeline sites are disturbed by land use effects (Schickhoff et al. 2015), the study slopes provide a unique research opportunity for detecting a climate change signal when assessing treeline dynamics.

We analysed the treeline ecotone at three north-facing slopes with deviating aspect (Northeast [NE]-Northwest [NW]) along transects from closed subalpine forest via treeline to lower alpine vegetation (3700 m to 4300 m a.s.l.) (Fig. 5.3). The slopes were stratified according to differences in elevational vegetation zonation. Upper subalpine forests (corresponding to elevational zones A and B in

Fig. 5.2) consist of tall upright growing individuals of *Betula utilis* and *Abies spectabilis*, with gnarled trunks of *Rhododendron campanulatum* and *Sorbus microphylla* forming a second tree layer. *Acer caudatum* and *Prunus rufa* occasionally occur as companions. Dense and nearly impenetrable *R. campanulatum* thickets form the krummholz belt in zone C, which contains the uppermost stunted individuals of *Abies spectabilis* and *Betula utilis* with dbh ≥7 cm. The uppermost trees with true tree habitus (upright stem with crown) grow in elevational zone B. Thus, the treeline coincides with the transition from zone B to C, which is located at 3910 m (NW-exp.)/4010 m (NE-exp.), as in zone C only a few outpost-treeline trees occur (Schwab et al. 2016). Zone D represents the alpine vegetation zone composed of dwarf shrub heaths (*Rhododendron* dwarf thickets) interspersed with few low-growing individuals of *Sorbus microphylla*. It is evident from the elevational zonation of vegetation within the treeline ecotone that the Rolwaling treeline has to be assigned physiognomically to treelines with a krummholz belt.

We randomly selected and sampled a total of 50 square plots of 20 m x 20 m in the pre-monsoon and monsoon seasons in 2013 and 2014. In accordance with standard forest inventory procedures (Van Laar and Akça 2007), we measured the diameter at breast height (dbh) at 130 cm above ground level, as well as the height of all vital tree individuals with dbh ≥7 cm. We identified and counted individuals of tree species with dbh <7 cm and assigned them to height classes. We termed individuals with dbh ≥7 cm 'adult trees', while smaller ones were categorized as 'juvenile individuals' or 'recruits' (cf. Bürzle et al. 2017; Schwab et al. 2017, 2020). Regeneration analyses included precisely characterizing microsite conditions on subplots and recording relevant variables such as number, height, apical growth rate, distribution, growth form, vitality, mortality rate, site-microtopography, proximity to potential shelter elements and elevational distance to treeline.

In order to analyse tree growth-climate relationships, we followed standard procedures for the collection, preparation and measurement of tree-ring samples (Speer 2010). To collect data on the species composition and ecology of treeline vegetation types and to differentiate plant communities along the treeline ecotone, we extended the number of randomly selected and sampled plots to 91. In each plot, we collected phytosociological data following the Braun-Blanquet approach (Braun-Blanquet 1964; Kent 2012). We recorded the height and cover of the separate vegetation layers, followed by detailed inventories of all vascular plants using the Braun-Blanquet cover-abundance scale. A voucher specimen of each species was collected for final identification in the laboratory (http://www.efloras.org; Grierson and Long 1983–2001; Watson et al. 2011) and in the herbarium of the Botanische Staatssammlung München (M).

To generate a climate database, we integrated available observational data resources and climate modelling approaches in a comprehensive analysis and modelling framework since the lack and unrepresentative distribution of climatic observations is still a major deficit in the High Himalayas. In order to evaluate the role of topoclimatic settings in treeline dynamics and for the elevational differentiation of vegetation structure along the treeline ecotone, we implemented eight automated meteorological stations at representative locations up to 5100 m, six of them running since April 2013 and two newly implemented since October 2017 (Gerlitz et al. 2016; Weidinger et al. 2021). Merged with the EVK2R instrumentation in the Khumbu Himal east of Rolwaling, a unique contiguous high-mountain meteorological network is provided for environmental analyses as well as for the validation and calibration of spatial high resolution (convection permitting) model simulations using the non-hydrostatic Weather Research and Forecast (WRF) model (for a comprehensive description of the model setup, its parameterization schemes, model physics and nesting architectures, see Karki et al. 2017a,b, 2018, 2020b). Sampling of climatic data included the measurement of soil temperatures and soil moisture (Müller et al. 2016b). Climate data from the Climatic Research Unit (CRU) TS4.00 dataset (Harris et al. 2014; CRU 2017), in particular the mean, maximum and minimum temperatures and precipitation for the period 1901–2012, were consulted to assess tree growth-climate relationships. In addition, we deployed the drought indices of the 1-month and 3-month Standardised Precipitation-Evapotranspiration Index (SPEI) (Vicente-Serrano et al. 2010) based on CRU data.

The sampling of abiotic data included the survey of several topographic variables characterizing aspect, slope, curvature, ground cover and microrelief in order to capture the variation in topography between and within plots (Schwab et al. 2017, 2020). In addition, all plots were sampled for Of, Ah and Ae soil horizons. Soil samples from all plots were analysed via standard methods at the University of Tübingen, Laboratory for Soil Science and Geoecology and at the University of Hamburg, Institute of Geography (cf. Müller et al. 2016b, 2017; Bürzle et al. 2017; Schwab et al. 2017).

In terms of data analyses, we performed computations and figure plotting using the functions of packages in the most recent versions of the program R (R Core Team 2018), unless stated otherwise. To analyse population structures, regeneration patterns and abruptness of transitions along the treeline ecotone, we first calculated stem numbers and stand densities per hectare. Population structures and species compositions were visualized by creating bar graphs. In order to describe and analyse variation in stand structures and to characterize abrupt or smooth and gradual transition patterns between elevational zones, the abruptness of transitions was calculated as the difference between the scaled values of successive zones (Wiegand et al. 2006; Batllori and Gutiérrez 2008; cf. Schwab et al. 2016). Abruptness was determined species-wise for the density of trees. In order to comparatively analyse the abruptness patterns of tree species, we calculated the abruptness of annual and seasonal soil temperatures. We applied functions of the R-packages 'plyr' (Wickham 2011) and 'vegan' (Oksanen et al. 2015).

Relationships between tree species population density and environmental variables were analysed through redundancy analyses (RDA) with backward elimination of explanatory variables as well as through variation partitioning (Legendre and Legendre 2012). To preselect important variables and apply variation partitioning, the explanatory dataset was split into soil, topography and climate variables (cf. Schwab et al. 2016). In order to avoid high multicollinearity, within-group correlations of $|r| > 0.7$ (Spearman, $p < 0.05$, adjusted according to Benjamini and Hochberg 1995) were removed by excluding variables. We included all variables selected by this procedure and all uncorrelated variables in the three final matrices of explanatory variables, further reduced by the RDA backward elimination process. Functions of the R-packages 'PCNM' (Legendre et al. 2013), 'pgirmess' (Giraudoux 2015), 'plyr' (Wickham 2011), 'psych' (Revelle 2015), 'rgdal', 'sp' (Bivand et al. 2013, 2015), 'vegan' (Oksanen et al. 2015), and 'xlsx' (Dragulescu 2014) were used in this data processing. Some missing values in variables such as soil temperature and soil moisture would have restricted the multivariate analyses of tree species density to some extent. To estimate the missing values based on the relationships between variables, we applied multivariate imputation by chained equations (MICE; Van Buuren 2012) based on a random forest classification (Doove et al. 2014) (cf. Schwab et al. 2017).

To analyse tree growth-climate relationships, we followed standard methods and used, inter alia, the programs CooRecorder and CDendro (Larsson 2016), COFECHA (Holmes 1983) and ARSTAN (Cook 1985) to cross-date the tree-ring width (TRW) chronology of *Abies spectabilis*, to remove age-related growth trends and to assess chronology quality. The relationships between the standardized TRW chronologies and the climate variables were analysed by calculating Pearson's correlation coefficients and confidence intervals through a stationary bootstrapped correlation function with optimal block-length selection (Politis and Romano 1994; Politis and White 2004) at a level of significance of $p < 0.05$. We calculated the correlations for the period of climate data availability, that is, 1901–2012. We computed static correlations of TRW with mean, mean minimum and mean maximum temperatures, as well as precipitation sums and SPEI for single months and seasons of the present year and the year before ring formation. We calculated 31-year moving window correlations (Biondi and Waikul 2004) with a 1-year offset between consecutive windows to examine the stability of correlations for the analysed period. We used backward evolving window correlations to determine the lengths of most recent periods with stable correlations, applying the R-packages car (Fox and Weisberg 2011) and treeclim (Zang and Biondi 2015).

Phytosociological data analyses included the use of standard phytosociological methods such as cluster analyses, table arrangement with iterative re-sorting of relevés, determination of diagnostic species by calculating the phi-coefficient and definition of character species in order to classify plant communities of the treeline ecotone. In view of the deficit state of knowledge, we consider the proposed classes as provisional (cf. Bürzle et al. 2017). Nomenclature of vascular plants follows Press et al. (2000) and Watson et al. (2011), with minor adaptations and alterations.

To analyse vegetation-environment relationships, significant differences in site conditions between phytosociological units were assessed with one-way analysis of variance (ANOVA) and post-hoc tests, followed by the visualization of compositional patterns of phytosociological units by means of detrended correspondence analysis (DCA). Corresponding scatter plots were restricted to two dimensions and showed the diversity in species composition along the first two axes, measured in standard deviation units of species turnover (SD) (Gauch 1982; Kent 2012). We interpreted visualized compositional patterns by applying post-hoc correlations of the first two ordination axes (with highest eigen value) with environmental variables (cf. Bürzle et al. 2017).

5.3 Results and Discussion

5.3.1 Regional Climate Inputs Triggering Treeline Response and Topoclimatic Variations

Time series analysis based on available station records and gridded climate data products (Hasson et al. 2016), as well as statistically downscaled reanalyses (Gerlitz et al. 2014, 2016), revealed large warming trends in the Hindu Kush Himalayas in the past 30–40 years, in line with the strong amplification of warming rates to 0.2 °C per decade over the period 1951–2014 assessed in recent reviews (Ren et al. 2017; Krishnan et al. 2019). Observed temperature trends, summarized in Schickhoff et al. (2022), substantially exceed the increase in global mean surface temperature, which was 1.09 °C higher in 2011–2020 than in 1850–1900 (IPCC 2021). Warming trends have accelerated in the past two decades and in recent years, connected with reduced frost days (up to −17 days per decade in the Nepal Himalayas) and a significant positive trend of growing degree days (Gerlitz et al. 2014; Schickhoff et al. 2016b). Extreme warm days and nights show an increasing trend of occurrence (nights by 2.54 days per decade), while occurrences of cold days and nights have declined (Hijioka et al. 2014; Krishnan et al. 2019). We conducted a comprehensive evaluation of temperature trends across Nepal over the period 1980–2016, and confirmed widespread significant warming, which is higher for maximum temperature (0.4 °C per decade) than for minimum temperature (0.2 °C per decade), higher in the mountainous region than in valleys and lowlands, and higher in the pre-monsoon season than in the rest of the year (Karki et al. 2020b). Shrestha et al. (2019) reported more or less equal magnitudes of warming, with a more pronounced rate of increase after 2005 (see also Dahal et al. 2019).

Maximum warming rates in the Hindu Kush Himalayas have been assessed between 4000 and 5000 m a.s.l., locally even at higher elevations (cf. Gao et al. 2018; Pepin et al. 2019; Rangwala et al. 2020). Patterns of elevation-dependent warming have been widely observed in the entire mountain system, in particular on the Tibetan Plateau (Hasson et al. 2016; Karki et al. 2020a; Krishnan et al. 2019; Dimri et al. 2020). High-resolution temperature trends over the Himalayas for the period since the 1980s show a clear elevational gradient in the pre-monsoon season with maximum values of up to 1.2 °C per decade at higher elevations (Gerlitz et al. 2014; Schickhoff et al. 2015). In general, mean annual and mean annual maximum temperatures have increased at rates between 0.6 and c. 1 °C per decade over the past 40 years at higher elevations (cf. Shrestha et al. 1999; Liu et al. 2006, 2009; Bhutiyani et al. 2007, 2010; Shrestha and Aryal 2011; Yang et al. 2011). Elevation-dependent warming signals are most distinct in winter and pre-monsoon seasons. During post-monsoon and monsoon seasons, warming trends were found to be rather reduced throughout the Himalayas (Hasson et al. 2016). Referring to modelling experiments of Karki et al. (2017b, 2018) proving the dominant role of large-scale moisture supply and associated latent heat release on the Himalayan slopes for thermal stratification, the dumped warming signal in summer (and eventually post-monsoon) suggests that present-day anthropogenic radiative forcing initially increases the enthalpy of the summer monsoon system and intensifies the latent heat release on the Himalayan slopes at the expense of direct sensible heating. In addition to the significant warming the Hindu Kush Himalayas have seen in the past, the climate is projected to change more dramatically in the coming decades, with warming to be at least 0.3 °C higher, and in the NW Himalaya and Karakoram at least 0.7 °C higher than the targeted 1.5 °C as a global mean (Dimri et al. 2018; Krishnan et al. 2019).

Trends in annual precipitation in the Hindu Kush Himalayas are difficult to derive due to the widespread non-availability of long-term observations and distinct variabilities prevalent in different subregions and seasons (Schickhoff et al. 2016a, 2022). Over the last 100-plus years, the trend of annual precipitation is characterized by a slight decrease (Ren and Shrestha 2017; Ren et al. 2017; Krishnan et al. 2019). Observations over recent decades show either slightly decreasing or slightly increasing trends, but trends are rarely significant. A clear shift in the temporal characteristics of precipitation variation has been assessed after 1990 with greater inter-annual variability and more frequent intense precipitation events and less frequent light precipitation events (Krishnan et al. 2019). In Nepal, most regions experienced a positive trend of annual precipitation, in particular of monsoonal precipitation, in the period 1979–2016, notably in the years after 2000 (Shrestha et al. 2019; see also Panthi et al. 2015 for the Kali Gandaki River Basin). We observed a significant increase in high-intensity precipitation extremes in some parts of Nepal over the period 1970–2012, at the same time the number of rainy days significantly decreased over the whole of Nepal, while the number of consecutive dry days significantly increased (Karki et al. 2017b).

Shifting the focus from large-scale atmospheric controls and their seasonal variations to the local scale of the Rolwaling Himal (Table 5.1 provides basic climatic parameters), it becomes obvious that the climate is strongly influenced by topoclimatic processes, which result in a strong spatial variability of temperature and precipitation (Böhner 2006; Gerlitz et al. 2016; Weidinger et al. 2021). Both point-source observations and numerical modelling studies generally prove the controlling influence of hydro-meteorological processes for the fine-scale spatial differentiation of radiation and energy exchange. In particular, the diurnal formation of convective clouds with a characteristic lull in the morning hours and secondary or even primary peaks at late afternoon, especially over ridges, alters the normal topographic irradiation pattern (cf. Böhner and Bechtel 2018) with higher radiation totals at NE-compared to NW-facing slopes (due to direct solar insolation during cloudless morning hours), well reflected in soil temperatures, stand structures and treeline position. The local wind systems within the valleys are particularly controlled by incoming solar shortwave radiation and thermal forcing, resulting in mountain-valley wind systems with characteristic nocturnal and diurnal air currents (Böhner et al. 2015; Gerlitz et al. 2016).

Table 5.1 Seasonal climatic averages and precipitation sums measured at NE top station at 4158 m a.s.l. (Dec. 2013–Sep. 2019); annual precipitation is 1225 mm (3*winter + 3*pre-monsoon + 4*monsoon + 2*post-monsoon months), without complete coverage of heavy snowfall events (after data in Weidinger et al. 2021)

	T_{min} (°C)		T_{mean} (°C)		T_{max} (°C)		RH%		Rad (W m^{-2})		$Prec_{sum}$ (mm)	
Season	Mean	SD	Mean	SD	Mean	SD	Mean	SD	Mean	SD	Mean	SD
Winter	−11.7	2.0	−2.3	2.2	8.7	3.3	40.5	10.7	149.8	22.4	6.0	6.3
Pre-monsoon	−5.4	3.0	1.6	2.4	11.7	2.5	73.0	10.6	228.7	20.3	59.8	29.3
Monsoon	3.1	2.0	8.0	1.0	15.7	1.5	97.5	2.6	150.4	32.8	248.4	113.3
Post-monsoon	−5.8	1.2	2.1	1.0	11.2	1.1	60.5	22.3	159.2	15.2	17.5	21.4

Strong topographic forcings of local-scale temperature variations affect the temperature patterns in the treeline ecotone. During clear nights, particularly in winter season (DJF), the mean nocturnal temperatures at valley stations in Rolwaling are substantially lower compared to treeline elevations at upper slopes. The occurrence of positive temperature lapse rates of up to +3 °C/100 m is accompanied by low values of relative humidity at all stations, clearly indicating the importance of radiative cooling at higher elevations and cold air drainage into the valleys for the nocturnal distribution of near-surface temperatures under dry conditions. The monsoon season (JJA) shows mainly negative lapse rates, which are, however, shallower compared to the commonly suggested environmental lapse rate of 0.65 °C/100 m (mean temperature lapse rate in summer c. 0.5 °C/100 m). Highly positive pre-monsoon lapse rates on northeast-facing slopes at noon, to be attributed to the diurnal variations of solar insolation, are a relevant topoclimatic phenomenon at treeline elevations. The east-facing slopes receive high insolation rates in the morning and thus are significantly warmer. The development of convective clouds after midday prevents the opposite effect of enhanced warming of west-facing slopes in the afternoon. Ecological effects of pre-monsoonal temperature variations on the local scale are well reflected in significantly higher elevations of the treeline and altitudinal vegetation zones at the northeast-facing slopes (Gerlitz et al. 2016).

5.3.2 Soil Physical and Chemical Conditions Affecting Treeline Response

Very little is still known about how soil properties control tree growth at its upper limit and affect treeline dynamics, notwithstanding the consensus that at a global scale and throughout diverse ecozones, growing season soil temperature is considered a key factor for tree growth (Müller et al. 2016a). Data evaluations at the Rolwaling treeline revealed higher soil temperatures than expected and soil moisture deficits in the pre-monsoon season, and shed new light on the role of soil and foliar macronutrient concentrations and the significance of soil C:N:P ratios and nutrient availability with soil depth.

We measured a growing season mean soil temperature of 7.5 ± 0.6 °C at 10 cm soil depth (mean of three growing seasons under uppermost forest stands of *Abies spectabilis* and *Betula utilis*) (Fig. 5.4), which is considerably higher than the mean air temperature during the growing season of around 6 °C under which natural, climatic treelines establish according to the global treeline isotherm concept of Körner (2012, 2020, 2021a). As mean soil temperatures were derived from data loggers buried in full shade largely excluding solar radiation-driven soil heat flux, they can be used as a proxy for mean air temperatures. Notably high air and soil temperatures at treeline elevations indicate that the actual treeline position is much lower than the climatic treeline elevation. Thus, the mean temperatures reflect a mismatch of treeline elevation with climate. Even without the temperature rise over the twentieth century, the potential treeline is located at a much higher elevation (Müller et al. 2016a, 2016b), suggesting that the Rolwaling treeline is either in a state

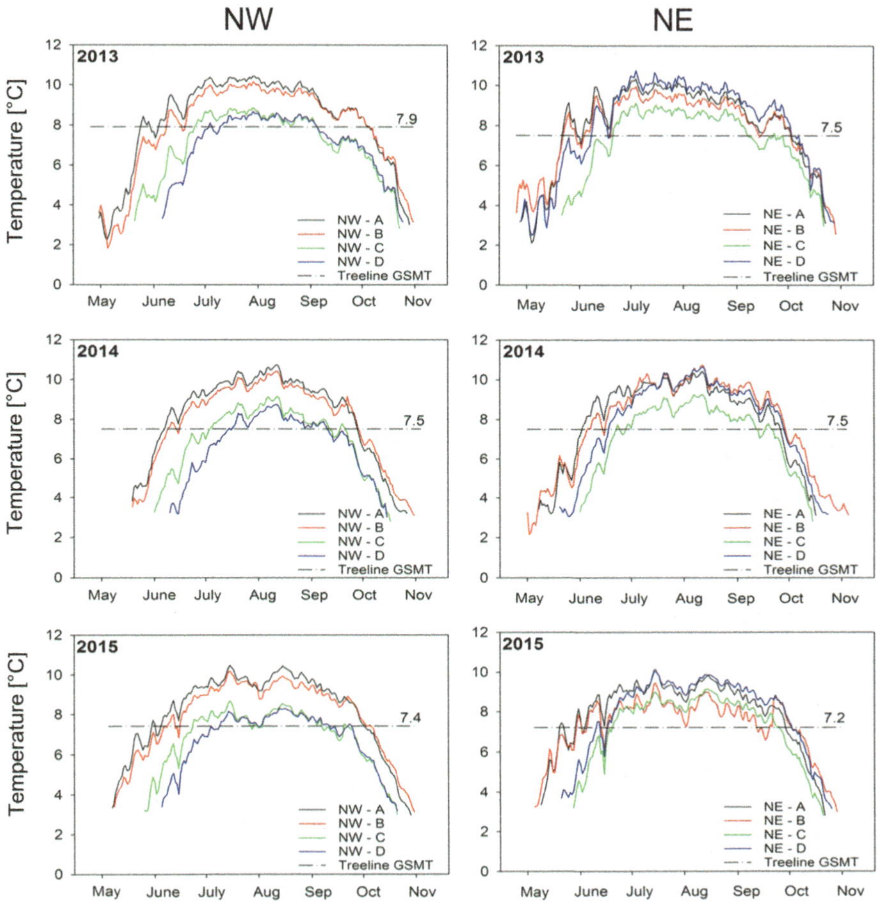

Fig. 5.4 Daily mean soil temperatures during growing season in 2013, 2014 and 2015 at 10 cm depth in altitudinal zones A (closed forest), B (uppermost closed forest), C (krummholz) and D (alpine tundra) on NW (bottom) and NE transect (top). Treeline GSMT = Growing season mean temperature at 10 cm soil depth at treeline (modified from Müller et al. 2016b)

of climatic tracking and lagging behind driving climatic signals since the Little Ice Age, or has not yet recovered from a local disturbance such as fire, avalanches, or other extreme events. However, we did not find any evidence for the latter.

We assessed a marked decline of soil temperatures with elevation at the transition from uppermost closed forests (zone B) to krummholz (zone C), followed upslope by a distinct increase in the NE-exposed alpine tundra (zone D), and a slight increase in the NW-exposed alpine tundra during the second half of the growing season (Aug-Oct) (cf. Figs. 5.4 and Fig. 5.5). These findings suggest that the maximum radiation energy consumption in the krummholz thickets shelters the surface from direct radiative heating, dampens the energy exchange at the surface-atmosphere interface, and efficiently prevents soil heat flux, resulting in cold soils with reduced

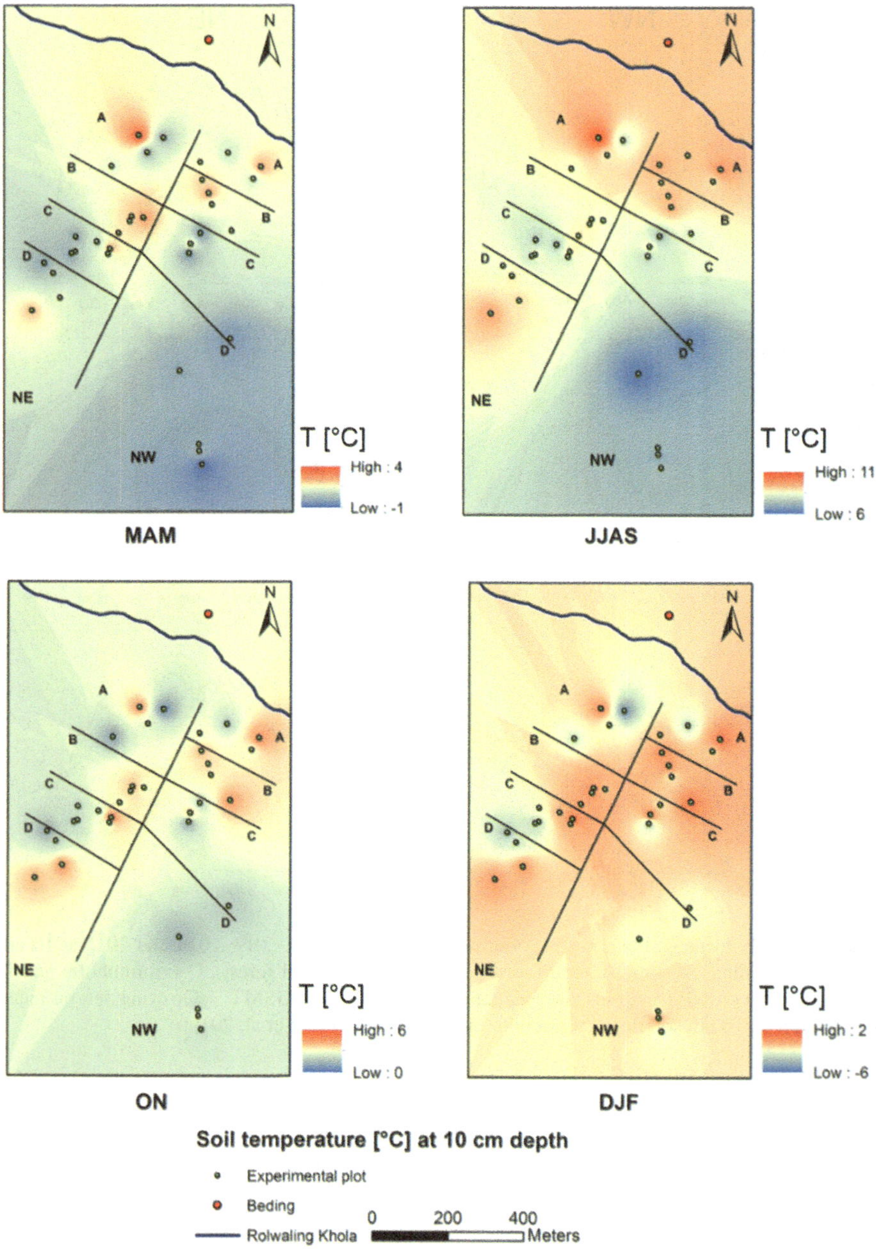

Fig. 5.5 Spatiotemporal distribution of soil temperatures at 10 cm depth. A, B, C, D = altitudinal zones. NW = northwest. NE = northeast. Seasons: MAM = pre-monsoon (March, April, May), JJAS = monsoon (June, July, August, September), ON = autumn (October, November), DJF = winter (December, January, February) (modified from Müller et al. 2016b)

root and microbial activity. The temperature distribution in the alpine tundra (zone D) with highest values at NE aspect corresponds to the asymmetric solar radiation load. However, relatively high soil temperatures relative to elevation and distinct exposure effects likely reflect the increasing duration of (isolating) snow cover at this elevation and its wind-driven distribution with snow driftage at NW and accumulation at NE aspects (cf. Weidinger et al. 2018), strengthening exposure effects.

In the uppermost closed forests (zone B), the seasonally higher solar radiation load on the NE slope is mirrored in both soil temperatures and forest structural parameters, that is, taller tree heights, larger diameters and higher leaf area indices (LAI). Despite a lower elevation of treeline of c. 100 m on the NW slope, mean soil temperatures at both NE- and NW-exposed treelines are almost equal, supporting the global treeline isotherm concept. Multivariate statistical analyses yield a significant relation between soil temperatures and forest structural parameters, that is, variability in tree height, crown length, crown width and leaf area index (LAI). In turn, soil temperature variability is controlled by the tree physiognomy itself (Müller et al. 2016b).

The results further suggest that, beyond the overriding influence of soil temperature, the importance of soil nutrients and soil moisture as factors affecting treeline response to climate warming should not be underestimated. Knowledge on soil nutrient availability (primarily N, P) in Himalayan treeline environments is still scarce. We classified the soil types as podzols, and assessed a sharp decline of nutrients (N, K, Mg) in soils, litter and foliar concentrations with elevation along the treeline ecotone. Soil C concentrations (Of, Ah) slightly decrease with elevation, resulting in a significantly widening C:N ratio and decreasing humus quality (Müller et al. 2016a, 2017). While cation exchange capacity decreases significantly with elevation in the Of layer, correlations between elevation and exchangeable cations are not significant in other horizons. Tree species show diverse relationships between foliar and soil stoichiometric ratios and soil nutrient availability. Significantly increasing foliar C:N and C:P ratios with elevation due to significantly decreasing foliar N and P concentrations suggest a limitation in N and P (Müller et al. 2017). Concentrations of foliar P significantly decrease with elevation, contrasting with extraordinarily high foliar Mn concentrations in the upper treeline ecotone (Drollinger et al. 2017). Nutrient minima in the upper treeline ecotone indicate that plant nutrition is most likely coupled to a recycling of litter nutrients and does not predominantly depend on subsequent supply of nutrients from the mineral soil. Thick leaching horizons with high skeleton fractions and highly acidic soils (pH 2.5–4.0), resulting in restricted nutrient uptake rates due to poor replacement of base cations, support this assumption. The results suggest that low soil-nutrient availability adversely affects establishment, persistence, fitness and survival of trees not only at treeline elevations, but even below in the uppermost subalpine forests, and that nutrient deficiency exerts a considerable influence on treeline ecotone species composition, community patterns, stand structures and treeline position, most likely retarding the attainment of a concordance between treeline position and current climatic conditions (Müller et al. 2016a, 2017; Drollinger et al. 2017). While tree growth at treeline is generally adversely affected

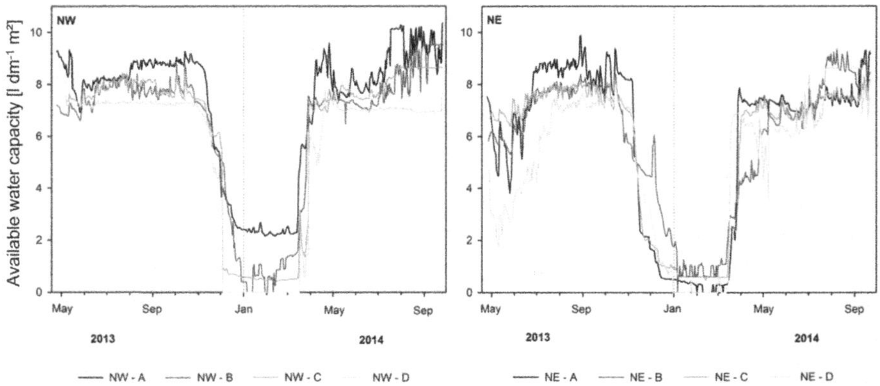

Fig. 5.6 Spatial and temporal variation in available water capacity (AWC; L/dm/m^2) in altitudinal zones A (closed forest), B (closed forest), C (krummholz zone) and D (dwarf shrub heath) on an NE- and an NW-exposed slope in Rolwaling (modified from Müller et al. 2016b)

by pedogenetic processes (podzolization) and extremely acidic soils, one key finding is that soil N and P availability as well as soil C:P and N:P ratios decrease with elevation, in particular at the transition from krummholz to the alpine tundra.

In principle, shallow and coarse-grained soils in the Rolwaling treeline ecotone with a considerable amount of water percolating through the sandy texture from top soils to subsoils induce low water-holding capacities. Soil moisture conditions (available water capacity, 0–10 cm) show a characteristic seasonal pattern (Fig. 5.6). After the monsoon season, soil moisture dramatically drops from November to April during a period of very low precipitation, resulting in soil moisture deficits in the pre-monsoon season (March, April, May) when the growing season starts (Fig. 5.7) (Müller et al. 2016a, 2016b). Warming-induced higher evapotranspiration and soil moisture deficits during dry spring months adversely affect tree growth at Himalayan treelines as well as seedling/sapling density since we found significantly positive correlations of seedling/sapling abundance with soil moisture (Schickhoff et al. 2015). Sites which are prone to drought stress are particularly affected. Moisture supply in the pre-monsoon season might become an effective control of future Himalayan treeline dynamics (Schickhoff et al. 2016b; Mishra and Mainali 2017; Sigdel et al. 2018; Lyu et al. 2019; Schwab et al. 2018, 2020). Drought stress may be reinforced after winter seasons with reduced snow cover. Along the elevational gradient we assessed all year round lowest soil moisture in the alpine tundra (zone D), to be attributed to high solar radiation load, wind, and high evapotranspiration (Müller et al. 2016b). We assume low soil moisture in zone D to be a crucial factor for tree seedling establishment and performance.

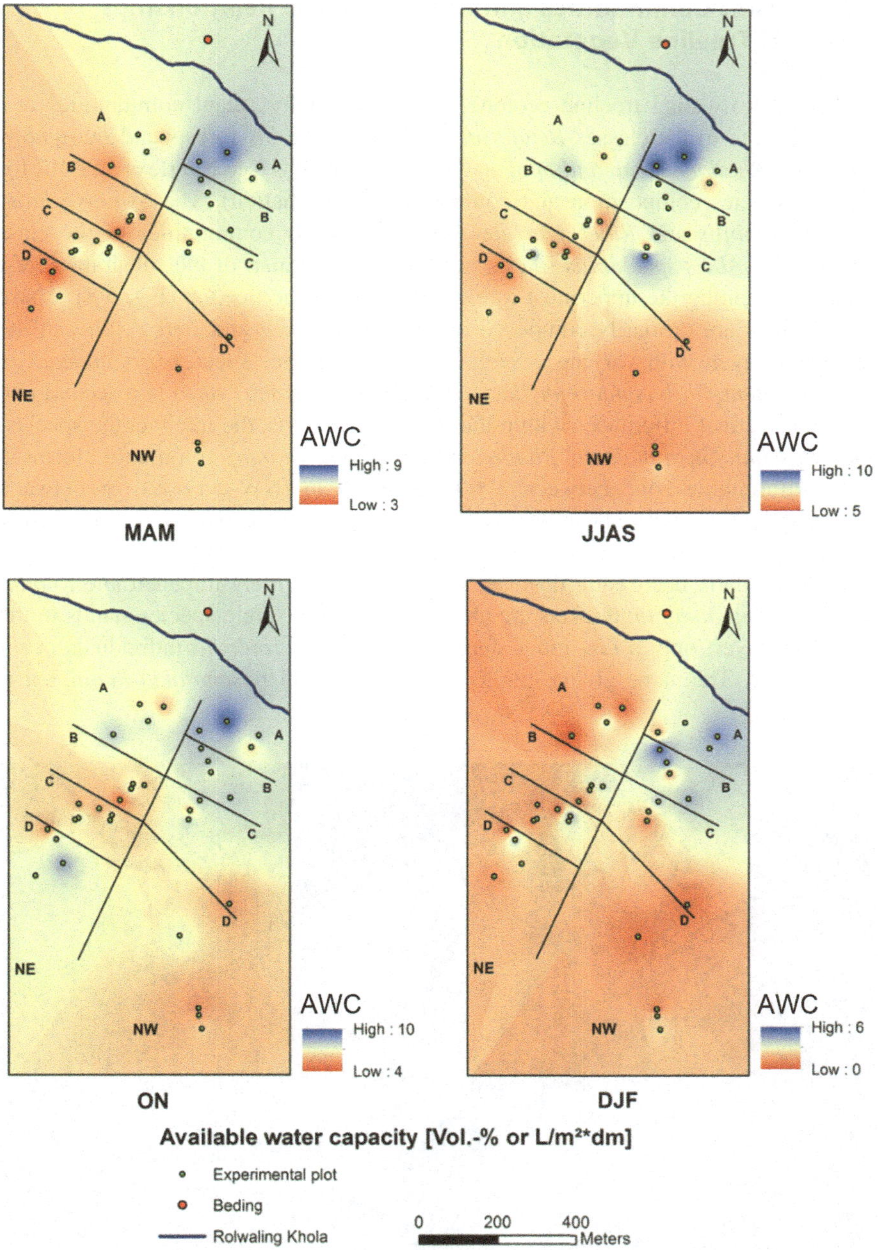

Fig. 5.7 Spatiotemporal variation in available water capacity (Vol.% or L/m^2/dm). A, B, C, D = altitudinal zones. NW = northwest. NE = northeast. Seasons: MAM = pre-monsoon (March, April, May), JJAS = monsoon (June, July, August, September), ON = autumn (October, November), DJF = winter (December, January, February) (modified from Müller et al. 2016b)

5.3.3 Plant Communities and Environmental Relationships of Treeline Vegetation

Along the Rolwaling treeline ecotone, we identified five plant communities and assigned them to the classes *Betula utilis-Abies spectabilis* forests and *Dasiphora arbuscula-Rhododendron anthopogon* dwarf shrub heaths (see Bürzle et al. 2017 for detailed species composition and community differentiation). The *Synotis alata–Abies spectabilis* and *Ribes glaciale–Abies spectabilis* communities of the class *Betula utilis-Abies spectabilis* forests occupy the upper part of the subalpine forest zone, representing the uppermost forest stands below the treeline (Fig. 5.8). These communities are primarily composed of *Abies spectabilis* and *Betula utilis* in the upper tree layer, with varying cover of several other tree species (*Rhododendron campanulatum*, *Sorbus microphylla*, *Acer caudatum*, *Prunus rufa*) in a second tree layer and distinct differences within the herb layer. Above the treeline, the species-poor *Boschniakia himalaica–Rhododendron campanulatum* community forms a dense krummholz zone between 3900 and 4000 m (NW-exposed) or between 4000 and 4100 m (NE-exposed) (Fig. 5.9). The layer (mean cover 85%) of severely gnarled and stunted krummholz trees or tall shrubs attains a height of two to three meters, gradually decreasing upslope. The dense and largely impenetrable *Rhododendron campanulatum* thickets are almost exclusively single-species stands in the tree/shrub layer; only a few multi-stemmed *Sorbus microphylla* individuals occur occasionally. The name-giving, highly diagnostic taxon *Boschniakia himalaica* is

Fig. 5.8 *Betula utilis-Abies spectabilis* forests represent the uppermost forest stands below the treeline (Photo © Udo Schickhoff, August 17, 2013)

Fig. 5.9 Above the treeline, a dense krummholz zone (*Boschniakia himalaica–Rhododendron campanulatum* community) is developed (NW-exposed: between 3900 and 4000 m; NE-exposed: between 4000 and 4100 m) (Photo © Udo Schickhoff, April 13, 2013)

parasitic on *Rhododendron* species, and regularly occurs in *Rhododendron* krummholz thickets in the Himalayas (Miehe 1990; Miehe et al. 2015). Below the *Rhododendron* canopy, shrub and herb layers have low productivity, indicated by small height, low cover (17.5% and 25%, respectively) and low numbers of vascular plant species. The krummholz community, also assigned to the class *Betula utilis–Abies spectabilis* forests, gives way at its upper boundary to communities assigned to the class *Dasiphora arbuscula–Rhododendron anthopogon* dwarf shrub heaths of the alpine zone. The species-poor *Pedicularis* cf. *microcalyx–Rhododendron anthopogon* community has a less developed herbaceous layer, whereas the *Anaphalis royleana–Rhododendron anthopogon* community is characterized by higher cover and higher species richness of herbs and grasses including tree seedlings.

The vegetation zonation and phytosociological differentiation assessed in this study for the Rolwaling Valley more or less corresponds to the patterns described in previous accounts of treeline ecotone vegetation on north-facing slopes in the West and Central Himalayas, emphasizing mixed forest stands dominated by conifers (mainly *Abies* species), with varying proportions of the Himalayan birch (*Betula utilis*) and other deciduous trees, followed upslope by a krummholz zone and alpine dwarf shrub heaths (Schweinfurth 1957; Miehe 1990; Schickhoff 1993; Miehe and Miehe 2000; Schickhoff 2005; Miehe et al. 2015). However, a closed belt of pure *Betula utilis* stands above mixed *Abies spectabilis-Betula utilis* forests, as described in Schickhoff (1993, 2005), Miehe (1990) and Miehe et al. (2015) for shady slopes in the western and central Himalayas, is not developed in Rolwaling. Here, subalpine mixed forests directly merge into the *Rhododendron campanulatum* krummholz zone, also described by Kikuchi and Ohba (1988), who found similar vegetation zonation in a southeastern range of the Rolwaling Himal. The species composition and community structure of subalpine forests in Rolwaling largely correspond to the *Abies spectabilis* forests of the Langtang/Helambu region, c. 100 km to the west, comprehensively analysed by Miehe (1990). Thus, we assigned the *Synotis alata-Abies spectabilis* and the *Ribes glaciale-Abies spectabilis* communities to a class of Himalayan fir forests (upper cloud forests), proposed by Miehe (1990) as the provisional class *Abietea himalayae*. We term this class '*Betula utilis-Abies spectabilis* forests' on the basis of our material. We also assigned the *Boschniakia himalaica-Rhododendron campanulatum* krummholz community to this class. To some extent, the krummholz zone has floristically a transitional character, with floristic affinities also to the dwarf shrub communities further upslope, including some diagnostic species (*Rhododendron anthopogon*, *Bistorta vacciniifolia*). However, a high number of common highly frequent species (inter alia *Galium paradoxum*, *Koenigia delicatula*, *Deparia allantodioides*) and in particular the predominantly forest-like physiognomy suggest a closer relation to the class *Betula utilis-Abies spectabilis* forests.

Himalayan dwarf shrub communities of the lower alpine zone are mainly composed of different small-leaved *Rhododendron* species (*Rhododendron anthopogon*, *Rhododendron setosum*, *Rhododendron nivale*) and are characterized by generally sparsely developed herbaceous layers (Schweinfurth 1957; Kikuchi and Ohba 1988; Miehe 1990; Schickhoff 1993; Miehe et al. 2015). We assessed species compositions that deviate to some extent from previous descriptions of *Rhododendron anthopogon* communities in SE Rolwaling (Kikuchi and Ohba 1988), Langtang Valley (Miehe 1990) and Kaghan Valley in the western Himalayas (Schickhoff 1993), indicating a considerable species turnover within these communities in different regions due to the variation of local site conditions and the severity of human impact (Bürzle et al. 2017). Human impact on upper subalpine forests and krummholz facilitates the downslope expansion of dwarf thickets (Schickhoff et al. 2015). According to our observations in western and central Himalayan valleys, *Rhododendron* dwarf shrub communities extend their range into areas where the natural krummholz belt has been destroyed, adding to their floristic heterogeneity. A syntaxonomical scheme for Himalayan vegetation units of the lower alpine zone has not been established to date. We proposed a syntaxonomical unit at class level incorporating Himalayan *Rhododendron* dwarf shrub heaths and suggested the term '*Rhododendretea anthopogonis*' (Bürzle et al. 2017).

In Rolwaling, the species composition of the identified communities is mainly differentiated by elevation, soil temperature, nitrogen supply and availability, and soil moisture content, as indicated by the DCA (Fig. 5.10) and the ANOVA (Table 5.2) (Bürzle et al. 2017; Schwab et al. 2022). Strongly acidic soils (mean pH c. 3.0) with a dominating sand fraction (sandy loams) along the treeline ecotone indicate rather unfavourable edaphic conditions for plant growth. Along the

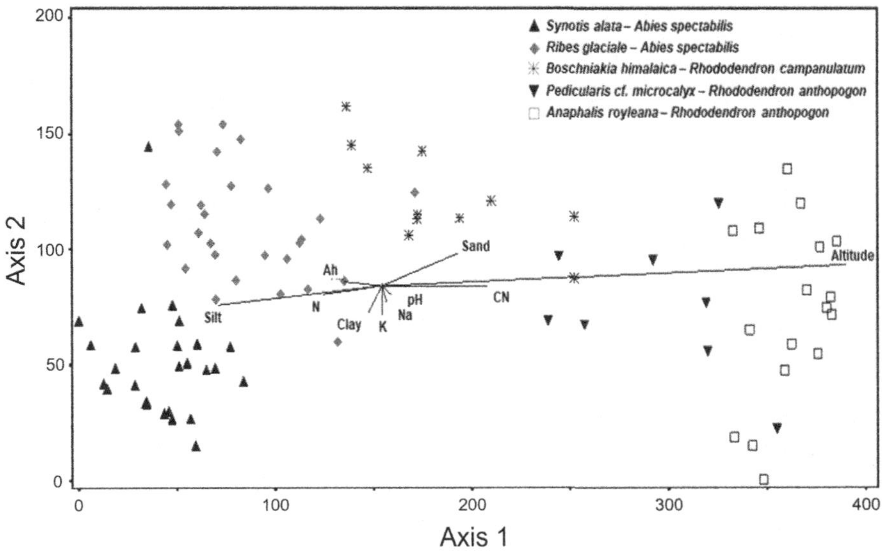

Fig. 5.10 Detrended correspondence analysis (DCA) with post-hoc correlation of environmental parameters. Relevés ($N = 91$) were classified into five plant communities (modified from Bürzle et al. 2017)

elevational gradient, the sand fraction slightly increases implying decreasing water-holding capacity and soil moisture as well as decreasing nutrient availability. Soil C:N ratios widen with elevation, most notably in the krummholz and alpine tundra zones, indicating increasing nutrient shortage above the treeline (Müller et al. 2017). Somewhat more favourable edaphic conditions with higher soil temperatures and soil moisture, as well as enhanced nutrient supply (nitrogen and potassium contents, C:N ratio), still prevail in the lower subalpine zone (3700–3900 m) where the *Synotis alata-Abies spectabilis* community occurs. In the upper subalpine zone (3800–4050 m), occupied by the *Ribes glaciale-Abies spectabilis* community, soil temperatures during summer months are already significantly lower, and the C:N ratio and nitrogen supply become more unfavourable. The *Boschniakia himalaica-Rhododendron campanulatum* krummholz community has to cope with further deteriorating growth conditions along the elevational gradient. Annual mean soil temperature and connected growing degree days further decrease, while the C:N ratio widens, indicating lower decomposition rates and lower availability of nitrogen. The mean pH value slightly decreases to 2.8 (Bürzle et al. 2017; Schwab et al. 2022).

The *Rhododendron* dwarf shrub heaths above the krummholz zone are distributed on podzols between 3950 and 4300 m and formed by *Rhododendron anthopogon*, *Rhododendron setosum* and *Rhododendron lepidotum* (Fig. 5.11). Sites of these *Rhododendron* dwarf thickets are characterized by lower soil temperatures, wider C:N ratios and lower nitrogen contents compared to the forest communities below. The *Pedicularis* cf. *microcalyx-Rhododendron anthopogon* community is developed as a

Table 5.2 Mean values and standard deviation (SD) of environmental parameters of plant communities. Parameters of communities with different letters in column 'Sig'. are significantly different at $p < 0.05$ (one-way-ANOVA, post-hoc pairwise t-test with Holm correction) (after Bürzle et al. 2017; Müller et al. 2016a, 2016b, 2017)

Communities	Synotis alata—A. spectabilis				Ribes glaciale—A. spectabilis				Boschniakia himalaica-R. campanulatum				Pedicularis cf. microcalyx—R. anthopogon				Anaphalis royleana—R. anthopogon			
Parameters	*N*	Mean	SD	Sig.	*N*	Mean	SD	Sig.	*N*	Mean	SD	Sig.	*N*	Mean	SD	Sig.	*N*	Mean	SD	Sig.
Landscape parameters																				
Altitude (m.a.s.l.)	27	3815.5	44.1	a	28	3928.5	58.0	b	11	4051.4	601	c	8	4098.4	72.8	c	17	4215.1	43.3	d
Inclination (°)	27	33.2	4.4	a	28	36.9	3.5	b	11	35.1	3.1	ab	8	37.1	3.6	ab	17	33.4	3.3	a
Exposition (°)	27	41.4	63.0		28	50.9	62.1		11	91.7	128.4		8	75.3	114.2		17	38.0	17.0	
Soil parameters																				
pH (KCl)	27	3.1	0.4	ab	28	3.0	0.3	a	11	2.8	0.3	a	8	3.0	0.2	ab	17	3.2	0.2	b
Ct (%)	27	29.6	12.3		28	25.9	13.6		11	34.4	12.4		8	26.6	13.7		17	23.6	11.8	
N (%)	27	1.5	0.6	a	28	1.3	0.6	ab	11	1.4	0.4	ab	8	1.1	0.5	ab	17	1.0	0.3	b
C:N ratio	27	19.6	3.2	a	28	20.4	2.6	a	11	24.1	4.6	b	8	23.9	3.1	b	17	23.9	4.8	b
CEC (μmolc/g)	27	188.0	135.1		28	188.6	116.4		11	234.3	117.4		8	183.7	120.3		17	151.0	68.0	
Base saturation (%)	27	61.1	22.7		28	57.0	24.6		11	66.8	19.7		8	40.5	17.7		17	67.0	20.5	
Exchange acidity (μmolc/g)	27	67.1	39.7	ab	28	68.2	43.2	ab	11	75.5	45.3	ab	8	106.0	53.2	a	17	45.2	18.6	b
Na (μmolc/g)	27	0.3	0.2		28	0.2	0.3		11	0.2	0.1		8	0.2	0.2		17	0.3	0.5	
K (μmolc/g)	27	10.8	7.5		28	8.2	5.3		11	8.9	4.4		8	7.8	4.3		17	10.5	5.4	
Mg (μmolc/g)	27	32.0	25.6		28	29.7	23.3		11	42.3	23.8		8	19.7	12.2		17	30.0	20.4	
Ca (μmolc/g)	27	109.6	119.3		28	107.1	99.1		11	142.0	122.8		8	58.3	71.9		17	86.8	60.8	
Mn (μmolc/g)	27	5.4	12.7	ab	28	3.7	7.3	ab	11	0.9	0.9	ab	8	0.5	0.3	a	17	4.4	4.8	b
Fe (μmolc/g)	27	7.7	8.9		28	7.3	7.5		11	11.4	10.0		8	9.3	2.7		17	7.4	5.3	
Al (μmolc/g)	27	35.4	29.8	ab	28	32.7	26.9	ab	11	34.7	25.5	ab	8	49.6	21.2	a	17	24.2	15.9	b
Ah (cm)	27	8.3	7.1	a	28	9.2	62	a	11	9.3	5.4	a	8	11.1	10.9	a	17	2.8	2.4	b
Organic (%)	27	51.7	21.6		28	44.3	23.4		11	59.2	21 .4		8	46.3	22.9		17	41.3	20.0	

Sand (%)	23	59.9	16.9		25	61.1	14.0		9	68.6	10.7		8	69.5	6.2		17	68.0	5.9	
Silt (%)	23	28.0	6.6	a	25	26.0	5.3	a	9	22.1	6.7	ab	8	22.1	5.7	ab	17	19.0	4.1	b
Clay (%)	23	15.0	11.3		25	12.9	11.0		9	9.3	5.2		8	8.4	3.6		17	13.0	5.4	
Climate parameters																				
Mean soil temp. (year) [°C]	6	3.5	0.6	ab	8	3.6	0.4	b	3	2.7	0.4		6	2.5	0.7	ac	3	2.5	0.8	c
Mean soil temp. (ON) [°C]	7	3.1	0.7		10	3.3	0.6		3	2.8	0.4		6	2.6	0.8		3	2.7	0.2	
Mean soil temp. (DJF) [°C]	7	-2,1	1,4		8	-0,9	0.7		3	-1.9	0.8		6	-2,9	1.5		3	-2,2	0.9	
Mean soil temp. (MAM) [°C]	6	1,3	0.6		8	1,3	0.8		3	0.3	0.2		6	0.3	0.5		3	0.0	0.4	
Mean soil temp. (JJAS) [°C]	7	9.5	0.2	a	10	8.8	0.4	b	3	7.8	0.6	ab	6	8,2	0.7	ab	3	7,8	0.7	ab
Growing degree days	7	169.6	8.3	a	8	169.3	15.0	ab	3	141.7	4.5	c	6	145.8	11.9	bc	3	137.7	10.7	ac
Mean Soil Moisture (year) [pF]	7	2.1	0.5		10	2.3	0.6		3	2.3	0.2		7	2.3	0.4		3	2.6	0.4	

Fig. 5.11 Above the krummholz belt, alpine dwarf shrub heaths (*Rhododendron anthopogon* communities) occupy the uppermost zone of the Rolwaling treeline ecotone (Photo © Udo Schickhoff, April 15, 2013)

dense dwarf thicket between 3950 and 4150 m, reaching a mean height of 0.8 m and a mean cover of 80%. Growth conditions further deteriorate, in particular in terms of reduced base saturation and nitrogen content. Different *Rhododendron* species (*Rhododendron anthopogon*, *Rhododendron lepidotum*, dwarf individuals of *Rhododendron campanulatum*) dominate this community, accompanied by *Bistorta vacciniifolia* shrubs and *Sorbus microphylla* shrubs or small trees. The herb layer is less developed (mean height 0.1 m, mean cover 45%) and is composed of fern species and species of the genera *Kobresia* and *Calamagrostis*. The *Anaphalis royleana*-*Rhododendron anthopogon* community occurs between 4100 and 4300 m, showing a reduced mean cover of the shrub layer (60%) and a reduced height of less than 50 cm. Site conditions are the most unfavourable along the elevational gradient, in particular in terms of mean soil temperatures during the vegetation period, number of growing degree days, soil moisture and thickness of the Ah-horizon, cation exchange capacity and nitrogen supply (cf. Fig. 5.10; Table 5.2). The shrub layer is dominated by *Rhododendron anthopogon*, *Bistorta vacciniifolia*, *Rhododendron setosum* and *Dasiphora arbuscula*. The herbaceous layer is not as sparse as in the *Pedicularis* cf. *microcalyx*-*Rhododendron anthopogon* community, reaching a mean cover of 50%. The comparatively species-rich herb layer is characterized by a high percentage of *Poaceae* and *Cyperaceae* (genera *Calamagrostis* and *Kobresia* and others) (Bürzle et al. 2017; Schwab et al. 2022).

In Rolwaling, both soil nutrients and foliar macronutrient concentrations decrease significantly with elevation (Drollinger et al. 2017), indicating that nutrient limitation is a potentially constraining factor for plant productivity and tree growth (cf. Larcher 2003; Körner 2021a), as well as for the treeline responsiveness to climate warming. Lower nitrogen availability and nutrient deficiency in the krummholz and dwarf shrub thickets is, inter alia, caused by the low leaf litter quality of *Rhododendron* species. Thick layers of recalcitrant *Rhododendron* litter, with leaves containing high amounts of flavonoids, diterpenoids, phenol derivates and other chemical compounds (Popescu and Kopp 2013), are associated with low N mineralization rates and N availability (cf. Northup et al. 1995; Maithani et al. 1998). In addition, chemical compounds of *Rhododendron* leaves are known to be potentially allelopathic (Li et al. 2010), negatively influencing seedling establishment and growth performance of other species (Nilsen et al. 1999; Chou et al. 2010). The impoverished flora and the extremely low productivity of the herbaceous layer in the krummholz zone suggest that allelopathic effects of *Rhododendron campanulatum* facilitate the development of monospecific krummholz thickets (Bürzle et al. 2017).

5.3.4 Seedling Establishment in the Treeline Ecotone

The establishment of seedlings and a successful performance during early life stages are the prior conditions for any treeline advance to higher elevations. A treeline shift requires successful seedling establishment in the contiguous treeline ecotone, that is, beyond the uppermost tree individuals or tree groups. The number of respective regeneration studies at treeline elevations in the Himalayas is limited, and the available studies are reviewed in Schickhoff et al. (2015, 2016a). High levels of recruitment in recent decades become apparent from these studies, as long as respective treeline ecotones are not too heavily disturbed by grazing and other human impact.

The results we obtained from regeneration studies in Rolwaling and Langtang support the emerging pattern of a generally intense regeneration at Himalayan treeline sites which are less or not disturbed by pastoral use (Schickhoff et al. 2015, 2016a; Bürzle et al. 2018; Schwab et al. 2016, 2017, 2020). We assessed prolific regeneration with seedling establishment of *Betula utilis*, *Abies spectabilis*, *Rhododendron campanulatum* and *Sorbus microphylla*, sometimes far above the upper limit of adult trees. Some individuals of more than 2 m height even grew vigorously above the krummholz zone, that is, 100–200 m above the current treeline. Seedling distribution was not uniform along the treeline ecotone (Table 5.3). We recorded the highest number of *Abies spectabilis* seedlings of all height classes (>39,000 N/ha) in the lower and upper subalpine forest (zones A and B). *Betula utilis* seedlings of all height classes were most frequent in the lower subalpine forest (zone A) (almost 10,000 N/ha). Maximum recruit density (>46,000 N/ha) occurred in zone B, in spite of competing ground vegetation of a fairly dense herbaceous layer including many tall forbs (average cover of 60%). Here, also small (0–5 cm) and medium-sized (5–15 cm) *Rhododendron campanulatum* seedlings showed most intense recruitment (>40,000 N/ha). Seedlings of *Rhododendron campanulatum* had a very high density in the krummholz zone (zone C) as well (>33,000 N/ha),

Table 5.3 Tree species density of seedlings/saplings 0–5 cm, 5–15 cm, 15–50 cm in the lower subalpine forest (Zone A), upper subalpine forest (Zone B), *Rhododendron* krummholz (Zone C) and alpine dwarf shrub heath (Zone D) (modified from Bürzle et al. 2018)

Height (cm)	Elevational zone	Seedling density (N/ha) *A. spectabilis*	Seedling density (N/ha) *B. utilis*	Seedling density (N/ha) *R. campanulatum*	Total (N/ha)
0–5	A	34.500	3.313	3.563	41.376
0–5	B	6.688	0	39.875	46.563
0–5	C	3.125	125	33.750	37.000
0–5	D	875	0	10.813	11.688
5–15	A	3.813	4.938	625	9.376
5–15	B	313	0	250	563
5–15	C	0	0	0	0
5–15	D	0	0	875	875
15–50	A	938	1.375	63	2.376
15–50	B	188	0	438	626
15–50	C	250	0	125	375
15–50	D	0	0	1.000	1.000

Fig. 5.12 *Rhododendron campanulatum* shows a high level of recruitment up to the alpine dwarf shrub heath (Photo © Udo Schickhoff, April 15, 2013)

where they already clearly predominate (cf. Table 5.3). The percentage of *Abies spectabilis* seedlings in zone C decreases to 9.0%, while *Betula utilis* seedlings have a share of 0.3% only (see also Rana et al. 2016 for Manaslu Himal). Seedling/sapling density sharply decreased in the alpine dwarf shrub heaths (zone D), where only

Rhododendron campanulatum recruits occurred with greater number of individuals (Fig. 5.12), also in the sapling size class of 15–50 cm (1000 N/ha). In zone D, the percentage of *Abies spectabilis* seedlings decreases to 6.5%; occasionally also some *Sorbus microphylla* seedlings were found (Schwab et al. 2016; Bürzle et al. 2018).

The dense krummholz thickets have been found to be an effective barrier for upslope migration of other tree species, reflected by a negative correlation between the abundance and density of *Rhododendron campanulatum* seedlings and recruitment of other tree species. Permanently dense foliation of evergreen *Rhododendron* and presumed allelopathic effects adversely affect the establishment of seedlings of competing tree species, in particular those of *Betula utilis* and the performance of taller recruits. Seedling/sapling density and size structure point to a much higher competitiveness of *Rhododendron* recruits above treeline compared to *Abies* and *Betula* recruits (Schwab et al. 2016, 2017; Bürzle et al. 2018).

Successful recruitment necessitates the availability of safe sites for seedling establishment. Safe sites of the tree species *Abies spectabilis*, *Betula utilis* and *Rhododendron campanulatum* were analysed by investigating the microhabitat ground cover of established seedlings, followed by examining protective elements that potentially ameliorate seedlings' growth conditions, and by quantifying the distance of seedlings to the nearest protective element. The results showed species-specific, distinct microhabitat preferences for germination and seedling establishment (Fig. 5.13) (Bürzle et al. 2018). Microsite characteristics differed significantly between *Abies spectabilis* on the one hand and *Betula utilis* and *Rhododendron campanulatum* on the other. *Abies spectabilis* seedlings predominantly establish on litter accumulations, while *Betula utilis* and *Rhododendron campanulatum* seedlings primarily emerge on bryophyte mats. These preferences are largely similar across all vegetation types, reflecting the species-specific requirements for safe sites and the importance of ground cover for safe site structure. Previous studies reported facilitative effects of a moss cover for seed germination and seedling establishment of other *Betula* species as well (Anschlag et al. 2008; Sanz et al. 2011; see also Hughes et al. 2009). The moss cover acts as a seed trap for the extremely light and wind-dispersed seeds. Likewise, the suitability of humid, low-shaded bryophyte mats as safe sites for *Rhododendron* seedling establishment was assessed elsewhere (Kameyama et al. 1999; Mejías et al. 2002). That *Abies spectabilis* seedlings predominantly occur in litter-covered microhabitats may be attributed to the high physiological shade-tolerance and the high amount of seed reserves that enable the species to cope with an impediment by litter (Walters and Reich 2000; Maher and Germino 2006). Field layer vegetation and the early successional ground cover types 'bare soil' and 'stones' played a negligible role in seedling establishment of all species with cover values less than 5%. We found little variation in microsite cover among older seedlings, similar to the regeneration of other subalpine tree species in other treeline environments (Jones and del Moral 2005; Maher and Germino 2006). We did not find any *Betula utilis* seedlings in microhabitats covered with *Rhododendron* leaf litter; the percentage of *Rhododendron* litter was also very low in microhabitats of *Rhododendron campanulatum* seedlings. Thus, *Rhododendron* litter is obviously an unsuitable substrate for

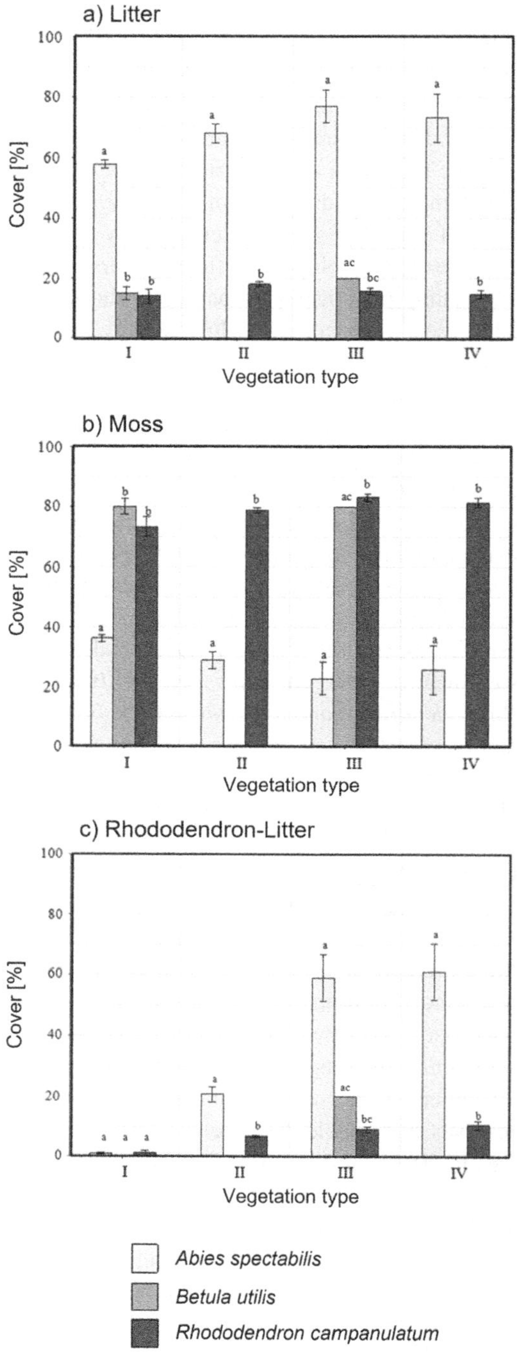

Fig. 5.13 Variations among tree species' seedlings (0–5 cm) in microhabitat cover of most important cover types; (**a**) litter, (**b**) moss and (**c**) *Rhododendron* litter in the (I) lower subalpine forest, (II) upper subalpine forest, (III) *Rhododendron* krummholz and (IV) alpine dwarf shrub heath; lower case letters indicate significant differences detected by Dunn's post-hoc test ($p < 0.01$) (modified from Bürzle et al. 2018)

germination and seedling establishment of *Betula utilis* and *Rhododendron campanulatum*. This applies only to a limited extent to *Abies* seedlings (Bürzle et al. 2018).

The vast majority of seedlings of all species are located close to at least one potentially protective element. Geomorphic elements (rocks, stones) and deadwood are the most important shelters for seedling establishment. The results emphasize the importance of safe sites for successful tree regeneration in treeline environments, and suggest an increasing significance of shelter elements under climate change conditions, in particular with regard to buffering potential extremes such as enhanced drought stress in the pre-monsoon season. The results also suggest that the availability of shelters for seedling establishment will play a major role for patterns of treeline dynamics (Bürzle et al. 2018). We assessed significantly positive correlations of seedling/sapling abundance with soil moisture for *Abies*, *Betula* and *Rhododendron*, and with soil temperature for *Abies*, *Betula* and *Sorbus*, in each case over almost all size classes. Thus, higher soil moisture and higher soil temperatures facilitate higher recruitment density for the majority of tree species in the Rolwaling treeline ecotone, in particular for *Abies spectabilis* and *Betula utilis* (Schickhoff et al. 2015, 2016a; Schwab et al. 2017).

5.3.5 Spatial and Temporal Patterns of Tree Species Population Density

Adult and juvenile tree species densities show major variations along the elevational zones within the treeline ecotone (Fig. 5.14) (Schwab et al. 2016, 2020, 2022). From the lower (zone A) to the upper subalpine forest (zone B), the population density of *Rhododendron campanulatum* considerably increases, both in adult trees and in juveniles, while that of *Abies spectabilis*, *Betula utilis* and *Sorbus microphylla* decreases (except for adult *Betula* trees). Striking contrasts are to be observed at the transition to the krummholz zone (zone C), with *Rhododendron campanulatum* becoming absolutely predominant (Fig. 5.15), and at the transition to the alpine dwarf shrub heath, where adult trees no longer occur and juveniles are almost exclusively composed of *Rhododendron campanulatum* and *Sorbus microphylla* individuals. The uppermost stunted individuals of adult *Abies spectabilis* and *Betula utilis* trees occur in the krummholz zone, while the uppermost trees with true tree habitus (upright stem with crown) were observed in the upper subalpine forest (zone B). Thus, we consider the treeline to coincide with the transition from the subalpine forest to the krummholz zone. At this transition, the most abrupt change in tree species composition occurs along the elevational gradient, associated with strongly modified tree physiognomy reflected in significantly declining values of stand structural parameters such as dbh, tree height and crown length. At this transition, we also assessed the most abrupt changes in annual and seasonal mean soil temperatures (Schwab et al. 2020). A distinctly higher number of recruits than trees (cf. Fig. 5.14), as well as the reverse J-shaped density-diameter distributions, indicates intense, sustainable regeneration and stable populations (Schwab et al.

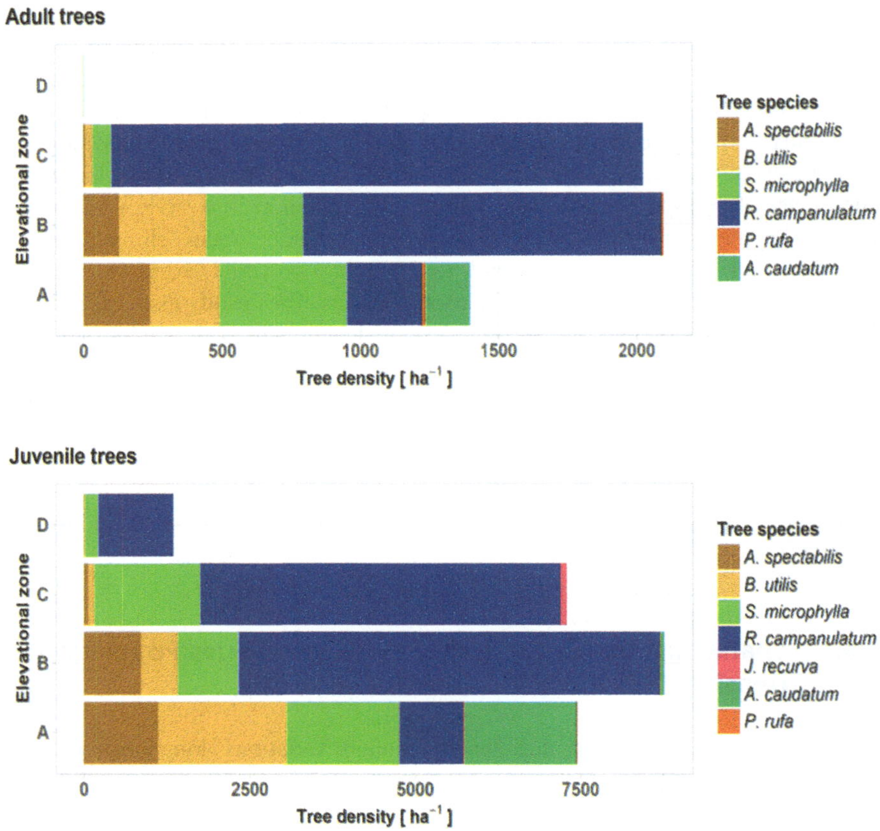

Fig. 5.14 Tree species population density of adult (≥7 cm dbh) and juvenile individuals (<7 cm dbh) across the elevational gradient; note the different scale of the *X* axis (modified from Schwab et al. 2020)

2016). Tree age data of *Abies spectabilis* confirm the high number of younger trees reflecting a maximum in stand density in recent decades. The majority of trees became established since the 1950s, coincident with the recent climate warming period. The occurrence of seedlings and juveniles of *Abies spectabilis*, *Betula utilis*, *Sorbus microphylla* and *Rhododendron campanulatum* in zone D (cf. Table 5.3; Fig. 5.16) indicates the potential of tree species to become established in the alpine dwarf shrub heath. These recruits are potential indicators of an upslope migration of tree species triggered by recent climate warming. The recent increase in recruit density could increase the number of trees if seedlings and juveniles survive subsequent life stages and result in stand densification.

In order to correlate modified population densities of tree species along the treeline ecotone to changing environmental conditions and altered constellations of site factors, we analysed tree species-environment relationships across the ecotone by means of redundancy analyses (RDA) and variation partitioning. Results indicate

Fig. 5.15 At the transition from upper subalpine forest to krummholz (treeline), *Rhododendron campanulatum* becomes absolutely predominant (NW-exposed: at 3900 m; NE-exposed: at 4000 m) (Photo © Udo Schickhoff, April 13, 2013)

Fig. 5.16 Recruits of tree species in the alpine dwarf shrub heath indicate a potential upslope migration under climate warming; this juvenile *Abies spectabilis* grows at 4200 m, c. 200 m above the current treeline (Photo © Udo Schickhoff, August 20, 2013)

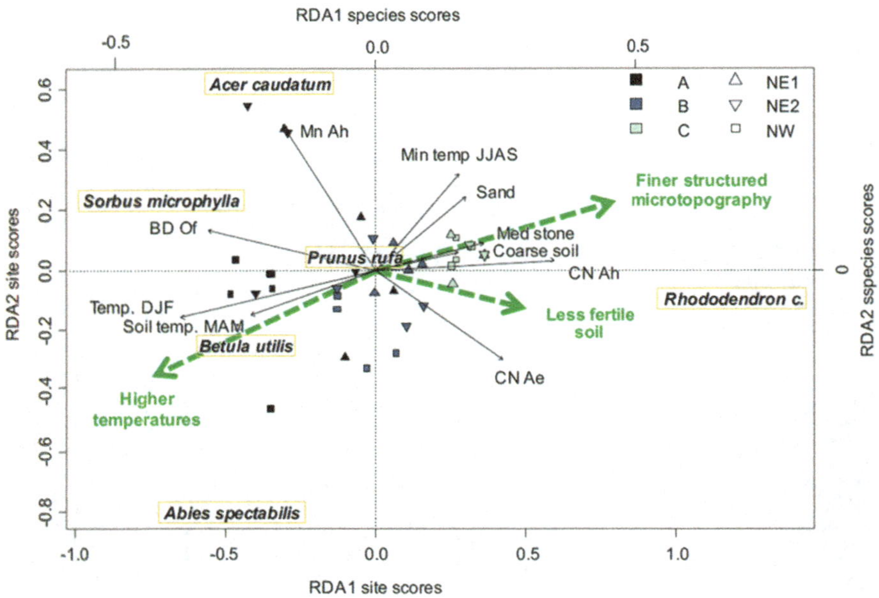

Fig. 5.17 Ordination diagram of RDA analysis for adult tree density: Vector directions show relation to environmental variables, vector lengths show their importance; only the most important environmental variables with a biplot score (axis one, axis two or both) ≥0.3 are depicted; dashed arrows summarize main gradients, with the length not true to scale (modified from Schwab et al. 2020)

that treeline tree species show species-specific, as well as life stage-specific, responses to gradients of environmental parameters (Schwab et al. 2017, 2020). The RDA ordination for adult trees (Fig. 5.17) shows a clear separation between plots of the *Rhododendron campanulatum* krummholz (zone C), plots of upper subalpine forest (zone B) admixed with some zone A plots, and a pure lower subalpine forest (zone A) group. These three groups are arranged mainly along the first RDA axis, while the plots spread to a lesser extent along the second RDA axis. The first RDA axis accounted for 58% of the variance, and the first and second axes together explained 68% of the variance (for explanatory variables and biplot scores of both axes, see Schwab et al. 2020). Plots of the krummholz zone show a tight cluster due to the strong dominance of *Rhododendron campanulatum*. In contrast, the plots of the subalpine forest (zones A and B) with higher number of tree species exhibit greater variation along both axes. The complex temperature-nutrient gradient along RDA axis 1 has a major influence on the tree species distribution in the ordination space, in particular on the separation of *Rhododendron campanulatum* from all other tree species, suggesting a species-specific response to environmental variables. Plots with a high density of *Rhododendron campanulatum* show lower soil and air temperatures, wider C:N ratios, and more distinctive nutrient deficiency compared to the subalpine forest plots, suggesting that low temperature and poor

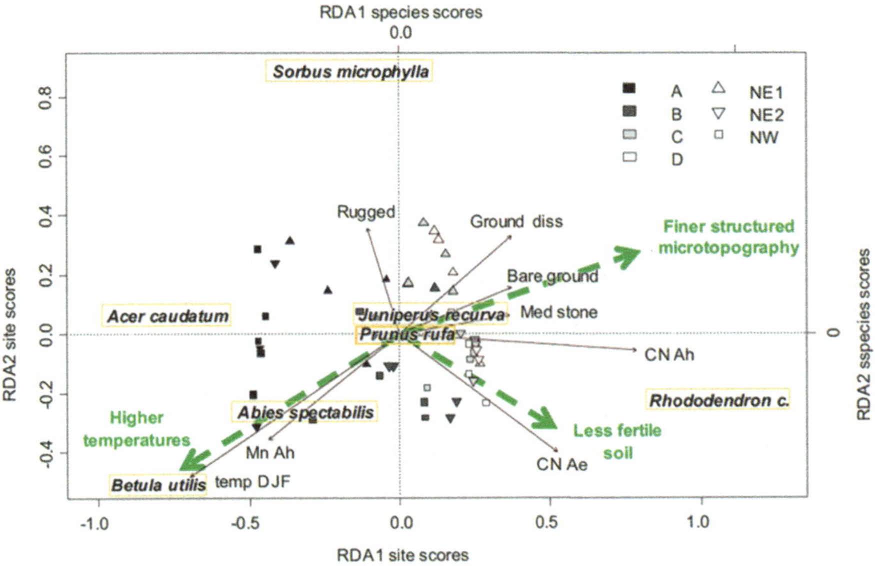

Fig. 5.18 Ordination diagram of RDA analysis for tree species density of juveniles: Vector directions show relation to environmental variables, vector lengths show their importance; only the most important environmental variables with a biplot score (axis one, axis two or both) ≥0.3 are depicted; dashed arrows summarize main gradients, with the length not true to scale (modified from Schwab et al. 2020)

nutrient availability accurately characterize the nature of the ecological niche that *Rhododendron campanulatum* occupies within the treeline ecotone.

In contrast to the adult tree ordination, sites and species are scattered to a higher degree in the ordination diagram of juveniles (<7 cm dbh) (Fig. 5.18). The first RDA axis accounts here for 51% of the variation, and the first and second axes explain a total of 60% of the variance. The ordination of juvenile individuals indicates a clear separation into a group of plots from the lower subalpine forest (zone A), a group from the upper subalpine forest (zone B), and a mixed group of plots from the krummholz zone and alpine dwarf shrub heaths (zones C and D) (cf. Fig. 5.18). The groups are separated from each other along the first RDA axis, and the plots are at the same time more widely scattered along the second RDA axis compared to the adult tree ordination. Differences in juvenile tree species composition between the slopes result in plots from slope NE1 forming a more heterogeneous cluster, in particular in case of krummholz and alpine dwarf shrub heath plots (cf. Schwab et al. 2017). A complex gradient of environmental variables, even more complex as in the adult tree ordination (cf. Schwab et al. 2020), causes the differentiations in plot groups, with differences in nitrogen availability (C:N ratios of the Ah and Ae horizons), the manganese content of the Ah horizon, the calcium content of the Ae horizon, the dissimilarity of surface structure within and between plots, and differences in mean and minimum air temperatures playing a major role.

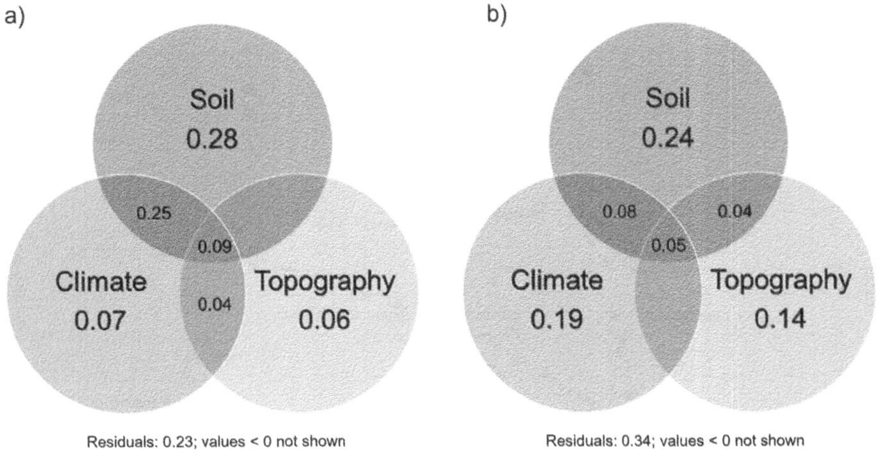

Fig. 5.19 Variation partitioning of (**a**) adult tree density and (**b**) juvenile density using the matrices of edaphic, climatic, and topographic variables; the numbers indicate the percentage of total explained variability, explained exclusively or jointly by the variable matrices (modified from Schwab et al. 2020)

However, as in the adult tree ordination, the main factors are related to nutrient availability, temperature and microtopography.

Variation partitioning indicated that the edaphic group of variables is the most important independent predictor of adult tree species distribution (28% explained variability), while climatic variables (including soil temperature) and topographic variables are of subordinate importance (7% and 6%, respectively) (Fig. 5.19) (Schwab et al. 2020). Of the shared variation of adult tree density, 25% is explained by soil and/or climate variables. The tree species density of juveniles revealed a more balanced distribution of explained variance per variable group when subjected to variation partitioning (cf. Fig. 5.19). However, as in the case of adult tree density, the edaphic group of variables was the key independent predictor of species distribution (24% explained variability), while climatic (19%) and topographic variation (14%) was of subordinate importance. The results show that edaphic variables, nutrient conditions in particular, significantly contribute to the total explained variation of both adult tree and juvenile density. The latter shows a stronger dependence on the climatic variable of temperature in comparison to adult tree density. In addition, differences in microrelief, surface structure and within-plot ground cover influence the occurrence of juvenile individuals to a larger extent compared to adult tree density, corroborating previous results concerning differences between juvenile and adult tree growth (e.g. Smith et al. 2003; Wieser et al. 2014).

The prominent role of edaphic variables in variation partitioning corresponds to the finding that soil temperature, nitrogen supply and availability, and soil moisture differentiated the species composition of plant communities throughout the Rolwaling treeline ecotone (see Sect. 5.3.3). The analysis of tree species population densities showed that the included environmental variables could not account for some of the variance of spatial patterns. Other factors such as species-specific migration potentials, seed-dispersal strategies, ecological drift and the like most likely contribute to this yet-unexplained variance, as relevant model-based results also suggest (Dullinger et al. 2004; Holtmeier and Broll 2010b; Nathan et al. 2011; Martínez et al. 2012; Johnson et al. 2017).

5.3.6 Tree Growth-Climate Relationships

Tree growth-climate relationships indicate that mature treeline trees are generally responsive to climate warming. In many mountain ranges, stimulating effects on tree radial growth have been assessed in the twentieth century due to rising temperatures. Growth responses to a warming climate, however, tend to be different where lack of available moisture becomes a controlling factor, for example, in arid or semiarid regions or in regions with seasonal drought periods. We analysed growth-climate relationships in the Rolwaling treeline ecotone for more than a century. We determined the relationships of *Abies spectabilis* radial tree growth to climate by correlating growth to temperature, precipitation and drought index data. The results showed a changing dendroclimatic signal over the past century (Schwab et al. 2018, 2020). During the period up to the 1970s, the dendroclimatic signal exhibited a fragmented pattern with longer insignificant phases, but indicated a predominantly positive relationship of tree growth and summer temperature (Fig. 5.20). A period of rather stable climate conditions during the mid-twentieth century coincided with a positive relationship of tree ring width (TRW) to summer temperature and a negative

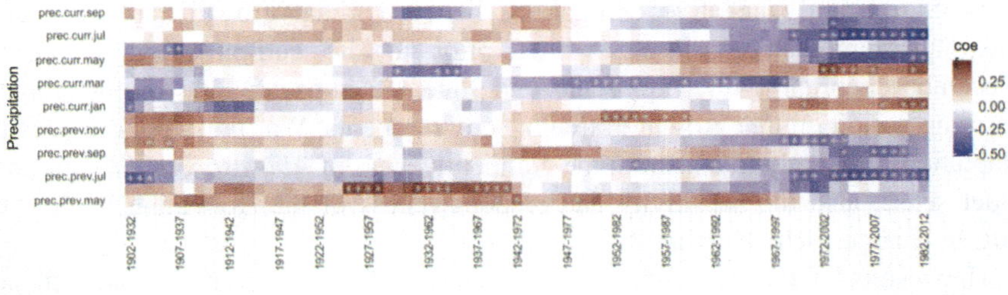

Fig. 5.20 Moving window correlations of the TRW chronology with mean temperatures and precipitation sums: 31-year windows were shifted in 1-year steps throughout the entire period of climate data availability (1901–2012); periods with asterisks indicate significant ($p < 0.05$) correlations (modified from Schwab et al. 2020)

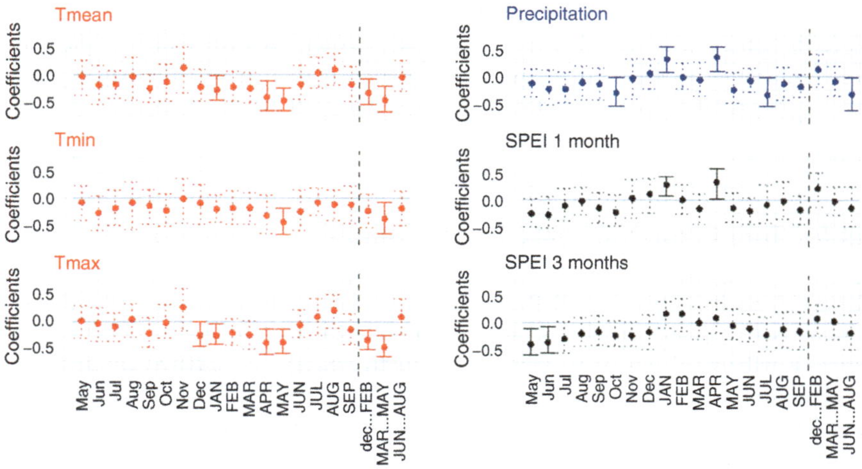

Fig. 5.21 Static correlations (1972–2012) of the TRW chronology with temperature, precipitation and drought indices (SPEI) for current and previous year's months and current year seasons; solid bars indicate significant correlations ($p < 0.05$) (modified from Schwab et al. 2018)

relationship of TRW to March precipitation, indicating sufficient moisture availability in spring, and growth limitation mostly due to low summer temperatures.

The dendroclimatic signal increased in strength in particular since the 1970s, showing a significantly negative relationship between radial growth and mean, minimum and maximum temperatures for the current year's pre-monsoon season (March–May) (Fig. 5.21). Since high warming rates imply increasing evapotranspiration, the negative relationship suggests that climate warming-induced moisture deficits during pre-monsoon seasons became a major factor limiting radial tree growth over recent decades. Significantly positive correlations of TRW to 1-month SPEI variables (cf. Fig. 5.21) corroborate the pre-monsoon moisture sensitivity, also highlighted by significantly positive correlations of precipitation sums and TRW during the current January and April. The current monsoon season (June–August) showed a significantly negative correlation between radial growth and precipitation sums. Significantly negative correlations between radial growth and mean and maximum temperatures during winter prior to the current growing season (December–February) (cf. Fig. 5.21) underline the impact of the recent climate warming period on tree growth in treeline environments. Warming-induced growth adaptations are likely to constrain the competitiveness of *Abies spectabilis* and, in a wider sense, also the sensitivity and responsiveness of the Rolwaling treeline to climate change (Schwab et al. 2018, 2020).

The results of the dendroecological studies in Rolwaling are in line with an increasing number of studies in the Himalayas detecting a widespread strong sensitivity of tree growth to pre-monsoon temperature and humidity conditions (Dawadi et al. 2013; Liang et al. 2014; Ram and Borgaonkar 2014; Bräuning et al. 2016; Panthi et al. 2017; Sohar et al. 2017; Tiwari et al. 2017; Gaire et al. 2017,

2020). Warming-induced higher evapotranspiration and soil moisture deficits during dry spring months will increasingly affect tree growth at Himalayan treelines (verified for *Abies spectabilis* and *Betula utilis*), in particular on sites which are prone to drought stress. Thus, moisture supply in the pre-monsoon season is likely to influence the pace of future treeline dynamics (Schickhoff et al. 2015, 2016b, 2022; Mishra and Mainali 2017; Sigdel et al. 2018; Lyu et al. 2019). Some other studies found enhanced tree growth at high-elevation sites, in particular at more humid sites (Fan et al. 2009; Huang et al. 2017; Thapa et al. 2017; Shi et al. 2020). Largely positive correlations between tree ring width and winter temperatures in treeline ecotones were detected in some Himalayan studies (e.g. Gaire et al. 2020). On the other hand, increasing winter temperatures were found to be detrimental to the growth of *Rhododendron* shrub species above the treeline (Bi et al. 2017).

5.3.7 Treeline Shift Potential

The Rolwaling treeline (at the transition from uppermost subalpine forest to krummholz) is at a position much warmer than a growing season mean temperature of c. 6 °C that still supports tree growth. The current treeline position obviously lags behind the driving climatic signal and reflects a legacy effect of the Little Ice Age. Even under moderate post-Little Ice Age warming up to the 1970s, the treeline position had been lagging behind climatic change. Over the recent decades with a persistent and increased warming trend, climatic change has been even more ahead of an assumed future treeline shift. The results of the treeline ecotone studies in Rolwaling suggest that the treeline trees are in a process of climatic tracking that is delayed due to several lag factors and feedback processes. Influential non-thermal drivers that modulate climate warming inputs need to be analysed since a warming-induced detectable treeline advance to higher elevation was not observed over recent decades. A recent study on the treeline history of the adjoining southeastern Tibetan Plateau (Li et al. 2019) supports the view that the treeline elevation has not yet recovered from downslope movements during cold phases of the Little Ice Age. In general, mean treeline shift rates are much lower than expected from climate warming (Harsch et al. 2009; Lu et al. 2021).

In Rolwaling, population processes at the front edge of the uppermost subalpine forest are primarily slowed down by the dense *Rhododendron campanulatum* krummholz zone. The self-persistent *Rhododendron* krummholz thickets (Fig. 5.22) constitute a very effective barrier that largely prevents the expected upslope migration of *Abies*, *Betula* and other tree species from the upper subalpine forests below, indicated by a negative correlation between the abundance and density of *Rhododendron campanulatum* and the recruitment of other tree species. *Rhododendron campanulatum* modifies site conditions in the krummholz zone to such an extent that the competitiveness of other treeline tree species under the dense krummholz canopy is adversely affected, and the krummholz' persistence and dominance is facilitated. Thus, positive feedback processes consolidate the *Rhododendron* krummholz belt (Schwab et al. 2020; Schickhoff et al. 2020). The nitrogen

Fig. 5.22 Dense *Rhododendron* krummholz thickets constitute a very effective barrier, largely preventing the upslope migration of other tree species (Photo © Udo Schickhoff, April 15, 2013)

mineralization loop is a major feedback process. Combined effects of slowly decomposing *Rhododendron* litter, causing low nitrogen mineralization rates, and high amounts of polyphenols in decaying *Rhododendron* leaves, enhancing nitrate and ammonium immobilization, result in reduced nitrogen availability. Nutrient deficiency, in turn, contributes to the high competitiveness of *Rhododendron campanulatum* in the krummholz zone since it is a relatively more serious physiological stressor for other tree species. *Rhododendron* species (ericaceous perennials in general) are known to be low nutrient users and to be very competitive under poor soil conditions (e.g. Ristvey et al. 2007). Evergreenness and high root-to-shoot ratio of *Rhododendron campanulatum* enhance its competitiveness due to optimized nutrient storage and increased photosynthetic capacity in spring and fall seasons. The competitiveness is further enhanced by the allelopathic effects of *Rhododendron* litter rich in flavonoids, diterpenoids, phenol derivates and other chemical compounds impeding seedling establishment and growth performance of other species and facilitating the development of monospecific krummholz thickets (see Sect. 5.3.3). In addition, other species have to cope with reduced light availability and a marked decline in soil temperatures under the dense krummholz canopy, further constraining their competitiveness. Thus, maintained by positive feedbacks, the dense krummholz thickets might represent an alternative stable state (not necessarily in equilibrium with climatic conditions) and to some extent decouple spatial patterns and processes from the regional climate change input. Similar feedback processes as in Rolwaling control spatial patterns and treeline dynamics, and result in a mismatch of treeline elevation with climate in other treeline environments as well (cf. Bader et al. 2008; Kim and Lee 2015; Camarero et al. 2017; Millar et al. 2020).

Given the strong competition and the barrier function of the firmly established krummholz zone, it is not surprising that the treeline position in Rolwaling is lagging far behind climatic changes. The Rolwaling treeline has a rather stable elevational position, characterized by a feedback-driven persistence. A considerable treeline advance is not to be expected in the short- to medium term (several years to a few decades). Nevertheless, the largely prolific regeneration with seedling establishment far above the upper limit of adult trees (even beyond the krummholz zone including *Betula*, *Abies* and *Sorbus* juveniles of up to more than 2 m height) indicates that diaspores occasionally succeed in overcoming the krummholz zone, become established, and survive early life stages. In addition, a process of stand densification was observed over recent decades. Thus, an upslope forest expansion seems possible, also suggested by an ecological niche modelling of Himalayan treeline species (Bobrowski et al. 2017; Lamsal et al. 2017; Chhetri et al. 2018). A treeline advance in Rolwaling would be even more likely if the krummholz zone were to develop into a forest with tall upright growing trees. A shift from a krummholz to a forest state could occur once threshold conditions are reached (e.g. through temperature forcing or improved nutrient availability) or as a resultant effect of a disturbance. In the medium term, however, cold, nutrient-poor conditions will most likely continue to be more beneficial to *Rhododendron campanulatum* and *Rhododendron* dwarf shrubs in krummholz and alpine tundra, as they have been in previous decades and centuries regardless of slightly increasing mean temperatures. It is also possible that the strong competition of *Rhododendron campanulatum* is even reinforced, at least in the medium term, by taking greater advantage from continued climate warming. For instance, *Rhododendron campanulatum* shows superior water use efficiency (cf. De Lillis et al. 2004) and may prove less susceptible to amplifying pre-monsoonal drought stress than *Abies* and *Betula*. The result could be the expansion and upslope migration of the krummholz zone.

5.4 Conclusions

Climatic treelines at the transition from subalpine to alpine zones in the world's high mountains are altitudinal limits caused by heat deficiency during the growing season. Above-average soil and air temperatures indicate that the Rolwaling treeline is located at an elevation below the potential treeline, suggesting that the current treeline position is lagging behind the driving climatic signal and that the treeline trees are in a state of climatic tracking. The potential timeframe for tracking, for reaching a new steady state at higher elevation is long-term and currently inestimable. Unless moving upslope is not prevented by controlling processes and dynamics involving non-thermal site factors and interrelationships, global high-elevation tree limits including the Rolwaling treeline will shift to higher elevations in the further course of climate change. Treeline elevations have responded to climate oscillations throughout the Holocene, and empirical studies from varied high mountain ranges reported treeline advance over the past century. However, the Rolwaling treeline has not shown a shift in elevation over recent decades. The process of climatic tracking is being retarded by diverse lag factors and feedback processes, resulting in current treeline persistence. The dense, self-sustaining *Rhododendron campanulatum* krummholz zone plays a crucial role in any treeline shift to higher elevations. The krummholz belt is being consolidated and sustained by positive feedback processes, and acts as a very effective barrier that largely prevents the upslope migration of *Abies spectabilis*, *Betula utilis*, and other treeline tree species. Major feedback processes include low nitrogen mineralization rates, allelopathic effects and stand structural development, resulting in site conditions such as nutrient deficiency, reduced light availability, and lower soil temperatures that severely restrict the competitiveness of other tree species. The results from Rolwaling support the view that krummholz treelines show rather low responsiveness to climate warming. The results are presumably transferable to other Himalayan treeline ecotones where the natural vegetation zonation is still intact and a krummholz zone of *Rhododendron campanulatum* is still developed.

Notwithstanding the persistent treeline position, the Rolwaling treeline ecotone undergoes substantial changes in environmental conditions and related pattern-process relationships. For instance, growth patterns of treeline trees have become particularly responsive to pre-monsoon temperature and humidity conditions. Ongoing climate change is affecting site factors such as air and soil temperature, soil moisture and nutrient availability that are significant for species distribution along the elevational gradient. Biotic responses include phenological shifts, changing species compositions, dominance patterns and community structures, and changes in primary production, affecting treeline spatial patterns and functioning of ecosystems. Stand densification and the intense recruitment with sufficient presence of seedlings and juveniles even in the alpine tundra indicate that the preconditions for a future treeline advance exist. Thus, the sensitivity of the Rolwaling treeline ecotone is clearly evident with respect to stand density, seed-based regeneration, and tree growth patterns. A treeline shift to higher elevation, however, is only to be expected in the medium to long term (decades to centuries).

Acknowledgements This research was funded by the German Research Foundation (DFG; SCHI436/14-1, BO 1333/4-1, SCHO 739/14-1). We are deeply grateful to Nepalese authorities for research permits and to the community in Rolwaling for friendly cooperation, assistance in fieldwork, and hospitality, in particular to Tenzing and Lakpa Sherpa from Beding, who provided lodging and support during field data collection.

References

Anschlag K, Broll G, Holtmeier FK (2008) Mountain birch seedlings in the treeline ecotone, subarctic Finland: variation in above- and below-ground growth depending on microtopography. Arct Antarct Alp Res 40:609–616

Bader MY, Rietkerk M, Bregt AK (2008) A simple spatial model exploring positive feedbacks at tropical alpine treelines. Arct Antarct Alp Res 40:269–278

Bader MY, Llambí LD, Case BS, Buckley HL, Toivonen JM et al (2021) A global framework for linking alpine-treeline ecotone patterns to underlying processes. Ecography 44:265–292

Baker BB, Moseley RK (2007) Advancing treeline and retreating glaciers: implications for conservation in Yunnan, P.R. China. Arct Antarct Alp Res 39:200–209

Batllori E, Gutiérrez E (2008) Regional tree line dynamics in response to global change in the Pyrenees. J Ecol 96:1275–1288

Baumgartner R (2015) Farewell to yak and yeti? The Sherpas of Rolwaling facing a globalised world. Vajra Books, Kathmandu

Benjamini Y, Hochberg Y (1995) Controlling the false discovery rate: a practical and powerful approach to multiple testing. J Roy Stat Soc Ser B 57:289–300

Bhutiyani MR, Kale VS, Pawar NJ (2007) Long-term trends in maximum, minimum and mean annual air temperatures across the northwestern Himalaya during the twentieth century. Clim Chang 85:159–177

Bhutiyani MR, Kale VS, Pawar NJ (2010) Climate change and the precipitation variations in the northwestern Himalaya: 1866-2006. Int J Climatol 30:535–548

Bi Y, Xu J, Yang J, Li Z, Gebrekirstos A et al (2017) Ring-widths of the above tree-line shrub *Rhododendron* reveal the change of minimum winter temperature over the past 211 years in southwestern China. Clim Dyn 48:3919–3933

Biondi F, Waikul K (2004) DENDROCLIM2002: a C++ program for statistical calibration of climate signals in tree-ringchronologies. Comput Geosci 30:303–311

Bivand RS, Pebesma E, Gómez-Rubio V (2013) Applied spatial data analysis with R. Springer, New York

Bivand RS, Keitt T, Rowlingson B (2015) rgdal: bindings for the geospatial data abstraction library. R package version 1.1-1. http://cran.r-project.org/package=rgdal

Bobrowski M, Schickhoff U (2017) Why input matters: selection of climate data sets for modelling the potential distribution of a treeline species in the Himalayan region. Ecol Model 359:92–102

Bobrowski M, Gerlitz L, Schickhoff U (2017) Modelling the potential distribution of *Betula utilis* in the Himalaya. Glob Ecol Conserv 11:69–83

Bobrowski M, Bechtel B, Böhner J, Oldeland J, Weidinger J, Schickhoff U (2018) Application of thermal and phenological land surface parameters for improving ecological niche models of *Betula utilis* in the Himalayan region. Remote Sens 10:814

Böhner J (2006) General climatic controls and topoclimatic variations in Central and High Asia. Boreas 35:279–295

Böhner J, Bechtel B (2018) GIS in climatology and meteorology. In: Huang B (ed) Comprehensive geographic information systems, vol 2. Elsevier, Oxford, pp 196–235

Böhner J, Miehe G, Miehe S, Nagy L (2015) Climate and weather. In: Miehe G, Pendry C, Chaudhary RP (eds) Nepal: an introduction to the natural history, ecology and human environment in the Himalayas. Royal Botanic Garden Edinburgh, Edinburgh, pp 23–90

Brandt JS, Haynes MA, Kuemmerle T, Waller DM, Radeloff VC (2013) Regime shift on the roof of the world: alpine meadows converting to shrublands in the southern Himalayas. Biol Conserv 158:116–127

Braun-Blanquet J (1964) Pflanzensoziologie. Springer, Wien

Bräuning A, Grießinger J, Hochreuther P, Wernicke J (2016) Dendroecological perspectives on climate change on the southern Tibetan Plateau. In: Singh RB, Schickhoff U, Mal S (eds) Climate change and dynamics of glaciers and vegetation in the Himalaya. Springer, Cham, pp 347–364

Bürzle B, Schickhoff U, Schwab N, Oldeland J, Müller M et al. (2017) Phytosociology and ecology of treeline ecotone vegetation in Rolwaling Himal, Nepal. Phytocoenologia 47:197–220

Bürzle B, Schickhoff U, Schwab N, Wernicke LM, Müller YK et al (2018) Seedling recruitment and facilitation dependence on safe site characteristics in a Himalayan treeline ecotone. Plant Ecol 219:115–132

Camarero JJ, García-Ruiz JM, Sangüesa-Barreda G, Galván JD, Alla AQ et al (2015) Recent and intense dynamics in a formerly static Pyrenean treeline. Arct Antarct Alp Res 47:773–783
Camarero JJ, Linares JC, García-Cervigón AI, Batllori E, Martínez I, Gutiérrez E (2017) Back to the future: the responses of alpine treelines to climate warming are constrained by the current ecotone structure. Ecosystems 20:683–700
Camarero JJ, Gazol A, Sánchez-Salguero R, Fajardo A, McIntire EJ et al (2021a) Global fading of the temperature–growth coupling at alpine and polar treelines. Glob Chang Biol 27:1879–1889
Camarero JJ, Gazol A, Sánchez-Salguero R, Fajardo A, McIntire EJ, Liang E (2021b) Tree growth and treeline responses to temperature: different questions and concepts. Glob Chang Biol 27. https://doi.org/10.1111/GCB.15728
Chhetri PK, Cairns DM (2018) Low recruitment above treeline indicates treeline stability under changing climate in Dhorpatan Hunting Reserve, western Nepal. Phys Geogr 39:329–342
Chhetri PK, Gaddis KD, Cairns DM (2018) Predicting the suitable habitat of treeline species in the Nepalese Himalayas under climate change. Mt Res Dev 38:153–164
Chhetri PK, Bista R, Shrestha KB (2020) How does the stand structure of treeline-forming species shape the treeline ecotone in different regions of the Nepal Himalayas? J Mt Sci 17:2354–2368
Chou SC, Huang CH, Hsu TW, Wu CC, Chou CH (2010) Allelopathic potential of *Rhododendron formosanum* Hemsl in Taiwan. Allelop J 25:73–92
Compostella C, Caccianiga M (2017) A comparison between different treeline types shows contrasting responses to climate fluctuations. Plant Biosyst 151:436–449
Cook ER (1985) A time series analysis approach to tree ring standardization. PhD Diss., University of Arizona
Core Team R (2018) R: alanguage and environment for statistical computing. R Foundation for Statistical Computing, Vienna
CRU (University of East Anglia Climatic Research Unit), Harris IC, Jones PD (2017) CRU TS4.00: Climatic Research Unit (CRU) Time-Series (TS) version 4.00 of high resolution gridded data of month-by-month variation in climate (Jan. 1901- Dec. 2015). https://doi.org/10.5285/edf8febfdaad48abb2cbaf7d7e846a86
Cudlín P, Klopčič M, Tognetti R, Máli F, Alados CL et al (2017) Drivers of treeline shift in different European mountains. Clim Res 73:135–150
Dahal N, Shrestha UB, Tuitui A, Ojha HR (2019) Temporal changes in precipitation and temperature and their implications on the streamflow of Rosi River, Central Nepal. Climate 7:3
Dawadi B, Liang E, Tian L, Devkota LP, Yao T (2013) Pre-monsoon precipitation signal in tree rings of timberline *Betula utilis* in the Central Himalayas. Quat Int 283:72–77
Dawes MA, Schleppi P, Hättenschwiler S, Rixen C, Hagedorn F (2017) Soil warming opens the nitrogen cycle at the alpine treeline. Glob Chang Biol 23:421–434
De Lillis M, Matteucci G, Valentini R (2004) Carbon assimilation, nitrogen, and photochemical efficiency of different Himalayan tree species along an altitudinal gradient. Photosynthetica 42: 597–605
Dimri AP, Kumar D, Choudhary A, Maharana P (2018) Future changes over the Himalayas: mean temperature. Glob Planet Chang 162:235–251
Dimri AP, Choudhary A, Kumar D (2020) Elevation dependent warming over Indian Himalayan region. In: Dimri AP, Bookhagen B, Stoffel M, Yasunari T (eds) Himalayan weather and climate and their impact on the environment. Springer, Cham, pp 141–156
Doove LL, Van Buuren S, Dusseldorp E (2014) Recursive partitioning for missing data imputation in the presence of interaction effects. Comput Stat Data Anal 72:92–104
Dragulescu AA (2014) xlsx: read, write, format Excel 2007 and Excel 97/2000/XP/2003 files. Rpackage version 0.5.7. http://CRAN.R-project.org/package=xlsx
Drollinger S, Müller M, Kobl T, Schwab N, Böhner J, Schickhoff U, Scholten T (2017) Decreasing nutrient concentrations in soils and trees with increasing elevation across a treeline ecotone in Rolwaling Himal, Nepal. J Mt Sci 14:843–858
Dullinger S, Dirnböck T, Grabherr G (2004) Modelling climate change-driven treeline shifts: relative effects of temperature increase, dispersal and invasibility. J Ecol 92:241–252

Elliott GP, Bailey SN, Cardinal SJ (2021) Hotter drought as a disturbance at upper treeline in the southern Rocky Mountains. Ann Am Assoc Geogr 111:756–770
Fajardo A, McIntire EJ (2012) Reversal of multicentury tree growth improvements and loss of synchrony at mountain tree lines point to changes in key drivers. J Ecol 100:782–794
Fan ZX, Bräuning A, Cao KF, Zhu SD (2009) Growth-climate responses of high-elevation conifers in the Central Hengduan Mountains, southwestern China. For Ecol Manage 258:306–313
Feuillet T, Birre D, Milian J, Godard V, Clauzel C, Serrano-Notivoli R (2020) Spatial dynamics of alpine tree lines under global warming: what explains the mismatch between tree densification and elevational upward shifts at the tree line ecotone? J Biogeogr 47:1056–1068
Fox J, Weisberg S (2011) An R companion to applied regression. Sage Publ, Thousand Oaks
Gaire NP, Koirala M, Bhuju DR, Borgaonkar HP (2014) Treeline dynamics with climate change at the Central Nepal Himalaya. Clim Past 10:1277–1290
Gaire NP, Koirala M, Bhuju DR, Carrer M (2017) Site- and species-specific treeline responses to climatic variability in eastern Nepal Himalaya. Dendrochronologia 41:44–56
Gaire NP, Fan ZX, Bräuning A, Panthi S, Rana P, Shrestha A, Bhuju DR (2020) *Abies spectabilis* shows stable growth relations to temperature, but changing response to moisture conditions along an elevation gradient in the centralHimalaya. Dendrochronologia 60:125675
Gao Y, Chen F, Lettenmaier DP, Xu J, Xiao L, Li X (2018) Does elevation-dependent warming hold true above 5000 m elevation? Lessons from the Tibetan Plateau. NPJ Clim Atmos Sci 1:19
Gauch HG (1982) Multivariate analysis in community ecology. Cambridge Univ Press, New York
Gerlitz L, Conrad O, Thomas A, Böhner J (2014) Warming patterns over the Tibetan Plateau and adjacent lowlands derived from elevation- and bias-corrected ERA-Interim data. Clim Res 58: 235–246
Gerlitz L, Bechtel B, Böhner J, Bobrowski M, Bürzle B et al (2016) Analytic comparison of temperature lapse rates and precipitation gradients in a Himalayan treeline environment: implications for statistical downscaling. In: Singh RB, Schickhoff U, Mal S (eds) Climate change, glacier response, and vegetation dynamics in the himalaya. Springer, Cham, pp 49–64
Giraudoux P (2015) pgirmess: data analysis in ecology. R package version 1.6.3. http://CRAN.R-project.org/package=pgirmess
Grierson AJC, Long DG (1983–2001) Flora of Bhutan. Including a record of plants from Sikkim, vol 1−3. Royal Botanic Garden, Edinburgh
Hagedorn F, Shiyatov SG, Mazepa VS, Devi NM, Grigor'ev AA et al (2014) Treeline advances along the Urals mountain range - driven by improved winter conditions? Glob Chang Biol 20: 3530–3543
Hansson A, Dargusch P, Shulmeister J (2020) A review of methods used to measure treeline migration and their application. J Environ Inform Lett 4:1–10
Harris I, Jones PD, Osborn TJ, Lister DH (2014) Updated high-resolution grids of monthly climatic observations – the CRU TS3.10 dataset. Int J Climatol 34:623–642
Harsch MA, Bader MY (2011) Treeline form - a potential key to understanding treeline dynamics. Glob Ecol Biogeogr 20:582–596
Harsch MA, Hulme PE, McGlone MS, Duncan RP (2009) Are treelines advancing? A global meta-analysis of treeline response to climate warming. Ecol Lett 12:1040–1049
Hasson S, Gerlitz L, Schickhoff U, Scholten T, Böhner J (2016) Recent climate change over High Asia. In: Singh RB, Schickhoff U, Mal S (eds) Climate change, glacier response, and vegetation dynamics in the Himalaya. Springer, Cham, pp 29–48
Hijioka Y, Lin E, Pereira JJ, Corlett RT, Cui X et al. (2014) Asia. In: IPCC (ed). Climate change 2014: impacts, adaptation, and vulnerability. Part B: regional aspects. Cambridge Univ Press, Cambridge, pp 1327–1370
Holmes RL (1983) Computer-assisted quality control in tree-ring dating and measurement. Tree-Ring Bull 43:69–78
Holtmeier FK, Broll G (2007) Treeline advance - driving processes and adverse factors. Landscape Online 1:1–21

Holtmeier KF, Broll G (2010a) Altitudinal and polar treelines in the northern hemisphere - causes and response to climate change. Polarforschung79:139–153
Holtmeier FK, Broll G (2010b) Wind as an ecological agent at treelines in North America, the Alps, and the European Subarctic. Phys Geogr 31:203–233
Holtmeier FK, Broll G (2017) Treelines - approaches at different scales. Sustainability 9:808
Holtmeier FK, Broll G (2018) Subalpine forest and treeline ecotone under the influence of disturbances: a review. J Environ Protect 9:815–845
Holtmeier FK, Broll G (2020) Treeline research — from the roots of the past to present time. A review. Forests 11:38
Huang R, Zhu H, Liu X, Liang E, Grießinger J et al (2017) Does increasing intrinsic water use efficiency (iWUE) stimulate tree growth at natural alpine timberline on the southeastern Tibetan Plateau? Glob Planetary Chang 148:217–226
Hughes NM, Johnson DM, Akhalkatsi M, Abdaladze O (2009) Characterizing *Betula litwinowii* seedling microsites at the alpine-treeline ecotone, Central Greater Caucasus Mountains, Georgia. Arct Antarct Alp Res 41:112–118
IPCC (ed) (2021) Climate change 2021: the physical science basis. Contribution of working group I to the sixth assessment report of the Intergovernmental Panel on Climate Change. Cambridge University Press, Cambridge
Johnson AC, Yeakley JA (2019) Microsites and climate zones: seedling regeneration in the alpine treeline ecotone worldwide. Forests 10:864
Johnson JS, Gaddis KD, Cairns DM, Krutovsky KV (2017) Seed dispersal at alpine treeline: an assessment of seed movement within the alpine treeline ecotone. Ecosphere 8:e01649
Jones CC, del Moral R (2005) Effects of microsite conditions on seedling establishment on the foreland of Coleman Glacier, Washington. J Veg Sci 16:293–300
Kameyama Y, Nehira NN, Nehira K (1999) Safe site for seedlings of *Rhododendron metternichii* var. *hondoense*. Plant Spec Biol 14:237–242
Karki R, Talchabhadel R, Aalto J, Baidya SK (2016) New climatic classification of Nepal. Theor Appl Climat 125:799–808
Karki R, Hasson S, Gerlitz L, Schickhoff U, Scholten T, Böhner J (2017a) Quantifying the added value of convection-permitting climate simulations in complex terrain: a systematic evaluation of WRF over the Himalayas. Earth Syst Dynam 8:507–528
Karki R, Schickhoff U, Scholten T, Böhner J (2017b) Rising precipitation extremes across Nepal. Climate 5:4
Karki R, Hasson S, Gerlitz L, Schenk E, Schickhoff U, Scholten T, Böhner J (2018) WRF-based simulation of an extreme precipitation event over the central Himalayas: atmospheric mechanisms and their representation by microphysics parameterization schemes. Atmos Res 214:21–35
Karki R, Hasson S, Gerlitz L, Talchabhadel R, Schickhoff U, Scholten T, Böhner J (2020a) Rising mean and extreme near-surface air temperature across Nepal. Int J Climat 40:2445–2463
Karki R, Hasson S, Schickhoff U, Scholten T, Böhner J, Gerlitz L (2020b) Near surface air temperature lapse rates over complex terrain: a WRF based analysis of controlling factors and processes for the central Himalayas. Clim Dyn 54:329–349
Kent M (2012) Vegetation description and data analysis: a practical approach, 2nd edn. Wiley-Blackwell, Chichester
Kikuchi T, Ohba H (1988) Preliminary study of alpine vegetation of the Himalayas, with special reference to the small-scale distribution patterns of plant communities. Himal Plants 1:47–70
Kim JW, Lee JS (2015) Dynamics of alpine treelines: positive feedbacks and global, regional and local controls. J Ecol Environ 38:1–14
Körner C (2012) Alpine treelines. Functional ecology of the global high elevation tree limits. Springer, Basel
Körner C (2020) Climatic controls of global high elevation treelines. In: Goldstein MI, DellaSala DA (eds) Encyclopedia of the world's biomes, vol 1. Elsevier, Amsterdam, pp 275–281

Körner C (2021a) Alpine treelines. In: Körner C (ed) Alpine plant life. Functional plant ecology of high mountain ecosystems, 3rd edn. Springer, Cham, pp 141–173

Körner C (2021b) 'Fading of the temperature-growth coupling' in treeline trees reflects a conceptual bias. Glob Change Biol. https://doi.org/10.1111/GCB.15730

Krishnan R, Shrestha AB, Ren G, Rajbhandari R, Saeed S et al (2019) Unravelling climate change in the Hindu Kush Himalaya: rapid warming in the mountains and increasing extremes. In: Wester P, Mishra A, Mukherji A, Shrestha AB (eds) The Hindu Kush Himalaya assessment. Springer, Cham, pp 57–96

Kullman L (2019) Early signs of a fundamental subalpine ecosystem shift in the Swedish Scandes - the case of the pine (*Pinus sylvestris* L.) treeline ecotone. Geo-Öko 40:122–175

Lamsal P, Kumar L, Shabani F, Atreya K (2017) The greening of the Himalayas and Tibetan Plateau under climate change. Glob Planet Chang 159:77–92

Larcher W (2003) Physiological plant ecology: ecophysiology and stress physiology of functional groups. Springer, Berlin-Heidelberg

Larsson LÅ (2016) CooRecorder and CDendro programs of the CooRecorder/CDendro package version 8.1. http://www.cybis.se/forfun/dendro

Legendre P, Legendre L (2012) Numerical ecology. Elsevier, Amsterdam

Legendre P, Borcard D, Blanchet FG, Dray S (2013) PCNM: MEM spatial eigenfunction and principal coordinate analyses. R package version 2.1-2/r109. http://R-Forge.Rproject.org/projects/sedar

Lett S, Dorrepaal E (2018) Global drivers of tree seedling establishment at alpine treelines in a changing climate. Funct Ecol 32:1666–1680

Li ZH, Wang Q, Ruan X, Pan CD, Jiang DA (2010) Phenolics and plant allelopathy. Molecules 15: 8933–8952

Li K, Liao M, Ni J, Liu X, Wang Y (2019) Treeline composition and biodiversity change on the southeastern Tibetan Plateau during the past millennium, inferred from a high-resolution alpine pollen record. Quat Sci Rev 206:44–55

Liang E, Dawadi B, Pederson N, Eckstein D (2014) Is the growth of birch at the upper timberline in the Himalayas limited by moisture or by temperature? Ecology 95:2453–2465

Liang E, Wang Y, Piao S, Lu X, Camarero JJ et al (2016) Species interactions slow warming-induced upward shifts of treelines on the Tibetan Plateau. Proc Natl Acad Sci 113:4380–4385

Liu X, Yin ZY, Shao X, Qin N (2006) Temporal trends and variability of daily maximum and minimum, extreme temperature events, and growing season length over the eastern and central Tibetan Plateau during 1961–2003. J Geophys Res Atmos 111:D19109

Liu X, Cheng Z, Yan L, Yin ZY (2009) Elevation dependency of recent and future minimum surface air temperature trends in the Tibetan Plateau and its surroundings. Glob Planet Chang 68:164–174

Lu X, Camarero JJ, Wang Y, Liang E, Eckstein D (2015) Up to 400-year-old *Rhododendron* shrubs on the southeastern Tibetan Plateau: prospects for shrub-based dendrochronology. Boreas 44: 760–768

Lu X, Liang E, Wang Y, Babst F, Camarero JJ (2021) Mountain treelines climb slowly despite rapid climate warming. Glob Ecol Biogeogr 30:305–315

Lyu L, Zhang QB, Pellatt MG, Büntgen U, Li MH, Cherubini P (2019) Drought limitation on tree growth at the northern hemisphere's highest tree line. Dendrochronologia 53:40–47

Maher EL, Germino MJ (2006) Microsite differentiation among conifer species during seedling establishment at alpine treeline. Ecoscience 13:334–341

Mainali K, Shrestha BB, Sharma RK, Adhikari A, Gurarie E, Singer M, Parmesan C (2020) Contrasting responses to climate change at Himalayan treelines revealed by population demographics of two dominant species. Ecol Evol 10:1209–1222

Maithani K, Arunachalam A, Tripathi RS, Pandey HN (1998) Influence of leaf litter quality on N mineralization in soils of subtropical humid forest regrowths. Biol Fertil Soils 27:44–50

Malanson GP, Resler LM, Butler DR, Fagre DB (2019) Mountain plant communities: uncertain sentinels? Progr Phys Geogr 43:521–543

Martínez I, González-Taboada F, Wiegand T, Camarero JJ, Gutiérrez E (2012) Dispersal limitation and spatial scale affect model based projections of *Pinus uncinata* response to climate change in the Pyrenees. Glob Change Biol 18:1714–1724

Mejías JA, Arroyo J, Ojeda F (2002) Reproductive ecology of *Rhododendron ponticum* (*Ericaceae*) in relict Mediterranean populations. Bot J Linn 140:297–311

Miehe G (1990) Langtang Himal: Flora und Vegetation als Klimazeiger und -zeugen im Himalaya. Cramer, Berlin

Miehe G, Miehe S (2000) Comparative high mountain research on the treeline ecotone under human impact: Carl Troll's "Asymmetrical zonation of the humid vegetation types of the world" of 1948 reconsidered. Erdkunde 54:34–50

Miehe G, Miehe S, Böhner J, Bäumler R, Ghimire SK et al (2015) Vegetation ecology. In: Miehe G, Pendry CA, Chaudhary RP (eds) Nepal: an introduction to the natural history, ecology and human environment of the Himalayas. Royal Botanic Garden Edinburgh, Edinburgh, pp 385–472

Millar CI, Delany DL, Westfall RD (2020) From treeline to species line: thermal patterns and growth relationships across the krummholz zone of whitebark pine, Sierra Nevada, California, USA. Arct Antarct Alp Res 52:390–407

Mishra NB, Mainali KP (2017) Greening and browning of the Himalaya: spatial patterns and the role of climatic change and human drivers. Sci Total Environ 587:326–339

Mohapatra J, Singh CP, Tripathi OP, Pandya HA (2019) Remote sensing of alpine treeline ecotone dynamics and phenology in Arunachal Pradesh Himalaya. Int J Rem Sens 40:7986–8009

Müller M, Schickhoff U, Scholten T, Drollinger S, Böhner J, Chaudhary RP (2016a) How do soil properties affect alpine treelines? General principles in a global perspective and novel findings from Rolwaling Himal, Nepal. Progr Phys Geogr 40:135–160

Müller M, Schwab N, Schickhoff U, Böhner J, Scholten T (2016b) Soil temperature and soil moisture patterns in a Himalayan alpine treeline ecotone. Arct Antarct Alp Res 48:501–521

Müller M, Oelmann Y, Schickhoff U, Böhner J, Scholten T (2017) Himalayan treeline soil and foliar C:N:P stoichiometry indicate nutrient shortage with elevation. Geoderma 291:21–32

Nathan R, Horvitz N, He Y, Kuparinen A, Schurr FM, Katul GG (2011) Spread of North American wind-dispersed trees in future environments. Ecol Lett 14:211–219

Nilsen ET, Walker JF, Miller OK, Semones SW, Lei TT, Clinton BD (1999) Inhibition of seedling survival under *Rhododendron maximum* (*Ericaceae*): could allelopathy be a cause? Am J Bot 86:1597–1605

Northup RR, Yu Z, Dahlgren RA, Vogt KA (1995) Polyphenol control of nitrogen release from pine litter. Nature 377:227–229

Oksanen J, Blanchet FG, Kindt R, Legendre P, Minchin PR et al. (2015) vegan: community ecology package. R package version 2.3-1. http://CRAN.R-project.org/package=vegan

Panthi J, Dahal P, Shrestha ML, Aryal S, Krakauer NY et al (2015) Spatial and temporal variability of rainfall in the Gandaki river basin of Nepal Himalaya. Climate 3:210–226

Panthi S, Bräuning A, Zhou ZK, Fan ZX (2017) Tree rings reveal recent intensified spring drought in the central Himalaya, Nepal. Glob Planet Chang 157:26–34

Panthi S, Fan ZX, Bräuning A (2021) Ring widths of *Rhododendron* shrubs reveal a persistent winter warming in the central Himalaya. Dendrochronologia 65:125799

Paulsen J, Körner C (2014) A climate-based model to predict potential treeline position around the globe. Alp Bot 124:1–12

Pepin N, Deng H, Zhang H, Zhang F, Kang S, Yao T (2019) An examination of temperature trends at high elevations across the Tibetan Plateau: The use of MODIS LST to understand patterns of elevation-dependent warming. J Geophys Res: Atmos 124:5738–5756

Politis DN, Romano JP (1994) The stationary bootstrap. J Am Stat Ass 89:1303–1313

Politis DN, White H (2004) Automatic block-length selection for the dependent bootstrap. Econ Rev 23:53–70

Popescu R, Kopp B (2013) The genus *Rhododendron*: an ethnopharmacological and toxicological review. J Ethnopharmacol 147:42–62

Press JR, Shrestha KK, Sutton DA (2000) Annotated checklist of the flowering plants of Nepal (updated online version 2014). Natural History Museum Publ, London

Ram S, Borgaonkar HP (2014) Tree-ring analysis over western Himalaya and its long-term association with vapor pressure and potential evapotranspiration. Dendrochronologia 32:32–38

Rana P, Koirala M, Bhuju DR, Boonchird C (2016) Population structure of *Rhododendron campanulatum* D. Don and associated tree species along the elevational gradient of Manaslu conservation area, Nepal. J Inst Sci Technol 21:95–102

Rangwala I, Palazzi E, Miller JR (2020) Projected climate change in the Himalayas during the twenty-first century. In: Dimri AP, Bookhagen B, Stoffel M, Yasunari T (eds) Himalayan weather and climate and their impact on the environment. Springer, Cham, pp 51–71

Ren GY, Shrestha AB (2017) Climate change in the Hindu Kush Himalaya. Adv Clim Chang Res 8: 137–140

Ren YY, Ren GY, Sun XB, Shrestha AB, You QL et al (2017) Observed changes in surface air temperature and precipitation in the Hindu Kush Himalayan region over the last 100-plus years. Adv Clim Chang Res 8:148–156

Revelle W (2015) psych: procedures for personality and psychological research. Northwestern University, Evanston. http://CRAN.Rproject.org/package=psych

Ristvey AG, Lea-Cox JD, Ross DS (2007) Nitrogen and phosphorus uptake efficiency and partitioning of container-grown *Azalea* during spring growth. J Am Soc Hortic Sci 132:563–571

Sacherer J (1979) The high altitude ethnobotany of the Rolwaling Sherpas. Contrib Nepal Stud 6: 45–64

Sanz R, Pulido F, Camarero JJ (2011) Boreal trees in the Mediterranean: recruitment of downy birch (*Betula alba*) at its southern range limit. Ann For Sci 68:793–802

Schickhoff U (1993) Das Kaghan-Tal im Westhimalaya (Pakistan): Studien zur landschaftsökologischen Differenzierung und zum Landschaftswandel mit vegetationskundlichem Ansatz. Bonner Geogr Abh 87, Dümmlers, Bonn

Schickhoff U (2005) The upper timberline in the Himalayas, Hindu Kush and Karakorum: a review of geographical and ecological aspects. In: Broll G, Keplin B (eds) Mountain ecosystems. Springer, Berlin, pp 275–354

Schickhoff U, Mal S (2020) Current changes in alpine ecosystems of Asia. In: Goldstein MI, DellaSala DA (eds) Encyclopedia of the world's biomes, vol 1. Elsevier, Amsterdam, pp 589–598

Schickhoff U, Bobrowski M, Böhner J, Bürzle B, Chaudhary RP et al (2015) Do Himalayan treelines respond to recent climate change? An evaluation of sensitivity indicators. Earth Syst Dynam 6:245–265

Schickhoff U, Bobrowski M, Böhner J, Bürzle B, Chaudhary RP et al (2016a) Climate change and treeline dynamics in the Himalaya. In: Singh RB, Schickhoff U, Mal S (eds) Climate change, glacier response, and vegetation dynamics in the Himalaya. Springer, Cham, pp 271–306

Schickhoff U, Singh RB, Mal S (2016b) Climate change and dynamics of glaciers and vegetation in the Himalaya: an overview. In: Singh RB, Schickhoff U, Mal S (eds) Climate change, glacier response, and vegetation dynamics in the Himalaya. Springer, Cham, pp 1–26

Schickhoff U, Bobrowski M, Schwab N (2020) Alpine Waldgrenzen im Klimawandel - Wie sind die heterogenen Reaktionsmuster zu erklären? In: Lozán JL, Breckle SW, Escher-Vetter H, Grassl H, Kasang D, Paul F, Schickhoff U (eds) Warnsignal Klima: Hochgebirge im Wandel. Wissenschaftliche Auswertungen, Hamburg, pp 232–238

Schickhoff U, Bobrowski M, Mal S, Schwab N, Singh RB (2022) The world's mountains in the Anthropocene. In: Schickhoff U, Singh RB, Mal S (eds) Mountain landscapes in transition: effects of land use and climate change. Springer, Cham, pp 1–144

Schwab N, Schickhoff U, Bobrowski M, Böhner J, Bürzle B et al (2016) Treeline responsiveness to climate warming: insights from a krummholz treeline in Rolwaling Himal, Nepal. In: Singh RB, Schickhoff U, Mal S (eds) Climate change, glacier response, and vegetation dynamics in the Himalaya. Springer, Cham, pp 307–345

Schwab N, Schickhoff U, Bürzle B, Müller M, Böhner J et al (2017) Implications of tree species-environment relationships for the responsiveness of Himalayan krummholz treelines to climate change. J Mt Sci 14:453–473

Schwab N, Kaczka RJ, Janecka K, Böhner J, Chaudhary RP, Scholten T, Schickhoff U (2018) Climate change-induced shift of tree growth sensitivity at a Central Himalayan treeline ecotone. Forests 9:267

Schwab N, Janecka K, Kaczka RJ, Böhner J, Chaudhary RP, Scholten T, Schickhoff U (2020) Ecological relationships at a near-natural treeline, Rolwaling Valley, Nepal Himalaya: implications for the sensitivity to climate change. Erdkunde 74:15–44

Schwab N, Bürzle B, Böhner J, Chaudhary RP, Scholten T, Schickhoff U (2022) Environmental drivers of species composition and tree species density of a near-natural central Himalayan treeline ecotone: consequences for the response to climate change. In: Schickhoff U, Singh RB, Mal S (eds) Mountain landscapes in transition: effects of land use and climate change. Springer, Cham, pp 349–370

Schweinfurth U (1957) Die horizontale und vertikale Verbreitung der Vegetation im Himalaya. Bonner Geogr Abh 20, Dümmlers, Bonn

Schwörer C, Kaltenrieder P, Glur L, Berlinger M, Elbert J et al (2014) Holocene climate, fire and vegetation dynamics at the treeline in the northwestern Swiss Alps. Veg Hist Archaeobot 23: 479–496

Sharma KP, Tiwari A, Shrestha BB (2020) Changes in regeneration and leaf traits of *Rhododendron campanulatum* along a treeline ecotone in central Nepal. J Mt Sci 17:602–613

Sherry J, Curtis A, Mendham E, Toman E (2018) Cultural landscapes at risk: exploring the meaning of place in a sacred valley of Nepal. Glob Environ Chang 52:190–200

Shi C, Schneider L, Hu Y, Shen M, Sun C et al (2020) Warming-induced unprecedented high-elevation forest growth over the monsoonal Tibetan Plateau. Environ Res Lett 15:054011

Shrestha AB, Aryal R (2011) Climate change in Nepal and its impact on Himalayan glaciers. Reg Environ Chang 11:S65–S77

Shrestha AB, Wake CP, Mayewski PA, Dibb JE (1999) Maximum temperature trends in the Himalayas and its vicinity: an analysis based on temperature records from Nepal for the period 1971-1994. J Clim 12:2775–2786

Shrestha UB, Shrestha AM, Aryal S, Shrestha S, Gautam MS, Ojha H (2019) Climate change in Nepal: a comprehensive analysis of instrumental data and people's perceptions. Clim Chang 154:315–334

Sigdel SR, Wang Y, Camarero JJ, Zhu H, Liang E, Peñuelas J (2018) Moisture-mediated responsiveness of treeline shifts to global warming in the Himalayas. Glob Chang Biol 24: 5549–5559

Sigdel SR, Liang E, Wang Y, Dawadi B, Camarero JJ (2020) Tree-to-tree interactions slow down Himalayan treeline shifts as inferred from tree spatial patterns. J Biogeogr 47:1816–1826

Singh SP (2018) Research on Indian Himalayan treeline ecotone: an overview. Trop Ecol 59:163–176

Singh P, Arya V, Negi GCS, Singh SP (2018) Expansion of *Rhododendron campanulatum* krummholz in the treeline ecotone in Tungnath, Garhwal Himalaya. Trop Ecol 59:287–295

Singh SP, Sharma S, Dhyani PP (2019) Himalayan arc and treeline: distribution, climate change responses and ecosystem properties. Biodivers Conserv 28:1997–2016

Singh CP, Mohapatra J, Pandya HA, Gajmer B, Sharma N, Shrestha DG (2020) Evaluating changes in treeline position and land surface phenology in Sikkim Himalaya. Geocarto Int 35:453–469

Smith WK, Germino MJ, Hancock TE, Johnson DM (2003) Another perspective on altitudinal limits of alpine timberlines. Tree Physiol 23:1101–1112

Sohar K, Altman J, Lehečková E, Doležal J (2017) Growth-climate relationships of Himalayan conifers along elevational and latitudinal gradients. Int J Climat 37:2593–2605

Speer JH (2010) Fundamentals of tree-ring research. Univ. of Arizona Press, Tucson

Suwal MK, Shrestha KB, Guragain L, Shakya R, Shrestha K, Bhuju DR, Vetaas OR (2016) Land-use change under a warming climate facilitated upslope expansion of Himalayan silver fir (*Abies spectabilis* (D. Don) Spach). Plant Ecol 217:993–1002
Thapa UK, St. George S, Kharal DK, Gaire NP (2017) Tree growth across the Nepal Himalaya during the last four centuries. Progr Phys Geogr 41:478–495
Tiwari A, Jha PK (2018) An overview of treeline response to environmental changes in Nepal Himalaya. Trop Ecol 59:273–285
Tiwari A, Fan ZX, Jump AS, Zhou ZK (2017) Warming induced growth decline of Himalayan birch at its lower range edge in a semi-arid region of Trans-Himalaya, Central Nepal. Plant Ecol 218: 621–633
Trant AJ, Hermanutz L (2014) Advancing towards novel tree lines? A multispecies approach to recent tree line dynamics in subarctic alpine Labrador, northern Canada. J Biogeogr 41:1115–1125
Treml V, Veblen TT (2017) Does tree growth sensitivity to warming trends vary according to treeline form? J Biogeogr 44:1469–1480
Treml V, Šenfeldr M, Chuman T, Ponocná T, Demková K (2016) Twentieth century treeline ecotone advance in the Sudetes Mountains (Central Europe) was induced by agricultural land abandonment rather than climate change. J Veg Sci 27:1209–1221
Van Buuren S (2012) Flexible imputation of missing data. CRC Press, Boca Raton
Van Laar A, Akça A (2007) Forest mensuration. Springer, Dordrecht
Vicente-Serrano SM, Beguería S, López-Moreno JI, Angulo M, El Kenawy A (2010) A new global 0.5° gridded dataset (1901–2006) of a multiscalar drought index: comparison with current drought index datasets based on the Palmer Drought Severity Index. J Hydrometeorol 11: 1033–1043
Vincze I, Orbán I, Birks HH, Pál I, Finsinger W et al (2017) Holocene treeline and timberline changes in the South Carpathians (Romania): climatic and anthropogenic drivers on the southern slopes of the Retezat Mountains. The Holocene 27:1613–1630
Walters MB, Reich PB (2000) Seed size, nitrogen supply, and growth rate affect tree seedling survival in deep shade. Ecology 81:1887–1901
Watson MF, Mkiyama S, Ikeda H, Pendry C, Rajbhandari KR, Shrestha KK (2011) Flora of Nepal, vol 3. Royal Botanic Garden, Edinburgh
Weidinger J, Gerlitz L, Bechtel B, Böhner J (2018) Statistical modelling of snow cover dynamics in the Central Himalaya Region, Nepal. ClimRes 75:181–199
Weidinger J, Gerlitz L, Bobrowski M, Böhner J, Chaudhary RP et al. (2021) TREELINE – Longterm atmospheric and pedo-climatic observations along an upper treeline ecotone in the Himalayas, Nepal. Open Access Dataset at CEN-ICDC. https://doi.org/10.25592/uhhfdm.9562
Weijers S, Myers-Smith IH, Loeffler J (2018) A warmer and greener cold world: summer warming increases shrub growth in the alpine and high Arctic tundra. Erdkunde 72:63–85
Wickham H (2011) The split-apply-combine strategy for data analysis. J Stat Softw 40:1–29
Wiegand T, Camarero JJ, Rüger N, Gutiérrez E (2006) Abrupt population changes in treeline ecotones along smooth gradients. J Ecol 94:880–892
Wieser G, Holtmeier F-K, Smith WK (2014) Treelines in a changing global environment. In: Tausz M, Grulke N (eds) Trees in a changing environment. Springer, Dordrecht, pp 221–263
Wieser G, Oberhuber W, Gruber A (2019) Effects of climate change at treeline: lessons from space-for-time studies, manipulative experiments, and long-term observational records in the Central Austrian Alps. Forests 10:508
Yang X, Zhang T, Qin D, Kang S, Qin X (2011) Characteristics and changes in air temperature and glacier's response on the north slope of Mt. Qomolangma (Mt. Everest). Arct Antarct Alp Res 43:147–160
Zang C, Biondi F (2015) treeclim: an R package for the numerical calibration of proxy-climate relationships. Ecography 38:431–436
Zhang F, Chen Q, Gou X, Du M, Wang F, Zhang J (2021) Climatic control on the growth and regeneration of *Juniperus przewalskii* at alpine treeline in the eastern Qilian Mountains, northwest China. Trees 35. https://doi.org/10.1007/s00468-021-02123-w

6 Challenges of Timberline Mapping in the Himalaya: A Case Study of the Sikkim Himalaya

Priyanka Sah, Avantika Latwal, and Subrat Sharma

Abstract

High-altitude Himalayan vegetation is considered a prominent indicator of climate change. Field-based observations are limited in coverage to understand regional patterns and to be used as inputs in ecological models for generalization. This study standardizes the terminology by defining the rules for the interpretation of timberline through auto-extraction methods and proposes a harmonized approach to regional-scale mapping of timberline by incorporating field-based observations. At rare locations in the state of Sikkim, high-altitude timberline may occur much below (~2620 m) or much higher altitude (~4390 m) than the normal range reported from field observations. Change analysis for timberline (upward or downward shift) in the Sikkim Himalaya indicates that the majority of the timberlines (76.5%) remained stationary in last three and half decades (1977–2015). The mean upward shift of timberline was 100 m ± 89 (@ 2.71 m/year) and the mean downward shift was 56 m ± 54 (@ 1.52 m/year). This study reports for the first time a stationary timberline in the Sikkim Himalaya, and the rate of change observed in this study is in tune with the previous studies. A well-harmonized approach, using satellite imagery in conjunction with field observations, can be useful for the regular monitoring of timberline change in order to study the impacts of global warming and biotic pressures at high altitudes.

Keywords

Change detection · Himalayas · Mapping · Timberline · Satellite Images

P. Sah (✉) · A. Latwal · S. Sharma
G.B. Pant National Institute of Himalayan Environment (NIHE), Ladakh Regional Centre, Leh, Ladakh (UT), India

S. P. Singh et al. (eds.), *Ecology of Himalayan Treeline Ecotone*,
https://doi.org/10.1007/978-981-19-4476-5_6

6.1 Introduction

High-altitude treelines and timberlines have gained global attention as a barometer for climate change in the last few decades. In high mountains, trees at their upper altitudinal limit make a conspicuous vegetation boundary. In this respect, two terms, timberline and treeline, are most commonly used. Timberline (or forest limit) is the maximum elevation of forest with cover at least 20–40%. Treeline refers to the transition from forest to treeless vegetation. It is defined as the connection among the highest-elevation trees (Singh and Rawal 2017). Studies from different parts of world show that warming is promoting vegetation or species richness in the alpines, which means that higher the temperature, greater is the richness (Hamid et al. 2020; Holzinger et al. 2008; Lenoir and Svenning 2015; Dolezal et al. 2016; Lamprecht et al. 2018; Steinbauer et al. 2018). Across the mountains on the Earth, altitudinal shift in the occurrence of tree species has been attributed to different ongoing natural or anthropogenic processes and interactions of landscape elements (Kullman 2007; Germino et al. 2002; Ives and Hansen-Bristow 1983; Leonelli et al. 2009, 2016; Motta et al. 2006; Treml and Banas 2008; Colombaroli et al. 2010; Smith et al. 2009; Holtmeier and Broll 2005, 2007; Vittoz et al. 2008; Baker and Moseley 2007; Sarmiento and Frolich 2002; Singh et al. 2019).

Observations of many field biologists have paved the way for regional- or global-scale generalization for the formation of treelines and timberlines (Körner and Paulsen 2004; Schickhoff 2005; Berdanier 2010), but the Himalayan ranges pose a challenge for such global generalization. For example, the globally latitudinal decrease in timberline elevation in the Northern Hemisphere is distributed between the high peaks and terrains of the Himalaya. Based on some studies (point dataset), models have been developed to explain the relationship of treeline/timberline occurrence with geography, climate and their combinations (Zhao et al. 2014). Such field point observations have limited coverage in developing an understanding of local/regional patterns. Even studies on extensive field monitoring to realize the responses of vegetation are few (Kullman 2007; Germino et al. 2002), and a regional-scale model for the Himalaya is not available to explain timberline elevation and influence of geography. Such limitation is largely due to the inaccessibility of high-altitude terrains where field observations are not possible, hence a complete picture is always missing to realize a synoptic view, and in developing a complete geo-spatial database of high-altitude treeline/timberline. Details are required to determine the timberline elevations (Beaman 1962) for better predictions and modelling. Accurate remote sensing of mountainous forest cover change is important for myriad social and ecological reasons (Van Den Hoek et al. 2021). Timberline mapping through remote sensing techniques (Klinge et al. 2015) is a better way to realize the actual position of timberline in the mountains.

Mapping of the Himalayan timberline has started recently with an aim to understand the rate of migration of timberline (Bharti et al. 2012; Juntunen et al. 2002; Panigrahy et al. 2010a, b; Singh et al. 2012, 2018; Latwal et al. 2018; Mohapatra et al. 2019), and attempts have been initiated to develop a geo-database of regional timberline and analyse its relationship with topographical features (Sah and Sharma

2018; Singh et al. 2021a, b). Variability in terms of scale and methods in different studies makes it incomparable, and the actual shift in timberline remains debatable (Bharti et al. 2011). An evaluation of different remote sensing approaches for the highly rugged terrains of the Himalaya reveals that all methods do not perform adequately, and results may include methodological errors (Bharti et al. 2012). Thus, there is a need to adopt a uniformly applicable remote sensing approach for developing an ecologically compatible database of the Indian Himalayan Region (IHR) to assess timberline attributes at the regional scale, and its response to climate change and other factors. With this rationale, the chapter attempts to (1) standardize terminology for a regional-scale remote sensing-based mapping of timberline, (2) assess the role of scale in timberline mapping at regional (entire Sikkim state) as well as local scale (a watershed of the state of Sikkim) and (3) develop a methodological framework for a remote sensing approach to be applied for creating a comparable timberline geospatial database over the heterogeneous Himalayan landscape. It is expected that such a harmonized geospatial database will be a useful input for the regional modelling of change detection, future prediction and geographical explanations of diverse Himalayan timberline spread in about 2000 km length of these mountain ranges.

6.2 Study Area and Field Observations

High-altitude timberline in the Indian Himalayan Region (IHR) exists in five states/ union territories, viz., Jammu & Kashmir, Himachal Pradesh, Uttarakhand, Sikkim and Arunachal Pradesh (arranged from west to east on the Himalayan Arc, Fig. 6.1). Sikkim and Arunachal Pradesh represent the north-eastern part and Uttarakhand, Himachal Pradesh and Jammu & Kashmir are in north-western part of the IHR. Because Sikkim is the smallest state, wide altitudinal variations, and a large area of protected area network in high altitudes were the criteria to take it as a test case to compare products (scale dependent) obtained from different satellite images and also to develop a methodological framework for evaluation and use. The state has four districts, namely East district, West district, North district and South district (Fig. 6.1), with district headquarters at Gangtok, Geyzing, Mangan and Namchi respectively.

High-resolution mapping was done in the Dzongri watershed where field surveys have been carried out for vegetation studies (Pandey et al. 2018a, b) in a part of the Khangchendzonga National Park. In the field study along the elevation gradient of 3000–4000 m, 23 tree species were recorded. In this altitudinal zone, only six species had a long elevation range (600 m or more, viz., *Abies densa, Pieris villosa, Acer caudatum, Sorbus thomsonii, Prunus rufa, Rhododendron hodgsonii*). Towards higher elevations (3800 m or above), different evergreen small tree/erect shrub species of rhododendrons were observed, viz., *R. hodgsonii* (3–7 m tall), *R. thomsonii* (3–5 m tall), *R. lanatum* (1–3 m tall), *R. wightii* (1–3 m tall) and *R. fulgens* (1–3 m tall). Above 3800 m, large trees of evergreen coniferous *Abies densa* (which can reach up to 60 m in height) along with deciduous tree elements, viz., *Acer*

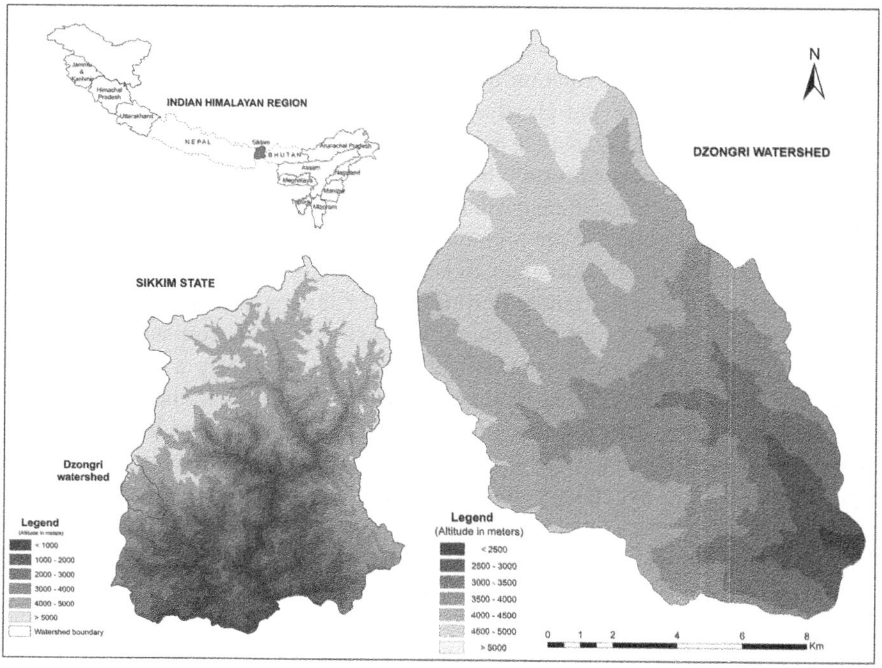

Fig. 6.1 Location of Sikkim State in the Indian Himalayan region and altitudinal profile of state and Dzongri watershed

caudatum (10 m tall), *Sorbus thomsonii* (8–10 m tall), *Prunus rufa* (5–6 m tall) and *Pieris villosa* (1–3 m tall) were observed. Information about plant height of tree species is based on description from e-flora (http://www.efloras.org). It appears that only a few large tree species occur in the high altitudes of Sikkim, and the total basal area of *Abies* forests or communities decreases with increase in altitude (52.5 m^2/ha at 3200 m to 15.06 m^2/ha at 3900 m; Pandey et al. 2018a).

In another study, Pandey et al. (2018b) reported that between 3790 m and 3990 m altitudes, the canopy tree layer was dominated by *Abies densa* and the under-canopy layer consisted of *Sorbus microphylla* and *Rhododendron lanatum*. At several sites, understory tree species were more prominent than *A. densa* or were sole tree species. At some sites, deciduous *Sorbus microphylla* and evergreen *Rhododendron lanatum* had higher densities than the canopy species *A. densa*. Interestingly this study in a narrow band of altitude (3790 m to 3990 m) recorded the presence of *Rhododendron arboreum* at this elevation, which was not documented in a previous study (Pandey et al. 2018a) at this high elevation, hence field-based studies may also provide different results. In these circumstances, diffused type of timberline exists, and the intermixing of different attributes of vegetation (short and tall height, coniferous and broadleaved canopy, evergreen and deciduous) may pose challenges in auto-extraction of information from satellite images.

6.3 Methodological Framework

Defining Timberline: Tree vegetation towards the high elevation is not a continuous feature and is disrupted by many factors. Realizing the heterogeneous appearance of vegetation in the landscape (Pandey et al. 2018a, b), the following rules were framed to draw a line on the map as timberline. (1) *Determining timberline*—up-slope termination of continuous forested landscape (canopy cover more than 30%; Singh and Rawal 2017) to alpine meadows (towards higher slopes) was considered timberline, (2) *Discontinuity in timberline*—disruption in the continuity of line due to large-scale geographical features and natural process (landslides, eroding soils, anthropogenic disturbances, rocks on higher slopes and their continuity towards permanent snow etc.) and (3) *Horizontal and vertical visualizations*—vertical sides of a forested landscape which faces the adjacent forested landscape on another slope were not considered as upper limit of trees (those drawn as limit of timberline in many studies). Otherwise, a lower altitude of timberline may be recorded than its normal occurrences. 3D visualization of the timberline was done by using DEM, so that the side edges of a forest were not confused with its topmost edge. This approach is much different from the indices-based mapping of timberline, and the line drawn in this study can be fairly considered the upper limit of the timberline in the study area (watershed and state).

6.3.1 Satellite-Based Observations

6.3.1.1 Selection of Appropriate Earth Observing Satellites

For developing a geo-database, mapping of timberline, and a regional synoptic perspective, satellite-based observation is most applicable technique. Satellite-based mapping in the higher Himalaya is a challenge where influence of local weather (local/monsoonal clouds or temporary snow cover over land surface) may obstruct the view of optical satellites and create confusion in the interpretation of various landscape features. Spatial resolutions in dominating 3D (folded) topography of mountains also pose challenges (hill shadows, steepness and analysis in 2D perspective) in realizing mapping objectives. High-resolution images are recent but consistency for comparison between decades is another issue. Use of high-resolution images to capture regional pictures of the object under study (timberline in this case) is also not economically feasible. Various satellite products which can be used in the Himalayan region are described by Ramachandran and Roy (2018). A high temporal resolution (few days) with technological consistency (instrument having same bands and resolutions) for a longer period (decades of operation) is a prerequisite in employing remote sensing approach towards (1) capturing various attributes of natural vegetation in the high altitudes and (2) analysing the impacts of natural factors and anthropogenic activities over a long period (change detection).

Keeping in view the tree heights and canopy dimensions of trees (Pandey et al. 2018a, b; www.eflora.org), use of satellite images with medium-range spatial resolution (20–30 m) appears suitable. Our search for instruments meeting all the above criteria ended with (i) Thematic Mapper and its enhanced version (LANDSAT Series, NASA) and (ii) Linear Imagine Self Scanning System (LISS; IRS, ISRO).

We started with the LISS sensor (Linear Imagine Self Scanning System) of ISRO for the current period (1) LISS-IV series—spatial resolution of 5.8 m for revisit time of 5 days on one platform and (2) LISS-III series—spatial resolution of 23.5 m. Other details are given in Table 6.1. The presence of instruments on different platforms increased the temporal resolution of LISS products during overlap periods of different functional satellites (IRS-1C- 1996–2007, IRS-1D- 1998–2009, Resourcesat-1or IRS-P6- since 2003, Resourcesat-2- since 2011, Resourcesat-2A- since 2016).

Historical baseline records were developed from LANDSAT series of satellites those have consistency with bands and spatial resolution of 30 m (Table 6.1). The identified sensors were Thematic Mapper and its improved versions: Enhanced Thematic Mapper+ and Operational Land Imager. These sensors were on board from 1982 on Landsat 4, and were/are available on different platforms of subsequent satellites of the Landsat series (Landsat 5, 7 and 8). The operational overlapping period of different satellite missions makes more products availability on temporal scale. The revisit time is 16 days on one platform. After a thorough revision of various features of both the satellite series (IRS and Landsat), we finalized the products of TM/ETM+/OLI sensors for current timberline mapping and also to document historical changes in Himalayan timberline across the region (1977–2015) (difference of nearly four decades by availability of satellite data products). To check the influence of high resolution on local/regional-scale mapping, we used LISS-4 satellite images (spatial resolution of 5.8 m in multispectral) of recent years.

6.3.1.2 Challenges in the Himalayan Landscape and Mapping Season

The growing season in the high altitudes of Himalaya is limited to about 4 months, from May to September. Snowfall in winters, which extends beyond the permanent snowline, and clouds during the rainy season (June to September) are barriers to getting useful satellite scenes from the optical sensors. Options for usable images are limited to a narrow passage of season in a year. This is further restricted by the limited pass of satellites over the area in that narrow temporal window. Lush growth after rains does not provide much separation (mostly saturated) between tree canopy species but the onset of autumn and following the period before snowfall give an opportunity to separate deciduous from evergreen vegetation. The interpretation using these images of the pre-snowfall period can further be verified by the ground situation in spring. Leaf emergence and canopy development in deciduous trees vary considerably at the beginning of the growing season (spring onwards) in deciduous species. This attains maximum greenness during the rainy season; hence, images from the onset of autumn to the time when the area is snow covered are more appropriate. Keeping in view these characteristics, satellite images from October (Year 1) to January (following Year 2) were obtained.

Table 6.1 Spectral reflectance and spatial resolution of different sensors in a range of 20-30 m spatial resolution

	Band				Wavelength (μm)				Pixel Size (m)			
Spectrum	LISS-IV	L1/L2/L3 (MSS)	L4/5 (TM)/L7 (ETM+)	L8 (OLI/TIRS)	LISS-IV	L1/L2/L3 (MSS)	L4/5(TM)/L7(ETM+)	L8 (OLI/TIRS)	LISS-IV	L1/L2/L3 (MSS)	L4/5 (TM)/L7 (ETM+)	L8 (OLI/TIRS)
Visible				B1	–		–	0.43–0.45	–		–	30
	B2		B1	B2	–		0.45–0.52	0.45–0.51	–		30	30
	B3	B4	B2	B3	0.52–0.59	0.5–0.6	0.52–0.60	0.52–0.60	5.8	60[a]	30	30
	B4	B5	B3	B4	0.62–0.68	0.6–0.7	0.63–0.69	0.63–0.68	5.8	60[a]	30	30
Infrared												
Near		B6/B7	B4	B5	0.77–0.86	0/7–0.8/ 0.8–1.1	0.76–0.90	0.84–0.88	5.8	60[a]	30	30
Short Wave			B5	B6	–		1.55–1.75	1.56–1.66	–	60[a]	30	30
Middle			B7	B7	–		2.08–2.35	2.10–2.30	–		30	30
Thermal			B6	–	–		10.40–12.50	–	–		120/ (30)[b]/ 60	–
TIRS 1				B10			10.6–11.19					30 (100)[c]
TIRS 2				B11			11.5–12.51					30 (100)[c]
Years	2003	1972/ 1975/ 1978	1982/ 1984/ 1999	2013								
								Other Sensors				
		Panchromatic		B8		0.5–0.75		0.52–0.90	0.50–0.68	5.8	15	15
		Cirrus		B9					1.36–1.39			30

[a] Original MSS pixel size was 79 × 57 m; production systems now resample the data to 60 m
[b] Band 6 was acquired at 120-m resolution, but products are resampled to 30-m pixels
[c] TIRS bands 10–11: collected at 100 m but resampled to 30 m to match OLI multispectral bands

6.3.2 Test Case

To map the current position of timberline, a thorough search was made to obtain appropriate satellite images covering the timberline zone of entire Indian Himalayan Region. We preferred the state of Sikkim because it is the smallest state and is usually covered in one satellite scene and therefore stitching of two images of varying atmospheric conditions is not required. Images were downloaded from the United States Geological Survey (USGS) portal (https://earthexplorer.usgs.gov/) for the year 2015, and the image of 28 December 2015 (OLI on Landsat, multi-spectral, spatial resolution of 30 m) was chosen due to maximum visual separability between different features and absence of fresh snow at high altitudes.

To compare with Indian satellite products, we also obtained high-resolution multispectral images of LISS-IV (spatial resolution 5.8 m, www1 n.d.) for the same year (2015) but could not get usable LISS-IV scenes covering the entire timberline zone of Sikkim. We were restricted to procure three different scenes of January (spread between 2012 and 2015) to cover the entire timberline zone, assuming that there will not be drastic changes in the vegetation. This comparison also provided an opportunity to realize improvement in mapping at different scales, that is, 1:50,000 (Landsat) and 1:12,500 (LISS-IV). A small watershed in the *Khangchendzonga* National Park (field studies on vegetation done by Pandey et al. 2018a, b) was also selected to compare various outputs of different satellite images at local scale. In addition to the timberline drawn from Landsat and LISS-IV images, additional timberline was drawn using high-resolution data of WorldView-3 (spatial resolution; 1.24 m) of 11 November 2017 for the same watershed. A small portion of high altitudes was not covered on this date so another image of 13 January 2015 from the same source was taken to complete the local-level timberline mapping. To make true colour images of WorldView-3, layers of three multispectral bands (band 5; red, band 3; green, band 2; blue) were stacked.

6.3.3 Image Interpretation and Mapping of Timberline

For realizing the timberline on the map, layer stacking (Band 2, 3, 4, 5 and 7) of Landsat data was done to develop a false colour composite. Image enhancement techniques were used to improve the interpretability of the images. Various tools of ERDAS IMAGINE 2016 image processing software were used.

Realizing the complex heterogeneous landscape features at high altitudes and limitations of auto-extraction methods (Bharti et al. 2012) in mountains, knowledge-based manual interpretation (visual) was preferred to delineate the timberline from the satellite scene. Working in Sikkim, Singh et al. (2018) realized that auto-extraction methods overlook/leave certain complex areas, which requires manual corrections. Visual interpretation appeared more appropriate in the mountains (rugged terrain), where complex topography (steep, including shadow) challenges auto extraction of features.

An interpreter was trained to read and correlate various features in satellite images (Landsat, LISS series and true colour images of WorldView-3). The interpreter visited the high-altitude field to understand the ground features, particularly the different types of vegetation and their appearance on satellite images. Inaccessibility in all the areas of high altitudes is a major issue in the Himalaya. Field-based researchers draw inferences on the basis of limited ground knowledge, while much more vegetation is present than what is seen/sampled during field visits. In this situation, true colour high-resolution satellite images (available on Google Earth) are a useful resource to support the interpretation. High-resolution natural colour satellite images of different seasons were used to validate inaccessible sites. Keeping in view the knowledge acquired from field visits and training on visual interpretation, a single person was assigned to work on all the images to visually develop the timberline and to maintain consistency in results. An isoline connecting the highest edge of the forests (highest edge of the forests towards the alpine, as defined in 6.3) was created as 'timberline'. This line breaks at various places due to various reasons. These broken lines were termed as segments/fragments of the timberline.

6.3.4 Geo-Spatial Attributes

Topography (elevation and aspects) influences the presence and distribution of vegetation (Klinge et al. 2015). Topography controls the overall presence of timberline; therefore, various attributes (slope—to measure the steepness of the terrain, which influences the recruitment of trees; aspect—a proxy for moisture and temperature gradient on a given altitude; elevation—a proxy for air temperature along the altitudinal gradient) were developed for the entire state of Sikkim with the help of a digital elevation model (DEM). The ASTER DEM of same spatial resolution as that of satellite images (30 m) was obtained from the USGS portal (https://earthexplorer.usgs.gov/) to develop the relationship between the spatial characteristics of the timberline and the topographical features (altitude, slope and aspect). ArcGIS was used for various spatial analyses and the extraction of attribute data of the timberline. The generated data on timberline, using satellite images, was subjected to various statistical treatments and correlated with various spatial attributes (latitude, longitude, altitude, slope, aspect).

6.3.5 Delineation of Timberline and Change Analysis

Remote sensing is of utmost importance in delineating timberline and demonstrating the changes occurring in the Himalayan landscape. The long-term availability of Landsat (since 1972) makes it possible to realize spatio-temporal variability of timberlines at a larger scale.

In order to map the longest spatio-temporal dynamics of the timberline in the Sikkim Himalaya, Landsat 8 (OLI, path/row 139/041 acquired on 28 December 2015) and Landsat-2 (MSS, path/row, 149/041 acquired on 23 January 1977) were used. A usable clear image of TM (Thematic Mapper) for Sikkim was available for 1989. Hence, for long-term change analysis, we also searched for MSS, where we faced again non-availability of suitable scenes for the Sikkim area from 1972 to 1976 (cloud cover more >50%, snow cover and stripping). The oldest required suitable scene was available in 1977. We opted to use this image, which was resampled at 30 m to make it comparable with the image of 2015. Different images were co-registered with the latest image of 2015 to do change analysis of the timberline. Spatial tools of ERDAS IMAGINE 2016 were used for these purposes (geometric correction, image enhancement, techniques etc.). The satellite images were then subjected to knowledge-based interpretation techniques and the timberline was delineated by applying visual interpretation. Changes in the timberline were recorded as a function of the shift in altitude from the past (1977) to the current (2015) position.

ArcGIS 10.5 was used for various spatial analyses and extraction of attribute data of the timberline. Following Sah and Sharma (2018), timberlines were further categorized on the basis of spatially distinct arrangements as (1) '*continuous*' type (parallel to permanent snow line having the alpine region in between these two lines mostly in the inner Himalayan ranges [hereafter CTL]), and (2) summits with high elevations, away from the snowy ranges, arise due to geological processes where the island-type shape of the alpine zone occurs around the summit. Such alpine is surrounded by the timberline towards lower elevations, these types of timberlines were termed as '*island*' type timberlines (hereafter ITL). These continuous line data (vector) were used to create point data (30 m separation between two points in a line) for the corresponding years (1977–2015) using the Pixel to ASCII Converter feature of ERDAS IMAGINE 16. Thirty-meter distance (spatial resolution) of points were in tune to resolution of ASTER DEM (30 m), which was used to extract altitudinal information. Points at every 30 m were generated over the entire timberline (past and present; ITL and CTL) to match the spatial attributes of DEM, and differences (elevation and distance) were recorded.

Temporal changes were marked as 'shift' (upward/downward) and 'no change' (stationary) in timberline position with respect to the base year (1977). The interaction of the line points and DEM provided the maximum altitude of timberline occurrence at each location. For each point of the timberline, spatial attributes (latitude, longitude and elevation) were extracted. These attributes were subjected to various statistical analysis.

6.4 Result and Discussion

6.4.1 Geo-Spatial Attributes of Sikkim

Sikkim is a typical mountainous state with preponderance of high elevation area (nearly 75% area above 2000 m amsl). Of the state's total area, 21.3% is above the permanent snow line (above 5000 m, Fig. 6.1). Field observations on treeline/ timberline in the eastern Himalaya illustrate its zone ranges from 3000 m to

4500 m (Singh et al. 2019; Schickhoff 2005; Chaudhuri 1992). Only 38.4% of the state falls in this altitude zone. Slopes are important features for the stabilization and recruitment of trees. Nearly 28% of the slopes in the state are gentle (<20°) but mostly in the higher altitudes of the trans-Himalayan region. The middle part of the state has more steepness than other parts. Nearly 30% area of the state has slopes steeper than 35°. A negligible portion of the state can be considered flat topography (0.01%). Distribution of different mountain aspects ranged between 10.8% (north) and 13.9% (south-east). Cooler aspects (NW-N-NE) occupy nearly 40% of the total area.

6.4.2 Timberline in Sikkim and Its Geo-Spatial Characters

The timberline in Sikkim was mapped on various scales from different spatial resolution satellite images. For example, 1:50,000 using 30 m spatial resolution and 1:12,500 using 5.8 m spatial resolution for regional (Himalayan) and local-scale mapping were used, respectively.

6.4.2.1 Regional-Scale Mapping Using Spatial Resolution of 30 m

The timberline drawn for 2015 (derived from Landsat; 1:50,000 scale) is presented in Fig. 6.2. The entire timberline of the state (730.42 km) is between nearly 1° spread of latitude (27.24 to 27.89 N) and longitude (88.04 to 88.90 E). At rare locations of these mountains, high-altitude timberline may occur at a much lower (i.e. 2620 m) or much higher altitude (i.e. 4390 m) than the range reported by field studies from Sikkim. The lowest point of high-altitude timberline is usually not described in the field-based studies. Working on images of near resolutions in the same state, Singh et al. (2018) observed the highest treeline elevations at 4804 m for 2013 and 4579 m for 1977, while the mean was at 3542 m (2013). Such differences could be due to (1) different methods employed (auto-extraction followed by manual corrections in a past study and visual interpretation in the present case), (2) detection capability of different spatial resolutions (23.5 m and 77 m in a previous study and 30 m in this study), (3) source of elevation (CartoDEM vs ASTER DEM) and (4) definition of treeline or timberline (synonymous or antonymous).

The mean elevation of the entire timberline length was 3620 m amsl. Distribution of timberline was more apparent from 3000 m onwards and scarcely reaches above 4200 m. Nearly 60% of the total timberline of the state occurred between 3400 m and 4000 m altitude, and 27.4% of the timberline is present below 3400 m (Fig. 6.3). Altitude above 4000 m accounted for minimal presence of timberline (12.8%) in the state. Such details were not available previously.

The majority edge of high-altitude forests in Sikkim, that is, timberlines, occurs on gentle slopes (14% of total timberline on slopes less than 20°) or moderate mountain slopes (nearly half of the timberline, 49.2%, on slopes between 20° and 35°). With increasing steepness, the occurrence of timberline decreased (24.8% in 35–45° and 12.2% in >45° slopes). Among the faces of a slope (i.e. aspects), the highest occurrence was observed on E-facing slopes (15% of the total timberline)

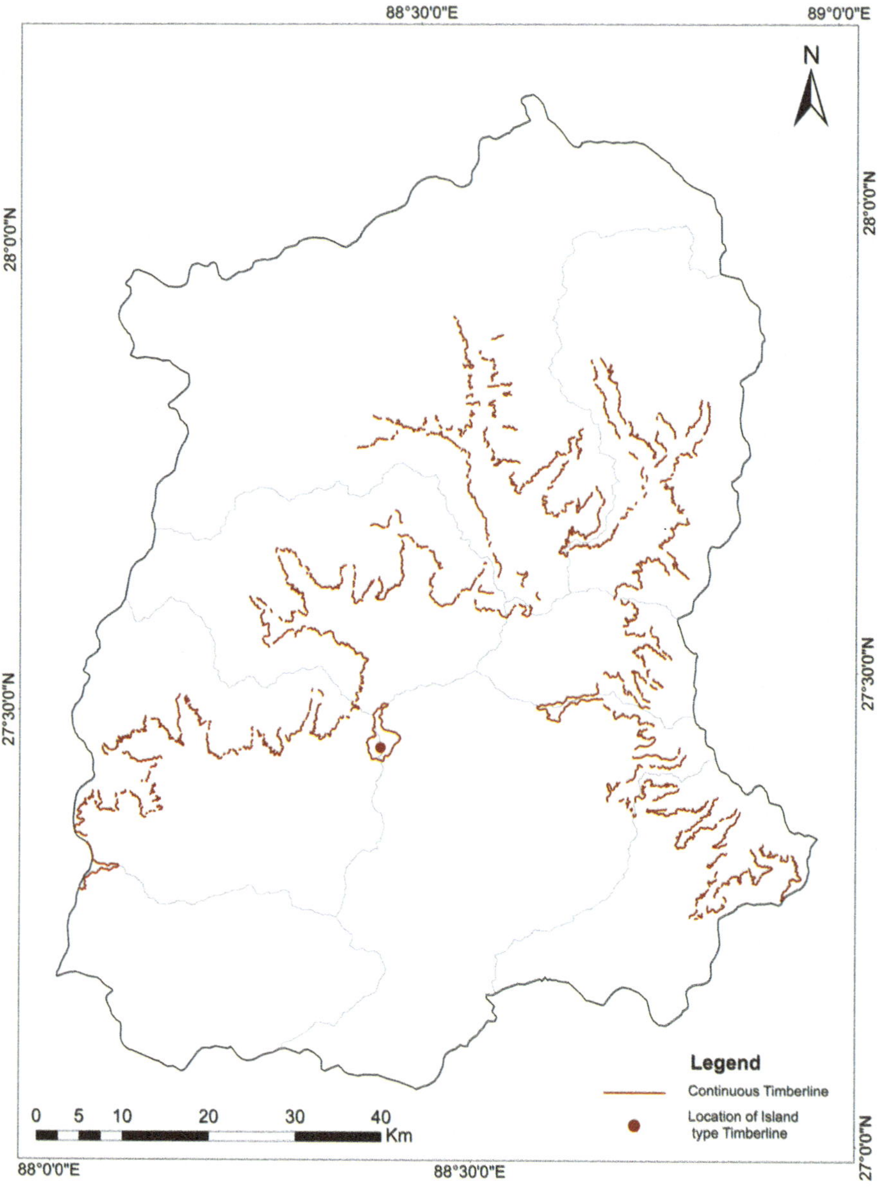

Fig. 6.2 Timberline drawn from Landsat image of 2015. Dot shows an isolated presence of timberline which is away from the main snow peaks

and the lowest on NW aspect (9.5%). On cooler aspects (NW-N-NE), 32.2% of the total timberline occurred, while the remaining was on warmer slopes.

The derived timberline was not a continuous line over its entire range. The timberline was fragmented due to various natural reasons (long stretches of rocks,

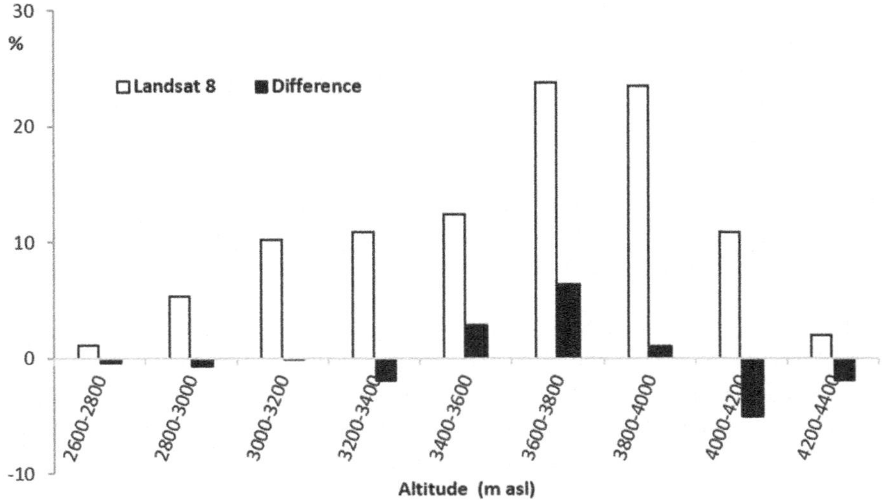

Fig. 6.3 Comparison of two spatial scales (Landsat 8 & LISS-4): Proportional distribution of total timberline length derived from two images of different resolutions in different elevational bands

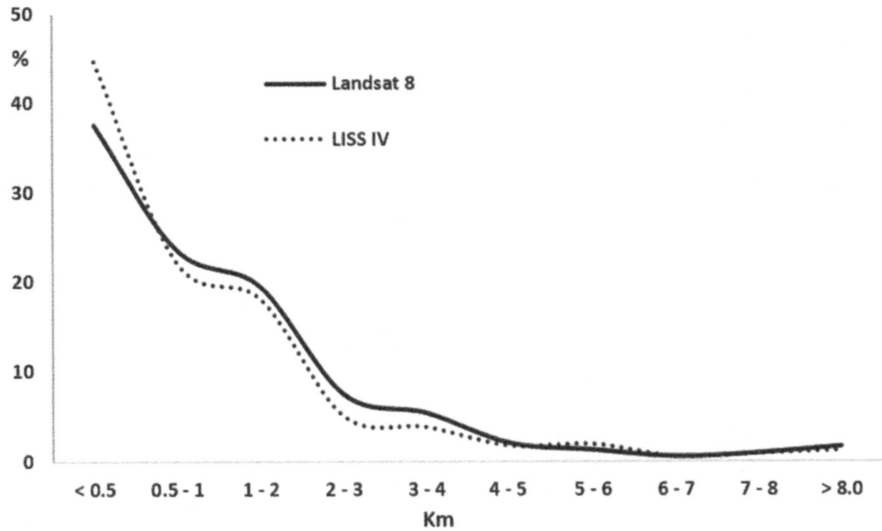

Fig. 6.4 Comparison of two spatial scales (Landsat 8 & LISS-4): Proportional distribution of number of timberline segments (arranged according to length) derived from two images of different resolutions

landslides etc.), and therefore, was divisible into 534 segments. The length of a timberline segment varied from 0.09 km to 19.98 km; however, more than one-third (37.6%) of the segments of timberline had a length below 500 m (Fig. 6.4). With increase in the length of timberline segments, the number of segments decreased, and 34 segments (6.4%) had length 5 km or more (Fig.6.4). Mapping fragments (gaps between the timberlines) is important for understanding movements/shifts in upfront vegetation. In the absence of such details, misleading results may be obtained using automated methods (e.g. band ratioing, supervised classification etc.) for timberline shift.

A common belief regarding timberline presence in the Himalayas is that it exists parallel to the permanent snow line in the Himalayas, and the alpine zone lies in between these lines. We observed that the timberline may also exist around isolated summits where appropriate high altitude exists, termed as island type (Sah and Sharma 2018).

For the first time for the Sikkim Himalaya, we have mapped one such timberline (Fig. 6.2) which is not parallel to the permanent snow line (disjointed landscape from the Himalayan snow peaks). About 20 km of timberline was mapped with a minimum elevation of 2806 m and a maximum of 3436 m. Nearly 70% of this timberline falls between 3000 and 3400 m altitude and 8% lies above 3400 m altitude. Such geo-spatial details have been created for the first time and using visual interpretation method.

6.4.2.2 Local-Scale Mapping Using Spatial Resolution of 5.8 m, and Comparison with Regional-Scale Mapping

The total length of timberline derived from the mosaic of LISS-IV images was ~690 km (1:12,500), which decreased by 5.5% than the 1:50,000 scale. Many long segments were further divided due to better resolution, which resulted in 40 new timberline fragments. Thus, at this scale, timberline length was composed of 574 segments. Matching resolution between DEM and LISS IV (5.8 m) is not available; therefore, an elevational analysis was done with the present available DEM of 30 m resolution. A decrease in the elevational position of the timberline was also observed for both points of timberline's occurrence—lowest (down by ~10 m) and highest point (down by ~115 m). This indicates a complex terrain of high altitude requires fine-resolution satellite images for clear differentiation between features/classes; however, at a regional scale, such deviations are minor (entire altitudinal range of timberline has elevation difference of 1.6 km, i.e. from 2600 m to 4200 m).

The same is also evident from the distribution of timberline length in different altitudinal classes. The distribution pattern remains the same as in the case of regional mapping (1:50 K scale) but gain/loss in a category was substantial (Fig. 6.3). Nearly 70% of the total derived timberline at fine resolution (1:12.5 K) occurred between 3400 m and 4000 m altitude, which was 60% at the 1:50 K scale. An increase in the proportion of timberline was observed in three altitudinal bands (1% to 6.3%); however, towards both the tail ends of elevation band, the proportion of timberline decreased at the 1:12.5 K scale (Fig. 6.3).

In agreement with field-based observations (Pandey et al. 2018a, b), the maximum expression of timberline at both the scales remained highest in the same range of altitude (3600 m–4000 m) and the highest portion of occurrence in an elevation band was from 3800–4000 m at both the scales of mapping.

Total 48% of the timberline occurred on moderate slopes. The patterns of distribution on different slope categories were also changed at the 1:12.5 K scale where the proportion of timberline declined in gentle (<20°; 13.2% of the total in comparison to 14% in 1:50 K) and moderate slope categories (20–35°; 48%, compared to 49.2% in 1:50 K). On the contrary, a rise was observed in the categories with more steepness (25.7% of the total in 35–45° and 13.2% of the total in >45° class) than 1:50 k. This further demonstrates that in rugged mountain terrains, finer spatial resolution images can perform better to separate diverse landscape features, which are confined in narrow zones. Among the slope aspects, the highest occurrence of timberline was continued on E-facing slopes (14.4% of the total timberline but lower than the 1:50 K), and warmer aspect of 'W' had almost similar proportion of timberline (13.9%, slightly higher than the 1:50 K). The NW aspect remained the lowest (10.6%) in terms of proportion of timberline, but slightly increased (1%) than the 1:50 K. On cooler aspects (NW-NNE), the proportion of timberline increased by 1% (33.9% of the total timberline).

6.4.2.3 Test Case of Watershed

At the watershed level (area 143.73 km^2, altitude range 2215–6084 m amsl; Fig. 6.1), the length of timberline mapped varied considerably between the source images, that is, 29.76 km (Landsat 8), 31.05 km (LISS-4) and 45.76 km (WorldView-3) (Table 6.2). Difference between 1:50 K and 1:12.5 K was only 4% increase in timberline length, while WorldView-3 images increased detailing of features and interpretation capability, hence timberline length increased 1.5 times over the length obtained from Landsat image. Using all the three satellite images

Table 6.2 Various attributes of timberline/treeline derived from different satellite images on watershed level and date generated from Field observations

Sl No.	Attribute	Present Study (Source Satellite Images)			Past Study (Singh et al. 2018)	Filed Study
		Landsat 8	LISS-4	WorldView-3	Landsat and LISS-3	
1	Maximum Altitude (m)	4095	4042	4073	4579 and 4804	4000 (Pandey et al. 2018a) 3787–3989 ± 4 (Pandey et al. 2018b)
2	Minimum Altitude (m)	3442	3347	3412	Not given	Not mentioned
3	Timberline Length (km)	29.76	31.05	45.76	Not given	Never measured
4	Segments (Number)	24	42	61	Not given	Not Known

maximum timberline elevation in this watershed increased between 40–95 m over the field studies conducted by Pandey et al. (2018a, b). Hence in inhospitable and inaccessible locations of Himalayan high altitudes, satellite image is useful tool to capture glimpses of tree vegetation. Minimum timberline elevation drawn from various satellite images was also comparable between Landsat (1:50 K) and WorldView-3 (1:3 K) image (3442 m *vs* 3412 m respectively), however due to hazy clouds in LISS-4 image interpretation was not clear at some places. Hence, elevation of timberline was decreased by 100 m over the other two images due to this misinterpretation. It appears that features of timberline elevations drawn at 1:50 K scale are comparable to finer scales if interpreted correctly with images of good quality. Interesting observation was that, with increasing finer spatial resolution of an image, number of timberline fragments was increasing (24 to 61) indicating that disruption in continuity in frontline tree vegetation towards alpine is frequent due to local features.

6.4.2.4 Changes in Timberline Elevations Between 1977 and 2015

The gain (32.7 km; increase) and loss (8.56 km; decrease) in different elevation bands were recorded since 1977, and the absolute increase in total length of timberline was about 23 km during the studied period. These changes occurred in less than one-fourth of the timberline length of 1977 (23.5% of total; 142.43 km upward and 23.8 km downward, Fig.6.5) while majority of the timberline (76.5%) remained stationary (i.e. no change) since 1977 (baseline data).

Mean elevation of the entire timberline in 2015 moved upward by 18 m since 1977; however, maximum elevation of occurrence remained same. Minimum elevation (lowest occurrence) increased by 79 m which indicates disappearance of lower end of timberline from unusual sites of occurrence.

We further analysed proportional distribution of timberline length in different elevational bands. It was realized that above 3600 m elevations there is gain in timberline length and below that elevation timberline is shrinking (3200–3400 m elevation). These finding suggests that high-altitude timberline may occur at any altitude in a given geography of the Earth, its response to natural factors is more visible in the favourable elevation zone in that regime. This zone is characterized by the location on the Earth (latitude, longitude and precipitation regime). For example, in Sikkim Mountains, maximum expression of timberline is between a zone of 400 m elevational difference (47.3% of the present timberline occurs between 3600 and 4000 m altitude) which has witnessed 75% of the total increase in timberline length in 37 years.

One more interesting observation was that entire timberline away from snowy ranges (Island type timberline, ITL) moved upwards where none of the segments remained stationary. It may be due to escape from more warming conditions in outer ranges in comparison to inner Himalayan ranges. ITL altitudes were warmer (average annual mean temperature 13.8 °C ± 0.7) than the same timberlines altitudes of CTL (8.6 °C ± 3.0) by Latwal et al. (2022). Thus, outer Himalayan sites appear more prone to global warming, and to disturbances caused by anthropogenic activities (decrease in maximum elevation in 2015 may be attributed to later one). This is also

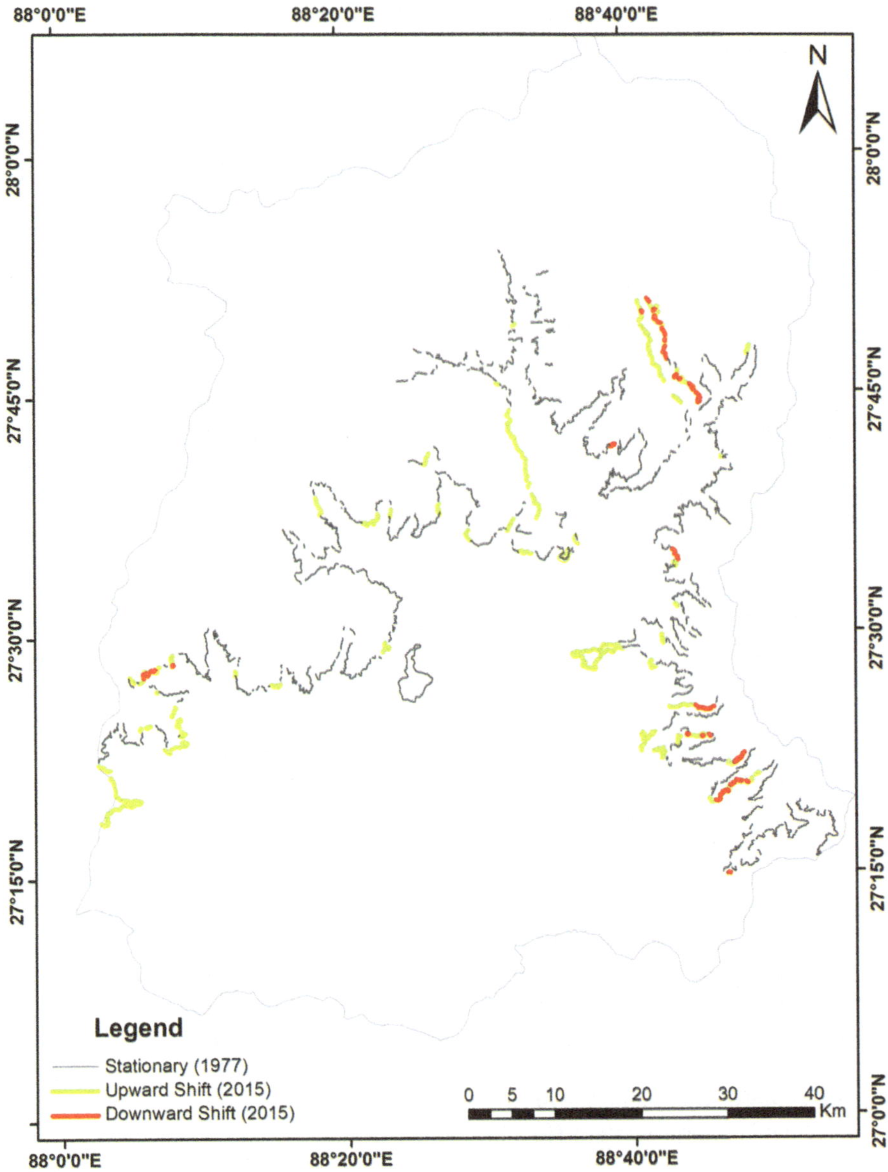

Fig. 6.5 The position of timberline in 1977 and 2015 in the state of Sikkim Himalaya

supported by the fact that minimum elevation of ITL increased by 43 m since 1977, and mean elevation increase of upwardly moved timberline was nearly 2.5 times greater in ITL (42 m) than the CTL (17 m). CTL increased by 22.9 km (3.3% of length in 1977) between 37 years of time frame with an increase of 79 m in minimum altitude while maximum elevation of occurrence remained same in this time period.

No expansion to upper limit may be attributed to non-availability of suitable sites in highest altitudes, which is characterized by poor or no soil cover and very low temperature, and occupied by largely permanent snow, glaciers, moraine and alpine meadows.

Detailed analysis of each timberline point (created at 30 m distance) shifted from its place (up or down) indicates that mean upward movement was 100 m ± 89 since 1977 with an average rate of 2.71 m/year in 130 segments, while mean downward movement of various points was 56 m ± 54 with an average rate of 1.52 m/year in 48 segments. More rainfall (high soil moisture) along with warmer conditions in Eastern Himalayan precipitation regime than the western Himalayan conditions increases probability of expansion of timberline ecotone due to developing warming conditions in the face of climate change, while landslide events due to same climatic conditions may erode timberline habitats at some locations bringing timberline elevation down.

6.4.2.5 Comparison with Previous Studies

Shift in treelines in the states of Uttarakhand, Sikkim and Arunachal Pradesh have been studied (Singh et al. 2012, 2018; Mohapatra et al. 2019). In nearly four decades (1977–2013) mean upward shift of treeline of Sikkim (eastern Himalaya) was recorded 301 ± 66 m (@ 81 m/decade), and in north-western Himalaya (Uttarakhand) it was recorded 388 + 60 m (@ 114 m/decade) in three decades time period (1976–2006) while this was 452 + 74 m (@ 113 m/decade) for alpine treeline ecotone of Arunachal Pradesh (1977–2013). Singh et al. (2021a, b) also reported a fundamental niche shift of c. 109.9 m/decade for Sikkim treeline (highest in Indian Himalaya) in response to climate change scenario (IPCC$_5$ RCP8.5 for year 2061–2080). For a part of Uttarakhand, Panigrahy et al. (2010a) reported an upward shift of timberline vegetation by 300 m between 1986 and 2004. On a repeat study of the same area on methodological issues, Bharti et al. (2011) concluded that remote sensing approaches require careful steps including selection of appropriate methodology and sufficient ground knowledge for such interpretations. From a dendrochronological study, in Himachal Pradesh, on a long-time scale (1860 to 2000), Dubey et al. (2003) concluded that the upward shift rate of treeline ecotone (dominated by blue pine, *Pinus wallichiana*) might vary from 14 m to 19 m decade^{-1} on different aspects. Our study observed a rate (27.1 m/decade) close to these values in moist wet conditions of Sikkim Himalaya than the dry conditions of blue pine habitats in north-western Himalaya.

6.5 Conclusions

Remote sensing method can help to derive spatial attributes of timberline at regional/micro scale. Satellite images of high spatial resolution (≤6 m) can improve timberline data (length and variation) by 10% over the medium spatial resolution (25–30 m) but availability of 'useful' high spatial resolution images is restricted by limited coverage, clouds etc., hence becomes a low priority in regional scale mapping (viz., entire Himalayan timberline) and further non-availability of historical data restricts studies on impacts of climate change. However, spatial resolution refines certain attributes of object under investigation (e.g. timberline); consistent availability of medium spatial resolution images is useful to create scenarios on temporal scale (T_1 & T_2, decadal etc.). Change analysis for timberline altitude (upward or downward shift) in the Sikkim Himalaya indicates that major portion of timberline (76.5%) remained stationary in last three and half decades (1977–2015) but mean elevation of entire timberline in the Sikkim Himalaya moved upward by 18 m due to changes in remaining 23.5% timberline. Both types of shifts (upward or downward movement) were observed at different locations with different rates. For example, mean upward shift was 100 m $\pm$ 89 (@ 2.71 m/year) and mean downward shift was 56 m $\pm$ 54 (@ 1.52 m/year). Rate of change is in tune with previous studies; however, stationary timberline from Sikkim is reported first time. A careful harmonized method, using satellite imagery, can be useful in regular monitoring of changes in Himalayan timberline to analyze impacts of climate change in high altitudes, particularly induced by global warming. Development of such harmonized geospatial database is useful input for regional modelling of change detection, future prediction and geographical explanations on the formation of diverse mountain timberline along the 2000 km long Himalayan arc.

This study highlights that there is need for spatial data-bank for the Himalayan region to monitor long term changes in biological processes (as timberline movement in this case) influenced by climate change or any other factors. Monitoring through satellites, having consistent and comparable products, may play a pivotal role in this regard.

Acknowledgement Authors are thankful to Prof. S. P. Singh for encouragement and guidance to conduct this research. Director, G. B. Pant 'National Institute of Himalayan Environment' (NIHE), Kosi-Katarmal, Almora for providing necessary facility. Financial grant for this study was supported by National Mission on Himalayan Studies, Ministry of Environment, Forest and Climate Change, Govt. of India. The LISS data provided by the National Remote Sensing Centre (NRSC) Hyderabad, and Landsat data from National Aeronautics and Space Administration (NASA) and the United States Geological Survey (USGS) are duly acknowledged.

References

Baker BB, Moseley RK (2007) Advancing Treeline and retreating glaciers: implications for conservation in Yunnan, P.R. China. Arct Antarct Alp Res 39(2):200–209

Beaman JH (1962) The timberlines of Iztaccihuatl and Popocatepetl, Mexico. Ecology 43:377–385

Berdanier AB (2010) Global treeline position. Nat Educ Knowl 3(10):11

Bharti RR, Rai ID, Adhikari BS et al (2011) Timberline change detection using topographic map and satellite imagery: a critique. Trop Ecol 52:133–137

Bharti RR, Adhikari BS, Rawat GS (2012) Assessing vegetation changes in timberline ecotone of Nanda Devi National Park, Uttarakhand. Int J Appl Earth Obs Geoinf 18:472–479

Chaudhuri AB (1992) Himalayan ecology and environment. New Delhi

Colombaroli D, Henne PD, Kaltenrieder P et al (2010) Species responses to fire, climate and human impact at tree line in the Alps as evidenced by palaeo-environmental records and a dynamic simulation model. J Ecol 98:1346–1357

Dolezal J, Dvorsky M, Kopecky M et al (2016) Vegetation dynamics at the upper elevational limit of vascular plants in Himalaya. Sci Rep 6:24881. https://doi.org/10.1038/srep24881

Dubey B, Yadav RR, Singh J, Chaturvedi R (2003) Upward Shift of Himalayan Pine in Western Himalaya, India. Curr Sci 85(8):1135–1136

Germino MJ, Smith WK, Resor AC (2002) Conifer seedling distribution and survival in an alpine-treeline ecotone. Plant Ecol 162(2):157–168

Hamid MK, Malik AA, Ahmad AH et al (2020) Early evidence of shifts in alpine summit vegetation: a case study from Kashmir Himalaya. Front Plant Sci 11:421

Holtmeier FK, Broll G (2005) Sensitivity and response of northern hemisphere altitudinal and polar treelines to environmental change at landscape and local scales. Glob Ecol Biogeogr 14:395–410

Holtmeier FK, Broll G (2007) Treeline advance– driving processes and adverse factors. Landscape Online 1:1–33

Holzinger B, Hülber K, Camenisch M et al (2008) Changes in plant species richness over the last century in the eastern Swiss Alps: elevational gradient, bedrock effects and migration rates. Plant Ecol 195:179–196. https://doi.org/10.1007/s11258-007-9314-9

Ives JD, Hansen-Bristow KJ (1983) Stability and instability of natural and modified upper timberline landscapes in the colorado rocky mountains. Mt Res Dev 3(2):149–155

Juntunen V, Neuvonen S, Norokorpi Y et al (2002) Potential for timberline advance in Northern Finland, as revealed by monitoring during 1983–99. Arctic. 55:348–361

Klinge M, Böhner J, Erasmi S (2015) Modeling forest lines and forest distribution patterns with remote-sensing data in a mountainous region of semiarid central Asia. Biogeosciences 12:2893–2905

Körner C, Paulsen J (2004) A world-wide study of high-altitude treeline temperatures. J Biogeogr 31:713–732

Kullman L (2007) Tree line population monitoring of *Pinus sylvestris* in the Swedish Scandes, 1973-2005: implications for tree line theory and climate change ecology. J Ecol 95:41–52

Lamprecht A, Semenchuk PR, Steinbauer K et al (2018) Climate change leads to accelerated transformation of high elevation vegetation in the central Alps. New Phytol 220:447–459. https://doi.org/10.1111/nph.15290

Latwal A, Sah P, Sharma S (2018) A cartographic representation of a timberline, treeline and wood vegetation around a Central Himalayan summit using remote sensing method. Trop Ecol 59(2): 177–187

Latwal A, Sah P, Sharma S et al (2022) Relationship between timberline elevation and climate in Sikkim Himalaya. In: Singh SP, Reshi ZA, Joshi RJ (eds) Ecology of Himalayan treeline ecotone. Springer Nature

Lenoir J, Svenning JC (2015) Climate related range shifts – a global multidimensional synthesis and new research directions. Ecography 38:15–28. https://doi.org/10.1111/ecog.00967

Leonelli G, Pelfini M, di Cella UM (2009) Detecting climatic treelines in the Italian alps: the influence of geomorphological factors and human impacts. Phys Geogr 30(4):338–352

Leonelli G, Masseroli A, Pelfini M (2016) The influence of topographic variables on treeline trees under different environmental conditions. Phys Geogr 37(1):56–72

Mohapatra J, Singh CP, Tripathi OP et al (2019) Remote sensing of alpine treeline ecotone dynamics and phenology in Arunachal Pradesh Himalaya. Int J Remote Sens 40(20):7986–8009

Motta R, Moralis M, Nola P (2006) Human land-use, forest dynamics and tree growth at the treeline in the Western Italian Alps. Ann For Sci 63:739–747

Pandey A, Rai S, Kumar D (2018a) Changes in vegetation attributes along an elevation gradient towards timberline in Khangchendzonga National Park, Sikkim. Trop Ecol 59(2):259–271

Pandey A, Badola HK, Rai S et al (2018b) Timberline structure and woody taxa regeneration towards treeline along latitudinal gradients in Khangchendzonga National Park, Eastern Himalaya. PLoS One 13(11):e0207762. https://doi.org/10.1371/journal.pone.0207762

Panigrahy S, Anitha D, Kimothi MM et al (2010a) Timberline change detection using topographic map and satellite imagery. Trop Ecol 51:87–91

Panigrahy S, Singh CP, Kimothi MM et al (2010b) Alpine Treeline Atlas of Indian Himalaya: Uttarakhand, India. Space Application Centre (ISRO), Ahmedabad

Ramachandran RM, Roy PS (2018) Vegetation response to climate change in Himalayan hill ranges: a remote sensing perspective. In: Das AP, Bera S (eds) Plant diversity in the Himalaya hotspot region, vol 1. M/s Bishen Singh Mahendra Pal Singh, Dehradun, pp 369–392

Sah P, Sharma S (2018) Topographical characterisation of high-altitude timberline in the Indian Central Himalayan region. Trop Ecol 59(2):187–196

Sarmiento FO, Frolich LM (2002) Andean cloud forest tree lines. Mt Res Dev 22(3):278–288

Schickhoff U (2005) The upper timberline in the Himalayas, Hindu Kush and Karakoram: a review of geographical and ecological aspects. In: Broll G, Keplin B (eds) Mountain ecosystems. Springer, Berlin, pp 275–354

Singh CP, Mohapatra J, Pandya HA et al (2018) Evaluating changes in treeline position and land surface phenology in Sikkim Himalaya. Geocarto Int 35(5):453–469. https://doi.org/10.1080/10106049.2018.1524513

Singh CP, Mohapatra J, Mathew JR et al (2021a) Long-term observation and modeling on the distribution and patterns of alpine treeline ecotone in Indian Himalaya. J Geom 15(1): 68–84. ISSN: 0976-1330

Singh CP, Panigrahy S, Thapliyal A et al (2012) Monitoring the alpine treeline shift in parts of the Indian Himalayas using remote sensing. Curr Sci 102:559–562

Singh SP, Rawal RS (2017) Manual of field methods- Indian Himalayan timberline. CHEA, Nainital

Singh SP, Sharma S, Dhyani PP (2019) Himalayan arc and treeline: distribution, climate change responses and ecosystem properties. Biodivers Conserv 28(8–9):1997–2016

Singh SP, Bhattacharyya A, Mittal A et al (2021b) Indian Himalayan timberline ecotone in response to climate change — initial findings. Curr Sci 120(5):859–871

Smith WK, Germino MJ, Johnson DM et al (2009) The altitude of alpine treeline: a bellwether of climate change effects. Bot Rev 75:163–190

Steinbauer MJ, Grytnes JA, Jurasinski G et al (2018) Accelerated increase in plant species richness on mountain summits is linked to warming. Nature 556:231–234. https://doi.org/10.1038/s41586-018-0005-6

Treml V, Banas M (2008) The effect of exposure on alpine treeline position: a case study from the high Sudetes, Czech republic. Arct Antarct Alp Res 40(4):751–760

Van Den Hoek J, Smith AC, Hurni K et al (2021) Shedding new light on mountainous forest growth: a cross-scale evaluation of the effects of topographic illumination correction on 25 years of forest cover change across Nepal. Remote Sens 13(11):2131

Vittoz P, Rulence B, Largey T et al (2008) Effects of Climate and Land-UseChange on the Establishment and Growth of Cembran Pine (*Pinuscembra* L.) over the Altitudinal Treeline Ecotone in the Central Swiss. Alps Arct Antarc Alp Res 40(1):225–232

www1 (n.d.) NRSC, ISRO. Satellite data product https://nrsc.gov.in/Satellite_Data_Products_Overview?q=Price_List. accessed 19 December 2015

Zhao F, Zhang B, Pang Y et al (2014) A study of the contribution of mass elevation effect to the altitudinal distribution of timberline in the Northern Hemisphere. J Geogr Sci 24(2):226–236

Mapping the Spatial Patterns of Biodiversity Along the Alpine Treeline Ecotone in the Eastern Himalaya Using Information Theory

7

Jakesh Mohapatra, C. P. Singh, Priyadarshini Gurung, Bandan Gajmer, Narpati Sharma, Dhiren G. Shrestha, and Himanshu A. Pandya

Abstract

Climatic variability causes loss of biodiversity in high-elevation ecosystems. Monitoring of biodiversity along the temperature-sensitive treeline ecotone in the Himalaya is necessary. The applications of information theory in understanding the patterns of biodiversity along the alpine treeline ecotone in the Himalaya are highlighted in the chapter. Spatially explicit monitoring of plant biodiversity using satellite remote sensing imagery and information theory was conducted. α-and β-diversity were estimated using high-resolution imagery of the MultiSpectral Instrument (MSI) onboard the Sentinel-2A satellite. The relationship between topography and biodiversity along the treeline ecotone in the Eastern Himalaya was studied. There was a statistically significant ($p < 0.001$) difference in the spectral Shannon's entropy along the gradients of elevation, slope, and summits'

J. Mohapatra (✉)
Earth, Ocean, Atmosphere, Planetary Sciences and Applications Area (EPSA), Space Applications Centre (SAC), Indian Space Research Organisation (ISRO), Ahmedabad, Gujarat, India

Department of Botany, Bioinformatics, and Climate Change Impacts Management, University School of Sciences, Gujarat University, Ahmedabad, Gujarat, India

North Eastern Space Applications Centre, Department of Space, Government of India, Umiam, Meghalaya, India

C. P. Singh
Earth, Ocean, Atmosphere, Planetary Sciences and Applications Area (EPSA), Space Applications Centre (SAC), Indian Space Research Organisation (ISRO), Ahmedabad, Gujarat, India

P. Gurung · B. Gajmer · N. Sharma · D. G. Shrestha
Sikkim State Council of Science and Technology, Vigyan Bhawan, Gangtok, Sikkim, India

H. A. Pandya
Department of Botany, Bioinformatics, and Climate Change Impacts Management, University School of Sciences, Gujarat University, Ahmedabad, Gujarat, India

S. P. Singh et al. (eds.), *Ecology of Himalayan Treeline Ecotone*,
https://doi.org/10.1007/978-981-19-4476-5_7

aspects. The highest biodiversity (Shannon's entropy, 3.04) was observed in the 4000 to 4500 m elevation range. The chapter concludes with the limitations and prospects of remote sensing biodiversity along the treeline ecotone in the Himalaya from space.

Keywords

Alpine ecosystems · Entropy · Geographic Information System · Macroecology · Remote sensing · Topography

7.1 Introduction

The alpine ecosystems of the Himalaya are vulnerable ecosystems mainly due to climate change. These ecosystems are mostly found in the mountain summit peaks bordered between the treeline and the permanent snowline. Climatic variability causes short- and long-term changes in the biodiversity in alpine ecosystems. One such climatic variability is increasing near-surface air temperature. Temperature-bounded ecosystems, like in the mountains, are sensitive to warming (Grabherr et al. 1994). In mountains, the temperature is increasing at the rate of 0.3 ± 0.2 °C/decade outpacing the global rate of 0.2 ± 0.1 °C/decade (IPCC 2018; Hock et al. 2019). Alpine ecosystem is found where along the elevation gradient the woody vegetated montane forest gives way to shrubs, grasses, mosses, and finally to open snow-covered ground (Nagy and Grabherr 2009). Alpine treeline ecotone (ATE) is an imaginary line, which is the lower boundary of alpine pasture in the mountain peaks. This line is the upper limit of tree growth along the elevation gradient. The ATE is ecologically important because it is the transition between the tree species down-slope, and the herbs and shrubs species in the alpine meadows. Tree species in the treeline ecotone are better competitors than those of the herbs and shrubs in the alpine tundra at higher-elevation zones (Mayor et al. 2017). Therefore, the plant communities in the alpine ecosystem are vulnerable to warming-induced treeline advancement.

The treeline ecotone in the Himalaya is sensitive to temperature change leading it to shift upwards. Higher temperatures lead to a shift in the treeline ecotone's potential niche to higher-elevation zones in the Himalaya, which is likely to continue in the future as per the climate projections (Mohapatra et al. 2019a). There is an upward shift of the treeline ecotone to the higher-elevation zones in Uttarakhand (*c.* 110 m/decade; Singh et al. 2011, 2012), Sikkim (*c.* 81 m/decade; Singh et al. 2020), and Arunachal Pradesh (*c.* 113 m/decade; Mohapatra et al. 2019b). Rapid shrub expansion has threatened the species biodiversity in the alpine meadows of the Himalaya (Brandt et al. 2013).

There are changes in the biodiversity in response to warming at the high-elevation mountain regions. In the alpine ecosystems of European mountains, there is a dominance of more warmth-adapted species and a decline of more cold-adapted species due to climate warming (Gottfried et al. 2012). This is significant at the continental scale and is likely to continue in the future as well. As per the modelled future climate, there would be an increase in the treeline species niche in the Himalaya (Mohapatra et al. 2019a), which could cause the decline of the alpine species biodiversity. Therefore, biodiversity monitoring at the landscape scale in the Himalaya is important.

Alpine plant communities in the Himalaya survive in complex topography. Variable abiotic factors determine their distribution and biodiversity. Local-level biodiversity studies are insufficient in the Himalaya. Landscape-level estimation of biodiversity is essential for understanding the overall effect of global climatic variability on the plant species biodiversity in the Himalaya. For this, landscape-level studies of biodiversity are essential.

The Himalaya is a large stretch of the physiographic region. To understand the overall patterns of alpine species biodiversity in such a large region, landscape-level studies are necessary. Landscape-level studies of ecological processes are essential for understanding the rapid and broad-scale changes in an ecosystem. Such studies show the interactions between spatial patterns and ecological processes, including the causative factors and consequences of spatial heterogeneity (Turner and Gardner 2015). The treeline ecotone is seen as a line when we observe it from a far distance, like from another nearby summit or below the summit. That means the treeline is distinctively seen at the landscape level. The "trees" in the treeline can be better monitored spatially through remote sensing at the landscape level. The spatial patterns arising due to the assemblage of biodiversity are closely linked to the niche characteristics (Goodall 1970). Biodiversity at different topographic gradients of elevation, summit slope, and summit aspects can be well-understood through landscape-level studies.

The relationship between biodiversity and the environment (like climatic variability) is scale dependent, more importantly in mountain ecosystems (Antonelli et al. 2018). Monitoring of plant species biodiversity in a spatially explicit way is necessary along the ATE in the Himalaya. Spatially explicit models fill the gap between the basic and applied ecology (DeAngelis and Yurek 2017). To understand the impacts of global climate change on ecosystems, studies at multiple scales, times, and organizations are necessary (Levin 1992). Studies on different scales provide unique causes and consequences (Levin 1992). Although the ecosystems in the Himalaya are more vulnerable to climate change, landscape-scale studies of plant species biodiversity in the Himalaya are rare. This gap can be filled using satellite remote sensing and information theory along the treeline ecotone in the Himalaya.

Conducting landscape-level studies in the remote, harsh, and difficult terrains of the Himalaya is a daunting task. Such studies can be better done through space-based remote sensing techniques with limited field studies. In field-based studies of biodiversity, there are scale and sampling limitations (Palmer 1990). Remote sensing of vegetation at the "crude" level, suppressing the field-level details, is essential for the overall understanding of the ecosystem processes (Levin 1992). Satellite remote sensing provides contiguous multi-temporal data with long-term records. Satellite imagery can be used to study any given area (including inaccessible areas) in the Himalaya continuously in time and space.

At the current pace of biodiversity loss, decisions for the conservation of species in large areas can be achieved through remote sensing measurements (Turner et al. 2003). Remote sensing of biodiversity is still a challenge and is critically necessary for the twenty-first century (Cavender-Bares et al. 2020). In the alpine ecosystem of the Western Himalaya, habitat diversity or spatial heterogeneity measured through satellite imagery and information theory has been attempted to link with community-based biodiversity measurements (Mohapatra et al. 2019c; Nayak et al. 2019). Likewise, information theory could be used for estimating the biodiversity along the treeline ecotone through satellite remote sensing in the entire Himalayan arc.

Information theory (Shannon 1948) is the measurement of information and the probability of uncertainty in the given information. Information theory and entropy are similar in concept. The word "entropy" in thermodynamics means disorders in a system. In ecology, the higher the entropy or disorder in the species distribution or their niche, the higher is the biodiversity. Biodiversity is a multidimensional entity. That is, it is measured in many dimensions. Measurement of biodiversity is more important than simple species counting (MacArthur and MacArthur 1961).

There are different metrics for the quantification of biodiversity. While α-diversity is the estimation of biodiversity at the local level, β-diversity measures biodiversity between sites (Whittaker 1972). β-diversity is the measure of community differentiation along the habitat gradients. The total or γ-diversity is the product of α-diversity and the degree of β differentiation among them (Whittaker 1972). Biodiversity patterns at the landscape scale can be better understood using information theory. Derivation of biodiversity maps at the landscape scale using basic information theory in plant communities is a spatially explicit model. In this chapter, an assessment of biodiversity at the landscape scale using information statistics along the treeline ecotone of the Himalaya has been brought out.

7.2 Materials and Methods

7.2.1 Study Area

The study of deriving biodiversity maps using information theory was conducted through remote sensing of the extent, 27.018° N to 28.021° N latitude, and 88.008° E to 89.133° E longitude. The region was over parts of Sikkim state, Eastern Himalaya (Fig. 7.1). There was a decrease in vegetation density from high to low, and finally to open frozen grounds with increase in elevation. The elevation ranged from *c*. 3000 to 7112 m above mean sea level (amsl). The slope ranged from 0° to *c*. 85°.

The climate of the study area was analyzed from the WorldClim dataset for the current time (1970–2000; version 2.1; Fick and Hijmans 2017). The monthly minimum temperature in the region ranges from −11.69 °C (in January) to 5.53 °C (in July). The monthly maximum temperature ranges from 2.07 °C (in January) to 13.09 °C (in July). The monthly precipitation ranges from 4.71 mm (in December) to 144.14 mm (in July).

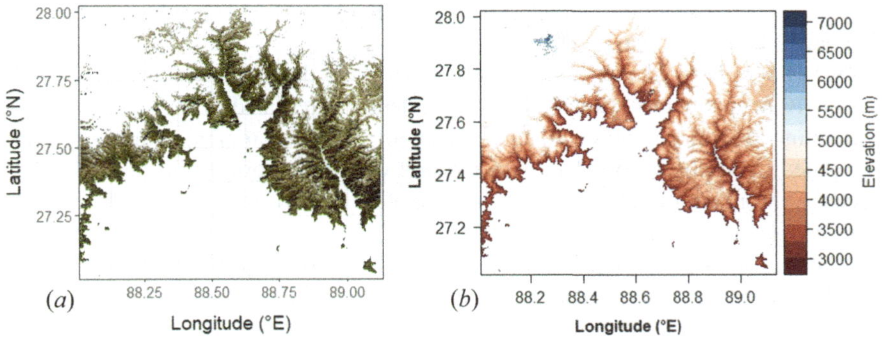

Fig. 7.1 (**a**) Natural color composite (NCC) imagery (Sentinel-2 MultiSpectral Instrument; Date of acquisition, 31 October 2020) of vegetation above 2999 m elevation; and (**b**) elevation gradient (from Cartosat-1 Digital Elevation Model) along the treeline ecotone in the Eastern Himalaya (over Sikkim). Clear continuous vegetation along the longitude is observed in the NCC

7.2.2 Dataset

The optical imagery of the European Space Agency's MultiSpectral Instrument (MSI) sensor onboard Sentinel-2A (ESA 2015) was used to derive biodiversity indices along the treeline ecotone in the Himalaya. The date of acquisition of the image was 15 November 2020. This is the time when the vegetation in the Sikkim Himalaya has peak vegetation growth just after the end of the monsoon (Singh et al. 2020). The dataset has spectral bands with the highest spatial resolution of 10 m × 10 m. Top Of Atmosphere (TOA) reflectance data (Level-1C) of MSI was acquired from Earth Explorer (https://earthexplorer.usgs.gov/). TOA data was atmospherically corrected to surface reflectance using the Sen2Cor v2.9 software (Mueller-Wilm 2020).

The Cartosat-1 Digital Elevation Model (CartoDEM Version-3 R1; NRSA 2006) with a spatial resolution of 30 m × 30 m was used for studying the topographic features along the treeline ecotone. The data was acquired from the *Bhuvan* web portal (https://bhuvan.nrsc.gov.in) of the National Remote Sensing Centre, Indian Space Research Organisation, Hyderabad, India.

7.2.3 Measurement of α-Diversity and β-Diversity

Habitat is the geographical area where a particular species survives. Species richness is the total number of species present in an area. Biodiversity is the measurement of species diversity in a region. Normalized Difference Vegetation Index (NDVI) values can be used to study these parameters in an ecosystem from space. NDVI differentiates vegetation based on spectral reflectance from the plant leaves. The difference in the internal plant leaf structure gives different NDVI values. Therefore,

NDVI was used to derive vegetation heterogeneity (diversity) along the treeline ecotone in the Himalaya.

NDVI was derived using the surface reflectance data of red and near-infrared bands (10 m spatial resolutions). The NDVI data was rescaled to an 8-bit integer for computing the information statistics indices of diversity. α- and β-diversity were computed using the NDVI data with moving window algorithm in the "rasterdiv" package (Marcantonio et al. 2021; Rocchini et al. 2021b) in R statistical programming language (R Core Team 2021).

Information theories that are simple to understand in quantifying biodiversity should be used (Hill 1973). The computed α-diversity indices were Shannon-Weiner entropy (*H'*; Eq. 7.1; Shannon 1948), and Pielou's evenness (*J'*; Eq. 7.2; Pielou 1966). Since the population of plant species considering the large spatial extent along the treeline ecotone in the Himalaya is indefinitely large, *H'* was calculated. *H'* considers the rare species in an ecosystem with the dominant ones. *H'* is calculated as,

$$H' = -\sum_{i=1}^{S} p_i \ln p_i \tag{7.1}$$

where, S is the total number of unique pixel values, and p_i is the proportion of each unique pixel value.

An ecosystem would be considered highly diverse when the species individuals are distributed evenly (Pielou 1966). For the calculation of evenness (*J'*), the species number or the unique digital numbers in the space-borne imagery of a community are necessary. *J'* is calculated as,

$$J' = \frac{H'}{\ln(S)} \tag{7.2}$$

Measurement of β-diversity is based on differences, and these differences are measured along with the differential habitat, niche, or coenoclines (Whittaker 1972, p. 232). There are different techniques in measuring β-diversity from space (Rocchini et al. 2018). β-diversity was computed using Rao's quadratic entropy (Rao's Q; Eq. 7.3; Rao 1982). Rao's Q entropy distinguishes not only unique pixel values (like Shannon's entropy) but also the magnitude of the values (Rocchini et al. 2017, 2018). This is useful in distinguishing the diversity of the vegetation's functional traits in an ecosystem. It takes into consideration the differences in the values along with proportions. The Rao's Q (*Q*) was calculated as,

$$Q = \sum_{i,j=1}^{S} p_i p_j d_{ij} \tag{7.3}$$

where, d_{ij} is the distances between the pixel pairs i and j.

Biodiversity maps were computed using different moving window sizes from three to 15. The higher the moving window size, the higher is the chance that the values get smoothed out, reducing heterogeneity. After analyzing the results generated from different window sizes, finally, a moving window size of nine was selected that showed clear distinctions in the diversity.

7.2.4 Analysis of the Relationship Between Biodiversity and Topography

Slope and aspect maps were generated from CartoDEM data using the "raster" package (Hijmans 2021) in the R language. Since the terrain is relatively rough in the Himalaya, the Horn (1981) method was used for the computation of slope and aspect maps.

Species diversity in the high elevation ecosystem of the Himalaya is majorly determined by topographic factors. α- and β-diversity values along the gradient of elevation, slope, and summits' aspect were estimated. The experimental design for the study was non-orthogonal, as the number of biodiversity pixel values differs in different topography ranges. The design was completely randomized with diversity as the dependent variable and topographic levels as classification variables.

A non-parametric one-way analysis of variance using the Kruskal-Wallis rank sum test (Kruskal and Wallis 1952) was conducted, followed by a post hoc test (Dunn's multiple comparisons). The tests were conducted to check the significant differences in the biodiversity values at different topography features and gradients. All the statistical analyses were done in the "stats" and "FSA" (Ogle et al. 2021) packages in the R programming language. Plots were done in "rasterVis" (Lamigueiro and Hijmans 2021) and "ggpubr" packages (Kassambara 2020) in R.

7.2.5 Field-Level Evaluation of Biodiversity Along the Treeline Ecotone in Sikkim Himalaya

Species biodiversity studies in the field were conducted in Kabi-Tingda, North Sikkim (sampled in 2016), and Kyongnosla Alpine Sanctuary, East Sikkim (sampled in 2021). Field-level evaluation of biodiversity was conducted in three mountain summits of each location. In Kabi-Tingda, the sites were KBT-3151 m (27° 24′ 41.5″ N, 88° 40′ 36.7″ E, 3151 m amsl), KBT-3701 m (27° 24′ 54.4″ N, 88° 41′ 43.0 E, 3701 m), and KBT-3926 m (27° 25′ 30.4″ N, 88° 42′ 59.5″ E, 3926 m). In the Kyongnosla Alpine Sanctuary, the sites along the elevation gradient were KAS-3749 m (27° 21′ 56.33″ N, 88° 43' 54.40″ E, 3749 m), KAS-3810 m (27° 21′ 56.24″ N, 88° 43' 50.59″ E, 3810 m), KAS-3846 m (27° 21′ 56.34″ N, 88° 44′ 24.97″ E, 3846 m), KAS-3896 m (27° 21′ 55.46″ N, 88° 44′ 49.55″ E, 3896 m), KAS-3913 m (27° 21′ 53.74″ N, 88° 44′ 27.37″ E, 3913 m), KAS-3923 m (27° 21′ 53.57″ N, 88° 44′ 26.92″ E, 3923 m), KAS-3979 m (27° 21′ 55.09″ N, 88° 44' 59.67″ E, 3979 m), and KAS-4046 m (27° 21′ 53.92″ N, 88° 45′ 2.95″ E, 4046 m).

Shannon-Weiner entropy and Pielou's evenness were measured using the species abundance data collected from quadrat sampling. For understanding the patterns of co-occurrence amongst multiple sites, averaging the Sørensen pairwise dissimilarities is not appropriate (Baselga 2013). Overall heterogeneity was quantified using Sørensen-based multiple-site dissimilarity (Baselga 2010, Eq. 7.4). The analyses were done using the "vegan" (Oksanen et al. 2020), and "betapart" (Baselga et al. 2021) packages in R.

$$\beta_{SOR} = \frac{\left[\sum_{i<j}\min(b_{ij}, b_{ji})\right] + \left[\sum_{i<j}\max(b_{ij}, b_{ji})\right]}{2\left[\sum_{i}S_i - S_T\right] + \left[\sum_{i<j}\min(b_{ij}, b_{ji})\right] + \left[\sum_{i<j}\max(b_{ij}, b_{ji})\right]} \tag{7.4}$$

Field-measured geolocations of the quadrats were converted to vector point format using QGIS software. The field-measured biodiversity values were then used to compare with the values of the diversity maps derived from the satellite imagery. The flowchart in Fig. 7.2 shows the overall methodology of the study.

7.3 Results

7.3.1 α-Diversity and β-Diversity Along the Treeline Ecotone

Distinct patterns of α- and β-diversity were observed along the treeline ecotone in the Himalaya. The highest biodiversity was in the ecotone region (Fig. 7.3). The higher NDVI values clearly distinguished the vegetation from the non-vegetation regions. The mean NDVI value in the region was 0.61. The mean and maximum values of Shannon's entropy over the region were 2.79 and 4.10, respectively. Pielou's evenness had similar patterns with a mean value of 0.92. The mean and maximum values of Rao's Q entropy along the treeline were 8.99 and 50.0, respectively.

The outlier values in Shannon's entropy were less than in Pielou's evenness and Rao's Q entropy (Fig. 7.4). Shannon's entropy had a higher range of values than Pielou's evenness and Rao's Q entropy.

7.3.2 Spectral Diversity of Vegetation Along the Topographic Gradients

There was a significant (Kruskal-Wallis test, $p < 0.001$) difference in Shannon's entropy and Rao's Q entropy along the gradients of elevation, slope, and summits' aspects along the treeline ecotone (Fig. 7.4 and Table 7.1). A pairwise significant difference (with Dunn's post hoc test, $p < 0.05$) in Shannon's entropy and Rao's Q entropy was between most of the paired elevation ranges. There was a significant difference (at $p < 0.001$) in the Pielou's evenness values along the gradients of

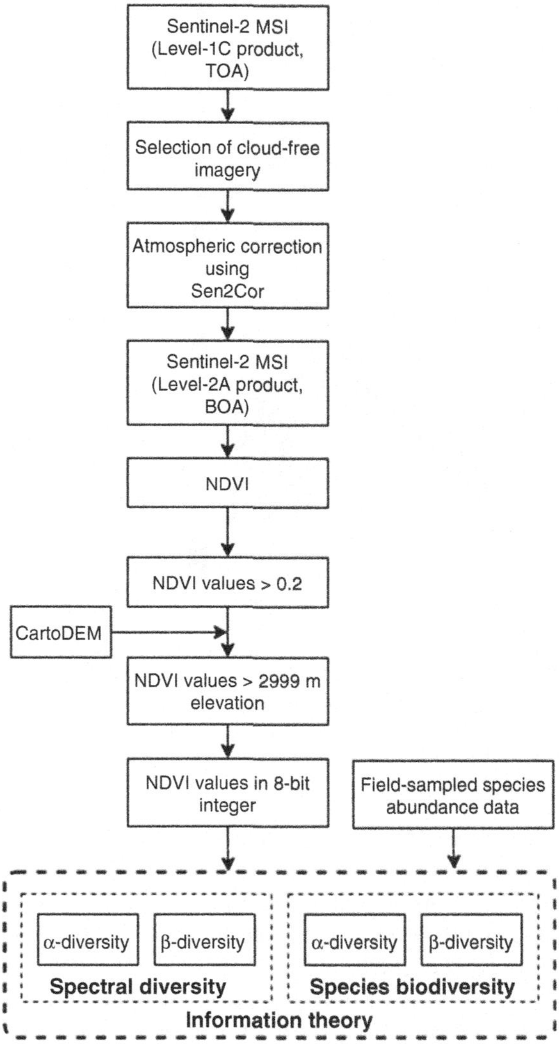

Fig. 7.2 Schematic representation of the overall methodology

elevation, while non-significant difference along the gradients of the slope, and aspects. The Shannon's entropy in the elevation ranges 4000–4500 m was more normally distributed (excluding the extreme values or outliers) than in the other elevation ranges (Fig. 7.4a). Pielou's evenness in different elevation ranges had a similar pattern as that of Shannon's entropy. The distribution of Rao's Q entropy was normally distributed in the elevation range 3500–4000 m (Fig. 7.4c).

The median values in Shannon's entropy were highest (3.12) in the elevation range 3500–4000 m. The lowest median value (2.32) was in the elevation range greater than 5000 m. The median values of Shannon's entropy in the elevation ranges 3000–3500 m, 4000–4500 m, and 4500–5000 m were 2.94, 3.09, and 2.90,

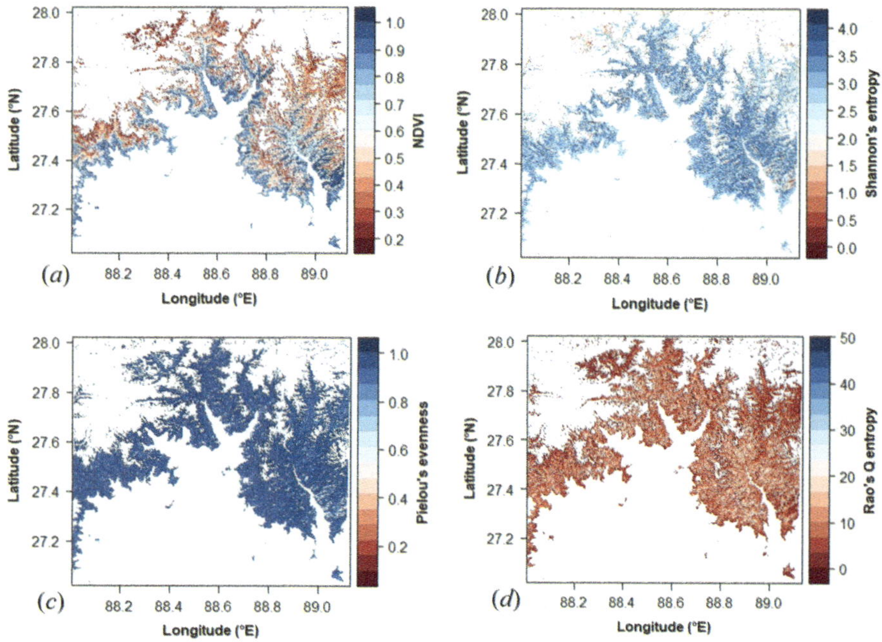

Fig. 7.3 (**a**) Normalized Difference Vegetation Index; and information theory-based α-diversity map along the alpine treeline ecotone in the Himalaya, viz. (**b**) Shannon-Weiner entropy, (**c**) Pielou's evenness, and (**d**) Rao's quadratic entropy. The higher the values, the higher is the biodiversity in the area

respectively. The median values in Pielou's evenness were highest (0.94) in the elevation range 4000–5000 m, followed by 3500–4000 m (0.93) and 3000–3500 m (0.92). The median values in Rao's Q entropy were highest (9.69) in the elevation range 3500–4000 m, followed by 4000–4500 m (8.78), 3000–3500 m (7.28), and 4500–5000 m (6.99).

The distribution of α- (Fig. 7.4d, e) values in the different summits' aspects were mostly negatively skewed or normally distributed. The β-diversity values had normal distribution and negative skew (Fig. 7.4f). The median values of Shannon's entropy and Rao's Q entropy in different summits' aspects were more distinctive than in Pielou's evenness. The median value of Shannon's entropy was highest in the west aspect (3.07), followed by the north (3.03), east (3.02), and south (2.93) aspects. A similar pattern emerged for Pielou's evenness and Rao's Q entropy as well. The median value of Pielou's evenness was highest in the north and east aspects (0.94), followed by the south and west aspects (0.93). The median value of Rao's Q entropy was highest in the north aspect (9.24), followed by the east (8.59), west (8.37), and south (6.86) aspects.

There was a progressive increase in the median value of Shannon's entropy along the increasing gradient of the summit's slope angle (Fig. 7.4g). Lower slope ranges had the highest dispersion. With the increase in slope ranges, the dispersion

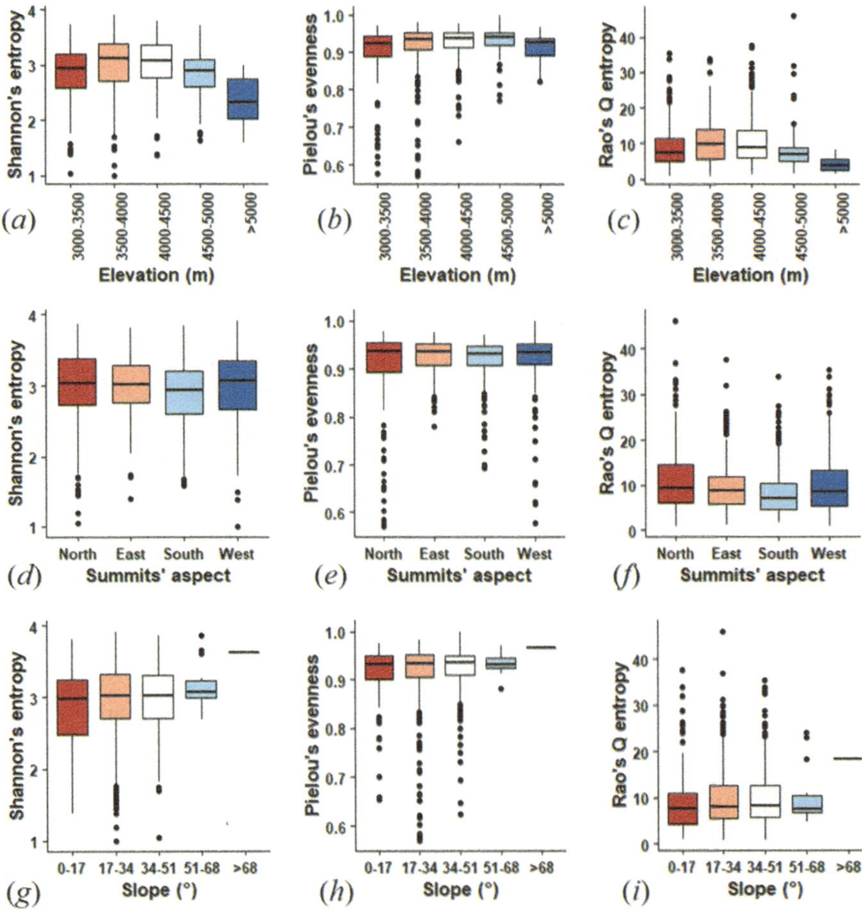

Fig. 7.4 Boxplot showing the variations in Shannon's entropy (α-diversity), Pielou's evenness (α-diversity), and Rao's quadratic entropy (β-diversity) along the elevation gradient (**a**–**c**), summits' aspects (**d**–**f**), and slope ranges (**g**–**i**) along the treeline ecotone in the Himalaya. The bold horizontal line in the box plot shows the median value of diversity

Table 7.1 Summary of Kruskal-Wallis test for diversity values in different topography ranges along the treeline ecotone in the Himalaya. Significant difference is considered at $p < 0.05$

	Elevation ranges			Summits' aspects			Slope ranges		
Biodiversity	χ^2	df	p	χ^2	df	p	χ^2	df	p
Shannon's entropy	75.31	4	< 0.001	12.72	3	0.005	11.21	4	0.024
Pielou's evenness	33.29	4	< 0.001	2.46	3	0.482	4.03	4	0.401
Rao's Q entropy	66.92	4	< 0.001	29.68	3	< 0.001	11.79	4	0.019

Table 7.2 Summary of mean α-diversity (Shannon's entropy and Pielou's evenness) and β-diversity (Rao's quadratic entropy) in different elevation ranges along the treeline ecotone in the Eastern Himalaya

	α-diversity		β-diversity
Elevation ranges (m)	Shannon's entropy	Pielou's evenness	Rao's Q entropy
3000–3500	2.87	0.91	9.11
3500–4000	3.03	0.92	10.74
4000–4500	3.04	0.93	10.49
4500–5000	2.83	0.93	7.82
>5000	2.33	0.91	3.95

Table 7.3 Summary of mean α-diversity (Shannon's entropy and Pielou's evenness) and β-diversity (Rao's quadratic entropy) in different summits' aspects along the treeline ecotone in the Eastern Himalaya

	α-diversity		β-diversity
Summits' aspect	Shannon's entropy	Pielou's evenness	Rao's Q entropy
North	2.97	0.90	11.52
East	2.99	0.93	9.56
South	2.88	0.92	8.22
West	2.98	0.92	10.27

decreased. The highest value (3.63) was in the slope range greater than 68°. The lowest value was 2.96. There was a progressive increase in the median value of Pielou's evenness along the slope gradient up to the 34°–51° range and further increase at a slope greater than 68°. The highest value (0.97) was in the slope range greater than 68°. The lowest value (0.93) was in the range 0°–17°. This trend of median values along the slope gradient was the same for Rao's Q entropy. The highest median value of Rao's Q entropy (18.30) was in the slope range greater than 68°. The lowest value (7.34) was in the slope range 0°–17°.

The trend in the mean values of spectral α- and β-diversity along the elevation gradient was similar. The mean Shannon's entropy was highest (3.04) in the elevation range 4000–4500 m, and lowest (2.33) in the elevation range greater than 5000 m (Table 7.2). There was an increase in the spectral diversity from the elevation ranges 3000 m–4500 m, and then a decrease at a range greater than 4500 m.

The mean Pielou's evenness was highest (0.93) in the elevation range 4000–5000 m and lowest (0.91) in the elevation range greater than 5000 m and below 3500 m. The mean Rao's Q entropy was highest (10.74) in the elevation range 3500–4000 m, and lowest (3.95) in the elevation range greater than 5000 m.

The highest mean Shannon's entropy was in the east aspect (2.99), followed by the west (2.98), north (2.97), and south aspects (2.88; Table 7.3). Pielou's evenness had the highest value of 0.93 in the east aspect, similar to Shannon's entropy. The lowest value (0.90) was in the north aspect. The mean β-diversity value was highest (11.52) in the north aspect, followed by the west (10.27), east (9.56), and south aspects (8.22).

Table 7.4 Summary of mean α-diversity (Shannon's entropy and Pielou's evenness) and β-diversity (Rao's quadratic entropy) along the slope gradient along the treeline ecotone in the Eastern Himalaya

	α-diversity		β-diversity
Slope ranges (°)	Shannon's entropy	Pielou's evenness	Rao's Q entropy
0–17	2.87	0.92	8.84
17–34	2.97	0.92	9.76
34–51	2.97	0.92	10.31
51–68	3.13	0.93	9.90
>68	3.63	0.97	18.30

The mean α-diversity values increased with an increase in the slope gradient from 0° to greater than 68° (Table 7.4). The mean β-diversity values increased with an increase in the slope gradient up to 34°–51°, and then further increased at slope ranges greater than 68°. The highest (10.31) and lowest (8.84) Rao's Q entropy were in the slope ranges 34°–51° and 0°–17°, respectively.

7.3.3 Field-Level Species Biodiversity Along the Elevation Gradient in the Alpine Ecosystems of the Sikkim Himalaya

α- and β-diversity at the community level (through field studies) was estimated in different mountain summits of the Sikkim Himalaya. The total species richness in the three summits was 63. KBT-3151 m, KBT-3701 m, and KBT-3926 m had species richness 30, 37, and 26, respectively. The overall Shannon's index and Pielou's evenness in the three summits were 3.18 and 0.96, respectively. In KBT-3151 m, Shannon's index and Pielou's evenness were 3.11 and 0.93, respectively. In KBT-3701 m, Shannon's index and Pielou's evenness were 3.90 and 0.93, respectively. In KBT-3926 m, Shannon's index and Pielou's evenness were 2.83 and 0.92, respectively. β_{SOR}in Kabi-Tingda was 0.65. The species richness in the region was 63. The major species were *Rhododendron thomsonii*, *Juniperus indica*, *Polygonum molle*, *Juncus thomsonii*, *Potentilla peduncularis*, *Iris clarkei*, *Selinum tenuifolium*, *Senecio diversifolius*, and *Usnea orientalis*.

In the Kyongnosla Alpine Sanctuary, the Shannon's entropy values in KAS-3749 m, KAS-3810 m, KAS-3846 m, KAS-3896 m, KAS-3913 m, KAS-3923 m, KAS-3979 m, and KAS-4046 m were 2.86, 2.03, 1.65, 2.73, 2.44, 2.51, 3.08, and 2.63, respectively. Pielou's evenness in KAS-3749 m, KAS-3810 m, KAS-3846 m, KAS-3896 m, KAS-3913 m, KAS-3923 m, KAS-3979 m, and KAS-4046 m was 0.85, 0.66, 0.54, 0.85, 0.72, 0.75, 0.87, and 0.82, respectively. The species richness in the region was 68. β_{SOR} in KAS-3749 m, KAS-3810 m, KAS-3846 m, KAS-3896 m, KAS-3913 m, KAS-3923 m, KAS-3979 m, and KAS-4046 m was 0.89, 0.81, 0.83, 0.85, 0.83, 0.93, 0.86, and 0.81. The dominant species were *Rhododendron lepidotum*, *Rhododendron anthopogon*, mosses, ferns,

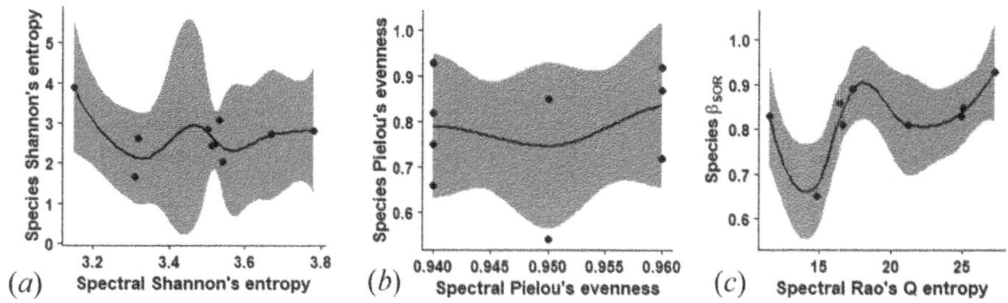

Fig. 7.5 Scatterplot showing the relationship between spectral diversity and species diversity in the alpine ecosystem of the Sikkim Himalaya measured through (**a**) Shannon's entropy, (**b**) Pielou's evenness, and (**c**) Rao's Q entropy information theories. The points are fitted with local regression (Cleveland et al. 1993)

Polygonum barbatum, *Salix calyculata*, *Rumex nepalensis*, *Rubus nepalensis*, *Himalaiella* sp., *Juncus* sp., and *Ligusticopsis* sp.

The field-based estimation of biodiversity was compared with the spectral diversity estimated through satellite imagery. The relationship between field-based measurement and satellite-derived measurement spectral diversity was curvilinear (Fig. 7.5). An increasing trend was observed between β_{SOR} and Rao's Q entropy.

7.4 Discussion

7.4.1 Patterns of α-Diversity and β-Diversity Along the Treeline Ecotone

There was a gradual decrease in the NDVI values after a certain elevation limit, indicating the transition from montane to treeline, alpine tundra, and open frozen ground. This was visible in the natural color composite image as well (Fig. 7.1a). The patterns of biodiversity along the treeline ecotone were well-distinguished with Shannon's entropy (Fig. 7.3b). As shown in the map, Shannon's entropy was lower in the region upslope (mostly dominated by snow), and downslope had higher dominance of a few montane tree species.

The ecotone region has higher functional diversity (Fig. 7.3d). Similar to the pattern of α-diversity, the β-diversity (Rao's Q entropy) was higher in the ecotone region compared to the vegetation downslope and areas upslope dominated with snow cover. The magnitude of the range of values in Rao's Q entropy was higher than Shannon's entropy because Rao's Q entropy considers differences in the magnitude of input values with proportions of each unique value (Rocchini et al. 2017), in this case, the NDVI values. The presence of different vegetation with highly distinct NDVI values caused the differences to be high in Rao's Q entropy in specific regions.

The valley region had higher entropy and was distinct in Rao's Q entropy. Valley regions have an interface of water, land, and vegetation that brings out higher habitat heterogeneity. Rocchini et al. (2021b) described and used the same technique to estimate the spatial heterogeneity of the alpine ecosystem in the Alps, Italy. Similar to this study, there are observations of low heterogeneity in the areas dominated by snow, and high in the ecotone regions, and the valleys of the Alps.

All the three information theory algorithms showed higher entropy (indicating higher biodiversity) along the treeline ecotone in the Himalaya compared to the areas far upslope and downslope the treeline. However, Shannon's entropy was better in showing distinct spatial heterogeneity along the treeline in the Himalaya (Fig. 7.3).

7.4.2 Biodiversity Along the Treeline Ecotone Varies with Topographic Factors

Along with the different habitat gradients, species evolve to occupy different positions in the gradient (Whittaker 1972). The observation of the highest biodiversity in this elevation range is similar to that reported in another region of the Eastern Himalaya (Salick et al. 2004; Brandt et al. 2013). Although the general pattern in the mean values of Shannon's entropy, Pielou's evenness, and Rao's Q entropy along the elevation gradient varied, mid-domain effect (MDE, Colwell and Lees 2000) was observed in all the measured indices. In this study, the MDE hypothesis was proved along the treeline ecotone at the landscape level through satellite imagery and information theory.

In the Himalaya, the treeline ecotone is more elevation-driven than the latitude and the longitude (Mohapatra et al. 2019a). In the Sikkim Himalaya, the mean elevation of the current treeline position is 3542 m (Singh et al. 2020). The modelled 'highly suitable' potential niche of *Betula utilis* (a dominant treeline species) for the current time in the Hindu-Kush Himalaya region is at the elevation range 3801–4000 m (Mohapatra et al. 2019a). Generally, it is expected that with the increase in elevation, the biodiversity would decrease due to cold stress, which is observed after a certain elevation in the Himalaya. In this study, the decreasing Shannon's entropy and Rao's Q entropy along the elevation gradient were observed in elevation ranges greater than 4500 m. The patterns of Shannon's entropy and Rao's Q entropy in the different aspects of mountain summits are similar. The south and east aspects have less α- and β-diversity than the north and west aspects. The south and east aspects in the Himalayan terrain get higher insolation and rainfall. This causes specific species to be more abundant than others. The higher the dominance of a few species in an ecosystem, the lower is the biodiversity. This is observed in spectral heterogeneity as well.

The lower slope had less spectral diversity than the higher slope ranges. Lower slope ranges hold more water and nutrient in the soil. The higher availability of water and nutrients makes a few species more dominant than the others, thus, lower biodiversity. With increase in the slope ranges, less water and nutrients remain in the soil, leading to lower selective species dominance, that is, higher biodiversity in the summits.

Different biodiversity indices have different values. No particular algorithm for estimating biodiversity has a pre-eminent advantage (Hill 1973). A particular diversity index favors a specific community structure, and the values could be high or low. Some algorithm assign more weightage to dominant species in the community, while others assign more to rare species.

A community could be considered highly diverse when it has higher values in all diversity measurements (multi-dimensional; Tóthmérész 1995). This is important for diversity comparison (Tóthmérész 1995). A particular community or habitat can be considered more diverse than the other if the diversity profile (a curve of biodiversity values against parametric indices) is entirely above the other. The curve may intersect, and if it does, then the communities are not directly comparable but could have important ecological phenomena (Tóthmérész 1995). A topography range with high values in all the parametric values (continuum of biodiversity) could be considered highly diverse.

Diversity ordering (measuring continuum of biodiversity) of α-diversity using parametric Hill's diversity (Hill 1973), and β-diversity using generalized Rao's quadratic entropy (Rocchini et al. 2021a) along the topography gradients was not studied for high computational demands. Rare species can be more efficiently mapped with lower parametric values and dominant species with higher parametric values of Hill's diversity and generalized Rao's Q entropy. Such studies in the alpine ecosystem of the Himalayan terrain can be attempted in the future.

Elevation, slope, and aspect are the primary determinants of species distribution in the alpine ecosystem of the Himalaya. Local topographic features determine the microclimatic conditions in the alpine ecosystems. Heterogeneity in the elevations, slopes, and aspect values (quantified from DEM) can be used with the NDVI to study the spatial heterogeneity along the treeline.

The alpine ecosystem is a short-area ecosystem in the Himalaya. Assessment of biodiversity through remote sensing is considered "too coarse" to identify most species (Turner et al. 2003). However, for an overall understanding of the ecological processes in an ecosystem, studies at different scales are necessary (Levin 1992; Turner et al. 2003).

There were no perfect monotonic trends in comparing the field-based indices with the satellite-based spectral diversity (Fig. 7.5). This is because, at the field level, sampling is possible at the community level only in the difficult terrain of the Himalayan ecosystem. Spectral diversity measured through satellite imagery represents landscape-level heterogeneity. Biodiversity estimates on the field from smaller quadrat sizes do not correlate well with the higher grain-sized diversity estimates derived through satellite imagery in the alpine ecosystem of the Himalaya (Mohapatra et al. 2019c). Nonetheless, in this study, the measurement of biodiversity in the field, and spectral heterogeneity through satellite imagery is based on information theory. Although difficult to conduct in the Himalayan terrain, an attempt in sampling species abundances in the field along the treeline using random macroplots would be useful for the validation of the satellite-derived biodiversity maps.

Monitoring of biodiversity in the alpine ecosystem of the Himalaya requires high-resolution imagery. This could make species diversity measurement more precise by differentiating individual species at the treeline ecotone. However, using some information theory algorithms on the higher spatial resolution imagery will be computationally intensive.

Estimation of actual species richness from multispectral satellite imagery is a challenge. However, species richness could be better estimated through hyperspectral imagery. Different plant species reflect a specific wavelength of incident light based on their cells' biochemical composition and leaf structure. Hyperspectral data would make biodiversity monitoring more species specific. Entropy in the spectral signatures of vegetation would provide a deeper insight into species distribution patterns along the treeline ecotone. Derivation of the biodiversity estimates along the treeline using the concept of "spectral species" (Féret and Asner 2014; Féret and de Boissieu 2019) is necessary.

The information theory algorithms of estimating biodiversity at the landscape level can be used in airborne imagery and imagery with multiple spectral bands at a time. Temporal studies of biodiversity along the treeline ecotone can be studied using the historical archived satellite imagery, like since the year 1972 of the Landsat series (Wulder et al. 2019). Other than the entropy in NDVI, entropy in other functional traits of vegetation like solar-induced chlorophyll-*a* fluorescence, leaf area index, and net primary productivity along the treeline ecotone in the Himalaya could be attempted using Rao's Q entropy.

The transition zones or the ecotones have the highest biodiversity. This is because an ecotone has a heterogeneous niche or habitat. The ecotone region is mostly affected by climatic variability and hence, the biodiversity of species. Niche overlap between the treeline and the alpine grassland species can lead to a decline of grassland species at the summit peak. It is necessary to monitor the biodiversity and habitat heterogeneity along the treeline ecotone. Assessment of biodiversity from space using remote sensing and information theory is necessary for the remote terrains of the Himalayan treeline. Such assessments would help in the rapid decision-making process in selecting areas for species conservation along the treeline ecotone in the Himalaya.

7.5 Conclusion

The alpine ecosystems in the summits' peak of the Himalaya are islands in the sea of forest ecosystem downslope. The treeline ecotone is the boundary for the alpine tundra and montane region downslope. As species in islands are vulnerable to extinction, so are species in the alpine ecosystem of the Himalaya. The alpine ecosystem in the Himalaya is one of the "difficult to assess" ecosystems on the globe. The remoteness, harsh weather condition, and undulating terrain of the Himalaya make biodiversity monitoring at the field level practically difficult. In this case, remote sensing is a useful tool in biodiversity assessment. Space-based studies capture mostly the habitat heterogeneity in an ecosystem that is a proxy of biodiversity. The higher the heterogeneity in the species' habitat, the higher is the chance of high species biodiversity in the ecosystem. Information statistics are useful for understanding the patterns of habitat heterogeneity.

The α- and β-diversity (in terms of habitat heterogeneity) in the Himalaya is higher along the treeline ecotone. There is a statistically significant ($p < 0.001$) difference in both α- and β-diversity along the gradients of elevation, slope, and summits' aspects. The highest diversity is in the 3500–4000 m a. m. s. l. elevation range. The lowest diversity is in the elevation range greater than 5000 m. Considering the diversity in different summits' aspects, the highest is in the south, and the lowest is in the north. There is increase in diversity with increase in slope gradients.

Monitoring biodiversity in the alpine ecosystem is important for species conservation. The areas in the high-elevation regions of the Himalaya with high habitat heterogeneity could be selected for conservation. Landscape-level study of biodiversity patterns along the treeline ecotone is important for alpine species conservation in the Himalaya.

In the majorly inaccessible alpine ecosystems of the Himalaya, satellite remote sensing of biodiversity is necessary. Satellite imagery acquired multiple times would help in investigating the temporal changes in species biodiversity. Such biodiversity studies can be conducted since year 1972 by using the freely available archived Landsat satellite data. Investigating further the mid-domain effect along the treeline at the landscape level through satellite remote sensing would provide important insights into the treeline dynamics and its effect on species biodiversity in the alpine ecosystem of the Himalaya.

Acknowledgments The authors gratefully acknowledge Director, SAC; Head of Agriculture and Land Ecosystem Division (AED); Group Director of Biological and Planetary Sciences and Applications Group (BPSG); and Deputy Director of EPSA, SAC, ISRO, Ahmedabad, India. We thank the Head of the Department, Department of Botany, Bioinformatics, and Climate Change Impacts Management, Gujarat University, Ahmedabad, for her kind support. We thank European Space Agency for providing Sentinel-2A MSI Level-2A optical imagery. CartoDEM Version-3 R1 data was used in this study for topographic analysis. We acknowledge National Remote Sensing Centre, ISRO, Hyderabad, for the CartoDEM data available in the *Bhuvan* web portal. We acknowledge the current time climate dataset (2.5 min) of WorldClim version 2.1. Field exploration of the alpine ecosystem in the Himalayan terrain is a hard task. We acknowledge all the team members. Our thanks are due to all the editors, and the anonymous reviewers for their constructive comments, contribution and support to the chapter.

References

Antonelli A, Kissling WD, Flantua SGA, Bermúdez MA, Mulch A, Muellner-Riehl AN et al (2018) Geological and climatic influences on mountain biodiversity. Nat Geosci 11:718–725

Baselga A (2010) Partitioning the turnover and nestedness components of beta diversity. Glob Ecol Biogeogr 19(1):134–143

Baselga A (2013) Multiple site dissimilarity quantifies compositional heterogeneity among several sites, while average pairwise dissimilarity may be misleading. Ecography 36:124–128

Baselga A, Orme D, Villeger S, Bortoli JD, Leprieur F, Logez M (2021) betapart: Partitioning beta diversity into turnover and nestedness components. R package version 1.5.4. https://CRAN.R-project.org/package=betapart

Brandt JS, Haynes MA, Kuemmerle T, Waller DM, Radeloff VC (2013) Regime shift on the roof of the world: alpine meadows converting to shrublands in the southern Himalayas. Biol Conserv 158:116–127

Cavender-Bares J, Gamon JA, Townsend PA (2020) The use of remote sensing to enhance biodiversity monitoring and detection: a critical challenge for the twenty-first century. In: Cavender-Bares J, Gamon JA, Townsend PA (eds) Remote sensing of plant biodiversity. Springer, Switzerland, pp 1–12. ISBN 978-3-030-33157-3

Cleveland WS, Grosse E, Shyu WM (1993) Local regression models. In: Chambers JM, Hastie TJ (eds) Statistical models in S. Chapman & Hall, London, pp 309–376

Colwell RK, Lees DC (2000) The mid-domain effect: geometric constraints on the geography of species richness. Trends Ecol Evol 15(2):70–76

DeAngelis DL, Yurek S (2017) Spatially explicit modeling in ecology: a review. Ecosystems 20: 284–300

ESA (European Space Agency) (2015) Sentinel-2 user handbook. Issue 1, Revision 1, pp. 1–64

Féret J-B, Asner GP (2014) Mapping tropical forest canopy diversity using high-fidelity imaging spectroscopy. Ecol Appl 24(6):1289–1296

Féret J-B, de Boissieu F (2019) biodivMapR: An R package for α- and β-diversity mapping using remotely sensed images. Methods Ecol Evol 11:64–70

Fick SE, Hijmans RJ (2017) WorldClim 2: New 1-km spatial resolution climate surfaces for global land areas. Int J Climatol 37(12):4302–4315

Goodall DW (1970) Statistical plant ecology. Annu Rev Ecol Syst 1:99–124

Gottfried M, Pauli H, Futschik A, Akhalkatsi M, Barančok P, Alonso JLB et al (2012) Continent-wide response of mountain vegetation to climate change. Nat Clim Chang 2:111–115

Grabherr G, Gottfried M, Pauli H (1994) Climate effects on mountain plants. Nature 369:448

Hijmans RJ (2021) raster: Geographic data analysis and modeling. R package version 3.4-10. https://CRAN.R-project.org/package=raster

Hill MO (1973) Diversity and evenness: a unifying notation and its consequences. Ecology 54(2): 427–432

Hock R, Rasul G, Adler C, Cáceres B, Gruber S, Hirabayashi Y et al (2019) High mountain areas. In: Pörtner H-O,Roberts DC, Masson-Delmotte V, Zhai P, Tignor M, Poloczanska E, Mintenbeck K, Alegría A, Nicolai M, Okem A, Petzold J, Rama B, Weyer NM, editors. IPCC Special Report on the Ocean and Cryosphere in a Changing Climate. In press

Horn BKP (1981) Hill shading and the reflectance map. P IEEE 69(1):14–47

IPCC (Intergovernmental Panel on Climate Change) (2018) Summary for Policymakers. In: Masson-DelmotteV, Zhai P, Pörtner H-O, Roberts D, Skea J, Shukla PR, Pirani A, Moufouma-Okia W, Péan C, Pidcock R, Connors S, Matthews JBR, Chen Y, Zhou X, Gomis MI, Lonnoy E, Maycock T, Tignor M,Waterfield T, editors. Global Warming of 1.5°C. An IPCC Special Report on the impacts of global warming of 1.5°C above pre-industrial levels and related global greenhouse gas emission pathways, in the context of strengthening the global response to the threat of climate change, sustainable development, and efforts to eradicate poverty. In Press

Kassambara A (2020) ggpubr: 'ggplot2' Based Publication Ready Plots. R package version 0.4.0. https://CRAN.R-project.org/package=ggpubr

Kruskal WH, Wallis WA (1952) Use of ranks in one-criterion variance analysis. J Am Stat Assoc 47(260):583–621

Lamigueiro OP, Hijmans R (2021) rasterVis. R package version 0.50.2

Levin SA (1992) The problem of pattern and scale in ecology. Ecology 73(6):1943–1967

MacArthur RH, MacArthur JW (1961) On bird species diversity. Ecology 42(3):594–598

Marcantonio M, Iannacito M, Thouverai E, Da Re D, Tattoni C, Bacaro G et al (2021) rasterdiv: Diversity indices for numerical matrices. R package version 0.2-3. https://CRAN.R-project.org/package=rasterdiv

Mayor JR, Sanders NJ, Classen AT, Bardgett RD, Clément J-C, Fajardo A et al (2017) Elevation alters ecosystem properties across temperate treelines globally. Nature 542:91–95

Mohapatra J, Singh CP, Hamid M, Verma A, Semwal SC, Gajmer B et al (2019a) Modelling *Betula utilis* distribution in response to climate-warming scenarios in Hindu-Kush Himalaya using Random Forest. Biodivers Conserv 28:2295–2317
Mohapatra J, Singh CP, Tripathi OP, Pandya HA (2019b) Remote sensing of alpine treeline ecotone dynamics and phenology in Arunachal Pradesh Himalaya. Int J Remote Sens 40(20):7986–8009
Mohapatra J, Singh CP, Hamid M, Khuroo AA, Malik AH, Pandya HA (2019c) Assessment of the alpine plant species biodiversity in the Western Himalaya using Resourcesat-2 imagery and field survey. J Earth Syst Sci 128(189):1–16
Mueller-Wilm U (2020) S2 MPC: Sen2Cor software release note. Ref. S2-PDGS-MPC-L2A-SRN-V2.9.0, Issue 01, European Space Agency
Nagy L, Grabherr G (2009) The biology of alpine habitats. Oxford University Press, New York
Nayak PP, Gita, Chandra GS, Mishra R, Pandian S, Deobhanj S et al (2019) Biodiversity from space: Conserving alpine habitats. In Science last fortnight, Curr Sci India 117(7): 1129
NRSA (National Remote Sensing Agency) (2006) Cartosat-1: Data user's handbook. NRSA Data Center, NRSA, Hyderabad, Indian Space Research Organisation, India. Report No. CARTOSAT-1/NRSA/NDC/HB-09/06
Ogle DH, Wheeler P, Dinno A (2021) FSA: Fisheries stock analysis. R package version 0.8.32, https://github.com/droglenc/FSA
Oksanen J, Blanchet FG, Friendly M, Kindt R, Legendre P, McGlinn D, Minchin PR, O'Hara RB, Simpson GL, Solymos P, Stevens MHH, Szoecs E, Wagner H (2020) vegan: Community ecology package. R package version 2.5-7. https://CRAN.R-project.org/package=vegan
Palmer MW (1990) The estimation of species richness by extrapolation. Ecology 71(3):1195–1198
Pielou EC (1966) The measurement of diversity in different types of biological collections. J Theor Biol 13:131–144
R Core Team (2021) R: A language and environment for statistical computing. R Foundation for Statistical Computing, Vienna, Austria. URL https://www.R-project.org/
Rao CR (1982) Diversity: Its measurement, decomposition, apportionment and analysis. Sankhya Ser A 44(1):1–22
Rocchini D, Marcantonio M, Ricotta C (2017) Measuring Rao's Q diversity index from remote sensing: an open source solution. Ecol Indic 72:234–238
Rocchini D, Luque S, Pettorelli N, Bastin L, Doktor D, Faedi N et al (2018) Measuring β-diversity by remote sensing: a challenge for biodiversity monitoring. Methods Ecol Evol 9(8):1787–1798
Rocchini D, Marcantonio M, Da Re D, Bacaro G, Feoli E, Foody GM et al (2021a) From zero to infinity: minimum to maximum diversity of the planet by spatio-parametric Rao's quadratic entropy. Global Ecol Biogeogr 30(5):1153–1162
Rocchini D, Thouverai E, Marcantonio M, Iannacito M, Da Re D, Torresani M et al (2021b) Rasterdiv-an information theory tailored R package for measuring ecosystem heterogeneity from space: to the origin and back. Methods Ecol Evol. https://doi.org/10.1111/2041-210X.13583
Salick J, Anderson D, Woo J, Sherman R, Cili N, Na A, Dorje S (2004) Tibetan ethnobotany and gradient analyses: Menri (Medicine Mountains), Eastern Himalayas. In: Millennium ecosystem assessment: Bridging Scales and Epistemologies, Alexandria, Egypt
Shannon CE (1948) A mathematical theory of communication. Bell Syst Tech J XXVII(3):379–423
Singh CP, Panigrahy S, Parihar JS (2011) Alpine vegetation ecotone dynamics in Gangotri catchment using remote sensing techniques. Int Arch Photogramm XXXVIII-8(W20):162–169
Singh CP, Panigrahy S, Thapliyal A, Kimothi MM, Soni P, Parihar JS (2012) Monitoring the alpine treeline shift in parts of the Indian Himalayas using remote sensing. Curr Sci India 102(4): 559–562
Singh CP, Mohapatra J, Pandya HA, Gajmer B, Sharma N, Shrestha DG (2020) Evaluating changes in treeline position and land surface phenology in Sikkim Himalaya. Geocarto Int 35(5): 453–469
Turner MG, Gardner RH (2015) Landscape ecology in theory and practice: pattern and process, 2nd edn. New York, Springer

Turner W, Spector S, Gardiner N, Fladeland M, Sterling E, Steininger M (2003) Remote sensing for biodiversity science and conservation. Trends Ecol Evol 18(6):306–314
Tóthmérész B (1995) Comparison of different methods for diversity ordering. J Veg Sci 6:283–290
Whittaker RH (1972) Evolution and measurement of species diversity. Taxon 21(2/3):213–251
Wulder MA, Loveland TR, Roy DP, Crawford CJ, Masek JG, Woodcock CE et al (2019) Current status of Landsat program, science, and applications. Remote Sens Environ 225:127–147

Quantifying Variation in Canopy Height from LiDAR Data as a Function of Altitude Along Alpine Treeline Ecotone in Indian Himalaya

8

Jincy Rachel Mathew, C. P. Singh, Jakesh Mohapatra, Ritesh Agrawal, Hitesh Solanki, Anzar A. Khuroo, Maroof Hamid, A. H. Malik, Rameez Ahmad, Amit Kumar, and Anirudh Verma

Abstract

Canopy height is a key physiognomic parameter of biodiversity, productivity and other ecosystem functions in high-elevation alpine ecosystems. However, little is known as to how altitude influences canopy height in these ecosystems. This study makes use of an open-access global forest canopy height map with a spatial resolution of 30 m that integrates Global Ecosystem Dynamics Investigation (GEDI)–Light Detection and Ranging (LiDAR) data (April–October 2019) and Landsat analysis-ready time-series data (year 2019). The variation in canopy height was quantified for each 100 m elevation band starting 500 metres below the alpine treeline ecotone at 3780 masl and extending up to 500 m above the alpine treeline ecotone. The global forest height map was compared to the in situ data (root-mean-square error [RMSE] = 6.6 m; mean absolute error [MAE] =

J. R. Mathew (✉)
EPSA, Space Applications Centre (SAC), ISRO, Ahmedabad, Gujarat, India

Department of Environmental Sciences, Gujarat University, Ahmedabad, Gujarat, India

C. P. Singh · J. Mohapatra · R. Agrawal
EPSA, Space Applications Centre (SAC), ISRO, Ahmedabad, Gujarat, India
e-mail: cpsingh@sac.isro.gov.in; ritesh_agrawal@sac.isro.gov.in

H. Solanki
Department of Environmental Sciences, Gujarat University, Ahmedabad, Gujarat, India
e-mail: hasolanki@gujaratuniversity.ac.in

A. A. Khuroo · M. Hamid · A. H. Malik · R. Ahmad
Centre for Biodiversity and Taxonomy, University of Kashmir, Srinagar, Jammu and Kashmir, India

A. Kumar · A. Verma
CSIR-Institute of Himalayan Bioresource Technology (IHBT), Palampur, Himachal Pradesh, India
e-mail: amitkr@ihbt.res.in

S. P. Singh et al. (eds.), *Ecology of Himalayan Treeline Ecotone*,
https://doi.org/10.1007/978-981-19-4476-5_8

4.45 m). We observed a strong negative correlation ($R^2 = 0.96$) between altitude and LiDAR-estimated canopy height. The altitude alone explained 96% of the variation in canopy height ($p < 0.001$). This chapter provides the first of its kind landscape-level quantification of the rate at which canopy height decreases with the increase in altitudinal gradient across the Indian Himalayan treeline.

Keywords

Alpine ecosystem · Canopy height · GEDI LiDAR · Indian Himalayan region · Treeline ecotone

8.1 Introduction

In recent decades, Himalaya has experienced the most rapid climate warming (Field et al. 2014) along with an increase in extreme climate events (Shrestha et al. 2012). The pristine environment with the least amount of anthropogenic activities makes it an ideal 'natural laboratory' to study tree growth responses in altitude-mediated thermal compression regime. Owing to heat deficiency, tree development at high altitude is predicted to be more vulnerable to climate change (Körner and Paulsen 2004), which in turn will alter the forest ecosystem composition and functioning (Körner 2012). As a result of the ongoing rapid warming, tree growth may respond by upward shifting of treeline (Grabherr et al. 1994) and enhanced tree growth (Shi et al. 2020).

Altitude is a strong limiting factor for the tree growth enhancement in mountainous areas since the height of trees is affected by decreasing temperatures with increasing altitude. This phenomenon results in shorter growth and peculiar formations at the treeline, i.e. the altitudinal limit of tree growth (Körner 2012). However, little is known about role of altitude in regulating canopy height in these ecosystems at a landscape scale. Therefore, it is critical to understand altitude-dependent canopy height responses and their spatial patterns at a landscape scale. Quantifying tree height along altitudinal gradients aids in understanding of climate-driven tree patterns and their climate sensitivity to continuing climate change. Quantifying canopy height along altitudinal gradients aids in understanding of tree growth sensitivity to ongoing climate change. Canopy height dynamics of the alpine ecosystem of Himalaya has rarely been studied, and the present study, therefore, addresses this research gap. We compiled canopy height data at treeline in Indian Himalaya to analyse spatial variations in climate-driven canopy height patterns.

Traditionally, canopy height is acquired directly through field measurements *or* indirectly through two-dimensional (2-D) image information. However, these methods are expensive and impractical for landscape-level studies. The new generation of active remote sensing technologies like Light Detection and Ranging (LiDAR) system provides direct, three-dimensional (3-D) measurement of the vertical structure of vegetation and terrain surface. The most recent space-borne LiDAR instrument is National Aeronautics and Space Administration's (NASA's) Global

Ecosystem Dynamics Investigation (GEDI). It is designed to give information about forest vertical structure through application of GEDI, and it is anticipated to acquire 4% of Earth's land surface over its 2-year nominal mission. It gives information about canopy height, canopy cover, plant area index and topography (Dubayah et al. 2020). For practical applications, standard GEDI data packages have significant restrictions. Since GEDI data provide footprint-level products (average footprint 25 m) represented by a point sample, leaving considerable land surface without observations, GEDI's transect sampling may not be appropriate for quantifying variation in canopy height over the study area. Potapov et al. (2021) developed the global forest height map by integrating GEDI-derived canopy height measurements with Landsat multi-temporal surface reflectance data. To achieve high-quality forest height estimation, this open-access global forest canopy height map was used for the study.

Here, we quantified the variation in canopy height for each 100 m elevation band starting from 500 m below the treeline to 500 m above the treeline. Continuous datasets at large geographic extents allowed us to identify the relationship between altitude and canopy height at a landscape level. Specifically, in this study, we addressed the following questions: (a) What is the canopy height of the alpine treeline ecotone along parts of the Indian Himalaya? (b) Are the variations in canopy height at treeline similar with respect to altitude from west to east? (c) Does other physiographic factors such as latitude have relationship with canopy height?

8.2 Study Area

The study area includes the part of Indian Himalayan region featuring highly heterogeneous terrain and dense forest (Fig. 8.1). This includes five Indian Himalayan states, viz. Jammu and Kashmir (J and K), Himachal Pradesh (H.P.), Uttarakhand, Sikkim, and Arunachal Pradesh (A.P.). Because of the complex three-dimensional geo-ecological variability with diverse thermal and biotic variables, Himalayan treeline ecotones display significant differences in physiognomy and altitudinal position (Schickhoff 2005). In the Himalaya, climate is strongly regulated by altitude, resulting in steep environmental gradients. The region's microclimatic variability leads to significant variations in vegetation over short distances. Heavy and long-duration precipitation in both winter and summer leads to luxuriant vegetation and regions of diverse biological communities with a high level of endemism. The competitive ability of evergreen trees and large shrubs of the genus *Rhododendron*, which have replaced the deciduous birch belt in the eastern Himalaya, increases along the northwest–southeast gradient due to strongly decreasing temperature during winter months and increasing humidity during monsoon months along the gradient (Schickhoff et al. 2015).

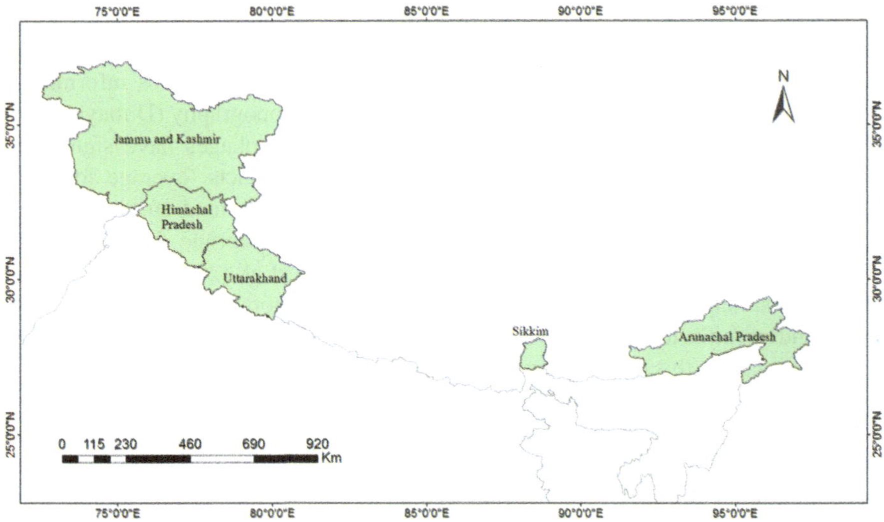

Fig. 8.1 Study area (shown in shade)

8.3 Data Used

8.3.1 Global Forest Canopy Height Map

The global forest canopy height map of 30 m spatial resolution was acquired from Global Land Analysis and Discover-University of Maryland (UMD-GLAD) website https://glad.umd.edu/dataset/gedi. The global forest height map was developed by integrating GEDI-derived canopy height measurements with Landsat multi-temporal surface reflectance data (Potapov et al. 2021). The dependent variable for model calibration was the relative height at 95% and the independent variable was the Landsat multi-temporal metrics obtained from the analysis-ready data (GLAD ARD) time-series. The global canopy height map for selected Himalayan states under the study is shown in Fig. 8.2.

8.3.2 Treeline Data

The treeline position delineation was acquired from the previous research carried on the alpine ecosystems by Singh et al. (2021), where the authors have used Cloud-free, terrain corrected, ortho-rectified (Universal Transverse Mercator World Geodetic System-1984 projection) Resourcesat-1/2 Linear Imaging Self Scanning Sensor (LISS-III) multispectral images to delineate alpine treeline over the Indian Himalayan region. The details of the data used are given in Table 8.1. First digital numbers (DNs) in the multispectral images were converted to radiance using the equation given by Chander and Markham (2003). This was atmospherically

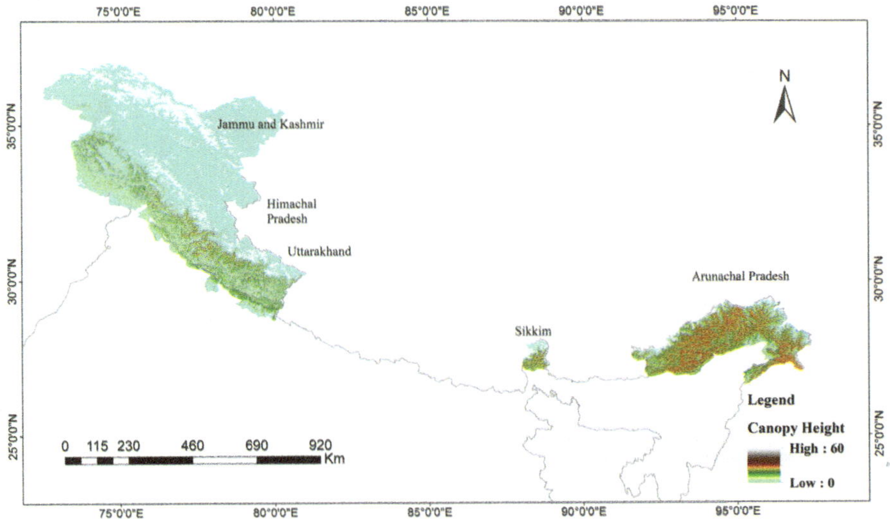

Fig. 8.2 Global forest canopy height map. Pixel values ranging from 0 to 60 represent forest canopy height in metres. (Downloaded from website https://glad.umd.edu/dataset/gedi)

Table 8.1 Satellite data used for treeline delineation

Sr. No	Himalayan state	No. of scenes	Month/year range
1	Jammu and Kashmir	31	August–December, 2012–2014
2	Himachal Pradesh	6	October–November, 2014
3	Uttarakhand	10	October–December, 2012–2014
4	Sikkim	4	November–December, 2012–2014
5	Arunachal Pradesh	13	October–December, 2012–2014

Source: Singh et al. (2021)

corrected using the Fast Line-of-Sight Atmospheric Analysis of Spectral Hypercube (FLAASH) technique in the Environment for Visualising Images (ENVI) software. The automatic contour delineation method in Aeronautical Reconnaissance Coverage Geographic Information System (ArcGIS) (ESRI 2016) was used to delineate the treeline. Figure 8.3 shows the treeline position delineation of Himalayan states under the study.

8.4 Methodology

8.4.1 Quantification of Canopy Height

The height map was downloaded in GeoTiff format. The pixel values representing canopy heights in the raster data were extracted to station points generated over the alpine treeline vector and buffer around it (100 m and 500 m) along with the information about geolocation as attributes.

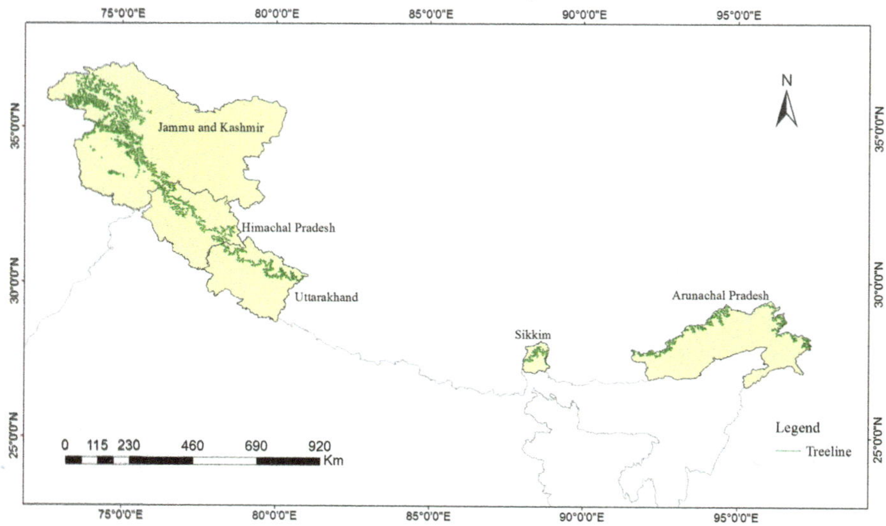

Fig. 8.3 Alpine treeline ecotone position in Indian Himalaya. (Adapted from Singh et al. 2021)

A shapefile of the buffer zone was created around the treeline and it extends between 500 m altitude below and above the treeline location. The shapefile was created in ArcGIS (ESRI 2016) in such a way that in every 100 m altitude a distinct buffer zone was formed. The average canopy height was calculated for each 100 m altitudinal group. The variation in canopy height as a function of altitude was quantified using Pearson's correlation coefficient (also known as Pearson *R*-test) represented by '*r*' and coefficient of determination (COD), i.e. square of '*r*', represented by 'R^2'. '*r*' was computed using the following formula:

$$r = \frac{n\left(\sum xy\right) - \left(\sum x\right)\left(\sum y\right)}{\sqrt{\left[n\sum x^2 - \left(\sum x\right)^2\right]\left[n\sum y^2 - \left(\sum y\right)^2\right]}}$$

where x and y are two different variables and n is the number of observations.

8.4.2 Ground Survey and Validation

Field validations were conducted in Jammu and Kashmir (74°20′ E; 34°01′ N) and Himachal Pradesh (77°57′ E; 31°14′ N). Endemic alpine tree species are abundant in these regions. The vegetation varies from closed montane forest to high-elevation alpine meadows across the elevation gradient. Anthropogenic activities have little influence on these areas. Canopy heights were collected from a sample plot of size 400 m^2 (20 m × 20 m) in Jammu and Kashmir and Himachal Pradesh in elevations ranging from 3345–3540 m and 2469–3719 m, respectively. The height of each tree was calculated with a clinometer using angles measured after walking 10 m from the base of the tree. The average canopy height of each sample plot (37 points) is treated as a training point for the validation of the satellite-derived treeline canopy height. The elevation and geolocations of the treeline species were noted along the altitudinal gradient.

The global forest canopy height map and ground measured canopy heights were compared using the statistical indices: mean absolute error (MAE), and root-mean-square error (RMSE). These indices, which are checked for goodness-of-fit, are defined as follows:

$$\mathrm{RMSE} = \sqrt{\frac{1}{N}\sum_{i=1}^{N}(\mathrm{S}-\mathrm{I})^2}$$

$$\mathrm{MAE} = \frac{1}{N}\sum_{i=1}^{N}|\mathrm{S}-\mathrm{I}|$$

where S stands for satellite observations, I for in situ observations and N is the number of points.

8.5 Results

Scatter plots of in situ observations of canopy height and satellite-derived canopy height from global forest height map is shown in Fig. 8.4. It is observed that MAE and RMSE for the global forest height map are 4.23 m and 5.23 m, respectively. Further, a significant correlation (correlation coefficient = 0.63) is also found

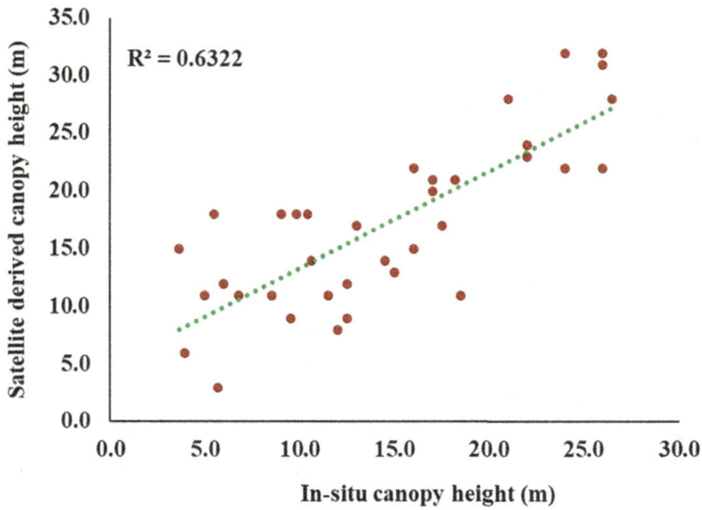

Fig. 8.4 Scatter plots of satellite-derived canopy height against in situ datasets

between ground measured and satellite-derived canopy heights ($p < 0.001$). The significant correlation and minimum MAE and RMSE provide high confidence for utilization of the global forest height map. Ground truth observations compared with satellite-derived canopy height are shown in Fig. 8.4.

8.5.1 Response of Canopy Height to Altitude

Estimation of canopy height varied between 3 m and 41 m along alpine treeline ecotone in Indian Himalaya. We observed a clear, strong negative response ($R^2 = 0.96$) of canopy height to increasing altitude and around 96% of the variability in mean canopy height is explained by altitude ($p < 0.001$). The correlation between canopy height and altitude is shown in Fig. 8.5. Canopy height decreased at a rate of 0.4 m per 100 m altitude. Along the altitudinal gradient, canopy height decreased from 12.42 ± 6.81 m to 8.84 ± 4.48 m.

8.5.2 Relationship Between Canopy Height and Treeline in Himalaya

Canopy height gets shortened at 0.3 m per 100 m starting from 500 m below the treeline to 500 m above the treeline in Jammu and Kashmir; 0.5 m in Himachal Pradesh; 0.7 m in Uttarakhand; and 0.4 m in Sikkim and Arunachal Pradesh. The results are shown in Fig. 8.6. As calculated from Cartosat-1 DEM, the mean

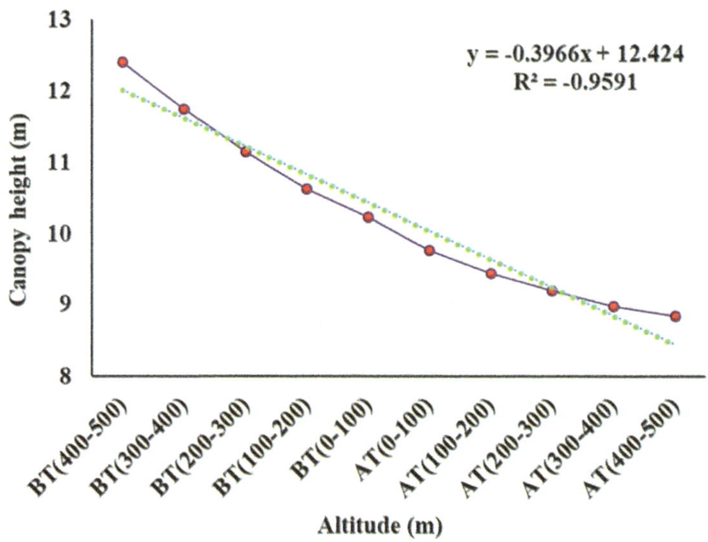

Fig. 8.5 Relationship between altitude and canopy height in the Indian Himalayan alpine treeline ecotone. The blue line represents the trend line. R^2 for the model is calculated as the coefficient of determination of the relationship between the altitude and canopy height

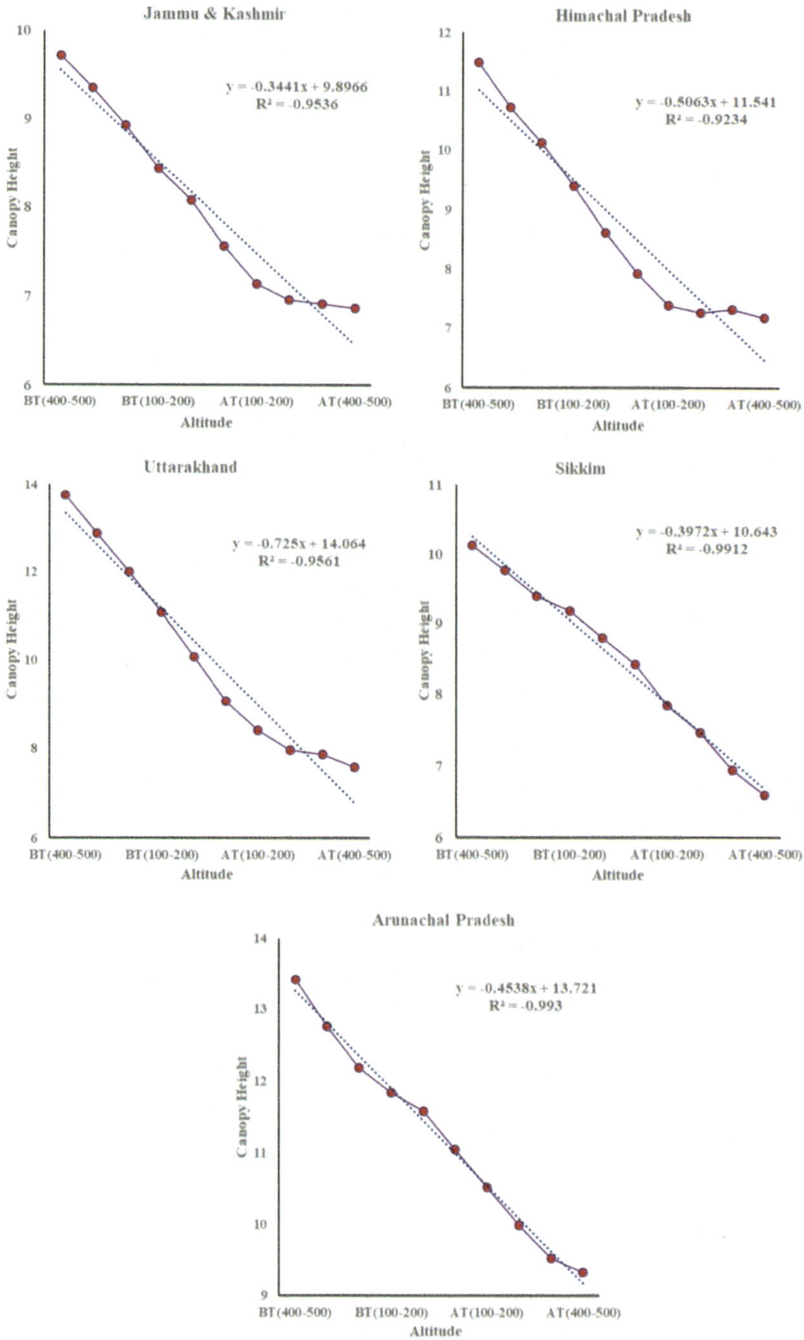

Fig. 8.6 Relationship between altitude and canopy height in Jammu and Kashmir, Himachal Pradesh, Uttarakhand, Sikkim and Arunachal Pradesh

elevation of the treeline in Jammu and Kashmir, Himachal Pradesh, Uttarakhand, Sikkim and Arunachal Pradesh Himalaya was 4121 m, 3520 m, 3615 m, 3542 m and 4136 m, respectively (Singh et al. 2021).

8.6 Discussion

8.6.1 Canopy Height Decreases with Altitude

The declining trend in canopy height with increasing altitude is in agreement with reports from other areas (Wang et al. 2012; Paulsen et al. 2000; Liang et al. 2011). The reduction in canopy height may be interpreted as direct feedback to decreasing temperature with increasing altitude (Körner 1998, 2012; Wieser and Tausz 2007; Holtmeier 2009). The variations in canopy height show how trees respond to climatic factors (temperature, wind and moisture) in a natural environment, and help in the study of processes that limit tree growth. The height of trees increases, where stresses are minimal and resources are abundant (King 1990; Cary and Pittermann 2018). The individual canopy height is known to be influenced by climatic and soil conditions, terrain, vertical vegetation structure and light competition (Zhang et al. 2016). The physiological processes essential for tissue formation, including photosynthesis, respiration, food allocation and shoot growth, are limited by the low temperature at high altitudes (Koch et al. 2004).

The reduction in canopy height points to the stress along the altitudinal transect, which is marked by a strong temperature decrement. Environmental factors, such as low temperature, caused water stress, and poor nutrient availability impacted the tree performance at high altitudes (Körner 2003). Previous studies on mountainous biomes present significant role of hydraulic limit in canopy height (Klein et al. 2015; Tao et al. 2016; Petit et al. 2011). Wang et al. (2012) present strong evidence of the dominant role of low temperature-induced water uptake, along with gravity, in limiting canopy height. In our case, the adiabatic gradient, i.e. the reduction in temperature with altitude, appears to be the major cause of the decrease in canopy height with altitude. In a modelling study, Ameztegui et al. (2021) reported that the relationship between elevation and maximum canopy height is not progressive; however, it begins at a specified location below the treeline but above the trailing limit of the species' range. Other characteristics such as slope, direction or dominant tree species showed no effect on the negative correlation.

8.6.2 Canopy Height Decrement Patterns Along Himalayan Arc

It is clearly observed that canopy height decrement with respect to altitude declines from west to east within Himalayan arc. Western Himalayan states (Himachal Pradesh and Uttarakhand) show a relatively higher rate of change in canopy height with respect to altitude than eastern Himalayan states (Sikkim and Arunachal Pradesh). It is apparent that ecological processes in mountains are influenced not by elevation alone, but by a variety of other variables too (Körner and Spehn 2019). The Himalayan arc takes a latitudinal dip in its middle part (Nepal) as it advances from northwest to southeast, and the extreme southeast end is at lower latitudes than the northwest end. A previous study conducted on the global variation of forest canopy height reported that maximum canopy height is inversely proportional to latitude (Zhang et al. 2016). The decreasing trend of canopy height decrement may be due to the decrease in latitude as we go from the northwest to the southeast. A 1° latitude increase results in a 0.55 °C drop in mean yearly temperature. As a result, temperatures at a given elevation are more likely to be warmer towards the southeast end than the northwest end.

Large amounts of rainfall are there on the windward slopes of the eastern Himalaya as a result of the south Asian monsoon, while staggered mountain ranges function as climate barriers, preventing moisture-bearing monsoonal air masses from reaching the western Himalaya, where moisture air masses are only brought in by minor monsoons from the Arabian Sea. The climate in the western Himalaya is drier with less precipitation, whereas the climate in the eastern Himalaya is dominated by a heavy moist climate with heavy precipitation (Schickhoff et al. 2015). From east to west and from low elevation to high elevation, moisture decreases (Singh et al. 2017). Studies have also reported that the temperature lapse rate declines from west to east for treeline transect, i.e. from less moist to high moist sites in Himalaya. A previous study carried out in treeline transect has reported lower temperature lapse rate value in Uttarakhand Himalaya (Joshi et al. 2018). Thus, along the northwest–southeast gradient, strongly decreasing winter cold and higher monsoonal humidity levels increase the maritime climatic conditions for trees to grow taller and result in lower canopy height reduction in eastern Himalayan states.

8.7 Conclusion

Our study is the first landscape analysis of the relationship between altitude and canopy height over the Indian Himalayan alpine ecotone. With the canopy height information, we were able to address fundamental questions about how canopy height varies with increase in altitude, and provide evidence of the existence of a significant negative response. The altitude–canopy height relationship has the prospect of becoming a fundamental tool in the study of responses of mountain trees to environmental changes. Our results reveal that the height response of treeline trees spatially varies along Indian Himalayan arc in response to varying environmental factors and this will have a bearing in future climate change scenarios as well. Finally, GEDI (Global Ecosystem Dynamics Investigation)-derived parameters such as total canopy cover, total plant area index, foliage height diversity index, foliage clumping index and volumetric scattering coefficient of the canopy open a promising future for evaluating the relationship between tree physiognomic parameters and climatic variables at the global scale.

The present study advocates the use of the canopy height–altitude model to further study the carbon sequestration patterns in the light of changing climate. The Dynamic Global Vegetation Model (DGVM) can be set up to simulate shifts in potential vegetation and its associated biogeochemical and hydrological cycles as a response to shifts in climate. DGVMs require realistic initial values and when we run it in a spatially distributed mode, we generally assume to have homogeneous (default) conditions within each cell as far as canopy height is concerned. Now, with the available canopy height–altitude model we can have dynamic heights for Indian Himalayan region as an input to the DGVM. However, more accurate and fine-resolution data will be required to achieve more consistent estimates of various ecosystem parameters.

References

Ameztegui A, Rodrigues M, Gelabert PJ, Lavaquiol B, Coll L (2021) Maximum height of mountain forests abruptly decreases above an elevation breakpoint. GIScі Remote Sens 58(3):442–454

Cary KL, Pittermann J (2018) Small trees, big problems: comparative leaf function under extreme edaphic stress. Am J Bot 105(1):50–59

Chander G, Markham B (2003) Revised Landsat-5 TM radiometric calibration procedures and postcalibration dynamic ranges. IEEE Trans Geosci Remote Sens 41(11):2674–2677

Dubayah R, Blair JB, Goetz S, Fatoyinbo L, Hansen M, Healey S, Hofton M, Hurtt G, Kellner J, Luthcke S et al (2020) The global ecosystem dynamics investigation: high-resolution laser ranging of the earth's forests and topography. Sci Remote Sens 1:100002

ESRI (2016) ArcGIS Desktop Environmental Systems Research Institute. https://desktoparcgiscom/en/arcmap

Field CB, Barros VR, Dokken DJ, Mach KJ, Mastrandrea MD, Bilir TE, Chatterjee M, Ebi KL, Estrada YO, Genova RC, Girma B, Kissel ES, Levy AN, MacCracken S, Mastrandrea PR, White LL (eds) (2014) Climate change 2014: impacts, adaptation and vulnerability. Part A: global and sectoral aspects. Contribution of working Group II to the fifth assessment report of the intergovernmental panel on climate change. Cambridge University Press, Cambridge

Grabherr G, Gottfried M, Pauli H (1994) Climate effects on mountain plants. Nature 369:448

Holtmeier FK (2009) Mountain timberlines. Ecology, patchiness, and dynamics. Springer, New York

Joshi R, Sambhav K, Singh SP (2018) Near surface temperature lapse rate for treeline environment in western Himalaya and possible impacts on ecotone vegetation. Trop Ecol 59(2):197–209

King DA (1990) Theadaptive significance of tree height. Am Nat 135(6):809–828

Klein T, Randin C, Körner C (2015) Water availability predicts forest canopy height at the global scale. Ecol Lett 18(12):1311–1320

Koch GW, Sillett SC, Jennings GM, Davis SD (2004) The limits to tree height. Nature 428(6985): 851–854

Körner C (1998) Worldwide positions of alpine treelines and their causes. In: Beniston M, Innes JL (eds) The impacts of climate variability on forests, Lecture notes in earth sciences, vol 74. Springer, New York, pp 221–229

Körner C (2003) Alpine plant life: functional plant ecology of high mountain ecosystems. Springer, Heidelberg

Körner C (2012) Treelines will be understood once the functional difference between a tree and a shrub is. Ambio 41:197–206

Körner C, Paulsen J (2004) A world-wide study of high altitude treeline temperatures. J Biogeogr 31:713–732

Körner C, Spehn E (2019) A Humboldtian view of mountains. Science 365(6458):1061

Liang EY, Liu B, Zhu LP, Yin ZY (2011) A short note on linkage of climatic records between a river valley and the upper timberline in the Sygera Mountains southeastern Tibetan Plateau. Glob Planet Chang 77:97–102

Paulsen J, Weber UM, Körner C (2000) Tree growth near treeline: Abrupt or gradual reduction with altitude? Arct Antarct Alp Res 32:14–20
Petit G, Anfodillo T, Carraro V, Grani F, Carrer M (2011) Hydraulic constraints limit height growth in trees at high altitude. New Phytol 189(1):241–252
Potapov P, Li X, Hernandez-Serna A, Tyukavina A, Hansen MC, Kommareddy A, Pickens A, Turubanova S, Tang H, Silva CE et al (2021) Mapping global forest canopy height through integration of GEDI and Landsat data. Remote Sens Environ 253:112165
Schickhoff U (2005) The upper timberline in the Himalayas, Hindu Kush and Karakorum: a review of geographical and ecological aspects. In: Broll G, Beate K (eds) Mountain ecosystems studies in treeline ecology. Springer, Berlin, pp 275–354
Schickhoff U, Bobrowski M, Boehner J et al (2015) Do Himalayan treelines respond to recent climate change? An evaluation of sensitivity indicators. Earth Syst Dynam 6(1):245–265
Shi C, Schneider L, Hu Y et al (2020) Warming-induced unprecedented high-elevation forest rowth over the monsoonal Tibetan Plateau. Environ Res Lett 15:054011
Shrestha UB, Gautam S, Bawa KS (2012) Widespread climate change in the Himalayas and associated changes in local ecosystems. PLoS One 7:e36741
Singh SP, Singh RD, Gumber S, Bhatt S (2017) Two principal precipitation regimes in Himalayas and their influence on tree distribution. Trop Ecol 58(4):679–691
Singh CP, Mohapatra J, Mathew JR et al (2021) Long-term observation and modelling on the distribution and patterns of alpine treeline ecotone in Indian Himalaya. J Geomat 15(1):68–84
Tao S, Guo Q, Li C, Wang Z, Fang J (2016) Global patterns and determinants of forest canopy height. Ecology 97(12):3265–3270
Wang Y, Čufar K, Eckstein D, Liang E (2012) Variation of maximum tree height and annual shoot growth of Smith fir at various elevations in the Sygera Mountains southeastern Tibetan Plateau. PLoS One 7(3):e31725
Wieser G, Tausz M (eds) (2007) Trees at their upper limit: treelife limitation at the alpine timberline, Plant ecophysiology series, vol 5. Springer, Dordrecht
Zhang J, Nielsen SE, Mao L, Chen S, Svenning JC (2016) Regional and historical factors supplement current climate in shaping global forest canopy height. J Ecol 104:469–478

Patterns of Plant Taxonomic, Life-Form and Phylogenetic Diversity at a Treeline Ecotone in Northwestern Himalaya: Role of Aspect and Elevation

9

Subzar Ahmad Nanda and Zafar A. Reshi

Abstract

Treeline ecotone, though studied the world over because of its sensitivity to changing climate, has received limited attention in the Himalaya. It is in this backdrop that an extensive study in the Daksum-Sinthan Top area of Kashmir Himalaya, India, was carried out to document the taxonomic, life-form and phylogenetic diversity of plant assemblages at the treeline ecotone in relation to elevation and aspect. A total of 235 species belonging to 168 genera and 71 families were recorded in the ecotone. Only 26% of species were common between the north-facing and south-facing aspects, and a decline in the total number of species with elevation was the general trend. Herbs were predominant at all the elevations on both aspects. Sørensen's dissimilarity across the elevations and aspects was low and the turnover component (β_{sim}) was the major contributor to the overall dissimilarity. Phylogenetic overdispersion was noticed at lower elevations and phylogenetic clustering was prevalent at higher elevations on both aspects. Diffuse-type treeline-form was more common on the north-facing slope and tree-island type on the south-facing slope. The dominant treeline species on the north-facing slope was *Betula utilis*, and *Pinus wallichiana* on the south-facing slope.

Keywords

α-Diversity · β-Diversity · Nestedness · Treeline form · Turnover

S. A. Nanda · Z. A. Reshi (✉)
Department of Botany, University of Kashmir, Srinagar, Jammu and Kashmir, India
e-mail: zreshi@uok.edu.in

S. P. Singh et al. (eds.), *Ecology of Himalayan Treeline Ecotone*,
https://doi.org/10.1007/978-981-19-4476-5_9

9.1 Introduction

Treeline ecotone, a vegetation zone between closed-canopy forests and alpine meadows (Holtmeier 2000; Holtmeier et al. 2003), is dynamic (Nagy and Grabherr 2009; Camarero et al. 2017), species-rich (Wielgolaski et al. 2017) and sensitive to environmental conditions (Nagy and Grabherr 2009; Camarero et al. 2017). Its dynamic nature is due to climatic factors, such as the growing season temperature (Körner 2012) and precipitation (Kharuk et al. 2010; Liang et al. 2014), and non-climatic factors, such as nutrient limitation (Li et al. 2008), biotic interactions (Hofgaard et al. 2010; Speed et al. 2011; Shrestha et al. 2015) and land-use change (Schickhoff 2005). Globally, treeline ecotones vary from abrupt lines to extended zones of increasingly small, stunted and/or dispersed trees (Bader et al. 2021). Spatial treeline patterns also differ greatly reflecting the underlying ecological processes that produce such patterns, and Holtmeier (2009) distinguished four types of treeline form, namely abrupt forest limit, transition zone (ecotone), 'true krummholz belt' and gradual transition from high-stemmed forest to crippled trees of the same species. 'True krummholz' was used for genetic krummholz, i.e. species that can grow only as shrubs, but now krummholz usually refers to environmentally induced stunted and crippled trees and genetic krummholz are referred to as shrubs (Bader et al. 2021). Subsequently, four treeline forms – diffuse, abrupt, island and krummholz—were recognized by Harsch and Bader (2011) with growth limitation being dominant only in the most common diffuse treeline, and dieback and seedling mortality in other treeline forms. Recently, Bader et al. (2021) proposed a framework of 12 forms at the hillslope scale based on the spatial pattern of tree cover as seen from above and changes in tree stature and physiognomy. The response of these treeline forms to climate change varies (Harsch and Bader 2011; Schickhoff et al. 2015) with diffuse treeline responding frequently to growing season warming and growth limitation is also higher in comparison to other treeline forms (Harsch and Bader 2011). Treeline shift in response to climate change has been a major focus of scientific investigation globally for more than a century (Bader et al. 2007), and a meta-analysis of 166 treeline sites revealed advancement in treeline at 52% of sites since 1900, recession in only 1% of sites and no change in treeline position at 47% sites (Harsch et al. 2009).

Himalaya, owing to its complexity, dynamic nature and greater sensitivity towards changing environmental conditions (Telwala et al. 2013; Aryal et al. 2014; Anup and Ghimire 2015), is likely to undergo changes, and more so in the treeline ecotone (Xu et al. 2009; Mishra and Mainali 2017) with serious consequences for invaluable goods and services that these ecosystems provide to the dependent humans. Pastoralism, wood logging, fuelwood gathering, excessive grazing and overexploitation of economically important species (Schickhoff 2002, 2005; Miehe et al. 2015) are the other factors that affect species composition, community structure (Lee et al. 2019; Ali et al. 2020), species distribution range (Pacifici et al. 2015; Kosanic et al. 2018) and more common upward shift in the species ranges in response to global climate warming (Chen et al. 2011; Lenoir and Svenning 2015). Many studies have been carried out in the Himalaya that deal with the taxonomic composition of vascular plants (Bharti et al. 2012; Pandita et al. 2019; Singh et al. 2019), phytosociology (Bürzle et al. 2017), dry matter dynamics (Rai et al. 2020), community structure and species regeneration (Shrestha et al. 2007; Gairola et al. 2008; Ghimire et al. 2010; Wong et al. 2010; Pandey et al. 2018) and community structure (Adhikari et al. 2012), but these studies do not focus specifically on treeline ecotone and, in particular, the influence of aspect and elevation on taxonomic, functional and phylogenetic diversity (PD) has not been studied.

Taking note of the above-mentioned ecological significance and lack of detailed ecological studies in the Himalaya, a comprehensive study was undertaken in a typical Kashmir Himalayan treeline ecotone to assess the complementary and interrelated components of taxonomic (species richness), functional (life-form) and phylogenetic diversity, and how these multiple measures of diversity vary in relation to elevation and aspect. While the documentation of species richness is the first essential step in biodiversity studies, functional diversity quantified based on species traits not only offers valuable insights into the functional uniqueness, redundancy and complementarity of communities but also provides clues to the resistance or resilience of communities. Likewise, phylogenetic diversity is a measure of the evolutionary relationships among the taxa, and species assemblage with greater phylogenetic diversity may provide cushion against long-term environmental change. We also examined how elevation and aspect influence α- and β-diversity patterns in the treeline ecotone. Such detailed studies are expected to expand our understanding of the structural organization of the treeline ecotone and could provide valuable insights into the response of communities to changing climate and other anthropogenic factors that are necessary for better management of ecosystems and ecotones.

9.2 Methodology

9.2.1 Study Area

This study was carried out in the Daksum-Sinthan Top area of Kashmir Himalaya, India (Fig. 9.1). Daksum lies within the geographical coordinates of 33°36′43″ N and 75°26′6″ E and is located at an altitude of 2438 m above sea level (masl). Sinthan Top lies within the geographical coordinates of 33°34′ N and 75°30′ E and a road (NH1B highway) traverses this top that connects Kashmir Valley with Kishtwar, which lies in Jammu province of the Union Territory and is at 12,500 ft. (3800 m) above sea level. Physiographically, the area is hilly with lush mixed coniferous forests and grassy meadows. The average annual precipitation is about 1000 mm, with November as the driest month (31 mm rainfall), and the average annual air temperature is about 12.8 °C, with the highest temperature of about 20 °C recorded in July (Nanda et al. 2018; Fayaz et al. 2019).

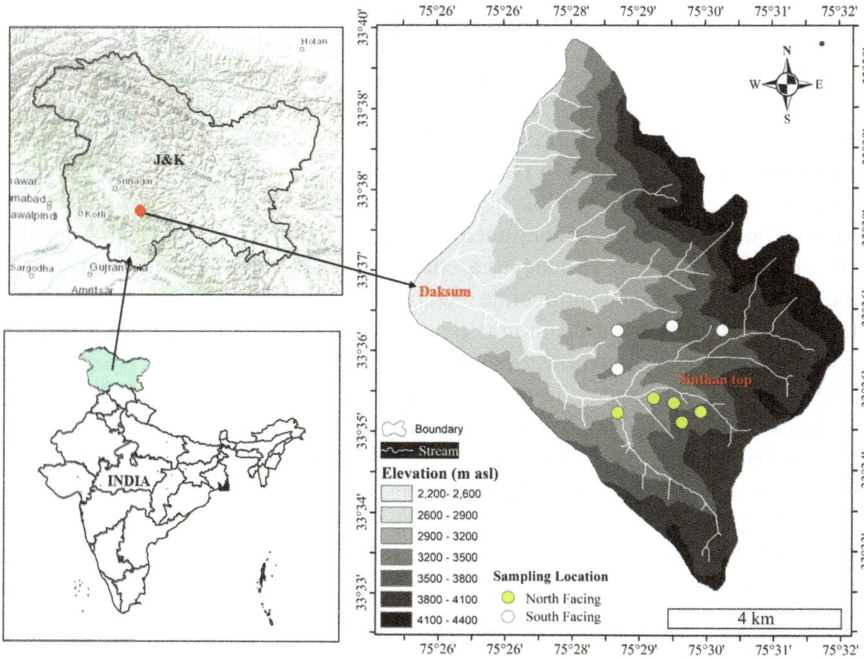

Fig. 9.1 Location map of the study area showing north-facing and south-facing slopes

9.2.2 Vegetation Sampling

Systematic sampling using the less biased quadrat method (Bellhouse 2005; Bhatta et al. 2012) was used to assess plant species composition in the treeline ecotone along the two elevation transects on the north-facing and south-facing aspects. Treeline ecotone in this study refers to the transition zone between the timberline to the treeless alpine vegetation (Fig. 9.2).

The treeline ecotone on the north-facing slope extended from 3200 to 3600 masl while on the south-facing slope it extended from 3300 to 3600 masl. The treeline ecotone on the north-facing and south-facing aspects was divided into elevation bands that were 100 m apart and hence we had five elevation bands on the north-facing slope and four bands on the south-facing slope. Three plots of 50×50 m^2 area were established in each elevational band and in each of these 3 plots, 10 quadrats of 10×10 m^2 area each were laid for sampling trees, 20 quadrats of 5×5 m^2 area each for shrub sampling and 40 quadrats of 1×1 m^2 area each for herbs were laid randomly during the growing season. Thus, in each elevational band, 30 quadrats were laid for tree sampling, 60 quadrats for the sampling of shrubs and 120 quadrats for herbs. In all, 1890 quadrats (270 for tress, 540 for shrubs and 1080 for herbs) were used for sampling vegetation in the treeline ecotone. During the sampling, if a rock of cover >10% was encountered along the transect or at places where the slope

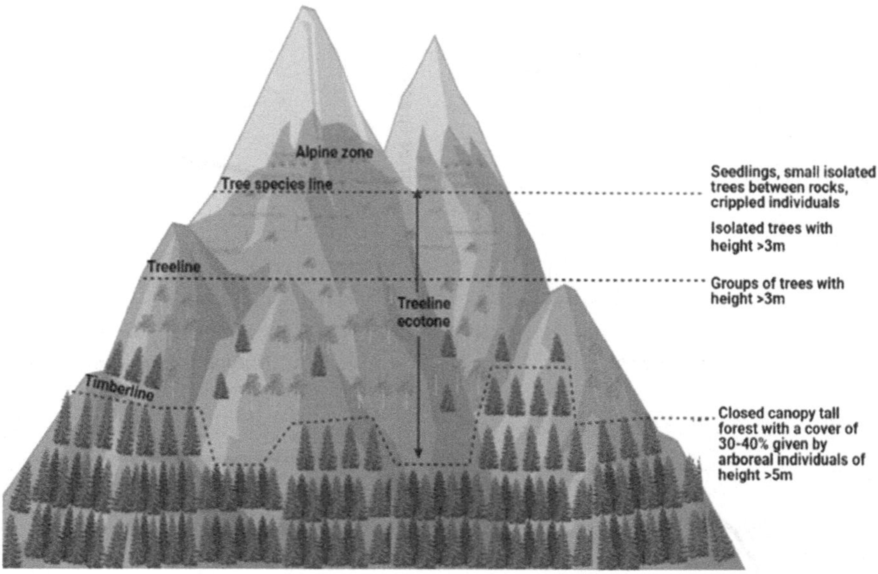

Fig. 9.2 A diagrammatic representation of treeline ecotone. (After Körner and Paulsen 2004 and Batllori et al. 2009)

was inclined over 30°, the distance between the adjacent plots was increased by 10–20 m (Mosley et al. 1989; Gómez-Díaz et al. 2017; Bhatta et al. 2018).

We constructed species accumulation curves using the *accumresult* function of the *BiodiversityR* package in R (version 3.6.2; R Core Team, 2019) with 1000 permutations and random method to evaluate sampling adequacy separately for trees, shrubs and herbs (Fig. 9.3), because of differing size and number of sampling units used for the three life-form categories.

9.2.3 Data Analysis

Vegetation data were analysed for taxonomic and life-form composition, and phytosociological characteristics such as frequency, density and abundance were computed following Mandal and Joshi (2014). While presence–absence data were recorded for all the species, density and abundance were recorded for trees and shrubs only. Non-metric multidimensional scaling (NMDS), based on presence–absence of species in the sampling units, was performed to visualize the differences in species composition between north- and south-facing slopes, and analysis of similarity (ANOSIM) based on Hellinger distance with 999 permutations was used to test whether the overall compositional difference between north- and south-facing slopes was significant (Yang et al. 2020). All analyses were performed using *vegan* package in R.

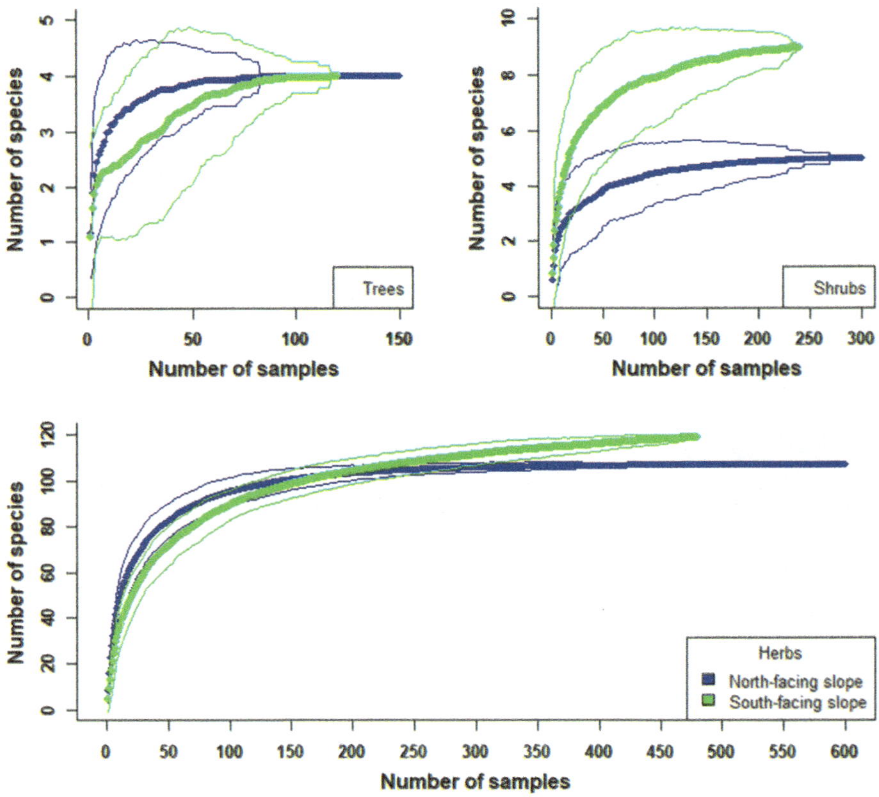

Fig. 9.3 Species accumulation curves for trees, shrubs and herbs on the north- and south-facing slopes with 95% confidence interval

Species richness was estimated from replicate samples laid across elevation bands on the two aspects. Since the total number of species observed is almost always an underestimate of the total number of species in the assemblage (Colwell and Elsensohn 2014), we used several incidence-based non-parametric estimators (Chao, Jack1, Jack2, Bootstrap) to correct this bias using *EstimateS* (9.1.0; Colwell 2013). Standard errors for all the estimators (except Jackknife2 for which it is not available as yet) were also computed. The performance of each non-parametric estimator was evaluated by quantifying bias, precision and accuracy (Walther and Moore 2005) at 25% and 50% sampling effort following Walther and Morand (1998) and Walther and Martin (2001) who suggested that the performance of species richness estimators should be evaluated at 'intermediate' sampling effort when the observed species richness has not reached the asymptote and is still increasing. Scaled performance measures, namely scaled mean error (SME) for bias, coefficient of variation (CV) for precision and scaled mean square error (SMSE) were used. Estimators that have smaller bias, higher precision and reduced SMSE were considered more accurate than others (Branco et al. 2018). To assess the changes in the richness and abundance of trees and shrubs with elevation, we computed Simpson' dominance, Shannon's diversity and Pielou's evenness (J-evenness) indices for each elevation band on the two aspects in the treeline ecotone.

Beta-diversity (β-diversity) was calculated on the basis of presence–absence data, which was then partitioned into turnover and nestedness components using function beta.pair in R package *betapart.* In all, three matrices were produced based on pairwise comparisons of all elevation bands: the Sørensen dissimilarity index (βsor) that expresses the total compositional variation with values ranging between 0 and 1, the Simpson dissimilarity index matrix (βsim) expressing compositional changes due to species turnover, and βsor minus βsim giving the resultant nestedness component (βsne) of β-diversity.

β-diversity data (βsor, βsim, βsne) were further processed to distinguish between (1) along-elevation beta-diversity accounting for pairs starting from the lowest elevation to all other elevations, and (2) stepwise beta-diversity including pairs from one elevation to the next neighbouring one (Fontana et al. 2020). The effect of elevation on both stepwise and along-gradient beta-diversity was tested with a linear mixed-effects model using the ordinary least squares regression (OLS) method in the Past software 4.02 (Hammer et al. 2001).

We constructed a phylogenetic tree for all the vascular plants using the *V. PhyloMaker* (Jin and Qian 2019) in the R platform. This package contains a megaphylogeny, which was a combination of GBOTB (i.e., GenBank taxa with a backbone provided by Open Tree of Life version 9.1) for seed plants (Smith and Brown 2018) and the clades in (Zanne et al. 2014) phylogeny. For the genera and species in our dataset that are absent from the megaphylogeny, we added them to their respective genera (in the case of species) and families (in the case of genera) using the Phylomatic and BLADJ approaches (Webb et al. 2008) implemented in *V. PhyloMaker* (scenario 3; Jin and Qian 2019). *V. PhyloMaker* sets branch lengths of added taxa of a family by placing the nodes evenly between dated nodes and tips within the family, and adds a missing genus at the mid-point of its family branch and a missing species at the mid-point of its genus branch.

Five metrics were computed to document the phylogenetic diversity. These included: phylogenetic diversity (PD) (Faith 1992), which is the sum of evolutionary history in millions of years; mean pairwise distance (MPD) (Webb 2000), which is one of the most popular measures for computing the phylogenetic distance between a given group of species; mean nearest taxon distance (MNTD) (Webb et al. 2002), which is the average distance between an individual and the most closely related (non-conspecific) individual; net relatedness index (NRI); and nearest taxon index (NTI). NRI and NTI compare the phylogenetic diversity in the dataset to a randomly generated null model, from the regional species pool, revealing either phylogenetic overdispersion or evenness (co-occurring species more distantly related than expected by chance) or phylogenetic clustering (species more closely related than expected by chance). A positive NRI/NTI indicates that co-occurring species are more phylogenetically related (phylogenetic clustering) than expected by chance, whereas the negative value indicates that species are more distantly related (phylogenetic overdispersion) than expected by chance (Webb et al. 2002). All the phylogenetic indices were obtained with the R package *picante* (Kembel et al. 2010) (Table 9.1).

9.3 Results

9.3.1 Taxonomic and Life-Form Diversity

A total of 235 species belonging to 168 genera and 71 families were recorded at the treeline ecotone. Angiosperms with 164 species (116 genera and 40 families) were predominant and dicots were more numerous with 150 species (108 genera and 35 families) than monocots, which were represented by 14 species (8 genera and 5 families). Gymnosperms were represented by 4 species belonging to 4 genera and 2 families. Pteridophytes included 24 species (13 genera and 9 families) and bryophytes were represented by 43 species belonging to 35 genera and 20 families (Table 9.2).

Dominant families at the ecotone among angiosperms were Asteraceae, which included 27 species, followed by Lamiaceae (12 spp.), Rosaceae, Poaceae and Caryophyllaceae each with 9 species, Polygonaceae (7 spp.), Ranunculaceae and Boraginaceae (6 spp.) and Crassulaceae and Gentianaceae (5 spp.). In gymnosperms, Pinaceae was the dominant family. In the case of pteridophytes, Dryopteridaceae and Athyriaceae were the dominant families, which included 7 and 5 species respectively. In bryophytes, Pottiaceae was the dominant family with 7 species followed by Polytrichaceae and Dicranaceae (6 spp.).

Among the life-form groups, herbs were predominant with 176 species belonging to 117 genera and 39 families followed by mosses, which included 42 species belonging to 34 genera and 19 species. Other life-form groups, such as trees, shrubs and liverworts, were not well represented in the ecotone (Table 9.2).

9.3.2 Effect of Aspect on Taxonomic and Life-Form Diversity

Aspect had a distinct influence on species richness as 132 species were recorded on the north-facing slope as against 164 species on the south-facing slope (Table 9.3). In particular, angiosperms and bryophytes were more on the south-facing slope than on the north-facing slope. In respect of life-forms, herbs and mosses were more on the south-facing slope compared to the north-facing slope (Table 9.3).

Non-metric multidimensional scaling (Fig. 9.4) showed that species composition on the north- and south-facing slopes were strongly separated in ordination space with no overlap (stress value = 0.028).

The analysis of similarity (ANOSIM; Fig. 9.5) showed significant compositional difference between the aspects ($R = 0.994$, $p = 0.007$).

Table 9.1 Formulae used for computation of various indices and metrics

Index	Data	Formula	Symbols and notations	Packages and software used	References
Species accumulation curve	Presence–absence	$S_{obs} = \sum_{i=1}^{H} Si$	Where S_{obs} is the number of species in the pooled samples (mean among runs), and Si is the number of species found in exactly i samples of the empirical sample set, which has a total of H samples	Vegan and BiodiversityR in R	Colwell (2006)
Chao2 (bias corrected)	Presence–absence	$\widehat{S} = S_{observed} + \left(\frac{m-1}{m}\right)\left(\frac{Q_1(Q_1-1)}{2(Q_2+1)}\right)$	Where $S_{observed}$ is the total number of species observed in all sampled pooled, m is the total number of samples, Q_1 is the frequency of uniques, Q_2 is the frequency of duplicates, and p_k is the proportion of samples that contain species k	EstimateS	Chao (1984)
First-order jacknife (Jack1)		$\widehat{S} = S_{observed} + Q_1\left(\frac{m-1}{m}\right)$			Burnham and Overton (1979)
Second-order jacknife (Jack2)		$\widehat{S} = S_{observed} + \left[\frac{Q_1(2m-3)}{m} - \frac{Q_2(m-2)^2}{m(m-1)}\right]$			
Bootstrap		$\widehat{S} = S_{observed} + \sum_{k=1}^{S_{observed}} (1-p_k)^m$			Smith and van Belle (1984)
Scaled mean error (SME)		$\frac{1}{An}\sum_{j=1}^{n}(E_j - A)$	Where A is the asymptotic or total species richness, E_j is the estimated species richness for jth sample, n is the number of samples, and $\tilde{E}$ is the mean or runs	–	Walther and Moore (2005)
Coefficient of variation		CV = 100 SD/$\tilde{E}$			
Scaled mean square error (SMSE)		$\frac{1}{A^2n}\sum_{j=1}^{n}(E_j - A)^2$			
Simpson diversity index (D)	Abundance	$D = 1 - \sum_{i=1}^{s} pi^2$	Where p is the proportion (n/N) of individuals of one particular species found (n) divided by the total number of individuals found (N), Σ is still the sum of the calculations, and s is the number of species	Vegan in R	Simpson (1949)

(continued)

Table 9.1 (continued)

Index	Data	Formula	Symbols and notations	Packages and software used	References
Shannon index (H')	Abundance	$H' = -\sum \frac{ni}{n} \ln \frac{ni}{n}$	Where ni = abundance of nth species and n = abundance of all species		Shannon and Weaver (1998)
J-evenness index (E)	Abundance	$E = \frac{H'}{\ln(S)}$	H′ = Shannon index, S = total number of species		Pielou (1975)
Sorensen dissimilarity index (βsor)	Presence–absence	$\beta_{\mathrm{sor}} = \frac{b+c}{2a+b+c}$	Where a = number of shared species between two sites, b = number of species unique to the poorest site, and c = number of species unique to the richest site	Betapart, R	Baselga (2010); Baselga and Orme (2012)
Simpson dissimilarity index (βsim)	Presence–absence	$\beta_{\mathrm{sim}} = \frac{\min(b,\ c)}{a+\min(b,\ c)}$			Baselga (2010); Baselga and Orme (2012)
Nestedness-resultant dissimilarity index (βsne)	Presence–absence	$\beta_{\mathrm{sne}} = \frac{\max(b,\ c)-\min(b,\ c)}{2a+\min(b,\ c)+\max(b,\ c)} \times \frac{a}{a+\min(b.c)}$			Baselga (2010); Baselgaand Orme (2012)
Phylogenetic diversity index (Faith)	Plant community	Sum of all branch lengths connecting the species in the community	–	Picante in R	Faith (1992)
Mean pairwise distance (MPD)	Plant community	The average evolutionary distance between all pairwise species	–		Webb (2000)
Mean nearest taxon distance (MNTD)	Plant community	Average branch lengths connecting each species to its nearest relative	–		Webb et al. (2002)
Net relatedness index (NRI)	Plant community	$\mathrm{NRI} = -1 \times \frac{\mathrm{MPD}_{\mathrm{sample}} - \mathrm{MPD}_{\mathrm{randomsample}}}{\mathrm{SD}(\mathrm{MPD}_{\mathrm{randomsample}})}$	MPD = mean pairwise distance, SD = standard deviation		
Nearest taxon index (NTI)	Plant community	$\mathrm{NTI} = -1 \times \frac{\mathrm{MNTD}_{\mathrm{sample}} - \mathrm{MNTD}_{\mathrm{randomsample}}}{\mathrm{SD}(\mathrm{MNTD}_{\mathrm{randomsample}})}$	MNTD = mean nearest taxon distance, SD = standard deviation		

Table 9.2 Number of species belonging to various taxonomic and life-form groups growing at the treeline ecotone

Plant group	Treeline ecotone		
	Species	Genera	Families
Taxonomic group			
Angiosperms	164	116	40
Dicots	150	108	35
Monocots	14	8	5
Gymnosperms	4	4	2
Pteridophytes	24	13	9
Bryophytes	43	35	20
Total	235	168	71
Life-form group			
Herbs	176	117	39
Shrubs	10	10	8
Trees	6	6	4
Liverworts	1	1	1
Mosses	42	34	19
Total	235	168	71

Table 9.3 Number of species belonging to various taxonomic and life-form groups growing on the two aspects at the treeline ecotone

Groups	North-facing slope			South-facing slope		
	Species	Genera	Families	Species	Genera	Families
Taxonomic group						
Angiosperms	96	80	34	115	87	36
Dicots	88	74	30	105	82	32
Monocots	8	6	4	10	5	4
Gymnosperms	2	2	2	4	4	2
Pteridophytes	20	12	8	13	8	6
Bryophytes	14	12	8	32	28	16
Total	132	106	52	164	127	60
Life-form group						
Herbs	109	85	35	119	86	35
Shrubs	5	5	5	9	9	7
Trees	4	4	4	4	4	2
Liverworts	1	1	1	0	0	0
Mosses	13	11	7	32	28	16
Total	132	106	52	164	127	60

9.3.3 Effect of Elevation on Taxonomic and Life-Form Diversity

Taxonomic composition and life-form diversity in different elevation bands on the two aspects are presented in Figs. 9.6 and 9.7. The overall trend was that the species number of all the taxonomic groups declined with elevation on both aspects (Fig. 9.6). Angiosperms were the predominant elements at all the elevations on the

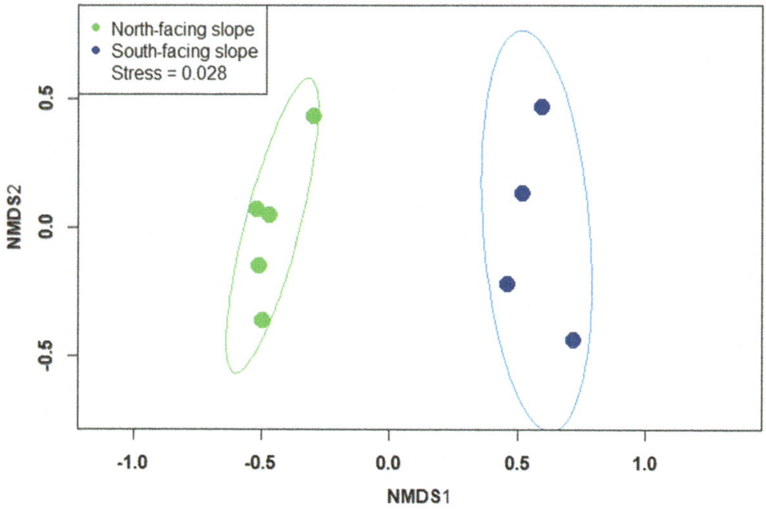

Fig. 9.4 Non-metric multidimensional scaling (NMDS) of north-facing and south-facing slopes based on Bray–Curtis dissimilarity

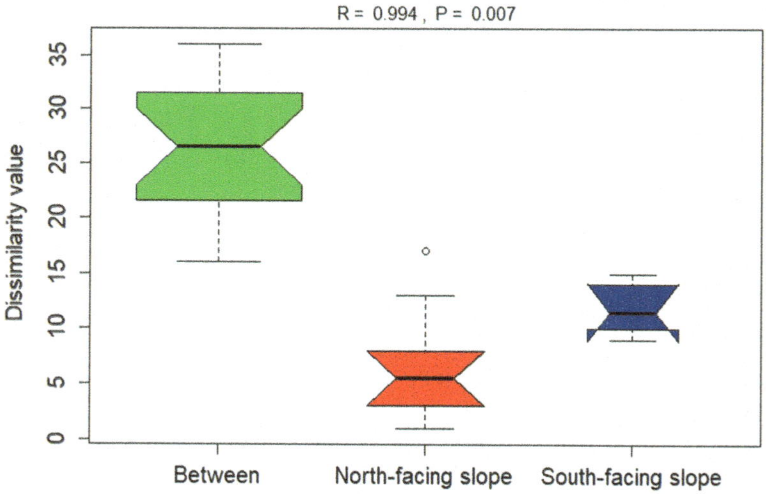

Fig. 9.5 Analysis of similarity (ANOSIM) plot showing dissimilarity between and within the two aspects. The bold horizontal bar in the box indicates median; bottom of the box indicates 25th percentile; top of the box indicates 75th percentile; and whiskers extend to the most extreme data point. Data points falling outside the range (o) are plotted as outliers

two aspects and their number ranged from 58 at the lowest elevation to 48 at the highest elevation in the treeline ecotone on the north-facing slope. Likewise, the number of angiosperms on the south-facing slope declined from 65 species at the lowest elevation to 61 species at the highest elevation (Fig. 9.6). Another notable

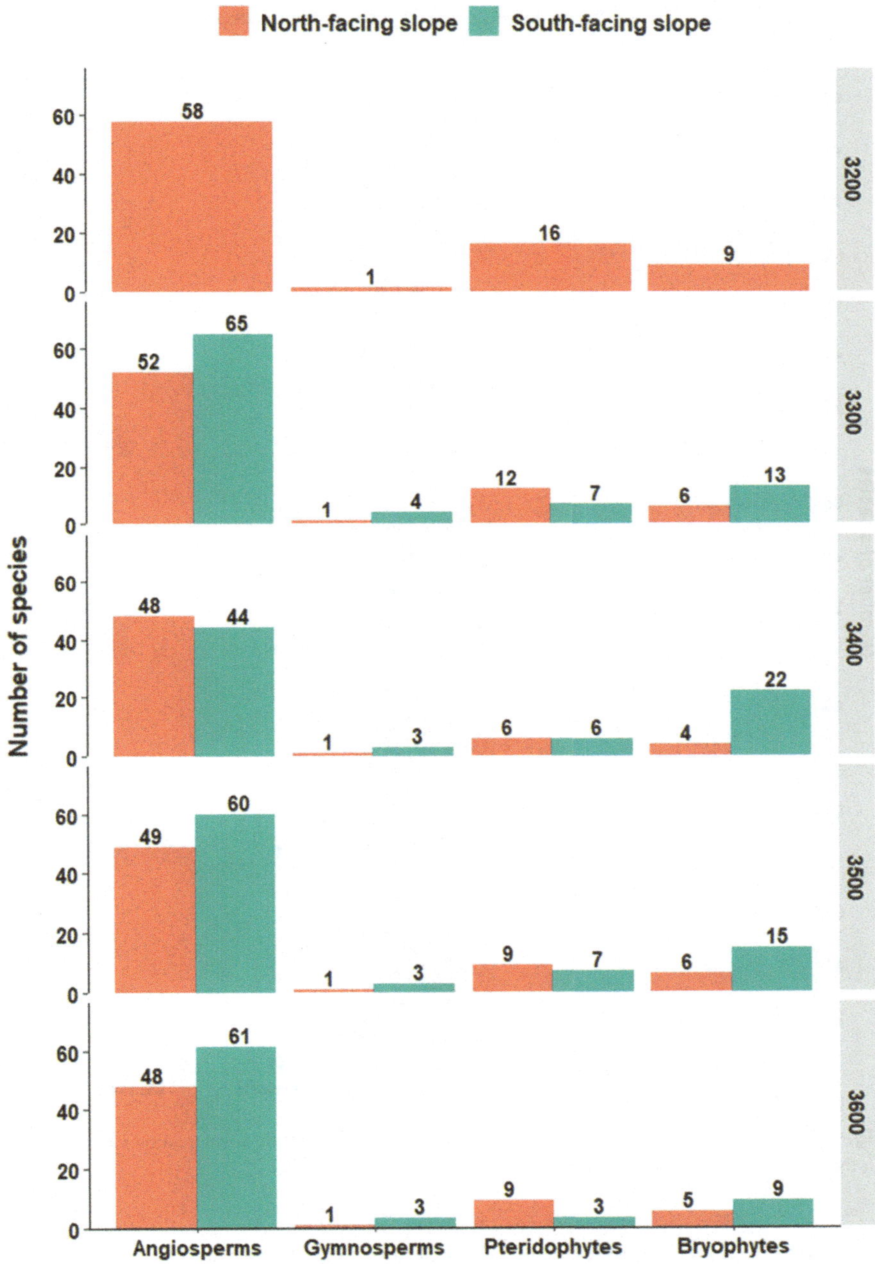

Fig. 9.6 Changes in the number of species belonging to different taxonomic groups with elevation on the south-facing and north-facing slopes

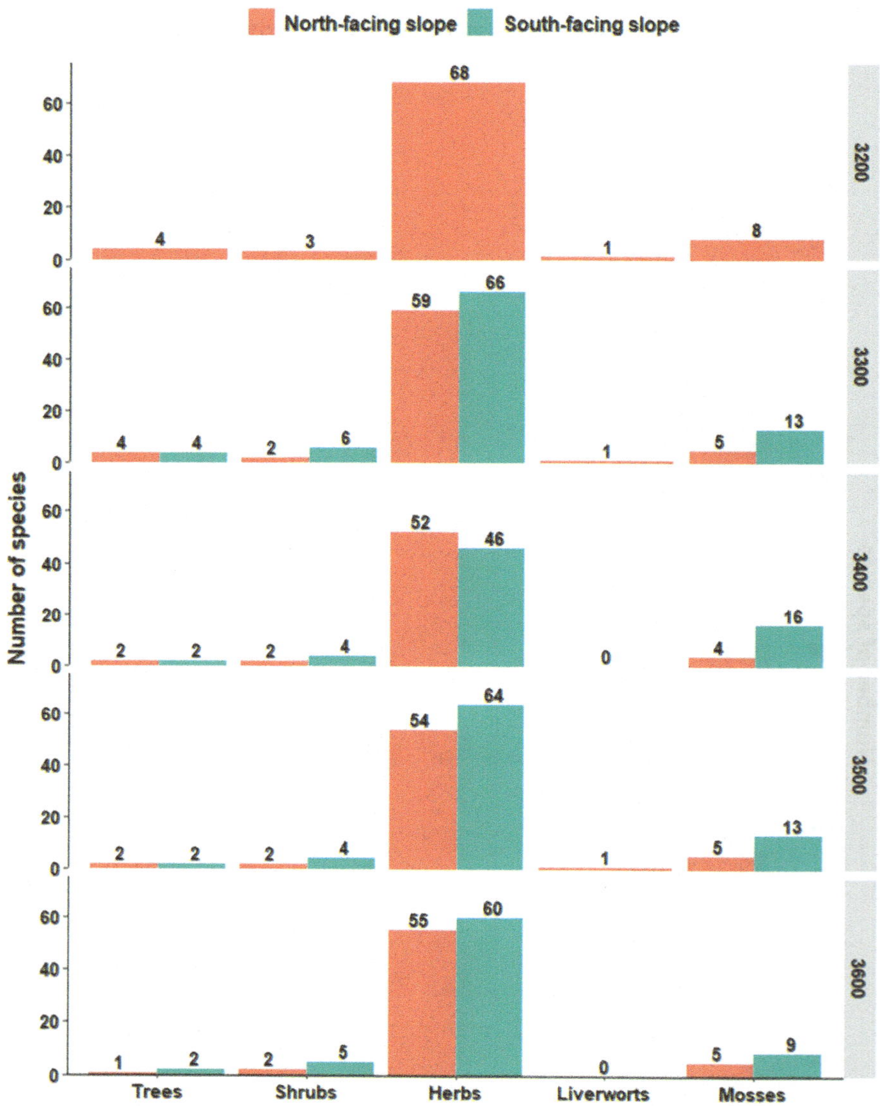

Fig. 9.7 Changes in the number of species belonging to different life-form groups with elevation on the south-facing and north-facing slopes

feature was that pteridophytes were more common on the north-facing slope than the south-facing slope. In contrast, bryophytes and gymnosperms were more represented on the south-facing slope than the north-facing slope.

In life-form groups, a declining trend in the number of species with elevation was also noticed on both aspects (Fig. 9.7). Herbs were predominant floristic elements on both the aspects and at all the elevations. The number of herbs declined from

Table 9.4 Number of species belonging to various taxonomic and life-form groups shared or exclusive to the two aspects (north- and south-facing slopes) at the treeline ecotone

	Number of shared or exclusive species	
Group	Occurring on both aspects	Occurring on only one aspect
Taxonomic groups		
Angiosperms	47	117
Dicots	43	107
Monocots	4	10
Gymnosperms	2	2
Pteridophytes	9	15
Bryophytes	3	40
Total	61	174
Life-form groups		
Herbs	52	124
Shrubs	4	6
Trees	2	4
Liverworts	0	1
Mosses	3	39
Total	61	174

68 species at the lowest treeline ecotone elevation to 55 species almost at the upper limit of treeline ecotone on the north-facing slope. However, the number of herb species on the south-facing slope declined from 66 to 46 in the adjacent elevation bands (Fig. 9.7).

Only 61 (26%) species occurred on both aspects and 174 (74%) species occurred in only one of the two aspects (Table 9.4). Angiosperms, particularly dicots, among the taxonomic groups and herbs and mosses among the life-form groups included most of the exclusive species.

9.3.4 Species Richness

The observed number of species belonging to various life-form groups based on sampling in different elevation bands and the estimated number of species using incidence-based non-parametric estimators are presented in Table 9.5. A perusal of the data reveals that estimated values of species richness based on all the estimators were more or less similar to the observed species richness on the north-facing slope. On the contrary, the richness estimators predicted higher values for herbs in comparison to observed richness across all elevations (Table 9.5).

Species richness estimated using various non-parametric indices across the entire ecotone on both the slopes is presented in Figs. 9.8 and 9.9. On the north-facing slope, species accumulation curves reached saturation with our sampling effort (Fig. 9.8). Bootstrap richness estimators approached the asymptote earlier than other estimators in respect of trees and shrubs while for herbs all the estimators

Table 9.5 Observed and estimated number of species across elevations and aspects

Elevation (m)	Life-form	Actual number of species	Estimated number of species (±standard error [SE])			
			Chao	Jack1	Jack2	Bootstrap
North-facing slope						
3200	Trees	4	4.97 ± 2.10	5.93 ± 1.37	7.80	4.72 ± 0.67
	Shrubs	3	3.00 ± 0.00	3.00 ± 0.00	3.00	3.03 ± 0.18
	Herbs	68	68.00 ± 0.00	68.00 ± 0.00	68.00	68.23 ± 0.48
3300	Trees	4	4.00 ± 0.00	4.00 ± 0.00	3.10	4.14 ± 0.36
	Shrubs	2	2.00 ± 0.00	2.00 ± 0.00	2.00	2.02 ± 0.13
	Herbs	59	61.98 ± 4.59	61.98 ± 1.72	64.93	60.63 ± 1.14
3400	Trees	2	2.00 ± 0.00	2.00 ± 0.00	2.00	2.00 ± 0.00
	Shrubs	2	2.00 ± 0.00	2.00 ± 0.00	2.00	2.00 ± 0.00
	Herbs	52	52.00 ± 0.00	52.00 ± 0.00	50.05	52.35 ± 0.61
3500	Trees	2	2.00 ± 0.00	2.00 ± 0.00	2.00	2.00 ± 0.00
	Shrubs	2	2.00 ± 0.00	2.00 ± 0.00	2.00	2.00 ± 0.01
	Herbs	54	54.50 ± 1.31	54.99 ± 0.99	55.00	54.56 ± 0.65
3600	Trees	1	2.00 ± 0.34	2.97 ± 0.97	3.90	2.36 ± 0.48
	Shrubs	2	2.00 ± 0.00	2.00 ± 0.00	1.05	2.13 ± 0.34
	Herbs	55	55.00 ± 0.00	55.00 ± 0.00	54.02	55.29 ± 0.53
South-facing slope						
3300	Trees	4	4.00 ± 0.00	4.00 ± 0.00	3.10	4.14 ± 0.36
	Shrubs	6	6.00 ± 0.00	6.00 ± 0.00	5.05	6.15 ± 0.37
	Herbs	66	76.47 ± 7.61	78.89 ± 3.58	83.87	72.17 ± 2.11
3400	Trees	2	2.00 ± 0.00	2.00 ± 0.00	2.00	2.00 ± 0.00
	Shrubs	4	4.00 ± 0.00	4.00 ± 0.00	4.00	4.05 ± 0.21
	Herbs	46	51.29 ± 4.89	53.93 ± 3.14	55.95	50.02 ± 1.94

(continued)

Table 9.5 (continued)

Elevation (m)	Life-form	Actual number of species	Estimated number of species (±standard error [SE]) Chao	Jack1	Jack2	Bootstrap
3500	Trees	2	2.00 ± 0.00	2.00 ± 0.00	2.00	2.04 ± 0.20
	Shrubs	4	4.00 ± 0.00	4.00 ± 0.00	3.05	4.13 ± 0.34
	Herbs	64	73.92 ± 8.30	73.92 ± 3.14	78.87	68.61 ± 1.79
3600	Trees	2	2.00 ± 0.00	2.00 ± 0.00	0.20	2.25 ± 0.46
	Shrubs	5	5.98 ± 2.17	6.97 ± 1.39	8.90	5.75 ± 0.73
	Herbs	60	63.47 ± 3.47	66.94 ± 2.98	67.00	63.84 ± 1.86

reached asymptote almost at the same level of sampling effort. Jack1 and Jack2 for trees and shrubs had an erratic behaviour at low sampling effort, but all the estimators tended to stabilize with increasing sampling effort. On the south-facing slope also, almost all the indices reached asymptote with our sampling effort for trees, shrubs and herbs (Fig. 9.9).

The species richness estimators had a different performance computed on the basis of bias, precision and accuracy (Tables 9.6 and 9.7). Invariably, Jack2 performed better across the groups and slopes and even at a low sampling effort.

9.3.5 Species Diversity

Species diversity estimated for each elevation band on the two slopes (Fig. 9.10) revealed that Simpson's index ranged from 0.72 in the lowest elevation band to 0.26 in the highest elevation band of the north-facing slope. Simpson's index on the south-facing slope ranged from 0.69 to 0.46 indicating a relatively modest decline with elevation and thus comparatively more uniform pattern across elevations. Shannon's index ranged from 1.44 to 0.44 on the north-facing slope and 1.29 to 0.88 on the south-facing slope. On both the slopes, the Shannon diversity index was lowest in the uppermost elevation band of the ecotones, but higher values were recorded in different elevation bands on the two slopes. J-evenness values also showed a pattern similar to other indices (Fig. 9.10) with values ranging from 0.93 to 0.40 on the north-facing slope and 0.93 to 0.55 on the south-facing slope. Thus, the plant assemblages had even distribution of individuals among the species at lower elevations than in the higher elevations.

Analysis of variance (ANOVA; Table 9.8) revealed that elevation and aspect had a significant effect on the three diversity indices but the significant interactive effects of elevation and aspect were noticed only for Simpson's and Shannon's indices.

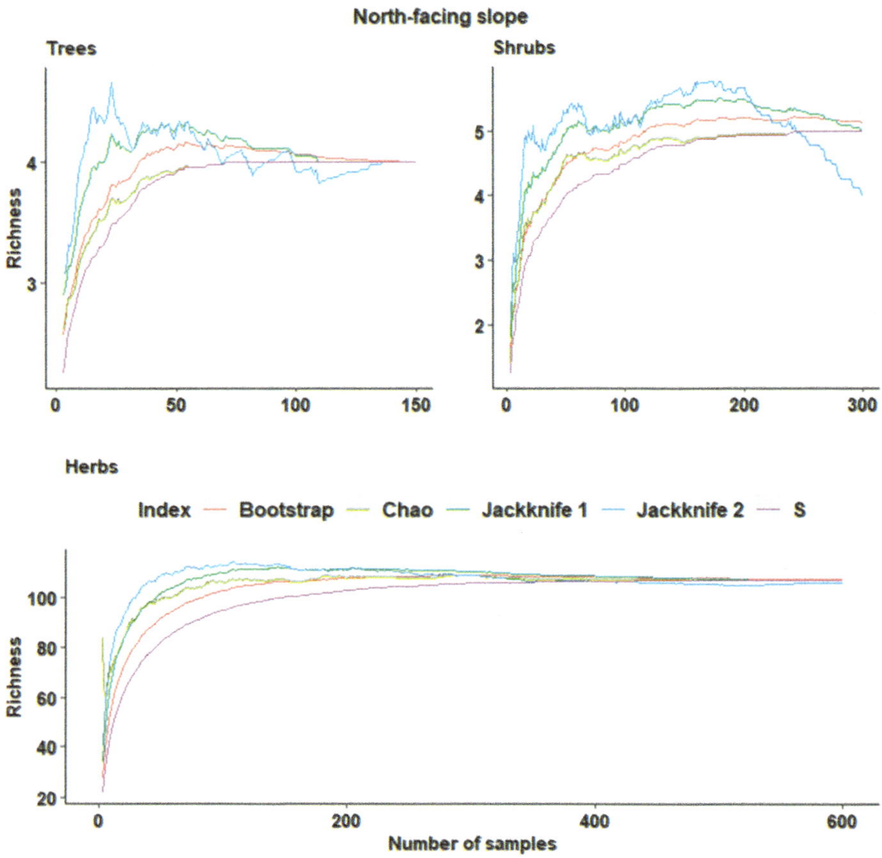

Fig. 9.8 Performance of species richness estimators as a function of sampling effort on the north-facing slope

9.3.6 β-Diversity

Data presented in Fig. 9.11a and b reveal that Sørensen's dissimilarity on both the slopes is low and the turnover component (β_{sim}) was the major contributor to the overall dissimilarity. Cluster analysis based on turnover (β_{sim}) revealed that the highest elevation band (3600 masl) in the treeline ecotone on the north-facing slope (Fig. 9.11c) was most dissimilar while as on the south-facing slope, the first two elevation bands (3300 and 3400 masl) represented one dissimilar pair and elevation bands at 3500 and 3600 masl represented the other dissimilar pair. Cluster analysis performed on the basis of nestedness (β_{sne}) revealed that the basal elevation band (3200 masl) was most dissimilar to the rest of the elevation bands on the north-facing slope (Fig. 9.11e) but on the south-facing slope two dissimilarity clusters were obtained (Fig. 9.11f). When turnover (β_{sim}) and nestedness (β_{sne}) components of β-diversity were computed for pairs starting from the lowest elevation to all other

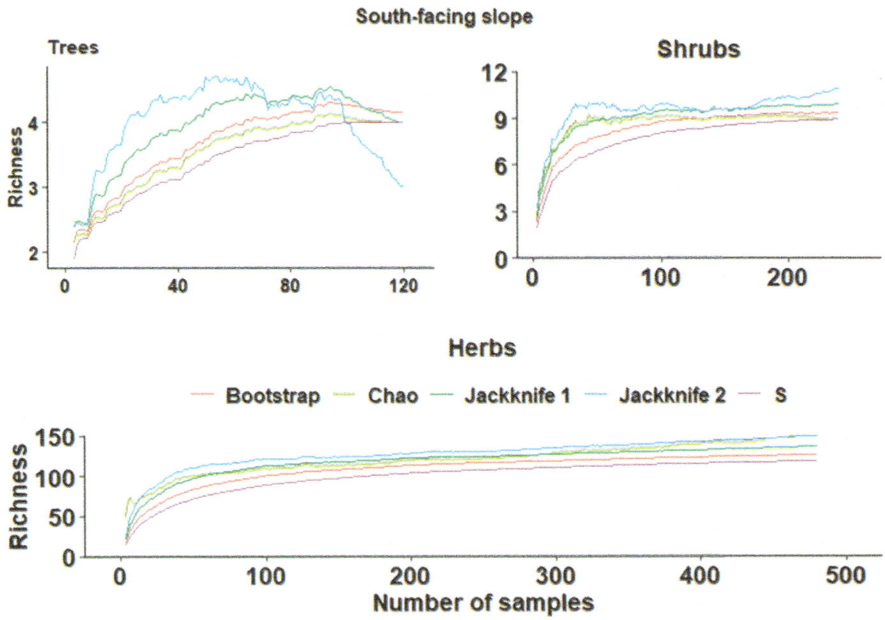

Fig. 9.9 Performance of species richness estimators as a function of sampling effort on the south-facing slope

elevations, and for pairs from one elevation to the next neighbouring elevation (Fig. 9.12), it again revealed that turnover contributed to the dissimilarity between the elevation bands.

Total β-diversity and turnover increased significantly with elevation (*p*-value = 0.05 or <0.05), but nestedness decreased insignificantly along the north-facing slope (Table 9.9, Fig. 9.13). Likewise, stepwise total β-diversity and turnover also increased at higher elevations but the increase was not statistically significant. Stepwise nestedness decreased significantly at higher elevation steps (*p*-value = 0.05) along the north-facing slope. On the south-facing slope, the total β-diversity did not show a significant increase but its turnover component showed a significant increase with elevation ($p < 0.05$). The stepwise total β-diversity and stepwise turnover also increased while stepwise nestedness decrease was not significant at higher elevations on the south-facing slope (Table 9.9, Fig. 9.14).

9.3.7 Phylogenetic Diversity

Phylogenetic diversity computed in terms of PD, MPD and MNTD, and NRI and NTI (Fig. 9.15) varied both in relation to elevation and aspect. Faith's phylogenetic diversity (FD) decreased with elevation on the north-facing slope but did not show such a declining trend with elevation on the south-facing slope. Such a pattern was

Table 9.6 Performance of non-parametric species richness estimators on the north-facing slope

	Non-parametric richness estimator (north-facing slope)							
	25% sample size				50% sample size			
Performance measure	Chao2	Jack1	Jack2	Bootstrap	Chao2	Jack1	Jack2	Bootstrap
Trees								
Bias SME	−0.057	−0.004	−0.024	−0.028	−0.029	−0.0011	−0.0133	−0.012
Precision CV	9.608	8.603	17.600	9.290	14.784	5.896	12.212	6.509
Accuracy SMSE	0.011	0.006	0.028	0.008	0.0053	0.0032	0.0141	0.0040
Shrubs								
Bias SME	−0.220	−0.118	−0.055	−0.186	−0.142	−0.036	0.023	−0.103
Precision CV	11.636	11.455	11.571	12.088	8.032	8.165	8.165	8.183
Accuracy SMSE	3.636	1.040	0.224	2.599	3.017	0.192	0.081	1.601
Herbs								
Bias SME	−0.109	−0.071	−0.026	−0.149	−0.055	−0.016	0.002	−0.070
Precision CV	14.480	17.32	15.29	19.930	11.261	12.85	10.89	15.469
Accuracy SMSE	0.029	0.031	0.023	0.051	0.014	0.016	0.012	0.026

CV: coefficient of variation; *SME*: scaled mean error; *SMSE*: scaled mean square error.

Table 9.7 Performance of non-parametric species richness estimators on the north-facing slope

	Non-parametric richness estimator (South-facing slope)							
	25% sample size				50% sample size			
Performance measure	Chao2	Jack1	Jack2	Bootstrap	Chao2	Jack1	Jack2	Bootstrap
Trees								
Bias SME	−0.313	−0.203	−0.131	−0.279	−0.204	−0.068	−0.004	−0.156
Precision CV	0.066	0.115	0.053	0.000	17.018	18.443	20.768	18.078
Accuracy SMSE	2.933	1.234	0.512	2.344	0.059	0.034	0.043	0.047
Shrubs								
Bias SME	−0.258	−0.179	−0.093	−0.270 s	−0.153	−0.063	0.0095	−0.153
Precision CV	0.019	0.017	0.016	0.000	17.833	17.381	16.830	18.765
Accuracy SMSE	3.993	1.917	0.521	4.388	0.046	0.031	0.029	0.049
Herbs								
Bias SME	−0.216	−0.217	−0.122	−0.323	−0.125	−0.101	−0.029	−0.195
Precision CV	21.478	26.022	23.967	28.195	17.162	20.597	18.068	23.237
Accuracy SMSE	0.075	0.089	0.060	0.141	0.038	0.045	0.032	0.073

CV: coefficient of variation; *SME*: scaled mean error; *SMSE*: scaled mean square error.

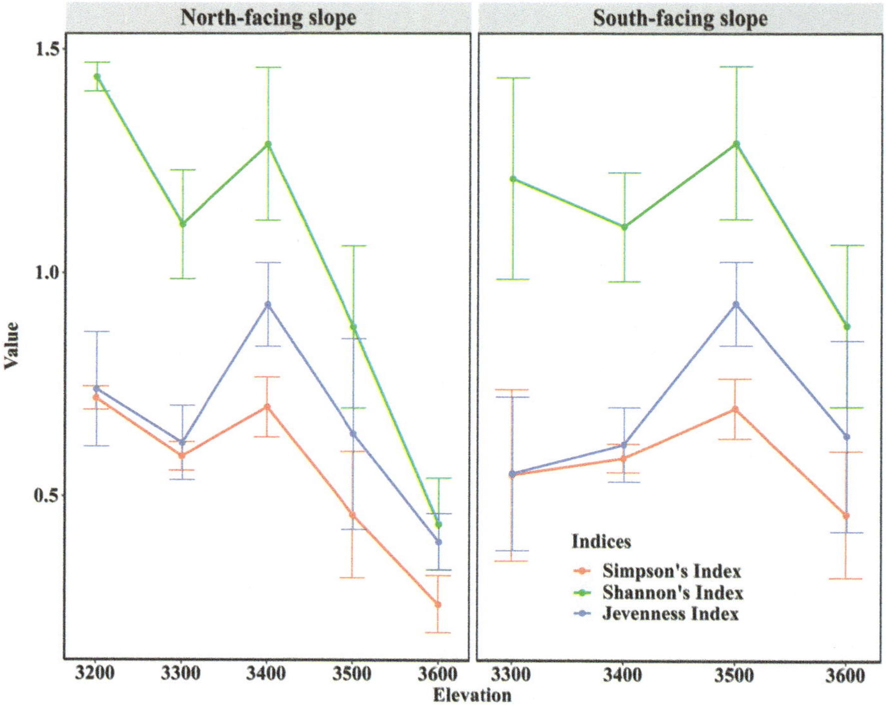

Fig. 9.10 Species diversity in relation to elevation and aspect (error bars represent standard deviation [SD])

also recorded for mean pairwise distance (MPD). The mean nearest taxon distance (MNTD) index increased initially with elevation on both the north-facing and south-facing slopes but thereafter declined (Fig. 9.15).

Further analysis of the phylogenetic pattern using the net relatedness index (NRI) revealed an increase from a negative value of −0.57 at 3200 masl to the highest value of 2.20 at 3400 masl followed by comparatively lower values at higher elevations on the north-facing slope. NRI on the south-facing slope showed an initial decline at 3400 masl to −0.64 but it subsequently increased with elevation reaching a maximum value of 2.19 at the highest elevation of treeline ecotone. Nearest taxon index (NTI), on the north-facing slope, showed a pattern that was almost reverse of NRI (Fig. 9.15) but it almost followed the pattern similar to NRI on the south-facing slope.

Based on the above-mentioned indices, phylogenetic diversity was compared between the two aspects (Fig. 9.16). Analysis of variance revealed that MNTD ($F = 43$, $p = 0.0003$) and NTI ($F = 7.26$, $p = 0.031$) were significantly different between the two aspects and the rest of the indices did vary significantly between the aspects. The phylogenetic distribution of vascular plants on the two slopes is shown in Fig. 9.17a and b.

Table 9.8 Two-way analysis of variance (ANOVA) results for various diversity indices, elevation and aspect

	DF	Sum of square	Mean square	*F*-value	*Pr* (>*F*)
Simpson's index					
Elevation	4	0.26	0.0649	4.73	0.00873
Aspect	1	0.315	0.3151	22.96	0.00015
Elevation × aspect	3	0.264	0.088	6.41	0.00382
Residuals	18	0.247	0.0137		
Shannon's index					
Elevation	4	0.733	0.183	4.09	0.01569
Aspect	1	0.832	0.832	18.56	0.00042
Elevation × aspect	3	0.601	0.2	4.47	0.01632
Residuals	18	0.807	0.045		
J-evenness index					
Elevation	4	0.373	0.093	3.120	0.04100
Aspect	1	1.021	1.021	34.120	0.00002
Elevation × aspect	3	0.257	0.086	2.870	0.06500
Residuals	18	0.539	0.030		

9.3.8 Treeline-Form

Two types of treeline forms were observed in the study area: diffuse type and tree-island type. The diffuse type was more common on the north-facing slope and tree-island type was more common on the south-facing slope (Fig. 9.18). The dominant treeline species on the north-facing slope was *Betula utilis* and on the south-facing slope, the treeline species was *Pinus wallichiana*. These treeline species were associated with thickets of shrubs, such as *Rhododendron campanulatum* and *Juniperus squamata*.

9.4 Discussion

A total of 235 species belonging to 168 genera and 71 families of various taxonomic and life-form groups were recorded at the treeline ecotone during this study, which is higher in comparison to many previous studies. For example, Junyan et al. (2014) reported 218 species, 164 genera and 67 families in the tropical coniferous broadleaved forest ecotone in China; Shrestha and Vetaas (2009) reported 84 species of vascular plants from 37 families and 55 genera in an ecotone in arid Trans-Himalayan Landscape of Nepal; Bürzle et al. (2017) reported 103 species of vascular plant species in treeline ecotone in Rolwaling Himal, Nepal; and Gaire et al. (2010) recorded just 30 plant species in the treeline ecotone of Langtang National Park, Central Nepal. Rai et al. (2012) reported 58–75 species in different communities in the Kedarnath Wildlife Sanctuary. Pandita and Dutt (2017) observed that herbaceous

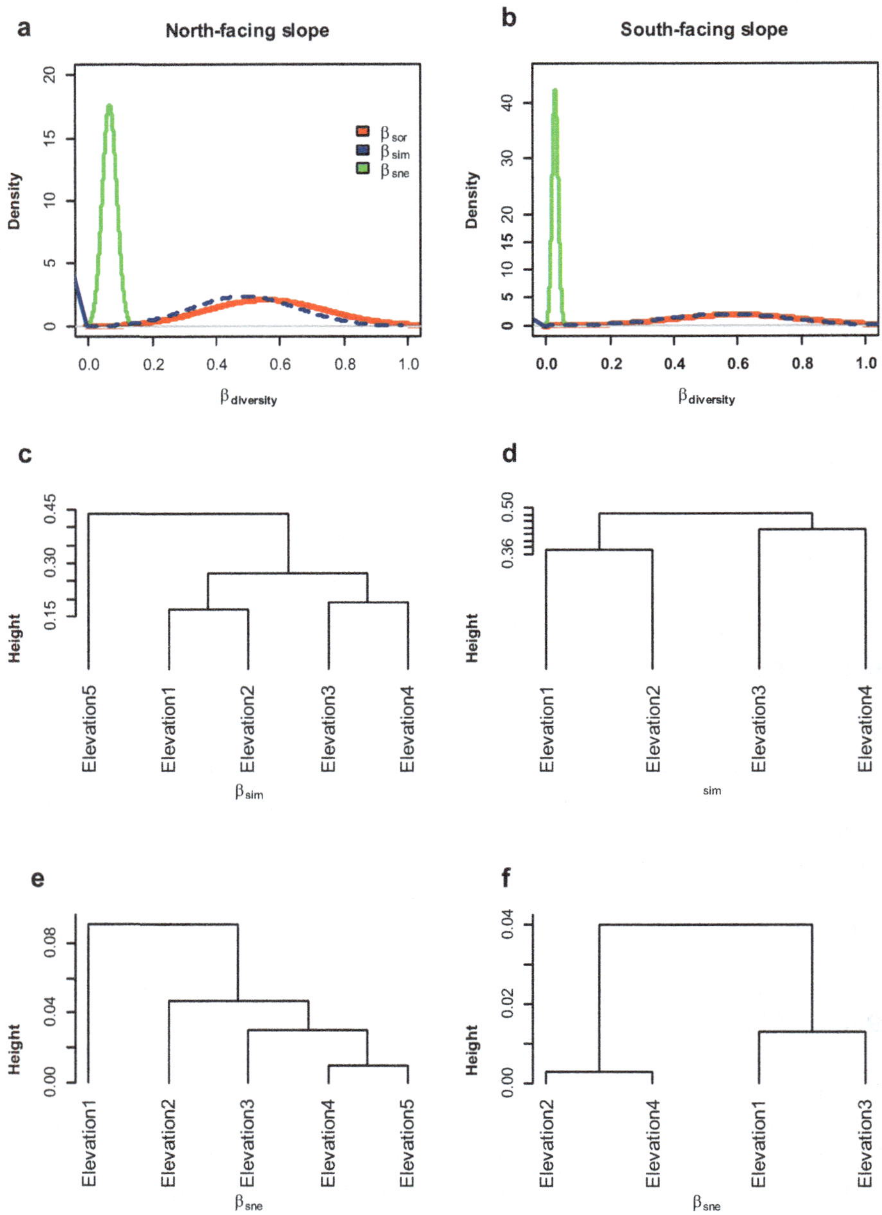

Fig. 9.11 Compositional dissimilarities across elevations and aspect. (**a** and **b**) β-diversity, turnover and nestedness on the north-facing slope (**a**) and south-facing slope (**b**). Pairwise average clustering of elevation bands based on β_{sim} dissimilarity on north-facing slope (**c**) and south-facing slope (**d**). Pairwise average clustering of elevation bands based on β_{sne} dissimilarity on north-facing slope (**e**) and south-facing slope (**f**)

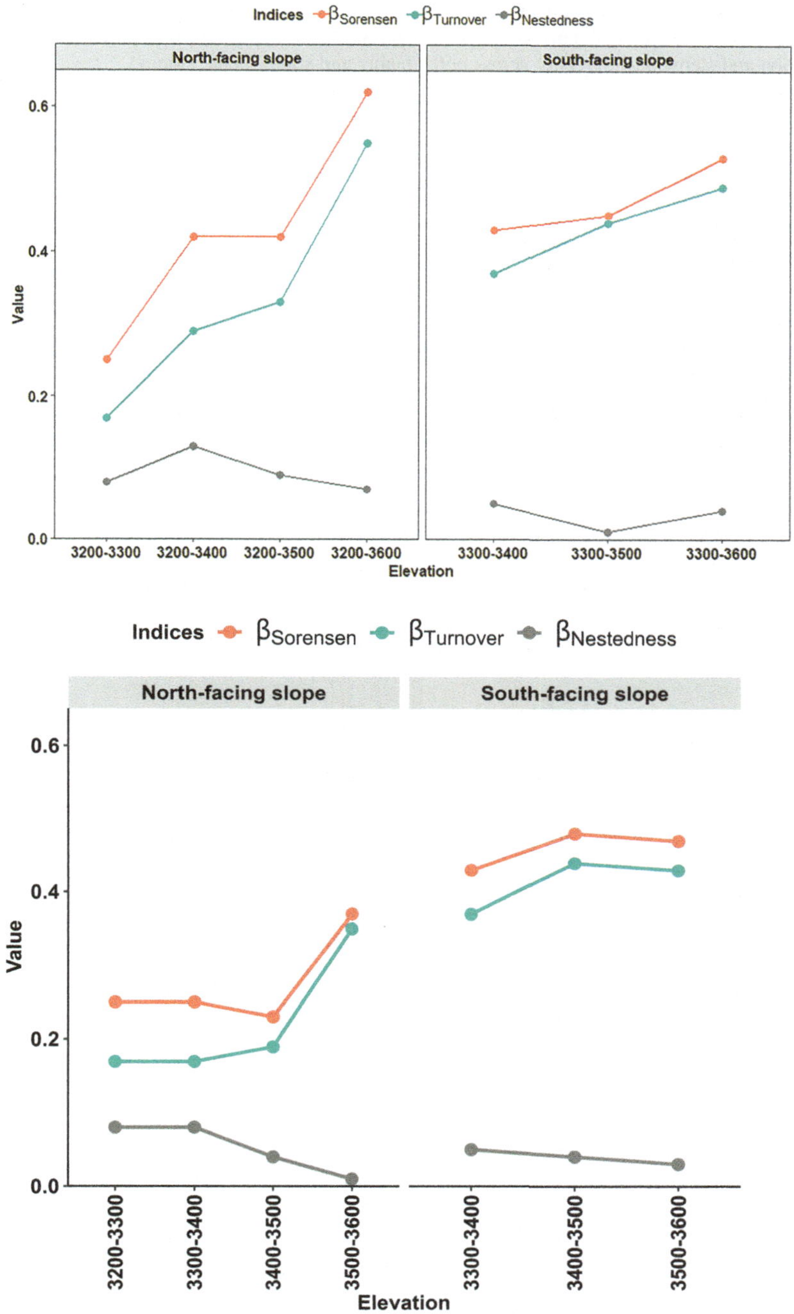

Fig. 9.12 Along-elevation (above) and stepwise (below) β-diversity (β_{Sor}), and its components turnover (β_{Sim}) and nestedness (β_{Sne}) across the north-facing and south-facing slopes. The effect of elevational distance on turnover, nestedness and total β-diversity was tested with a linear mixed-effects model

Table 9.9 Results of linear effect models testing the effect of elevational distance for overall β-diversity and the components of turnover and nestedness performed separately for along-elevation and stepwise β-diversity across north-facing and south-facing slopes

	North-facing slope				South-facing slope			
	r	r^2	t-Value	p-Value	r	r^2	t-Value	p-Value
Along-elevation diversity								
β-Sorensen	0.947	0.897	4.175	0.053	0.949	0.901	3.023	0.203
Turnover	0.960	0.922	4.866	0.040	0.999	0.997	18.881	0.034
Nestedness	−0.344	0.118	−0.517	0.656	−0.316	0.100	−0.333	0.795
Stepwise diversity								
β-Sorensen	0.949	0.901	3.023	0.203	0.756	0.571	1.155	0.454
Turnover	0.829	0.688	2.099	0.171	0.824	0.679	1.455	0.383
Nestedness	−0.948	0.899	−4.226	0.052	−0.924	0.855	−2.425	0.249

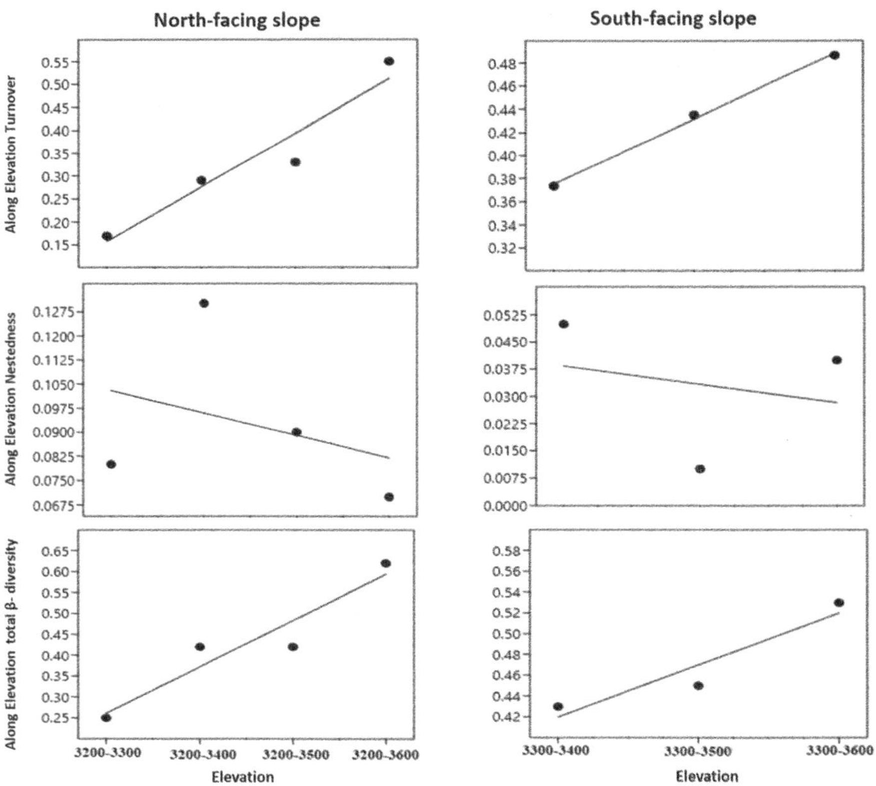

Fig. 9.13 Effect of elevational distance on along-elevation turnover, nestedness and total β-diversity

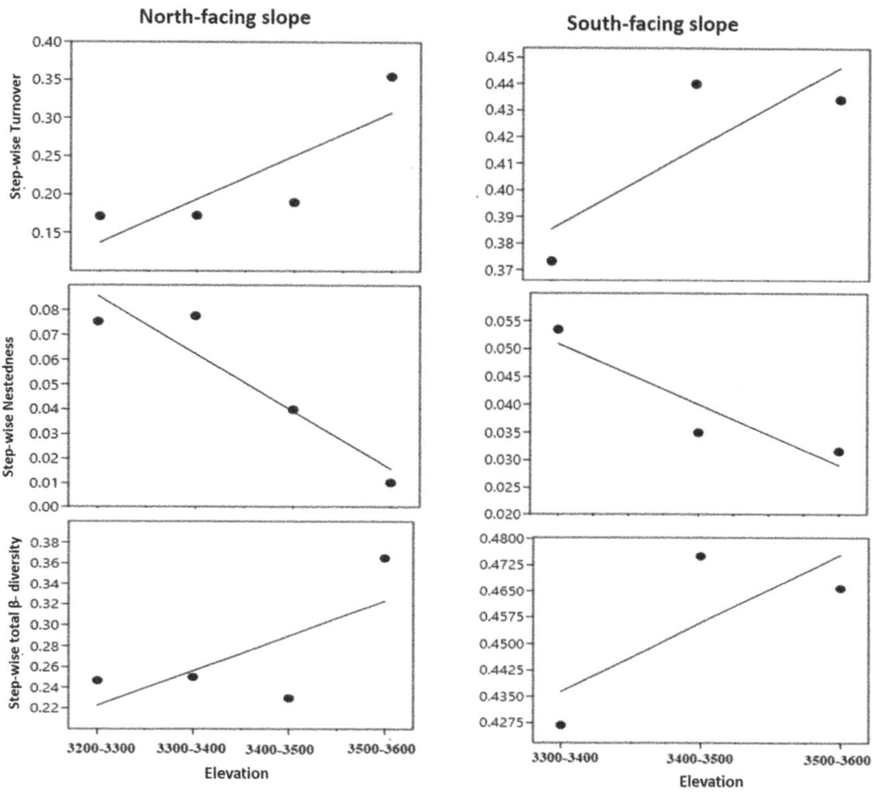

Fig. 9.14 Effect of elevational distance on stepwise turnover, nestedness and overall β-diversity

species in the northwest Himalayan ecotone range from 26 to 42 species. Mir et al. (2017) reported 48 and 54 plant species from Sonamarg and Gulmarg areas respectively of Kashmir Himalaya. However, a higher number of species (253 species) compared to the present study was reported by Singh et al. (2019) at the semi-disturbed treeline ecotone in Bhaderwah, Jammu and Kashmir. Such differences in species composition of treeline ecotones at regional and smaller spatial scales have been reported previously as well (Wielgolaski et al. 2017) and attributed to differences in the biotic and abiotic factors, such as human activities, soil conditions (Callaghan et al. 2004; Vittoz et al. 2010), temperature and precipitation (including snow cover) (Kullman 1995, 2003; Mathisen et al. 2014; Schwörer et al. 2014), micro-topography and disturbances (Rai et al. 2012), habitat heterogeneity (Junyan et al. 2014) and evolutionary processes (Schilthuizen 2000).

Among the life-forms, herbs were predominant in the treeline ecotone on both the slopes. A multitude of factors could be responsible for high herb richness, such as wide range of life-history adaptations that allow the herbs to persist and flourish (Whigham 2004), forest stand characteristics (Petersson et al. 2019) like spaced-out

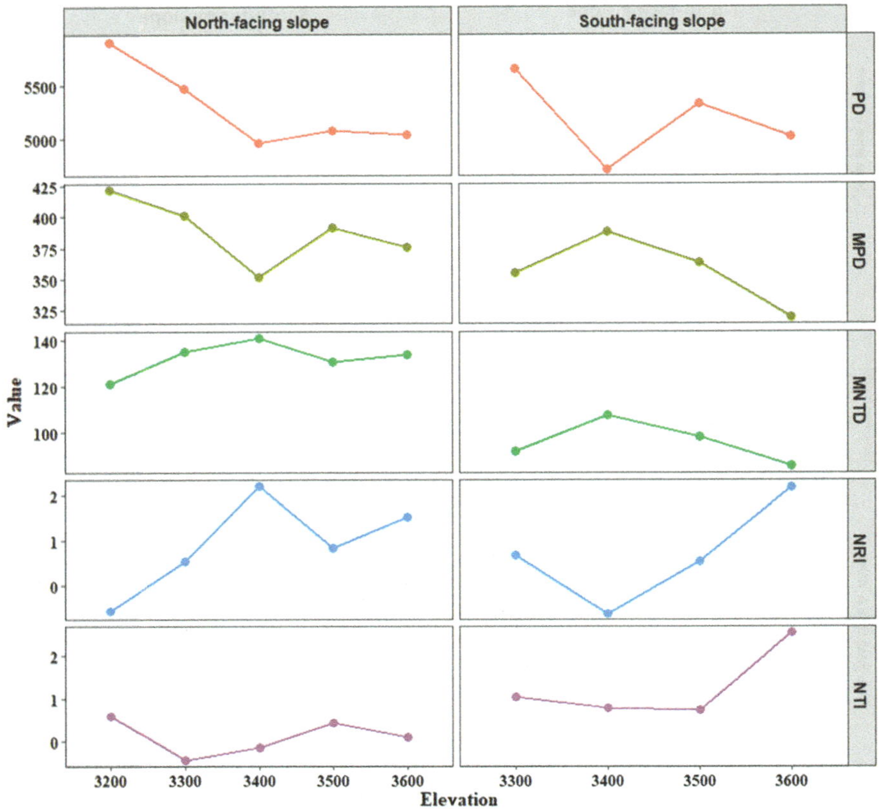

Fig. 9.15 Variation in phylogenetic indices along the elevation and across the aspects

trees in the treeline ecotone that increase light availability (Dormann et al. 2020), soil physical and chemical properties (Hulshof and Spasojevic 2020), topography (Tardella et al. 2019) and topography-related microclimatic conditions (Macek et al. 2019). In addition to these present-day conditions, historical processes like climate fluctuations over millions of years (glacial events) could be also responsible for the present structure of treeline ecotone in Kashmir Himalaya with a very less number of tree species. Post-glacial low recolonization has been attributed to dispersal limitations and a majority of European tree species are reported to be filling less than 50% of their potentially climatically suitable range (Svenning and Skov 2004).

In this study, we also noticed that the number of species in the treeline ecotone was higher compared to the alpine meadow but lower than the sub-alpine closed-canopy forest (unpublished personal data). Such an observation is consistent with studies that have found species diversity at ecotones to be intermediate between the two bounded communities (Harper 1995; Turton and Duff 1992; Mészáros 1990; Meiners et al. 2000). Thus, the enhanced diversity at ecotones as reported in many

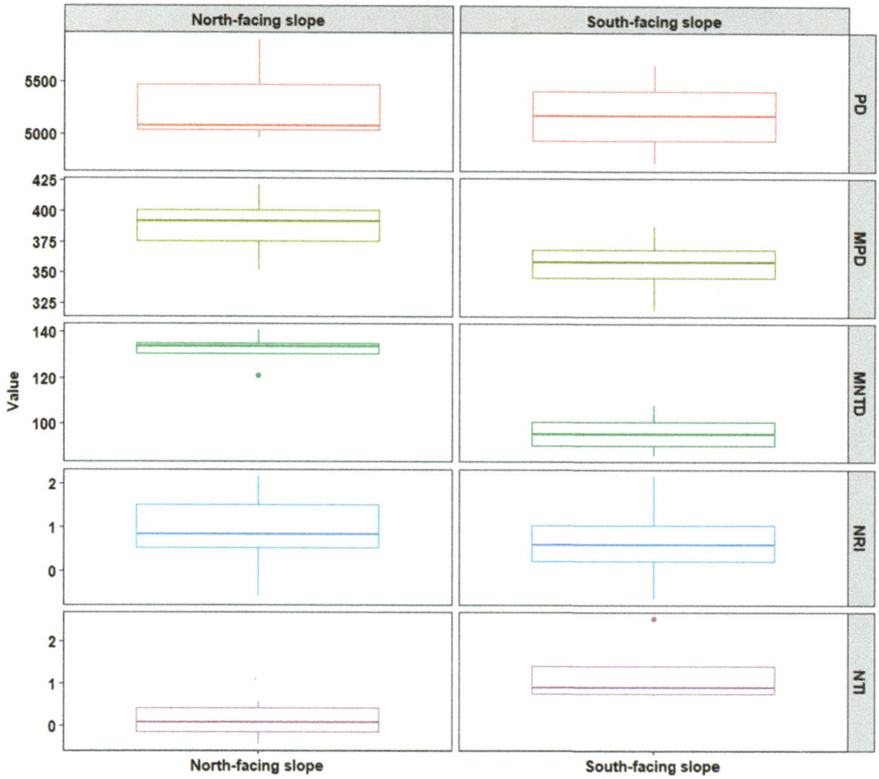

Fig. 9.16 Boxplots of phylogenetic diversity indices

studies (Shmida and Wilson 1985; Wolf 1993; Kernaghan and Harper 2001) is not a universal property but depends on the attributes of an ecotone in question. Further, it was also observed that a small fraction of species was ecotonal in nature with 16 (12.12%) and 14 (8.83%) species in the north-facing and south-facing slopes respectively growing exclusively in the ecotone. Most of the species in the treeline ecotone were either from the sub-alpine closed-canopy forest below the treeline ecotone or from the alpine meadow above the treeline ecotone, which indicates additive blending, instead of ecotonal specialization (Pandita and Dutt 2017; Pandita et al. 2019), is responsible for the species composition in the treeline ecotone. The two aspects did not differ significantly in terms of the number of species ($F = 0.209$, $p = 0.653$) but the species composition was different, with only 61 (26%) species being shared between the two aspects. Such a variation could be attributed to the modification of the local environment by aspect, slope and other topographic elements. Previous studies have also reported a significant influence of aspect on vegetation distribution (Moeslund et al. 2013; Hamid et al. 2020a, 2020b; Yang et al. 2020). This study also brought out that the south-facing (drier) slope in the ecotone was relatively species-rich compared to the north-facing (wetter) slope, which could

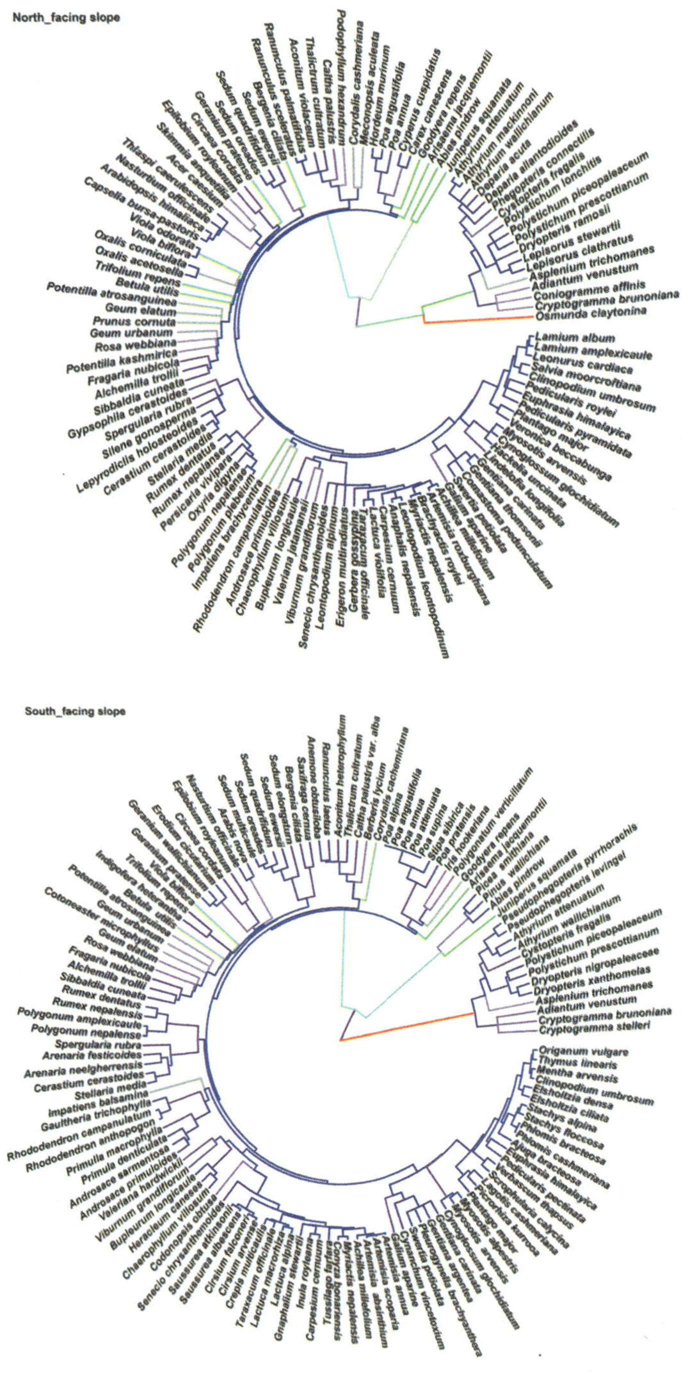

Fig. 9.17 Phylogenetic distribution of vascular plants on the north-facing slope (above) and south-facing slope (below)

a

Fig. 9.18 Treeline-forms in the treeline ecotone. (**a**) Diffuse type on the north-facing slope and (**b**) diffuse as well as tree-island treeline types on the south-facing slope

be due to micro-relief spatial heterogeneity and favourable temperature on the south-facing slope. While the role of soil temperature in species compositional differences between aspects has been invoked by Dearborn and Danby (2017) and Hamid et al. (2020a; b), the influence of micro-relief heterogeneity in species richness has been established by Leutner et al. (2012). In fact, a higher number of mosses on the drier slope as observed in this study was also reported by Leutner et al. (2012) who attributed it to heterogeneity caused by rocks and boulders, which we also noticed during sampling in the study area. Species diversity computed as Simpson's diversity decreased at the highest elevations on both aspects. Thus, it indicates that species diversity is relatively higher at lower elevations and lower at higher elevations. Lower diversity at higher elevations has been reported by other workers as well (Ahmad et al. 2020) and has been attributed to hard climatic conditions in the higher elevations (Lee et al. 2013; Gómez-Díaz et al. 2017). Evenness, measured as the relative abundance of different species in different elevation bands, revealed more or less similar pattern on the two slopes with higher values of evenness in the mid-elevation bands of the treeline ecotone and low evenness in the higher elevations. Our results are in contrast to the findings of Dar and Sundarapandian (2016) who observed an increase in evenness with elevation in Kashmir Himalaya. A decrease in Shannon index with elevation as recorded in this study is supported by similar observations of other workers (Gracia et al. 2007; Zhang et al. 2016). However, an increase in Shannon index with elevation has been also reported (Cui and Zheng 2016) with the highest in middle elevations (Kessler 2001; Grytnes and Vetaas 2002; Zhang et al. 2013). Such differences in species richness and diversity with elevation have been attributed to variation in climate, spatial habitat heterogeneity and anthropogenic factors (Cui and Zheng 2016). Mean β-diversity based on all possible pairs of comparisons of the sampled elevation bands did not vary much between the aspects, and turnover was the major contributor to the dissimilarity between the elevation bands irrespective of the aspect, which points towards the role of historical events, geographical isolation, habitat specialization, and environmental filtering and dispersal processes in shaping the plant assemblages (Baselga 2010; Pinto-Ledezma et al. 2018) and not due to patterns of colonization and extinction (Fontana et al. 2020). Along-elevation and stepwise β-diversity (Fig. 9.12) further reveals that species turnover increased with elevation and the increase was more distinct on the north-facing slope while nestedness decreased with elevation. Fontana et al. (2020) also reported an increase in turnover and a decrease in nestedness for vascular plants with elevation. Our results indicate that species sorting by environmental filtering and dispersal limitations seems to prevail, leading to higher species turnover and thereby reducing the probability of species-poor plant assemblages being a subset of species richness assemblages (nestedness) in the ecotone. On the contrary, Hamid et al. (2020a, b) reported that nestedness-resultant dissimilarity contributed more to β-diversity among the summits than the species turnover. Such variations could be because of differences in the elevation ranges considered in these studies for the study of β-diversity and its components.

Faith's phylogenetic diversity index (FD) declined with elevation on the north-facing slope but not on the south-facing slope (Fig. 9.15). We also noticed a significant positive correlation between PD and species richness on both north-facing ($r = +0.995$, $p < 0.001$) and south-facing ($r = +0.953$, $p = 0.047$) slopes, which is consistent with the findings of Cheng et al. (2018) and Zhu et al. (2019). It is quite likely that species richness may be the dominant factor driving the phylogenetic diversity across elevations and aspect. Our data also revealed higher mean pairwise distance (MPD) values than the mean nearest taxon distance (MNTD) index (Fig. 9.15). It is expected because MPD is the average of all pairwise species and MNTD is the average of only the nearest neighbour distances and such observations have been made also by Cadotte et al. (2010) and Kellar et al. (2015). MPD was higher at lower elevations in the ecotone and then declined sharply with elevation on the north-facing slope but the decline was steady on the south-facing slope (Fig. 9.15). Higher values of MPD indicate the presence of distantly related species co-occurring in the lower elevations of treeline ecotone particularly on the north-facing slope where it declined with elevation. On the south-facing slope, MPD did not show a sharp decline with elevation (Fig. 9.15). With respect to MNTD, the pattern was different between the two aspects and on the north-facing slope, it increased with elevation and then declined. A small mid-elevation increase in MNTD was also noticed on the south-facing slope. It indicates that mid-elevations in the ecotone have witnessed recent diversification compared to lower elevations where diversification might have taken place in the past. It implies that there are more closely related terminal species at higher elevations compared to lower elevations, which has also been reported by Zhang et al. (2021). Net relatedness index (NRI) and nearest taxon index (NTI) revealed phylogenetic overdispersion at lower elevations on both aspects but phylogenetic clustering at higher elevations, particularly on the south-facing slope. Such a pattern has also been reported by Manish and Pandit (2018) in the Himalaya, and ecological filtering at higher elevations could be responsible for phylogenetic clustering while interspecific competition may be determining phylogenetic overdispersion at low and mid-elevations of the treeline ecotone.

Two types of treeline forms, namely diffuse type and tree-island type, were common in the study area (Fig. 9.18). The diffuse type was more common on the north-facing slope and the tree-island type on the south-facing slope. Diffuse type gets formed due to slow density change with elevation in the treeline species (Bader et al. 2021) with mortality within the ecotone coupled with environmental heterogeneity or stochasticity and/or seed-limited colonization. Tree-island type treeline form is also a consequence of mortality within the ecotone, environmental heterogeneity, clustered seed dispersal, clonal reproduction and microclimate alterations due to wind or snow distribution, abrasion etc. (Bader et al. 2021). Others (Schickhoff et al. 2016; Harsch and Bader 2011; Körner 2012) have also reported diffuse, abrupt, island and krummholz treeline forms in different parts of Himalaya with the diffuse form being more vulnerable to climate change (Harsch and Bader 2011). In this study, we noticed that the treeline species differ on the two aspects. On the north-facing aspect, *Betula utilis* (tree) and *Rhododendron campanulatum* (shrub) were the dominant treeline species. On the contrary, the dominant treeline species on the south-facing slope was *Pinus wallichiana*. As many as 58 treeline species belonging to 14 genera and 8 families have been reported by Singh et al. (2020). The most common treeline species in the Himalaya belong to genera such as *Abies*, *Betula*, *Picea* (Table 9.10).

9.5 Concluding Remarks

While compiling this chapter, it became apparent that many knowledge gaps exist about community structure, species composition and turnover, and ecological and evolutionary diversification of species in the treeline ecotones across the Himalaya. Thus, it is high time that detailed and holistic studies are undertaken related to species diversity with a focus not only on vascular plants but also on other plant groups, animals and even so far neglected microbes as well and all other interacting factors at varied spatial scales. Particularly interesting would be the studies explicating the role of mutualists (mycorrhizas and dark septate hyphae etc.) in the growth of certain treeline species and their interaction under various climate change scenarios. Equally valuable would be to discern the relative role of inorganic and organic forms of nutrients, particularly nitrogen, in the growth and development of species in the treeline ecotone. It is also necessary to establish sites for long-term monitoring of treeline ecotone across Himalaya for the study of temporal changes in the structural organization and functional integrity of treeline ecotone essentially in

Table 9.10 Treeline species in different Himalayan regions

Himalayan region	Characteristic treeline species	Location	Reference
Indian Himalayan region	*Abies pindrow*	Daksum-Sinthan top (Kashmir); Sind Forest division (Sonamarg); Tangmarg Forest division (Gulmarg)	Nanda et al. (2018); Mir et al. (2017)
	Betula utilis	Daksum-Sinthan top (Kashmir); Sind Forest division (Sonamarg); Tangmarg Forest division (Gulmarg); Kedarnath wildlife sanctuary (Uttarakhand)	Nanda et al. (2018); Mir et al. (2017); Rai et al. (2013)
	Pinus wallichiana	Daksum-Sinthan top (Kashmir); Ngawal, Manang, Lauribinayak, Rasuwa (Central Nepal); Ngawal, Manang (Central Nepal)	Nanda et al. (2018); Shrestha et al. (2015); Chhetri et al. (2020)
	Abies densa	Khangchendzonga National Park (Sikkim)	Pandey et al. (2018)
	Abies spectabilis	Tungnath (Uttarakhand); Kedarnath wildlife sanctuary (Uttarakhand)	Adhikari and Kumar (2020); Rai et al. (2013)
	Quercus semecarpifolia	Chaudas Valley (Uttarakhand); Bhaderwah (Jammu and Kashmir); Tungnath (Uttarakhand); Kedarnath wildlife sanctuary (Uttarakhand)	Singh (2018); Singh et al. (2019); Adhikari and Kumar (2020); Rai et al. (2013)
	Rhododendron wightii	Sikkim	Pradhan and Lachungpa (1990)
	Rhododendron arboreum	Kedarnath wildlife sanctuary (Uttarakhand)	Rai et al. (2013)
Nepal Himalaya	*Abies spectabilis*	Langtang National Park, Rolwaling Valley (Dolakha District); Yangle, Barun (eastern Nepal)	Gaire et al. (2010); Schwab et al. (2018); Chhetri et al. (2020); Chhetri and Cairns (2015)
	Betula utilis	Rugakharka, Dhorpatan (Western Nepal); Langtang, Himalaya (Nepal)	Chhetri et al. (2020); Körner and Paulsen (2004)
China Himalaya	*Abies georgei, Larix potaninii*	Yunnan, P.R. China	Baker and Moseley (2007)
	Picea asperata	Sichuan (China)	Körner and Paulsen (2004)
	Picea crassifolia	Qilian Mountains	He et al. (2013)
	Picea schrenkiana	Tianchi nature reserve, Central Xinjiang, Uygur autonomous region	Wang et al. (2006)

(continued)

Table 9.10 (continued)

Himalayan region	Characteristic treeline species	Location	Reference
	Abies faxoniana	Balang Mountain, Wolong nature reserve, Sichuan Province	Shi et al. (2006)
	Betula ermanii	Changbai Mountains National Nature Reserve (CMNNR) in Jilin province	Zong et al. (2014)
	Betula platyphylla	Mozhugongka County, Tibet	Shi et al. (2008)
	Juniperus tibetica	Nagarze County town, Tibet	Shi et al. (2008)

relation to climate change and other anthropogenic pressures. Well-planned long-term spatial and temporal studies are likely to open new vistas in our understanding of the seasonal, annual and more long-term changes in the treeline ecotone and its response to changing environment, which is of pivotal importance in formulating effective strategies for conservation of such fragile ecosystems in the Himalaya for larger and longer benefit to the dependent and marginalized human population.

Acknowledgements Thanks are due to Prof. S.P. Singh, Chairman, Central Himalayan Environment Association, Nanital, for his valuable intellectual inputs. Financial support under the Timberline project supported by the Ministry of Environment, Forest and Climate change (New Delhi) through G.B. Pant National Institute of Himalayan Environment and Sustainable Development (Almora), Uttarakhand, is gratefully acknowledged. We are thankful to the Head, Department of Botany, for facilities and encouragement.

References

Adhikari BS, Kumar R (2020) Effect of snowmelt regime on phenology of herbaceous species at and around treeline in Western Himalaya, India. Not Sci Biol 12(4):901–919

Adhikari BS, Rawat GS, Bargal K (2012) Community structure along timberline ecotone in relation to micro-topography and disturbances in Western Himalaya. Not Sci Biol 4(2):41–52

Ahmad M, Uniyal SK, Batish DR, Singh HP, Jaryan V, Rathee S, Sharma P, Kohli RK (2020) Patterns of plant communities along vertical gradient in Dhauladhar Mountains in lesser Himalayas in North-Western India. Sci Total Environ 716:136919

Ali A, Sanaei A, Li M, Nalivan OA, Ahmadaali K, Pour MJ, Valipour A, Karami J, Aminpour M, Kaboli H (2020) Impacts of climatic and edaphic factors on the diversity, structure and biomass of species-poor and structurally-complex forests. Sci Total Environ 706:135719

Anup K, Ghimire A (2015) High-altitude plants in era of climate change: a case of Nepal Himalayas. In: Climate change impacts on high-altitude ecosystems. Springer, Cham, pp 177–187

Aryal A, Brunton D, Raubenheimer D (2014) Impact of climate change on human-wildlife-ecosystem interactions in the trans-Himalaya region of Nepal. Theor Appl Climatol 115(3): 517–529

Bader MY, Rietkerk M, Bregt AK (2007) Vegetation structure and temperature regimes of tropical alpine treelines. Arct Antarct Alp Res 39(3):353–364

Bader MY, Llambí LD, Case BS, Buckley HL, Toivonen JM, Camarero JJ, Cairns DM, Brown CD, Wiegand T, Resler LM (2021) A global framework for linking alpine-treeline ecotone patterns to underlying processes. Ecography 44(2):265–292

Baker B, Moseley R (2007) Advancing treeline and retreating glaciers: implications for conservation in Yunnan, PR China. Arct Antarct Alp Res 39(2):200–209

Baselga A (2010) Partitioning the turnover and nestedness components of beta diversity. Glob Ecol Biogeogr 19(1):134–143

Baselga A, Orme CDL (2012) Betapart: an R package for the study of beta diversity. Methods Ecol Evol 3(5):808–812

Batllori E, Blanco-Moreno J, Ninot J, Gutiérrez E, Carrillo E (2009) Vegetation patterns at the alpine treeline ecotone: the influence of tree cover on abrupt change in species composition of alpine communities. J Veg Sci 20(5):814–825

Bellhouse D (2005) Systematic sampling methods. Encyclopedia Biostat. https://doi.org/10.1002/9781118445112.stat05723

Bharti RR, Adhikari BS, Rawat GS (2012) Assessing vegetation changes in timberline ecotone of Nanda Devi National Park, Uttarakhand. Int J Appl Earth Obs Geoinf 18:472–479

Bhatta KP, Chaudhary RP, Vetaas OR (2012) A comparison of systematic versus stratified-random sampling design for gradient analyses: a case study in subalpine Himalaya. Nepal Phytocoenologia 42(3–4):191–202

Bhatta KP, Grytnes JA, Vetaas OR (2018) Scale sensitivity of the relationship between alpha and gamma diversity along an alpine elevation gradient in Central Nepal. J Biogeogr 45(4):804–814

Branco M, Figueiras FG, Cermeño P (2018) Assessing the efficiency of non-parametric estimators of species richness for marine microplankton. J Plankton Res 40(3):230–243

Burnham KP, Overton WS (1979) Robust estimation of population size when capture probabilities vary among animals. Ecology 60(5):927–936

Bürzle B, Schickhoff U, Schwab N, Oldeland J, Müller M, Böhner J, Chaudhary RP, Scholten T, Dickoré WB (2017) Phytosociology and ecology of treeline ecotone vegetation in RolwalingHimal, Nepal. Phytocoenologia 47(2):197–220

Cadotte MW, Jonathan Davies T, Regetz J, Kembel SW, Cleland E, Oakley TH (2010) Phylogenetic diversity metrics for ecological communities: integrating species richness, abundance and evolutionary history. Ecol Lett 13(1):96–105

Callaghan TV, Björn LO, Chernov Y, Chapin T, Christensen TR, Huntley B, Ims RA, Johansson M, Jolly D, Jonasson S (2004) Biodiversity, distributions and adaptations of Arctic species in the context of environmental change. AMBIO: A J Human Environ 33(7):404–417

Camarero JJ, Linares JC, García-Cervigón AI, Batllori E, Martínez I, Gutiérrez E (2017) Back to the future: the responses of alpine treelines to climate warming are constrained by the current ecotone structure. Ecosystems 20(4):683–700

Chao A (1984) Nonparametric estimation of the number of classes in a population. Scand J Stat. 11: 265–270

Chen I-C, Hill JK, Ohlemüller R, Roy DB, Thomas CD (2011) Rapid range shifts of species associated with high levels of climate warming. Science 333(6045):1024–1026

Cheng X-L, Yuan L-X, Nizamani MM, Zhu Z-X, Friedman CR, Wang H-F (2018) Taxonomic and phylogenetic diversity of vascular plants at Ma'anling volcano urban park in tropical Haikou, China: Reponses to soil properties. PLoS One 13(6):e0198517

Chhetri PK, Cairns DM (2015) Contemporary and historic population structure of Abies spectabilis at treeline in Barun valley, eastern Nepal Himalaya. J Mt Sci 12(3):558–570

Chhetri PK, Bista R, Shrestha KB (2020) How does the stand structure of treeline-forming species shape the treeline ecotone in different regions of the Nepal Himalayas? J Mt Sci 17(10): 2354–2368

Colwell RK (2006) EstimateS: statistical estimation of species richness and shared species from simples, version 8.0. http://purl oclc org/estimates

Colwell R (2013) EstimateS: statistical estimation of species richness and shared species from samples. Version 9.1.0, User's Guide and Application http://purl.oclc.org/estimates

Colwell RK, Elsensohn JE (2014) EstimateS turns 20: statistical estimation of species richness and shared species from samples, with non-parametric extrapolation. Ecography 37(6):609–613
Cui W, Zheng X-X (2016) Spatial heterogeneity in tree diversity and forest structure of evergreen broadleaf forests in southern China along an altitudinal gradient. Forests 7(10):216
Dar JA, Sundarapandian S (2016) Patterns of plant diversity in seven temperate forest types of Western Himalaya, India. J Asia-Pac Biodivers 9(3):280–292
Dearborn KD, Danby RK (2017) Aspect and slope influence plant community composition more than elevation across forest–tundra ecotones in subarctic Canada. J Veg Sci 28(3):595–604
Dormann CF, Bagnara M, Boch S, Hinderling J, Schäfer D, Schall P, Hartig F (2020) Plant species richness increases with light availability, but not variability, in temperate forests understorey. BMC Ecol 20:43. https://doi.org/10.1186/s12898-020-00311-9
Faith DP (1992) Conservation evaluation and phylogenetic diversity. Biol Conserv 61(1):1–10
Fayaz M, Jain AK, Bhat MH, Kumar A (2019) Ethnobotanical survey of Daksum forest range of Anantnag District, Jammu and Kashmir, India. J Herbs Spices Med Plants 25(1):55–67
Fontana V, Guariento E, Hilpold A, Niedrist G, Steinwandter M, Spitale D, Nascimbene J, Tappeiner U, Seeber J (2020) Species richness and beta diversity patterns of multiple taxa along an elevational gradient in pastured grasslands in the European Alps. Sci Rep 10(1):1–11
Gaire N, Dhakal Y, Lekhak H, Bhuju D, Shah S (2010) Vegetation dynamics in treeline ecotone of Langtang National Park, Central Nepal. Nepal J Sci Technol 11:107–114
Gairola S, Rawal R, Todaria N (2008) Forest vegetation patterns along an altitudinal gradient in sub-alpine zone of west Himalaya, India. Afr J Plant Sci 2(6):042–048
Ghimire B, Mainali KP, Lekhak HD, Chaudhary RP, Ghimeray AK (2010) Regeneration of Pinus wallichiana AB Jackson in a trans-Himalayan dry valley of north-Central Nepal. Himal J Sci 6(8):19–26
Gómez-Díaz JA, Krömer T, Carvajal-Hernández CI, Gerold G, Heitkamp F (2017) Richness and distribution of herbaceous angiosperms along gradients of elevation and forest disturbance in Central Veracruz, Mexico. Bot Sci 95(2):307–328
Gracia M, Montané F, Piqué J, Retana J (2007) Overstory structure and topographic gradients determining diversity and abundance of understory shrub species in temperate forests in Central Pyrenees (NE Spain). For Ecol Manag 242(2–3):391–397
Grytnes JA, Vetaas OR (2002) Species richness and altitude: a comparison between null models and interpolated plant species richness along the Himalayan altitudinal gradient, Nepal. Am Nat 159(3):294–304
Hamid M, Khuroo AA, Malik AH, Ahmad R, Singh CP (2020a) Assessment of alpine summit flora in Kashmir Himalaya and its implications for long-term monitoring of climate change impacts. J Mt Sci 17(8):1974–1988
Hamid M, Khuroo AA, Malik AH, Ahmad R, Singh CP, Dolezal J, Haq SM (2020b) Early evidence of shifts in alpine summit vegetation: a case study from Kashmir Himalaya. Front Plant Sci 11: 421
Hammer Ø, Harper DA, Ryan PD (2001) PAST: paleontological statistics software package for education and data analysis. Palaeontol Electron 4(1):9
Harper KA (1995) Effect of expanding clones of Gaylussacia baccata (black huckleberry) on species composition in sandplain grassland on Nantucket Island, Massachusetts. Bull Torrey Bot Club 122:124–133
Harsch MA, Bader MY (2011) Treeline form–a potential key to understanding treeline dynamics. Glob Ecol Biogeogr 20(4):582–596
Harsch MA, Hulme PE, McGlone MS, Duncan RP (2009) Are treelines advancing? A global meta-analysis of treeline response to climate warming. Ecol Lett 12(10):1040–1049
He Z, Zhao W, Zhang L, Liu H (2013) Response of tree recruitment to climatic variability in the alpine treeline ecotone of the Qilian Mountains, northwestern China. For Sci 59(1):118–126
Hofgaard A, Løkken JO, Dalen L, Hytteborn H (2010) Comparing warming and grazing effects on birch growth in an alpine environment–a 10-year experiment. Plant Ecol Divers 3(1):19–27
Holtmeier F (2000) Die Höhengrenze der Gebirgswälder (ArbeitenausdemInstitut für Landschaftsökologie 8). WestfälischeWilhelms-Universität, Münster

Holtmeier F-K (2009) Mountain timberlines: ecology, patchiness, and dynamics, vol 36. Springer, Cham

Holtmeier F-K, Broll G, Müterthies A, Anschlag K (2003) Regeneration of trees in the treeline ecotone: northern Finnish Lapland. Fennia 181(2):103–128

Hulshof CM, Spasojevic MJ (2020) The edaphic control of plant diversity. Glob Ecol Biogeogr 29(10):1634–1165

Jin Y, Qian H (2019) V. PhyloMaker: an R package that can generate very large phylogenies for vascular plants. Ecography 42(8):1353–1359

Junyan Z, Kewu C, Runguo Z, Yi D (2014) Changes in floristic composition, community structure and species diversity across a tropical coniferous-broadleaved forest ecotone. Trop Conserv Sci 7(1):126–144

Kellar PR, Ahrendsen DL, Aust SK, Jones AR, Pires JC (2015) Biodiversity comparison among phylogenetic diversity metrics and between three north American prairies. Appl Plant Sci 3(7): 1400108

Kembel S, Cowan P, Helmus M, Cornwell W, Morlon H, Ackerly D, Blomberg S, Webb C (2010) Picante: R tools for integrating phylogenies and ecology. Bioinformatics 26:1463–1464

Kernaghan G, Harper K (2001) Community structure of ectomycorrhizal fungi across an alpine/ subalpine ecotone. Ecography 24(2):181–188

Kessler M (2001) Patterns of diversity and range size of selected plant groups along an elevational transect in the Bolivian Andes. Biodivers Conserv 10(11):1897–1921

Kharuk VI, Im ST, Dvinskaya ML, Ranson KJ (2010) Climate-induced mountain tree-line evolution in southern Siberia. Scand J For Res 25(5):446–454

Körner C (2012) Alpine treelines: functional ecology of the global high elevation tree limits. Springer, Cham

Körner C, Paulsen J (2004) A world-wide study of high altitude treeline temperatures. J Biogeogr 31(5):713–732

Kosanic A, Anderson K, Harrison S, Turkington T, Bennie J (2018) Changes in the geographical distribution of plant species and climatic variables on the West Cornwall peninsula (south West UK). PLoS One 13(2):e0191021

Kullman L (1995) Holocene tree-limit and climate history from the Scandes Mountains, Sweden. Ecology 76(8):2490–2502

Kullman L (2003) Recent reversal of Neoglacial climate cooling trend in the Swedish Scandes as evidenced by mountain birch tree-limit rise. Glob Planet Chang 36(1–2):77–88

Lee C-B, Chun J-H, Song H-K, Cho H-J (2013) Altitudinal patterns of plant species richness on the Baekdudaegan Mountains, South Korea: mid-domain effect, area, climate, and Rapoport's rule. Ecol Res 28(1):67–79

Lee CK, Laughlin DC, Bottos EM, Caruso T, Joy K, Barrett JE, Brabyn L, Nielsen UN, Adams BJ, Wall DH (2019) Biotic interactions are an unexpected yet critical control on the complexity of an abiotically driven polar ecosystem. Comm Biol 2(1):1–10

Lenoir J, Svenning JC (2015) Climate-related range shifts–a global multidimensional synthesis and new research directions. Ecography 38(1):15–28

Leutner BF, Steinbauer MJ, Müller CM, Früh AJ, Irl S, Jentsch A, Beierkuhnlein C (2012) Mosses like it rough—growth form specific responses of mosses, herbaceous and woody plants to micro-relief heterogeneity. Diversity 4(1):59–73

Li MH, Xiao WF, Shi P, Wang SG, Zhong YD, Liu XL, Wang XD, Cai XH, Shi ZM (2008) Nitrogen and carbon source–sink relationships in trees at the Himalayan treelines compared with lower elevations. Plant Cell Environ 31(10):1377–1387

Liang E, Dawadi B, Pederson N, Eckstein D (2014) Is the growth of birch at the upper timberline in the Himalayas limited by moisture or by temperature? Ecology 95(9):2453–2465

Macek M, Kopecký M, Wild J (2019) Maximum air temperature controlled by landscape topography affects plant species composition in temperate forests. Landsc Ecol 34(11):2541–2556

Mandal G, Joshi SP (2014) Analysis of vegetation dynamics and phytodiversity from three dry deciduous forests of Doon Valley, Western Himalaya, India. J Asia-Pac Biodivers 7(3):292–304

Manish K, Pandit MK (2018) Phylogenetic diversity, structure and diversification patterns of endemic plants along the elevational gradient in the eastern Himalaya. Plant Ecol Divers 11(4):501–513
Mathisen IE, Mikheeva A, Tutubalina OV, Aune S, Hofgaard A (2014) Fifty years of tree line change in the K hibiny M ountains, R ussia: advantages of combined remote sensing and dendroecological approaches. Appl Veg Sci 17(1):6–16
Meiners S, Handel S, Pickett S (2000) Tree seedling establishment under insect herbivory: edge effects and inter-annual variation. Plant Ecol 151(2):161–170
Mészáros I (1990) Spatial changes in herb layer in a beech forest/clear cut area ecotone from northern Hungary. Spatial processes in plant communities Academia, Prague, pp 59–71
Miehe G, Gravendeel B, Kluge J, Kuss P, Long D, Opgenoorth L, Pendry C, Rajbhandari K, Rajbhandary S, Ree R (2015) Flora. Royal Botanic Garden Edinburgh, Edinburgh
Mir NA, Masood T, Sofi P, Husain M, Rather T (2017) Life form spectrum of vegetation in Betula dominant tree stands along the available altitudinal gradient in north western Himalayas of Kashmir. J Pharmacogn Phytochem 6(4):267–272
Mishra NB, Mainali KP (2017) Greening and browning of the Himalaya: spatial patterns and the role of climatic change and human drivers. Sci Total Environ 587:326–339
Moeslund JE, Arge L, Bøcher PK, Dalgaard T, Svenning JC (2013) Topography as a driver of local terrestrial vascular plant diversity patterns. Nord J Bot 31(2):129–144
Mosley JC, Bunting SC, Hironaka M (1989) Quadrat and sample sizes for frequency sampling mountain meadow vegetation. Great Basin Nat 49:241–248
Nagy L, Grabherr G (2009) The biology of alpine habitats. Oxford University Press on Demand, Oxford
Nanda SA, Reshi ZA, Ul-haq M, Lone A, Mir SA (2018) Taxonomic and functional plant diversity patterns along an elevational gradient through treeline ecotone in Kashmir. Trop Ecol 59(2): 211–224
Pacifici M, Foden WB, Visconti P, Watson JE, Butchart SH, Kovacs KM, Scheffers BR, Hole DG, Martin TG, Akçakaya HR (2015) Assessing species vulnerability to climate change. Nat Clim Chang 5(3):215–224
Pandey A, Badola HK, Rai S, Singh SP (2018) Timberline structure and woody taxa regeneration towards treeline along latitudinal gradients in Khangchendzonga National Park, Eastern Himalaya. PLoS One 13(11):e0207762
Pandita S, Dutt HC (2017) Herbaceous species in structuring the ecotone between forest-grassland in lesser stratum of north-west (NW) Himalaya. NeBIO 8:287â
Pandita S, Kumar V, Dutt HC (2019) Environmental variables Vis-a-Vis distribution of herbaceous tracheophytes on northern sub-slopes in Western Himalayan ecotone. Ecol Process 8(1):1–9
Petersson L, Holmström E, Lindbladh M, Felton A (2019) Tree species impact on understory vegetation: vascular plant communities of scots pine and Norway spruce managed stands in northern Europe. For Ecol Manag 448:330–345
Pielou F (1975) Ecological diversity. wileyIntescieco, New York
Pinto-Ledezma JN, Larkin DJ, Cavender-Bares J (2018) Patterns of beta diversity of vascular plants and their correspondence with biome boundaries across North America. Front Ecol Evol 6:194
Pradhan UC, Lachungpa ST (1990) Sikkim-Himalayan Rhododendrons. Primulaceae Books
Rai ID, Adhikari BS, Rawat GS (2012) Floral diversity along sub-alpine and alpine ecosystems in Tungnath area of Kedarnath wildlife sanctuary, Uttarakhand. Indian For 138(10):927–940
Rai I, Bharti R, Adhikari B, Rawat G (2013) Structure and functioning of timberline vegetation in the western Himalaya: a case study. In: Wu N, Rawat GS, Joshi S, Ismail M, Sharma E (eds) High-altitude rangelands and their interfaces in the Hindu Kush Himalayas. International Centre for Integrated Mountain Development, Nepal, p 91107
Rai ID, Padalia H, Singh G, Adhikari BS, Rawat GS (2020) Vegetation dry matter dynamics along treeline ecotone in Western Himalaya, India. Trop Ecol 61(1):116–127
Schickhoff U (2002) Die Degradierung der GebirgswälderNordpakistans. Faktoren, Prozesse und Wirkungszusammenhänge in einemregionalen Mensch-Umwelt-System ErdwissForsch 41

Schickhoff U (2005) The upper timberline in the Himalayas, Hindu Kush and Karakorum: a review of geographical and ecological aspects. In: Mountain ecosystems. Springer, Cham, pp 275–354
Schickhoff U, Bobrowski M, Böhner J, Bürzle B, Chaudhary R, Gerlitz L, Heyken H, Lange J, Müller M, Scholten T (2015) Do Himalayan treelines respond to recent climate change? An evaluation of sensitivity indicators. Earth Syst Dynam 6(1):245–265
Schickhoff U, Bobrowski M, Böhner J, Bürzle B, Chaudhary RP, Gerlitz L, Lange J, Müller M, Scholten T, Schwab N (2016) Climate change and treeline dynamics in the Himalaya. In: Climate change, glacier response, and vegetation dynamics in the Himalaya. Springer, Cham, pp 271–306
Schilthuizen M (2000) Ecotone: speciation-prone. Trends Ecol Evol 15(4):130–131
Schwab N, Kaczka RJ, Janecka K, Böhner J, Chaudhary RP, Scholten T, Schickhoff U (2018) Climate change-induced shift of tree growth sensitivity at a central Himalayan treeline ecotone. Forests 9(5):267
Schwörer C, Henne PD, Tinner W (2014) A model-data comparison of Holocene timberline changes in the Swiss Alps reveals past and future drivers of mountain forest dynamics. Glob Chang Biol 20(5):1512–1526
Shannon C, Weaver W (1949) rep. 1998, The mathematical theory of communication (Urbana. University of Illinois Press). Foreword by Richard E. Blahut and Bruce Hajek
Shi P, Körner C, Hoch G (2006) End of season carbon supply status of woody species near the treeline in western China. Basic Appl Ecol 7(4):370–377
Shi P, Körner C, Hoch G (2008) A test of the growth-limitation theory for alpine tree line formation in evergreen and deciduous taxa of the eastern Himalayas. Funct Ecol 22(2):213–220
Shmida A, Wilson MV (1985) Biological determinants of species diversity. J Biogeogr 12(1):1–20
Shrestha KB, Vetaas OR (2009) The forest ecotone effect on species richness in an arid trans-Himalayan landscape of Nepal. Folia Geobot 44(3):247–262
Shrestha BB, Ghimire B, Lekhak HD, Jha PK (2007) Regeneration of treeline birch (Betula utilis D. Don) forest in a trans-Himalayan dry valley in Central Nepal. Mt Res Dev 27(3):259–267
Shrestha KB, Hofgaard A, Vandvik V (2015) Recent treeline dynamics are similar between dry and Mesic areas of Nepal, central Himalaya. J Plant Ecol 8(4):347–358
Simpson E (1949) Measurement of diversity. Nature 163:688. Simpson688163Nature1949
Singh SP (2018) Research on Indian Himalayan treeline ecotone: an overview. Trop Ecol 59(2): 163–176
Singh D, Sharma A, Sharma N (2019) Composition, richness and floristic diversity along an elevational gradient in a semi-disturbed treeline ecotone, Bhaderwah, Jammu and Kashmir. J Appl Nat Sci 11(1):23–34
Singh SP, Gumber S, Singh RD, Singh G (2020) How many tree species are in the Himalayan treelines and how are they distributed? Trop Ecol 61(3):317–327
Smith SA, Brown JW (2018) Constructing a broadly inclusive seed plant phylogeny. Am J Bot 105(3):302–314
Smith EP, van Belle G (1984) Nonparametric estimation of species richness. Biometrics. 40:119–129
Speed JD, Austrheim G, Hester AJ, Mysterud A (2011) Growth limitation of mountain birch caused by sheep browsing at the altitudinal treeline. For Ecol Manag 261(7):1344–1352
Svenning JC, Skov F (2004) Limited filling of the potential range in European tree species. Ecol Lett 7(7):565–573
Tardella FM, Postiglione N, Bricca A, Cutini M, Catorci A (2019) Altitude and aspect filter the herb layer functional structure of sub-Mediterranean forests. Phytocoenologia 49(2):185–198
Telwala Y, Brook BW, Manish K, Pandit MK (2013) Climate-induced elevational range shifts and increase in plant species richness in a Himalayan biodiversity epicentre. PLoS One 8(2):e57103
Turton S, Duff G (1992) Light environments and floristic composition across an open forest-rainforest boundary in northeastern Queensland. Aust J Ecol 17(4):415–423
Vittoz P, Camenisch M, Mayor R, Miserere L, Vust M, Theurillat J-P (2010) Subalpine-nival gradient of species richness for vascular plants, bryophytes and lichens in the Swiss inner Alps. Bot Helv 120(2):139–149

Walther BA, Martin JL (2001) Species richness estimation of bird communities: how to control for sampling effort? Ibis 143(4):413–419
Walther BA, Moore JL (2005) The concepts of bias, precision and accuracy, and their use in testing the performance of species richness estimators, with a literature review of estimator performance. Ecography 28(6):815–829
Walther B, Morand S (1998) Comparative performance of species richness estimation methods. Parasitology 116(4):395–405
Wang T, Zhang QB, Ma K (2006) Treeline dynamics in relation to climatic variability in the central Tianshan Mountains, northwestern China. Glob Ecol Biogeogr 15(4):406–415
Webb CO (2000) Exploring the phylogenetic structure of ecological communities: an example for rain forest trees. Am Nat 156(2):145–155
Webb CO, Ackerly DD, McPeek MA, Donoghue MJ (2002) Phylogenies and community ecology. Annu Rev Ecol Syst 33(1):475–505
Webb CO, Ackerly DD, Kembel SW (2008) Phylocom: software for the analysis of phylogenetic community structure and trait evolution. Bioinformatics 24(18):2098–2100
Whigham DF (2004) Ecology of woodland herbs in temperate deciduous forests. Annu Rev Ecol Evol Syst 35:583–621
Wielgolaski FE, Hofgaard A, Holtmeier F-K (2017) Sensitivity to environmental change of the treeline ecotone and its associated biodiversity in European mountains. Clim Res 73(1–2): 151–166
Wolf JH (1993) Diversity patterns and biomass of epiphytic bryophytes and lichens along an altitudinal gradient in the northern Andes. Ann Mo Bot Gard 80:928–960
Wong MH-G, Duan C-Q, Long Y-C, Luo Y, Xie G-Q (2010) How will the distribution and size of subalpine Abies georgei forest respond to climate change? A study in Northwest Yunnan, China. Phys Geogr 31(4):319–335
Xu J, Grumbine RE, Shrestha A, Eriksson M, Yang X, Wang Y, Wilkes A (2009) The melting Himalayas: cascading effects of climate change on water, biodiversity, and livelihoods. Conserv Biol 23(3):520–530
Yang J, El-Kassaby YA, Guan W (2020) The effect of slope aspect on vegetation attributes in a mountainous dry valley, Southwest China. Sci Rep 10(1):1–11
Zanne AE, Tank DC, Cornwell WK, Eastman JM, Smith SA, FitzJohn RG et al (2014) Three keys to the radiation of angiosperms into freezing environments. Nature 506(7486):89–92
Zhang B, Liang C, He H, Zhang X (2013) Variations in soil microbial communities and residues along an altitude gradient on the northern slope of Changbai Mountain, China. PLoS One 8(6): e66184
Zhang C, Liu G, Xue S, Wang G (2016) Soil bacterial community dynamics reflect changes in plant community and soil properties during the secondary succession of abandoned farmland in the loess plateau. Soil Biol Biochem 97:40–49
Zhang R, Zhang Z, Shang K, Zhao M, Kong J, Wang X, Wang Y, Song H, Zhang O, Lv X (2021) A taxonomic and phylogenetic perspective on plant community assembly along an elevational gradient in subtropical forests. J Plant Ecol 14(4):702–716
Zhu Z-X, Nizamani MM, Sahu SK, Kunasingam A, Wang H-F (2019) Tree abundance, richness, and phylogenetic diversity along an elevation gradient in the tropical forest of Diaoluo Mountain in Hainan, China. Acta Oecol 101:103481
Zong S, Wu Z, Xu J, Li M, Gao X, He H, Du H, Wang L (2014) Current and potential tree locations in tree line ecotone of Changbai Mountains, Northeast China: the controlling effects of topography. PLoS One 9(8):e106114

Vegetation Structure Along an Elevation Gradient at the Treeline Ecotone of Eastern Himalayan Forests in Sikkim

10

Devendra Kumar and Aseesh Pandey

Abstract

This chapter presents an extensive vegetation analysis for Yuksam-Dzongri transect in Sikkim (in the eastern Himalaya). The investigation was conducted along an elevation gradient between 1700 m and 4000 m at 100 m elevation steps and considered its physiognomy life form, species-area relationships and species compositional changes. The plant species assemblages in the transect was represented by 267 species belonging to 174 genera and 81 families. The familial composition showed the predominance of Ericaceae, Fagaceae, and Lauraceae in tree layer; Ericaceae, Rosaceae, Berberidaceae, and Rubiaceae in shrub layer, and Compositae, Polygonaceae, Rosaceae, and Primulaceae in the herb layer. The species richness (tree, shrub, and herb) and total basal area (TBA) of trees declined monotonically along the elevation gradient with peaks at 1700–1900 m. The tree density in present studied transect was significantly higher than in its western Himalayan counter parts of Indian Himalayan region. The presence of 77 tree species in the high forested zone highlights the high tree species richness of the eastern Himalaya to which the contribution of *Rhododendron* was particularly high. Sparse canopy layer and dense undercanopy tree layer were the characteristic features of the forests. Further, inclusion of biotic interactions and disturbance may improve ecological understanding of these patterns of plant species richness across the Indian Himalayan region.

D. Kumar (✉) · A. Pandey
G.B. Pant National Institute of Himalayan Environment, Sikkim Regional Centre, Gangtok, Sikkim, India

S. P. Singh et al. (eds.), *Ecology of Himalayan Treeline Ecotone*, https://doi.org/10.1007/978-981-19-4476-5_10

Keywords

Eastern Himalaya · Elevation gradient · Sikkim · Species richness · Treeline

10.1 Introduction

Variability in environmental factors along the elevation gradient plays an important role in shaping the distribution patterns of plant species in mountainous landscapes (Hunter and Yonzon 1993; Vetaas and Grytnes 2002; Baniya et al. 2012; Rai et al. 2018). An elevation transect in the Himalaya supports diverse vegetation types from tropical monsoon forest to moist alpine scrub and meadows (Singh and Singh 1992). Diversity and distribution of species are influenced by their physiological tolerance and competitive ability in environmental variables, such as elevation, exposure to radiation, moisture, precipitation, substrate attributes, temperature, and topography (John and Dale 1990; Belnap and Gillette 1998; Ponzetti and McCune 2001). Therefore, elevation gradients are particularly suitable for detecting climate change and assessing climate change impacts (Nogués-Bravo et al. 2008).

Although many theoretical explanations are available, still patterns of plant species distribution along the elevation gradient are debatable (Sharma et al. 2019). Several postulations about species richness patterns along the elevation gradients have been proposed. Among them, monotonous decline in species richness with elevation gain (Odland and Birks 1999; Körner 2004; Saikia et al. 2017; Shooner et al. 2018) and mid elevation peak showing hump-shaped pattern of species richness (Carpenter 2005; Manish et al. 2016) are most prominent. In the Himalaya, the hump-shaped pattern of species richness is most common, (Grytnes and Vetaas 2002; Bhattarai and Vetaas 2006; Acharya et al. 2011; Pandey et al. 2018b). Reports on monotonic decline in tree species richness with elevational gain are also available (Sharma et al. 2009; Bhattarai and Vetaas 2006).

The East Himalayan state, Sikkim offers an ideal research area for studying elevational influence on phytodiversity (Pandey et al. 2018b). Acharya et al. (2011) have analyzed the tree richness pattern involving data points from 23 sample locations at every 100 m interval between 300 and 4700 m. Manish et al. (2016) have studied life forms (herb, shrub, and tree), richness of angiospermic plants using 51 elevational bands of 100 m each along 300–5300 m elevation gradient. Pandey et al. (2018b) have studied changes in vegetation attributes using 11 elevational bands of 100 m each along 3000–4000 m elevation gradient in Khangchendzonga National Park. Sharma et al. (2019) studied richness pattern of life forms (herb, shrub, and tree) in an elevation gradient 500–3300 m using 28 elevation bands of 100 m each. Besides elevation, the ecotone effect in transitional zones is also reported in eastern Himalaya (Oommen and Shanker 2005; Behera and Kushwaha 2007). Further, the eastern Himalaya possesses an ultra-varied topography that fosters species diversity and endemism (Myers 1988). Thus, for better understanding of phytodiversity patterns, there is a need of detailed study focusing on various ecological attributes along elevation gradient. This study focuses on analyzing the patterns of the life-form richness of vascular plants, vegetation structure, species composition, and distribution along an elevation gradient from mixed broad leaved forest to treeline ecotone between 1700 and 4000 m in Sikkim.

10.2 Materials and Methods

10.2.1 Study Area

This study considers a Yuksam-Dzongri transect located in West district of Sikkim state in the Eastern Himalaya (Fig. 10.1). The Yuksam-Dzongri transact covers an elevation gradient of 2400 m starting from mixed broad-leaved forest at 1700 m to moist alpine scrub at 4000 m. Nestled in Khangchendzonga Biosphere Reserve (KBR), the transect runs across the undisturbed forest ridge supporting *Prek Chu-Rathong Chu* Catchment. Most of the area of Yuksam-Dzongri transect falls within the Khangchendzonga National Park (KNP) the first "Mixed World Heritage Site" of

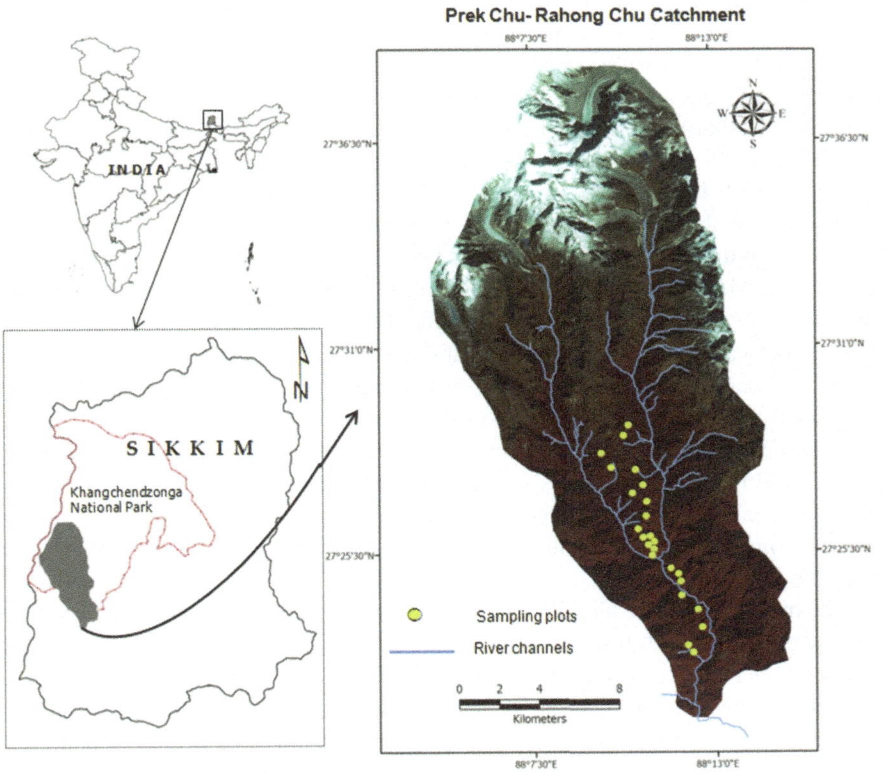

Fig. 10.1 Map showing location of Khangchendzonga National Park (KNP), Sikkim in northeast India and various sampling location distributed over different elevation in Prekchu-Rathongchu Catchment

India on UNESCO World Heritage List under both cultural and natural heritage category (Pandey et al. 2018a, b). The KNP covers an area of 1784 km^2 (approx. 25% of the geographical area of Sikkim state) and located between 27°03′41″ and 28°7′34″ N latitude and, 88°03'40″ and 88°57′19″ E longitude. The area under KNP varies from 1220 m to 8586 m elevation (Mount Khangchendzonga peak) includes a wide range of vegetation with a considerable area under subalpine forests and alpine meadows. The climate is monsoonal with annual precipitation of 1021.0 ± 157.01 mm, and mean annual temperature 5.68 ± 1.79 °C with maximum temperature of warmest month 15.93 ± 1.64 °C and, minimum temperature of coldest month −8.36 ± 3.47 °C (Hijmans et al. 2005; http://www.worldclim.org).

10.2.1.1 Vegetation Sampling and Data Analysis

Three 0.1 ha plots (50 m × 20 m) were sampled in every 100 m elevational the mixed-broad-leaved forest to moist alpine scrub (1700–4000 m). In each 0.1 ha plot, five 10 m × 10 m quadrats were laid randomly for enumerating tree species, and in every 10 m × 10 m quadrats one 5 m × 5 m sub-quadrat was laid randomly for enumerating shrub species, and four 1 m × 1 m random quadrat to sample herb species (Pandey et al. 2018b). All individuals of species present within the quadrats were enumerated. Woody stems of size ≥10 cm girth at breast height (1.37 m from the ground) were measured and the height of each stem was recorded. The voucher specimens of sampled species were put up on herbarium sheet and identified by consulting herbaria and regional flora (Hooker 1872–1897; Maity et al. 2018; Ghosh and Mallick 2014). The accepted plant species names and families were adapted from The Plant List (2013).

The data collected in the field along elevation gradient were analyzed to obtain the values of density, frequency, and total basal area of the species following Misra (1968) and Mueller-Dombois and Ellenberg (1974). A frequency distribution was developed in five frequency classes of Raunkiaer (1934). The importance value index (IVI) was calculated by the summation of relative values of frequency, density, and basal area (Curtis and McIntosh 1950). The Shannon's diversity index (H') was calculated following Shannon (1948), Simpson's dominance index was calculated following Simpson (1949) and Pielou's index of evenness was calculated following Pielou (1975). Species richness was determined as the total number of species in sampled area for different life forms. Linear regression analysis was used to understand the linear/quadratic relationship between tree diversity and tree richness with elevational gradients.

The tree species were classified in to following growth forms: large tree, medium tree, and small tree following Whitehurst et al. (2013) and Kumar (2014). The species which occupied upper canopy and emergent strata (≥15 m) were large species, species which occurred in middle canopy (−15 m) were moderate size species, and the species which were restricted to understory (<5 m) were small trees. For phanerophytic life form, each tree species was classified into following life forms; megaphanerophyte (>25 m), mesophanerophyte (>8–25 m), and microphanerophyte (2–8 m) following Raunkiaer (1934).

10.3 Results

A total of 5959 (≥10 cm) tree individuals were recorded in 3.6 ha area that was sampled along the 1700–4000 m elevation gradient for tree layer. They belonged to 32 families, 48 genera, and 77 species, and were distributed across five forests types. In shrub layers, a total of 7862 individuals belonging to 56 species, 42 genera, and 21 families were recorded, and in herbaceous layer, a total of 35,935 individuals of 134 species belonging to 95 genera and 50 families were recorded. Thus about 50% species were woody, which is close to global scale percentage of woody species (43%).

10.3.1 Forest Type Distribution

The elevation transect was divisible into five forest types based on dominance of trees: (1) mixed-broad leaved forests between 1700 and 2000 m, (2) oak-dominated forests between 2300 and 2700 m, (3) rhododendron mixed forests between 2800 and 3000 m, (4) conifer mixed forests between 3100 and 3900 m, and (5) moist alpine scrub, in and around treeline at 4000 m (Table 10.1).

10.3.2 Family Dominance

In tree layer, out of 32 families, 18 had a single species each, 4 families had two species each, and 6 families had three species each. The most speciose families were Ericaceae (14 species), followed by Fagaceae (7), Lauraceae (7), and Rosaceae (5). At generic level, Lauraceae topped the list with five genera and Fagaceae and Araliaceae followed with three genera each. As many as 21 families had single genus each and 8 families had two genera each (Table 10.2). Of the 21 shrub families, seven families had a single species each, another seven had two species each, and three families had three species each. The most speciose families in shrub layer were Ericaceae (11), Rosaceae (7), Berberidaceae (4), and Rubiaceae (4). At generic level Rosaceae with five genera were most important families, and Ericaceae and Rubiaceae followed with four genera each (Table 10.3). In herbaceous layer, out of 50 families, 25 had a single species each, 6 had two species each, and 10 had three species each. The most speciose families in herb layer were Compositae (17), Polygonaceae (10), Rosaceae (8), and Primulaceae (7). At generic level Compositae were topped with 12 genera, followed by Ranunculaceae with five genera, Polygonaceae and Rosaceae with four genera in each (Table 10.3).

10.3.3 Physiognomy, Life Form, and Frequency Distribution

The tree height was between 25 and 28 m for emergent tree species, between 15 and 25 m for upper-canopy tree species, between 5 and 15 m for middle canopy tree

Table 10.1 Vegetation attributes of sampling plots along the elevation gradient (1700–4000 m) in Yuksam-Dzongri transect, Khangchendzonga National Park (KNP), Sikkim

Elevation (m)	Forest types	Dominant tree species
1700	Mixed-broad leaved forest	*Schima wallichii, Macaranga indica, Engelhardtia spicata, Castanopsis hystrix, Castanopsis tribuloides, Litsea cubeba, Leptobotia elongata, Lyonia ovalifolia, Eurya* spp., *Prunus cerasoides, Symplocos glomerata., Alnus nepalensis*
1800		
1900		
2000		
2100		
2200		
2300	Oak-dominated forest	*Quercus lamellose, Quercus lineata, Lithocarpus pachyphyllus, Laniisoma elegans, Alnus nepalensis, Cinnamomum impressinervium, Castanopsis hystrix, Prunus bracteopadus*
2400		
2500		
2600		
2700		
2800	Rhododendron mixed forest	*Rhododendron arboreum, Rhododendron falconeri, Rhododendron grande, Magnolia hodgsonii, Tsuga dumosa*
2900		
3000		
3100	Conifer mixed forest	*Abies densa, Tsuga dumosa, Rhododendron hodgsonii, Viburnum nervosum, Sorbus microphylla*
3200		
3300		
3400		
3500		
3600		
3700		
3800		
3900		
4000	Moist alpine scrub	*Rhododendron wigthii, Rhododendron lanatum, Rhododendron thomsonii, Sorbus microphylla, Symplocos dryophila*

species and for below 5 m for understory species. The majority of tree species along the elevation gradient were large tree (29), followed by medium (28) and small trees (20) (Table 10.4). The number of individuals of tree species was high for small trees followed by medium and large trees. In term of IVI, large trees shared the higher value followed by small and medium trees (Fig. 10.2). The forests of Sikkim Himalaya exhibited a phanerophytic life form, where mega- and meso-phanerophytes were dominant in number of species, number of individuals, basal area, and importance value index (Fig. 10.3). The proportion of evergreen species was much higher than the proportion of deciduous tree species in all phanerophytic life form.

In Raunkiaer's frequency occupancy, 82.1% of tree species belonged to class "A" with ≤20% frequency of occurrence (Fig. 10.4), 12.3% of species belonged to class

Table 10.2 Familial composition (number of species and genera, and different growth form) of tree layer in different forest types along the elevation gradient in Khangchendzonga National Park, Sikkim. The total number of family was 32

Family	Number of genera	Number of species	LT	MT	ST
Actinidiaceae	1	1	1		
Adoxaceae	1	1			1
Anacardiaceae	1	1		1	
Aquifoliaceae	1	1	1		
Araliaceae	3	3		2	1
Betulaceae	2	3	3		
Cornaceae	1	1		1	
Elaeocarpaceae	1	1	1		
Ericaceae	2	14		5	9
Euphorbiaceae	1	1	1		
Fagaceae	3	7	7		
Garryaceae	1	1		1	
Hamamelidaceae	1	1	1		
Juglandaceae	2	2	2		
Lamiaceae	2	2		2	
Lauraceae	5	7	2	5	
Leguminosae	1	1		1	
Magnoliaceae	1	1	1		
Moraceae	1	1		1	
Myrtaceae	1	1		1	
Pentaphylacaceae	1	3		1	2
Phyllanthaceae	1	1			1
Pinaceae	2	2	2		
Primulaceae	1	1			1
Rosaceae	2	5	2	2	1
Rutaceae	2	2	1	1	
Salicaceae	1	1	1		
Sapindaceae	1	3	2	1	
Simaroubaceae	1	1			1
Symplocaceae	1	3		3	
Theaceae	1	1	1		
Thymelaeaceae	2	3			3
Total	48	77	29	28	20

LT: Large tree; *MT*: Medium tree; *ST*: Small tree

"B" (>20–≤40% frequency of occurrence), 3.1% belonged to class "C" (>40–≤60%), 2.1% belonged to class "D" (>60–≤80%), and 0.4% belonged to class "E" (>80–100%).

Table 10.3 Familial composition (count of species and genera) of herb and shrub layer in different forest types along the elevation gradient in Khangchendzonga National Park, Sikkim. The number of families was 50 for herbs, and 21 for shrubs

Family	Number of species	Number of genus
Herb layer		
Acanthaceae	1	1
Amaranthaceae	1	1
Apiaceae	3	3
Araceae	3	1
Araliaceae	3	3
Asparagaceae	4	3
Balsaminaceae	3	1
Begoniaceae	1	1
Berberidaceae	1	1
Boraginaceae	2	2
Brassicaceae	2	2
Campanulaceae	1	1
Caprifoliaceae	1	1
Caryophyllaceae	2	1
Commelinaceae	1	1
Compositae	17	12
Convolvulaceae	1	1
Cyperaceae	3	3
Dryopteridaceae	3	2
Equisetaceae	1	1
Euphorbiaceae	1	1
Gentianaceae	3	2
Geraniaceae	3	1
Hypericaceae	1	1
Hypoxidaceae	1	1
Juncaceae	1	1
Lamiaceae	3	3
Leguminosae	1	1
Liliaceae	1	1
Linaceae	1	1
Melanthiaceae	1	1
Melastomataceae	1	1
Nartheciaceae	1	1
Orchidaceae	1	1
Orobanchaceae	1	1
Oxalidaceae	2	1
Papaveraceae	1	1
Phrymaceae	1	1
Plantaginaceae	4	3
Poaceae	6	4

(continued)

Table 10.3 (continued)

Family	Number of species	Number of genus
Polygonaceae	10	4
Primulaceae	7	2
Pteridaceae	1	1
Ranunculaceae	6	5
Rosaceae	8	4
Rubiaceae	2	1
Saxifragaceae	3	3
Urticaceae	5	3
Violaceae	2	1
Zingiberaceae	1	1
Shrub layer		
Adoxaceae	2	2
Anacardiaceae	1	1
Berberidaceae	4	2
Cupressaceae	1	1
Ericaceae	11	4
Gleicheniaceae	2	2
Grossulariaceae	1	1
Hydrangeaceae	2	2
Lamiaceae	2	2
Leguminosae	1	1
Melastomataceae	3	3
Pentaphylacaceae	1	1
Poaceae	2	2
Polygonaceae	1	1
Primulaceae	2	1
Ranunculaceae	1	1
Rosaceae	7	5
Rubiaceae	4	4
Rutaceae	2	1
Thymelaeaceae	3	2
Urticaceae	3	3

10.3.4 Elevation Pattern of Species Richness

A significant decline in total number of species (including tree, shrub, and herb) was observed toward the higher end of elevation gradient (Fig. 10.5). Along the elevation gradient the tree species richness followed a reverse J-shaped curve and peaked at lower elevation (1700 m). The shrub and species richness followed exponentially random decline curve and peaked at 1900 m. Herb species richness followed random distribution with richness declined through the gradient and peaked at lower (1700 m) and mid (2900 m) elevation.

Table 10.4 Species richness, density, basal, area, and importance value index IVI of trees by different growth form in Khangchendzonga National Park, Sikkim

	Species		Density		Basal area		IVI	
Growth form	Number	%	ha^{-1}	%	m^2ha^{-1}	%	Value	%
Large tree	29	37.7	339.33	20.5	13.33	46.7	114.2	38.07
Medium tree	28	36.4	576.04	34.8	9.19	34.8	83.1	27.70
Small tree	20	26.0	739.91	44.7	4.88	18.5	102.7	34.23
Total	77	100	1655.28	100	27.4	100	300	100

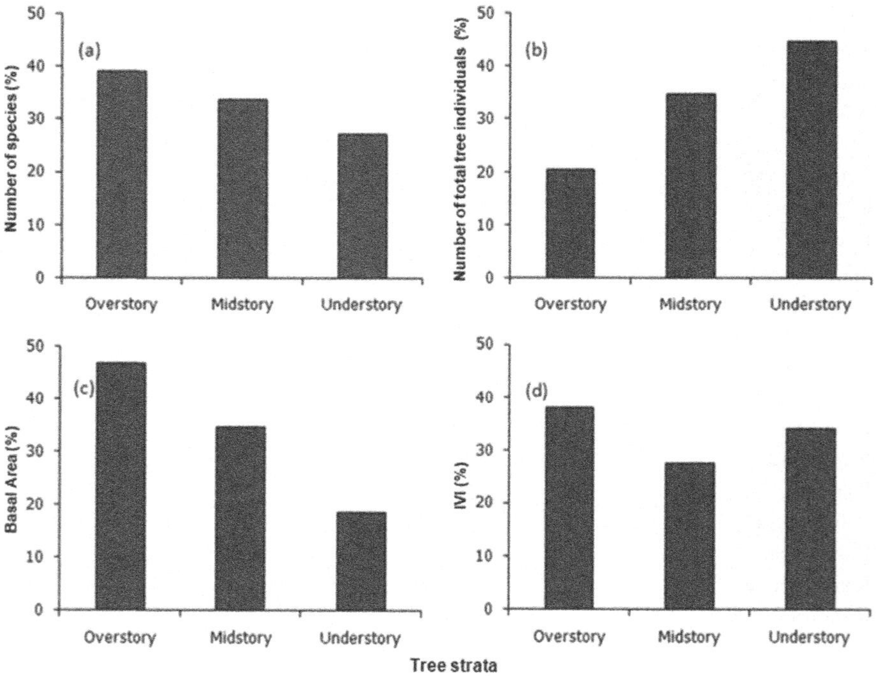

Fig. 10.2 Percentage of number of tree species, density, basal area, and IVI in different canopy layer structure in eastern Himalayan forests of Sikkim

10.3.5 Abundance (Tree Density and Basal Area) along the Elevation Gradient

The stand tree density (mean ± SD; 1655.2 ± 309 individuals/ha) varied significantly ($P < 0.05$) along the studied elevation gradient, with significantly higher stand density (2615.6 individuals/ha; $P < 0.05$) at 1800 m in mixed-broad leaved forests and minimum (303.7 individuals/ha; $P < 0.05$) at treeline (4000 m) (Fig. 10.6). Total Basal Area (TBA) values (mean ± SD; 27.4 ± 4.52 m^2/ha) differed significantly ($P < 0.05$) along the elevation gradient ranging from 0.90 ± 0.51 m^2/ha

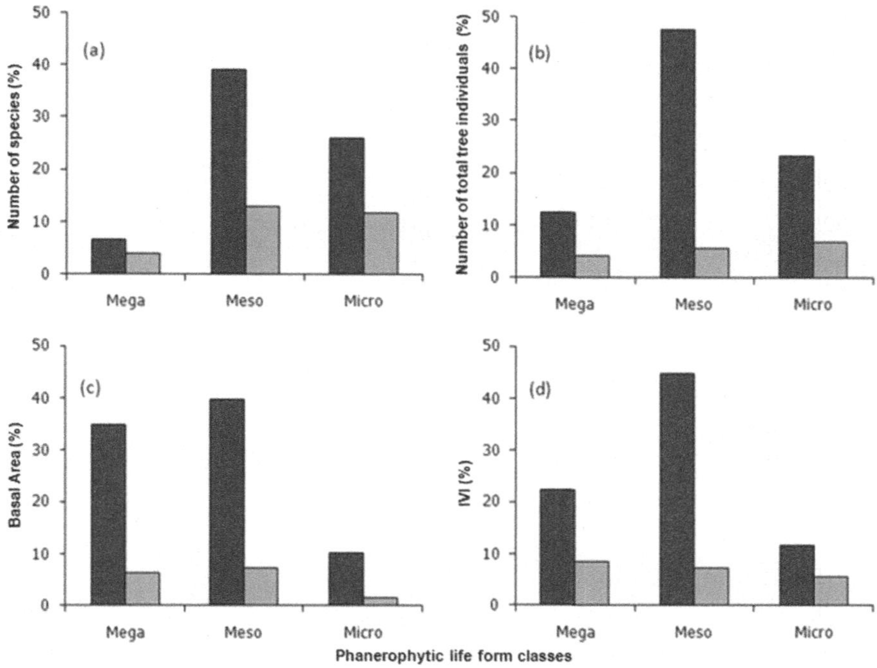

Fig. 10.3 Life-form spectrum of eastern Himalayan forests of Sikkim. The woody species were categorized into three phanerophytic classes: megaphanerophyte, mesophanerophyte, and microphanerophyte. The black bars indicate evergreen species and gray bars indicate deciduous species

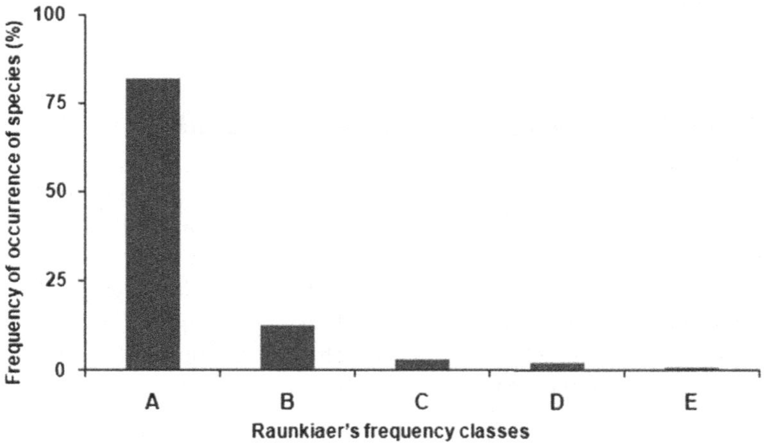

Fig. 10.4 Frequency of occurrence of species in subplots of 0.15 ha size, modeled on Raunkiaer's frequency classes in eastern Himalayan forests of Sikkim. Class "A" with ≤20% frequency of occurrence "B" (>20–≤40%), "C" (>40–≤60), "D" (>60–≤80), and "E" (>80–100%)

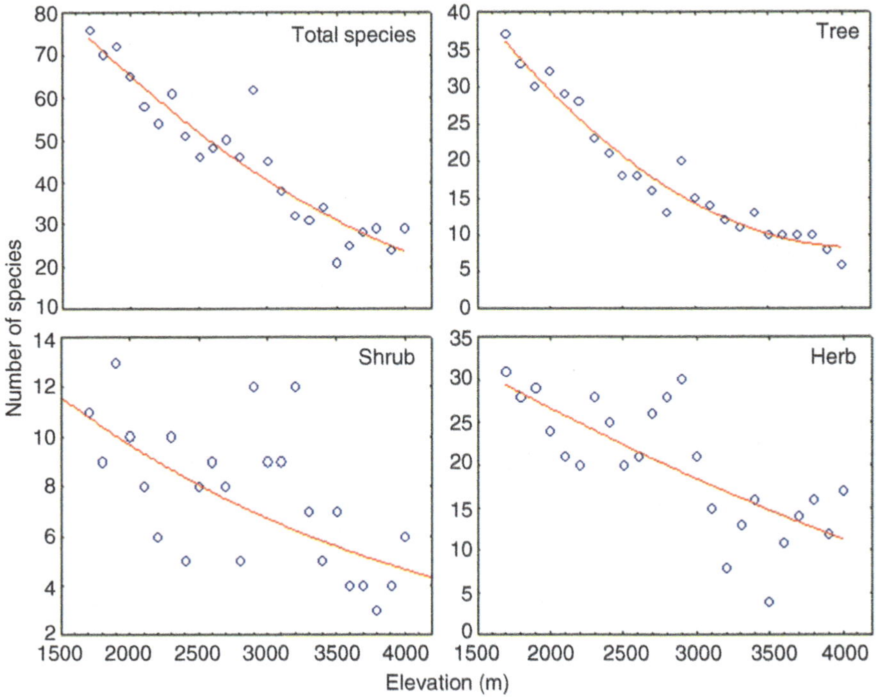

Fig. 10.5 Trends of species richness along the elevation gradient (1700–4000 m) in different growth form, in Khangchendzonga National Park, Sikkim

at treeline (4000 m) to 52.5 m^2/ha at 3200 m in *Abies* and *Tsuga* mixed communities (Fig. 10.6).

10.3.6 Species Diversity along the Elevation Gradient

The species diversity and dominance showed significant quadratic relation with elevation. The tree species diversity declined monotonically toward the higher elevation. The Shannon's diversity index (H') was 3.18 at 1700 m, whereafter it decreased continuing and was up to 1.21 at the treeline at 4000 m (Fig. 10.7a). The Pielou's evenness or homogeneity index (E) decreased toward the higher elevation, and ranged from 0.671 to 0.882 (Fig. 10.7b). The species dominance increased with increase in elevation and the Simpson's dominance index (D) ranged from 0.011 at 1700 m to 0.882 at 4000 m.

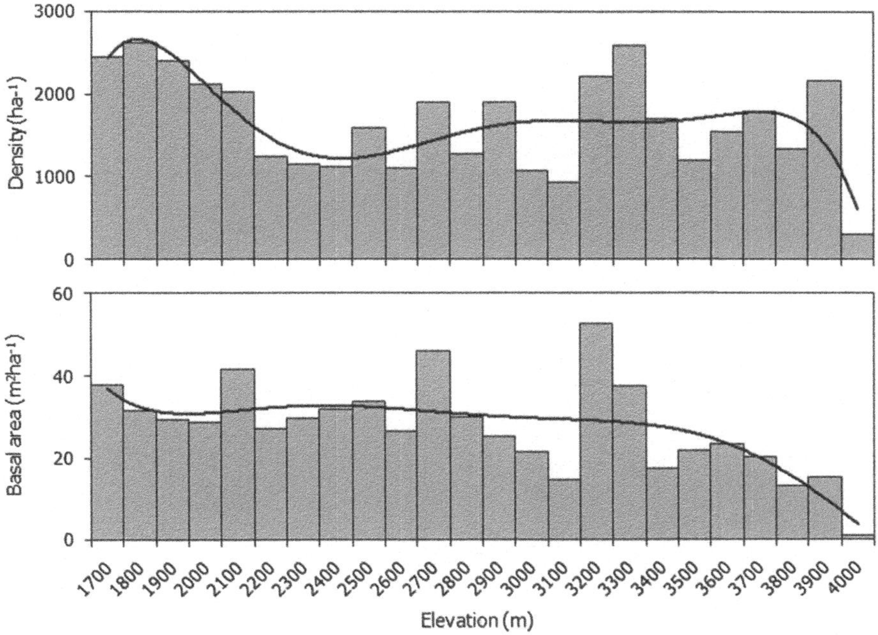

Fig. 10.6 Changes in density and basal area of tree species along the elevation gradient (1700–4000 m) in Khangchendzonga National Park, Sikkim

10.4 Discussion

Understanding species diversity and distribution patterns along the elevation gradient is important for helping managers to evaluate the complexity as well as the prospects of Himalayan forest ecosystems. In this study, we recorded 77 tree species, 56 shrub species, and 134 herb species in Sikkim, indicating the richness of the region. The abundance of herb species has also reported by Manish et al. (2016) and Sharma et al. (2019) for Sikkim Himalaya. Overall, this study recorded (including tree, shrub, and herb species) 81 families belonging to 267 species and 174 genera, in which Ericaceae was the most speciose family with 25 species. Out of the total 81 recorded families 37 were monotypic. Whereas Asteraceae (49 species) is reported to be the most speciose family with 28 monotypic family recorded in the Western Himalayan forests of India (Sharma et al. 2014), and Fagaceae (27) was the most speciose family with 49 monotypic family in the Eastern Himalayan forest of Arunachal Pradesh (Saikia et al. 2017). This difference can be related to the variation in terms of growth forms, species composition, climatic condition, and study area, as trees and shrubs are more dominant in the Eastern Himalayan region, whereas herbaceous species are dominant in the Western Himalayan region of forests.

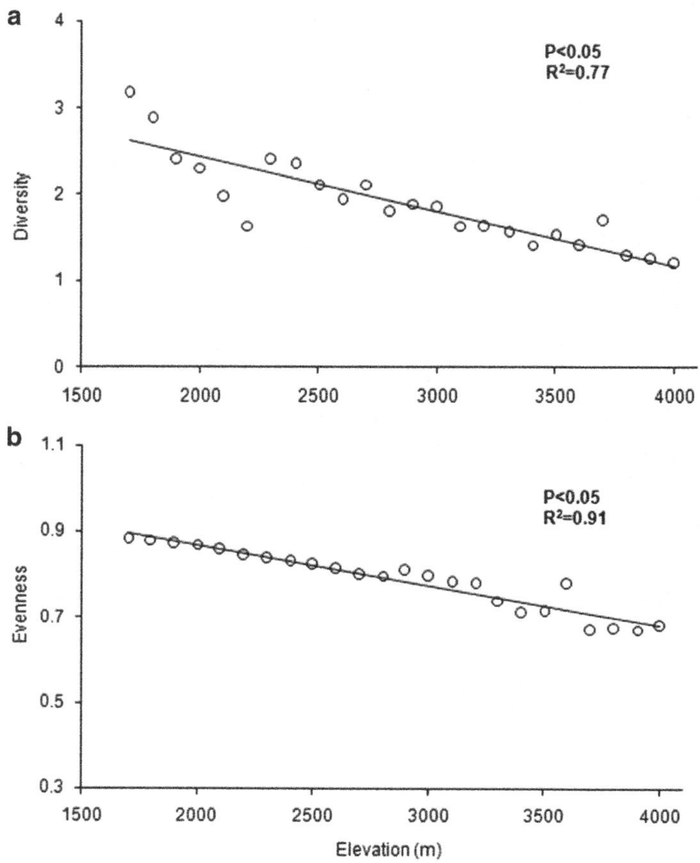

Fig. 10.7 Relationship between elevation and tree species diversity were: (**a**) Shannon index of diversity (*H′*); and (**b**) Pielou's evenness or homogeneity index (*E*)

In Sikkim, Acharya et al. (2011) reported the tree density of 965 individuals ha^{-1}) at elevation ranging from 300 to 4700 m, which is lower with the recorded tree density (1655.2 individuals/ha) in this study from mixed-broad leaved forest to most alpine scrub (1700–4000 m). The tree density values for Sikkim, are comparatively higher than that of Arunachal Pradesh, (456 individuals/ha), ranging 87–4161 m elevation (Saikia et al. 2017), western Himalayan (728 individuals/ha) subalpine forests (Gairola et al. 2015), and (546–616 individuals/ha) different ridge top forests of Uttarakhand west Himalayan region (Sharma et al. 2017). This difference in tree density can be attributed to the forest community structure, age class, site history, on site condition, and other factors (Parthasarathy 2001; Kumar et al. 2006). In the present study, the tree density showed irregular trend along the elevation gradient and higher value was recorded at 1800 m (2615.6 individuals/ha) in mixed-broad leaved forests and at 3300 m (2586.7 individuals/ha) in conifer dominated forests. This may be due to high regeneration potential of mixed-broad leaved forests (Paul et al. 2018) and less vulnerability to herbivory for conifer dominated forests (Begon et al. 2006). The mean basal area (27.4 ± 4.52 m^2/ha) of tree species in our study is lower than the reported basal area (72.1 ± 69.8 m^2/ha) in other parts of Sikkim (Acharya et al. 2011). These differences in the basal area may be attributed to altitudinal variations, species composition, population structure, and successional stage of the forests (Swamy et al. 2000). The maximum basal area was observed between mid elevation oak-dominated (45.9 m^2/ha at 2700 m) and conifer mixed (52.5 m^2/ha at 3000 m) forests, and it's attributed to the presence of large sized tree species in these forests.

Along the elevation gradient, we documented 77 tree species and the number of tree species declined with increasing girth size. Similar trends are exhibited by various taxa along the elevation gradients in the mountain regimes studied elsewhere (Graham 1990; Gaston 1996; Cardelus et al. 2006; Acharya et al. 2011; Saikia et al. 2017; Pandey et al. 2018b). Because of restricted resilience to climatic variations, many tree species are unable to expand their ranges beyond certain elevation range (Jetz and Rahbek 2002). Further, factors such as habitat availability, competition, dispersal and establishment ability, abundance, climatic tolerance, and historical incidences are the determinant factors of species range limit (Gaston 1996). The tree species richness along the elevation gradient exhibited a hump-shaped distribution pattern, which is usual pattern in the Indian Himalayan region (Behera and Kushwaha 2007; Acharya et al. 2011; Saikia et al. 2017; Pandey et al. 2018b). Our study showed the similar trend, thus confirming hump-shaped distribution pattern of tree species richness along the elevation gradient in Sikkim Himalaya region. This finding is supported by other studies in the forests of eastern Himalayan region of India (Behera and Kushwaha 2007; Acharya et al. 2011; Saikia et al. 2017; Sharma et al. 2019). Acharya et al. (2011) reported peak in tree species richness in Sikkim, at 1500 m and then abrupt decrease in richness till 3800 m. The Shannon's diversity index (H') is generally high in tropical forests of the Indian subcontinent and ranges from 0.81 to 4.1 (Singh et al. 1984; Parthasarathy et al. 1992; Bhuyan et al. 2003). The Shannon's diversity values for tree species in this study ranged between 1.2 and 3.2, which is slightly higher than that of the temperate forests of western Himalaya (Gairola et al. 2011; Dar and Sundarapandian 2016; Singh et al. 2016). The tree species diversity decreased with increasing elevations, which was in accordance with the earlier reports in the western Himalaya (Sharma et al. 2009; Singh and Kaushal 2006) and eastern Himalaya (Saikia et al. 2017).

One of the notable features of the present transect of Sikkim was that canopy was broken and undercanopy layers were continuous. On an average canopy tree density was 20.51% and undercanopy density was 79.51% of total density. It seems that the well-developed undercanopy did not allow canopy species to regenerate adequately. It is also possible that once canopy trees are uprooted or damaged the undercanopy layer thickens, making canopy regeneration difficult. The subtropical to temperate forests of Sikkim Himalaya exhibited a phanerophytic life-form spectrum. Among the woody life forms (≥10 gbh), the mega- and mesophanerophytes are predominant, and share up to 83% value of IVI, whereas the microphanerophytes contribute more species richness than IVI. These patterns of life-form spectrum substantiate with the other observation by Champion and Seth (1968), Whitmore (1984), and Shankar and Tripathi (2017) in various part of Indian subcontinent.

10.5 Conclusion

This study analyzes the floristics and phytosociological attributes along an elevation range of 1700–4000 m at 100 m elevation steps in eastern Himalaya forests of Sikkim. Species richness patterns of various growth forms were studied in the forests communities, ranging from broad-leaved forests (1700 m) to treeline ecotone (4000 m), which include 267 species from 174 genera and 81 families. The three layers of trees: (1) understory layer of low girth size trees, (2) canopy layer of *Betula* spp., *Vitex quinata*, *Rhododendron* spp., at higher elevation and *Prunus* spp., *Alnus nepalensis*, and *Eurya* spp. at middle elevation and (3) emergent layer of *Abies densa* and *Tsuga dumosa* at higher elevations and *Engelhardtia spicata, Castanopsis* spp., and *Lithocarpus* spp. at middle elevations. The total basal area was highest in middle elevations. The overall species richness declined monotonically with elevation. The reduction in tree height and richness along the elevation gradient is due its greater sensitivity to low temperature, dispersal and establishment abilities, and liquid water availability. Climatic and anthropogenic factors, which are not evaluated in this study, might influence for species diversity pattern along the elevation gradients in the forests of eastern Himalaya. A sparse canopy layer and continuously dense undercanopy layer and overall high tree density are the characteristics of the study forests.

Acknowledgements This study is part of the *Timberline and Altitudinal Gradient Ecology of Himalayas, and Human Use Sustenance in a Warming Climate* project funded by the Ministry of Environment, Forest & Climate Change, Govt. of India, under National Mission on Himalayan Studies (NMHS) program. We thank the Department of Forests and Environment and Department of Home, Government of Sikkim for permission to carry out research work in protected areas. We thank Director G.B. Pant National Institute of Himalayan Environment for providing necessary facilities to carry out the research work. The support of Mr. Bhim Pandey, Mr. Anoop Raj Singh, and Mr. Nimesh Timsina during field survey is duly acknowledged.

References

Acharya BK, Chettri B, Vijayan L (2011) Distribution pattern of trees along an elevation gradient of eastern Himalaya, India. Acta Oecol 37:329–336

Baniya C, Solhøy T, Gauslaa Y, Palmer MW (2012) Richness and composition of vascular plants and cryptogams along a high elevational gradient on Buddha mountain, Central Tibet. Folia Geobot 47:135–151

Begon M, Townsend CR, Harper JL (2006) Ecology: from individuals to ecosystems. Blackwell Scientific Publications, Oxford

Behera MD, Kushwaha SPS (2007) An analysis of altitudinal behavior of tree species in Subansiri district, eastern Himalaya. Biodivers Conserv 16:1851–1865

Belnap J, Gillette DA (1998) Vulnerability of desert biological soil crusts to wind erosion: the influence of crust development, soil texture and disturbance. J Arid Environ 39:133–142

Bhattarai KR, Vetaas OR (2006) Can Rapoport's rule explain tree species richness along the Himalayan elevation gradient, Nepal? Divers Distrib 12(4):373–378

Bhuyan P, Khan ML, Tripathi RS (2003) Tree diversity and population structure in undisturbed and human-impacted stands of tropical wet evergreen forest in Arunachal Pradesh, eastern Himalayas, India. Biodivers Conserv 12:1753–1773

Cardelus CL, Colwell RK, Watkins JE (2006) Vascular epiphyte distribution patterns: explaining the mid elevation richness peak. J Ecol 94:144–156

Carpenter C (2005) The environmental control of plant species density on a Himalayan elevation gradient. J Biogeogr 32:999–1018

Champion HG, Seth SK (1968) A revised survey of the Forest types of India. Government of India, New Delhi

Curtis JT, McIntosh RP (1950) The interrelations of certain analytic and synthetic phytosociological characters. Ecology 31:435–455

Dar JA, Sundarapandian S (2016) Patterns of plant diversity in seven temperate forest types of Western Himalaya, India. J Asia-Pac Biodivers 9:280–292

Gairola S, Sharma CM, Suyal S, Ghildiyal SK (2011) Species composition and diversity in mid-altitudinal moist temperate forests of the Western Himalaya. J For Sci 27(1):1–15

Gairola S, Rawal RS, Todaria NP (2015) Effect of anthropogenic disturbance on vegetation characterristics of sub-alpine forests in and around Valley of Flowers National Park, a world heritage site of India. Trop Ecol 56:357–365

Gaston KJ (1996) Species-range-size distributions: patterns, mechanisms and implications. Trends Ecol Evol 11:197–201

Ghosh DK, Mallick JK (2014) Flora of Darjeeling Himalayas and foothills (angiosperms). Research Circle, Forest Directorate, Govt. of West Bengal

Graham GL (1990) Bats versus birds: comparisons among Peruvian volant vertebrate faunas along an elevational gradient. J Biogeogr 17:657–668

Grytnes JA, Vetaas OR (2002) Species richness and altitude: a comparison between null models and interpolated plant species richness along the Himalayan altitudinal gradient, Nepal. Am Nat 159: 294–304

Hijmans RJ, Cameron SE, Parra JL, Jones PG, Jarvis A (2005) Very high resolution interpolated climate surfaces for global land areas. Int J Climatol 25:1965–1978. https://doi.org/10.1002/joc.1276

Hunter ML, Yonzon P (1993) Altitudinal distributions of birds, mammals, people, forests, andparks in Nepal. Conserv Biol 7:420–423

Jetz W, Rahbek C (2002) Geographic range size and determinants of avian species richness. Science 297:1548–1551

John E, Dale MRT (1990) Environmental correlates of species distributions in a saxicolous lichen community. J Veg Sci 1:385–392

Körner C (2004) Individuals have limitations, not communities – A response to Marrs, Weiher and Lortie et al. J Veg Sci 15(4):581–582

Kumar D (2014) Mapping and distribution of selected tree species in lower Assam region of India. Ph.D. thesis, North-Eastern Hill University, Shillong, Meghalaya

Kumar A, Marcot BG, Saxena A (2006) Tree species diversity and distribution patterns in tropical forests of Garo Hills. Curr Sci 91:1370–1381

Maity D, Mait GG, Chauhan AS (2018) Flora of Kanchenjunga biosphere reserve Sikkim. Bot Survey India, Kolkata

Manish K, Telwala Y, Nautiyal DC, Pandit MK (2016) Modelling the impacts of future climate change on plant communities in the Himalaya: a case study from eastern Himalaya, India. Model Earth Syst Environ 2(2):1–12

Misra R (1968) Ecology work book. Oxford and IBH Publishing Co., Calcutta

Mueller-Dombois D, Ellenberg H (1974) Aims and methods of vegetation ecology. Wiley, Hoboken, NJ

Myers N (1988) Threatened biotas: 'hotspots' in tropical forestry. Environmentalist 8:1–20

Nogués-Bravo D, Araújo MB, Romdal T, Rahbek C (2008) Scale effects and human impact on the elevational species richness gradients. Nature 453(7192):216
Odland A, Birks HJB (1999) The altitudinal gradient of vascular plant richness in Aurland, western Norway. Ecogeg 22:548–566
Oommen MA, Shanker K (2005) Elevational species richness patterns emerge from multiple local mechanisms in Himalayan woody plants. Ecology 86(11):3039–3047
Pandey A, Badola HK, Rai S, Singh SP (2018a) Timberline structure and woody taxa regeneration towards treeline along latitudinal gradients in Khangchendzonga National Park, Eastern Himalaya. PLoS One 13(11):e0207762
Pandey A, Rai S, Kumar D (2018b) Changes in vegetation attributes along an elevation gradient towards timberline in Khangchendzonga National Park, Sikkim. Trop Ecol 59(2):259–271
Parthasarathy N (2001) Changes in forest composition and structure in three sites of tropical evergreen forest around Sengaltheri, Western Ghats. Curr Sci 80:389–393
Parthasarathy N, Kinhal V, Kumar LP (1992) Plant species diversity and human impacts in the tropical wet evergreen forests of southern western Ghats. Indo-French workshop on tropical Forest ecosystems: natural functioning and anthropogenic impact. French Institute, Pondicherry
Paul A, Khan ML, Das AK (2018) Population structure and regeneration status of rhododendrons in temperate mixed broad-leaved forests of western Arunachal Pradesh, India. Geol Ecol Landsc 3(3):168–186
Pielou EC (1975) Ecological diversity. Wiley, New York
Ponzetti JM, McCune BP (2001) Biotic soil crusts of Oregon's shrub steppe: community composition in relation to soil chemistry, climate, and livestock activity. Bryologist 104:212–225
Rai S, Pandey A, Badola HK (2018) Biomass and carbon stock estimation across the timberline of Khangchendzonga National Park, eastern Himalaya, India. Taiwania 63(4):311–320
Raunkiaer C (1934) The life forms of plants and statistical plant geography. Clarendon Press, Oxford
Saikia P, Deka J, Bharali S, Kumar A, Tripathi OP, Singha LB, Khan ML (2017) Plant diversity patterns and conservation status of eastern Himalayan forests in Arunachal Pradesh, Northeast India. For Ecosyst 4(1):28
Shankar U, Tripathi AK (2017) Rainforests north of the tropic of cancer: physiognomy, floristics and diversity in 'lowland rainforests' of Meghalaya, India. Plant Divers 39:20–36
Shannon CE (1948) A mathematical theory of communication. Bell Syst Tech J 27:379–423. 623-656
Sharma CM, Suyal S, Gairola S, Ghildiyal SK (2009) Species richness and diversity along an altitudinal gradient in moist temperate forest of Garhwal Himalaya. J Am Sci 5(5):119–128
Sharma P, Rana JC, Devi U, Randhawa SS, Kumar R (2014) Floristic diversity and distribution pattern of plant communities along altitudinal gradient in Sangla Valley, northwest Himalaya. Sci World J 2014:1–11
Sharma CM, Mishra AK, Tiwari OP, Krishan R, Rana YS (2017) Effect of altitudinal gradients on forest structure and composition on ridge tops in Garhwal Himalaya. Energy Ecol Environ 2: 404–417
Sharma N, Behera MD, Das AP, Panda RM (2019) Plant richness pattern in an elevation gradient in the eastern Himalaya. Biodivers Conserv 28:2085–2104
Shooner S, Davies TJ, Saikia P, Deka J, Bharali S, Tripathi OP, Dayanandan S (2018) Phylogenetic diversity patterns in Himalayan forests reveal evidence for environmental fltering of distinct lineages. Ecosphere 9(5):e02157
Simpson EH (1949) Measurement of diversity. Nature 163:688
Singh Y, Kaushal A (2006) Extraction of geomorphological features using Radarsat 318 data. J Indian Soc Remote 34:300–307
Singh JS, Singh SP (1992) Forests of Himalaya: structure, functioning and impact of man. GyanodayaPrakashan, Nainital
Singh JS, Singh SP, Saxena AK, Ravat YS (1984) The forest vegetation of Silent Valley. Tropical rain forests – the Leeds symposium, pp. 25–52

Singh S, Malik ZA, Sharma CM (2016) Tree species richness, diversity, and regeneration status in different oak (*Quercus* spp.) dominated forests of Garhwal Himalaya, India. J Asia-Pac Biodivers 9:293–300
Swamy PS, Sundarapandian SM, Chandrasekar P, Chandrasekaran S (2000) Plant species diversity and tree population structure of a humid tropical forest in Tamil Nadu, India. Biodivers Conserv 9:1643–1669
The Plant List (2013) Version 1.1. Published on the Internet; http://www.theplantlist.org/
Vetaas OR, Grytnes JA (2002) Distribution of vascular plant species richness and endemic richness along the Himalayan elevation gradient in Nepal. Glob Ecol Biogeogr 11:291–301
Whitehurst AS, Swatantran A, Blair JB, Hofton MA, Dubayah A (2013) Characterization of canopy layering in forested ecosystems using full waveform Lidar. Remote Sens 5:2014–2036
Whitmore TC (1984) Tropical rain forests of the Far East, 2nd edn. Clarendon Press, Oxford

Patterns of Plant Species Richness Across the Himalayan Treeline Ecotone

11

Anzar Ahmad Khuroo, Firdous A. Dar, Maroof Hamid, Rameez Ahmad, Sajad A. Wani, Aadil Gulzar, A. H. Malik, and C. P. Singh

Abstract

Treeline ecotones, the transition zone between the upper closed forest limit (timberline) and treeless alpine vegetation, are the most conspicuous features of mountain ecosystems around the world. Understanding the patterns of plant species richness in the treeline ecotones is crucial in accurately assessing and monitoring the treeline shifts and vegetation dynamics in the face of climate warming. This chapter, based on a quantitative systematic review of the scientific literature published over the last 30 years (1991–2021), unravels the taxonomic and distributional patterns of plant species richness in the Himalayan treeline ecotone. We found a total of 593 vascular plant species belonging to 232 genera in 70 families reported in the literature on treeline ecotones of the Himalaya. Asteraceae, Rosaceae, Ericaceae, Ranunculaceae and Pinaceae were found to be species-rich families. Overall, the treeline ecotone plant species pool was higher in Western Himalaya as compared to that of Eastern Himalaya and only a small proportion of species was common to both the regions. The majority of the species were of herbaceous growth form and the trend was consistent across the treeline ecotones of Western and Eastern Himalaya. In the ecotone flora of the Himalaya, there are 2 plant species currently recognized by IUCN as near

A. A. Khuroo (✉) · F. A. Dar · M. Hamid · R. Ahmad · S. A. Wani · A. H. Malik
Centre for Biodiversity & Taxonomy, Department of Botany, University of Kashmir, Srinagar, Jammu & Kashmir, India
e-mail: anzarak@uok.edu.in

A. Gulzar
Department of Environmental Science, University of Kashmir, Srinagar, Jammu & Kashmir, India

C. P. Singh
Space Applications Centre, Indian Space Research Organization, Ahmedabad, Gujarat, India
e-mail: cpsingh@sac.isro.gov.in

S. P. Singh et al. (eds.), *Ecology of Himalayan Treeline Ecotone*,
https://doi.org/10.1007/978-981-19-4476-5_11

threatened, 4 vulnerable, 7 endangered, and 2 as critically endangered. Based on the insights gained from the present empirical synthesis, we highlight the research knowledge gaps and suggest way forward for better understanding of patterns of treeline ecotone flora in the Himalaya.

Keywords

Climate change · Flora · Himalaya · Species richness · Treeline ecotone

11.1 Introduction

Globally, mountains are considered as rich repositories of biodiversity, with a relatively high endemic species richness compared to low-land areas (Perrigo et al. 2020). Along the elevation gradient on mountains, the compression of life zones brings different biomes into close proximity within a short distance (Testolin et al. 2021). The treeline ecotone, occupying an intermediate position between the forests below and the treeless alpine vegetation above, represents one of the most conspicuous vegetation zones in the mountain landscapes (Körner 2021; Plate 11.1). These ecotones have attracted the research attention since quite earlier times and received much attention in recent times (Körner 1998; Schickhoff 2005; Körner and Paulsen 2004; Holtmeier and Broll 2005). The reasons for such an increasing scientific interest are manifold. They provide life-supporting ecosystem services, are critical for maintaining global biodiversity patterns, and act as an ecological indicator of global environmental change (Mayor et al. 2017; Skre 2019). These ecosystems are reported to be susceptible to climate warming and results have been widely reported from various treeline ecotones of the world (Körner and Paulsen 2004; Feuillet et al. 2020; Hamid et al. 2019, 2020a, b; Bader et al. 2021). Despite their immense ecological importance and susceptibility to climate warming, we still lack a true picture of the patterns of species richness of these transition zones, especially from the developing world.

Changes in the elevational position and species composition of treeline ecotones, for example in response to recent climate warming, have implications for global carbon dynamics (Cairns and Malanson 1998), local climates as well as on the ecosystem goods and services they provide (Palomo 2017). The treeline ecotones may show gradual or abrupt transitions in several variables (such as tree density, height, age), thereby manifesting in many forms (Plate 11.1). The most important is the formation of an abrupt line or sometimes fuzzy or diffuse boundary, where the highest groups of trees >3 m height in closed forest stands end. Termed as 'treeline' or 'tree limit', it is widely considered as temperature-sensitive and expected to respond to global warming (Holtmeier and Broll 2005). One widespread hypothesis is that global warming will shift the climatic ranges of the treeline and subalpine/alpine species upward along the elevation gradients (Gottfried et al. 2012; Steinbauer et al. 2018). These species' range shifts or treeline expansion are often modulated by local-scale drivers such as historical land-use changes, neighbouring vegetation and

Plate 11.1 Glimpses of treeline ecotone vegetation in Kashmir Himalaya (Picture credits: A. A. Khuroo, M. Hamid)

biotic interactions, which in turn will directly depend on the magnitude and patterns of local species pool. Warming induced treeline expansion also depends on the successful establishment of viable seeds and seedlings beyond the treeline (Lett and Dorrepaal 2018). In addition to other biotic and abiotic factors, neighbouring vegetation and their pattern of richness or distribution play an important role in determining the failure or successful establishment of seedlings at and beyond treelines. For example, neighbouring vegetation can facilitate this establishment by sheltering seedlings from adverse climate effects and herbivores (Batllori et al. 2009a; b). On the other hand, dense vegetation can impede seedling emergence and establishment by competing for light, water and nutrients (Thrippleton et al. 2016). Therefore, understanding the patterns of species richness at the treeline ecotone will indirectly help in determining how ongoing climate warming shifts the species composition and distributions, treeline expansion. It will be also crucial in recognizing the combination of species that will face one another in the future or have potential to colonize alpine landscapes at higher elevations. Moreover, determining the species richness patterns at treeline ecotones may be a key to understanding the dynamic processes that determine the current and future distribution of species and vegetation dynamics (Bader et al. 2021).

The Himalaya, recognized globally as a biodiversity hotspot (Mittermeier et al. 2004), is experiencing rapid changes in land use and climate with temperature increasing faster than the global average at higher elevations (Zaz et al. 2019; Shugar et al. 2021). It holds the world's highest mountain peaks and the highest treelines in the Northern Hemisphere (Miehe et al. 2007). *Abies, Betula, Larix, Picea, Pinus* and *Rhododendron* are among the most conspicuous treeline taxa in the Himalayan mountains. However, because of the remoteness and inaccessibility, the patterns of species richness studies of the Himalayan treeline ecotones have been largely understudied and therefore underrepresented in the scientific literature. Besides, a lack of appreciation of its significance from the developing world is also apparent (Singh 2018; Singh et al. 2020). Recently, studies on treeline-forming tree species, dendroecology, plant physiology, remote sensing have become available (Behera and Kushwaha 2006; Acharya et al. 2011; Liang et al. 2014; Singh et al. 2020). Some studies have focused on treeline shift and expansion as a consequence of rapid warming rates (Schickhoff et al. 2015; Bobrowski et al. 2017; Hamid et al. 2019; Mainali et al. 2020; Singh et al. 2021). Despite these studies, a synthesis of the taxonomic and distributional patterns of plant species richness including all the life forms (trees, shrubs and herbs) that occur typically across the treeline ecotone in the Himalaya is still lacking (see Singh et al. 2020 for the trees).

To fill this knowledge gap, this chapter provides a quantitative systematic review of scientific literature published over the last 30 years (1991–2021), so as to unravel the taxonomic and distributional patterns of plant species richness in the Himalayan treeline ecotone. In this study, we specifically aim to address the following research questions: (i) what is the taxonomic composition of the vegetation forming the Himalayan treeline ecotone? (ii) do the distribution patterns of plant species richness in the Himalayan treeline ecotone differ between Western and Eastern Himalaya? (iii) what are patterns of growth forms (i.e. tree, shrubs and herbs) in the Himalayan treeline ecotone? (iv) which are the currently threatened plant species in the Himalayan treeline ecotone?

11.2 Materials and Methods

11.2.1 Data Sources and Search Tools Used

We performed comprehensive literature searches using the Web of Science (https://www.webofknowledge.com) and Google Scholar (https://scholar.google.com/) databases on 10 May 2021. We used the following keyword combinations to search the relevant publications on treeline ecotone in the Himalaya: 'Himalaya and treeline or tree line', 'alpine and treeline or timberline', 'treeline ecotone and vegetation and elevational gradient'. The search generated over 27,000 publications which were refined to 4147 publications. In comparison to other mountain regions in the world, a lesser number of publications were available specifically focusing on the Himalayan treeline. Based on the explicit mention of 'Himalaya treeline' or 'Himalaya timberline' either in title or abstract, we finally retained 80 publications (Appendix 1) published over the last three decades (1991–2021) for plant species' data extraction. One of the common causes of arbitrary inflation of species numbers in biodiversity research is synonymy (Khuroo et al. 2007). Therefore, we undertook rigorous taxonomic scrutiny to avoid the multiple species entries on account of synonymy. The number of species name records that were extracted from the literature was 670, which was finally refined to 593 valid names after taxonomic curation. The scientific nomenclature of all the species was updated using the recently available WorldFlora R package (https://cran.r-project.org/package=WorldFlora; Kindt 2020). There were few species with ambiguous names which were further resolved by Kew-based Plants of the World Online (https://wwww.plantsoftheworldonline.org). We excluded the taxa mentioning only the genus name. Several species (e.g. *Meconopsis latifolia*, *Iris hookeriana*) were added to the database based on our recent field records.

The species' distribution data were recognized under three categories: (1) distributed across the entire Himalaya, (2) restricted to the Eastern Himalaya and (3) restricted to the Western Himalaya. In this study, the Eastern Himalaya comprises North-Eastern states of India, Eastern Nepal, Bhutan, Tibet autonomous Region and Sichuan Province of China, while the Western Himalaya includes Hindu Kush, Karakoram, Jammu and Kashmir, Himachal Pradesh, Uttarakhand and Western Nepal. The occurrence of each species was recorded in the form of a presence–absence data matrix, wherein the presence of each species was represented with the symbol '1' and absence with the symbol '0'.

The plant species were categorized into 3 growth forms: tree, shrub and herb (Plate 11.1). However, in the case of some species such as *Salix lindleyana*, *Rhododendron barbatum* and *Juniperus indica* which show the tree as well as shrub growth forms were recognized under a different category: shrub/tree. To evaluate the conservation status of the treeline ecotone flora, we used the IUCN (International Union for Conservation of Nature) Red List (IUCN Red List 2017, Version 3, https://www.iucnredlist.org). Moreover, the distribution of each species was checked by using the Plants of the World Online database (https://wwww.

plantsoftheworldonline.org), supplemented with local floras (Dhar and Kachroo 1983; Polunin and Stainton 1984).

11.2.2 Data Analysis

To test whether there are differences in the observed numbers of plant species among the two Himalayan regions (i.e. Eastern and Western Himalaya) in terms of their representation within different growth form categories, we used the chi-square test on Pearson's contingency table. The observed number within each combination of categories was compared with the counts expected based on equal distribution across the categories. The deviation of observed values from expected values within individual categories was expressed using Pearson's residuals in the R package 'vcd 1.4-8' (Meyer et al. 2020).

11.3 Results

11.3.1 Taxonomic Diversity and Distribution Patterns

The systematic review revealed that a total of 593 plant species belonging to 232 genera in 70 families have been reported in the scientific literature published on the Himalayan treeline ecotone (Table 11.1). Region-wise, 423 plant species belonging to 184 genera and 63 families have been reported from the Eastern Himalaya, while 449 plant species belonging to 211 genera and 66 families reported from the Western Himalaya (Fig. 11.1). While 144 and 170 plant species show restricted distribution to Eastern and Western Himalaya, respectively, there were 279 plant species which are common to both the regions (Fig. 11.2).

The top ten species-rich families in the Himalayan treeline ecotone in a decreasing order are Asteraceae (64 species), Rosaceae (45 species), Ericaceae (41 species), Ranunculaceae (34 species), Pinaceae (24 species), Primulaceae (22 species), Poaceae (21 species), Gentianaceae and Polygonaceae (18 species each) and Caryophyllaceae (17 species) (Fig. 11.3a). These ten families contribute ca. 52% (309) of the total plant species present in the region and thus resulting into a higher average family:species ratio of ca. 1:31. The same families with the exception of Caprifoliaceae again comprise the species-rich families in the Eastern Himalaya, although not in the same order and showing an average family:species ratio of ca. 1: 23 (Fig. 11.3b). In contrast, the top ten species-rich families for the Western Himalaya in decreasing order include Asteraceae (57 species), Rosaceae (37 species), Ranunculaceae (31 species), Poaceae (17 species), Polygonaceae (16 species), Apiaceae (14 species), Caprifoliaceae and Caryophyllaceae (13 species each), Primulaceae (12 species) and Ericaceae (11 species) (Fig. 11.3c). These families, contributing ca. 49% (221) of the total plant species in the Western Himalaya, had an average family:species ratio of ca. 1:22. In terms of monotypic families (i.e. families

Table 11.1 Taxonomic inventory of treeline ecotone flora of the Himalaya (The inventory is arranged alphabetically by scientific name, family and growth form)

S. No	Scientific name	Family	Growth form
1	*Abies delavayi* Franch.	Pinaceae	Tree
2	*Abies delavayi* var. *motuoensis* W.C. Cheng & L.K. Fu	Pinaceae	Tree
3	*Abies densa* Griff.	Pinaceae	Tree
4	*Abies fabri* (Mast.) Craib	Pinaceae	Tree
5	*Abies fargesii* var. *faxoniana* (Rehder & E.H. Wilson) Tang S. Liu	Pinaceae	Tree
6	*Abies forrestii* Coltm.-Rog.	Pinaceae	Tree
7	*Abies forrestii* var. *smithii* R. Vig. & Gaussen	Pinaceae	Tree
8	*Abies pindrow* (Royle ex D. Don) Royle	Pinaceae	Tree
9	*Abies spectabilis* (D. Don) Mirb.	Pinaceae	Tree
10	*Abies squamata* Mast.	Pinaceae	Tree
11	*Acer acuminatum* Wall. ex D.Don	Sapindaceae	Tree
12	*Acer caesium* Wall. ex Brandis	Sapindaceae	Tree
13	*Acer campbellii* Hook.f. & Thomson ex Hiern	Sapindaceae	Tree
14	*Acer caudatum* Wall.	Sapindaceae	Tree
15	*Acer pictum* Thunb.	Sapindaceae	Tree
16	*Acer sterculiaceum* Wall.	Sapindaceae	Tree
17	*Achillea millefolium* L.	Asteraceae	Herb
18	*Aconitum ferox* Wall. ex Ser.	Ranunculaceae	Herb
19	*Aconitum fletcherianum* G. Taylor	Ranunculaceae	Herb
20	*Aconitum heterophyllum* Wall. ex Royle	Ranunculaceae	Herb
21	*Aconitum hookeri* Stapf	Ranunculaceae	Herb
22	*Aconitum leave* Royle	Ranunculaceae	Herb
23	*Aconitum lethale* Griff.	Ranunculaceae	Herb
24	*Aconitum rotundifolium* Kar. & Kir.	Ranunculaceae	Herb
25	*Aconitum violaceum* Jacquem. ex Stapf	Ranunculaceae	Herb
26	*Actaea kashmiriana* (J. Compton & Hedd.) J. Compton	Ranunculaceae	Herb
27	*Adiantum venustum* D. Don	Pteridaceae	Herb
28	*Agrostis munroana* Aitch & Hemsl.	Poaceae	Herb
29	*Ainsliaea aptera* DC.	Asteraceae	Herb
30	*Ajaniatibetica* (Hook.f. & Thomson ex C. B. Clarke) Tzvelev	Asteraceae	Herb
31	*Alchemilla trollii* Rothm.	Rosaceae	Herb
32	*Aletris pauciflora* (Klotzsch) Hand.-Mazz.	Nartheciaceae	Herb
33	*Allium wallichii* Kunth	Amaryllidaceae	Herb
34	*Anaphalis busua* (Buch.-Ham.) DC.	Asteraceae	Herb
35	*Anaphalis contorta* (D. Don) Hook.f.	Asteraceae	Herb
36	*Anaphalis nepalensis* (Spreng.) Hand.-Mazz.	Asteraceae	Herb
37	*Anaphalis nubigena* DC.	Asteraceae	Herb
38	*Anaphalis royleana* DC.	Asteraceae	Herb
39	*Anaphalis triplinervis* (Sims) Sims ex C.B. Clarke	Asteraceae	Herb

(continued)

Table 11.1 (continued)

S. No	Scientific name	Family	Growth form
40	*Androsace rotundifolia* Hardw.	Primulaceae	Herb
41	*Androsace sarmentosa* Wall.	Primulaceae	Herb
42	*Anemone elongata* D. Don	Ranunculaceae	Herb
43	*Anemone obtusiloba* D. Don	Ranunculaceae	Herb
44	*Anemone polyanthes* D. Don	Ranunculaceae	Herb
45	*Anemone rivularis* Buch.-Ham. ex DC.	Ranunculaceae	Herb
46	*Anemone rupestris* Wall. ex Hook.f. & Thomson	Ranunculaceae	Herb
47	*Angelica archangelica* L.	Apiaceae	Herb
48	*Angelica glauca* Edgew.	Apiaceae	Herb
49	*Aphragmus hobsonii* (H. Pearson) Al-Shehbaz & Warwick	Brassicaceae	Herb
50	*Aquilegia fragrans* Benth.	Ranunculaceae	Herb
51	*Aquilegia nivalis* (Baker) Falc. ex B.D. Jacks.	Ranunculaceae	Herb
52	*Aquilegia pubiflora* Wall. ex Royle	Ranunculaceae	Herb
53	*Aralia cachemirica* Decne.	Araliaceae	Herb
54	*Arenaria ciliolata* Edgew. & Hook.f.	Caryophyllaceae	Herb
55	*Arenaria densissima* Wall. ex Edgew. & Hook.f.	Caryophyllaceae	Herb
56	*Arenaria festucoides* Benth.	Caryophyllaceae	Herb
57	*Arenaria kansuensis* Maxim.	Caryophyllaceae	Herb
58	*Arenaria orbiculata* Royle ex Edgew. & Hook.f.	Caryophyllaceae	Herb
59	*Argentina polyphylla* (Wall. ex Lehm.) Soják	Rosaceae	Herb
60	*Aria lanata* (D. Don) Decne.	Rosaceae	Tree
61	*Arisaema elephas* Buchet	Araceae	Herb
62	*Arisaema jacquemontii* Blume	Araceae	Herb
63	*Arisaema propinquum* Schott	Araceae	Herb
64	*Arnebia benthamii* (Wall. ex G. Don) I.M. Johnst.	Boraginaceae	Herb
65	*Arnebia euchroma* (Royle) I.M. Johnst.	Boraginaceae	Herb
66	*Artemisia absinthium* L	Asteraceae	Herb
67	*Artemisia brevifolia* Wall. ex DC.	Asteraceae	Shrub
68	*Artemisia roxburghiana* Wall. ex Besser	Asteraceae	Herb
69	*Aruncus dioicus* (Walter) Fernald	Rosaceae	Herb
70	*Aster albescens* (DC.) Wall. ex Hand.-Mazz.	Asteraceae	Herb
71	*Aster falconeri* (C.B. Clarke) Hutch.	Asteraceae	Herb
72	*Aster methodorus* (Benth.) Govaerts	Asteraceae	Herb
73	*Aster thomsonii* C.B. Clarke	Asteraceae	Herb
74	*Astragalus chlorostachys* Lindl.	Fabaceae	Shrub
75	*Astragalus himalayanus* Klotzsch	Fabaceae	Shrub
76	*Astragalus lessertioides* Bunge	Fabaceae	Shrub
77	*Astragalus pycnorhizus* Benth.	Fabaceae	Shrub
78	*Athanasia linifolia* Burm.f.	Asteraceae	Herb
79	*Berberis angulosa* Wall. ex Hook.f. & Thomson	Berberidaceae	Shrub
80	*Berberis concinna* Hook.f.	Berberidaceae	Shrub

(continued)

Table 11.1 (continued)

S. No	Scientific name	Family	Growth form
81	*Berberis erythroclada* Ahrendt	Berberidaceae	Shrub
82	*Berberis jaeschkeana* C.K. Schneid.	Berberidaceae	Shrub
83	*Berberis koehneana* C.K. Schneid.	Berberidaceae	Shrub
84	*Berberis kumaonensis* C.K. Schneid.	Berberidaceae	Shrub
85	*Berberis mucrifolia* Ahrendt	Berberidaceae	Shrub
86	*Berberis pachyacantha* Bien. ex Koehne	Berberidaceae	Shrub
87	*Berberis umbellata* Wall. ex G. Don	Berberidaceae	Shrub
88	*Bergenia ciliata* (Haw.) Sternb.	Saxifragaceae	Herb
89	*Bergenia purpurascens* (Hook.f. & Thomson) Engl.	Saxifragaceae	Herb
90	*Bergenia stracheyi* (Hook.f. & Thomson) Engl.	Saxifragaceae	Herb
91	*Betula alnoides* Buch.-Ham. ex D. Don	Betulaceae	Tree
92	*Betula cylindrostachya* Lindl. ex Wall.	Betulaceae	Tree
93	*Betula platyphylla* Sukaczev	Betulaceae	Tree
94	*Betula utilis* D. Don	Betulaceae	Tree
95	*Bistorta affinis* (D. Don) Greene	Polygonaceae	Herb
96	*Bistorta griffithii* (Hook.f.) Grierson	Polygonaceae	Herb
97	*Bistorta macrophylla* (D. Don) Soják	Polygonaceae	Herb
98	*Bistorta vacciniifolia* (Wall. ex Meisn.) Greene	Polygonaceae	Herb
99	*Bupleurum candollei* Wall. ex DC.	Apiaceae	Herb
100	*Bupleurum longicaule* Wall. ex DC.	Apiaceae	Herb
101	*Bupleurum longifolium* L.	Apiaceae	Herb
102	*Calamagrostis emodensis* Griseb.	Poaceae	Herb
103	*Calamagrostis zenkeri* (Trin.) Davidse	Poaceae	Herb
104	*Caltha palustris* L.	Ranunculaceae	Herb
105	*Campanula argyrotricha* Wall. ex A.DC.	Campanulaceae	Herb
106	*Campanula aristata* Wall.	Campanulaceae	Herb
107	*Caragana gerardiana* Benth.	Fabaceae	Shrub
108	*Caragana sukiensis* C.K. Schneid.	Fabaceae	Shrub
109	*Cardamine trifoliolata* Hook.f. & Thomson	Brassicaceae	Herb
110	*Carex haematostoma* Nees	Cyperaceae	Herb
111	*Carex lemanniana* Boott	Cyperaceae	Herb
112	*Carex munda* Boott	Cyperaceae	Herb
113	*Carex nubigena* D. Don ex Tilloch & Taylor	Cyperaceae	Herb
114	*Carex obscura* Nees	Cyperaceae	Herb
115	*Carex setosa* Boott	Cyperaceae	Herb
116	*Carex stracheyi* Boott ex C.B. Clarke	Cyperaceae	Herb
117	*Cassiope fastigiata* (Wall.) D. Don	Ericaceae	Shrub
118	*Cassiope selaginoides* Hook.f. & Thomson	Ericaceae	Shrub
119	*Cedrus deodara* (Roxb. ex D. Don) G. Don	Pinaceae	Tree
120	*Cenchrus ciliaris* L.	Poaceae	Herb
121	*Cerastium cerastoides* (L.) Britton	Caryophyllaceae	Herb
122	*Cerastium glomeratum* Thuill.	Caryophyllaceae	Herb

(continued)

Table 11.1 (continued)

S. No	Scientific name	Family	Growth form
123	*Chrysopogon aciculatus* (Retz.) Trin.	Poaceae	Herb
124	*Chrysopogon gryllus* (L.) Trin.	Poaceae	Herb
125	*Chrysosplenium carnosum* Hook.f. & Thomson	Saxifragaceae	Herb
126	*Chrysosplenium tenellum* Hook.f. & Thomson	Saxifragaceae	Herb
127	*Circaea alpina* subsp. *micrantha* (Skvortsov) Boufford	Onagraceae	Herb
128	*Cirsium falconeri* (Hook.f.) Petr.	Asteraceae	Herb
129	*Cirsium griffithii* Boiss.	Asteraceae	Herb
130	*Cirsium wallichii* DC.	Asteraceae	Herb
131	*Clematis barbellata* Edgew.	Ranunculaceae	Herb
132	*Clematis montana* Buch.-Ham. ex DC.	Ranunculaceae	Herb
133	*Clinopodium umbrosum* (M. Bieb.) Kuntze	Laimaceae	Herb
134	*Clinopodium vulgare* L.	Lamiaceae	Herb
135	*Clintonia udensis* Trautv. & C.A. Mey.	Liliaceae	Herb
136	*Codonopsis ovata* Benth.	Campanulaceae	Herb
137	*Codonopsis rotundifolia* Benth.	Campanulaceae	Herb
138	*Cortia depressa* (D. Don) C. Norman	Apiaceae	Herb
139	*Corydalis cashmeriana* Royle	Papaveraceae	Herb
140	*Corydalis chaerophylla* DC.	Papaveraceae	Herb
141	*Corydalis cornuta* Royle	Papaveraceae	Herb
142	*Corydalis crithmifolia* Royle	Papaveraceae	Herb
143	*Corydalis diphylla* Wall.	Papaveraceae	Herb
144	*Corydalis govaniana* Wall.	Papaveraceae	Herb
145	*Corydalis juncea* Wall.	Papaveraceae	Herb
146	*Corydalis thyrsiflora* Prain	Papaveraceae	Herb
147	*Cotoneaster acuminatus* Wall. ex Lindl.	Rosaceae	Shrub
148	*Cotoneaster bacillaris* Wall. ex Lindl.	Rosaceae	Shrub
149	*Cotoneaster frigidus* Wall. ex Lindl.	Rosaceae	Shrub
150	*Cotoneaster microphyllus* Wall. ex Lindl.	Rosaceae	Shrub
151	*Cotoneaster rotundifolius* Wall. ex Lindl.	Rosaceae	Shrub
152	*Cotoneaster thymifolius* hort. ex Baker	Rosaceae	Shrub
153	*Crassula campestris* (Eckl. & Zeyh.) Endl.	Crassulaceae	Herb
154	*Cremanthodium oblongatum* C.B. Clarke	Asteraceae	Herb
155	*Cremanthodium purpureifolium* Kitam.	Asteraceae	Herb
156	*Cremanthodium reniforme* (DC.) Benth.	Asteraceae	Herb
157	*Crepis sancta* (L.) Bornm.	Asteraceae	Herb
158	*Cryptogramma stelleri* (S.G. Gmel.) Prantl	Pteridaceae	Herb
159	*Cyananthus incanus* Hook.f. & Thomson	Campanulaceae	Herb
160	*Cyananthus inflatus* Hook.f. & Thomson	Campanulaceae	Herb
161	*Cyananthus lobatus* Wall. ex Benth.	Campanulaceae	Herb
162	*Cyananthus macrocalyx* subsp. *spathulifolius* (Nannf.) K.K. Shrestha	Campanulaceae	Herb

(continued)

Table 11.1 (continued)

S. No	Scientific name	Family	Growth form
163	*Cyananthus microphyllus* Edgew.	Campanulaceae	Herb
164	*Cynoglossum wallichii* G. Don	Boraginaceae	Herb
165	*Cynoglossum wallichii* var. *glochidiatum* (Wall. ex Benth.) Kazmi	Boraginaceae	Herb
166	*Cypripedium cordigerum* D. Don	Orchidaceae	Herb
167	*Cypripedium elegans* Rchb.f.	Orchidaceae	Herb
168	*Cypripedium himalaicum* Rolfe	Orchidaceae	Herb
169	*Dactylorhiza hatagirea* (D. Don) Soó	Orchidaceae	Herb
170	*Delphinium brunonianum* Royle	Ranunculaceae	Herb
171	*Delphinium cashmerianum* Royle	Ranunculaceae	Herb
172	*Delphinium elatum* L.	Ranunculaceae	Herb
173	*Delphinium vestitum* Wall. ex Royle	Ranunculaceae	Herb
174	*Deparia allantodioides* (Bedd.) M. Kato	Athyriaceae	Herb
175	*Diapensia himalaica* Hook.f. & Thomson	Diapensiaceae	Herb
176	*Dipsacus inermis* Wall.	Caprifoliaceae	Herb
177	*Dolomiaea macrocephala* DC. ex Royle	Asteraceae	Herb
178	*Doronicum kamaonense* (DC.) Alv.Fern.	Asteraceae	Herb
179	*Dryopteris stewartii* Fraser-Jenk.	Dryopteridaceae	Herb
180	*Dryopteris wallichiana* (Spreng.) Hyl.	Dryopteridaceae	Herb
181	*Elaeagnus rhamnoides* (L.) A. Nelson	Elaeagnaceae	Herb
182	*Elymus lanceolatus* (Scribn. & J.G. Sm.) Gould	Poaceae	Herb
183	*Ephedra gerardiana* Wall. ex Stapf	Ephedraceae	Shrub
184	*Epilobium amurense* Hausskn.	Onagraceae	Herb
185	*Epilobium brevifolium* D. Don	Onagraceae	Herb
186	*Epilobium latifolium* L.	Onagraceae	Herb
187	*Epilobium laxum* Royle	Onagraceae	Herb
188	*Epilobium parviflorum* Schreb.	Onagraceae	Herb
189	*Epilobium royleanum* Hausskn.	Onagraceae	Herb
190	*Epilobium tetragonum* L.	Onagraceae	Herb
191	*Eremopoa persica* (Trin.) Roshev.	Poaceae	Herb
192	*Eremurus himalaicus* Baker	Asphodelaceae	Herb
193	*Erigeron acris* var. *multicaulis* (Wall. ex DC.) C.B. Clarke	Asteraceae	Herb
194	*Erigeron multiradiatus* (Lindl. ex DC.)	Asteraceae	Herb
195	*Euonymus fimbriatus* Wall.	Celastraceae	Tree
196	*Euphorbia wallichii* Hook.f.	Euphorbiaceae	Herb
197	*Euphrasia himalayica* Wettst.	Orobanchaceae	Herb
198	*Fargesia melanostachys* (Hand.-Mazz.) T.P. Yi	Poaceae	Herb
199	*Ferula jaeschkeana* Vatke	Apiaceae	Herb
200	*Fragaria daltoniana* J. Gay	Rosaceae	Herb
201	*Fragaria nubicola* (Lindl. ex Hook.f.) Lacaita	Rosaceae	Herb
202	*Fritillaria cirrhosa* D. Don	Liliaceae	Herb

(continued)

Table 11.1 (continued)

S. No	Scientific name	Family	Growth form
203	*Galium aparine* L.	Rubiaceae	Herb
204	*Galium asperifolium* Wall.	Rubiaceae	Herb
205	*Galium paradoxum* Maxim	Rubiaceae	Herb
206	*Gaultheria nummularioides* D. Don	Ericaceae	Herb
207	*Gaultheria pyroloides* Hook.f. & Thomson ex Miq.	Ericaceae	Herb
208	*Gaultheria trichophylla* Royle	Ericaceae	Herb
209	*Gentiana argentea* (Royle ex D. Don) Royle ex D. Don	Gentianaceae	Herb
210	*Gentiana carinata* (D. Don) Griseb.	Gentianaceae	Herb
211	*Gentiana cephalodes* Edgew.	Gentianaceae	Herb
212	*Gentiana elwesii* C.B. Clarke	Gentianaceae	Herb
213	*Gentiana phyllocalyx* C.B. Clarke	Gentianaceae	Herb
214	*Gentiana prostrata* var. *karelinii* (Griseb.) Kusn.	Gentianaceae	Herb
215	*Gentiana sikkimensis* C.B. Clarke	Gentianaceae	Herb
216	*Gentiana stylophora* C.B. Clarke	Gentianaceae	Herb
217	*Gentiana tubiflora* (G. Don) Griseb.	Gentianaceae	Herb
218	*Geranium donianum* Sweet	Geraniaceae	Herb
219	*Geranium lambertii* Sweet	Geraniaceae	Herb
220	*Geranium nepalense* Sweet	Geraniaceae	Herb
221	*Geranium polyanthes* Edgew. & Hook.f.	Geraniaceae	Herb
222	*Geranium pratense* L.	Geraniaceae	Herb
223	*Geranium wallichianum* D. Don ex Sweet	Geraniaceae	Herb
224	*Gerbera gossypina* (Royle) Beauverd	Asteraceae	Herb
225	*Geum elatum* Wall. ex G. Don	Rosaceae	Herb
226	*Geum roylei* Wall.	Rosaceae	Herb
227	*Goodyera fusca* (Lindl.) Hook.f.	Orchidaceae	Herb
228	*Goodyera repens* (L.) R.Br.	Orchidaceae	Herb
229	*Gypsophila sedifolia* Kurz	Caryophyllaceae	Herb
230	*Halenia elliptica* D. Don	Gentianaceae	Herb
231	*Hedysarum kumaonense* Baker	Fabaceae	Herb
232	*Hemiphragma heterophyllum* Wall.	Plantaginaceae	Herb
233	*Heracleum candicans* Wall. ex DC	Apiaceae	Herb
234	*Hieracium umbellatum* L.	Asteraceae	Herb
235	*Himalaiella deltoidea* (DC.) Raab-Straube	Asteraceae	Herb
236	*Hippophae salicifolia* D. Don	Elaeagnaceae	Shrub
237	*Hippophae tibetana* Schltdl.	Elaeagnaceae	Shrub
238	*Hymenolaena candollei* DC.	Apiaceae	Herb
239	*Hyoscyamus niger* L.	Solanaceae	Herb
240	*Hypericum perforatum* L.	Hypericaceae	Herb
241	*Impatiens amplexicaulis* Edgew.	Balsaminaceae	Herb
242	*Impatiens bicolor* Royle	Balsaminaceae	Herb
243	*Impatiens biflora* var. *ecalcarata* M.E. Jones	Balsaminaceae	Herb
244	*Impatiens brachycentra* Kar. & Kir.	Balsaminaceae	Herb

(continued)

Table 11.1 (continued)

S. No	Scientific name	Family	Growth form
245	*Impatiens glandulifera* Royle	Balsaminaceae	Herb
246	*Impatiens laxiflora* Edgew.	Balsaminaceae	Herb
247	*Impatiens sulcata* Wall.	Balsaminaceae	Herb
248	*Impatiens thomsonii* Hook.f.	Balsaminaceae	Herb
249	*Inula orientalis* Lam.	Asteraceae	Herb
250	*Inula royleana* DC.	Asteraceae	Herb
251	*Iris hookeriana* Foster	Iridaceae	Herb
252	*Iris kemaonensis* Wall. ex D. Don	Iridaceae	Herb
253	*Jaeschkea canaliculata* (Royle ex G. Don) Knobl.	Gentianaceae	Herb
254	*Juncus concinnus* D. Don	Juncaceae	Herb
255	*Juncus himalensis* Klotzsch	Juncaceae	Herb
256	*Juncus leucanthus* Royle ex D.Don	Juncaceae	Herb
257	*Juncus membranaceus* Royle	Juncaceae	Herb
258	*Juncus thomsonii* Buchenau	Juncaceae	Herb
259	*Juniperus communis* L.	Cupressaceae	Shrub
260	*Juniperus convallium* Rehder & E.H. Wilson	Cupressaceae	Tree
261	*Juniperus excelsa* M. Bieb.	Cupressaceae	Tree
262	*Juniperus indica* Bertol.	Cupressaceae	Tree/ Shrub
263	*Juniperus komarovii* Florin	Cupressaceae	Tree
264	*Juniperus polycarpos* K. Koch	Cupressaceae	Tree
265	*Juniperus polycarpos* var. *seravschanica* (Kom.) Kitam.	Cupressaceae	Shrub
266	*Juniperus przewalskii* Kom.	Cupressaceae	Tree
267	*Juniperus pseudosabina* Fisch. & C.A. Mey.	Cupressaceae	Tree/ Shrub
268	*Juniperus recurva* Buch.-Ham. ex D. Don	Cupressaceae	Tree/ Shrub
269	*Juniperus saltuaria* Rehder & E.H. Wilson	Cupressaceae	Tree
270	*Juniperus semiglobosa* Regel	Cupressaceae	Tree
271	*Juniperus squamata* Buch.-Ham. ex D. Don	Cupressaceae	Tree/ Shrub
272	*Juniperus tibetica* Kom.	Cupressaceae	Tree
273	*Jurinea macrocephala* DC.	Asteraceae	Herb
274	*Kobresia duthiei* C.B. Clarke	Cyperaceae	Herb
275	*Kobresia esenbeckii* (Kunth) Noltie	Cyperaceae	Herb
276	*Kobresia royleana* (Nees) Boeckeler	Cyperaceae	Herb
277	*Koeleria pyramidata* (Lam.) P.Beauv.	Poaceae	Herb
278	*Koenigia polystachya* (Wall. ex Meisn.) T.M. Schust. & Reveal	Polygonaceae	Herb
279	*Kyllinga odorata* Vahl	Cyperaceae	Herb
280	*Lactuca lessertiana* (Wall. ex DC) Wall. ex C. B. Clarke	Asteraceae	Herb

(continued)

Table 11.1 (continued)

S. No	Scientific name	Family	Growth form
281	*Lactuca macrorhiza* (Royle) Hook.f.	Asteraceae	Herb
282	*Lactuca saligna* L.	Asteraceae	Herb
283	*Lamium album* L.	Lamiaceae	Herb
284	*Lancea tibetica* Hook.f. & Thomson	Mazaceae	Herb
285	*Larix griffithii* Hook.f.	Pinaceae	Tree
286	*Larix potaninii* Batalin	Pinaceae	Tree
287	*Lathyrus emodi* (Fritsch) Ali	Fabaceae	Herb
288	*Leontopodium jacotianum* Beauverd	Asteraceae	Herb
289	*Leontopodium stracheyi* (Hook.f.) C.B. Clarke ex Hemsl.	Asteraceae	Herb
290	*Lepidium capitatum* Hook.f. & Thomson	Brassicaceae	Herb
291	*Lepisorus clathratus* (C.B. Clarke) Ching	Polypodiaceae	Herb
292	*Ligularia amplexicaulis* DC.	Asteraceae	Herb
293	*Ligularia fischeri* (Ledeb.) Turcz.	Asteraceae	Herb
294	*Lilium nanum* Klotzsch	Liliaceae	Herb
295	*Lomatogonium caeruleum* (Royle) Harry Sm. ex B.L. Burtt	Gentianaceae	Herb
296	*Lonicera angustifolia* var. *myrtillus* (Hook.f. & Thomson) Q.E. Yang, Landrein, Borosova & J. Osborne	Caprifoliaceae	Shrub
297	*Lonicera asperifolia* Hook.f. & Thomson	Caprifoliaceae	Shrub
298	*Lonicera obovata* Royle ex Hook.f. & Thomson	Caprifoliaceae	Shrub
299	*Lonicera quinquelocularis* Hard.	Caprifoliaceae	Shrub
300	*Lonicera spinosa* (Decne.) Jacq. ex Walp.	Caprifoliaceae	Shrub
301	*Lonicera tomentella* Hook. f. & Thomson	Caprifoliaceae	Shrub
302	*Lonicera webbiana* Wall. ex DC.	Caprifoliaceae	Shrub
303	*Lyonia villosa* (Wall. ex C.B.Clarke) Hand.-Mazz.	Ericaceae	Tree
304	*Lysimachia prolifera* Klatt	Primulaceae	Herb
305	*Maianthemum oleraceum* (Baker) LaFrankie	Asparagaceae	Herb
306	*Maianthemum purpureum* (Wall.) LaFrankie	Asparagaceae	Herb
307	*Malaxis muscifera* (Lindl.) Kuntze	Orchidaceae	Herb
308	*Malva neglecta* Wallr.	Malvaceae	Herb
309	*Mazus pumilus* (Burm.f.) Steenis	Mazaceae	Herb
310	*Meconopsis aculeata* Royle	Papaveraceae	Herb
311	*Meconopsis bella* Prain	Papaveraceae	Herb
312	*Meconopsis latifolia* (Prain) Prain	Papaveraceae	Herb
313	*Meconopsis paniculata* (D. Don) Prain	Papaveraceae	Herb
314	*Meconopsis robusta* Hook.f. & Thomson	Papaveraceae	Herb
315	*Megacarpaea polyandra* Benth. ex Madden	Brassicaceae	Herb
316	*Melica persica* Kunth	Poaceae	Herb
317	*Morina coulteriana* Royle	Caprifoliaceae	Herb
318	*Morina longifolia* Wall. ex DC.	Caprifoliaceae	Herb
319	*Morina nepalensis* D. Don	Caprifoliaceae	Herb

(continued)

Table 11.1 (continued)

S. No	Scientific name	Family	Growth form
320	*Morina polyphylla* Wall. ex DC.	Caprifoliaceae	Herb
321	*Myosotis alpestris* F.W. Schmidt	Boraginaceae	Herb
322	*Myosotis sylvatica* Hoffm.	Boraginaceae	Herb
323	*Myriactis wallichii* Less.	Asteraceae	Herb
324	*Myricaria elegans* Royle	Tamaricaceae	Shrub
325	*Myricaria rosea* W.W. Sm.	Tamaricaceae	Shrub
326	*Nardostachys jatamansi* (D. Don) DC.	Caprifoliaceae	Herb
327	*Nepeta connata* Royle ex Benth.	Lamiaceae	Herb
328	*Nepeta laevigata* (D. Don) Hand.-Mazz.	Lamiaceae	Herb
329	*Onopordum acanthium* L.	Asteraceae	Herb
330	*Onosma hispida* Wall. ex G. Don	Boraginaceae	Herb
331	*Ophiopogon intermedius* D. Don	Asparagaceae	Herb
332	*Origanum vulgare* L.	Lamiaceae	Herb
333	*Osmunda claytoniana* L.	Osmundaceae	Herb
334	*Oxalis leucolepis* Diels	Oxalidaceae	Herb
335	*Oxygraphis polypetala* (Raf.) Hook.f. & Thomson	Ranunculaceae	Herb
336	*Oxyria digyna* (L.) Hill	Polygonaceae	Herb
337	*Panax pseudoginseng* Wall.	Araliaceae	Herb
338	*Parasenecio quinquelobus* (Wall. ex DC.) Y.L. Chen	Asteraceae	Herb
339	*Parnassia chinensis* Franch.	Celastraceae	Herb
340	*Parnassia nubicola* Wall. ex Royle	Celastraceae	Herb
341	*Parnassia pusilla* Wall. ex Arn.	Celastraceae	Herb
342	*Parochetus communis* D. Don	Fabaceae	Herb
343	*Pedicularis alaschanica* Maxim.	Orobanchaceae	Herb
344	*Pedicularis brevifolia* D. Don	Orobanchaceae	Herb
345	*Pedicularis flexuosa* Hook.f.	Orobanchaceae	Herb
346	*Pedicularis hookeriana* Wall. ex Benn.	Orobanchaceae	Herb
347	*Pedicularis lachnoglossa* Hook.f.	Orobanchaceae	Herb
348	*Pedicularis megalantha* D. Don	Orobanchaceae	Herb
349	*Pedicularis mollis* Wall. Ex Benth.	Orobanchaceae	Herb
350	*Pedicularis pectinata* Wall. ex Benn.	Orobanchaceae	Herb
351	*Pedicularis punctata* Decne.	Orobanchaceae	Herb
352	*Pedicularis pyramidata* Royle ex Benth.	Orobanchaceae	Herb
353	*Pedicularis roylei* Maxim.	Orobanchaceae	Herb
354	*Pedicularis siphonantha* D Don	Orobanchaceae	Herb
355	*Peristylus fallax* Lindl.	Orchidaceae	Herb
356	*Persicaria amplexicaulis* (D. Don) RonseDecr.	Polygonaceae	Herb
357	*Persicaria nepalensis* (Meisn.) Miyabe	Polygonaceae	Herb
358	*Persicaria vivipara* (L.) RonseDecr.	Polygonaceae	Herb
359	*Persicaria wallichii* Greuter & Burdet	Polygonaceae	Herb
360	*Phleum alpinum* L.	Poaceae	Herb

(continued)

Table 11.1 (continued)

S. No	Scientific name	Family	Growth form
361	*Phlomoides bracteosa* (Royle ex Benth.) Kamelin & Makhm.	Lamiaceae	Herb
362	*Phymatopteris malacodon* (Hook.) Pic. Serm.	Polypodiaceae	Herb
363	*Picea crassifolia* Kom.	Pinaceae	Tree
364	*Picea likiangensis* (Franch.) E. Pritz.	Pinaceae	Tree
365	*Picea likiangensis* var. *hirtella* W.C. Cheng	Pinaceae	Tree
366	*Picea likiangensis* var. *linzhiensis* W.C. Cheng & L.K. Fu	Pinaceae	Tree
367	*Picea likiangensis* var. *rubescens* Rehder & E.H. Wilson	Pinaceae	Tree
368	*Picea retroflexa* Mast.	Pinaceae	Tree
369	*Picea schrenkiana* Fisch. & C.A. Mey.	Pinaceae	Tree
370	*Picea smithiana* (Wall.) Boiss.	Pinaceae	Tree
371	*Picea spinulosa* (Griff.) A. Henry	Pinaceae	Tree
372	*Picris hieracioides* Sibth. & Sm.	Asteraceae	Herb
373	*Picrorhiza kurroa* Royle ex Benth.	Plantaginaceae	Herb
374	*Pilea racemosa* (Royle) Tuyama	Urticaceae	Herb
375	*Pinus gerardiana* Wall. Ex D. Don	Pinaceae	Tree
376	*Pinus wallichiana* A.B. Jacks.	Pinaceae	Tree
377	*Plantago himalaica* Pilg.	Plantaginaceae	Herb
378	*Platanthera urceolata* (Hook.f.) R.M. Bateman	Orchidaceae	Herb
379	*Pleurospermum angelicoides* (Wall. ex DC.) Benth. ex C.B. Clarke	Apiaceae	Herb
380	*Poa alpina* L.	Poaceae	Herb
381	*Poa angustifolia* L.	Poaceae	Herb
382	*Poa annua* L.	Poaceae	Herb
383	*Polemonium caeruleum* L.	Polemoniaceae	Herb
384	*Polygonatum cirrhifolium* (Wall.) Royle	Asparagaceae	Herb
385	*Polygonatum multiflorum* (L.) All.	Asparagaceae	Herb
386	*Polygonatum odoratum* (Mill.) Druce	Asparagaceae	Herb
387	*Polygonatum verticillatum* (L.) All.	Asparagaceae	Herb
388	*Polygonum delicatulum* Meisn.	Polygonaceae	Herb
389	*Polygonum filicaule* Wall. ex Meisn.	Polygonaceae	Herb
390	*Polypogon fugax* Nees ex Steud.	Poaceae	Herb
391	*Potentilla atrosanguinea* G. Lodd. ex D. Don	Rosaceae	Herb
392	*Potentilla coriandrifolia* G. Don	Rosaceae	Herb
393	*Potentilla eriocarpa* Wall. ex Lehm.	Rosaceae	Herb
394	*Potentilla fruticosa* Hook.f.	Rosaceae	Shrub
395	*Potentilla fruticosa* var. *arbuscula* (D. Don) Maxim.	Rosaceae	Shrub
396	*Potentilla fulgens* Wall. ex Sims	Rosaceae	Herb
397	*Potentilla lineata* Trevir.	Rosaceae	Herb
398	*Potentilla microphylla* D. Don	Rosaceae	Herb
399	*Potentilla multifida* L.	Rosaceae	Herb

(continued)

Table 11.1 (continued)

S. No	Scientific name	Family	Growth form
400	*Potentilla peduncularis* D. Don	Rosaceae	Herb
401	*Potentilla saundersiana* var. *caespitosa* Lehm. ex Th. Wolf	Rosaceae	Herb
402	*Primula capitata* Hook.	Primulaceae	Herb
403	*Primula denticulata* Sm.	Primulaceae	Herb
404	*Primula deuteronana* Craib	Primulaceae	Herb
405	*Primula dickieana* Watt	Primulaceae	Herb
406	*Primula elliptica* Royle	Primulaceae	Herb
407	*Primula elongata* Watt	Primulaceae	Herb
408	*Primula gambeliana* Watt	Primulaceae	Herb
409	*Primula glabra* Klatt	Primulaceae	Herb
410	*Primula glomerata* Pax	Primulaceae	Herb
411	*Primula macrophylla* D. Don	Primulaceae	Herb
412	*Primula megalocarpa* H. Hara	Primulaceae	Herb
413	*Primula nana* Wall.	Primulaceae	Herb
414	*Primula obliqua* W.W. Sm.	Primulaceae	Herb
415	*Primula poluninii* H.R. Fletcher	Primulaceae	Herb
416	*Primula primulina* (Spreng.) H. Hara	Primulaceae	Herb
417	*Primula reidii* Duthie	Primulaceae	Herb
418	*Primula reptans* Hook.f. ex Watt	Primulaceae	Herb
419	*Primula rosea* Royle	Primulaceae	Herb
420	*Primula sikkimensis* Hook.	Primulaceae	Herb
421	*Prunella vulgaris* L.	Lamiaceae	Herb
422	*Prunus cornuta* (Wall. ex Royle) Steud.	Rosaceae	Tree
423	*Prunus napaulensis* (Ser.) Steud.	Rosaceae	Tree
424	*Prunus rufa* Wall. ex Hook.f.	Rosaceae	Tree
425	*Pycnoplinthopsis bhutanica* Jafri	Brassicaceae	Herb
426	*Quercus aquifolioides* Rehder & E.H. Wilson	Fagaceae	Tree
427	*Quercus lineata* Blume	Fagaceae	Tree
428	*Quercus semecarpifolia* Sm.	Fagaceae	Tree
429	*Ranunculus brotherusii* Freyn	Ranunculaceae	Herb
430	*Ranunculus hirtellus* Royle	Ranunculaceae	Herb
431	*Ranunculus palmatifidus* Riedl	Ranunculaceae	Herb
432	*Ranunculus trivedii* Aswal & Mehrotra	Ranunculaceae	Herb
433	*Rhamnus procumbens* Edgew.	Rhamnaceae	Shrub
434	*Rheum acuminatum* Hook.f. & Thomson	Polygonaceae	Herb
435	*Rheum moorcroftianum* Royle	Polygonaceae	Herb
436	*Rheum nobile* Hook.f. & Thomson	Polygonaceae	Herb
437	*Rheum webbianum* Royle	Polygonaceae	Herb
438	*Rhodiola himalensis* (D. Don) S.H. Fu	Crassulaceae	Herb
439	*Rhodiola imbricata* Edgew.	Crassulaceae	Herb
440	*Rhodiola wallichiana* (Hook.) S.H. Fu	Crassulaceae	Herb

(continued)

Table 11.1 (continued)

S. No	Scientific name	Family	Growth form
441	*Rhododendron candelabrum* Hook.f.	Ericaceae	Shrub
442	*Rhododendron adenogynum* Diels	Ericaceae	Shrub
443	*Rhododendron anthopogon* D. Don	Ericaceae	Shrub
444	*Rhododendron arboreum* Sm.	Ericaceae	Tree
445	*Rhododendron argipeplum* Balf.f. & R.E. Cooper	Ericaceae	Tree/ Shrub
446	*Rhododendron barbatum* Wall. ex G. Don	Ericaceae	Tree/ Shrub
447	*Rhododendron bhutanense* D.G. Long & Bowes Lyon	Ericaceae	Tree/ Shrub
448	*Rhododendron camelliiflorum* Hook.f.	Ericaceae	Shrub
449	*Rhododendron campanulatum* D. Don	Ericaceae	Tree/ Shrub
450	*Rhododendron campanulatum* subsp. *aeruginosum* (Hook.f.) D.F. Chamb.	Ericaceae	Tree/ Shrub
451	*Rhododendron campylocarpum* Hook.f.	Ericaceae	Shrub
452	*Rhododendron cerasinum* Tagg	Ericaceae	Shrub
453	*Rhododendron cinnabarinum* Hook.f.	Ericaceae	Shrub
454	*Rhododendron decorum* Franch.	Ericaceae	Tree/ Shrub
455	*Rhododendron falconeri* Hook.f.	Ericaceae	Tree
456	*Rhododendron fulgens* Hook.f.	Ericaceae	Shrub
457	*Rhododendron fulvum* subsp. *fulvoides* (Balf.f. & Forrest) D.F. Chamb.	Ericaceae	Shrub
458	*Rhododendron glaucophyllum* Rehder	Ericaceae	Shrub
459	*Rhododendron grande* Wight	Ericaceae	Tree
460	*Rhododendron hodgsonii* Hook.f.	Ericaceae	Tree/ Shrub
461	*Rhododendron kesangiae* D.G. Long & Rushforth	Ericaceae	Tree
462	*Rhododendron lanatum* Hook.f.	Ericaceae	Tree/ Shrub
463	*Rhododendron lepidotum* Wall. ex G. Don	Ericaceae	Shrub
464	*Rhododendron mole* G. Don	Ericaceae	Shrub
465	*Rhododendron nivale* Hook.f.	Ericaceae	Shrub
466	*Rhododendron niveum* Hook.f.	Ericaceae	Shrub
467	*Rhododendron praestans* Balf.f. & W.W. Sm.	Ericaceae	Tree/ shrub
468	*Rhododendron rawatii* I. D. Rai & B. S. Adhikari	Ericaceae	Shrub
469	*Rhododendron rex* subsp. *fictolacteum* (Balf.f.) D.F. Chamb.	Ericaceae	Shrub
470	*Rhododendron selense* Franch.	Ericaceae	Shrub
471	*Rhododendron setosum* D. Don	Ericaceae	Shrub
472	*Rhododendron sikkimense* Pradhan & Lachungpa	Ericaceae	Shrub

(continued)

Table 11.1 (continued)

S. No	Scientific name	Family	Growth form
473	*Rhododendron thomsonii* Hook.f.	Ericaceae	Tree/ Shrub
474	*Rhododendron thymifolium* Maxim.	Ericaceae	Shrub
475	*Rhododendron wightii* Hook.f.	Ericaceae	Tree/ Shrub
476	*Ribes glaciale* Wall.	Grossulariaceae	Shrub
477	*Ribes griffithii* Hook. f. & Thomson	Grossulariaceae	Shrub
478	*Ribes orientale* Desf.	Grossulariaceae	Shrub
479	*Ribes takare* D. Don	Grossulariaceae	Shrub
480	*Ribes uva-crispa* L.	Grossulariaceae	Shrub
481	*Rosa macrophylla* Lindl.	Rosaceae	Shrub
482	*Rosa moschata* Herrm.	Rosaceae	Shrub
483	*Rosa sericea* Wall. ex Lindl.	Rosaceae	Shrub
484	*Rosa webbiana* Wall. ex Royle	Rosaceae	Shrub
485	*Rosularia alpestris* (Kar. & Kir.) Boriss.	Crassulaceae	Herb
486	*Rubus anteiafera* L.	Rosaceae	Herb
487	*Rubus fragarioides* Bertol.	Rosaceae	Herb
488	*Rubus niveus* Thunb.	Rosaceae	Shrub
489	*Rumex dentatus* L.	Polygonaceae	Herb
490	*Rumex nepalensis* Spreng.	Polygonaceae	Herb
491	*Salix daphnoides* Vill.	Salicaceae	Tree
492	*Salix denticulata* Andersson	Salicaceae	Tree/ Shrub
493	*Salix disperma* Roxb. ex D. Don.	Salicaceae	Tree
494	*Salix flabellaris* Andersson	Salicaceae	Shrub
495	*Salix karelinii* Turcz. ex Stschegl.	Salicaceae	Tree
496	*Salix lindleyana* Wall. ex Andersson	Salicaceae	Tree/ Shrub
497	*Salix microphylla* Schltdl. & Cham.	Salicaceae	Tree
498	*Salvia hians* Royle ex Benth.	Lamiaceae	Herb
499	*Satyrium nepalense* D. Don	Orchidaceae	Herb
500	*Saussurea costus* (Falc.) Lipsch.	Asteraceae	Herb
501	*Saussurea fastuosa* (Decne.) Sch.Bip.	Asteraceae	Herb
502	*Saussurea globosa* F.H. Chen	Asteraceae	Herb
503	*Saussurea gossypiphora* D. Don	Asteraceae	Herb
504	*Saussurea nepalensis* Spreng.	Asteraceae	Herb
505	*Saussurea obvallata* (DC.) Edgew.	Asteraceae	Herb
506	*Saussurea piptathera* Edgew.	Asteraceae	Herb
507	*Saussurea yakla* C.B. Clarke	Asteraceae	Herb
508	*Saxifraga brunonis* Wall. ex Ser.	Saxifragaceae	Herb
509	*Saxifraga hispidula* D. Don	Saxifragaceae	Herb
510	*Saxifraga jacquemontiana* Decne.	Saxifragaceae	Herb
511	*Saxifraga pallida* Wall. ex Ser.	Saxifragaceae	Herb

(continued)

Table 11.1 (continued)

S. No	Scientific name	Family	Growth form
512	*Scrophularia calycina* Benth.	Scrophulariaceae	Herb
513	*Scrophularia lucida* L.	Scrophulariaceae	Herb
514	*Scrophularia pauciflora* Benth.	Scrophulariaceae	Herb
515	*Scutellaria prostrata* Jacquem. ex Benth.	Lamiaceae	Herb
516	*Sedum ewersii* Ledeb.	Crassulaceae	Herb
517	*Sedum multicaule* Wall. ex Lindl.	Crassulaceae	Herb
518	*Sedum oreades* (Decne.) Raym.-Hamet	Crassulaceae	Herb
519	*Sedum quadrifidum* Pall.	Crassulaceae	Herb
520	*Selinum vaginatum* C.B.Clarke	Apiaceae	Herb
521	*Selinum wallichianum* (DC.) Raizada & H.O. Saxena	Apiaceae	Herb
522	*Senecio candolleanus* Wall. ex DC.	Asteraceae	Herb
523	*Senecio chrysanthemoides* DC.	Asteraceae	Herb
524	*Senecio graciliflorus* (Wall.) DC.	Asteraceae	Herb
525	*Senecio laetus* Edgew.	Asteraceae	Herb
526	*Sibbaldia cuneata* Schouw ex Kunze	Rosaceae	Herb
527	*Silene graminifolia* Otth	Caryophyllaceae	Herb
528	*Silene indica* (Roxb.) Roxb. ex Otth	Caryophyllaceae	Herb
529	*Silene moorcroftiana* Wall. ex Benth.	Caryophyllaceae	Herb
530	*Silene nigrescens* (Edgew.) Majumdar	Caryophyllaceae	Herb
531	*Sinopodophyllum hexandrum* (Royle) T.S. Ying	Berberidaceae	Herb
532	*Smilax vaginata* Decne.	Smilacaceae	Shrub
533	*Solidago virgaurea* L.	Asteraceae	Herb
534	*Sorbus aucuparia* L.	Rosaceae	Tree
535	*Sorbus cuspidata* (Spach) Hedl.	Rosaceae	Tree
536	*Sorbus foliolosa* (Wall.) Spach	Rosaceae	Tree
537	*Sorbus microphylla* (Wall. ex Hook.f.) Wenz.	Rosaceae	Tree
538	*Sorbus tianschanica* Rupr.	Rosaceae	Tree
539	*Sorbus ursina* (Wall. ex D. Don) Decne.	Rosaceae	Tree
540	*Sorbus vilmorinii* C.K. Schneid.	Rosaceae	Tree
541	*Soroseris hookeriana* (C.B. Clarke) Stebbins	Asteraceae	Herb
542	*Soroseris pumila* Stebbins	Asteraceae	Herb
543	*Spiraea arcuate* Hook.f.	Rosaceae	Shrub
544	*Spiraea canescens* D. Don	Rosaceae	Shrub
545	*Stellaria decumbens* Edgew.	Caryophyllaceae	Herb
546	*Stellaria monosperma* Buch.-Ham. ex D. Don	Caryophyllaceae	Herb
547	*Stellaria patens* D. Don	Caryophyllaceae	Herb
548	*Stellaria sikkimensis* Hook.f.	Caryophyllaceae	Herb
549	*Stellaria subumbellata* Edgew.	Caryophyllaceae	Herb
550	*Stipa roylei* (Nees) Duthie	Poaceae	Herb
551	*Stipa sibirica* (L.) Lam	Poaceae	Herb
552	*Streptopus simplex* D. Don	Liliaceae	Herb
553	*Swertia assamensis* Harry Sm.	Gentianaceae	Herb

(continued)

Table 11.1 (continued)

S. No	Scientific name	Family	Growth form
554	*Swertia ciliata* (D. Don ex G. Don) B.L. Burtt	Gentianaceae	Herb
555	*Swertia hookeri* C.B. Clarke	Gentianaceae	Herb
556	*Swertia multicaulis* D. Don	Gentianaceae	Herb
557	*Swertia petiolata* D. Don	Gentianaceae	Herb
558	*Swertia speciosa* Wall.	Gentianaceae	Herb
559	*Synotis alata* C. Jeffrey & Y.L. Chen	Asteraceae	Herb
560	*Synotis rufinervis* (DC.) C. Jeffrey & Y.L. Chen	Asteraceae	Herb
561	*Syringa emodi* Wall. ex Royle	Oleaceae	Tree
562	*Tanacetum dolichophyllum* (Kitam.) Kitam.	Asteraceae	Herb
563	*Taraxacum campylodes* G.E. Haglund	Asteraceae	Herb
564	*Taxus wallichiana* Zucc.	Taxaceae	Tree
565	*Tenaxia cachemyriana* (Jaub. & Spach) N.P. Barker & H.P. Linder	Poaceae	Herb
566	*Thalictrum alpinum* L.	Ranunculaceae	Herb
567	*Thalictrum chelidonii* DC.	Ranunculaceae	Herb
568	*Thalictrum cultratum* Wall.	Ranunculaceae	Herb
569	*Thalictrum foliolosum* DC.	Ranunculaceae	Herb
570	*Thalictrum secundum* Edgew.	Ranunculaceae	Herb
571	*Thamnocalamus spathiflorus* (Trin.) Munro	Poaceae	Shrub
572	*Themeda anathera* (Nees ex Steud.) Hack.	Poaceae	Herb
573	*Thermopsis barbata* Benth.	Fabaceae	Herb
574	*Thymus linearis* Benth	Lamiaceae	Herb
575	*Tordyliopsis brunonis* DC.	Apiaceae	Herb
576	*Trachydium roylei* Lindl.	Apiaceae	Herb
577	*Trifolium pratense* L.	Fabaceae	Herb
578	*Trifolium repens* L.	Fabaceae	Herb
579	*Trillium govanianum* Wall. ex D. Don	Melanthiaceae	Herb
580	*Triplostegia glandulifera* Wall. ex DC.	Caprifoliaceae	Herb
581	*Urtica dioica* L.	Urticaceae	Herb
582	*Valeriana jatamansi* Jones	Caprifoliaceae	Herb
583	*Valeriana pyrolifolia* Decne	Caprifoliaceae	Herb
584	*Verbascum thapsus* L.	Scrophulariaceae	Herb
585	*Veronica cana* Wall. ex Benth.	Plantaginaceae	Herb
586	*Veronica laxa* Benth.	Plantaginaceae	Herb
587	*Viburnum cotinifolium* D. Don	Adoxaceae	Shrub
588	*Viburnum grandiflorum* Wall. ex DC.	Adoxaceae	Shrub
589	*Viburnum mullaha* Buch.-Ham. ex D. Don	Adoxaceae	Shrub
590	*Viburnum nervosum* D. Don	Adoxaceae	Tree
591	*Viola biflora* L.	Violaceae	Herb
592	*Viola pilosa* Blume	Violaceae	Herb
593	*Viola reichenbachiana* Jord. ex Boreau	Violaceae	Herb

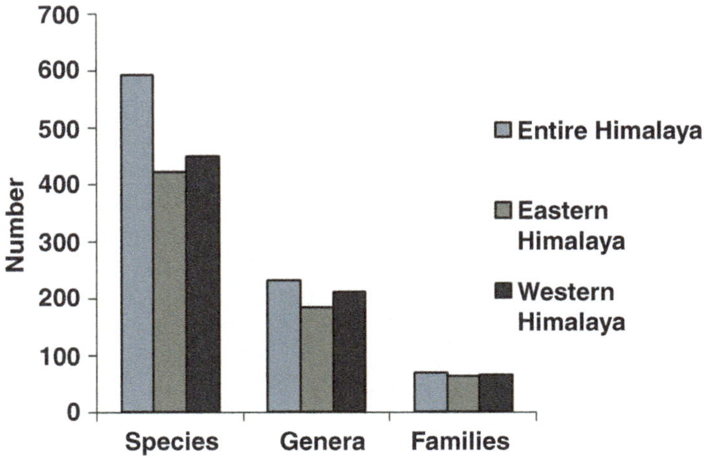

Fig. 11.1 A taxonomic summary of the treeline ecotone flora of the Himalaya

Fig. 11.2 Number of plant species in the treeline ecotone restricted only to Eastern Himalaya, Western Himalaya, and those common to both the Himalayan regions

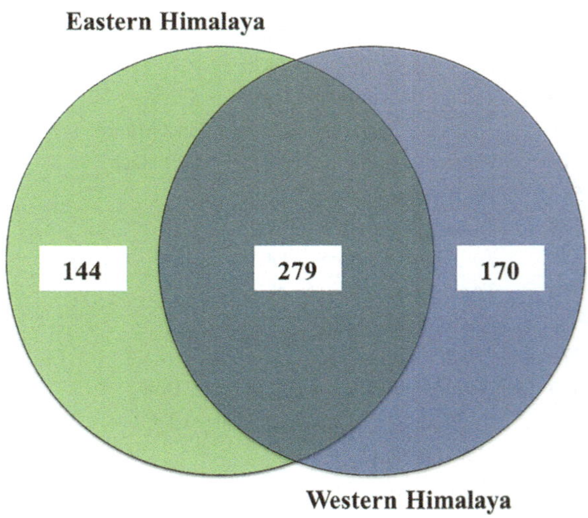

represented by single species), there were 18, 15 and 18 families in the entire, Eastern and Western Himalayan treeline ecotone respectively.

The top ten species-rich genera for the entire Himalayan treeline ecotone in a decreasing order are *Rhododendron* (35 species), *Primula* (19 species), *Juniperus* (14 species), *Pedicularis* (12 species), *Potentilla* (11 species), *Abies* (10 species), *Berberis* and *Gentiana* (9 species each), *Picea* (8 species) and *Aconitum* (7 species) (Fig. 11.4a). These ten genera contributed 136 species (ca. 23%) to the treeline ecotone flora of Himalayan region, and show a higher average genus:species ratio of ca. 1:14. Region-wise, the ten species-rich genera for the Eastern Himalaya in descending order include *Rhododendron* (34 species), *Primula* (14 species),

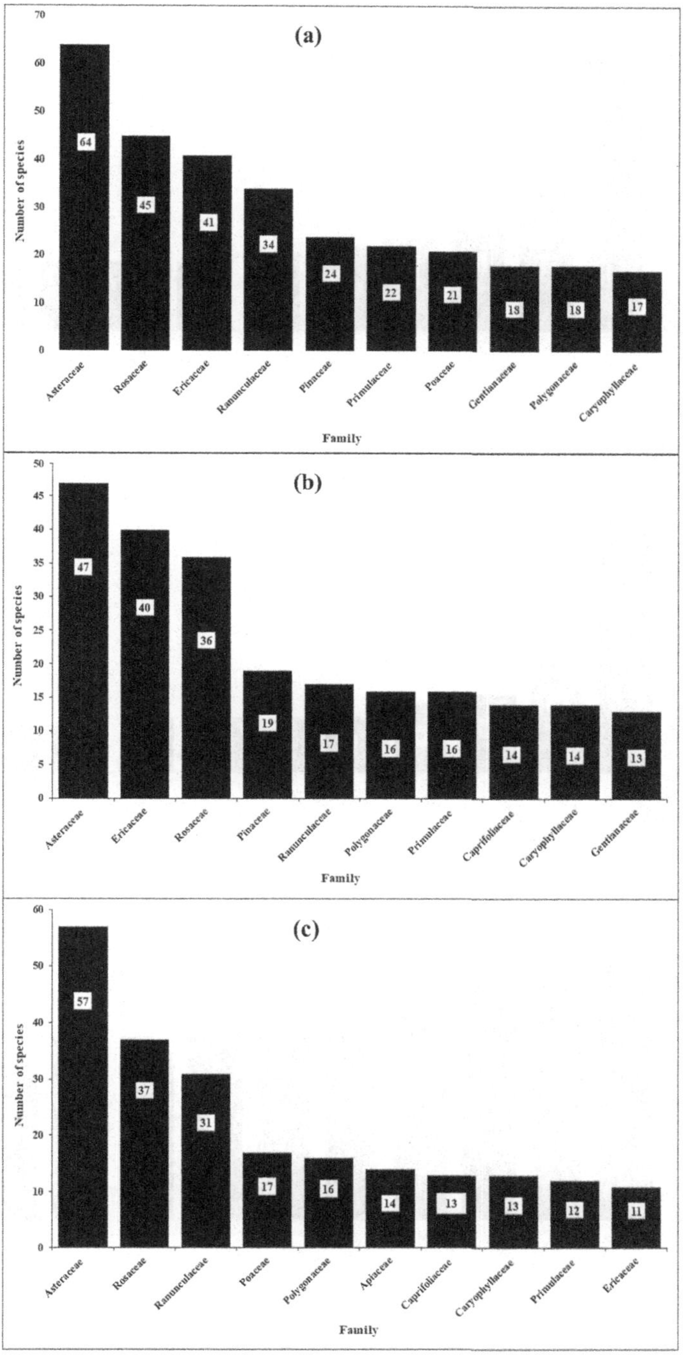

Fig. 11.3 Top ten species-rich families in the treeline ecotone flora of (**a**) the Entire Himalaya (**b**) Eastern Himalaya and (**c**) Western Himalaya

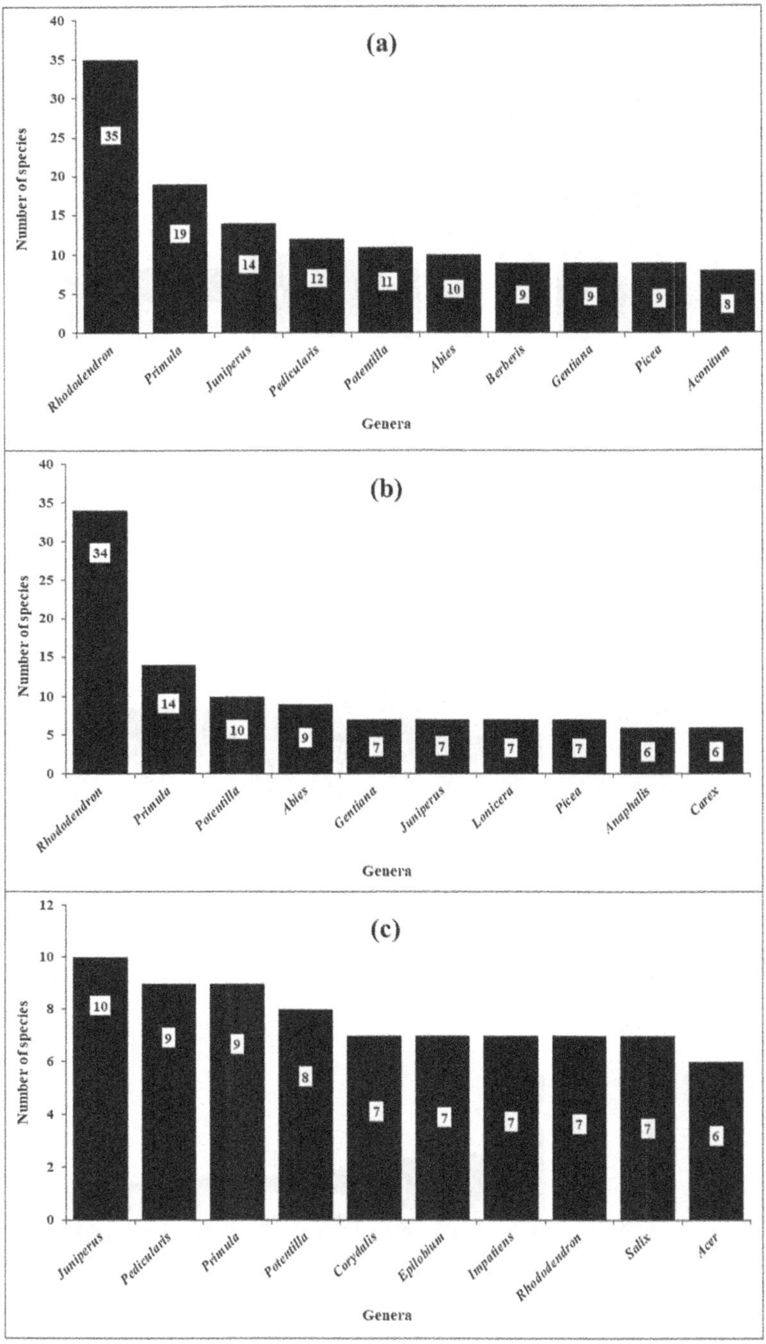

Fig. 11.4 Top ten species-rich genera in the treeline ecotone flora of (**a**) Entire Himalaya, (**b**) Eastern Himalaya and (**c**) Western Himalaya

Potentilla (10 species), *Abies* (9 species), *Gentiana, Juniperus, Lonicera, Picea* (7 species each), *Anaphalis* and *Carex* (6 species each), with a higher average genus:species ratio of ca. 1:11 (Fig. 11.4b). Contrary to this, the ten species-rich genera for the Western Himalaya in decreasing order are *Juniperus* (10 species), *Pedicularis* and *Primula* (9 species each), *Potentilla* (8 species), *Corydalis, Epilobium, Impatiens, Rhododendron, Salix* (7 species each) and *Acer* (6 species) (Fig. 11.4c). These genera contributed 77 species (ca. 17%) in the Western Himalayan region with an average genus:species ratio of ca. 1:8. Further, there were 125, 104 and 113 genera with a single species (i.e. monotypic genera) in the entire, Eastern and Western Himalayan regions respectively.

11.3.2 Growth Form Patterns

In the treeline ecotone flora of the entire Himalaya, herbs with 428 species (72%) were the highest in number, followed by shrubs with 79 species (13%) and trees with 69 species (12%) and 17 species (3%) growing both as shrub and tree (Fig. 11.5a). As expected, at the regional level also, herbs numerically represent the highest growth form category in the treeline ecotone flora of both Eastern and Western Himalaya (Fig. 11.5b, c). Further, the results of the chi-square analysis revealed that the growth form categories differed significantly in terms of their representation within the Eastern and Western Himalayan regions ($x^2 = 11.05$, df $= 3$, $p = 0.01145$). More specifically, an over-representation of shrubs and trees but under-representation of herbs than expected was observed for the Eastern Himalaya, whereas the opposite trend was found for the Western Himalaya (Fig. 11.6 and Table 11.2).

11.3.3 Threat Status of Himalayan Treeline Ecotone Flora

Of the total 593 plant species growing in the Himalayan treeline ecotones, we observed only 56 species that are currently assessed by IUCN (International Union for Conservation of Nature) under different conservation categories (Table 11.3). There are 41 species which are recognized as least concern, 2 near threatened, 4 vulnerable, 7 endangered and 2 as critically endangered.

11.4 Discussion

The substantial warming in the Himalaya, coupled with the high sensitivity of treeline ecotone vegetation to climate change, has severe implications for regional biodiversity and ecological integrity (Schickhoff et al. 2015). At a broad spatial scale, the upward shift of treeline ecotone as a result of climate warming would reduce alpine habitats and possibly will result in the reduction of high elevation alpine biodiversity. In this regard, availability of taxonomic and distributional data

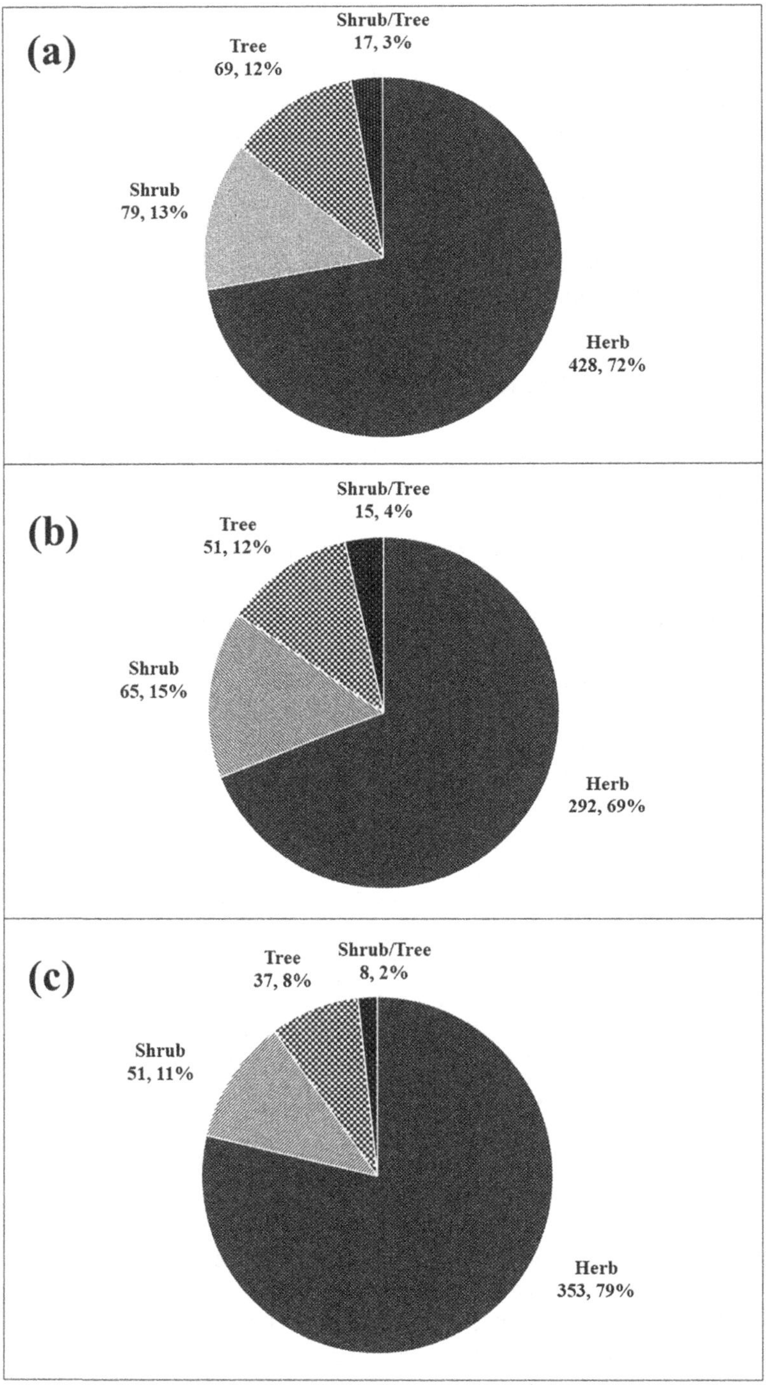

Fig. 11.5 Contribution of different growth form categories to the treeline ecotone flora of (**a**) Entire Himalaya, (**b**) Eastern Himalaya and (**c**) Western Himalaya

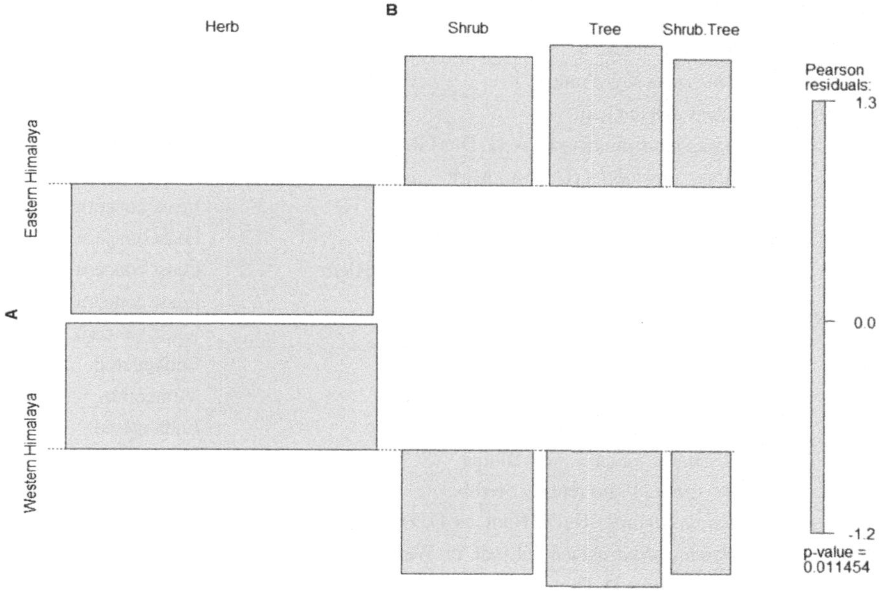

Fig. 11.6 Results of the chi-square test showing the comparison between observed and expected numbers of treeline ecotone species according to their distribution in the Eastern and Western Himalayan regions and growth form categories. Shown are Pearson's residuals derived from the contingency table. Positive values of residuals indicate higher observed values and negative residuals denote lower observed values than expected

Table 11.2 Summarized results of the Pearson's contingency table of treeline ecotone flora according to their regional distribution in the Himalaya and growth form. Shown are observed values and expected values (in brackets)

Region	Herb	Shrub	Tree	Shrub/Tree
Eastern Himalaya	292 (312.88)	65 (56.27)	51 (42.69)	15 (11.16)
Western Himalaya	353 (332.12)	51 (59.73)	37 (45.31)	8 (11.84)

on treeline ecotone flora will be the first step for informing the conservation and management of these fragile ecosystems in the Himalaya. This chapter, based on quantitative literature synthesis, provides a database on the treeline ecotone flora of the Himalaya with particular focus on understanding the taxonomic and distributional patterns of plant species richness. A total of 593 vascular plant species belonging to 232 genera in 70 families were reported in the scientific literature published on the Himalayan treeline ecotone. This high species richness corroborates with other floristic studies from the Himalayan treeline ecotones (Schickhoff 2005; Rawal et al. 2018) or elsewhere in the world (Batllori et al. 2009a, b). Such a high species number may be ascribed to widest bioclimatic gradients (Grytnes and Vetaas 2002) and high habitat herterogenity at regional scale across the Himalayan treeline ecotones (Schickhoff et al. 2015). Our results also reveal that a higher number of plant species in Western Himalaya as compared

Table 11.3 The IUCN threat status of the treeline ecotone flora of the Himalaya

S. No.	Scientific name	Threat status
1.	*Abies delavayi* Franch.	Least concern
2.	*Abies densa* Griff.	Least concern
3.	*Abies pindrow* (Royle ex D. Don) Royle	Least concern
4.	*Abies spectabilis* (D. Don) Mirb.	Near threatened
5.	*Acer acuminatum* Wall. ex D. Don	Least concern
6.	*Acer caesium* Wall. ex Brandis	Least concern
7.	*Acer campbellii* Hook.f. & Thomson ex Hiern	Least concern
8.	*Acer caudatum* Wall.	Least concern
9.	*Acer sterculiaceum* Wall.	Least concern
10.	*Aconitum heterophyllum* Wall. ex Royle	Endangered
11.	*Aconitum violaceum* Jacquem. ex Stapf	Vulnerable
12.	*Angelica glauca* Edgew.	Endangered
13.	*Arisaema jacquemontii* Blume	Least concern
14.	*Bergenia ciliata* (Haw.) Sternb.	Least concern
15.	*Betula alnoides* Buch.-Ham. ex D. Don	Least concern
16.	*Betula cylindrostachya* Lindl. ex Wall.	Least concern
17.	*Betula utilis* D. Don	Least concern
18.	*Cedrus deodara* (Roxb. ex D. Don) G. Don	Least concern
19.	*Cenchrus ciliaris* L.	Least concern
20.	*Cypripedium cordigerum* D. Don	Vulnerable
21.	*Cypripedium elegans* Rchb. f.	Endangered
22.	*Cypripedium himalaicum* Rolfe	Endangered
23.	*Elaeagnus rhamnoides* (L.) A. Nelson	Least concern
24.	*Ephedra gerardiana* Wall. ex Stapf	Vulnerable
25.	*Epilobium latifolium* L.	Least concern
26.	*Epilobium parviflorum* Schreb.	Least concern
27.	*Geranium wallichianum* D. Don ex Sweet	Least concern
28.	*Juniperus communis* L.	Least concern
29.	*Juniperus excelsa* M. Bieb.	Least concern
30.	*Juniperus indica* Bertol.	Least concern
31.	*Juniperus recurva* Buch.-Ham. ex D. Don	Least concern
32.	*Juniperus semiglobosa* Regel	Least concern
33.	*Juniperus squamata* Buch.-Ham. ex D. Don	Least concern
34.	*Larix griffithii* Hook. f.	Least concern
35.	*Lyonia villosa* (Wall. ex C.B.Clarke) Hand.-Mazz.	Least concern
36.	*Malaxis muscifera* (Lindl.) Kuntze	Vulnerable
37.	*Nardostachys jatamansi* (D. Don) DC.	Critically endangered
38.	*Parochetus communis* D. Don	Least concern
39.	*Phleum alpinum* L.	Least concern
40.	*Picea smithiana* (Wall.) Boiss.	Least concern
41.	*Picea spinulosa* (Griff.) A. Henry	Least concern
42.	*Picrorhiza kurroa* Royle ex Benth.	Endangered
43.	*Pinus gerardiana* Wall. ex D. Don	Near threatened

(continued)

Table 11.3 (continued)

S. No.	Scientific name	Threat status
44.	*Pinus wallichiana* A.B. Jacks.	Least concern
45.	*Poa angustifolia* L.	Least concern
46.	*Poa annua* L.	Least concern
47.	*Prunella vulgaris* L.	Least concern
48.	*Quercus lineata* Blume	Least concern
49.	*Quercus semecarpifolia* Sm.	Least concern
50.	*Rhododendron arboreum* Sm.	Least concern
51.	*Saussurea costus* (Falc.) Lipsch.	Critically endangered
52.	*Sorbus foliolosa* (Wall.) Spach	Least concern
53.	*Taxus wallichiana* Zucc.	Endangered
54.	*Trifolium pratense* L.	Least concern
55.	*Trillium govanianum* Wall. ex D. Don	Endangered
56.	*Viburnum mullaha* Buch.-Ham. ex D. Don	Least concern

to that of Eastern Himalaya and only a small proportion of species common to both regions. These findings highlight that treeline ecotones of these two main regions of the Himalaya are unique and/or dissimilar in terms of their species composition and richness. Such a dissimilarity in plant diversity could be due to marked differences in the environmental conditions (e.g. temperature, precipitation, snow cover) across the Himalaya, which have a substantial effect on the pool of species with optimal performance in that specific region (Schickhoff et al. 2015).

We observed a relatively disproportional number of plant species belonging to few families in the Himalayan treeline ecotones, rather than having a fair share of representation by many families. In this study, Asteraceae, Rosaceae, Ericaceae, Ranunculaceae and Pinaceae were the species-rich families and represented about 52% of the total species that were reported across the Himalaya. Since treeline ecotones globally experience severe environmental conditions such as low temperature, long snow cover duration and short growing season (Körner 2021), it is therefore likely that plant species belonging to these families are adapted to such specialized environmental conditions. Moreover, similar to species richness, both the regions of the Himalaya were unique in terms of the representation of families. These results indicate that treeline ecotones of the Himalaya are taxonomically varied and therefore limit our ability to draw broad generalizations about the treeline sensitivity and response to climate change (Schickhoff et al. 2015; Mainali et al. 2020; Singh et al. 2020).

In the treeline ecotone flora, the top species-rich genera were *Rhododendron, Primula Juniperus, Pedicularis* and *Potentilla*. *Rhododendron* was found to be the most speciose genus in Eastern Himalaya, while as same holds true for *Juniperus* in Western Himalaya. These results are more or less consistent with the previous floristic studies from the treeline ecotones of the same region (e.g. Schickhoff 2005; Behera and Kushwaha 2006; Rawal et al. 2018; Mainali et al. 2020; Singh et al. 2020; Gulzar et al. 2022). Along the Eastern Himalaya, increasingly more tropical climatic conditions, higher monsoonal humidity probably increases the competitiveness and dominance of large evergreen shrubs of genus *Rhododendron* (Schickhoff et al. 2015). Moreover, this study also observed noticeable differences in the physiognomy of the treeline forming tree species between Eastern and Western Himalaya. In the Western Himalaya, the treeline ecotone primarily consist of *Abies pindrow*, *Betula utilis*, *Pinus wallichiana* and *Picea smithiana* which ultimately merge into species of *Betula, Rhododendron, Juniperus* and *Salix* krummholz belt, until its final transition to treeless alpine habitat at higher elevations. In Eastern Himalaya, *Abies densa*, *Picea spinulosa*, *Larix griffithii* and Rhododendrons represent treeline forming species, with dwarf shrubs of *Rhododendron* transitioning into krummholz further upslope at higher elevations (Schickhoff et al. 2015).

This study observed higher proportion of herbaceous growth form in the treeline ecotones of the Himalaya and this trend was consistent across Eastern and Western Himalaya. The large number of herbs is the common feature of most of the high elevation communities (Körner 2016). Since treeline ecotones experience extreme environmental conditions and shorter growing seasons (Körner and Paulsen 2004; Körner 2021), therefore rapidly colonizing herbaceous species which can exploit resources quickly with ease get successfully established and predominate these ecotones.

This chapter observed only 10% of the plant species that are currently assessed under IUCN scheme of threat categories. Although this number is small, still there were some important treeline ecotone species such as *Aconitum heterophyllum, A. violaceum, Angelica glauca, Cypripedium cordigerum, C. elegans, C. himalaicum, Ephedra gerardiana, Malaxis muscifera, Nardostachys jatamansi, Picrorhiza kurroa, Saussurea costus, Trillium govanianum* and some treeline forming tree species such as *Pinus gerardiana, Taxus wallichiana* which are recognized under different threatened categories. Such a scenario calls for taking necessary steps toward formulating target-oriented conservation policies in the treeline ecotones of the Himalaya.

11.5 Conclusions and Way Forward

The treeline ecotones of the Himalaya are reported to be highly sensitive to climate change. However, a complete synthesis about the floristic composition and species richness patterns of this ecologically fragile transition zone in the Himalaya was lacking. Therefore, this chapter for the first time, to our best knowledge, provides a synthesis of the floristic composition, species richness and distributional patterns across treeline ecotones of the Himalaya. Our literature synthesis reveals that the treeline ecotones of the Himalaya harbour reasonably high species richness, and many species are currently threatened as well. We also found considerable differences in species composition and richness of particular sets of genera and families as well as on treeline forming tree species along the Western and Eastern Himalayan treeline ecotones. These results indicate that treeline ecotones of the Himalaya are quite heterogeneous not only in terms of climate, altitudinal location and topography but also in terms of species composition, richness and distribution patterns. In future, research studies on ecological and geospatial drivers of species richness and composition in the treeline ecotone of the Himalaya will allow more effective monitoring and advance our understanding of treeline shift and vegetation dynamics in response to climate warming.

Despite providing a comprehensive synthesis of taxonomic and distributional patterns across the treeline ecotones of the Himalaya, the results of this chapter should be taken with some reasonable caveats. In our literature review, we have strictly included only those studies which have explicitly mentioned 'treeline' or 'timberline' in the title, abstract, or keywords. There is possibility of some studies might have been missed, which although conducted on the treeline ecotone of the Himalaya; however, they have failed to explicitly mention the 'treeline' or 'timberline'. We highlight the major knowledge gaps as well as identify topics that can guide the future research.

1. This chapter observed several research-focus biases and geographical knowledge gaps. For example, the majority of the studies we reviewed were published or available from the Western Himalaya compared to Eastern Himalaya and only a few studies were covering the entire Himalaya. Therefore, in future, more information on treeline ecotone vegetation is needed from these data-deficient regions of the Himalaya. This challenge can be achieved through the integration of herbarium reconstruction studies as well as the collection of high-quality field data.
2. Furthermore, by targeting 'treeline ecotone' literature, ambiguity might have arisen as authors can differ in what they consider or refer to as 'treeline ecotone'. While reviewing the published literature, we observed that researchers hardly adhere to a universally accepted terminology in defining the spatial dimensions of a particular treeline environment. Therefore, in future, there is a need for accurately demarcating spatial dimensions and marking of treeline ecotones in the Himalaya. Integration of emerging remote sensing techniques can help address such challenges.
3. We also observed that taxon concepts were not applied consistently and problems with species identifications and detectability were more common. Although, by correcting misspelt and merging synonyms, we assume that the main sources of error relevant to the estimation of species richness were removed. Still, there is a need to integrate additional taxonomic studies such as using modern molecular and geo-spatial tools to provide a solution to the misidentified or erroneous distribution records in the future.
4. This chapter observed only 10% of plant species that were categorized under different IUCN threat categories. However, knowing that both land use and climate change impacts have recently intensified in the Himalaya compared to other regions, we believe that this number is an underestimate, which calls for priority conservation assessment in future studies. Such inconsistencies and knowledge gaps may be due to insufficient field data available on treeline ecotones of the Himalaya. Therefore, there is an immediate need to generate high-quality field data for accurate IUCN threat assessment in the future.

Acknowledgements We acknowledge the kind help provided by the research scholars and supporting staff of Centre for Biodiversity and Taxonomy, Department of Botany, University of Kashmir during the course of present study. Anzar A. Khuroo acknowledges Space Applications Centre (SAC) ISRO Ahmedabad, India, for providing financial support under HIMADRI project (SAC/EPSA/ABHG/ALPINE/PRACRITI-II/01/13), SHRESTI project (SAC/EPSA/BPSG/ALPINE/SHRESTI/01/2019) and MoEFCC, New Delhi, India, under AICOPTAX project (F No. 22018/12/2015/RE(Tax). We are highly thankful to the esteemed reviewer for critical and constructive comments, which has improved the quality of the chapter.

Appendix 1: List of Data Sources Used for the Literature Review During This Study

Adhikari BS, Kumar R, Singh, SP (2018) Early snowmelt impact on herb species composition, diversity and phenology in a western Himalayan treeline ecotone. Trop Ecol 59:365–382

Asad F, Zhu H (2018) Climate–growth relationships of dominant tree species in Karakoram region, northern pakistan. Silva 19:2

Bharti RR, Adhikari BS, Rawat, GS (2012) Assessing vegetation changes in timberline ecotone of Nanda Devi National Park, Uttarakhand. Int J Appl Earth Obs Geoinf 18:472–479

Bhattacharyya A, Chaudhary V (2003) Late-summer temperature reconstruction of the eastern Himalayan region based on tree-ring data of *Abies densa*. Arct Antarct Alp Res Research 35:196–202

Bhattacharyya S, Rawat GS, Adhikari BS (2019) Faunal abundance along timberline ecotone in western Himalaya with reference to anthropogenic pressure and season: a case study. Trop Ecol 60:288–298

Bhuju DR, Koirala M, Boonchird C (2017) Dendroecological studies of Rhododendron campanulatum D. Don along the elevational gradient of Manaslu conservation area, Nepal Himalaya. Pak J Bot 49:1749–1755

Bobrowski M, Gerlitz L, Schickhoff U (2017) Modelling the potential distribution of *Betula utilis* in the Himalaya. Glob Ecol Conserv 11:69–83

Bürzle B, Schickhoff U, Schwab N et al. (2017) Phytosociology and ecology of treeline ecotone vegetation in RolwalingHimal, Nepal. Phytocoenologia 47:197–220

Bürzle B, Schickhoff U, Schwab N, Wernicke LM, Müller YK, Böhner J, Oldeland J (2018) Seedling recruitment and facilitation dependence on safe site characteristics in a Himalayan treeline ecotone. Plant Ecol 219:115–132

Chandra S, Singh A, Singh CP, Nautiyal MC, Rawat LS (2018) Vascular plants distribution in relation to topography and environmental variables in alpine zone of Kedarnath Wild Life Sanctuary, West Himalaya. J Mt Sci 15:1936–1949

Chawla A, Yadav PK, Uniyal SK, Kumar A, Vats SK, Kumar S, Ahuja PS (2012) Long-term ecological and biodiversity monitoring in the western Himalaya using satellite remote sensing. Curr Sci 102:1143–1156

Chhetri PK, Gaddis KD, Cairns DM (2018) Predicting the suitable habitat of treeline species in the Nepalese Himalayas under climate change. Mt. Res Dev 38: 153–163

Chhetri PK, Shrestha KB, Cairns DM (2017). Topography and human disturbances are major controlling factors in treeline pattern at Barun and Manang area in the Nepal Himalaya. J Mt. Sci 14:119–127

Dasila K, Samant SS, Pandey A (2021) Studies on subalpine forests of Hamta Pass area in Himachal Pradesh, India with a focus on Betula utilis populations. Curr Sci 120:872–882

Dawadi B, Liang E, Tian L, Devkota LP, & Yao T (2013) Pre-monsoon precipitation signal in tree rings of timberline *Betula utilis* in the central Himalayas. Quat Int 283:72–77

Dharaiya N (2013) Modeling environmental niche of Himalayan birch and remote sensing based vicarious validation. Trop Ecol 54:321–329

Dolezal J, Kopecky M, Dvorsky M, Macek M, Rehakova K, Capkova K, Altman J (2019) Sink limitation of plant growth determines tree line in the arid Himalayas. Funct Ecol 33:553–565

Drollinger S, Müller M, Kobl T, Schwab N, Böhner J, Schickhoff U, Scholten T (2017) Decreasing nutrient concentrations in soils and trees with increasing elevation across a treeline ecotone in RolwalingHimal, Nepal. J Mt Sci 14:843–858

Dutta PK, Das AK (2014) Alpine Timberline Research Gap in Himalaya. Indian For 140:419–427

Dutta PK, Dutta BK, Sundriyal RC, Das AK (2013) Diversity, representativeness and biotic pressure on plant species along alpine timberline of Western Arunachal Pradesh in the Eastern Himalaya, India. Curr Sci 701–708

Dutta PK, Sundriyal, RC (2018) The easternmost timberline of the Indian Himalayan region: A socio-ecological assessment. Trop Ecol 59:241–257

Gaire NP, Dhakal YR, Lekhak HC, Bhuju DR, Shah SK (2010) Vegetation dynamics in treeline ecotone of Langtang National Park, Central Nepal. Nepal J Sci Technol11:107–114

Gaire NP, Fan ZX, Bräuning A, Panthi S, Rana P, Shrestha A, Bhuju DR (2020) Abies spectabilis shows stable growth relations to temperature, but changing response to moisture conditions along an elevation gradient in the central Himalaya. Dendrochronologia 60:125675

Garkoti SC (2011) Fine root dynamics in three central Himalayan high elevation forests ranging from closed canopied to open-canopied treeline vegetation. J For Res 16:136–143

Garkoti SC, Singh SP (1995) Variation in net primary productivity and biomass of forests in the high mountains of Central Himalaya. J Veg Sci 6:23–28

Ghimire BK, Lekhak HD, Chaudhary RP, Vetaas OR (2008) Vegetation analysis along an altitudinal gradient of Juniperus indica forest in Southern Manang Valley, Nepal. Int J Ecol Dev 9:20–29

Hamid M, Khuroo AA, Charles B, Ahmad R, Singh CP, Aravind NA (2019). Impact of climate change on the distribution range and niche dynamics of Himalayan birch, a typical treeline species in Himalayas. Biodivers Conserv 28:2345 2370

Hamid M, Khuroo AA, Malik AH, Ahmad R, Singh CP (2020) Assessment of alpine summit flora in Kashmir Himalaya and its implications for long-term monitoring of climate change impacts. J Mt. Sci 17:1974–1988

Jamtsho K, Sridith K (2015) Exploring the patterns of alpine vegetation of Eastern Bhutan: a case study from the Merak Himalaya. SpringerPlus 4:1–11

Joshi R, Sambhav K, Singh SP (2018) Near surface temperature lapse rate for treeline environment in western Himalaya and possible impacts on ecotone vegetation. Trop Ecol 59:197–209

Kala CP (2004) Pastoralism, plant conservation, and conflicts on proliferation of Himalayan knotweed in high altitude protected areas of the Western Himalaya, India. Biodivers Conserv 13:985–995

Khan SM, Page S, Ahmad H, Harper D (2013) Identifying plant species and communities across environmental gradients in the Western Himalayas: Method development and conservation use. Ecol Inform 14:99–103

Kosi-Katarmal A (2018) Static tree line of Himalayan silver fir since last several decades at Tungnath, western Himalaya. Trop Ecol 59:351–363

Latwal A, Sah P, Sharma S (2018) A cartographic representation of a timberline, treeline and woody vegetation around a Central Himalayan summit using remote sensing method. Trop Ecol 59:177–186

Li MH, Xiao WF, Shi P, Wang SG, Zhong YD, Liu XL, Shi ZM (2008). Nitrogen and carbon source–sink relationships in trees at the Himalayan treelines compared with lower elevations. Plant Cell Environ 31:1377–1387

Li MH, Xiao WF, Wang SG, Cheng GW, Cherubini P, Cai XH, Zhu WZ (2008). Mobile carbohydrates in Himalayan treeline trees. Evidence for carbon gain limitation but not for growth limitation. Tree Physiol 28:1287–1296

Liang E, Dawadi B, Pederson N, Eckstein D (2014) Is the growth of birch at the upper timberline in the Himalayas limited by moisture or by temperature? Ecology 95:2453–2465

Lv LX, Zhang QB (2012) Asynchronous recruitment history of Abies spectabilis along an altitudinal gradient in the Mt. Everest region. J Plant Ecol 5:147–156

Mainali K, Shrestha BB, Sharma RK, Adhikari A, Gurarie E, Singer M, Parmesan C (2020) Contrasting responses to climate change at Himalayan treelines revealed by population demographics of two dominant species. Ecol Evol 10:1209–1222

Miehe G, Miehe S, Vogel J, La D (2007) Highest treeline in the northern hemisphere found in southern Tibet. Mt. Res Dev 27:169–173

Mir NA, Masood TH, Sofi PA, Husain M, Rather TA (2017) Life form spectrum of vegetation in Betula dominant tree stands along the available altitudinal gradient in north western Himalayas of Kashmir. J Pharmacogn Phytochem 6:267–272

Misra KG, Singh V, Yadava AK, Misra S, Yadav RR (2020) Treeline migration and settlement recorded by Himalayan pencil cedar tree-rings in the highest alpine zone of western Himalaya, India. Curr Sci 118:192–195

Mohapatra J, Singh CP, Hamid M, Verma A, Semwal SC, Gajmer B, Pandya HA (2019) Modelling *Betula utilis* distribution in response to climate-warming scenarios in Hindu-Kush Himalaya using random forest. Biodivers Conserv 28:2295–2317

Mohapatra J, Singh CP, Tripathi OP, Pandya HA (2019) Remote sensing of alpine treeline ecotone dynamics and phenology in Arunachal Pradesh Himalaya. International Journal of Remote Sensing 40:7986–8009

Müller M, Schwab N, Schickhoff U, Böhner J, Scholten T (2016) Soil temperature and soil moisture patterns in a Himalayan alpine treeline ecotone. Arct Antarct Alp Res Research 48:501–521

Negi VS, Giri L, Sekar KC (2018) Floristic diversity, community composition and structure in Nanda Devi National Park after prohibition of human activities, Western Himalaya, India. Curr Sci 115:1056

Pandey A, Badola HK, Rai S, Singh SP (2018) Timberline structure and woody taxa regeneration towards treeline along latitudinal gradients in Khangchendzonga National Park, Eastern Himalaya. PloS one 13:e0207762

Pandey A, Rai S, Kumar D (2018) Changes in vegetation attributes along an elevation gradient towards timberline in Khangchendzonga National Park, Sikkim. Trop Ecol 59:259–271

Pandey S, Carrer M, Castagneri D, Petit G (2018) Xylem anatomical responses to climate variability in Himalayan birch trees at one of the world's highest forest limit. Plant Ecol Evol Syst 33:34 41

Pandey S, Cherubini P, Saurer M, Carrer M, Petit G (2020) Effects of climate change on treeline trees in Sagarmatha (Mt. Everest, Central Himalaya). J Veg Sci 31:1144–1153

Panigrahy S, Anitha D, Kimothi MM, Singh SP (2010) Timberline change detection using topographic map and satellite imagery. Trop Ecol 51:87–91

Rai ID, Adhikari BS (2012) *Rhododendron rawatii* (Ericaceae), a new species from the Western Himalaya, India. Phytotaxa, 71:10–16

Rai ID, Padalia H, Singh G, Adhikari BS, Rawat GS (2020) Vegetation dry matter dynamics along treeline ecotone in Western Himalaya, India. Trop Ecol 1–12

Rai S, Pandey A, Badola HK (2018) Biomass and carbon stock estimation across the timberline of Khangchendzonga National Park, Eastern Himalaya, India. Taiwania 63:311–320

Rawal RS, Bankoti NS, Samant SS, Pangtey YPS (1991) Phenology of tree layer species from the timber line around Kumaun in Central Himalaya, India. Vegetatio 93:108–118

Rawal RS, Dhar U (1997) Sensitivity of timberline flora in Kumaun Himalaya, India: conservation implications. Arct Antarct Alp Res 29:112–121

Rawal RS, Pangtey YPS (1994) Distribution and structural-functional attributes of trees in the high altitude zone of Central Himalaya, India. Vegetatio 112:29–34

Rawal RS, Rawal R, Rawat B, Negi VS, Pathak R (2018) Plant species diversity and rarity patterns along altitude range covering treeline ecotone in Uttarakhand: conservation implications. Trop Ecol 59:225–239

Sah P, Sharma S (2018) Topographical characterisation of high altitude timberline in the Indian Central Himalayan region. Trop Ecol 59:187–196

Sano M, Sheshshayee MS, Managave S, Ramesh R, Sukumar R, Sweda T (2010) Climatic potential of δ18O of *Abies spectabilis* from the Nepal Himalaya. Dendrochronologia 28:93–98

Schickhoff U (2005) The Upper Timberline in the Himalayas, Hindu Kush and Karakorum: a Review of Geographical and Ecological Aspects. In: Broll G, Keplin B (eds) Mountain Ecosystems. Springer, Berlin

Schickhoff U, Bobrowski M, Böhner J, Bürzle B, Chaudhary RP, Gerlitz L, Schwab N (2016) Climate change and treeline dynamics in the Himalaya. Climate change, glacier response, and vegetation dynamics in the Himalaya 271–306

Schwab N, Janecka K, Kaczka RJ, Boehner J, Chaudhary RP, Scholten T, Schickhoff U (2020) Ecological relationships at a near-natural treeline, Rolwaling valley, Nepal Himalaya: implications for the sensitivity to climate change. *Erdkunde* 74:15–44

Schwab N, Kaczka RJ, Janecka K, Böhner J, Chaudhary RP, Scholten T, Schickhoff U (2018) Climate change-induced shift of tree growth sensitivity at a central Himalayan treeline ecotone. Forests 9:267

Schwab N, Schickhoff U, Bürzle B, Müller M, Böhner J, Chaudhary RP, Oldeland J (2017) Implications of tree species–environment relationships for the responsiveness of Himalayan krummholz treelines to climate change. J Mt Sci 14: 453–473

Sharma KP, Tiwari A, Shrestha BB (2020) Changes in regeneration and leaf traits of Rhododendron campanulatum along a treeline ecotone in central Nepal. J Mt. Sci 17:602–613

Shrestha BB, Ghimire B, Lekhak HD, Jha PK (2007) Regeneration of treeline birch (*Betula utilis* D. Don) forest in a trans-Himalayan dry valley in central Nepal. Mt. Res Dev 27:259–267

Shrestha KB, Hofgaard A, Vandvik V (2015) Recent treeline dynamics are similar between dry and mesic areas of Nepal, central Himalaya. J Plant Ecol 8: 347–358

Sigdel SR, Liang E, Wang Y, Dawadi B, Camarero JJ (2020). Tree to tree interactions slow down Himalayan treeline shifts as inferred from tree spatial patterns. J Biogeogr 47:1816–1826

Sigdel SR, Wang Y, Camarero JJ, Zhu H, Liang E, Peñuelas J (2018) Moisture mediated responsiveness of treeline shifts to global warming in the Himalayas. Glob Chang Biol 24:5549–5559

Singh SP (2018) Research on Indian Himalayan treeline ecotone: an overview. Trop Ecol 59:163–176

Singh SP, Bhattacharyya A, Mittal A et al. (2021) Indian Himalayan timberline ecotone in response to climate change-initial findings. Curr Sci 120:859–871

Singh SP, Gumber S, Singh RD, Singh G (2020). How many tree species are in the Himalayan treelines and how are they distributed? Trop Ecol 61:317–327

Suwal MK, Shrestha KB, Guragain L, Shakya R, Shrestha K, Bhuju DR, Vetaas OR (2016) Land-use change under a warming climate facilitated upslope expansion of Himalayan silver fir (*Abies spectabilis* (D. Don) Spach). Plant Ecol 217:993–1002

Tiwari A, Fan ZX, Jump AS, Li SF, Zhou ZK (2017) Gradual expansion of moisture sensitive *Abies spectabilis* forest in the Trans-Himalayan zone of central Nepal associated with climate change. Dendrochronologia 41:34–43

Tiwari A, Fan ZX, Jump AS, Zhou ZK (2017) Warming induced growth decline of Himalayan birch at its lower range edge in a semi-arid region of Trans-Himalaya, central Nepal. Plant Ecol 218:621–633

Tiwari A, Jha PK (2018) An overview of treeline response to environmental changes in Nepal Himalaya. Trop Ecol 59:273–285

Wani ZA, Pant S (2021) *Betula utilis* D. Don: an Ecologically and Economically Important Timberline Species of Himalayan Region in Jeopardy. Bot Rev 1–15

Yadav RR, Singh J, Dubey B, Chaturvedi R (2004) Varying strength of relationship between temperature and growth of high-level fir at marginal ecosystems in western Himalaya, India. Curr Sci 1152–1156

Yadava AK, Sharma YK, Dubey B, Singh J, Singh V, Bhutiyani MR, Misra KG (2017) Altitudinal treeline dynamics of Himalayan pine in western Himalaya, India. Quat Int 444:44–52

References

Acharya BK, Chettri B, Vijayan L (2011) Distribution pattern of trees along an elevation gradient of eastern Himalaya, India. Acta Oecol 37:329–336

Bader MY, Llambí LD, Case BS et al (2021) A global framework for linking alpine-treeline ecotone patterns to underlying processes. Ecography 44:265–292

Batllori E, Blanco-Moreno JM, Ninot JM, Gutiérrez E, Carrillo E (2009a) Vegetation patterns at the alpine treeline ecotone: the influence of tree cover on abrupt change in species composition of alpine communities. J Veg Sci 20:814–825

Batllori E, Camarero JJ, Ninot JM, Gutierrez E (2009b) Seedling recruitment, survival and facilitation in alpine Pinus uncinata tree line ecotones. Implications and potential responses to climate warming. Glob Ecol Biogeogr 18:460–472

Behera MD, Kushwaha SPS (2006) An analysis of altitudinal behaviour of tree species in Subansiri district, Eastern Himalaya. In: Plant conservation and biodiversity. Springer, Dordrecht

Bobrowski M, Gerlitz L, Schickhoff U (2017) Modelling the potential distribution of *Betula utilis* in the Himalaya. Glob Ecol Conserv 11:69–83

Cairns DM, Malanson G (1998) Environmental variables influencing the carbon balance at the alpine Treeline: a modeling approach. J Veg Sci 9:679–692

Dhar U, Kachroo P (1983) Alpine Flora of Kashmir Himalaya. Scientific Publishers, Jodhpur

Feuillet T, Birre D, Milian J, Godard V, Clauzel C, Serrano-Notivoli R (2020) Spatial dynamics of alpine tree lines under global warming: what explains the mismatch between tree densification and elevational upward shifts at the tree line ecotone? J Biogeogr 47:1056–1068

Gottfried M, Hantel M, Maurer C, Toechterle R, Pauli H, Grabherr G (2012) Coincidence of the alpine–nival ecotone with the summer snowline. Environ Res Lett 6:014013

Grytnes JA, Vetaas OR (2002) Species richness and altitude: a comparison between null models and interpolated plant species richness along the Himalayan altitudinal gradient, Nepal. Am Nat 159: 294–304

Gulzar A, Hamid M, Dar FA, Wani SA, Malik AH, Kamili AN, Singh CP, Khuroo AA (2022) Patterns of floristic and functional diversity in two treeline ecotone sites of Kashmir Himalaya. Environ Monit Assess 194:420

Hamid M, Khuroo AA, Charles B, Ahmad R, Singh CP, Aravind NA (2019) Impact of climate change on the distribution range and niche dynamics of Himalayan birch, a typical treeline species in Himalayas. Biodivers Conserv 28:2345–2370

Hamid M, Khuroo AA, Malik AH, Ahmad R, Singh CP (2020a) Assessment of alpine summit flora in Kashmir Himalaya and its implications for long-term monitoring of climate change impacts. J Mt Sci 17:1974–1988

Hamid M, Khuroo AA, Malik AH, Ahmad R, Singh CP, Dolezal J, Haq SM (2020b) Early evidence of shifts in alpine summit vegetation: a case study from Kashmir Himalaya. Front Plant Sci 11: 421

Holtmeier FK, Broll G (2005) Sensitivity and response of northern hemisphere altitudinal and polar treelines to environmental change at landscape and local scales. Glob Eco lBiogeogr 14:395–410

IUCN (2017) The IUCN red list of threatened species version 3. http://www.iucnredlist.org. Accessed 22 June 2021

Khuroo AA, Dar GH, Khan ZS, Malik AH (2007) Exploring an inherent interface between taxonomy and biodiversity: current problems and future challenges. J Nat Conserv 15:256–261

Kindt R (2020) WorldFlora: an R package for exact and fuzzy matching of plant names against the world Flora online taxonomic backbone data. Appl Plant Sci 8:e11388. Accessed 15 June 2021

Körner C (1998) A re-assessment of high elevation treeline positions and their explanation. Oecologia 115:445–459

Körner C (2016) Plant adaptation to cold climates. F1000research 5:2769

Körner C (2021) Alpine treelines. In: Alpine plant life. Springer, Cham

Körner C, Paulsen J (2004) A worldwide study of high altitude treeline temperatures: study of high altitude treeline temperatures. J Biogeogr 31:713–732

Lett S, Dorrepaal E (2018) Global drivers of tree seedling establishment at alpine treelines in a changing climate. Funct Ecol 32:1666–1680

Liang E, Dawadi B, Pederson N, Eckstein D (2014) Is the growth of birch at the upper timberline in the Himalayas limited by moisture or by temperature? Ecology 95:2453–2465

Mainali K, Shrestha BB, Sharma RK, Adhikari A, Gurarie E, Singer M, Parmesan C (2020) Contrasting responses to climate change at Himalayan treelines revealed by population demographics of two dominant species. Ecol Evol 10:1209–1222

Mayor JR, Sanders NJ, Classen AT et al (2017) Elevation alters ecosystem properties across temperate treelines globally. Nature 542:91–95

Meyer D, Zeileis A, Hornik K (2020) vcd: Visualizing Categorical Data. R package version 1.4-8. Accessed 30 June 2021

Miehe G, Miehe S, Vogel J, Co S, La D (2007) Highest treeline in the northern hemisphere found in southern Tibet. Mt Res Dev 27:169–173

Mittermeier RA, Gil PR, Hoffmann M et al (2004) Hotspots revisited: Earth's biologically richest and Most endangered terrestrial ecoregions, Revised edn. University of Chicago Press, Chicago

Palomo I (2017) Climate change impacts on ecosystem services in high mountain areas: a literature review. Mt Res Dev 37(2):179–187

Perrigo A, Hoorn C, Antonelli A (2020) Why mountains matter for biodiversity. J Biogeogr 47: 315–325

Polunin O, Stainton A (1984) Flowers of the Himalaya. Oxford University Press, Delhi

Rawal RS, Rawal R, Rawat B, Negi VS, Pathak R (2018) Plant species diversity and rarity patterns along altitude range covering treeline ecotone in Uttarakhand: conservation implications. Trop Ecol 59:225–239

Schickhoff U (2005) The upper timberline in the Himalayas, Hindu Kush and Karakorum: a review of geographical and ecological aspects. In: Broll G, Keplin B (eds) Mountain ecosystems: studies in treeline ecology. Springer, Berlin

Schickhoff U, Bobrowski M, Böhner J et al (2015) Do Himalayan treelines respond to recent climate change? An evaluation of sensitivity indicators. Earth Syst Dyn 6:245–265

Shugar DH, Jacquemart M, Shean D et al (2021) A massive rock and ice avalanche caused the 2021 disaster at Chamoli, Indian Himalaya. Science 373:eabh4455

Singh SP (2018) Research on Indian Himalayan treeline ecotone: an overview. Trop Ecol 59:163–176

Singh SP, Gumber S, Singh RD, Singh G (2020) How many tree species are in the Himalayan treelines and how are they distributed? Trop Ecol 61:317–327

Singh CP, Mohapatra J, Mathew JR et al (2021) Long-term observation and modelling on the distribution and patterns of alpine treeline ecotone in Indian Himalaya. J Geom 15:68–84

Skre O (2019) Northern treelines as indicators of climate and land use changes—a literature review. Agrotechnology 9:190

Steinbauer MJ, Grytnes JA, Jurasinski G, Kulonen A, Lenoir J, Pauli H, Bjorkman AD (2018) Accelerated increase in plant species richness on mountain summits is linked to warming. Nature 556:231–234

Testolin R, Attorre F, Borchardt P et al (2021) Global patterns and drivers of alpine plant species richness. Glob Ecol Biogeogr 30:1218–1231

Thrippleton T, Bugmann H, Kramer-Priewasser K, Snell RS (2016) Herbaceous understorey: an overlooked player in forest landscape dynamics? Ecosystems 19:1240–1254

Zaz SN, Romshoo SA, Krishnamoorthy RT, Viswanadhapalli Y (2019) Analyses of temperature and precipitation in the Indian Jammu and Kashmir region for the 1980–2016 period: implications for remote influence and extreme events. Atmos Chem Phys 19:15–37

12 Lichen Diversity in High Elevations of Western Himalaya with Special Reference to Treeline Ecotone: Conservation and Indicator Value

Kapil Bisht, Shashi Upadhyay, Ranbeer S. Rawal, and Yogesh Joshi

Abstract

The treeline ecotone, which lies between subalpine forests and the alpine grasslands, offers special habitats and niches for several unique, representative, and sensitive biodiversity elements. Present study, with a focus on distribution patterns of macrolichens, attempts to describe such elements in high-altitude landscape associated with Tungnath, west Himalayan treeline ecotone. Two representative elevation transects [Kalsir–Chandrashila (KCT), 2080–3677 m asl and Pothibhasha–Chandrashila (PCT), 2120–3677 m asl] were investigated. Together these transects harbor 108 macrolichen species (41 genera, 15 families). PCT with 104 species (40 genera, 15 families) was species rich compared to KCT (73 species, 34 genera, 15 families). While family Parmeliaceae (46), Physciaceae (18), and Cladoniaceae (14) were species rich, genera like *Parmotrema* (12), *Heterodermia* (11), and *Cladonia* (8 species) had maximum species. Among families Candelariaceae, Coccocarpaceae, Nephromataceae, and Teloschistaceae had one species only. The species richness peaked at 2500 and 2600 m altitude bands (62 spp.). The presence of two foliicolous and caliciale lichen communities was interesting in view of their indicator value. Existence of the foliicolous lichens is indicative of human disturbance and increasing subtropical elements in high-elevation forests. Occurrence of caliciale lichens is associated with old growth forests. The distribution of macrolichens has implications for management of high–altitude forests including treelines.

K. Bisht (✉) · S. Upadhyay · R. S. Rawal
G.B. Pant National Institute of Himalayan Environment, Almora, Uttarakhand, India

Y. Joshi
Department of Botany, University of Rajasthan, Jaipur, Rajasthan, India

S. P. Singh et al. (eds.), *Ecology of Himalayan Treeline Ecotone*,
https://doi.org/10.1007/978-981-19-4476-5_12

Keywords

Conservation · Diversity · Macrolichen · Timberline ecotone · West Himalaya

12.1 Introduction

Treeline ecotone, a transition between alpine grasslands and subalpine forests, provides habitats and niches for unique and distinct biodiversity from both these ecosystems with many species sensitive to change in the climate (Rawal and Dhar 1997). Being highest in the world, treeline in the Himalaya is considered a special indicator of climate change (Singh 2018). Some studies in the region have established its uniqueness, and floristic sensitivity (Rawal and Dhar 1997). Few other studies have described responses of treeline ecotone vegetation to anthropogenic disturbances (Gairola et al. 2009, 2015; Rai et al. 2012), and its conservation in Western Himalaya (Dhar 2000; Singh 2018; Rawal et al. 2018). More recently, a comprehensive compilation of diverse studies on timberline of Indian Himalayan Region (IHR) has been a major outcome (Singh 2018). The study has pointed out that the Himalaya is warming with a rapid rate which has made timberlines more vulnerable. Despite being identified as conservation entities, the studies conducted on these regions are meager.

Among others, owing to their high-altitude location and relatively less anthropogenic pressure, the Himalayan treeline provides a suitable habitat with congenial environment for growth of different lichen species. As the lichens are well recognized for their sensitivity toward pollution (Vokou et al. 1999), they thrive very well in higher altitude areas where the pollution level remains low as compared to low-land areas. Lichens are the most successful symbiotic organisms in nature which dominate 8% or more of the Earth's terrestrial area (Ahmadjian 1995). They are better indicators of diverse environmental conditions (Will-Wolf et al. 2002; Shukla et al. 2014). High sensitivity of lichens to changing climate events is well recognized (Gauslaa 2014). Their slow growth rate allows the lichens to integrate the climatic conditions and modify their habitat as per the requirements (Aptroot 2009).

While lichens contribute to richness of forest biodiversity (Pharo et al. 1999), they form an important forage for several animals (Rosentreter et al. 1997), nesting material for birds (Starkey and Hagar 1999), special habitats for many invertebrates (Pettersson et al. 1995), and are also involved in nutrient cycling (Essen et al. 1996). However, being habitat specific, presence of some particular lichen communities in a forest indicates its status, and their diversity on a given site also reflects on habitat heterogeneity (McCune 2000). During recent years some lichenological investigations were conducted in high-altitude zones of Kumaun, west Himalaya, including glaciers (Bisht 2018; Bisht et al. 2018a, b; 2019a), enumeration of medicinally important lichens of high-altitude regions (Bisht et al. 2019b), discoveries of new species (Joshi et al. 2016, 2018a), new records of lichens (Joshi et al. 2018b), lichen growth rates (Bisht et al. 2020). However, in spite of all these studies, lichens in Himalaya, in general, and high-altitude regions, in particular (especially the timberline ecotone of IHR), have remained poorly investigated. Keeping this lacuna in view and considering the overall importance of lichens as components of biodiversity, the present study was conducted to: (i) enumerate the lichen diversity along the representative elevation transects [i.e., Kalsir–Chandrashila transect (KCT), and Pothibhasha–Chandrashila transect (PCT)], (ii) identify the indicator lichen communities, and (iii) understand the status of lichen diversity across the elevation bands.

12.2 Material and Methods

12.2.1 Study Area

The study was conducted in two representative elevation transects [Kalsir–Chandrashila (KCT), and Pothibhasha–Chandrashila (PCT)] in Tungnath area of Indian Himalayan region of Uttarakhand. Considering its association with treeline ecotone, we have termed the study landscape as Treeline Landscape (TL). Major parts of these transects fall in Rudraprayag district and a small part in Chamoli district. Elevation range of study transects varied from 2082 to 3677 m asl for KCT and 2120 to 3677 m asl for PCT. The Tungnath area is of great religious importance due to the presence of famous Lord Shiva shrine, which is recognized among famous Five Kedars of Garhwal region. The vegetation of the region broadly represents temperate mixed broad leaf and coniferous forests.

12.2.2 Sample Collection, Identification, and Herbarium Documentation

Several field expeditions were conducted in both the transects between the years 2016 and 2018 and an exhaustive search for lichen species was undertaken. Each transect was divided into 18 elevation bands with an interval of 100 m between each band. In each elevation band, 3 plots (50 m × 50 m) were laid and the representative lichen specimens were collected from the major available substrates (i.e., wood, soil, and rock). The collected specimens were packed in poly bags and then air dried. The air-dried specimens were then brought to the laboratory and the morphological characteristics of the species were examined under a stereo-zoom dissecting microscope (Olympus SZ2-ILST). The anatomical studies of handmade sections were done using a research microscope (CX21iLEDFS1). Spot test reactions on thalli, medulla, and fruiting bodies were tested with the standard reagents: 10% potassium hydroxide (K), sodium hypochlorite (C), and *para*-phenylenediamine (Pd). The specimens were identified using the available literature (Awasthi 2007; Divakar and Upreti 2005; Singh and Sinha 2010). The identified specimens have been deposited in the herbarium of G. B. Pant Institute (GBP).

Species richness in each elevation band was determined by pooling all species identified in different plots in that band. Whittaker's beta diversity across the elevation bands was analyzed by the following method mentioned in Rawal et al. (2018). Similarity coefficient between different selected elevations was calculated as per Sorenson's Similarity Index (Sorenson 1948).

12.3 Results

12.3.1 Macrolichen Species Diversity and Composition

Collectively in both the transects, the study revealed occurrence of 108 macrolichen species (41 genera and 15 families). The detailed inventory across transects and elevation bands of TL is presented in Table 12.1. Distribution of macrolichen species rich families is presented in Fig. 12.1. While the KCT harbored 73 species (belonging to 15 families and 34 genera), PCT was comparatively species rich with 104 species (belonging to 15 families and 40 genera). Family Parmeliaceae (46 species; 18 genera) was the dominant family followed by Physciaceae (18 species; 6 genera) and Cladoniaceae (14 species; 2 genera). In case of genera, *Parmotrema* (12 spp.) followed by *Heterodermia* (11 spp.) and *Cladonia* (8 spp.) were species rich. Among the families, Candelariaceae, Coccocarpaceae, Nephromataceae, and Teloschistaceae were species poor, represented by only one species. Likewise, a total of 18 (43.9%) genera, viz. *Candelaria*, *Canoparmelia*, *Cetraria*, *Coccocarpia*, *Collema*, *Dolichousnea*, *Flavoparmelia*, *Leucodermia*, *Melanelia*, *Nephroma*, *Parmelinella*, *Physconia*, *Protoparmeliopsis*, *Rhizoplaca*, *Sulcaria*, *Sticta*, *Xanthoparmelia,* and *Xanthoria* were represented by only one species. The species rich PCT had seven genera (viz *Collema*, *Melanelia*, *Myelochroa*, *Parmelinella*, *Physconia*, *Remototrachyna*, and *Sticta*) restricted to this transect whereas, transect Kalsir–Chandrashila had only one genus, that is, *Sulcaria* restricted to this transect. Considering the diverse substrates, maximum macrolichen species (41; 37.9%) were recorded exclusively on woody (corticolous) substrate followed by 15 (13.9%) species on rock (saxicolous) and 15 (13.9%) on soil (terricolous) substrate. While 10 spp. (9.2%) were common to all the substrates, wood and rock shared 19 (17.5%), wood and soil shared 5 spp. (4.6%), and soil and rock substrate share 3 spp. (2.7%). The distribution of species by substratum is presented (Fig. 12.2).

12.3.2 Diversity of Macrolichen in Different Elevation Bands

While considering elevation patterns of species richness, number of macrolichen species peaked at 2500 and 2600 m elevation bands (62 spp.) and thereafter a gradual decline was noticed with minimum diversity at 3700 m elevation (16 spp.). In general, the elevation pattern of lichens showed a hump shaped distribution pattern (Fig. 12.3). One-fourth of the total species were restricted to 2500 m altitude. Whereas, 8 species, viz. *Cetraria islandica*, *Cladonia pyxidata*, *Capitites ramulosa*, *Cladonia rangiferina*, *Hypotrachyna pindarensis*, *Parmelinella wallichiana*, *Stereocaulon alpinum*, and *Stereocaulon massartianum* were recorded only above

Table 12.1 Macrolichen diversity in Tungnath Timberline Landscape (TTL)

S. No.	Lichen taxa	Families	Transects		Altitude bands (m asl)										
			A	B	2000	2100	2200	2300	2400	2500	2600	2700	2800	2900	3000
1	*Bulbothrix isidiza* (Nyl.) Hale	Parmeliaceae	+	+	+	+	+	+	+	–	–	–	–	–	–
2	*Bulbothrix meizospora* (Nyl.) Hale	Parmeliaceae	+	–	+	+	+	+	+	–	–	–	–	–	–
3	*Bulbothrix setschwanensis* (Zahlbr.) Hale	Parmeliaceae	+	+	+	+	+	+	+	+	+	+	+	–	–
4	*Candelaria concolor* (dicks.) stein	Candelariaceae	+	+	+	+	+	+	+	+	+	+	+	+	+
5	*Canoparmelia texana* (Tuck.) Elix & hale	Parmeliaceae	+	+	+	+	+	+	+	–	–	–	–	–	–
6	*Cetraria islandica* (L.) Ach.	Parmeliaceae	+	+	–	–	–	–	–	–	–	–	–	–	–
7	*Cetrelia braunsiana* (Müll. Arg.) W.L. Culb. & C.F. Culb.	Parmeliaceae	+	+	+	+	+	+	+	+	+	–	–	–	–
8	*Cetrelia olivetorum* (Nyl.) W.L. Culb. & C.F. Culb.	Parmeliaceae	–	+	–	–	–	–	–	+	+	+	–	–	–
9	*Cladonia coccifera* (L.) Willd.	Cladoniaceae	+	+	+	+	+	+	+	+	+	+	+	–	–
10	*Cladonia coniocraea* (Flörke) Spreng.	Cladoniaceae	+	+	+	+	+	+	+	+	+	+	+	+	+
11	*Cladonia fimbriata* (L.) Fr.	Cladoniaceae	+	+	+	+	+	+	+	+	+	+	+	+	+
12	*Cladonia furcata* (Huds.) Schrad.	Cladoniaceae	+	+	+	+	+	+	+	+	+	+	+	–	–
13	*Cladonia pyxidata* (L.) Hoffm.	Cladoniaceae	+	+	–	–	–	–	–	–	–	–	–	–	+
14	*Cladonia ramulosa* (With.) J.R. Laundon	Cladoniaceae	+	+	–	–	–	–	–	–	–	–	–	–	+

(continued)

Table 12.1 (continued)

S. No.	Lichen taxa	Families	Transects A	B	Altitude bands (m asl) 2000	2100	2200	2300	2400	2500	2600	2700	2800	2900	3000
15	*Cladonia rangiferina* (L.) Weber ex F.H. Wigg.	Cladoniaceae	+	+	–	–	–	–	–	–	–	–	–	–	–
16	*Cladonia subulata* (L.) Weber ex F.H. Wigg.	Cladoniaceae	–	+	–	–	–	–	–	–	+	+	+	+	–
17	*Coccocarpia erythroxyli* (Spreng.) Swinscow & Krog	Coccocarpaceae	+	+	–	–	–	+	+	+	+	+	+	+	–
18	*Collema* sp.	Collemataceae	–	+	+	+	+	+	+	–	–	–	–	–	–
19	*Dermatocarpon miniatum* (L.) W. Mann	Verrucariaceae	+	+	–	–	–	–	+	+	+	+	+	–	–
20	*Dermatocarpon vellereum* Zschacke	Verrucariaceae	+	+	+	+	+	+	+	–	–	–	–	–	–
21	*Dolichousnea longissima* (ach.) Articus	Parmeliaceae	+	+	–	–	–	–	–	+	+	+	+	+	+
22	*Flavoparmelia caperata* (L.) Hale	Parmeliaceae	+	+	+	+	+	+	+	+	+	+	+	+	–
23	*Flavopunctelia flaventior* (Stirt.) Hale	Parmeliaceae	+	+	+	+	+	+	+	+	+	+	+	–	–
24	*Flavopunctelia soredica* (Nyl.) Hale	Parmeliaceae	–	+	–	–	–	–	–	–	+	+	+	+	+
25	*Heterodermia albidiflava* (Kurok.) D.D. Awasthi	Physciaceae	+	+	+	+	+	–	–	–	–	–	–	–	–
26	*Heterodermia comosa* (Eschw.) Follmann & Redón	Physciaceae	+	+	+	+	+	+	–	–	–	–	–	–	–
27	*Heterodermia diademata* (Taylor) D.D. Awasthi	Physciaceae	+	+	+	+	+	+	+	+	+	+	+	+	+

28	*Heterodermia dissecta* (Kurok.) D.D. Awasthi	Physciaceae	–	+	–	–	–	–	–	+	–	–	–	–	–
29	*Heterodermia incana* (Stirt.) D.D. Awasthi	Physciaceae	–	+	+	+	+	–	–	–	–	–	–	–	–
30	*Heterodermia indica* (H. Magn.) D.D. Awasthi	Physciaceae	–	+	–	–	–	–	–	–	+	–	–	–	–
31	*Heterodermia leucomela* (L.) Poelt	Physciaceae	–	+	–	–	–	–	–	–	–	–	+	–	–
32	*Heterodermia obscurata* (Nyl.) Trevis.	Physciaceae	–	+	+	–	–	–	–	–	–	–	–	–	–
33	*Heterodermia podocarpa* (Bél.) D.D. Awasthi	Physciaceae	+	+	+	+	+	+	+	+	+	–	–	–	–
34	*Heterodermia pseudospeciosa* (Kurok.) W.L. Culb.	Physciaceae	+	+	+	+	+	+	+	+	+	+	+	+	+
35	*Heterodermia speciosa* (Wulfen) Trevis.	Physciaceae	+	+	+	+	+	+	–	–	–	–	–	–	–
36	*Hypotrachyna cirrhata* (Fr.) Divakar, A. Crespo, Sipman, Elix & Lumbsch	Parmeliaceae	+	+	+	+	+	+	+	+	+	+	–	–	–
37	*Hypotrachyna nepalensis* (Taylor) Divakar, A. Crespo, Sipman, Elix & Lumbsch	Parmeliaceae	+	+	+	+	+	+	+	+	+	+	+	+	+
38	*Hypotrachyna pindarensis* (D.D. Awasthi & Suj. Singh) D.D. Awasthi	Parmeliaceae	–	+	–	–	–	–	–	–	–	–	–	–	–
39	*Hypotrachyna pseudosinuosa* (Asahina) Hale	Parmeliaceae	–	+	–	–	–	–	+	+	–	–	–	–	–

(continued)

Table 12.1 (continued)

S. No.	Lichen taxa	Families	Transects A	B	Altitude bands (m asl) 2000	2100	2200	2300	2400	2500	2600	2700	2800	2900	3000
40	*Hypotrachyna scytodes* (Kurok.) Hale	Parmeliaceae	–	+	–	–	+	+	+	+	–	–	–	–	–
41	*Leptogium askotense* D.D. Awasthi	Collemataceae	+	+	+	+	+	+	+	+	+	+	+	–	–
42	*Leptogium burnetiae* C.W. Dodge	Collemataceae	+	+	+	+	+	+	+	+	+	+	+	+	+
43	*Leptogium delavayi* Hue	Collemataceae	+	+	+	+	+	+	+	+	+	+	+	+	+
44	*Leptogium javanicum* (Mont. & Bosch) Mont.	Collemataceae	–	+	+	+	+	+	+	+	+	+	+	+	+
45	*Leptogium pedicellatum* P. M. Jørg.	Collemataceae	+	+	+	+	+	–	–	–	–	–	–	–	–
46	*Leptogium trichophorum* Müll. Arg.	Collemataceae	–	+	–	–	–	–	+	–	–	–	–	–	–
47	*Leucodermia boryi* (Fée) Kalb	Physciaceae	+	+	+	+	+	+	+	+	+	+	+	+	+
48	*Lobaria isidiosa* (Müll. Arg.) Vain.	Lobariaceae	–	+	–	–	+	+	+	+	+	+	–	–	–
49	*Lobaria pseudopulmonaria* Gyeln.	Lobariaceae	+	+	–	–	+	+	+	+	+	+	+	+	–
50	*Lobaria pulmonaria* (L.) Hoffm.	Lobariaceae	–	+	–	–	–	–	–	–	–	–	–	+	+
51	*Lobaria retigera* (Bory) Trevis.	Lobariaceae	+	+	+	+	+	+	+	+	+	+	+	+	+
52	*Melanelia stygia* (L.) Essl.	Parmeliaceae	–	+	–	–	–	–	–	+	+	–	–	–	–
53	*Myelochroa perisidians* (Nyl.) Elix & Hale	Parmeliaceae	–	+	–	–	–	+	+	–	–	–	–	–	–

54	*Myelochroa rhytidodes* (Hale) Elix & Hale	Parmeliaceae	–	+	–	–	–	–	–	+	+	+	–	–	–
55	*Nephroma helveticum* Ach.	Nephromataceae	+	+	–	–	–	–	–	–	–	–	+	+	+
56	*Parmelia meiophora* Nyl.	Parmeliaceae	+	+	–	–	–	–	–	–	+	+	+	+	+
57	*Parmelia saxatilis* (L.) Ach.	Parmeliaceae	–	+	–	–	–	–	–	–	+	+	+	+	+
58	*Parmelia sulcata* Taylor	Parmeliaceae	–	+	–	–	–	–	–	–	–	–	–	+	+
59	*Parmelinella wallichiana* (Taylor) Elix & hale	Parmeliaceae	–	+	–	–	–	–	–	–	–	–	–	–	–
60	*Parmotrema andinum* (Müll. Arg.) Hale	Parmeliaceae	–	+	–	–	+	+	–	–	–	–	–	–	–
61	*Parmotrema austrosinense* (Zahlbr.) Hale	Parmeliaceae	+	+	+	+	+	+	+	–	–	–	–	–	–
62	*Parmotrema crinitum* (Ach.) M. Choisy	Parmeliaceae	+	+	+	+	+	+	+	+	+	–	–	–	–
63	*Parmotrema grayanum* (hue) Hale	Parmeliaceae	+	+	+	+	+	+	+	+	+	+	–	–	–
64	*Parmotrema hababianum* (Gyeln.) Hale	Parmeliaceae	+	+	+	+	+	+	+	+	+	+	–	–	–
65	*Parmotrema melanothrix* (Mont.) Hale	Parmeliaceae	+	+	+	+	+	+	+	+	+	+	–	–	–
66	*Parmotrema nilgherrense* (Nyl.) Hale	Parmeliaceae	–	+	+	+	+	–	–	–	–	–	–	–	–
67	*Parmotrema praesorediosum* (Nyl.) Hale	Parmeliaceae	+	+	+	+	+	+	+	+	–	–	–	–	–
68	*Parmotrema reticulatum* (Taylor) M. Choisy	Parmeliaceae	+	+	+	+	+	+	+	+	+	+	+	+	+
69	*Parmotrema robustum* (Degel.) Hale	Parmeliaceae	–	+	–	–	–	–	–	+	+	–	–	–	–
70	*Parmotrema saccatilobum* (Taylor) Hale	Parmeliaceae	+	–	–	–	–	–	–	+	+	–	–	–	–

(continued)

Table 12.1 (continued)

S. No.	Lichen taxa	Families	Transects A	B	Altitude bands (m asl) 2000	2100	2200	2300	2400	2500	2600	2700	2800	2900	3000
71	*Parmotrema tinctorum* (Despr. ex Nyl.) Hale	Parmeliaceae	+	+	+	+	+	+	+	+	+	+	+	–	–
72	*Peltigera canina* (L.) Willd.	Peltigeraceae	–	+	–	–	–	–	+	+	+	+	+	–	–
73	*Peltigera polydactylon* (Neck.) Hoffm.	Peltigeraceae	+	+	+	+	+	+	+	–	–	–	–	–	–
74	*Peltigera praetextata* (Flörke ex Sommerf.) Zopf	Peltigeraceae	–	+	–	–	–	–	–	+	+	+	+	+	+
75	*Phaeophyscia endococcina* (Körb.) Moberg	Physciaceae	+	+	+	+	+	+	+	+	+	–	–	–	–
76	*Phaeophyscia hispidula* (Ach.) Essl.	Physciaceae	+	+	+	+	+	+	+	+	+	+	+	+	+
77	*Physcia dilatata* Nyl.	Physciaceae	+	+	–	–	–	+	+	+	–	–	–	–	–
78	*Polyblastidium hypocaesium* (Yasuda ex Räsänen) Kalb	Physciaceae	–	+	–	–	+	+	+	–	–	–	–	–	–
79	*Physconia* sp.	Physciaceae	–	+	–	–	–	–	–	+	+	+	+	–	–
80	*Polyblastidium japonicum* (M. Satô) Kalb	Physciaceae	+	+	–	–	+	+	+	+	+	+	+	+	+
81	*Protoparmeliopsis muralis* (Schreb.) M. Choisy	Lecanoraceae	+	+	–	–	–	–	–	–	–	–	+	+	+
82	*Punctelia rudecta* (ach.) Krog	Parmeliaceae	+	+	+	+	+	+	+	+	+	–	–	–	–
83	*Punctelia subrudecta* (Nyl.) Krog	Parmeliaceae	+	+	+	+	+	+	+	+	+	–	–	–	–
84	*Pyxine berteriana* (Fée) Imshaug	Caliciaceae	+	+	+	+	+	+	+	+	–	–	–	–	–
85	*Pyxine sorediata* (Ach.) Mont.	Caliciaceae	+	+	+	+	+	+	–	–	–	–	–	–	–

86	*Ramalina conduplicans* Vain.	Ramalinaceae	+	+	+	+	+	+	+	+	+	+	+	+	+
87	*Ramalina himalayensis* Räsänen	Ramalinaceae	–	+	–	–	–	–	+	+	+	+	+	+	+
88	*Ramalina sinensis* Jatta	Ramalinaceae	+	+	–	–	–	–	+	+	+	+	+	+	+
89	*Remototrachyna adducta* (Nyl.) Divakar, Lumbsch, Ferencová, Prado & A. Crespo	Parmeliaceae	–	+	–	+	–	–	–	–	–	–	–	–	–
90	*Remototrachyna crenata* (Kurok.) Divakar & A. Crespo	Parmeliaceae	–	+	–	–	–	+	–	–	–	–	–	–	–
91	*Rhizoplaca chrysoleuca* (Sm.) Zopf	Lecanoraceae	+	+	–	–	–	–	–	–	–	–	+	+	+
92	*Stereocaulon alpinum* Laurer	Cladoniaceae	+	+	–	–	–	–	–	–	–	–	–	–	–
93	*Stereocaulon foliolosum* Nyl.	Cladoniaceae	+	+	–	–	–	–	+	+	+	+	+	+	+
94	*Stereocaulon himalayense* D. D. Awasthi & I. M. Lamb	Cladoniaceae	+	+	–	–	–	–	+	+	+	+	+	+	+
95	*Stereocaulon macrocephalum* Müll. Arg.	Cladoniaceae	+	+	–	–	–	–	–	–	+	+	+	+	+
96	*Stereocaulon massartianum* Hue	Cladoniaceae	+	+	–	–	–	–	–	–	–	–	–	–	–
97	*Stereocaulon pomiferum* P. A. Duvign.	Cladoniaceae	+	+	–	–	–	–	–	–	–	–	–	+	+
98	*Sticta henryana* Müll. Arg.	Lobariaceae	–	+	–	–	–	–	–	+	+	+	+	+	–
99	*Sulcaria sulcata* (lev.) Bystrek ex Brodo and D. hawks.	Parmeliaceae	+	–	–	–	–	–	–	+	+	+	+	–	–

(continued)

Table 12.1 (continued)

S. No.	Lichen taxa	Families	Transects A	B	Altitude bands (m asl) 2000	2100	2200	2300	2400	2500	2600	2700	2800	2900	3000
100	*Umbilicaria indica* Frey	Umbilicariaceae	+	+	–	–	–	–	–	–	–	+	+	+	+
101	*Umbilicaria vellea* (L.) Ach.	Umbilicariaceae	+	+	–	–	–	–	–	–	–	+	+	+	+
102	*Usnea himalayana* C. Bab.	Parmeliaceae	+	–	–	–	–	+	+	+	+	+	–	–	–
103	*Usnea orientalis* Motyka	Parmeliaceae	+	+	+	+	+	+	+	+	+	+	+	+	+
104	*Usnea pseudosinensis* Asahina	Parmeliaceae	–	+	–	–	–	–	–	+	+	+	–	–	–
105	*Usnea stigmatoides* G. Awasthi	Parmeliaceae	–	+	–	–	–	–	–	–	–	–	–	+	–
106	*Usnea subfloridana* Stirt.	Parmeliaceae	+	+	+	+	+	+	+	+	+	+	+	+	+
107	*Xanthoparmelia mexicana* (Gyeln.) Hale	Parmeliaceae	+	+	–	–	–	–	+	+	+	+	+	+	+
108	*Xanthoria elegans* (Link) Th. Fr.	Teloschistaceae	+	+	–	–	–	–	–	–	–	–	+	+	+
Total species			73	104	50	50	55	56	60	62	62	54	50	43	39

(continued)

S.			Transects		Altitude bands (m asl)							Habitat		
No.	Lichen taxa	Families	A	B	3100	3200	3300	3400	3500	3600	3700	Corticolous	Saxicolous	Terricolous
1	*Bulbothrix isidiza* (Nyl.) Hale	Parmeliaceae	+	+	–	–	–	–	–	–	–	+	–	–
2	*Bulbothrix meizospora* (Nyl.) Hale	Parmeliaceae	+	–	–	–	–	–	–	–	–	+	–	–
3	*Bulbothrix setschwanensis* (Zahlbr.) Hale	Parmeliaceae	+	+	–	–	–	–	–	–	–	+	+	–
4	*Candelaria concolor* (dicks.) stein	Candelariaceae	+	+	+	+	–	–	–	–	–	+	–	–
5	*Canoparmelia texana* (Tuck.) Elix & hale	Parmeliaceae	+	+	–	–	–	–	–	–	–	+	–	–
6	*Cetraria islandica* (L.) Ach.	Parmeliaceae	+	+	–	+	+	+	+	+	–	–	–	+
7	*Cetrelia braunsiana* (Müll. Arg.) W.L. Culb. & C.F. Culb.	Parmeliaceae	+	+	–	–	–	–	–	–	–	+	–	–
8	*Cetrelia olivetorum* (Nyl.) W.L. Culb. & C.F. Culb.	Parmeliaceae	–	+	–	–	–	–	–	–	–	–	–	+
9	*Cladonia coccifera* (L.) Willd.	Cladoniaceae	+	+	–	–	–	–	–	–	–	–	–	+
10	*Cladonia coniocraea* (Flörke) Spreng.	Cladoniaceae	+	+	+	–	–	–	–	–	–	+	+	+
11	*Cladonia fimbriata* (L.) Fr.	Cladoniaceae	+	+	+	–	–	–	–	–	–	+	+	+
12	*Cladonia furcata* (Huds.) Schrad.	Cladoniaceae	+	+	–	–	–	–	–	–	–	+	+	+

(continued)

Table 12.1 (continued)

S. No.	Lichen taxa	Families	Transects		Altitude bands (m asl)							Habitat		
			A	B	3100	3200	3300	3400	3500	3600	3700	Corticolous	Saxicolous	Terricolous
13	*Cladonia pyxidata* (L.) Hoffm.	Cladoniaceae	+	+	+	+	+	+	+	+	+	–	+	+
14	*Cladonia ramulosa* (With.) J.R. Laundon	Cladoniaceae	+	+	+	+						–	–	+
15	*Cladonia rangiferina* (L.) Weber ex F.H. Wigg.	Cladoniaceae	+	+	–	–	–	–	+	+	+	–	–	+
16	*Cladonia subulata* (L.) Weber ex F.H. Wigg.	Cladoniaceae	–	+	–	–	–	–	–	–	–	–	–	+
17	*Coccocarpia erythroxyli* (Spreng.) Swinscow & Krog	Coccocarpaceae	+	+	–	–	–	–	–	–	–	–	+	–
18	*Collema* sp.	Collemataceae	–	+	–	–	–	–	–	–	–	+	–	–
19	*Dermatocarpon miniatum* (L.) W. Mann	Verrucariaceae	+	+	–	–	–	–	–	–	–	–	+	–
20	*Dermatocarpon vellereum* Zschacke	Verrucariaceae	+	+	–	–	–	–	–	–	–	–	+	–
21	*Dolichousnea longissima* (ach.) Articus	Parmeliaceae	+	+	+	+	–	–	–	–	–	+	–	–
22	*Flavoparmelia caperata* (L.) Hale	Parmeliaceae	+	+	–	–	–	–	–	–	–	+	+	+

23	*Flavopunctelia flaventior* (Stirt.) Hale	Parmeliaceae	+	+	–	–	–	–	–	–	–	+	–	+
24	*Flavopunctelia soredica* (Nyl.) Hale	Parmeliaceae	–	+	+	+	–	–	–	–	–	+	–	–
25	*Heterodermia albidiflava* (Kurok.) D.D. Awasthi	Physciaceae	+	+	–	–	–	–	–	–	–	+	+	–
26	*Heterodermia comosa* (Eschw.) Follmann & Redón	Physciaceae	+	+	–	–	–	–	–	–	–	+	–	–
27	*Heterodermia diademata* (Taylor) D.D. Awasthi	Physciaceae	+	+	+	+	+	+	+	+	+	+	+	+
28	*Heterodermia dissecta* (Kurok.) D.D. Awasthi	Physciaceae	–	+	–	–	–	–	–	–	–	+	+	–
29	*Heterodermia incana* (Stirt.) D.D. Awasthi	Physciaceae	–	+	–	–	–	–	–	–	–	+	–	–
30	*Heterodermia indica* (H. Magn.) D.D. Awasthi	Physciaceae	–	+	–	–	–	–	–	–	–	–	+	–
31	*Heterodermia leucomela* (L.) Poelt	Physciaceae	–	+	–	–	–	–	–	–	–	+	–	+
32	*Heterodermia obscurata* (Nyl.) Trevis.	Physciaceae	–	+	–	–	–	–	–	–	–	–	–	+

(continued)

Table 12.1 (continued)

S. No.	Lichen taxa	Families	Transects A	B	Altitude bands (m asl) 3100	3200	3300	3400	3500	3600	3700	Habitat Corticolous	Saxicolous	Terricolous
33	*Heterodermia podocarpa* (Bél.) D.D. Awasthi	Physciaceae	+	+	–	–	–	–	–	–	–	+	–	–
34	*Heterodermia pseudospeciosa* (Kurok.) W.L. Culb.	Physciaceae	+	+	+	–	–	–	–	–	–	+	+	–
35	*Heterodermia speciosa* (Wulfen) Trevis.	Physciaceae	+	+	–	–	–	–	–	–	–	+	+	+
36	*Hypotrachyna cirrhata* (Fr.) Divakar, A. Crespo, Sipman, Elix & Lumbsch	Parmeliaceae	+	+	–	–	–	–	–	–	–	+	+	+
37	*Hypotrachyna nepalensis* (Taylor) Divakar, A. Crespo, Sipman, Elix & Lumbsch	Parmeliaceae	+	+	+	+	+	+	–	–	–	+	+	+
38	*Hypotrachyna pindarensis* (D.D. Awasthi & Suj. Singh) D.D. Awasthi	Parmeliaceae	–	+	+	+	–	–	–	–	–	+	–	–
39	*Hypotrachyna pseudosinuosa* (Asahina) Hale	Parmeliaceae	–	+	–	–	–	–	–	–	–	+	–	–

40	*Hypotrachyna scytodes* (Kurok.) Hale	Parmeliaceae	–	+	–	–	–	–	–	–	–	+	–	–
41	*Leptogium askotense* D.D. Awasthi	Collemataceae	+	+	–	–	–	–	–	–	–	+	–	–
42	*Leptogium burnetiae* C.W. Dodge	Collemataceae	+	+	+	+	–	–	–	–	–	–	+	–
43	*Leptogium delavayi* Hue	Collemataceae	+	+	+	+	+	–	–	–	–	+	+	–
44	*Leptogium javanicum* (Mont. & Bosch) Mont.	Collemataceae	–	+	+	+	+	–	–	–	–	+	–	–
45	*Leptogium pedicellatum* P.M. Jørg.	Collemataceae	+	+	–	–	–	–	–	–	–	+	+	–
46	*Leptogium trichophorum* Müll. Arg.	Collemataceae	–	+	–	–	–	–	–	–	–	+	–	–
47	*Leucodermia boryi* (Fée) Kalb	Physciaceae	+	+	+	+	+	+	+	+	+	+	–	+
48	*Lobaria isidiosa* (Müll. Arg.) Vain.	Lobariaceae	–	+	–	–	–	–	–	–	–	+	+	–
49	*Lobaria pseudopulmonaria* Gyeln.	Lobariaceae	+	+	–	–	–	–	–	–	–	–	–	+
50	*Lobaria pulmonaria* (L.) Hoffm.	Lobariaceae	–	+	+	+	–	–	–	–	–	+	–	–
51	*Lobaria retigera* (Bory) Trevis.	Lobariaceae	+	+	+	+	+	–	–	–	–	+	+	+

(continued)

Table 12.1 (continued)

S. No.	Lichen taxa	Families	Transects		Altitude bands (m asl)							Habitat		
			A	B	3100	3200	3300	3400	3500	3600	3700	Corticolous	Saxicolous	Terricolous
52	*Melanelia stygia* (L.) Essl.	Parmeliaceae	–	+	–	–	–	–	–	–	–	–	+	–
53	*Myelochroa perisidians* (Nyl.) Elix & Hale	Parmeliaceae	–	+	–	–	–	–	–	–	–	+	+	–
54	*Myelochroa rhytidodes* (Hale) Elix & Hale	Parmeliaceae	–	+	–	–	–	–	–	–	–	+	+	–
55	*Nephroma helveticum* Ach.	Nephromataceae	+	+	+	+	+	–	–	–	–	+	–	–
56	*Parmelia meiophora* Nyl.	Parmeliaceae	+	+	–	–	–	–	–	–	–	+	–	–
57	*Parmelia saxatilis* (L.) Ach.	Parmeliaceae	–	+	–	–	–	–	–	–	–	+	+	–
58	*Parmelia sulcata* Taylor	Parmeliaceae	–	+	+	+	–	–	–	–	–	+	+	–
59	*Parmelinella wallichiana* (Taylor) Elix & hale	Parmeliaceae	–	+	+	+	–	–	–	–	–	+	+	–
60	*Parmotrema andinum* (Müll. Arg.) Hale	Parmeliaceae	–	+	–	–	–	–	–	–	–	+	+	–
61	*Parmotrema austrosinense* (Zahlbr.) Hale	Parmeliaceae	+	+	–	–	–	–	–	–	–	+	–	–
62	*Parmotrema crinitum* (Ach.) M. Choisy	Parmeliaceae	+	+	–	–	–	–	–	–	–	+	–	–

63	*Parmotrema grayanum* (hue) Hale	Parmeliaceae	+	+	–	–	–	–	–	–	–	+	+	–
64	*Parmotrema hababianum* (Gyeln.) Hale	Parmeliaceae	+	+	–	–	–	–	–	–	–	+	–	–
65	*Parmotrema melanothrix* (Mont.) Hale	Parmeliaceae	+	+	–	–	–	–	–	–	–	+	–	–
66	*Parmotrema nilgherrense* (Nyl.) Hale	Parmeliaceae	–	+	–	–	–	–	–	–	–	+	+	–
67	*Parmotrema praesorediosum* (Nyl.) Hale	Parmeliaceae	+	+	–	–	–	–	–	–	–	+	–	–
68	*Parmotrema reticulatum* (Taylor) M. Choisy	Parmeliaceae	+	+	+	+	+	–	–	–	–	+	–	–
69	*Parmotrema robustum* (Degel.) Hale	Parmeliaceae	–	+	–	–	–	–	–	–	–	+	–	–
70	*Parmotrema saccatilobum* (Taylor) Hale	Parmeliaceae	+	–	–	–	–	–	–	–	–	+	–	–
71	*Parmotrema tinctorum* (Despr. ex Nyl.) Hale	Parmeliaceae	+	+	–	–	–	–	–	–	–	+	+	–
72	*Peltigera canina* (L.) Willd.	Peltigeraceae	–	+	–	–	–	–	–	–	–	–	–	+
73	*Peltigera polydactylon* (Neck.) Hoffm.	Peltigeraceae	+	+	–	–	–	–	–	–	–	–	–	+
74	*Peltigera praetextata* (Flörke ex Sommerf.) Zopf	Peltigeraceae	–	+	–	–	–	–	–	–	–	–	–	+

(continued)

Table 12.1 (continued)

S. No.	Lichen taxa	Families	Transects		Altitude bands (m asl)							Habitat		
			A	B	3100	3200	3300	3400	3500	3600	3700	Corticolous	Saxicolous	Terricolous
75	*Phaeophyscia endococcina* (Körb.) Moberg	Physciaceae	+	+	–	–	–	–	–	–	–	+	–	–
76	*Phaeophyscia hispidula* (Ach.) Essl.	Physciaceae	+	+	+	+	+	+	+	+	+	+	–	–
77	*Physcia dilatata* Nyl.	Physciaceae	+	+	–	–	–	–	–	–	–	–	+	–
78	*Polyblastidium hypocaesium* (Yasuda ex Räsänen) Kalb	Physciaceae	–	+	–	–	–	–	–	–	–	–	–	+
79	*Physconia* sp.	Physciaceae	–	+	–	–	–	–	–	–	–	+	–	+
80	*Polyblastidium japonicum* (M. Satô) Kalb	Physciaceae	+	+	–	–	–	–	–	–	–	–	+	–
81	*Protoparmeliopsis muralis* (Schreb.) M. Choisy	Lecanoraceae	+	+	+	+	+	+	+	+	+	–	+	–
82	*Punctelia rudecta* (ach.) Krog	Parmeliaceae	+	+	–	–	–	–	–	–	–	+	+	–
83	*Punctelia subrudecta* (Nyl.) Krog	Parmeliaceae	+	+	–	–	–	–	–	–	–	+	+	–
84	*Pyxine berteriana* (Fée) Imshaug	Caliciaceae	+	+	–	–	–	–	–	–	–	+	–	+
85	*Pyxine sorediata* (Ach.) Mont.	Caliciaceae	+	+	–	–	–	–	–	–	–	+	+	+
86	*Ramalina conduplicans* Vain.	Ramalinaceae	+	+	+	+	–	–	–	–	–	+	+	–

87	*Ramalina himalayensis* Räsänen	Ramalinaceae	–	+	–	–	–	–	–	–	–	+	–	–
88	*Ramalina sinensis* Jatta	Ramalinaceae	+	+	+	+	+	+	–	–	–	+	–	–
89	*Remototrachyna adducta* (Nyl.) Divakar, Lumbsch, Ferencová, Prado & A. Crespo	Parmeliaceae	–	+	–	–	–	–	–	–	–	+	–	–
90	*Remototrachyna crenata* (Kurok.) Divakar & A. Crespo	Parmeliaceae	–	+	–	–	–	–	–	–	–	+	–	–
91	*Rhizoplaca chrysoleuca* (Sm.) Zopf	Lecanoraceae	+	+	+	+	+	+	+	+	+	–	+	–
92	*Stereocaulon alpinum* Laurer	Cladoniaceae	+	+	+	+	+	+	+	+	+	–	–	+
93	*Stereocaulon foliolosum* Nyl.	Cladoniaceae	+	+	+	+	+	+	+	+	+	–	+	+
94	*Stereocaulon himalayense* D. D. Awasthi & I. M. Lamb	Cladoniaceae	+	+	+	+	+	+	+	+	+	–	+	+
95	*Stereocaulon macrocephalum* Müll. Arg.	Cladoniaceae	+	+	+	+	+	+	+	+	+	–	+	–
96	*Stereocaulon massartianum* Hue	Cladoniaceae	+	+	–	+	+	+	+	+	+	–	–	+
97	*Stereocaulon pomiferum* P.A. Duvign.	Cladoniaceae	+	+	+	+	+	+	–	–	–	–	–	+
98	*Sticta henryana* Müll. Arg.	Lobariaceae	–	+	–	–	–	–	–	–	–	+	–	–

(continued)

Table 12.1 (continued)

S. No.	Lichen taxa	Families	Transects A	B	Altitude bands (m asl) 3100	3200	3300	3400	3500	3600	3700	Habitat Corticolous	Saxicolous	Terricolous
99	*Sulcaria sulcata* (lev.) Bystrek ex Brodo and D. hawks.	Parmeliaceae	+	–	–	–	–	–	–	–	–	+	–	–
100	*Umbilicaria indica* Frey	Umbilicariaceae	+	+	+	+	+	+	+	+	+	–	+	–
101	*Umbilicaria vellea* (L.) Ach.	Umbilicariaceae	+	+	+	+	+	+	+	+	+	–	+	–
102	*Usnea himalayana* C. Bab.	Parmeliaceae	+	–	–	–	–	–	–	–	–	+	–	–
103	*Usnea orientalis* Motyka	Parmeliaceae	+	+	–	–	–	–	–	–	–	+	–	–
104	*Usnea pseudosinensis* Asahina	Parmeliaceae	–	+	–	–	–	–	–	–	–	+	–	–
105	*Usnea stigmatoides* G. Awasthi	Parmeliaceae	–	+	–	–	–	–	–	–	–	+	–	–
106	*Usnea subfloridana* Stirt.	Parmeliaceae	+	+	+	–	–	–	–	–	–	+	–	–
107	*Xanthoparmelia mexicana* (Gyeln.) Hale	Parmeliaceae	+	+	+	+	+	+	+	+	+	–	+	–
108	*Xanthoria elegans* (Link) Th. Fr.	Teloschistaceae	+	+	+	+	+	+	+	+	+	–	+	–
Total species			73	104	36	34	24	19	17	17	16			

Note: A—Kalsir to Chandrashila, **B**—Pothibhasha to Chandrashila

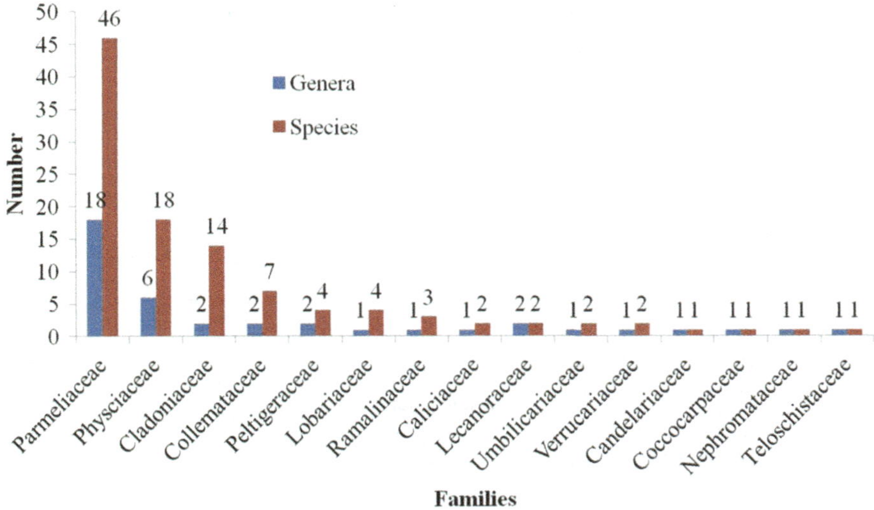

Fig. 12.1 Distribution of macrolichen species within genera and families

Fig. 12.2 Venn diagram showing distribution of lichen species on different substrates

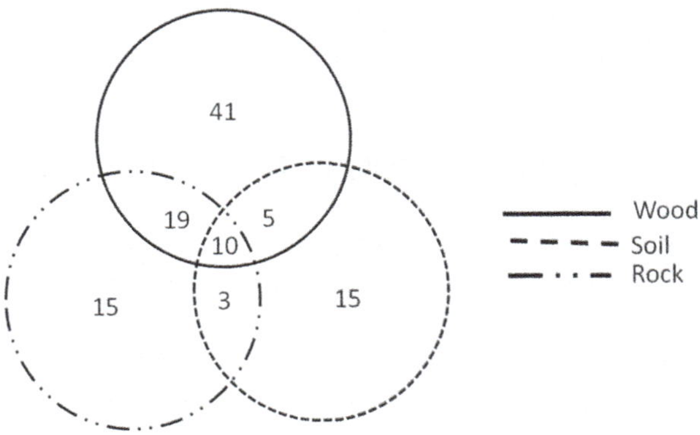

3000 m elevation. Fifteen species were confined to mid elevation zone, that is, 2400–3200 m altitude. Three species *Heterodermia diademata*, *Leucodermia boryi*, and *Phaeophyscia hispidula*, were, however, present in all the elevation bands. Other 8 species, that is, *Heterodermia dissecta* (2500 m), *Heterodermia indica* (2600 m), *Heterodermia leucomela* (2800 m), *Heterodermia obscurata* (2000 m), *Leptogium trichophorum* (2400 m), *Remototrachyna adducta* (2100 m), *Rinorea crenata* (2300 m) and *Usnea stigmatoides* (2900 m) were recorded from a single elevation band only. Fifteen species were frequently distributed (found in every elevation band) between 2000 and 2800–3400 m altitude. The species to family ratio (SR/F) across the elevation ranged from 2.4 at 3300 m to 6.2 at 2600 m. The species richness and genera ratio (SR/G) was maximum at 2200 m (2.5) and was minimum at 3300 m (1.41).

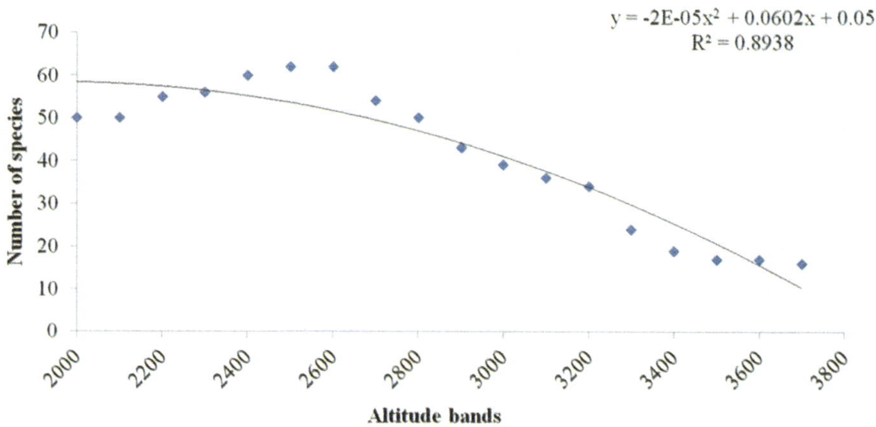

Fig. 12.3 Distribution of macrolichen species within altitude bands

12.3.3 Lichen Diversity in Closed Canopy Forests

The moisture and shade loving species of lichen families like Collemataceae (*Collema* spp. and *Leptogium* spp.), Lobariaceae (*Lobaria* spp.), and Peltigeraceae (*Peltigera* spp.) were found growing under closed canopy forests.

12.3.4 Lichen Diversity at Treeline Ecotone

Members of lichen families such as Caliciaceae (*Pyxine* spp.), Parmeliaceae (*Hypotrachyna* spp., *Nephroma* spp., *Parmelia* spp., *Parmotrema* spp., and *Usnea* spp.), Physciaceae (*Heterodermia* spp.), and Ramalinaceae (*Ramalina* spp.) were dominant at the treeline ecotone of the studied landscape. The species of these families grow well in open places with direct sunlight.

12.3.5 Lichen Diversity in Alpine Tundra

The alpine meadows above treeline were dominated by the members of families Cladoniaceae (*Cladonia* spp. and *Stereocaulon* spp.) on soil, while the rocks above treeline were colonized by the species of families Lecanoraceae (*Protoparmeliopsis* spp. and *Rhizoplaca* spp.), Umbilicariaceae (*Umbilicaria* spp.), and Teloschistaceae (*Xanthoria* spp.).

Table 12.2 Macrolichen diversity and distribution across the altitude bands in TTL

S. No.	Studies transects/ Altitude bands (m asl)	Species richness (SR)	Number of families (F)	Number of genera (G)	SR/ F	SR/ G	Whittaker's β diversity
1	Transect A	73	15	34	2.14	4.86	0.47
2	Transect B	104	15	40	6.93	2.60	0.03
3	2000	50	10	21	5.00	2.38	1.16
4	2100	50	10	22	5.00	2.27	1.16
5	2200	55	10	22	5.50	2.50	0.96
6	2300	56	11	26	5.09	2.15	0.92
7	2400	60	11	27	5.45	2.22	0.80
8	2500	62	11	30	5.63	2.06	0.74
9	2600	62	10	29	6.20	2.13	0.74
10	2700	54	11	28	4.90	1.92	1.00
11	2800	50	14	30	3.57	1.66	1.16
12	2900	43	13	26	3.30	1.65	1.51
13	3000	39	12	23	3.25	1.69	1.76
14	3100	36	11	21	3.27	1.71	2.00
15	3200	34	11	22	3.09	1.54	2.17
16	3300	24	10	17	2.40	1.41	3.50
17	3400	19	7	13	2.71	1.46	4.68
18	3500	17	6	11	2.83	1.54	5.35
19	3600	17	6	11	2.83	1.54	5.35
20	3700	16	5	10	3.20	1.60	5.75

12.3.6 Beta (β) Diversity across Elevation Bands

Whittaker's β diversity values, calculated for the studied transects and selected 18 altitude bands, are presented in Table 12.2. Perusal of the data revealed variations in macrolichen species across the elevations and transects. The maximum β diversity (5.75) was recorded at the 3700 m elevation. Across the elevation bands β diversity mostly increased with increase in elevation.

12.3.7 Similarity Coefficient between the Elevation Bands

Considerable variation in similarity coefficient was apparent across elevation bands (Table 12.3). Among all the bands, highest similarity was recorded between 3500 and 3600 m (100%) followed by 2000 and 2100 m, 3500 and 3700 m, and 3600 and 3700 m (98%, 97%, and 97%) respectively. The distantly located elevation bands (i.e., 2200 and 3500 m, 2200 and 3600 m, 2200 and 3700 m, 2300 and 3500 m, 2300 and 3600 m, and 2300 and 3700 m) exhibited low similarity (8% each).

Table 12.3 Similarity coefficient (in %) between different studied altitude bands in TTL

	2000	2100	2200	2300	2400	2500	2600	2700	2800	2900	3000	3100	3200	3300	3400	3500	3600	3700
2000	100	98	93	85	76	63	59	52	46	37	34	35	26	22	12	9	9	9
2100		100	93	85	76	63	59	52	46	37	36	35	26	22	12	9	9	9
2200			100	92	82	67	62	55	48	39	36	33	25	20	11	8	8	8
2300				100	88	71	64	58	49	40	36	33	24	20	11	8	8	8
2400					100	82	74	68	60	49	44	42	32	29	20	16	16	16
2500						100	90	81	68	53	48	41	33	28	20	15	15	15
2600							100	90	77	63	55	45	38	30	22	18	18	18
2700								100	87	72	65	53	45	38	30	25	25	26
2800									100	84	76	65	57	51	41	36	36	36
2900										100	90	70	70	60	48	40	40	41
3000											100	88	79	67	55	46	46	47
3100												100	91	73	62	53	53	54
3200													100	83	72	63	63	60
3300														100	88	78	78	75
3400															100	89	48	86
3500																100	100	97
3600																	100	97
3700																		100

Fig. 12.4 (**a**) Foliicolous lichen (*Strigula complanata*) on leaves of *Persea duthiei*, (**b**) Caliciale lichens on bark of *Abies spectabilis*

12.3.8 Indicator Lichen Communities

Besides the composition and diversity of macorlichens across the elevation bands, two special indicator lichen groups, viz. foliicolous (leaf colonizing lichens) and caliciales (pin head lichens), were encountered during the field survey. Leaves of *Persea duthiei* from Pothibhasha (2100 m) to Dugalbhitta (2300 m) were heavily colonized by a foliicolous lichen, that is, *Strigula complanata* (Fig. 12.4a). Similarly, most of the trees of *Abies spectabilis* were found colonized by the caliciale lichens (Fig. 12.4b).

12.4 Discussion

The elevation transects of present study represented about 4% and 10% of the total lichen diversity reported from India and Uttarakhand, respectively. In a previous study, a total of 85 species of macrolichens covering an elevation range of 1500–3700 m in a similar study region were reported (Negi 2000). While comparing these species with the species observed along the elevation transects of present study, a total of 69 species (81%) were common in both the studies. This suggests that most of the macrolichens have wide distribution in the study area. However, 16 species (viz *Cetreliopsis rhytidocarpa* (*Cetraria rhytidocarpa*), *Cladonia chlorophaea*, *Cladonia gymnopoda*, *Cladonia parasitica*, *Cladonia pleurota*, *Cladonia scabriuscula*, *Heterodermia angustiloba*, *Heterodermia punctifera*, *Hypotrachyna awasthii* (*Parmelia awasthi*), *Hypotrachyna crenata* (*Parmelia crenata*), *Parmelinella simplicior* (*Parmelia simplicior*), *Peltigera rufescens*, *Ramalina roesleri*, *Usnea baileyi*, *Usnea eumitrioides*, and *Usnea perplexans*) were reported in previous report only. Likewise 39 species have been reported in the present study only. However, putting both studies together it is reasonable to suggest the existence of 124 species of macrolichens in Tungnath area. Despite the maximum number of species were growing only on woody substrates, these species are at risk of disappearance from the region. It was observed that most of the trees were colonized by mosses and bryophytes and were devoid of lichens. In due course of time, with expected colonization of bryophyte species on whole trunk would lead to exclusion of lichens from them. In this case host specific (only corticolous) species are likely to suffer and struggle for their substrate requirements.

Elevation in Himalayan mountains remains an important factor for determining the diversity and distribution of biodiversity elements (Upadhya 2017; Rawal et al. 2018; Upadhyay et al. 2018). Along the elevation range, largely three types of distribution patterns of organisms exist: (1) monotonic decline, (2) increase with altitude, and (3) hump-shaped or intermediate relationship. Our study on macrolichens suggests, by and large, prevalence of monotonic decline. Although, there is an indication of increase in species number around 2600–2700 m altitude, thereby forming a small hump in species distribution. The study, however, established that the number of macrolichen species decline considerably with increasing elevation, specifically above 2600 m asl. In this context, species richness pattern of Tungnath differs from hump-shaped species relationship, with maximum richness at approximately 4000 m asl, reported in neighboring Nepalese Himalaya (Baniya et al. 2010). Considerable difference in richness of species in the two study transects (i.e., 73 species in KCT; 104 species in PCT) indicates that macrolichen diversity varied due to diversity in microclimatic conditions. Presence of maximum macrolichen species at an elevation of 2600 m asl may be due to the high atmospheric moisture and cooler summer temperature which are favorable for the growth and development of the macrolichens. Occurrence of 8 species of macrolichens (*Cetraria islandica*, *Cladonia pyxidata*, *Cladonia ramulosa*, *Cladonia rangiferina*, *Hypotrachyna pindarensis*, *Parmelinella wallichiana*, *Stereocaulon alpinum*, and *Stereocaulon massartianum*) only above 3000 m asl suggests that they are the ecotonal species. Three species, viz. *Heterodermia diademata*, *Leucodermia boryi*, and *Phaeophyscia hispidula*, with existence in all altitude bands, exhibited their wide ecological amplitude and resistance for growing in any environmental condition.

Whittaker's β diversity revealed that the high-altitude bands exhibit higher heterogeneity with respect to macrolichens. The higher beta diversity in high-altitude bands presumably reflects greater spatial variation in substrate and microclimatic conditions. This is obvious as <3200 m asl altitude bands are trees that are dominating elements, whereas >3200 m altitude bands are mostly small shrubs and herbaceous elements that dominate the landscape.

Overall heterogeneity in macrolichen species distribution across the elevation range is further reflected with lower similarity among low- and high-altitude bands. For instance, if the low-altitude bands (2000–2500 m) are compared with high-altitude bands (3100–3700 m), similarity in most cases remains less than 25% and even goes down to 8–9% in certain cases.

The lichens are considered bioindicators of forest health and ecological continuity, as well as atmospheric pollution in different regions of the world (McCune 2000; Kricke and Loppi 2002; Brunialti and Giordani 2003; Wolseley et al. 2006; Upadhya 2017). In this context, two interesting indicator communities (i.e., foliicolous and caliciales) were present in the study area. Existence of foliicolous lichens (lichens colonizing leaves of vascular plants), which are among the most abundant epiphytes in tropical rain forests (Lücking 2001; Anthony et al. 2002), indicates existence of higher human disturbance and microclimatic conditions of the studied forests (Lücking 1997). In India the distribution of foliicolous lichens is mostly confined to tropical regions of Eastern Himalaya and Western Ghats and over 135 species have been documented (Singh and Sinha 2010). Recently, Upadhyay et al. (2015) reported 6 species of foliicolous lichens from Nandhaur wildlife sanctuary of Uttarakhand. Presence of foliicolous lichens in some subtropical areas (Vězda 1983; Sérusiaux 1996; Puntillo and Ottonello 1997) and temperate regions (Malcolm and Galloway 1997; Lücking et al. 2003) has also been reported. Therefore, frequent occurrence of these lichens in studied transects that represent temperate climatic conditions is of special interest.

The lower parts of PCT, Pothibhasha (2100 m) to Dugalbhitta (2300 m), are broad leaved mixed forests. The leaves of *Persea duthiei* were heavily colonized by foliicolous lichens (Fig. 12.4a). This suggests that these lichens of tropical or subtropical climates are finding suitable habitats even at these elevations. In other words, it can be concluded that these lichens are extending their distributional range to temperate areas. These evidences correspond well with the reports of greater warming rate in higher elevations, in general, and the present study area, in particular (Singh 2018). However, more evidences from other transects would be required for a better generalization. Also, this will establish bio-indicator potential of these lichens to monitor the climatic conditions of various forest types across the Himalayan landscape.

The caliciales (Pin lichens) are the tiny unremarkable lichens which resemble with small pins that arise from a bed of green algae. Their association with late-successional and old-growth forests is, however, well documented (Rose 1992). They are substrate specific. Some of these lichen assemblages tend to be found on gymnosperms generally and on angiosperms rarely and vice versa (Selva 1994). Most of the *Abies spectabilis* trees in the areas of studied transects were colonized by caliciale lichens (Fig. 12.4b). The presence of these lichens on the bark of *Abies* trees is indicative of old growth of these forests that have reached the climax stage of succession. Such old growth forests that have taken a longer period of time to acquire the characteristics of microhabitats suitable for establishment of the rare caliciales have been described as "ancient forests" by Selva (1994).

The study suggests that there is an opportunity to harness the full bioindicator potential of different lichen communities to assess the ecosystem health in the study area. The identification of "indicator" lichens can provide a basis for appropriate management prescriptions and can effectively be used to assess climatic changes and potential forest recovery in areas where deforestation has caused changes in microclimate and thereby in the phanerogamic communities as well (Wolseley and Aguirre-Hudson 2007). The study also provides baseline information about the macrolichen community structure which can be used for long-term monitoring, temporal changes taking place in community structure, and migration of taxa from one habitat to another, which may be the consequence of change in climatic conditions. Any alteration in the lichen species and community structure may be further used for predicting future climate scenarios.

Acknowledgements Financial support from Indian Himalayan Timberline Project (IHTP) under the National Mission on Himalayan Studies (NMHS) of Govt. of India is gratefully acknowledged. Coordinator IHTP, Prof SP Singh is thanked for his encouragement. Thanks are due to Head, Department of Botany, S.S.J. Campus, Kumaun University, Almora and Director NIHE for laboratory facilities. We extend our thanks to Mr. Sunil Joshi, Ms. Renu Rawal, Ms. Poonam Mehta, Ms. Kamini Durgapal, and Ms. Medha Durgapal, NIHE, for supporting the authors (KB and SU) in the field.

References

Ahmadjian V (1995) Lichens are more important than you think. Bio Sci 45:124

Anthony PA, Holtum JAM, Jackes BR (2002) Shade acclimation of rainforest leaves to colonization by lichens. Funct Ecol 16:808–816

Aptroot A (2009) Lichens as an indicator of climate and global change. In: Letcher TM (ed) Climate change: observed impacts on planet earth. Elsevier, Oxford, p 444

Awasthi DD (2007) A compendium of the macrolichens from India, Nepal and Sri Lanka. Bishen Singh Mahendra Pal Singh Publication, Dehra Dun; pp. 580

Baniya CB, Solhøy T, Gauslaa Y, Palmer MW (2010) The elevation gradient of lichen species richness in Nepal. Lichenologist 42(1):83–96

Bisht K (2018) Impact of climate change on glaciers of Kumaun Himalaya: a lichenometric approach. Ph.D. thesis, Kumaun University, Nainital, Uttarakhand, India. http://hdl.handle.net/10603/257037

Bisht K, Joshi Y, Upadhyay S, Mehta P (2018a) Recession of Milam Glacier, Kumaun Himalaya, observed via lichenometric dating of moraines. J Geol Soc India 92(2):173–176

Bisht K, Joshi Y, Upadhyay S, Chandra K (2018b) Assessment of climate change impact on recession of Adi Kailash glacier, Kumaun Himalaya: a lichenometric observation. ENVIS Bull Himal Ecol 25:24–27

Bisht K, Upadhyay S, Joshi Y (2019a) Cairns as hurdles for lichenometric studies on Himalayan glaciers. J Geol Soc India 94(5):545–546

Bisht K, Upadhyay S, Joshi Y (2019b) Timberline forests: potential habitats for conserving Himalayan medicinal lichen diversity in Kailash Sacred Landscape part of India. ENVIS Bull Himal Ecol 26:24–27

Bisht K, Upadhyay S, Joshi Y (2020) Higher growth rates of saxicolous lichens in alpine regions of western Himalaya, a consequence of warming climate? ENVIS Bull Himal Ecol 27:69–72

Brunialti G, Giordani P (2003) Variability of lichen diversity in a climatically heterogeneous area (Liguria, NW Italy). Lichenologist 35:55–69

Dhar U (2000) Prioritization of conservation sites in the timberline zone of west Himalaya. In: Singh S, Sastry ARK, Mehta R, Uppal V (eds.) Setting biodiversity conservation priorities for India; pp. 193 – 211

Divakar PK, Upreti DK (2005) Parmelioid lichens in India. A revisionary study. Bishen Singh Mahendra Pal Singh Publicaton, Dehradun, p 488

Essen PA, Renhorn KE, Pettersson RB (1996) Epiphytic lichen biomass in managed and old-growth boreal forests: effects of branch quality. Ecol Appl 6:228–238

Gairola S, Rawal RS, Dhar U (2009) Patterns of litter fall and return of nutrients across anthropogenic disturbance gradients in three subalpine forests of West Himalaya, India. J For Res 14:73–80

Gairola S, Rawal RS, Todaria NP (2015) Effect of anthropogenic disturbance on vegetation characteristics of sub-alpine forests in and around valley of flowers National Park, a world heritage site of India. Trop Ecol 56(3):357–365

Gauslaa Y (2014) Rain, dew and humid air as drivers of morphology, function and spatial distribution in epiphytic lichens. Lichenologist 46:1–16

Joshi Y, Upadhyay S, Chandra K, Bisht K, Falswal A (2016) A new species of *Plectocarpon* (Roccellaceae, Lichenised Ascomycetes) from India. Act Bot Hung 58(3–4):257–264

Joshi Y, Tripathi M, Bisht K, Upadhyay S, Kumar V, Pal N, Gaira A, Pant S, Rawat KS, Bisht S, Bajpai R, Halda JP (2018a) Further contributions to the documentation of lichenicolous fungi from India. Kavaka 50:26–33

Joshi Y, Bisht K, Upadhyay S, Chandra K (2018b) Three new records of lichens from India. Nelumbo 60(1):90–94

Kricke R, Loppi S (2002) Bioindication: the I.a.P. approach. In: Nimis PL, Scheidegger C, Wolseley PA (eds) Monitoring with lichens – monitoring lichens. Kluwer Academic Press, Dordrecht, pp 21–37

Lücking R (1997) The use of foliicolous lichens as bioindicators in the tropics, with special reference to the microclimate. Abst Bot 21:99–116

Lücking R (2001) Lichens on leaves in tropical rainforests: life in a permanently ephemerous environment. In: Gottsberger G, Liede S (eds.), Life forms and dynamics in tropical forests; pp. 41–77

Lücking R, Wirth V, Ferrarom CMES (2003) Foliicolous lichens from Valdivian temperate rain forest of Chile and Argentina: evidence of an austral element, with the description of seven new taxa. Glob Ecol Biogeogr 12:21–36

Malcolm WM, Galloway DJ (1997) New Zealand Lichens. In: Checklists, key and glossary. Wellington, Museum of New Zealand Te Papa Tongarewa, p 192

McCune B (2000) Lichen communities as indicators of forest health. Bryologist 103(2):353–356

Negi HR (2000) On the patterns of abundance and diversity of macrolichens of Chopta-Tungnath in the Garhwal Himalaya. J Biosci 25(4):367–378

Pettersson RB, Ball JP, Renhorn KE, Essen PA, Sjberg K (1995) Invertebrate communities in boreal forest canopies as influenced by forestry and lichens with implications for passerine birds. Biol Conserv 74:57–63

Pharo EJ, Beattie AJ, Binns D (1999) Vascular plant diversity as a surrogate for bryophyte and lichen diversity. Conserv Biol 3:242–262

Puntillo D, Ottonello D (1997) A new foliicolous lichen station in Italy. Lichenologist 29:388–390
Rai ID, Adhikari BS, Rawat GS, Bargali K (2012) Community structure along timberline ecotones in relation to micro-topography and disturbance in West Himalaya. Not Sci Biol 4:41–52
Rawal RS, Dhar U (1997) Sensitivity of timberline flora in Kumaun Himalaya, India: conservation implications. Arct Alp Res 29:112–121
Rawal RS, Rawal R, Rawat B, Negi VS, Pathak R (2018) Plant species diversity and rarity patterns along altitude range covering treeline ecotones in Uttarakhand: conservation implications. Trop Ecol 59(2):225–239
Rose F (1992) Temperate forest management: its effects on bryophyte and lichen floras and habitats. In: Bates JW, Farmer AM (eds) Bryophytes and lichens in a changing environment. Clarendon, Oxford, pp 211 —21233
Rosentreter R, Hayward GH, Wicklow-Howard M (1997) Northern flying squirrel seasonal food habits in the interior conifer forests of Central Idaho, USA. Northwest Sci 71:97–101
Selva SB (1994) Lichen diversity and stand continuity in the Northern Hardwoods and Spruce-Fir forests of Northern New England and Western New Brunswick. Bryologist 97:424–429
Sérusiaux E (1996) Foliicolous lichens from Madeira, with the description of a new genus and two new species and a world-wide key of foliicolous *Fellhanera*. Lichenologist 28:197–227
Shukla V, Upreti DK, Bajpai R (2014) Lichen diversity in different lichenogeographical regions of India. In: Shukla V, Upreti DK, Bajpai R (eds) Lichen to biomonitor the environment. Springer, India, pp 61–96
Singh SP (2018) Research on Indian Himalayan treeline ecotone: an overview. Trop Ecol 59(2): 163–176
Singh KP, Sinha GP (2010) Indian lichens: an annotated checklist. Botanical Survey of India, Kolkata, p 571
Sorenson T (1948) A method of establishing groups of equal amplitude in plant sociology based on similarity of species content and its application to analysis of the vegetation on Danish commons. Biol Skrifter Kongelige Danske videnskabernes Selskab 5:1–34
Starkey E, Hagar JC (1999) Biodiversity of young Forest-role of birds. Final Report to US Department of Interior, Blue River District
Upadhya S (2017) Distribution and diversity pattern of macrolichens along an altitudinal gradient in Kumaun Himalaya with an emphasis to their indicator value. Ph.D. thesis. Kumaun University, Nainital
Upadhyay S, Shukla S, Joshi Y, Pinokiyo A (2015) Range extension of foliicolous lichens in India: a case study from Nandhaur Forest range, Lakhan Mandi, Haldwani, Uttarakhand, India. G - J Env Sci Tech 3:101–104
Upadhyay S, Jugran AK, Joshi Y, Suyal R, Rawal RS (2018) Ecological variables influencing the diversity and distribution of epiphytic macrolichens colonizing *Quercus leucotrichophora* in Thal Ke Dhar forest, Pithoragarh district, Uttarakhand. J Mt Sci 15:307. https://doi.org/10.1007/s11629-017-4397-9
Vězda A (1983) Foliicole Flechtenaus der Kolchis (West-Transkaukasien, UdSSR). Fol Geobot Phyto Praha 18:45–70
Vokou D, Pirintsos SA, Lopppi S (1999) Lichens as bioindicators of temporal variation in air quality around Thessaloniki, northern Greece. Ecol Res 14:89–96
Will-Wolf S, Neitlich P, Essen PA (2002) Monitoring biodiversity and ecosystem functions: forest. In: Nimis PL, Scheidegger C, Wolseley P (eds) Monitoring with lichens monitoring lichens, NATO science series. Kluwer Academic Publishers, The Hague, pp 203–222
Wolseley PA, Aguirre-Hudson B (2007) Lichens as indicators of environmental changes in the tropical forests of Thailand. Glob Ecol Biogeogr Lett 1(6):170–175
Wolseley PA, Stofer S, Mitchell R, Truscott AM, Vanbergen A, Chimonides J, Scheidegger C (2006) Variation of the lichen communities with landuse in Aberdeenshire, UK. Lichenologist 38(4):307–322

Community-Level Lichen Diversity Assessment in Alpine Zone of Indian Himalaya: Climate Change Implications

13

Rajesh Bajpai, C. P. Singh, and Dalip Kumar Upreti

Abstract

Lichens have prolonged history as excellent biomonitors of air pollutants and their responses are documented in climate change assessment. Lichen species composition within a community and change in composition are powerful tools to retrieve information about changes in climate, air quality and biological processes in the area. In this study, the composition of lichen communities in sub-alpine (3000–3500 m), moist alpine scrub (3500–4000 m) and dry alpine scrub (4000–4500 m) of Indian alpine Himalayan regions has been documented. A total of 18 bioindicator lichen communities are recorded which comprised 732 species under 148 genera and 47 families. Among different indicator communities, Parmelioid dominates the areas followed by Lecanorioid, Dimorphic and Physcioid communities represented by 185, 87, 71 and 64 species, respectively. It is assumed that Cyanophyceans are the most sensitive communities distributed at the middle range of altitude between 3000 and 4000 m in the studied areas. Furthermore, substratum and growth form also

Supplementary Information The online version contains supplementary material available at https://doi.org/10.1007/978-981-19-4476-5_13.

R. Bajpai (✉)
Environment, Agricultural & Education Society, Bareilly, Uttar Pradesh, India

Lichenology Laboratory, CSIR-National Botanical Research Institute, Lucknow, Uttar Pradesh, India

C. P. Singh
AED/BPSG/EPSA, Space Applications Centre, ISRO, Ahmedabad, Gujarat, India

D. K. Upreti
Lichenology Laboratory, CSIR-National Botanical Research Institute, Lucknow, Uttar Pradesh, India

S. P. Singh et al. (eds.), *Ecology of Himalayan Treeline Ecotone*,
https://doi.org/10.1007/978-981-19-4476-5_13

played a significant role in the assessment of varied environmental conditions. This study provides a baseline data about potential taxa having higher sensitivity to change in climate and which may exhibit migration, adaptation, and acclimation, in the near future.

Keywords

Altitudinal range · Biomonitoring · Habitats · Indicator community · Lichen

13.1 Introduction

The alpine zone in the Himalaya has attracted the attention of a large number of plant explorers, phyto-geographers, ecologist and naturalists since a very long time (Dickore and Nusser 2000). However, few studies of lichens are available in alpine areas of the Himalaya in the context of their relation to ecology, climate change and pollution. The lichens may allow a rapid assessment of forest health and status in regions where other forms of environmental monitoring are expensive or impractical. The identification of 'indicator communities' amongst lichens can provide a basis for management recommendations and can also be used to assess climatic change and potential forest recovery in areas where deforestation has caused a change in local climate and phanerogamic as well as cryptogrammic communities (Wolseley and Aguirre-Hudson 2007). However, lichens are valuable bio-indicators for evaluating the consequences of human activities that are progressively changing the Earth's ecosystems. Lichens have been referred as extremophiles because they have the ability to endure conditions ranging from Arctic/Antarctic to hot desert and high alpine regions (Hill and Hawksworth 1984). Lichens are often abundant in areas where vascular plants do not have ability to grow such as Antarctica and mountain peaks (Brodo et al. 2001).

The lichens survive in diverse climatic conditions, due to their peculiar poikilohydric nature, where they cannot regulate their own water content and depend on the immediate atmosphere for moisture mainly in the form of fog, dew and vapour. The poikilohydric nature enables lichens to tolerate extreme desiccation and hydration (Smith and Molesworth 1973). Owing to the mutualistic symbiotic nature, the lichens have an ability to sustain extreme environment and colonise on all available perennial substrata as well as man-made artefacts. On the bare substratum, the fungus provides a home for the alga, the alga supplies photosynthetic energy to the fungus. The unique morphology enables lichens to colonize a vast variety of substrates, such as bark of trees (Corticolous), rock (Saxicolous), leaves (Foliicolous), soil (Terricolous) and mosses (Muscicolous). The substratum plays an important role in colonization by selective taxa. The lichen growth forms depend upon arrangement of cortical algal and medullary tissue and mode of attachment to the substratum (Hale 1983). Lichens have the ability to produce vast array of different secondary metabolites as defence from high UV light and temperature stress in alpine areas (Elix and Stocker-Wörgötter 2008).

The increasing UV light, anthropogenic activities, deforestation and species' range extension have increased the necessity to understand ecosystem in which lichen communities are growing (Ellis and Coppins 2006; Hylander and Jonsson 2007). Lichens are involved in ecological processes from the canopy to the ground of an ecosystem, where they store, intercept and redistribute water and nutrients. In addition, lichens are a food source and habitat for various fauna and flora. Lichens play an important role in successional stages of plant communities, where certain lichen species are primary colonizers that create suitable conditions for later successional stages. Furthermore, in forest ecosystems, lichens represent a major part of the species diversity, and their diversity can exceed that of vascular plants (McCune et al. 1997). Consequently, there is high demand for lichenologists to contribute to the knowledge of forest ecosystems (Hylander and Jonsson 2007).

The composition of lichen communities based on different species of lichens can be used as lichen bioindicator communities for assessment of environmental disturbances in an area. Lichens are ideal systems for exploring the patterns and processes of community structure across space and time. For more than a decade, lichens have been recognised as an integral component of ecosystems and knowledge about the ecology of epiphytic lichens has enabled scientists to answer questions about environmental changes, sustainable forestry practices and forest health (McCune et al. 2000; Sillett et al. 2000). Furthermore, epiphytic lichens have been used as old-growth forest indicator species, such as, their presence or absence may suggest the present ecological conditions of the area, and thus they are used as an indicator of climate change (Boudreault et al. 2002).

The present studies discuss variation in lichen diversity in different altitudinal gradients (3000–4500 m) and their potential use as bioindicator of environmental alterations in alpine range. Further, the study examines the community patterns in Western Himalaya (WH) and Eastern Himalaya (EH) for recommending management practices to maintain and enhance cryptogamic diversity in different vegetation types in the area. The available information will provide baseline information on lichen communities in alpine areas and would also enhance our understanding of the mountain ecosystem which is vital for future management and restoration approaches in terms of climate change mitigation.

13.2 Material and Methods

13.2.1 Study Area and Sample Collection

The study is based on the specimens of lichens deposited in the herbarium of CSIR-National Botanical Research Institute, Lucknow (LWG) and collection from high alpine areas (altitudes between 3000 and 4500 m) in Indian Himalayan region (IHR) during the period 2013–2019 under the aegis of (HIMalayan Alpine Dynamics Research Initiative (HIMADRI). The data were collected from Jammu & Kashmir

(N 34° 5′ 1.1616″; E 74° 47′ 50.5356″), Himachal Pradesh (N 31° 6′ 16.5780″; E 77° 10′ 24.3264″), Uttarakhand (N 30° 19′ 18.8940″; E 78° 1′ 35.8284″), Sikkim (N 27° 20′ 20.1696″; E 88° 36′ 23.4216″), Arunachal Pradesh (N 27° 50′ 42.5328″; E 95° 14′ 50.4420″). The secondary lichen diversity data were generated from the past published literatures available in different monographs, revisionary, floristic studies on lichens of IHRs (Upreti and Nayaka 2000; Srivastava 2005; Prasher and Chander 2005; Yadav 2005; Thakur and Chander 2018; Bajpai et al. 2018). The collected specimens were identified using routine microscopic and laboratory techniques based on the following keys and taxonomic treatments (Awasthi 1991, 2007; Orange et al. 2001; Elix 2014). The lichen communities were grouped according to James et al. (1977) and Alatalo et al. (2017). The graphs were prepared using MS Excel.

13.3 Results

The variation in the diversity and composition of lichen communities along the altitudinal gradient (3000–4500 m) resulted in occurrence of 732 species of lichens belonging to 148 genera and 47 families of alpine areas of Indian Himalayan regions (Table 13.1). The area is dominated by the members of Parmeliaceae followed by Physciaceae and Cladoniaceae including 254, 64 and 48 species respectively. The dominant genera include *Parmotrema*, *Usnea*, *Bryoria*, *Heterodermia*, *Cladonia* and *Lecanora*. Out of the 47 families known from the area, 14 families are represented by single species within single genus (Table 13.1).

In the area, 133 species are crustose, 226 species are foliose with broad lobes, 72 species are with foliose narrow lobes, 69 species are fruticose filamentous and remaining are intermediates reported from the area. In this study a total of 40 broad lobe lichens with gelatinous nature are reported from the area. The gelatinous nature thalli are well known for their soil binding nature and capacity for production of soil humus as well as good indicator of moisture in the soil (Wade 1959). The species exhibit wide substratum preference, as 48 species prefer to grow on soil, 148 species on rocks, 233 species on bark and twigs, 4 on mosses and 7 on different plant leaves and remaining species grown on mixed type of substratum.

13.4 Discussion

A total of 18 lichen indicator communities were encountered in the area, among them Parmelioid community was the most dominant followed by lecanorioid, dimorphic and physcioid represented with 185, 87, 71 and 64 species respectively in the alpine areas. Lichens also have important biological roles in forests, including nitrogen fixation, and we found 56 species of cyanophycean community followed by Usnioid community with 61 species indicating good air quality. However, Arthonoid, Leprarioid, Pertusorioid and Lichenoid communities are in infirmity situations. Such findings on lichen community assemblage contribute to the investigation of several key forest ecosystem issues like contamination of natural resources, biodiversity status, productivity and sustainability.

Table 13.1 Lichen families represented by number of genera and species

S. No.	Families	Genera	Species
1	Acarosporaceae	3	19
2	Arthoniaceae	1	1
3	Baeomycetaceae	1	2
4	Caliciaceae	6	7
5	Candelariaceae	2	6
6	Catillariaceae	1	1
7	Cladoniaceae	2	48
8	Coccocarpiaceae	1	4
9	Coenogoniaceae	1	2
10	Collemataceae	2	37
11	Coniocybaceae	1	1
12	Gomphillaceae	1	1
13	Graphidaceae	3	12
14	Haematommataceae	1	1
15	Icmadophilaceae	4	6
16	Lecanoraceae	6	46
17	Lecideaceae	3	10
18	Leprocaulaceae	1	1
19	Lichinaceae	2	4
20	Lobariaceae	4	23
21	Megalariaceae	1	1
22	Megalosporaceae	1	1
23	Megasporaceae	2	11
24	Monoblastiaceae	1	1
25	Mycoblastaceae	1	1
26	Nephromataceae	1	5
27	Ochrolechiaceae	1	3
28	Pannariaceae	2	5
29	Parmeliaceae	44	254
30	Peltigeraceae	2	21
31	Peltulaceae	1	2
32	Pertusariaceae	1	6
33	Physciaceae	8	64
34	Pilocarpaceae	2	4
35	Porinaceae	1	1
36	Psoraceae	1	1
37	Pyrenulaceae	1	1
38	Ramalinaceae	4	19
39	Rhizocarpaceae	1	12
40	Roccellaceae	1	1
41	Sphaerophoraceae	1	2
42	Stereocaulaceae	2	18
43	Teloschistaceae	10	27

(continued)

Table 13.1 (continued)

S. No.	Families	Genera	Species
44	Tephromelataceae	1	2
45	Trapeliaceae	3	4
46	Umbilicariaceae	2	18
47	Verrucariaceae	6	15

13.4.1 Lichen Diversity along the Altitudinal Gradients

In all, 732 species of lichens were found throughout the altitudinal gradients, with highest richness observed between 3000 and 4000 m. The lowest values were recorded at an altitude of 4500 m. It appears that, the high altitudes bear the scanty substratum and have harsh climatic conditions which reduce the diversity of lichens but support the growth of some exclusive lichen taxa. The most recent account of lichens available from India revealed that more than 80% of species are from the Himalayan area (Singh and Sinha 2010; Singh et al. 2018). Among the study states, Uttarakhand, Sikkim and Arunachal Pradesh showed higher dominance of species than that in Himachal Pradesh and Jammu & Kashmir (Fig. 13.1).

However, Throughout the world, the lichen families Parmeliaceae, Physciaceae and Cladoniaceae showed highest species dominance (Lücking et al. 2016). The family Parmeliaceae, a widely distributed lichen family with cosmopolitan distribution, has higher abundance and wider distribution in high-altitude ecosystems (Giordani and Incerti 2008; MacDonald et al. 2013). Graphidaceae, Lecanoraceae, Arthoniaceae, Pyrenulaceae, and Trypetheliaceae are the dominant elements in low elevation forests with dry and humid areas, whereas the families namely Lobariaceae, Collemataceae, Peltigeraceae, Ramalinaceae are more frequent at middle altitudes with shaded conditions (Hekking and Sipman 1988). Foliose or fruticose lichens are not favoured in the areas with higher pollution load and higher humidity found mainly at lower elevations as compared to pristine higher elevation areas, because lichen thallus gets supersaturated with water, inhibiting photosynthesis. The crustose lichens exhibit strategies to avoid super saturation and can survive in high humidity environments (Lakatos et al. 2006). Low diversity is generally recorded when the natural vegetation is removed and replaced by exotic species, limiting the source area of lichen propagules or suitable substrata for colonization. Stumpy diversity in lower altitudes may have resulted because the recovery of specialist lichen species (mostly foliose and fruticose) is slow due to poor dispersal (Wolseley et al. 2007) and absence of specialised habitats in forest stands.

In this study, we found that, the crustose lichen families dominate mostly at lower altitudes. To withstand, adapt to the high UV radiation and bright sunlight the lichens develop bright pigments in their upper cortical regions while some secondary metabolites are produced in medulla to protect the photobiont against light stress conditions. Waring (2008) reported increase in depside concentrations in response to increasing light exposure, suggesting that compounds may have a photo protective role in higher altitudes. Shukla et al. (2017) studied the diversity of lichens with

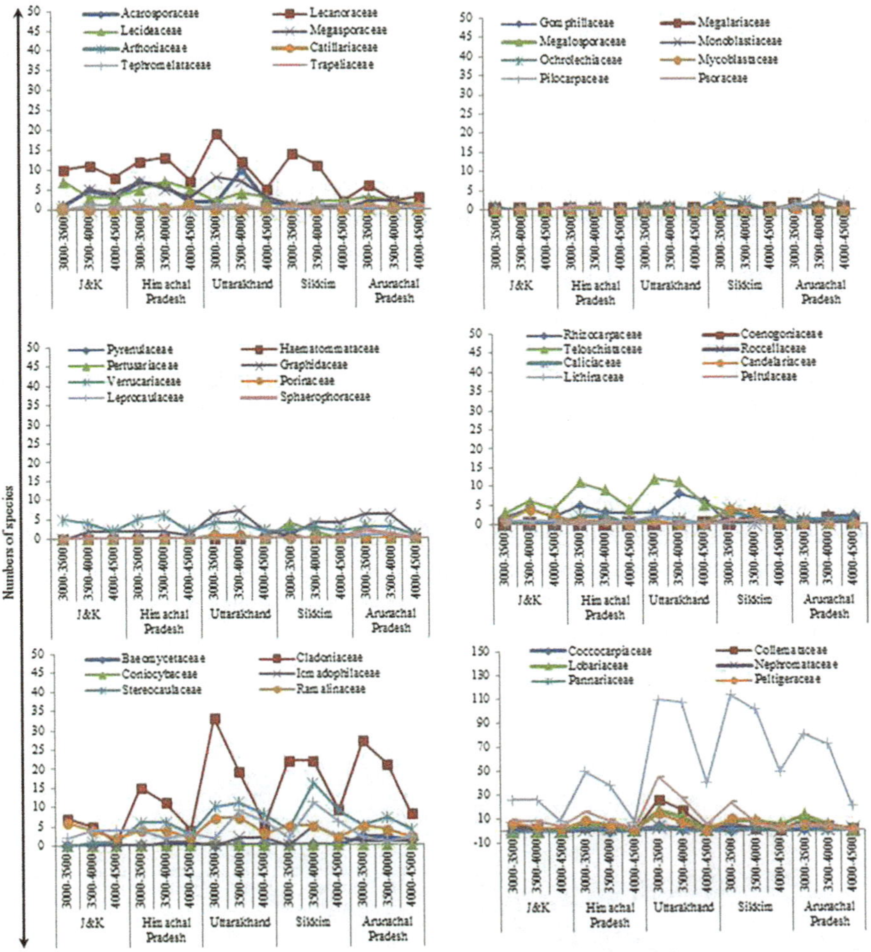

Fig. 13.1 Dominance of lichen families along altitudinal gradients of respective sites

qualitative assessment of secondary metabolites and suggested that secondary metabolites confer resistance to lichens against biotic and abiotic stresses.

The lichen families containing bigger thallus exhibit dominance in middle altitudes of 3000–3500 m. As compared to higher altitudes, the lower altitudes are drier and more disturbed and thus have poor lichen diversity, while higher altitudes being pristine correspond to rich lichen diversity. At higher altitudes, climate seems to favour growth of lichens but the incident solar radiation, low precipitation, low temperature acts as limiting factors and hence lead to decline in the diversity as the altitude increases. The similar findings have been reported by Grytnes et al. (2006) as the species diversity exhibits gradually increasing trend with altitude up to 2100 m indicating the impact of microclimatic conditions and forest cover on species composition.

13.4.2 Species Composition

It was found that the foliose and crustose growth forms dominated the area followed by fruticose and squamulose lichens. Further, the growth forms can be subdivided into different intermediate growth forms such as foliose bigger lobes (FB) dominated by 31%, foliose narrow lobate (FN) represented by 10% whereas crustose growth forms governs 18% of the total diversity (Fig. 13.2).

In lower altitudes, crustose lichens shared their dominance, which gradually decreased at higher elevation. The dominance at lower altitudes may be due to the survival potential of crusotse forms against various anthropogenic disturbances and by developing potential tolerance. As mentioned previously, crustose lichens have adaptations to the high humidity conditions at low elevations and can photosynthesize under low light intensity conditions (Lakatos et al. 2006). On the other hand, foliose forms dominate in higher altitude, because they no longer have problems of water supersaturation and can outcompete crustose lichens by virtue of their larger thalli (Sipman 1996; Lakatos et al. 2006). In addition, the increase of mist favours the growth of foliose lichens because of their three-dimensional morphology

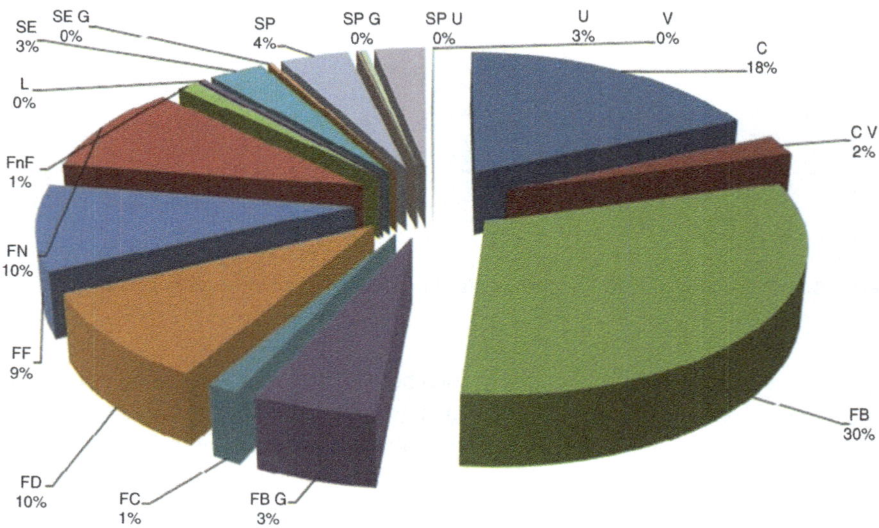

Fig. 13.2 Growth form of lichens represents the number of species found growing in Indian Alpine Himalayan areas. *C*: Crustose; *CV*: Crustose Verricouse; *FB*: Foliose with broad lobes; *FB G*: Foliose with broad lobes Gelatinous; *FC*:Foliose with canaliculate; *FD*: Fruticose dimorphic; *FF*: Fruticose filamentous; *FN*: Foliose with narrow lobes; *FnF*: Fruticose non-filamentous; *L*: Leprose; *SE*: Squamulose effigurate; *SE G*: Squamulose effigurate Gelatinous; *SP*: Squamulose peltate; *SP G*: Squamulose peltate Gelatinous; *U*: Umbilicate; *V*: Verricouse

(Stanton and Horn 2013; Stanton 2015). Fruticose growth forms, being pendulous from twigs in well-lit higher elevations have the advantage of being able to use light from all directions better than foliose lichens which can maximize the harvest of more or less unidirectional light (Gauslaa et al. 2009). Moreover, the dissected fruticose lichens have a high surface area to volume ratio, making them more closely coupled to ambient atmosphere than flat foliose lichens, and absorb moisture more readily from air, that results in their greater frequency in open sites (Baniya et al. 2009).

Lichens growing at higher elevations have abundance of hair like structures (rhizines) on their lower side. Rhizines act as reservoir and trap the atmospheric moisture for longer period. As altitude increases, the precipitation decreases thus the presence of rhizines helps in obtaining the scarce water that lichens need for photosynthesis (Stanton and Horn 2013). Lichens growing in open sites, where there is a higher intensity of light and low humidity, rhizines play vital role in absorbing moisture for lichens to survive in microhabitats. The higher frequency of cyanolichens was reported between the altitudes 2500–3000 m. The occurrence of cyanolichens related to the fact that these lichens require high humidity and low luminosity conditions in montane forests, especially in the forest interior (Soto-Medina and Londoño 2015). On the other hand, at low elevations, higher temperature reduces the presence of cyanolichens, whereas at high elevations there is a high intensity of light and low forest cover, which is a limiting factor for cyanolichens in the area.

13.4.3 Substrate Specificity

The colonization of lichen thalli is certainly one of the more apparent effects of difference in substrate texture. Lichen spores can get trapped and begin to develop on rough surface more easily than smooth surface. The types of substrata play an important role in distribution of different lichen species in the area. In this study, lichens growing on barks dominate with 233 species followed by 184 species growing over rocks and 48 species on soils. However, 190 species share two different substratum (Bark and Rock), 58 species share three types of substrates (Bark, Rock and Soil) while 8 species commonly grow on all available substrates. The study area revealed dominance of corticolous (32%) followed by saxicolous, terricolous and muscicolous with 25%, 7% and >1%, respectively (Fig. 13.3). The lower altitudes around <3000 m are dominated by a variety of phorophytes, which provide excellent substratum to a number of lichen taxa to colonize. However, in higher altitudes (>3000 m) mostly devoid of phorophytes and presence of exposed rocks represented the lesser number of epiphytic lichens in the area. In this study, most of the corticolous lichens reported are from lower and middle altitudes as compared to higher elevation.

All the IHR states exhibit the dominance of corticolous lichens in lower and middle altitudes region followed by foliiocolous and muscicolous. However, saxicolous lichens dominated at higher altitudes because of the presence of exposed

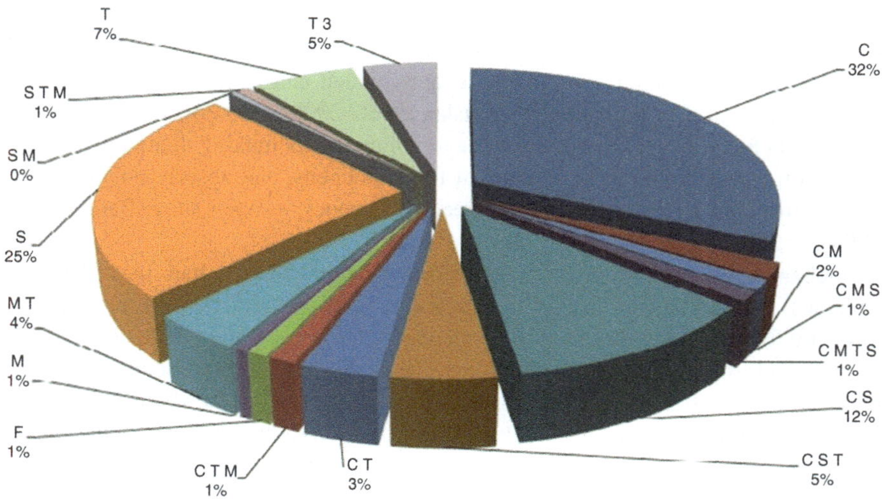

Fig. 13.3 Substratum of lichens represents the numbers of species found growing in Indian Alpine Himalayan regions. *C*: Corticolous; *CM*: Corticolous, muscicolous; *CMS*: Corticolous, muscicolous, saxicolous; *CMTS*: Corticolous, muscicolous, terricolous, saxicolous; *CS*: Corticolous, saxicolous; *CST*: Corticolous, saxicolous, terricolous; *CT*: Corticolous, terricolous; *CTM*: Corticolous, terricolous, muscicolous; *F*: Foliicolous; *M*: muscicolous; *MT:* Muscicolous, terricolous; *S*: Saxicolous; *SM:* Saxicolous, muscicolous; *STM:* Saxicolous, terricolous, muscicolous; *T*: Terricolous; *TS:* Terricolous, saxicolous

rocks, stones, boulders and lack of big trees. Members of lichen families Teloschistaceae, Cladoinaceae, Stereocauloiaceae, and some species of Lecanoraceae colonize most of the exposed rocks and boulders and soil over rocks or soil in open places at high altitudes. It is reported that at higher altitudes, climate seems to favour growth of lichens but the incident solar radiation acts as a stress factor leading to decline in the diversity as the altitude increases. Similar to the finding of Grytnes et al. (2006), the species diversity exhibited gradual increasing trend with increasing altitude, indicating the impact of microclimatic conditions and forest cover on species composition, which declined with further increase in altitudes.

13.4.4 Community Characterization

The increasing altitude, microclimate and/or air quality changes determine lichen community structure in the area. Both higher and lower altitudes of the study area exhibit distinct lichen community composition. Parmeliod community with 25% exhibits dominance in the area followed by Lecanorioid, Dimorphic, Physcioid and Usnioid represented by 12%, 10%, 9% and 8% respectively (Fig. 13.4).

The communities at lower altitudes comprised of species of the genera *Candelaria*, *Chrysothrix*, *Physcia* and *Phaeophyscia*, which indicates nutrient

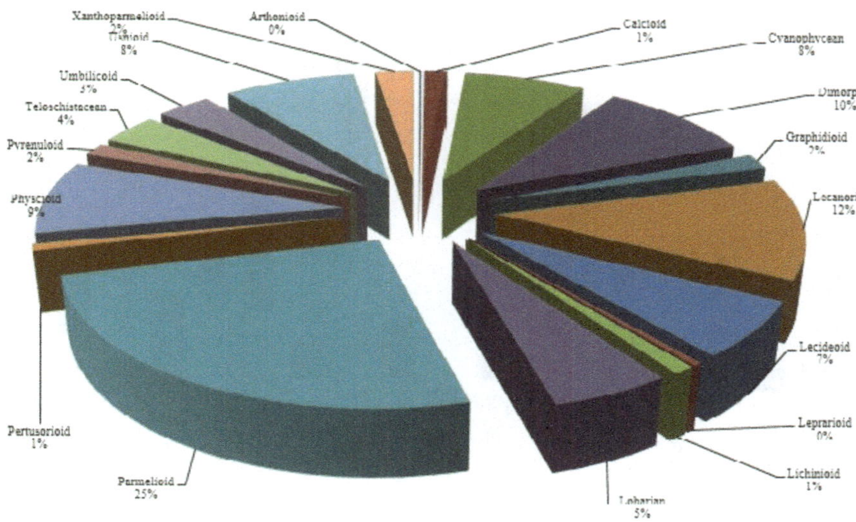

Fig. 13.4 Cumulative lichen communities representation in Indian Alpine Himalayan regions

enriched habitat whereas species of *Graphis* indicate an exposed and regenerated forest. As for as characterization of lichen indicator communities in both the Himalaya regions is concerned we have fascinating results. Most of the lichen communities were found dominating in Western Himalaya (WH) as compared to alpines of Eastern Himalaya (EH). The Parmelioid, Dimorphic, Graphidioid, Usnioid and Pyrenulioid were predominant in EH whereas, Lecanorioid, Cyanophycan, Lobarian, Lecidioid, Teloschistacean, Umbilicoid, Physcioid and Xanthoparmelioid were dominating in WH (Fig. 13.5). The lower altitudes usually have few lichens except in moist places. The lichen vegetation changes as the altitude increases, and the smooth bark mostly supports the crustose lichens belonging to species of lichen genera *Pyrenula, Calcioid, Graphis, Lecanora, Pertusaria, Verrucaria* and *Arthonia* but do not support the growth of foliose lichens as spores of bigger lichens get washed off from smooth surfaces in the EH (Sinha and Jagadeesh Ram 2014).

However, the alpine zone (3200–4500 m) between the treeline and permanent snow line is dominated by grassy meadows, locally called 'Bugyals' in the Himalaya. In terms of herbaceous species diversity, this is the richest zone, harbouring a number of plants. This vegetation is composed of tall forbs, mixed herbaceous formations, *Danthonia* grasslands and *Kobresia* sedge, with several intermixed cushionoid species, and is often lined by scrubs. Majority of the lichen species of genera *Cladonia, Fuscopannaria, Canoparmelia, Heterodermia, Hypogymnia, Lethariella, Melanelia, Melanelixia, Melanohalea, Myelochroa, Nephromopsis, Parmotrema, Remototrachyna, Xanthoparmelia* and some members of family Teloschistaceae prefer to grow at base of Bugyal vegetation along with mosses and small *Rhododendron* trees. The species of the above genera, not

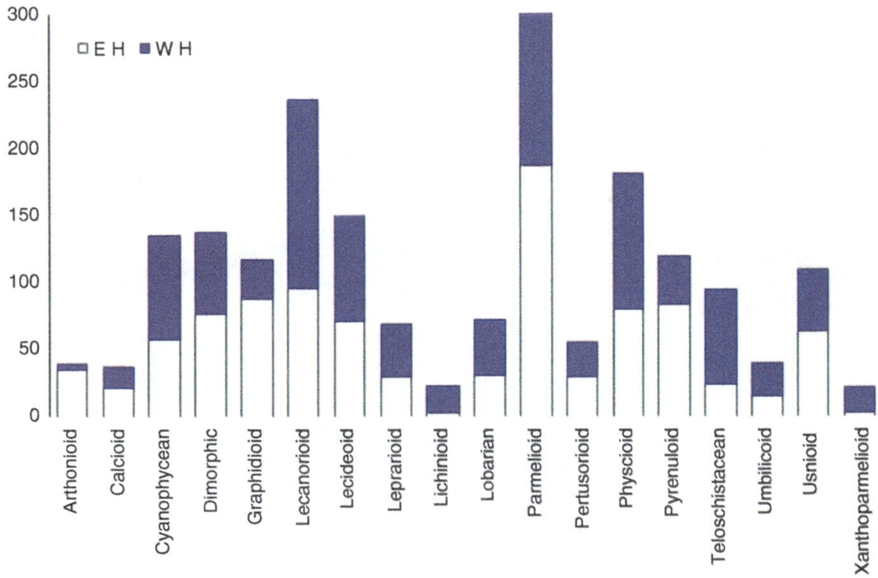

Fig. 13.5 Comparative assessment of lichen communities in Eastern Himalaya (EH) and Western Himalaya (WH)

supported by the exposed rock beyond treeline hence these grow luxuriantly in Bugyals. Rawat (2005) indicated that the Bugyal vegetation is high in the Western Himalaya. The Western Himalaya largely supports drought-resistant and cold-loving plants belonging mostly to Coniferae, Fabaceae, Asteraceae, Poaceae, Rosaceae, while in high altitudes, the landscape is dominated by species of *Abies, Cedrus, Picea, Pinus, Quercus, Rhododendron* and the treeline formed by *Betula utilis*. The landscape tree species, mostly support the growth of foliose and fruticose lichens. Those tree species have bark shedding nature do not favour the lichen growth, such as *Betula utilis*, which favours only the growth of some limited species of *Parmotrema, Flavoparmelia* at base and *Usnea* at twigs.

It is well known that the changes in temperature and precipitation have influenced phenology as well as diversity of plants in different parts of the Himalaya, and climate change in Himalaya may alter phenology, diversity at both individual species and community levels (Xu et al. 2009). According to Shrestha et al. (2012), the Himalaya has warmed by 1.5 °C from 1982 to 2006, at an average rate of 0.06 °C/year but the rate of warming varies across seasons and ecoregions. The greatest increase of 1.75 °C is observed in winter with an average increase of 0.07 °C/year whereas the least increase of 0.75 °C, is in summer, and an average increase is 0.03 °C/year. Average annual precipitation has increased by 163 mm (6.52 mm/year) during the 25 years period. However, the warming trend reported previously in the Himalaya was 0.04 °C/year in the Western Himalaya (Kumar et al. 2009) and 0.03 °C/year in the Eastern Himalaya (Sharma and Tsering 2009). Furthermore, both increasing and decreasing trends in Eastern Himalaya (Sharma et al. 2000), and decreasing trends in Western Himalaya (Kumar et al. 2009) have been reported for precipitation. The variability in abiotic components may lead to qualitative and quantitative variations in diversity and luxuriance of lichen species in the area.

The Eastern Himalaya, on the contrary, has cool, humid and semi-oceanic climate, with the vegetation characterized by the abundance of *Rhododendrons* and the remnants of *Podocarpus*, followed by tree ferns, *terminalias*, oaks, laurels and bamboos to support the growth of epiphytes orchids, ferns, bryophytes and lichens (Chowdhery 2008). Most of the tree species have smooth bark, favour only crustose growth form of lichens in the area (Sinha and Jagadeesh Ram 2014). Crustose lichen distribution patterns have their affinity for smoother bark. The bark condition of younger trees is often smoother which may allow these lichens to successfully attach to the substrate and grow compared with older trees bark which are rougher with numerous cracks, largely preventing these species from expanding. Whereas the foliose lichen species adhere closely to loosely to the substratum which may be adapted to survive in diverse water moisture stresses in different microhabitats and mostly prefer to grow on rough surface. Furthermore, the fruticose lichens grow up to trunk and twig in open canopy and adapt to survive in low precipitation (McCune et al. 1997).

In this study, following communities occur indicating the surrounding environment.

1. **Arthonioid community:** The species of lichen genera *Arthonia* and *Cryoptothecia* represent the Arthonioid community. Arthonioid community is present only at lower altitudes and prefers to grow on bark and leaves. The Arthonioid community is predominant in WH and shows increasing trend up to 3000 m altitudes (Fig. 13.6a). It clearly indicates that the area is moist with high humidity, especially at lower altitudes.
2. **Calcioid community:** The Calcioid community is composed of the species of lichen genera *Acroscyphus, Baculiferea, Calicium, Dibaeis, Diplotomma, Chaenotheca, Aderkomyces, Icmadophila* and *Bunodophoron*. The community prefers to grow on dead wood, bark, twigs at an altitudes up to 3000 m in both the Himalayan regions but the dominance is observed in EH (Fig. 13.6a). It indicates that the ecological contiguity of the aged forests tend to provide a wide variety of suitable substrates and microenvironments that provide favourable illumination conditions with high atmospheric humidity.
3. **Cyanophycean community:** The species of lichens genera *Coccocarpia, Collema, Leptogium, Pseudocyphellaria, Nephroma, Peltigera, Fuscopannaria, Pannaria* and *Solorina* represent the Cyanophycean community. The Cyanophycean lichens constitute a separate group of lichens having Cyanophycean algae either alone (bipartite) or along with a green primary photobiont (tripartite) and prefers to grow between lower altitudes and shows decreasing trend with increasing altitudes. The maximum dominance is observed in WH in comparisons to EH (Fig. 13.6b) as they are an exclusive group of lichens having capability of nitrogen fixation, enriching the substrate and playing an important role in nutrient cycling of the ecosystem. It indicates

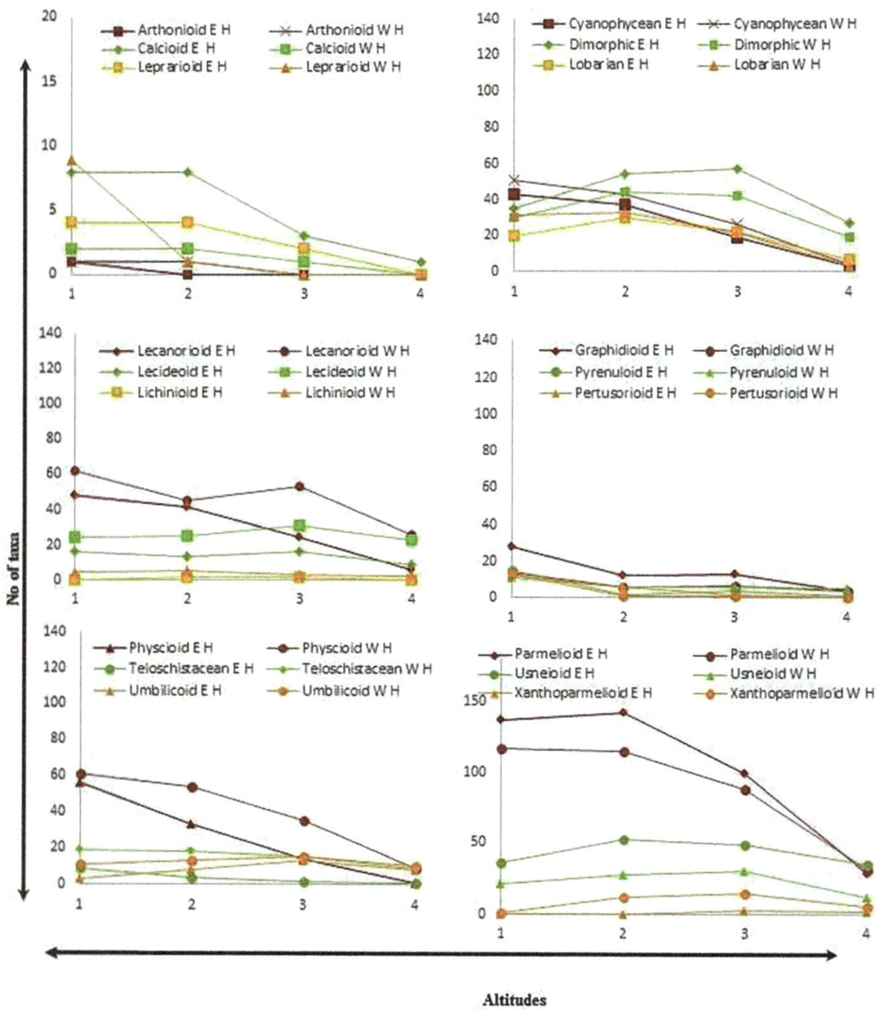

Fig. 13.6 Individual community assessment in EH and WH at different altitudinal gradients 1. 2500–3000 m, 2. 3000–3500 m, 3. 3500–4000 m, 4. 4000–4500 m

the forest age and continuity of forest ecosystem function. However, presence of the community indicates the humus rich soil, shift of the species habitat and loss of moisture in the respective substratum.

4. **Dimorphic community:** The Dimorphic community is composed of the species of lichen genera *Baeomyces, Cladia, Cladonia, Dibaeis, Siphula, Thamnolia* and *Stereocaulon*. The species found growing on soil, rock and sporadically along with mosses at an altitude between 3000 and 4000 m in both EH and WH

(Fig. 13.6b). However, the maximum dominance is observed in EH. The presence of dimorphic community in the area indicates undisturbed soil ecosystem.

5. **Graphidioid community:** The Graphidioid community is made up of the species of lichen genera *Diploschistes, Graphis, Phaeographis, Thecaria* and *Opegrapha.* The community growing at lower altitudes indicates open, thinned-out forest, presence of smooth bark trees and the area devoid of moisture. This community is dominat in EH and shows dominance with decreasing altitudes in both the Himalayan regions (Fig. 13.6d).
6. **Lecanorioid community:** The community is composed of the species of the lichen genera *Acarospora, Pleopsidium, Dimelaena, Candelaria, Candelariella, Coenogonium, Lecanora, Lecidea, Protoparmeliopsis, Rhizoplaca, Aspicilia, Lobothallia, Mycoblastus, Ochrolechia, Phyllopsora, Tephromela, Placopsis, Trapelia* and *Trapeliopsis.* This community prefers to grow on bark and rock at lower altitudes but *Protoparmeliopsis, Rhizoplaca* and *Aspicilia* are found growing over rocks at higher altitudes as well. The community indicates windy and well-illuminated environmental conditions. The Lecanorioid community is more common in WH and showing increasing trend from 4000 m altitudes (Fig. 13.6c).
7. **Lecideoid community:** This community is composed of species of the lichen genera *Buellia, Catillaria, Haematomma, Lecidea, Lecidella, Lecidoma, Porpidia, Megalaria, Megalospora, Lasioloma, Lopadium, Psora, Mycobilimbia, Toninia* and *Rhizocarpon.* The community members show their dominance in WH and abundance increases with increasing altitude (Fig. 13.6c). Some of the species of this community have bright coloured thallus indicating exposed and well illuminated area. The community members are more resistant against coldness and they are widely used in the lichenometric studies in alpine area.
8. **Leprarioid community:** The community represents the species of *Chrysothrix, Leprocaulon, Lepraria*and found growing at lower altitudes, and the dominance of this community decreases with increasing altitude. The maximum diversity is observed in EH (Fig. 13.6a). The community appears first after forest fires.
9. **Lichinioid community:** The community is found growing mostly on rock, lime plaster at lower altitudes. Several species of lichen genera *Lichinella, Peccania, Peltula* and *Endocarpon* represent the community. The Lichinioid community indicates dry desert habitat (devoid of moisture) and presence of medium to higher temperature in the area. The members of Lichinioid community show their dominance in WH and their dominance increases with increasing altitude (Fig. 13.6c).
10. **Lobarian community:** The community is composed of species of lichen genera *Dendriscosticta, Lobaria, Sticta, Peltigera* which are found growing at an altitude of 3000–4000 m in both EH and WH, but dominance is maximum in WH (Fig. 13.6b). The community indicates moist shady forest, productive forest and long forest continuity. The community is a good indicator of anthropogenic disturbance in the forest. The members of this community clearly indicate the presence of moisture in the area. If species of this community were found

growing on soil in past and currently growing on tree trunks or rocks, this escape indicates drying of soil (loss of moisture in the substratum). The periodical data about change of substrate by community are a good indicator of climate change.

11. **Parmelioid community:** The parmelioid community is composed of species of lichen genera *Allocetraria, Arctoparmelia, Bulbothrix, Canomaculina, Canoparmelia, Cetraria, Cetrariopsis, Cetrelia, Cetreliopsis, Evernia, Flavocetraria, Flavocetrariella, Flavoparmelia, Flavopunctelia, Hypogymnia, Hypotrachyna, Lethariella, Melanelia, Melanelixia, Melanohalea, Menegazzia, Montanelia, Myelochroa, Nephromopsis, Nipponoparmelia, Oropogon, Parmelia, Parmelina, Parmelinella, Parmelinopsis, Parmotrema, Platismatia, Punctelia, Relicinopsis, Remototrachyna, Rimelia, Tuckermannopsis, Tuckneraria* and *Vulpicida.* The community is found growing on diverse substrata and shows increasing trend with decreasing altitude. The maximum dominance is observed in EH as compared to WH (Fig. 13.6f). The community indicates thinned-out forest with more light, moist condition and lower to medium precipitation in the area.
12. **Pertusorioid community:** Mostly the species of *Pertusaria* represent the community and prefers to grow on rough bark and rarely on rock at lower latitudes. This community indicates well-illuminated environmental conditions, old tree forest, exposed to good winds. The members of Pertusorioid community were found dominating in EH and showed increasing trend with decreasing altitudes in both the Himalayan regions (Fig. 13.6d).
13. **Physcioid community:** The physcioid community is composed of the species of lichen genera *Rinodina, Anaptychia, Heterodermia, Hyperphyscia, Phaeophyscia, Physcia, Physconia* and *Pyxine*. The community members prefer to grow on bark, twigs and some time on rocks at lower to middle altitudes. This community is well known for pollution tolerance and their presence indicates a nitrophilous environment. The members of Physcioid community are dominant in WH but as far as the altitudinal distribution is concerned, it decreases with increasing altitude (Fig. 13.6e).
14. **Pyrenuloid community:** The community is composed of species of the lichen genera *Anisomeridium, Porina, Pyrenula, Lithothelium, Staurothele* and *Verrucaria*. This community prefers to grow on smooth bark at lower latitude and some species are found growing on higher altitudes on stable rock also. At lower altitudes, the community indicates the young and regenerated forest. However, the presence of *Staurothele, Verrucaria* on higher altitudes indicates the stable substratum and open area. Mostly this community is found in EH as compared to WH. However the community shows its dominance with decreasing altitudes (Fig. 13.6d). The community members prefer to grow on smooth bark substratum in comparison to rough bark.
15. **Teloschistacean community:** The community composed of the species of lichen genera *Calogaya, Caloplaca, Flavoplaca, Ioplaca, Oxneria, Rusavskia, Squamulea, Teloschistes, Variospora* and *Xanthoria* found growing on barks and rocks. Members of the Teloschistacean community is more common in WH as compared to EH (Fig. 13.6e). Dominance of this community increases with

increasing altitude. The colour of the thallus is mostly orange, brown and yellow due to presence of cortical pigment to protect algae inside against stress. The community indicates high UV irradiance in the area.

16. **Umbilicoid community:** The community grows in higher altitudes and shows increasing dominance with increasing altitude. This community is composed of species of lichen genera *Glypholecia, Lasallia, Umbilicaria, Catapyrenium, Dermatocarpon* and *Normandina*. This community is mostly found in WH as compared to EH (Fig. 13.6e). The presence of members of this community indicates the stable rock substratum along with high UV irradiance and low temperature.
17. **Usneioid, Alectoroid community:** The community is composed of the species of lichen genera *Alectoria, Bryoria, Sulcaria, Usnea* and *Ramalina*. This community is found growing on bark, twigs and rarely on rocks and shows increasing dominance with increasing altitude (Fig. 13.6f). The maximum dominance is reported in EH in comparison to WH. Most of the species of this community are pendulous thread like and indicate old forest and better air quality. This community is most sensitive to anthropogenic activities. However, increase in the pollution level causes disappearance of the species of this community.
18. **Xanthoparmelioid community:** Community is a group of species of a single lichen genus *Xanthoparmelia* found growing on higher altitudes between 3500 and 4500 m in both the Himalayan regions (Fig. 13.6f). The maximum dominance is observed in WH in comparison to EH. They grow luxuriantly over rocks along with mosses and rarely on barks. The presence of the community in the area indicate stable productive landscape and high UV radiation and the area being devoid of pollution.

The high altitude exhibits dominance of Parmelioid, Usnioid and Cyanophycean communities, which indicates fairly good air quality and minimum human disturbance. The change in microclimatic condition and pollution load across an altitudinal gradient determines the distribution and diversity of lichen communities with a specific pattern. The lower altitudes that are more easily accessible to the humans have poor lichen diversity, while higher altitudes being pristine exhibit rich lichen diversity. At higher altitude climate seems to favour growth of lichens but the higher UV radiation acts as a stress factor and may lead to decline in the diversity. The results correspond to the findings of Colwell and Lees (2000); Zapata et al. (2005) who also suggested that as we go up in altitudinal range there are strong barriers and consequently peak richness occurs in the middle (hump shaped curve). It is also reported that lichen species that are bright in colour belong to Xanthoparmeliod, Teloschistacean communities which dominate in the higher altitudes.

The fruticose and dimorphic growth form communities, in well-lit higher elevations, have the advantage to use light from all directions better than foliose lichens, thus maximizes the harvest of more or less unidirectional light (Gauslaa et al. 2009). The environmental filtering processes are occurring at low and high elevations, which can be explained by the extreme climatic conditions of the sites: high altitudes have low temperature, a strong temperature variation (high temperatures in day and low temperatures at night) and low average precipitation, which favours the presence of macro-lichen species (mainly fruticose) with few common features. On the other hand, lower altitudes with high temperatures and precipitation, which are a favourable environment only for crustose or filamentous lichens and some micro-foliose and squamulose that can survive in high humidity conditions. In this way, altitudinal extremes seem to be strong barriers for functional variation of lichens. The middle elevations have has balanced abiotic/biotic factors, which explains the greater diversity of lichens in this zone (Mayfield and Levine 2010; Prieto et al. 2017).

According to Ellis (2019), the species shows migration when anthropogenic activities increase in the area. The species can shift to a suitable climate space through lichen dispersal and establishment dynamics. On one hand, species may adapt to climate change and survive against anthropogenic disturbance by developing several functional traits. On other hand, lichens may acclimate to the changing climate space physiologically. The species produces some cortical pigments to acclimatize the species in the stressed environment. However, the habitat quality also plays an important role in relation to lichen diversity. The microclimate of the area also plays an an important factor to maintain habitat quality for survival of the species. These underlying factors *vis-à-vis* community study are important to understand the climate-driven fate of sensitive alpine ecosystems of Himalaya.

13.5 Conclusion

The study clearly brings out that Western Himalaya (WH) is richer as compared to Eastern Himalaya (EH) in terms of lichen indicator communities. The lichen community as well as the species show climate-related thresholds and is likely to be useful for targeting monitoring efforts. The status of present lichen bioindiactor communities (genus/species) at different altitudinal gradient can be used as the baseline data for carrying out future climatic change and environmental monitoring studies in different states of Indian Himalayan regions. More researches are required to be carried out to get a better understanding of community dynamics in response to ecological condition of the area. It is further emphasized that there is a need to develop a biomonitoirng programme of the country in terms of data repository for future studies.

Acknowledgements The authors are thankful to Prof. S. K. Barik, Director, CSIR-National Botanical Research Institute, Lucknow, for the laboratory facilities and Director, SAC-ISRO, Ahmedabad, also acknowledged for selecting CSIR-NBRI as partner of HIMADRI project of climate change. One of the authors RB like to thank to the Council of Scientific and Industrial Research, New Delhi for award of Scientist Pool fellowship (8909-A) and Director EAES, Bareilly India for kind support. The authors declare that they have no conflict of interests.

References

Alatalo JM, Jägerbrand AK, Chen S, Molau U (2017) Response of lichen communities to 18 years natural and experimental warning. Ann Bot 120(1):159–170

Awasthi DD (1991) A key to the microlichens of India, Nepal and Sri Lanka. Bibl Lichenol 40: 1–337

Awasthi DD (2007) A compendium of the macrolichens from India, Nepal and Sri Lanka. Bishen Singh and Mahendera Pal Singh, Dehradun

Bajpai R, Upreti DK, Nayaka S (2018) The lichen genera *Lepraria* (Stereocaulaceae) and *Leprocaulon* (Leprocaulaceae) in India. Phytotaxa 356(2):101–116

Baniya CB, Solhøy Y, Gauslaa M, Palmer W (2009) The elevation gradient of lichen species richness in Nepal. Lichenologist 42(1):83–96

Boudreault C, Bergeron Y, Gauthier S, Drapeau P (2002) Bryophyte and lichen communities in mature to old-growth stands in eastern boreal forests of Canada. Can J For Res 32:1080–1093

Brodo IM, Sharnoff SD, Sharnoff S (2001) Lichens of North America, vol 60. Yale University Press, New Haven

Chowdhery HJ (2008) Arunachal Pradesh- the cradle of flowering plants. In: Rawat GS (ed) ENVIS bulletin – special habitats and threatened plants of India. Wildlife Institute of India, Dehradun, pp 77–82

Colwell RK, Lees DC (2000) The mid-domain effect: geometric constraints on the geography of species richness. Trends Ecol Evolu 15:70–76

Dickore WB, Nusser M (2000) Flora of Nanga Parbat (NW Himalaya, Pakistan): an annotated inventory of vascular plants with remarks on vegetation dynamics. Englera (19). https://doi.org/10.2307/3776769

Elix JA (2014) A catalogue of standardized chromatographic data and biosynthetic relationship for lichen substances. 3rd edn. Published by the RSC, Australian National University, Canberra, ACT 0200, Australia

Elix JA, Stocker-Wörgötter E (2008) Biochemistry and secondary metabolites. In: NashTH III (ed) Lichen Biology, 2nd edn. Cambridge University Press, Cambridge, pp 104–133

Ellis CJ (2019) Climate change, bioclimatic models and the risk toLichen diversity. Diversity 11:54

Ellis CJ, Coppins BJ (2006) Contrasting functional traits maintain lichen epiphyte diversity in response to climate and autogenic succession. J Biogeogr 33:1643–1656

Gauslaa Y, Palmqvist K, Solhaug KA, Hilmo O, Holien H, Nybakken L, Ohlson M (2009) Size dependent growth in two old-growth associated macrolichen species. New Phytol 18:683–692

Giordani P, Incerti G (2008) The influence of climate on the distribution of lichens: a case study in a borderline area (Liguria, NW Italy). Plant Ecol 195:257–272

Grytnes JA, Heegaard E, Ihlen PG (2006) Species richness of vascular plants, bryophytes, and lichens along an altitudinal gradient in western Norway. Acta Oecol 29(3):241–246

Hale M (1983) The biology of lichens, 3rd edn. Edward Arnold Ltd, London

Hekking WHA, Sipman HJM (1988) The lichens reported from the Guianas before 1897. Willdenowia 17:193–228

Hill DJ, Hawksworth DL (1984) The Lichen forming fungi. Chapman and Hall, New York

Hylander K, Jonsson BG (2007) The conservation ecology of cryptogams. Biol Conserv 135:311–314

James PW, Hawaksworth JD, Rose F (1977) Lichen community in the British Isles: a primary conspectus. In: Seaward MRD (ed) Lichen ecology. Academic Press, London, pp 295–419

Kumar P, Yadav S, Sharma S, Lal S, Jha D (2009) Impact of climate change on seed production of cabbage in north western Himalayas. World J Agric Sci 5:18–26

Lakatos M, Rascher U, Büdel B (2006) Functional characteristics of corticolous lichens in the understory of a tropical low land rain forest. New Phytol 172:679–695

Lücking R, Hodkinson BP, Steven DL (2016) The 2016 classification of lichenized fungi in the Ascomycota and Basidiomycota - approaching one thousand genera. Bryologist 119(4): 361–416

MacDonald AMD, Coxson D, Björk C (2013) A framework for climate biomonitoring with lichens in British Columbia's inland temperate rainforest. J Ecosyst Manag 114:1–13
Mayfield MM, Levine JM (2010) Opposing effects of competitive exclusion on the phylogenetic structure of communities. Ecol Lett 13:1085–1093
McCune B, Dey J, Peck J, Heiman K, Will-Wolf S (1997) Regional gradients in lichen communities of the South-Eastern United States. Bryologist 100:145–158
McCune B, Rosentreter R, Ponzetti JM, Shaw DC (2000) Epiphyte habitats in an old conifer forest in western Washington, U.S.A. Bryologist 103:417–427
Orange A, James PW, White FJ (2001) Microchemical methods for the identification of lichens. British Lichen Society, London, p 101
Prasher IB, Chander H (2005) Lichens of Himachal Pradesh – I. Pb Univ Res J Sci 55(1 & 2): 109–129
Prieto M, Martinez I, Aragon G, Verdú M (2017) Phylogenetic and functional structure of lichen communities under contrasting environmental conditions. J Veg Sci 28(4):871–881
Rawat GS (2005) Alpine meadows of Uttaranchal- ecology, land use and status of medicinalandaromatic plants. Bishen Singh Mahendra Pal Singh, Dehradun
Sharma E, Tsering K (2009) Climate change in the Himalayas: the vulnerability of biodiversity. Sustain Mt Dev 55:10–12
Sharma K, Moore B, Vorosmarty C (2000) Anthropogenic, climatic, and hydrologic trends in the Koshi basin, Himalaya. Climate Change 47:141–165
Shrestha UB, Gautam S, Bawa KS (2012) Widespread climate change in the Himalayas and associated changes in local ecosystems. PLoS One 7(5):e36741
Shukla V, Bajpai R, Semwal M, Upreti DK (2017) Influence of chemical diversity in determining lichen communities structure along an altitudinal gradients in the Chopta, Tungnath, Western Himalaya. Int J Plant Environ 3(1):7–13
Sillett SC, McCune B, Peck JE, Rambo TR, Ruchty A (2000) Dispersal limitations of epiphytic lichens result in species dependent on old-growth forests. Ecol Appl 10:789–799
Singh KP, Sinha GP (2010) Indian lichens: an annotated checklist. Published by Botanical Survey of India (MoEF), Kolkata
Singh KP, Singh P, Sinha GP (2018 Lichen diversity in the eastern Himalaya biodiversity hotspot region, India Cryptogam Biodiversity and Assessment (Sp vol); pp 71–114
Sinha GP, Jagadeesh Ram TAM (2014) Lichen diversity in Sikkim. Botanical Survey of India report 2014, Ministry of Environmental and forest, India
Sipman HJM (1996) Corticolous lichens. In: Gradstein SR, Hietz P, Lücking R, Lücking A, Sipman HJM, Vesterh FM, Wolf JHD, Gardette E (eds) How to sample the epiphytic diversity of tropical rainforests, Ecotropica, vol 2, pp 66–67
Smith DC, Molesworth S (1973) Lichen physiology: XIII. Effects of rewetting dry lichens. New Phytol 72:525–533
Soto-Medina E, Londoño V, Díaz Escandón D (2015) Epiphytes from a forest type transition zone in the Chocó biogeographic region, Valle del Cauca, Colombia. Rev Biol Trop 63(4):915–926
Srivastava R (2005) Distribution, diversity and pollution status of lichen in great Himalayan National Park, Kullu district, Himachal Pradesh. PhD thesis submitted to Dr. Ram Manohar Lohia Avadh University, Faizabad, U.P., India
Stanton D (2015) Small scale fog gradients change epiphytic lichen shape and distribution. Bryologist 118(3):241–244
Stanton DE, Horn HS (2013) Epiphytes as 'filter-drinkers': life-form changes across a fog gradient. Bryologist 116:34–42
Thakur M, Chander H (2018) Common foliose macrolichens of Sikander Dhar, North Western Himalaya. CPUH Res J 3(2):179–186
Upreti DK, Nayaka S (2000) An enumeration of lichens from Himachal Pradesh. Professor D.D. Nautiyal Commemoration Volume Recent Trends In Botanical Researches, pp. 15–31. Botany Department. Allahabad University, Allahabad, India (D.K. Chauhan ed.)
Wade AE (1959) Soil binder for crustaceous ground lichens. Lichenologist 1(2):87–88

Waring B (2008) Light exposure affects secondary compound diversity in lichen communities in Monteverde, Costa Rica. Penn Sci 6:11–13

Wolseley PA Aguirre-Hudson B (2007). Lichens as Indicators of Environmental Changes in theTropical forests of Thailand [Online]. Available from: http://www.jstor.org/locate/envpol. (Accessed 03 May 2021)

Wolseley P, Ellis L, Chimonides J (2007) Corticolous lichen and moss communities in lowlanddipterocarp forests under differing management regimes. Bibl Lichenol 95:583–603

Xu J, Grumbine R, Shrestha A, Eriksson M, Yang X et al (2009) The melting Himalayas: cascading effects of climate change on water, biodiversity, and livelihoods. Conserv Biol 23:520–530

Yadav V (2005) Lichen flora of Himachal Pradesh. PhD thesis submitted to the University of Lucknow

Zapata FA, Gaston KJ, Chown SL (2005) The mid-domain effect revisited. Am Nat 166:144–148

Lower Plants of Tungnath-Chopta Timberline Zone, Garhwal Himalaya

14

Pradeep Singh, Vinod Joshi, Renu Rawal, Manisha Bhandari, Manish Tripathi, B. S. Kholia, and G. C. S. Negi

Abstract

This study was carried out in Pothivasa-Tungnath-Chopta area of Garhwal Himalaya, India to enumerate the occurrence of lower plants (Pteridophytes, Bryophytes and Lichens) along an elevation belt of 2100–2850 masl, encompassing NW and SW aspects and habitat types ranging from dense forests to timberline and alpine meadow in peak growing season in September 2020. We recorded a total of 54 species of Pteridophytes (species richness varied from 9 to 25 among these sites), 27 Lichen species (species richness 5–14) and 51species of Bryophytes, including 17 liverworts, 2 hornworts and 32 mosses. Across these sites, beta diversity was the highest for Pteridophytes (3.2), followed by Lichens (1.85) and Bryophytes (1.43). Along the elevation gradient species richness was almost invariant for bryophytes, it tended to be lower in the upper part of the altitudinal transect than the lower one for lichens, and it was highest for Pteridophytes at the highest altitude across the altitudinal transect. The entire elevational segment is rich in lower plant taxa and needs to be protected and intensively explored.

P. Singh · V. Joshi · R. Rawal · G. C. S. Negi (✉)
G.B. Pant National Institute of Himalayan Environment, Almora, Uttarakhand, India

M. Bhandari
Mahila Degree College, Haldwani, Uttarakhand, India

M. Tripathi
Soban Singh Jeena Campus, Kumaun University, Almora, Uttarakhand, India

B. S. Kholia
Botanical Survey of India, Northern Circle, Dehradun, Uttarakhand, India

S. P. Singh et al. (eds.), *Ecology of Himalayan Treeline Ecotone*,
https://doi.org/10.1007/978-981-19-4476-5_14

Keywords

Bryophytes · Lichens · Phytodiversity · Pteridophytes · Sub-timberline Garhwal Himalaya

14.1 Introduction

The lower group of plants (i.e. ferns, mosses, liverworts and lichens) are the oldest organisms on earth, and they play important role in ecosystems as primary producers and as nutrient and water recyclers. These plants resemble as well as differ from others with respect to several ecological and physiological features including dispersal by spores rather than seeds, mobile male gametes (ferns, bryophytes) and poikilohydry (lichens, bryophytes and filmy ferns). Because of this, these taxa often have similar abiotic requirements, and usually require high air humidity, and may abound in the same habitat, such as humid montane forests. They modify the physical and chemical weathering processes of the land surface and facilitate the establishment of higher plants. Uttarakhand is a storehouse of lower plants including 320 Pteridophytes (Joshi et al. 2009; Kholia 2010), 560 Bryophytes (Dandotiya et al. 2011), 1295 mosses (Sanjappa and Singh 2009; Singh and Dash 2018) and 560 Lichens species (Nayaka and Upreti 2011). Bryophytes are the second main group of green plants after the angiosperms. These workers have found that the abundance and diversity of lichens are maximum between 1000 and 3000 m elevations and their diversity increases from west to east. Both moist and humid habitats provide varied climatic conditions for foliose and other lichen taxa on rock and soil (Tewari and Pant 1994; Upreti and Chatterjee 1999a; b). Pteridophytes usually favour shady and moist conditions and grow across xerophytic, mesophytic and hydrophytic habitats. Commonly, the migration of fern from east to west is more than the west to east, and this rate is enhanced numerous times owing to recent climate change (Kholia et al. 2013). Bruun et al. (2006) studying the species richness of vascular plants, bryophytes and macro-lichens along the gradients of altitude and local topography in the alpine environment found that the richness of mosses and liverworts increases with altitude. They concluded that patterns in local species richness result from the action of two opposing forces: declining species pool and decreasing intensity of competition with altitude. Studies in Nepal (Grau et al. 2007) along the wide altitudinal transect,100–5000 masl reported a humped unimodal relationship between species richness and altitude for both liverworts and mosses with maximum richness at 2800 m and 2500 m, respectively. This peak was observed at 1900 m for ferns. It can be pointed out that different demands such as for water and temperature may be the main reason for this difference in patterns of occurrence of these plant groups. In this context of altitudinal and topographic patterns of occurrence, lower plants of the west Himalayan region remain yet to be explored and understood. It may be pointed out that the overexploitation of lichens is quite common and may affect not only lichen species richness but also of others plant groups in various ways including competition.

Here we give a preliminary assessment of species stock of lower plants, that is, Pteridophytes, Bryophytes and Lichens in an elevation belt of Tungnath, Garhwal Himalaya. It was based on a thorough survey of the elevational belt by a group of taxonomists and ecologists.

14.2 Study Area

This study was carried out in Pothivasa -Tungnath belt of Garhwal Himalaya, India. Total six sites across a gradient of altitude (2100–2850 masl), aspect (NW and SW) and habitat type (dense forests to timberline and alpine meadow) were surveyed for this study in September 2020 (Table 14.1). The rocks around Tungnath are mainly mylonite gneisses, augen gneisses, schists and granites (Agrawala 1973). The soil texture is sandy loam, light grey to brown in colour and acidic in nature with a pH range between 4 and 5 (Rai et al. 2012a). In the higher reaches of Tungnath, the snow cover lasts for about 4–5 months and melts during April–May that marks the arrival of favourable conditions for plant growth, which lasts for about 5–7 months. Annual precipitation was 2410.5 ± 432.2 mm, of which 89.5% was recorded during June–September (Adhikari et al. 2011). Forest types and vegetation according to Champion and Seth's (1968) classification in the study area fall in sub-alpine forest and alpine scrubs. Vegetation of the sub-alpine forests is formed by *Abies pindrow, Aglaia spectabilis, Acer spp.*, *Betula utilis*, *Quercus semecarpifolia*, *Quercus floribunda*, *Sorbus* spp. etc. (Table 14.1). Timberline in the study area ranges between 3250 and 3350 m, which is formed by *B. utilis* and *A. spectabilis* in the north to north-west facing slopes, while south to south west facing slopes and dominated by *Q. semecarpifolia* and *Rhododendron arboreum* (Rai et al. 2012b).

Table 14.1 Location and vegetation types of studied elevation transect in Chopta area

Site	Location	Elevation (masl.)	Aspect	Dominant forest trees
I	Pothiwasa	2107	NW	*Betula alnoides, Alnus nepalnesis, Quercus floribunda*
II	Dugalbitta	2370	NW	*B. alnoides, Abies pindrow, Aesculus indica, Acer sp., Machilus duthiei*
III	Mandal	2414	SE	*Q. semecarpfolia, Q. floribunda, Lyonia ovalifolia, Rhododendron arboreum, Ilex dipyrena, M. duthiei*
IV	Kachulakharak	2428	SE	*Q. semecarpifolia, A. nepalnesis, R. arboreum, Litsea umbrosa*
V	Dhotidhar	2718	SE	*R. arboreum, Q. semecarpifolia, A. nepalnesis, L. ovalifolia, Q. floribunda*
VI	Chopta	2844	NW	*Q. semecarpifolia, R. arboreum*

14.3 Material and Methods

Survey and assessment of the lower group of plants (Pteridophytes, Bryophytes and Lichens) were carried out by a group of 20 researchers from four regional organizations guided by experts from Botanical Survey of India, Dehradun, Kumaun University, Nainital and GBPNIHE, Almora in September 2020. The belt transect method (50 m long and 10 m wide) was used to determine the richness of these plant groups in the representative locations across six sites covering altitudinal intervals (200 m each), aspect (SE and NW) and forest vegetation type in Chopta-Tungnath timberline zone (Table 14.1).In each location, species richness, density and habitat-specific details of Pteridophytes, Bryophytes, and Lichens were recorded, and plant specimens were collected and preserved. The Bryophytes and Lichens were identified on the basis of morphological characteristics of the thallus, reproductive structure, colour, size and shape under the supervision of experts.

14.4 Results

The lower plant species stock of about 700 m elevation range was 132 species with about 50 each of pteridophytes and bryophytes and about half of that of lichens.

14.4.1 Pteridophytes

Across the six sites, a total of 54 species of Pteridophytes were recorded and the species richness varied from 9 to 25 among these sites (Table 14.2). The maximum species richness (25) was found in site I and the minimum (9) at site II (Dugalbitta 2370 m). *Selaginella chrysocaulos* was the only species that occurred in all six sites. Next to the site I, site VI was the highest in species richness. It had eight species not found at lower elevation sites. The site I had seven unique species. Thus, the species-rich sites also had high species uniqueness. *Dryopteris panda, Oleandra wallichii, Lycopodium japonicum, Diplazium maximum, Aleuritopteris leptolepis, Araiostegia beddomei, Athyrium atkinsonii, Dryopteris khullarii, Drynaria mollis, Pichisermollia quasidivaricata, Pichisermollia stewartii* and *Polystichum nepalense* were present only at higher elevation sites (Dhotidhar and Chopta, 2718–2844 m). Among all species, *Dryopteris chrysocoma* and *Dryopteris juxtaposita* were present in five sites across the studied transects. Some of the prominent Pteridophytes are given in Fig. 14.1.

14.4.2 Lichens

Across the six sites, a total of 27 Lichen species were recorded. The species richness of Lichen varied from 5 to 14 among the sites, being maximum (14) in *Q. semecarpifolia* dominated mixed broadleaf forest (Kachulakharak; Site IV).

Table 14.2 Pteridophytes recorded from study area (Sites I–VI arranged by increasing altitude)

S. no	Species	Site I 2107	Site II 2370	Site III 2414	Site IV 2428	Site V 2718	Site VI 2844
1	*Adiantum venustum*	–	+	–	–	–	–
2	*Aleuritopteris leptolepis*	–	–	–	–	–	+
3	*Araiostegia beddomei*	–	–	–	–	–	+
4	*Araiostegia pulchra*	+	–	–	+	+	+
5	*Arthromeris wallichiana*	–	–	–	+	–	–
6	*Asplenium ensiforme*	–	–	–	–	+	–
7	*Asplenium trichomanes*	–	–	–	–	+	–
8	*Asplenium laciniatum*	+	–	–	–	–	–
9	*Asplenium yoshinagae subsp. Indicum*	–	–	–	+	+	–
10	*Athyrium atkinsoni*	–	–	–	–	–	+
11	*Athyrium foliolosum*	+	–	+	–	–	–
12	*Athyrium pectinatum*	+	–	–	–	–	–
13	*Athyrium micropterum*	+	–	–	–	–	–
14	*Athyrium schimperi*	–	–	–	+	–	–
15	*Botrychiuml anuginosum*	+	–	–	–	–	–
16	*Coniogramme affinis*	+	–	+	–	+	+
17	*Coniogramme intermedia*	–	–	–	+	–	–
18	*Deparia japonica*	+	–	–	+	–	–
19	*Deparia subsimilis*	–	+	–	–	+	–
20	*Diplazium maximum*	+	–	+	–	–	–
21	*Dryopteris chrysocoma*	+	+	+	–	+	+
22	*Dryopteris juxtaposita*	+	+	+	+	–	+
23	*Dryopteris panda*	–	–	+	–	–	+
24	*Dryopteris khullarii*	–	–	+	–	–	–
25	*Dryopteris wallichiana*	+	–	+	–	+	+
26	*Drynaria mollis*	–	–	–	–	–	+
27	*Goniophlebiu margutum*	–	–	–	+	–	–
28	*Lepisorus mehrae*	+	+	–	–	+	–
29	*Lepisorus sesquipedalis*	+	–	–	+	+	–
30	*Lepisorusthunbergianus*	–	–	–	+	–	–
31	*Loxogramme involuta*	–	–	–	+	–	–
32	*Lycopodium japonica*	–	–	+	–	–	+
33	*Oleandra wallichii*	–	–	+	–	–	+
34	*Osmunda claytoniana*	–	–	–	–	+	–
35	*Onychium crypto grammoides*	–	+	–	–	–	+
36	*Pichi sermolli aquasi divaricate*	–	–	–	–	–	+
37	*Pichisermollia stewartii*	–	–	–	–	–	+
38	*Polypodiodeslachnopus*	+	–	–	+	–	+
39	*Polypodiodes amoena*	–	–	–	+	+	–

(continued)

Table 14.2 (continued)

S. no	Species	Site I 2107	Site II 2370	Site III 2414	Site IV 2428	Site V 2718	Site VI 2844
40	*Polystichum nepalense*	–	–	–	–	–	+
41	*Polystichum squarrosum*	+	–	–	+	–	–
42	*Polystichum stimulans*	+	–	–		–	–
43	*Polystichum piceopaleaceum*	–	–	–	+	–	–
44	*Polystichum manmeiense*	+	–	–	–	–	–
45	*Pteris aspericaulis*	+	–	–	–	–	+
46	*Pteris cretica*	+	–	–	–	–	+
47	*Pteris terminalis*	–	–	–	–	+	–
48	*Pteris wallichiana*	+	–	–	–	–	+
49	*Pyrrosia porosa*	+	–	–	–	+	–
50	*Pyrrosia flocculosa*	–	–	–	+	–	–
51	*Selaginella chrysocaulos*	+	+	+	+	+	+
52	*Thelypteris erubescens*	+	–	–	–	–	+
53	*Thelypteris pyrrhorhachis*	+	+	+	–	+	–
54	*Woodsia lanosa*	–	+	–	–	–	–
Total		25	9	12	17	16	22

Most of the species belong to the foliose category of lichens. *Heterodermia diademata*, *Lepraria lobificans Cladonia* sp. were recorded across all sites, whereas species such as *Bulbothrix* sp., *Dermatocarpon* sp., *Porpidia* sp., and *Stereocaulon* sp. were confined to one or two higher elevation sites. Similarly, *Phaeophyscia hispidula* and *Usnea longissima* were confined to low elevation sites only (Table 14.3). Site IV (*Q. semecarpifolia* forest site) was exceptionally rich in lichens, as it had seven lichen species (mostly Corticolous) not occurring at any other sites. More investigations are required to analyse its lichen and the sharp drop that followed in species number subsequently.

14.4.3 Bryophytes

Across the six sites, a total of 51 species of Bryophytes, including 17 liverworts, 2 hornworts and 32 mosses were recorded during the field work (Table 14.4). As many as 19 of 51 species were ubiquitous. Maximum species number (43 species) occurred at Kachulakharak site (*Q. semecarpifolia* dominated forest) that included all but one liverworts. Twenty species (*Atrichum sp., Bryum sp., Dicranum sp., Entodon sp., Funaria sp., Herpetineuron sp., Mnium sp., Plagiothecium sp., Pogonatum sp., Thuidium sp., Anomodon sp., Chiloscyphus sp., Frullania sp., Heteroscyphus sp., Plagiochilla sp., Scapania sp., Pellia sp., Cyathodium sp., Conocephalum conicum* and *Marchantia polymorpha*) were found common almost in all the six study sites. Bryotaxa restricted to higher elevation sites were species of

Fig. 14.1 Some prominent species of Bryophytes, Lichens and Pteridophytes in Chopta-Tungnath area

mosses viz., *Actinothuidium*, *Bryum*, *Rhacomitrium*, *Grimmia*, *Pogonatum* and hepatic like *Sauchia*, *Scapania*, *Bazzania*, *Plagiochilla* and *Herbertus spp.* Some photographs of prominent Bryophytes are given in Fig. 14.1.

Table 14.3 Lichen species recorded from study area (Sites I–VI arranged by increasing altitude)

S. no	Species	Growth forms	Substrate	Site I	Site II	Site III	Site IV	Site V	Site VI
1	*Heterodermia sp.*	Foliose	Saxicolous	–	–	–	–	+	–
2	*Heterodermia diademata*	Foliose	Saxicolous	+	+	+	+	+	–
3	*Heterodermia boryi*	Foliose	Saxicolous	–	+	–	–	+	–
4	*Parmotrema sp.*	Foliose	Saxicolous	–	+	+	+	–	–
5	*Parmotrema reticulatum*	Foliose	Saxicolous	–	–	–	–	+	+
6	*Caloplaca sp.*	Crustose	Saxicolous	–	+	+	+	–	–
7	*Lepraria sp.*	Leprose	Saxicolous	+	+	+	+	+	–
8	*Pyxine sp.*	Foliose	Saxicolous	–	+	–	–	–	–
9	*Pertusaria sp.*	Crustose	Saxicolous	–	+	–	–	–	–
10	*Bulbothrix sp.*	Foliose	Saxicolous	–	–	+	–	–	–
11	*Dermatocarpon sp.*	Crustose	Saxicolous	–	–	+	–	–	–
12	*Phaeophyscia hispidula*	Foliose	Saxicolous	–	–	–	+	–	–
13	*Porpidia sp.*	Crustose	Saxicolous	–	–	–	–	–	+
14	*Usnea sp.*	Fruticose	Corticolous	–	+	–	–	–	–
15	*Usnea longissima*	Fruticose	Corticolous	–	–	–	+	–	–
16	*Everniastrum sp.*	Foliose	Corticolous	–	+	–	+	+	+
17	*Ramalina conduplicans*	Fruticose	Corticolous	–	+	–	–	–	–
18	*Flavoparmelia caperata*	Fruticose	Corticolous	–	–	–	+	–	–
19	*Punctelia sp.*	Foliose	Corticolous	–	–	–	+	–	–
20	*Chrysothrix sp.*	Foliose	Corticolous	–	–	–	+	–	–
21	*Flavopuncteia sp.*	Foliose	Corticolous	–	–	–	+	–	–
22	*Canoparmelia sp.*	Foliose	Corticolous	–	–	–	+	+	–
23	*Parmotrema tinctorum*	Foliose	Corticolous	+	–	–	–	–	–
24	*Phaeophyscia sp.*	Foliose	Corticolous	–	–	–	+	+	+
25	*Cladonia sp.*	Foliose	Corticolous	+	+	–	+	+	–

26	*Stereocaulon sp.*	Foliose	Corticolous	–	–	–	–	–	+
27	*Graphis scripta*	Foliose	Corticolous	+	–	–	–	–	–
	Total			5	11	6	14	9	5

Table 14.4 Bryophytes with habitat specificity recorded from study area (Sites I–VI arranged by increasing altitude)

S. No	Species name	Habitat	Site I	Site II	Site III	Site IV	Site V	Site VI
1	*Actino thuidium hookerii* (Mitt.) Broth.	On moist slopes	–	–	+	–	–	–
2	*Anomodon minor* (Hedw.) Lindb.	On tree trunk	+	+	+	+	+	+
3	*Atrichum undulatum* (Mull. Hal.) A. Jaeger	On moist soil	+	+	+	+	+	+
4	*Brachythecium buchananii* (Hook.) A. Jaeger	On stone	+	+	–	+	+	+
5	*Brotheraleana* (Sull.) Mull. Hal	On rotten fallen log	–	+	+	+	+	+
6	*Bryumargenteum* Hedw.	On slopes soil, stone	+	+	+	+	+	+
7	*Bryum capillare* Hedw.	On tree bark	–	+	–	–	–	–
8	*Dicranum sp.*	On stone & bark	+	+	+	+	+	+
9	*Ditrichum heteromallum* (Hedw.) Britt.	Fallen logs	+	+	–	+	+	+
10	*Ectropothecium cyperoides* (Hook. & Harv.) A. Jaeger	Fallen logs	+	+	–	–	–	–
11	*Entodon plicatus* Mull. Hal.	On stone & bark	+	+	+	+	+	+
12	*Fissidens bryoides* Hedw.	Fallen logs	+	+	+	+	+	+
13	*Fissidens grandiforms* Brid.	On sloppy soil	–	–	+	–	–	–
14	*Funaria hygrometrica* Hedw.	Retaining walls	+	+	+	+	+	+
15	*Grimmia sp.*	On rocks	–	–	+	+	+	+
16	*Herpetineuron toccoae* (Sull. & Lesq.) Cardot	On bark, tree trunk	+	+	+	+	+	+
17	*Neckera crenulata* Harv.	On tree bark	–	+	–	–	–	–
18	*Leucodon secundus* (Harv.) Mitt.	On tree bark	+	+	–	+	+	+
19	*Macromitrium moorcroftii* (Hook. & Grev.) Schwagr.	On fallen twigs & tree branches	+	+	–	+	+	+
20	*Meteorium buchananii* (Brid.) Broth.	On fallen twigs & tree branches	+	+	–	+	+	+
21	*Mniumintegrum* Bosch & Sande Lac.	On tree base	+	+	+	+	+	+
22	*Philonotis sp.*	On moist rock	–	–	+	–	–	–

23	*Plagiothecium denticulatum* (Hedw.) Schimp.	On moist stone and slopes' soil	+	+	+	+	+	+
24	*Pogonatum microstomum* (R.Br. & Schwagr.) Brid.	On rocks boulder	+	+	+	+	+	+
25	*Pohliacamptotrachela* (Renauld & Cardot) Broth.	On slopes soil	–	–	+	–	+	–
26	*Racomitirium himalayanum* (Mitt.) A. Jaeger.	On rocks	–	–	+	+	+	+
27	*Rhodobryum roseum* (Hedw.) Limpr.	Tree bark	–	–	+	–	–	–
28	*Taxiphyllum taxirameum* (Mitt.) M. Fleisch.	On stone	+	+	–	+	+	+
29	*Thamnobryum fruticosum* (Mitt.) Gangulee	On tree bark	+	+	–	–	–	–
30	*Thuidium cymbifolium* (Dozy & Molk.) Dozy & Molk.	On forest floor, stone	+	+	+	+	+	+
31	*Trachypodopsis crispatula* (P. Beauv.) M. Fleisch	Tree bark	+	+	–	+	+	+
32	*Trematodon longicollis* Michx.	On rocks	–	+	–	+	+	+
Hornwort								
33	*Anthoceros himalayensis* Kashyap	Moist soil	–	+	+	+	–	–
34	*Anthoceros erectus* Kashyap	Moist soil	–	–	+	–	–	–
Liverworts								
35	*Asterella wallichiana* Beauv.	On rock & soil	–	+	–	+	+	+
36	*Cyathodium tuberosum* Kash.	On tree bark	+	+	+	+	+	+
37	*Conocephalum conicum* (L.) Lindb.	On sloppy soil	–	+	+	+	+	+
38	*Marchantia polymorpha* L.	Retaining walls	+	+	+	+	+	+
39	*Plagiochasma appendiculatum* Lehm. & Lindenb.	Retaining walls	+	+	+	–	–	–
40	*Reboulia hemisphaerica* (L.) Raddi.	On sloppy soil	+	+	+	–	–	–
41	*Stephnensoniella brevipedunculata* Kash.	On rocks	–	–	+	–	–	–
42	*Targionia hypophylla* L.	On slopes soil	+	+	+	–	–	–
43	*Apometzeria pubescens* (Schrank) Kuwah.	On moist slopes soil	+	+	–	–	–	–
44	*Pellia endiviifolia* (Dicks) Dum.	On moist slopes soil	+	+	+	+	+	+
45	*Chiloscyphus polyanthus* (L.) Corda.	Moist soil	+	+	+	+	+	+
46	*Frullania dilatata* (L.) Dumort.	On bark & rocks	+	+	+	+	+	+

(continued)

Table 14.4 (continued)

S. No	Species name	Habitat	Site I	Site II	Site III	Site IV	Site V	Site VI
47	*Heteroscyphus sp.*	On soil	+	+	+	+	+	+
48	*Plagiochilla spinulosa* (Dicks) Dumort.	On sloppy soil, bark & boulder	+	+	+	–	+	+
49	*Porella denticulata* (Kashyap & R.S. Chopra) J.X. Luo	On tree bark	–	+	–	–	+	+
50	*Scapania sp.*	On tree bark & tree base	+	+	+	+	+	+
51	*Radulla complanata* (L.) Dumort.	On bushy branches	+	+	–	+	+	–
Total			34	43	34	34	34	35

14.5 Discussion

This study has added to the otherwise unexplored region for the lower plants. In this altitudinal transect of about 700 m species richness of the lower plants did not show a definite trend with altitude ($R^2 = 0.2926$, $p = 0.267$; NS); the lower elevations recorded a greater number of species than the higher elevations (60 vs. 50) (Fig. 14.2; Table 14.5). However, a negative correlation was found for Lichens with altitude ($R^2 = 0.640$; $p = 0.056$; NS; Fig. 14.3). Across these sites, beta diversity was the highest for Pteridophytes (3.2), followed by Lichen (1.85) and Bryophytes (1.43). Along the elevation gradient species richness was almost invariant for bryophytes, it tended to be lower in the upper part of the altitudinal transect than the lower one for lichens, and it was highest for Pteridophytes at the highest altitude across the transect (Fig. 14.3). With increasing elevation up to timberline, the Bryophytes' richness and luxuriance continuously increased. Thalloid liverworts population occupies open exposed sites. However, with increasing altitude leafy liverworts and mosses dominate on diverse habitats. Due to the prevalence of more

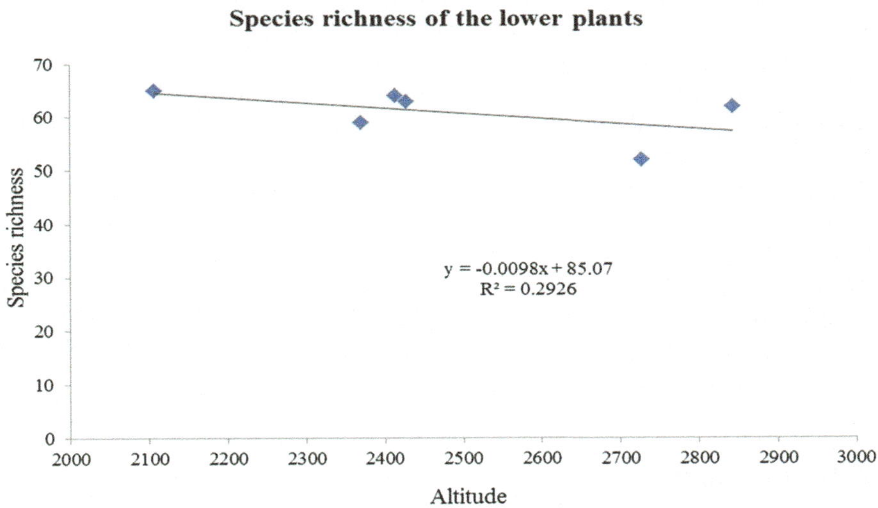

Fig. 14.2 Correlation between altitude and species richness of lower plants

Table 14.5 Pattern of lower plant species richness across altitudinal gradients in the Tungnath area

Site	Altitude (m. asl)	Pteridophytes	Lichens	Bryophytes	Total
Pothivasa	2107	17	14	34	65
Dugalbitta	2370	16	9	34	59
Mandal	2414	25	5	34	64
Kachulakharak	2428	9	11	43	63
Dhotidhar	2718	12	6	34	52
Chopta	2844	22	5	35	62

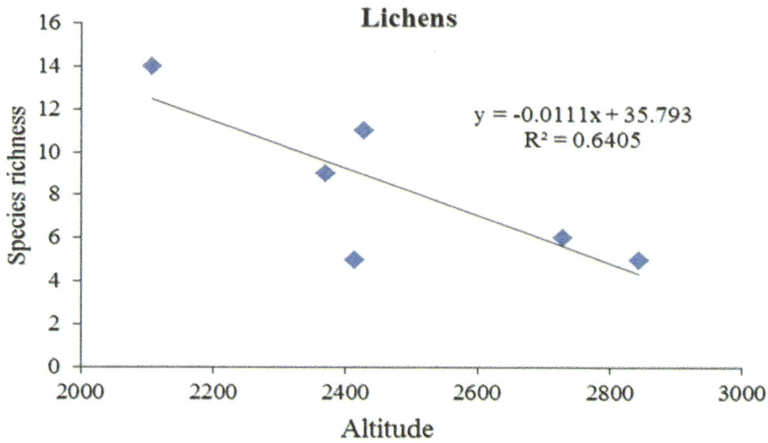

Fig. 14.3 Negative correlation between altitude and species richness (Lichen)

or less similar habitat conditions above timberline, only alpine-adapted bryotaxa dominated. Certain observations of Bryoflora made during the fieldwork were: (i) The tree trunk and branches were loaded with the epiphytic hanging mosses; (ii) Fruiting population of a hair cap moss (*Pogonatum* sp.) adorned the rocky ledges almost in all the study sites; (iii) Fruiting population of one of the unique rare, monotypic, endemic, liverwort *Stephensoniella brevipedunculata* Kash. was recorded from limestone-dominated rocky ledges between 2718 and 3025 m. Based on our current observations, it is interesting to highlight the occurrence and spread of *Sophronitis brevipenduculata* at higher altitudinal range up to 3000 m are a sharp indication of changing climatic conditions as this ecosensitive red list taxa was earlier known to grow at an altitudinal range of 2000–2400 m (Kashyap 1929; Pant and Tiwari 1995). The entire elevational segment is rich in lower plant taxa and needs to be protected (Fig. 14.4).

14.6 Conclusion

In this study, species richness of lower plants (Pteridophytes, Bryophytes and Lichens) along an elevation belt of 2100–2850 masl and NW and SW aspects within dense forests to timberline and alpine meadow during rainy season was found. Occurrence of a total of 54 species of Pteridophytes, 27 species of Lichen and 51species of Bryophytes revealed that across these sites beta diversity was the highest for Pteridophytes (3.2), followed by Lichens (1.85) and Bryophytes (1.43). Trend of species richness of lower plants along the elevation gradient was found almost invariant for bryophytes, it tended to be lower in the upper part of the altitudinal transect than the lower one for lichens, and it was highest for Pteridophytes at the highest altitude. Further exploration across different seasons is

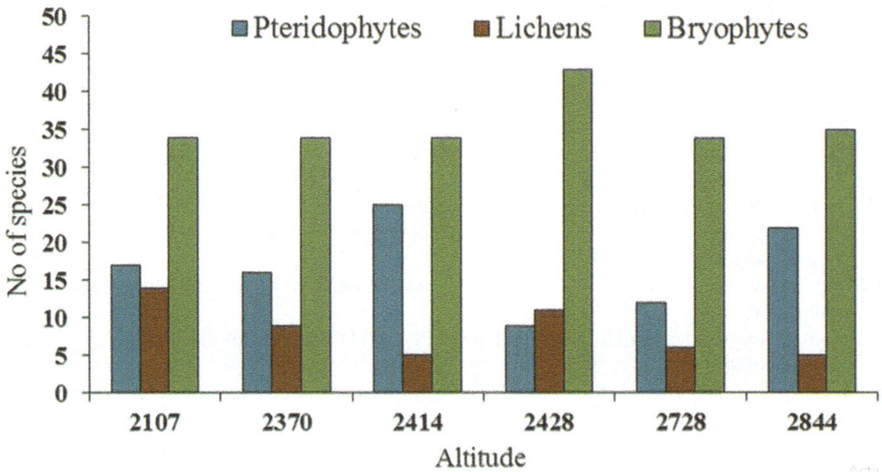

Fig. 14.4 Pattern of lower plants across altitudinal gradients

required to relate any relationship of occurrence of these lower plants with climate change.

In the high-altitude forests (encompassing temperate-subalpine-treeline zone) owing to the occurrence of a variety of microsites species richness of lower plants is quite high as found in this study (54 species of Pteridophytes, 51 species of Bryophytes and 27 species of Lichens) during the rainy season within an altitudinal span of 700 m. Also, the lower altitude sites were characterized with higher species richness than the high altitude sites. With increasing elevation up to timberline, the Bryophytes' richness and luxuriance continuously increased as opposed to richness of Lichens. The lower plants are considered the best elements to study the impact of climate change that has implications for survival and extinction of these species. For example, the occurrence and spread of *Stephensoniella brevipeduculata* at a higher altitudinal range up to 3000 m are a sharp indication of changing climatic condition as this ecosensitive red list taxa was earlier reported within an altitudinal range of 2000–2400 m in this region. Therefore, the high-altitude forests offer a great scope for investigation of the seasonal dynamics of occurrence of the hitherto neglected lower plant groups with a particular focus on their adaptation to atmospheric warming and changes in moisture regime.

Acknowledgements Authors gratefully acknowledge Drs. R.S. Rawal, Director, GBPNIHE, Kosi-Katarmal, and Almora for organizing a field workshop at Chopta-Tungnath in September 2020 during which over a dozen researchers of Kumaun University, Nainital, WII, Dehradun and CHEA, Nainital participated in field work to collect these lower plants. We also acknowledge Dr. S.D. Tewari, Head, Botany Department, Mahila Degree College Haldwani, Kumaun University, Nainital for identification of Bryoflora.

References

Adhikari BS, Rawat GS, Rai ID, Bharti RR, Bhattacharyya S (2011) Ecological assessment of timberline ecotone in western Himalaya with special reference to climate change and anthropogenic pressures. IV Annual Report Wildlife Institute of India, Dehradun, India

Agrawala NK (1973) Working plan for the Kedarnath Forest division 1972–73 to 1981–82. Working plan circle, Nainital, Uttar Pradesh

Bruun HH, Moen J, Virtanen R, Grytnes JA, Oksanen L, Angerbjörn A (2006) Effects of altitude and topography on species richness of vascular plants, bryophytes and lichens in alpine communities. J Veg Sci 17(1):37–46

Champion HG, Seth SK (1968) A revised survey of the forest types of India. Manager of publications

Dandotiya D, Govindapyari H, Suman S, Uniyal PL (2011) Checklist of the bryophytes of India. Arch Bryol 88:1–126

Grau O, Grytnes JA, Birks HJB (2007) A comparison of altitudinal species richness patterns of bryophytes with other plant groups in Nepal, Central Himalaya. J Biogeogr 34(11):1907–1915

Joshi P, Pande C, Pande PC (2009) Ferns of Central Himalayas-(Chamoli and Rudraprayag). Bishen Singh Mahendra Pal Singh

Kashyap SR (1929) Liverworts of Western Himalayas and Panjab plains, The Chronica Botanica Delhi, India; Vol. I

Kholia BS (2010) Ferns and fern allies of Sikkim: a pictorial handbook. In: Sikkim state biodiversity board. Botanical Survey of India, Salt Lake City, Kolkata

Kholia BS, Joshi R, Punetha R (2013) Extended distribution of *Cyathea spinulosa* wall. Ex hook.In Uttarakhand Himalaya with a note on distribution and diversification of Himalayan ferns in relation to recent climate change. NeBIO 4(2):40–45

Nayaka S, Upreti DK (2011) Lichen diversity in Western Ghats: Need for quantitative assessment and conservation. Commissioned Paper, WGEEP Report 2011, Ministry of Environment, Forest, and Climate Change, Government of India

Pant G, Tiwari SD (1995) Additional, upto date notes on red list monotypic endemic liverwort taxa of Kumaon (Western Himalayas). The Bryolog. Times 83–84:7

Rai ID, Adhikari BS, Rawat GS, Bargali K (2012a) Community structure along timberline ecotone in relation to micro-topography and disturbances in Western Himalaya. Not Sci Biol 4:41–52

Rai ID, Adhikari BS, Rawat GS (2012b) Floral diversity along sub-alpine and alpine ecosystems in Tungnath area of Kedarnath wildlife sanctuary, Uttarakhand. Indian For 138(10):927–940

Sanjappa M, Singh P (2009) Plant discoveries 2008. Govt. of India, Ministry of Environment and Forests, and Climate Change, Government of India, Botanical Survey of India

Singh P, Dash SS (eds) (2018) Plant discoveries 2017. BSI, Ministry of Environment and Forests, Kolkata, pp 1–150

Tewari SD, Pant GB (1994) Bryophytes of Kumaun Himalaya. Bishan Singh Mahendra Pal Singh, Shiva Offset Press, Dehradun

Upreti DK, Chatterjee S (1999a) A preliminary survey of lichens from Corbett National Park. JBNHS 96(1):88–92

Upreti DK, Chatterjee S (1999b) Epiphytic lichens on *Quercus* and *Pinus* trees in the three forest stands in Pithoragarh district, Kumaon Himalayas, India. Trop Ecol 40(1):41–49

Phenological Response of Treeline Ecotone Tree Species to Global Warming in Western Himalaya

15

Pradeep Singh and G. C. S. Negi

Abstract

Phenological behaviour of five treeline species (viz *Abies spectabilis, Betula utilis, Quercus semecarpifolia, Rhododendron arboreum* and *Rhododendron campanulatum*) was studied in Tungnath (Uttarakhand), West Himalaya. These species show annual variations in initiation and culmination of various phenophases, which were significantly correlated ($P < 0.001$) with mean annual temperature. Growth starts in early-April in *A. spectabilis* and *B. utilis* and in mid-May in the remaining species. In the slightly warmer months (April–June 2017, mean = 8.33 °C) bud-break and flowering occurred earlier by about 2–3 weeks in all the species than in the 1.8 °C cooler 2020 (April–June 2020, mean = 6.59 °C). Leaf drop was an extended activity in *A. spectabilis*, *R. arboreum* and *R. campanulatum* (5–7 months) and a rapid activity (~ 2 months) in *B. utilis* (deciduous) with a stable leaf population varying from 2 to 13 months across the species. Mean leaf longevity was recorded lowest for *B. utilis* (30 weeks) and highest for *A. spectabilis* (193 weeks). All these species flower before the onset of monsoon followed by fruiting and fruit maturation. However, fruit maturation and seed dispersal period varied considerably across these species. This study will serve as a benchmark for similar studies in future to ascertain the impact of CC on timberline vegetation.

Keywords

Bud-break and leafing · Climate change · Himalaya · Phenology · Treeline

P. Singh · G. C. S. Negi (✉)
G.B. Pant National Institute of Himalayan Environment, Almora, Uttarakhand, India

S. P. Singh et al. (eds.), *Ecology of Himalayan Treeline Ecotone*,
https://doi.org/10.1007/978-981-19-4476-5_15

15.1 Introduction

Alpine treeline ecotones across most of the world's mountains are being studied because of their potential for monitoring the effects of climate warming on forest ecosystems (Holtmeier and Broll 2005). Past studies have shown that treeline ecotones are good indicators of climate change (CC), where trees often respond to climatic warming with population structures, increase in seedling recruitment or tree density, as well as upward shift of the treeline (e.g. Camarero and Gutiérrez 2004). Treeline ecotones are sensitive bio-monitors of past and recent CC and variability (Kullman 1998), and thus well suited for monitoring CC impact (Becker et al. 2007; Kullman 1998). Treelines are considered to be constrained primarily by heat deficiency (Holtmeier 2009; Körner and Paulsen 2004), and the position of a treeline is determined mainly by low-temperature conditions (Körner and Paulsen 2004; Holtmeier 2009). Worldwide, high-altitude climatic treelines are associated with a seasonal mean ground temperature of 6.7 °C ± 0.8 during the growing period (Körner and Paulsen 2004), which has been computed 9.6 °C in our treeline study sites at Tungntah (Western Himalaya) (Joshi et al. this volume). The distribution and phenology of alpine plants are markedly controlled by snow, which produces various effects on the plants. In winter, snow cover protects the plants from freezing injury and from desiccation, especially in windy areas, and in summer, the lingering snow constantly provides water to the plants (Billings and Bliss 1959; Isard 1986). Other than the low air and soil temperature, the short growing season, snowfall, high exposure to wind, increased exposure to frost and lower availability of nutrients are some of the common features of high-altitude regions, which greatly influence plant phenophases and vegetation growth (Germino 2014; Korner 2012).

Among the studies on biological impacts of climate warming, those on phenological changes are the most numerous, probably since they are much easier to detect than changes in populations, distribution and community composition (Rosenzweig et al. 2008; Miller-Rushing and Inouye 2010). Both winter and summer temperatures are considered as accurate predictors of occurrence of phenophases (Galán et al. 2005; Schwartz 1999). Advances in leafing and flowering influenced by rise in atmospheric temperature have been frequently reported across Europe and North America (Menzel et al. 2006; Schwartz et al. 2006), and also in some parts of the Himalayas (e.g. Gaire et al. 2014). Phenological shifts (e.g. germination, bud-break, flowering time, its duration and synchrony) may affect the interactions of plants with their pollinators and seed dispersers, hence reproductive success and growth of plants (Elzinga et al. 2007; Singh and Kushwaha 2006). Of particular interest is leaf longevity, which has been evaluated in terms of nutrient-use efficiency, carbon balance and the ratio of cost to carbon gain, successional stage of plant, micro-environmental conditions, as well as the emergence times and types of survivorship curves of different leaf cohorts (Chabot and Hicks 1982; Kikuzawa 1995).

Schickhoff et al. (2015) have pointed out that for the Himalayan treeline there is limited knowledge on meteorological conditions, freezing and frost, drought, soil temperature, wind, snow cover, soil quality, regeneration and other aspects. However, for the timberline of the western Himalayan region, a few such studies are available (Mir et al. 2016; Rai et al. 2012; Rawal et al. 1991; Singh et al. 2018a; b), and a few past studies in the alpine zone of this region have shown that plant growth and phenophases here are triggered with the occurrence of favourable temperature, snow melt and consequent soil water availability (Nautiyal et al. 2001; Negi and Singh 1992; Sundriyal et al. 1987). In recent years, Himalayan treelines have also been studied from the standpoint of global warming (Gaire et al. 2014; Shrestha et al. 2012; Chhetri et al. 2018; Singh et al. 2018a; b). In Tungnath, the site of present investigation, Rai et al. (2013) reported that *Betula utilis* is highly sensitive to inter-annual climatic variations and early snowmelt. Liang et al. (2014) reported that Himalayan birch at its upper distribution boundary is increasingly at risk of survival. In the Tibetan plateau, Yu et al. (2010) reported that strong winter warming could slow the fulfilment of chilling requirements, which may delay spring phenology. In the mid-elevation forests of this region and other parts of the Himalaya CC induced changes in leafing and flowering phenology of trees (viz *Rhododendron* spp.) (e.g. Gaira et al. 2014), and alpine herbs (Telwala et al. 2013) have also been reported.

This study was carried out to answer the following research questions: (i) Does the dominant treeline ecotone tree species varying in growth forms differ in periodicity of major phenophases (bud-break, leafing, flowering, fruiting and leaf drop), leaf characteristics (leaf life-span, leaf area and leaf mass) and shoot growth phenology (shoot extension and radial growth and leaf population dynamics); and (ii) Whether there exist intra-specific and inter-annual changes in phenological responses of treeline species to global warming.

15.2 Material and Methods

15.2.1 Study Area

The study site lies in Tungnath, an alpine timberline forest in district Rudraprayag (Uttarakhand) at an elevation ranging from 2955 to 3334 m asl. In this region, December to March are intense cold months with heavy snowfall, while, during May and June, pre-monsoon showers and occasional hailstorms are common, and from July to early September heavy and regular rainfall is a characteristic feature. Rainfall at Tungnath (3350 m asl) during the study period ranged from 4217 to 5002 mm/year; and the mean annual temperature ranged from 5.1 to 5.9 °C. The mean temperature of the growing season (April–September) at this site varied from 9.52 to 9.74 °C (Joshi et al. this volume).

The five treeline species we selected for phenological observations in Tungnath were highly diverse in growth forms ranging from *Abies spectabilis* (evergreen conifer with multi-year leaf life-span), *Betula utilis* (winter deciduous broadleaf), *Quercus semecarpifolia* (semi-evergreen broadleaf) and *Rhododendron arboreum* (evergreen broadleaf) and *R. campanulatum* (evergreen krummholz species) distributed between 2955 and 3334 m asl (the upper limit distribution of these species). Thus our study site had the advantage of having both the set of species

Fig. 15.1 Phenological observations on *R. campanulatum* in Tungnath treeline

with a mixture of deciduous (*B. utilis, Sorbus foliolosa)* trees with temperate affinities and evergreen (broadleaf species with sub-tropical affinities (e.g. *Q. semecarpifolia*), which co-occur in the same environmental conditions.

For phenological records, a representative forest stand of 1 ha for each of the five species was selected and 100 mature individuals of each species were marked (for *B. utilis* only 70 trees were available at the study site) at the start of this study in May 2016, and phenological records on these marked tree populations were maintained at biweekly/monthly interval, except for January– March every year when the study area remains covered under snow (Fig. 15.1). Shoot extension and radial growth were measured at the monthly interval on 25 marked current year shoots for each of the tree species. To determine the leaf life-span leaf census on these 25 marked shoots was carried out at monthly intervals from leaf anthesis to leaf drop for each of the five tree species. Leaf characters (leaf area and leaf mass) were based on 100 randomly plucked current year leaves from the marked forest stands at monthly intervals. Phenological data thus collected over 5 years were treated statistically (Statistica 8.0). To determine the impact of temperature correlation analysis was performed. The phenological events were converted into binary variables (0 = phenological event absent; 1 = phenological event present) and a logistic regression model was applied with mean annual temperature (MAT) and mean temperature of the growing season (April–September) taken as independent variable and proportion of tree population in a given phenophase as a response variable (Menard 2002).

15.3 Results

15.3.1 Periodicity of Major Phenophases

This study carried out continuously for 5 years between May 2016 and April 2021 provided a strong database to decipher the inter-annual and intra-species variations in the periodicity of major phenophases. All the species showed significant ($P < 0.05$) annual variations both in initiation and culmination of the various phenophases (Fig. 15.2). On average for the entire study period of 5 years onset of

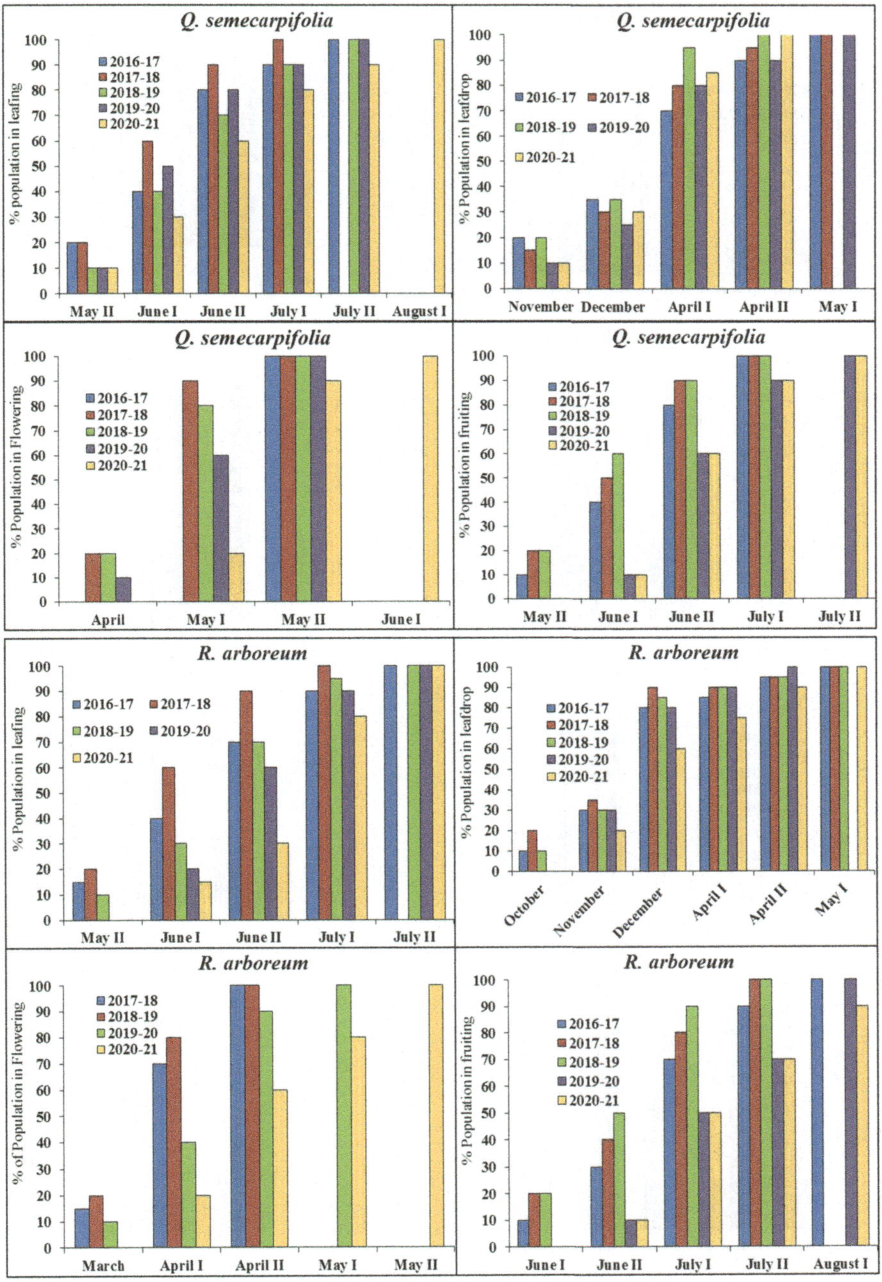

Fig. 15.2 Proportion of marked tree population in different phenological activity (mean across five study years)

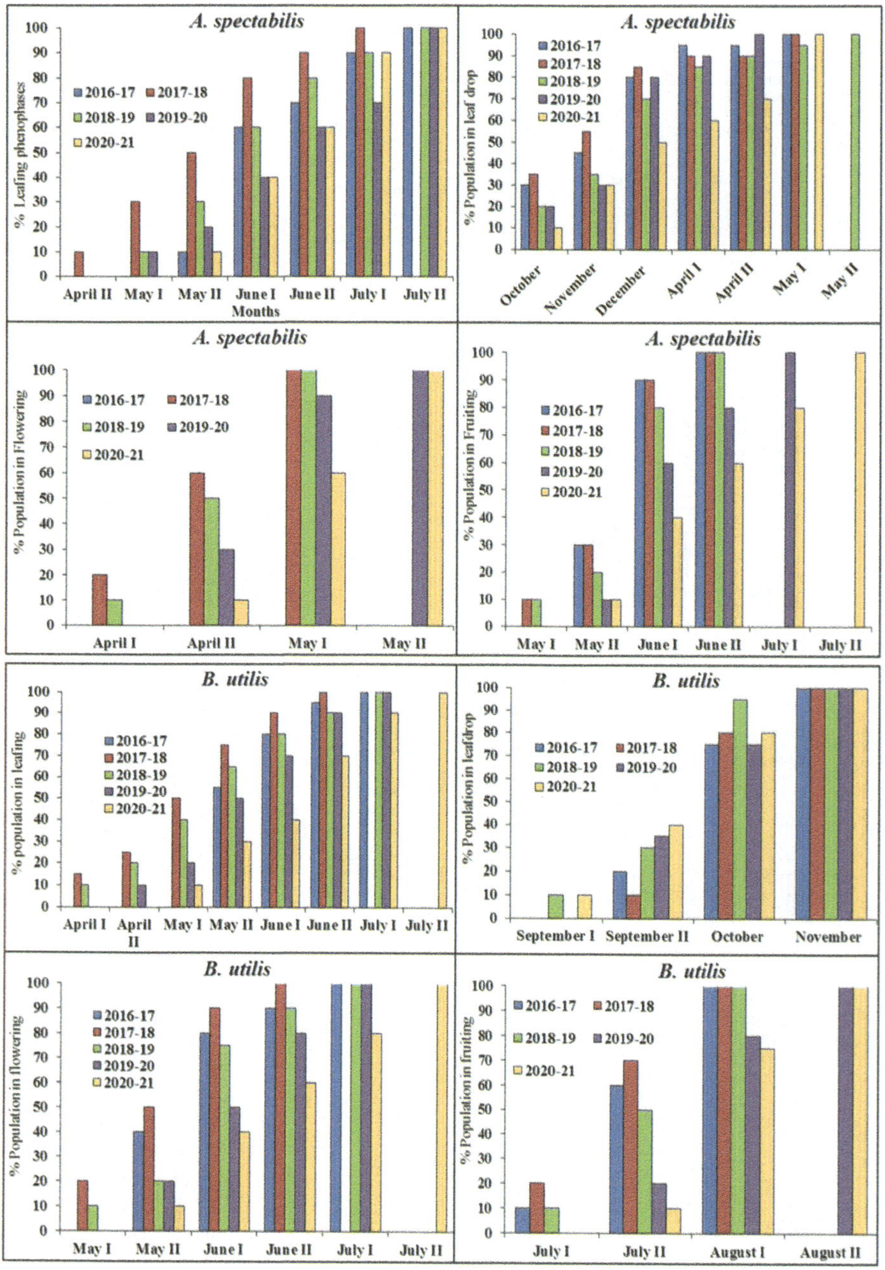

Fig. 15.2 (continued)

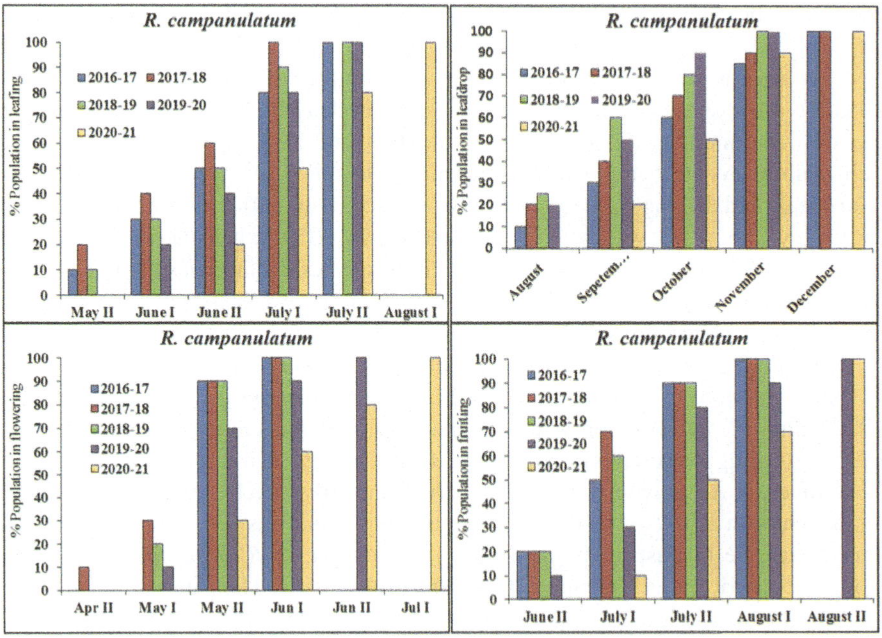

Fig. 15.2 (continued)

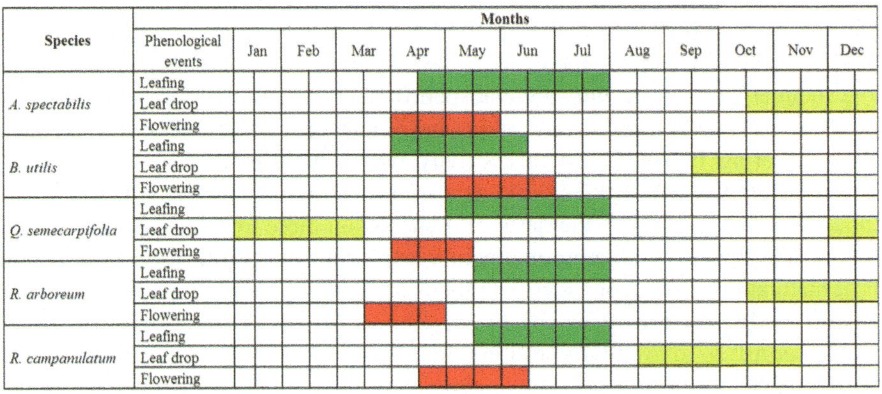

Fig. 15.3 Phenological calendar of various phenophases for the five treeline species (mean across 2016–2017, 2017–2018, 2018–2019, 2019–2020 and 2020–2021 phenological cycles)

growth signalled by bud-break was recorded as early as in mid-April in *A. spectabilis* and *B. utilis*, and in mid-May in the remaining three species (*Q. semecarpifolia*, *R. arboreum* and *Rhododendron campanulatum*) (Fig. 15.3). In *R. campanulatum*, which grows at the uppermost altitudinal limit of the treeline, leafing got delayed by about one month than all other treeline species studied by us. Bud-break and leafing activity in >90% of marked individuals were completed by July-end in all the species, except for *B. utilis* in which it was completed about 2–3 weeks earlier than the other species. Thus, in these species, the length of bud-break and leafing period was quite long (approximately 3 months), and all these species had attained full canopy by the commencement of monsoon rains. Inter-annual variations in bud-break and leafing activity were also observed. In 2017, bud-break occurred earlier by about 2–3 weeks in all the species. In 2020, when the bud-break and leafing were delayed by 2–3 weeks in all these species, leafing was continued for a longer period. The season of bud-break and leafing (April–June) was warmer in 2017 (mean temperature = 8.33 °C) and cooler by 1.8 °C in 2020 (mean = 6.59 °C).

Leaf drop was an extended activity in *A. spectabilis*, *R. arboreum* and *R. campanulatum* (5–7 months) and in *B. utilis* (deciduous) it was a rapid activity (2 months). *R. campanulatum* was characterized by the initiation of leaf drop in August (during peak rains), and in *Q. semecarpifolia* leaf drop was initiated in November. In *A. spectabilis* leaf drop starts from early October and continues until April. Thus, in *B. utilis* the stable leaf population period was the shortest (3 months) and it was the longest in *A. spectabilis* (12 months). In *R. arboreum*, stable leaf population period was computed 10 months (Table 15.1). Thus, except *B. utilis,* all the species were in foliage year round except for a few weeks during April end-May in *Q. semecarpifolia*. In *A. spectabilis*, *R. arboreum* and *R. campanulatum,* no leafless stage was observed.

As recorded for bud-break and leafing, flowering activity was also advanced by about 2–3 weeks in 2017 than in 2020. *R. arboreum* (growing at the lower extremity of the treeline ecotone) was the earliest flowering species (March–April) among all the five species. In *R. campanulatum* initiation of flowering occurred in April that continued till May-end. On the whole, flowering in the tree population continues for 1–2 months. In *A. spectabilis* (Gymnosperm) male (pollen) and female (cone) and *Q. semecarpifolia* (male flowers, catkins and female parts acorns) reproductive parts start appearing in April, and this activity was completed in May. In *B. utilis,* flowering (spikes and catkins) start appearing in May and this activity continues till June-end. Thus, before the onset of monsoon all these treeline species bear full foliage and flowers facilitating pollination and fertilization so that the fruiting and fruit maturation season coincides with the warm and wet season for optimal growth of fruits and seeds (Figs. 15.2 and 15.3).

Flowering is followed by fruiting in all the species. In *A. spectabilis* and *Q. semecarpifolia,* fruiting occurs in early-May and continues till June-end. In the other three species, fruiting starts in mid-June and continues till July-end. Thus, fruiting occurs rapidly in all these species during May—July, and the entire reproductive phase of these species (initiation of flowering to fruit set) lasted for a short duration of 3 months in *B. utilis* and longest for *R. arboreum* (12 months) and *A. spectabilis* (20 months). Thus, the fruit maturation and seed dispersal phases varied considerably across these species. The earliest fruit maturation and seed dispersal take place in *Q. semecarpifolia* (July–August; viviparous seeds), followed by *R. campanulatum* (September–October) and *R. arboreum* (October–November).

Table 15.1 Leaf population and shoot growth characteristics in the treeline ecotone tree species at Tungnath

Species	Leaf and shoot growth parameters								
	Mature shoot length (cm)	Leaf number/ shoot in mature shoot	Mean leaf life span (weeks)	Duration of leafing in shoots (months)	Duration of stable leaf population in shoots (months)	Mature leaf area (cm^2 $leaf^{-1}$)	Mature leaf mass (g $leaf^{-1}$)	Mature shoot diameter (mm)	Decline in shoot diameter (%)
A. spectabilis	10.2 ± 1.7	148.3 ± 10.2	193	May–July (3)	Aug–Sept (12)	0.33 ± 0.005	0.0083 ± 0.003	6.9 ± 0.2	0.75
B. utilis	13.3 ± 2.7	5.5 ± 1.8	30	April–June (3)	July–Aug (2)	24.2 ± 3.65	0.31 ± 0.01	2.9 ± 0.12	1.5
Q. semecarpifolia	5.7 ± 1.6	6.5 ± 1.7	53	May–July (3)	Aug–Nov (4)	26.9 ± 4.41	0.37 ± 0.01	3.47 ± 0.1	1.3
R. arboreum	5.6 ± 0.9	10.6 ± 2.2	94	June–Aug (2)	July–April (10)	41.20 ± 5.89	0.81 ± 0.033	6.5 ± 0.14	0.15
R. campanulatum	4.5 ± 0.7	4.7 ± 1.2	77	June–Aug (2)	Aug– March (8)	54.9 ± 9.33	1.17 ± 0.41	5.87 ± 0.2	0.32

In *B. utilis* seed maturation occurs in next March–April (6–7 months after fruit set), and in *A. spectabilis* the cones mature and disperse seeds after 17–18 months in November–December.

15.3.2 Leaf Population Dynamics & Shoot Growth

Vegetative bud-break gives rise to unfolding of leaves. Leaves are born gradually on the shoots during the onset of growth season (April–May), and the leaf recruitment rate rises rapidly during May–June, and culminates in July-end. The leaf recruitment period in the current year shoots was computed 8–12 weeks for these species. In the case of *A. spectabilis,* the shoot bears leaves (needles) of over 3 years of age, and the annual growth of shoot is divisible by a node and an internode. Leaves are produced only in the current year terminal shoots. In this species, leaf flushing is a brief activity (~3 months), and the peak leaf number in current year shoots was recorded in August (mean across five study years = 148/shoot). Peak leaf population in current year shoots remains stable for approximately 12 months. Leaf drop begins in year II in October and continues slowly, but concentrated leaf drop takes place in III year December onward, and by the IV year December the leaf population of I year is almost dropped. In *B. utilis,* peak leaf number per shoot (mean = 5.5) remains stable only for 2 months between July and August, which rapidly declines in September, and by October the trees become leafless. In *Q. semecarpifolia,* peak leaf number per shoot (mean = 6.5) was attained in July–August, and the leaf population remains stable until November (>4 months), and by the end of next year March-end and early-April, trees become leafless. In this species, the new leaves appear in April-end or early-May, thus the trees remain leafless for 2–3 weeks. In *R. arboreum*, peak leaf population (mean = 10.6/shoot) was attained in July-end to early-August and the leaf population remain more or less stable for 9 months until next year July, and leaf drop takes place gradually from September, and by next year March all the old leaves are dropped. In *R. campanulatum,* leaf recruitment in current year shoots continues from June to August and the peak leaf number per shoot was recorded 4.7. Peak leaf population remains stable from August–March (8 months), and leaf drop starts from next year in August and by November all the old leaf population is dropped (Table 15.1).

In all these species, the shoot extension growth continued for 4–5 months till September. The peak shoot length attained by these species was lowest in *R. campanulatum* (mean = 4.5 cm) and highest for *B. utilis* (mean = 13.3 cm). The shortest shoot growth period was recorded for *B. utilis* (4 months) and the longest for *A. spectabilis* (5–6 months). Leaf density (number of leaves/10 cm shoot length) at the mature stage of the shoot was lowest for *B. utilis* (4.0) and highest for *A. spectabilis* (133.0). Diameter growth of mature shoot was recorded lowest for *B. utilis* (2.9 mm) and highest for *A. spectabilis* (6.9 mm) (Table 15.1). Shoot grows in thickness in all the species for 4–5 months until October, except for *B. utilis* (August). A marginal shrinkage in shoot diameter was recorded in these species (range = 0.2–1.5%) during November–December due to frost and chilling conditions.

15.3.3 Leaf Life Span, Leaf Area and Leaf Mass Dynamics

Across the five species mean leaf longevity was recorded lowest for *B. utilis* (30 weeks) and highest for *A. spectabilis* (193 weeks) (Table 15.1). *A. spectabilis* was the only conifer species with multi-year leaf-life span (3.7 years). In this species, leaves are produced only in current year terminal shoots, and the old shoots have leaves of over 3 years of age. Leaf population of first and second year shoots remains stable while leaf senescence and leaf drop are exhibited by third and fourth year shoots, and the leaf crops of each year are divisible by nodes and internodes.

Immediately after bud-break and leaf unfolding leaf expansion take place rapidly in the initial months of leaf formation that attained peak in July (*B. utilis*), and in August–September in rest of the species. Peak leaf area (cm^2 $leaf^{-1}$) across the species ranged from 0.33 cm^2 (*A. spectabilis*; a conifer) to 54.9 cm^2 (*R. campanulatum*) (Table 15.1). Time taken to attain peak leaf area was 3–4 months (except for 2 months in *B. utilis*). Corresponding to leaf area, gain in leaf mass was rapid in the initial months of leaf formation. The time when full leaf mass was attained ranged from July in *B. utilis* and October in *R. arboreum* and *R. campanulatum*. At the mature leaf stage, the leaf dry weight (g/leaf) was measured lowest for *A. spectabilis* (0.0083 g) and highest for *R. campanulatum* (1.17 g). Leaf mass remains stable for varying periods in these species; *B. utilis* (1 month) and *A. spectabilis* (8–9 months). In most of the species, leaf mass loss started as early as in August–September in *R. campanulatum* and in December in *Q. semecarpifolia*. Leaf mass loss was computed in the order: *A. sepctabilis* (27.8%) <*Q. semecarpifolia* (31.4%) <*R. arboreum* (35.6%) <*R. campanulatum* (38.0%) and <42.5% (*B. utilis*). Here it is notable that leaf mass loss values are based on mean values of 4 years for *B. utilis*, 3 years for (*Q. semecarpifolia*), 2 years for *R. arboreum* and *R. campanulatum* and only leaf cohort for *A. spectabilis*.

15.4 Discussion

This study on phenological behaviour (periodicity of major phenophases, leaf population, leaf life-span and shoot growth) for major tree species over a sufficiently long period (5 years.; May 2016–April 2021) has generated some insights into intra-annual and intra-species patterns among these species as influenced by the climatic condition, particularly atmospheric temperature. Mean annual temperature of 2017–2020 (MAT = 5.5 °C) and growing season mean (April–September = 9.6 ° C) recorded for the study site (Joshi et al. this volume) was much higher than the seasonal mean ground temperature of 6.7 °C recorded for high altitude climatic treelines (Körner and Paulsen 2004). Rainfall at this site was recorded unusually high (mean of 2018 and 2019 = 4632 mm). Thus our site falling in the sub-tropical latitudes was not limited by temperature as well as moisture to support a long growing season (5–6 months) as opposed to 2–3 months for some temperate areas (Montgomery et al. 2020). Logistic regression test showed that in the warmer year (MAT as well as mean growing season temperature) probability of bud-break and leafing increased significantly ($P < 0.001$) for all the species as was recorded for 2017 at our study site. However, the other growth parameters of these species such as shoot length, leaf population, leaf area and leaf mass were found unaffected ($r = <$ 0.05) with annual variations in temperature and rainfall. It may be mainly due to the insignificant difference both in MAT for 2017 (5.9 °C), 2018 (5.6 °C) and 2019 (5.1 °C), and also for the growing season mean temperature (9.7 °C in 2017; 9.52 °C in 2018; 9.66 °C in 2019) at this site (Joshi et al. this volume). It is also quite likely that soil temperature may have a greater control on the periodicity of the phenophases that requires further studies. Our limited microclimatic data set collected for each of the forest stands of these species only day time during the field visits (approximately 1 week per month across the 4 years) (range = 8.3 °C to 10.3 ° C; mean = 9.8 °C at 30 cm depth for the forest stands) are higher than for alpine treelines of 6.7 °C (Korner 2012) that may have favoured nutrient mineralization and lengthening of leafing and shoot growth period. At a global scale, an average growing season soil temperature below 6.4 ± 0.7 °C in 10 cm soil depth is supposed to limit alpine tree growth (Korner 2012), whereas the growing season mean soil temperature at a local scale exhibits a large range from 5 °C to 12 °C (Müller et al. 2016). In a year (2010) with low soil temperature in Tungnath, a slow rate of leaf expansion, shoot growth and fewer leaves in shoot of *B. utilis* were recorded (Rai et al. 2012). Rai et al. reported that the development of leaves in a shoot and leaf fall at the end of the growth period were highly correlated with soil and air temperature, precipitation and relative humidity. They found that leaf fall was delayed in years with a higher temperature and vice versa, and higher temperatures and delayed monsoon extended the overall growth period of trees. In Nepal Himalaya, growth of *B. utilis*, reckoned by tree-ring width site chronologies was found limited by moisture availability (Liang et al. 2014), which was not the case for Tungnath, a north-west facing slope that had enough snowmelt and soil moisture due to unusually high annual rainfall (mean = 4217 mm). It has been pointed out that in exceptional cases timberline trees can suffer from drought stress during the growing season (Yang et al. 2013). In the Tibetan Plateau, Yu et al. (2010) reported that winter and spring warming has resulted in delayed spring phenology. The favourable condition of rising atmospheric temperature and snowmelt resulting in enough soil moisture at our study site must have supported leafing for a longer period, which in the low-altitude forests for these conspecifics is shorter (~2 months) due to seasonal drought and diminished soil moisture (Negi and Singh 1992).

In all the five species studied the onset of growth (vegetative bud-break and leafing) occurred in the first week of May (except for *B. utilis* in April); however, the number of trees (out of 100 marked trees of each species) in the bud-break phase varied from one species to other. This phase was particularly rapid in *B. utilis* (deciduous species) and rather slow in *R. campanulatum* growing at the similar altitudes in Tungnath. However, it was quite possible that bud-break in some species might have initiated before May when the site was inaccessible because of snow cover. For this study site, Rai et al. (2013) have reported 102 ± 11.5 Julian day (April first week), the average date of bud-break, and the appearance of first leaf on 111 ± 14 JD (mid-April) in *B. utilis*. We computed the mean date of bud-break (134 ± 7.8 JD), flowering (109 ± 8.3) and leaf drop (285 ± 12.5) across 5 years of observation for the studied species in Tungnath. In the dominant trees (both evergreen and deciduous species including *Quercus* spp. and *Rhododendron arboreum*) of mid-altitude (2000–2200 m asl), zone of this region peak leafing (~90% new leaf crop in trees) is accomplished by the end of April–early May (Negi 2006). Inter-annual trend of leaf recruitment in shoots reveals that in some years it was slow following bud-break, but it gains rapidly in later stages of growth and attain leaf crop and shoot length just comparable to other years. Thus, the dependence of leaf and shoot growth parameters on low magnitude of warming seems less related that needs to be studied in the year of extreme events of warming, drought or snowfall for better understanding of phenological response of treeline ecotone tree species.

The mid-altitude tree species are also characterized by concentrated leaf drop during March–April. Thus, in the treeline ecotone species leafing and accomplishment of full leaf canopy were delayed by about 2 months as compared to the mid-altitude tree species. Also, in the evergreen treeline species peak leaf drop occurred during May–June. However, during October–November due to heavy frost and cold winds, some leaf drop was also recorded in these species. Our study site had the advantage of having both the set of species with a mixture of deciduous trees with temperate affinities (e.g. *A. spectabilis* and *Betula utilis*) and broadleaf evergreen species with sub-tropical affinities (e.g. *Q. semecarpifolia* and *R. arboreum*), which co-occur in the same environmental conditions. The former initiated growth about 1 month earlier than the latter. Thus, the particular floristic composition of the present study site allowed the comparison of tree species with distinctly different leaf habits (deciduous vs. evergreen), and phyto-geographical affinities (temperate vs. sub-tropical), which has been considered ideal for examining questions related to differential patterns between functional group of trees, particularly between deciduous and evergreen broadleaf tree species. These observations may have ecological and phylogenetic significance and require further detailed investigations.

Leaf life-span we reported here (7–45 months) presents a wide range; the minimum recorded for *B. utilis* and maximum for *A. spectabilis*. Whole-canopy mean leaf life-span and associated leaf traits that include different age-group leaves from all tree species in a forest stand might be useful for indicating adaptations of natural forests to climatic gradients (Luo et al. 2005). To explain the pattern of leaf longevity in different forests and groups of species, a number of climatic, ecological eco-physiological, bio-geographical and evolutionary hypotheses have been proposed (Williams 2000). Leaf life-span appears to have taxonomic significance, with the more primitive taxa of flowering plants tending to have longer-lived leaves than the more advanced taxa (Rogers and Clifford 1993). Considering this, *B. utilis* is more advanced taxa than *Rhododendrons* in our study area. Short leaf life-span and a greater turnover of nutrients (N resorption = 58.5%) might have supported vigorous shoot growth in *B. utilis* to maximize the photosynthetic gains in leaves within a limited life span of 7 months (Singh and Negi 2018).

Long-term observations on inter-annual changes in growth phenology of the climatically sensitive treeline are also required to supplement ground truth data to verify the satellite-derived data in view of the reports on the advancement of the average starts of growing season (SOS), the length of growing season (LOS) and the end of the growing season (EOS). For example, in Eastern Himalaya, Mohapatra et al. (2019) reported that phenology has changed significantly from the year 1982 to 2015 and the past alpine treeline ecotone is getting greener and staying greener for a longer period than it was four decades ago in response to high altitude warming, which has an implication on treeline upward shift. Also, warming-induced favourable conditions for regeneration, seedling survival and upward shift of treeline ecotone need to be studied that will have conservation implications for the timberline ecosystems (Singh et al. 2018a; b). Also, a major contemporary research theme asks whether chilling at alpine-treeline affects the ability of trees to assimilate the growth resources and particularly carbon needed for growth or whether the growth itself is limited by the alpine environment (Germino 2014).

15.5 Conclusion

This study undertaken to determine the impact of climate change on the periodicity of major phenophases and leaf and shoot growth dynamics over a period of 5 years has concluded that the major treeline species of Tungnath (west Himalaya) respond to increase in atmospheric temperature with respect to earlier onset of bud-break and leafing. In the warmer year 2017 (MAT as well as growing season temperature), probability of earlier bud-break and leafing increased significantly ($P < 0.001$) for all the species. However, differences in the inter-annual variations in the leaf and shoot growth characteristics of these species were insignificant as the MAT and growing season temperature did not vary markedly across the four study years. Thus, it can be stated that the onset of growth signal is temperature dependent even if the mild increase in MAT occurs between 1 year and the other. Growing period at our study site was recorded greater (4–5 months) than the reported (2–3 months) for the timberline in the temperate climates of the world. This study has generated baseline data on phenology and leaf characteristics of the major treeline species that will serve as a benchmark for similar studies in future to ascertain the impact of CC. However, the implications of advancement of growing period induced by the rise in MAT and growing season temperature on the ecosystem level structure and functional aspects need to be better understood involving long-term observations to ascertain which species will outcompete others and expand its population and move up under global warming. Further, coordinated studies on soil moisture and temperature regime, tree-water relations and other eco-physiological aspects need to be undertaken to precisely explain the phenological behaviour and leaf and shoot growth traits of these species to decipher the impact of climate change.

This study spanning over a sufficiently long period (5 years) points out that treeline species respond to the annual variations both in initiation and culmination of the various phenophases (particularly, leafing and flowering), which was significantly correlated ($P < 0.001$) with MAT. However, the implications of large differences during the growing season (April–August = 4.3 °C) between air (14.1 °C) and soil temperature (9.74 °C) need to be further investigated on the occurrence of phenophases of the treeline species. Also, cooler soil associated low soil nutrient mineralization must have impeded the growth and development of plants, which is still not known from the Himalayan treeline. Our preliminary study has revealed that both C and N decrease with increasing altitude and the treeline is a nutrient-limited system (mean annual soil N = 0.44%; C = 4.3%). Therefore, eco-physiological investigations on soil–plant interaction (such mismatch between flowering and pollinators, seed set and desiccation of seed and seedling mortality due to freezing of soil), coupled with loss due to leaching of nutrients and seedling removal through grazing in the inhospitable conditions of Himalayan treeline seems to be a major area of research that may be a controlling factor in establishment and survival of new recruitment and upward movement of treeline.

Acknowledgements Authors thankfully acknowledge Prof. S.P. Singh, FNA and Coordinator of the IHTP project for guiding this research and reviewing this manuscript, Dr. Rajesh Joshi, Scientist, GBPNIHE for providing meteorological data of the study site, and Mr. Shail Joshi for assistance in fieldwork. This research was funded by National Mission on Himalayan Studies, MoEF&CC, New Delhi.

References

Becker A, Korner C, Brun JJ, Gusain A, Tappeiner U (2007) Ecological and land use studies along elevational gradients. Mt Res Dev 27:59–65

Billings WD, Bliss LC (1959) An alpine snowbank environment and its effects on vegetation, plant development, and productivity. Ecology 40:388–397

Camarero JJ, Gutiérrez E (2004) Pace and pattern of recent treeline dynamics: response of ecotones to climatic variability in the Spanish Pyrenees. Clim Chang 63(1–2):181–200

Chabot BF, Hicks DJ (1982) The ecology of leaf life spans. Annu Rev Ecol Syst 13:229–259

Chhetri PK, Gaddis KD, Cairns DM (2018) Predicting the suitable habitat of treeline species in the Nepalese Himalayas under climate change. Mt Res Dev 38(2):153–163

Elzinga JA, Atlan A, Biere A, Gigord L, Weis AE, Bernasconi G (2007) Time after time: flowering phenology and biotic interactions. Trends Ecol Evol 22(8):432–439

Gaira KS, Rawal RS, Rawat B, Bhatt ID (2014) Impact of climate change on the flowering of *Rhododendron arboreum* in Central Himalaya, India. Curr Sci 106(12):1735–1738

Gaire NP, Koirala M, Bhuju DR, Borgaonkar HP (2014) Treeline dynamics with climate change at the Central Nepal Himalaya. Clim Past 10(4):1277–1290

Galán C, García-Mozo H, Vázquez L, Ruiz L, De La Guardia CD, Trigo MM (2005) Heat requirement for the onset of the *Olea europaea* L. pollen season in several sites in Andalusia and the effect of the expected future climate change. Int J Biometeorol 49(3):184–188

Germino MJ (2014) Plants in alpine environments. In: Monson R (ed) Ecology and the environment, The plant sciences, vol 8. Springer, New York, NY

Holtmeier FK (2009) Mountain timberlines, ecology, patchiness, and dynamics, 2nd edn. Springer, New York

Holtmeier FK, Broll G (2005) Sensitivity and response of northern hemisphere altitudinal and polar treelines to environmental change at landscape and local scales. Glob Ecol Biogeogr 14(5): 395–410

Isard SA (1986) Factors influencing soil moisture and plant community distribution on Niwot ridge, front range, Colorado, U.S.A. Arct Antarc Alp Res 18:83–96

Kikuzawa K (1995) Leaf phenology as an optimal strategy for carbon gain in plants. Can J Bot 73: 158–163

Korner C (2012) Alpine Treelines. Functional Ecology of the global high elevation tree limits. Springer, Basel

Körner C, Paulsen J (2004) A world-wide study of high altitude tree-line temperatures. J Biogeogr 31:713–732

Kullman L (1998) Tree-limits and montane forests in the Swedish Scandes: sensitive biomonitors of climate change and variability. Ambio 27:312–321

Liang E, Dawadi B, Pederson N, Eckstein D (2014) Is the growth of birch at the upper timberline in the Himalayas limited by moisture or by temperature? Ecology 95(9):2453–2465

Luo T, Luo J, Pan Y (2005) Leaf traits and associated ecosystem characteristics across subtropical and timberline forests in the Gongga Mountains, eastern Tibetan plateau. Oecologia 142(2): 261–273

Menard S (2002) Applied logistic regression analysis, vol 106. Sage Publications, London

Menzel A, Sparks T, Estrella N, Koch E, Aasa A, Ahas R, Chmielewski FM (2006) European phenological response to climate change matches the warming pattern. Glob Chang Biol 12(10): 1969–1976

Miller-Rushing AJ, Inouye DW (2010) Variation in the impact of climate change on flowering phenology and abundance: an examination of two pairs of closely related wildflower species. Am J Bot 96(10):1821–1829

Mir NA, Masodi TH, Mir AA, Khan H, Sofi PA, Wani FJ, Hameed OB (2016) Phenology and growth performance of Himalayan birch (*Betula utilis*) in Kashmir west Himalaya along the different altitudinal gradients. Indian J Agric Sci 86:1086–1091

Mohapatra J, Singh CP, Tripathi OP, Pandya HA (2019) Remote sensing of alpine treeline ecotone dynamics and phenology in Arunachal Pradesh Himalaya. Int J Remote Sens 40(20):7986–8009

Montgomery RA, Rice, KE, Stefanski A, Rich RL, Reich PB (2020) Phenological responses of temperate and boreal trees to warming depend on ambient spring temperatures, leaf habit, and geographic range. Proc Natl Acad Sci 117(19):10397–10405

Müller M, Schwab N, Schickhoff U, Böhner J, Scholten T (2016) Soil temperature and soil moisture patterns in a Himalayan alpine treeline ecotone. Arct Antarc Alp Res 48:501–521

Nautiyal MC, Nautiyal BP, Prakash V (2001) Phenology and growth form distribution in an alpine pasture at Tungnath, Garhwal, Himalaya. Mt Res Dev 21:168–174

Negi GCS (2006) Leaf and bud demography and shoot growth in evergreen and deciduous trees of central Himalaya, India. Trees- Struct Funct 20:416–429

Negi GCS, Singh SP (1992) Leaf growth pattern in evergreen and deciduous species of the central Himalaya, India. Int J Biometeorol 36(4):233–242

Rai ID, Adhikari BS, Rawat GS, Bargali K (2012) Community structure along timberline ecotone in relation to micro-topography and disturbances in Western Himalaya. Not Sci Biol 4:41–52

Rai ID, Bharti RR, Adhikari BS, Rawat GS (2013) Structure and functioning of timberline vegetation in the Western Himalaya: a case study. In: high altitude rangelands and their interfaces in the Hindu Kush Himalayas. ICIMOD, Nepal, pp 91–107

Rawal RS, Bankoti NS, Samant SS, Pangety YPS (1991) Phenology of tree layer species from the timberline around Kumaun in central Himalaya, India. Vegetatio 93:109–118

Rogers RW, Clifford HT (1993) The taxonomic and evolutionary significance of leaf longevity. New Phytol 123:811–821

Rosenzweig C, Karoly D, Vicarelli M, Neofotis P, Wu Q, Casassa G, Tryjanowski P (2008) Attributing physical and biological impacts to anthropogenic climate change. Nature 453(7193):353–357

Schickhoff U, Bobrowski M, Böhner J, Bürzle B, Chaudhary RP, Gerlitz L, Heyken H, Lange J, Müller M, Scholten T, Schwab N, Wedegärtner R (2015) Do Himalayan treelines respond to recent climate change? An evaluation of sensitivity indicators. Earth Syst Dyn 6:245–265

Schwartz MD (1999) Advancing to full bloom: planning phenological research for the 21st century'. Int J Biometeorol 42:113–118

Schwartz MD, Ahas R, Aasa A (2006) Onset of spring starting earlier across the northern hemisphere. Glob Chang Biol 12(2):343–351

Shrestha UB, Gautam S, Bawa KS (2012) Widespread climate change in the Himalayas and associated changes in local ecosystems. PLoS One 7(5):e36741

Singh KP, Kushwaha CP (2006) Diversity of flowering and fruiting phenology of trees in a tropical deciduous forest in India. Ann Bot 97(2):265–276

Singh P, Negi GCS (2018) Seasonal changes in leaf area and leaf mass of dominant trees of Western Himalayan forests. Indian For 144(2):182–190

Singh P, Arya V, Negi GCS, Singh SP (2018a) Expansion of *Rhododendron campanulatum* krummholz in the treelineecotone in Tungnath, Garhwal Himalaya. Trop Ecol 59(2):287–295

Singh P, Negi GCS, Singh SP (2018b) Treeline species phenology: shoot growth, leaf characteristics and nutrient dynamics. Trop Ecol 59(2):297–311

Sundriyal RC, Joshi AP, Dhasmana R (1987) Phenology of high altitude plants at Tungnath in the Garhwal Himalaya. Trop Ecol 28:289–299

Telwala Y, Book WB, Kumar M, Pandit MK (2013) Climate induced elevational range shifts and increase in plant species richness in a Himalayan biodiversity epicenter. PLoS One 8:e57103

Williams LG (2000) Leaf demography and leaf traits of temperate-deciduous and tropical evergreen-broadleaved trees in a Mexican montane cloud forest. Plant Ecol 149(2):233–244

Yang B, He M, Melvin TM, Zhao Y, Briffa KR (2013) Climate control on tree growth at the upper and lower treelines: a case study in the Qilian Mountains, Tibetan plateau. PLoS One 8(7): e69065

Yu H, Luedeling E, Xu J (2010) Winter and spring warming result in delayed spring phenology on the Tibetan plateau. Proc Natl Acad Sci U S A 107:22151–22156

Water Relations of the Indian Himalayan Treeline Species

16

Ashish Tewari, Shruti Shah, Nandan Singh, Amit Mittal, and Krishna Kumar Tamta

Abstract

This chapter deals with study the variation in water relation and drought-adaptive mechanisms of five Himalayan treeline species *Quercus semecarpifolia* Sm., *Abies spectabilis* D. Don., *Betula utilis* D. Don and *Rhododendron campanulatum* D. Don and *R. arboreum* Wall. growing at elevation ranging from 3200 to 3450 masl. Soil water potential, tree water potential, water potential components and leaf conductance were measured for 3 years (2016 to 2018). Across all seasons and years, the soils were mostly moist. Moderate stress was observed in summer and winter seasons. The treeline species were never severely stressed during the study period and the Ψ_{PD} was generally above −0.99 MPa. The magnitude of diurnal change across all species was maximum during the rainy season (−0.72 MPa to −0.82 MPa) except in *Q. semecarpifolia*. The diurnal pattern of water potential during winter seasons indicated that water potential can become severely low during 8.30 a.m. and 10.30 a.m. and ranged between −1.62 ± 0.09 MPa in *R. campanulatum and* − 2.10 ± 0.03 MPa in *R. arboreum.* All treeline species showed osmotic adjustment between winters and spring season which coincided with the commencement of the phenological activities in majority of tree species. A smaller adjustment occurred in all species during rainy to winter season also.

A. Tewari (✉) · S. Shah · N. Singh
Department of Forestry & Environmental Science, D.S.B. Campus, Kumaun University, Nainital, Uttarakhand, India

A. Mittal
Department of Allied Sciences, Graphic Era Hill University, Bhimtal Campus, Nainital, Uttarakhand, India

K. K. Tamta
Central Himalayan Environment Association (CHEA), Nainital, Uttarakhand, India

S. P. Singh et al. (eds.), *Ecology of Himalayan Treeline Ecotone*,
https://doi.org/10.1007/978-981-19-4476-5_16

Keywords

Leaf conductance · Osmotic adjustment · Osmotic potential · Water potential

16.1 Introduction

Distribution of plants is strongly influenced by climate (Houle 1994; Walck et al. 2011). Precipitation and temperature are critical drivers for plant distribution at the global scale (Woodward 1987). The treeline marks the altitudinal limit of the tree life form. Trees growing at the treeline are exposed to various stress factors like short growing season, high wind speed, intense radiation and high snow load (Mayr et al. 2006a; b). Plant water relations at the treeline reflect the special climatic conditions of this alpine ecotone. There is evidence that water deficits can reach lethal levels in the northern hemisphere treeline in winters (Tranquillini 1976) and become a major cause of shoot die back. The Himalayas are warming 2–5 times more than global average rate and the degrees of temperature rise increase with altitude (Yao and Zhang 2012). Rising temperature and enhanced evapo-transpiration rates might diminish annual precipitation and result in more rain than snow and result in earlier snowmelts (Kapnick and Hall 2010) and increase in proportion of rain to snow means that tree will probably become more drought- stressed during the growing season. Drought can cause major damage to plant communities (Adams et al. 2010) although species-specific damage thresholds are not yet fully understood (Anderegg and Callaway 2012). Contrary to several researches, there are studies to indicate that trees growing at the alpine treeline very seldom undergo severe water stress because of high precipitation during the vegetative period (Anfodillo et al. 1998). However, since trees are adapted to moist conditions, even moderate water deficit may lead to a strong reduction in transpiration at the treeline.

Tree water potential and its components play an important role in the physiology and metabolism of plants (Kramer and Boyer 1995). Predawn water potential (Ψ_{PD}) varies along environmental gradients (Korner and Cochrane 1985), indicating how plants integrate soil water availability; hence it is a useful measure of plant water status which can be correlated with leaf conductance (Poudyal et al. 2004). The relation between water availability and sensitivity to water stress greatly affects the phenological activities of plants (Holbrook et al. 1995). The Himalayan treeline species have generally not been studied for the impact of drought and its possible repercussions on forest regeneration tree phenology and physiology. The data available on this aspect are from European temperate region which are very different from the Himalayan region.

In this study, we collected data to assess the level of stress faced by the major treeline species of Himalayan region and also tried to understand the adaptation mechanism of these species against the stress. We hypothesize that treeline species will be less affected by water stress because of low evapo-transpiration losses in treeline area. The tree ring width studies indicate that pre-monsoon drought affects growth in treeline areas (Singh et al. 2018). This study is the first of its kind in the

Indian Himalayan treeline region to develop understanding of tree water relations in treeline areas and provide baseline data to determine the impact of global warming on Himalayan treeline.

16.2 Material and Methods

16.2.1 Study Site

This study was carried out at Tungnath, situated between 30°49′22″ N latitude and $79^0 21'47''$ E longitude at an altitude between 3200 and 3500 m. The treeline species investigated were *Quercus semecarpifolia*, *Abies spectabilis*, *Betula utilis*, *Rhododendron campanulatum* and *R. arboreum*. The altitudinal range covered represents a transition from closed canopy temperate forest to open canopy sub-alpine forests. In this site the anthropogenic disturbance is very high in the form of grazing (including migratory grazing). Lopping of trees for fuel wood and fodder, removal of litter and tourist activities during summer season are the other factors. Besides Tungnath, two seasons' data (summer and autumn) were also collected from Chitkul, Himachal Pradesh which represents dry inner valleys of the region with annual precipitation about 2500 mm. The Chitkul site is situated between 3390 and 3460 m at 31°20′965″ N latitude and 78°26′243″ E longitude.

16.2.2 Climate

The climate of Tungnath site is characterized by short cool summers and long severe winters. Severe frosts, blizzards and hailstorms are common for most of the year (autumn, winter and spring season). The precipitation occurs as snow, hail, heavy rain and showers over the year. Snowfall occurs during December to March and snowmelt starts from April and May providing abundance of soil water prior to the monsoon period. Maximum rainfall occurs in July–August (Nautiyal et al. 2001). At the study site mean annual temperature was 6.65 °C and average annual precipitation was about 2410 mm of which 90% was received from June to September (Rai et al. 2012). However, air temperature as low as −7.0 to −10.5 °C was common on certain days at Tungnath site during end of December to mid of February.

16.3 Methods

16.3.1 Soil Water Potential

Soil water potential was measured at 3500 m (Site A) and 3300 m (Site B) with Psypro water potential system at 10 cm, 30 cm and 60 cm depths following Singh et al. (2006).

16.3.2 Tree Water Potential

Twig water potential was measured using a pressure chamber (PMS Instrument Co. model 1000, range 70 bars) at predawn (Ψ_{PD}) (5.30–6.30 a.m.) and in the midday (1.30–2.30 p.m.) (Ψ_{MD}) following Tewari (1999) and Zobel et al. (2001). Tree water potential was measured on five representative trees of each of the selected species across the seasons. Sampled trees were 15–25 m tall and measurements were made on twigs located 2–3 m from the ground. However, *R. campanulatum* individuals were 1–2 m tall.

16.3.3 Diurnal Change in Water Potential

Diurnal changes in water potential of trees and seedlings were measured during winter season on the lower side of treeline. The diurnal changes in twig water potential were measured at every 2 h interval. The first reading was taken at 6.30 a.m. in the morning and last at 2.30 p.m. in the afternoon.

16.3.4 Pressure–Volume Curves (P–V Curves)

P–V curves were prepared following the bench drying method from overnight saturated twigs. Osmotic potential at full turgor (OP_f), the osmotic potential at zero turgor (OP_z) and RWC% at turgor loss point (RWC_z) were calculated from P–V curves following Pallardy et al. (1991), Tewari (1998) and Tewari et al. (2018).

16.3.5 Leaf Conductance

Leaf conductance measurements were made seasonally, using AP_4 porometer (Delta-T Devices). The porometer was calibrated against a perforated plastic plate of known conductance (Poudyal et al. 2004). Three leaves were marked on each sample tree and leaf conductance was measured at morning (10.30–11.30 a.m.) and afternoon (1.30–2.30 p.m.) following Tewari (1998). The instrument AP_4 type diffusion porometer could not measure the leaf conductance of conifers hence the conductance of only broad-leaved species was measured. As the deciduous *B. utilis* was leafless during autumn and winter season, leaf conductance in these two seasons were not estimated.

The data were subjected to analysis of variance with a 95% confidence level using SPSS version 2016. Species, seasons and year were the factors used for ANOVA.

16.4 Result

16.4.1 Soil Water Potential

Soil water potential varied significantly across seasons and depths ($P < 0.01$). Across all seasons and years, the soils were mostly moist, and only during summer and winter some moderate stress was observed. At site A the most negative water potential was in the top layer during summer season (−3.05 ± 0.25Mpa and −2.8 ± 0.08 MPa) whereas at site B it was in winter (−3.40 ± 0.15 MPa) and summer season (−2.5 ± 0.02 MPa). Across the two sites and seasons at 30 cm depth the soil water potential ranged between −0.67 ± 0.03 MPa (rainy season) and −2.79 ± 0.13 MPa (winter season) (Fig.16.1). At 60 cm depth soil water potential was between −0.58 ± 0.02 MPa (rainy season) and − 2.59 ± 0.15 MPa (winter season).

16.4.2 Tree Water Potential

The tree water potential (Ψ) varied significantly across species and seasons ($P < 0.01$). The timberline species were never severely stressed during the study period and the Ψ_{PD} was generally above −0.99 MPa. In *Q. semecarpifolia* (−0.99 ± 0.05 MPa), *A. spectabilis* (−0.89 ± 0.03 MPa) and *B. utilis* (−0.76 ± 0.02 MPa) minimum Ψ_{PD} was during the summer season and in *R. arboreum* (−0.86 ± 0.03 MPa) and *R. campanulatum* (−0.80 ± 0.03 MPa) during the winter season (Table 16.1). The highest Ψ_{PD} in all species was observed during rainy season and the minimum generally during summer or winter seasons, and the minimum Ψ_{MD} values in all species except *B. utilis* were encountered during winter season (−1.24 ± 0.02 MPa and − 1.50 ± 0.06 MPa). *B. utilis* had the most negative water potential in autumn (−1.06 ± 0.03 MPa) just before the commencement of leaf fall (mid-September).

The daily change ($\Delta\Psi = \Psi_{MD} - \Psi_{PD}$) differed significantly among species and seasons ($P < 0.01$). The magnitude of diurnal change across all species was

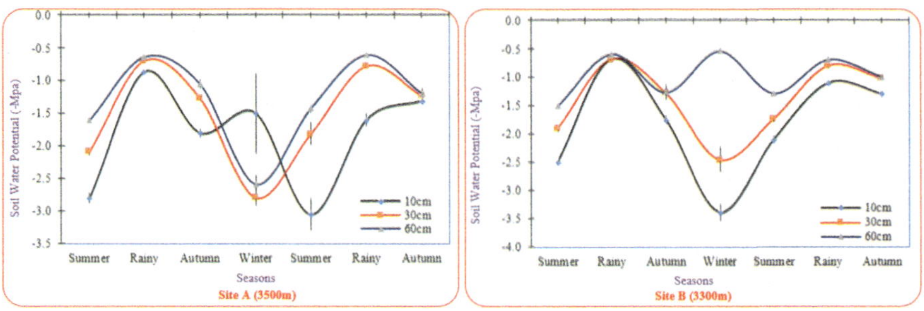

Fig. 16.1 Soil water potential at two different elevations at Tungnath site across different seasons and soil depths

Table. 16.1 Variation in predawn and midday water potential (-MPa) and daily change (-MPa) in treeline species (The values represented are mean of 3 years)

Species	Season	Predawn water Potential (-MPa)	Midday water Potential (-MPa)	Daily change (MPa)
Q. Semecarpifolia	Summer	−0.99 ± 0.05	−1.21 ± 0.05	0.22
	Rainy	−0.25 ± 0.02	−1.02 ± 0.03	0.78
	Autumn	−0.69 ± 0.02	−1.09 ± 0.03	0.40
	Winter	−0.57 ± 0.01	−1.37 ± 0.04	0.80
	Spring	−0.34 ± 0.01	−1.15 ± 0.02	0.81
A. spectabilis	Summer	−0.89 ± 0.03	−1.09 ± 0.03	0.20
	Rainy	−0.16 ± 0.03	−0.99 ± 0.03	0.82
	Autumn	−0.51 ± 0.03	−1.16 ± 0.03	0.65
	Winter	−0.69 ± 0.02	−1.24 ± 0.02	0.55
	Spring	−0.43 ± 0.02	−1.03 ± 0.04	0.60
R. campanulatum	Summer	−0.72 ± 0.03	−1.15 ± 0.04	0.43
	Rainy	−0.22 ± 0.01	−0.99 ± 0.04	0.77
	Autumn	−0.60 ± 0.02	−1.25 ± 0.03	0.65
	Winter	−0.80 ± 0.03	−1.50 ± 0.06	0.70
	Spring	−0.50 ± 0.02	−0.99 ± 0.05	0.49
R. arboreum	Summer	−0.80 ± 0.03	−1.00 ± 0.06	0.19
	Rainy	−0.31 ± 0.03	−1.11 ± 0.03	0.80
	Autumn	−0.47 ± 0.03	−0.92 ± 0.03	0.48
	Winter	−0.86 ± 0.03	−1.35 ± 0.02	0.49
	Spring	−0.32 ± 0.02	−1.11 ± 0.02	0.79
B. utilis	Summer	−0.76 ± 0.02	−0.88 ± 0.02	0.12
	Rainy	−0.18 ± 0.01	−0.90 ± 0.03	0.72
	Autumn	−0.58 ± 0.04	−1.06 ± 0.03	0.47
	Winter	−0.34 ± 0.02	−0.63 ± 0.02	0.29
	Spring	−0.33 ± 0.02	−0.91 ± 0.04	0.58

maximum during the rainy season except in *Q. semecarpifolia*, in which it was during the spring season 0.81 MPa (Table 16.1). The species varied in seasons and magnitude of diurnal change. *B. utilis*, the only deciduous species, had the highest daily change during the rainy season (0.72 MPa).

Data for tree water potential at Chitkul treeline were collected for two seasons, summer and autumn. During summer season, the Ψ_{PD} of trees ranged between −0.94 ± 0.05 MPa and −1.1 ± 0.02 MPa and of seedlings between −0.80 ± 0.04 MPa and −1.4 ± 0.01 MPa (Table 16.2). The Ψ_{PD} of trees at this site during summers were relatively more negative than Tungnath treeline. During autumn season, the Ψ_{PD} of trees ranged between −0.15 ± 0.01 MPa and −0.26 ± 0.02 MPa and of seedlings between −0.13 ± 0.01 MPa and −0.21 ± 0.01 MPa (Table 16.2). During autumn the Ψ_{PD} was more negative in Tungnath treeline than Chitkul treeline.

Table 16.2 Variation in predawn and midday water potential (-MPa) in trees and seedlings of *A. spectabilis*, *R. campanulatum* and *B. utilis* during summer and autumn seasons at Chitkul treeline

		Water Potential (-MPa)			
		Trees		Seedlings	
Species	Season	Predawn	Midday	Predawn	Midday
A. spectabilis,	Summer	−0.94 ± 0.05	−1.3 ± 0.03	−1.4 ± 0.01	−1.6 ± 0.05
	Autumn	−0.26 ± 0.02	−1.06 ± 0.01	−0.21 ± 0.01	−1.15 ± 0.04
R. campanulatum	Summer	−0.86 ± 0.04	−1.6 ± 0.05	−0.80 ± 0.04	−1.4 ± 0.04
	Autumn	−0.15 ± 0.01	−1.27 ± 0.04	−0.13 ± 0.01	−1.16 ± 0.02
B. utilis.	Summer	−1.1 ± 0.02	−1.9 ± 0.01	−1.3 ± 0.03	−1.8 ± 0.03
	Autumn	−0.19 ± 0.01	−0.96 ± 0.03	−0.18 ± 0.01	−1.44 ± 0.03

16.4.3 Winter Soil Temperature and Diurnal Pattern of Water Potential

Below zero-degree air temperatures are common during the winter seasons in the treelines around the world and frozen soil is a likely outcome. The soil temperature at 30 cm soil depth during the winter season generally ranged between 1.2 °C to −3.8 °C. The temperatures were lower at microsites where snow was less or absent. The diurnal pattern of water stress in trees during winter followed a pattern where Ψ_{PD} values between 6 and 6.30 a.m. were maximum and then declined significantly between 8 and 10 a.m. Thereafter, there was a recovery and Ψ_{MD} in the late afternoon was similar to Ψ_{PD}. In all the species except *B. utilis* the minimum water potential was during 8.30 and 10.30 a.m. and in *B. utilis* at 12.30 p.m.; thereafter there was recovery in water potential. The minimum water potential was −2.13 ± 0.03 MPa in *Q. semecarpifolia*, −1.94 ± 0.05 MPa in *A. spectabilis*, −2.25 ± 0.07 MPa in *R. campanulatum*, −2.10 ± 0.03 MPa in *R. arboreum* and −0.71 ± 0.02 MPa in *B. utilis* during winters of 2 years (Fig. 16.2). However, in the other seasons diurnal water potential pattern showed the normal trend and water potential rose gradually from pre-dawn to midday being least negative in pre-dawn and most negative during midday.

16.4.4 Water Potential Components

The water potential parameters (osmotic potential at full turgor and zero turgor and relative water content %) differed significantly across species and seasons ($P < 0.01$). All the treeline species showed a small osmotic adjustment between the rainy and autumn season and a larger adjustment between winter and spring season. The magnitude of adjustment in *B. utilis* at zero turgor was −0.17 MPa and −0.57 at full turgor. In all other species the osmotic adjustments between rainy and autumn season both at zero and full turgor was above −0.72 MPa. A large decline in osmotic potential was evident during winter to spring being the maximum in *B. utilis* (−1.24 MPa) followed by *A. spectabilis* (−1.22 MPa), *R. arboreum* (−0.82 MPa)

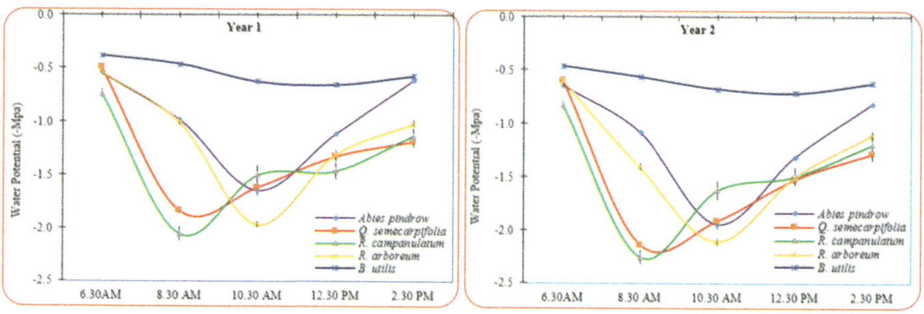

Fig. 16.2 Diurnal pattern of water potential of treeline species in winter season during Year 1 and Year 2

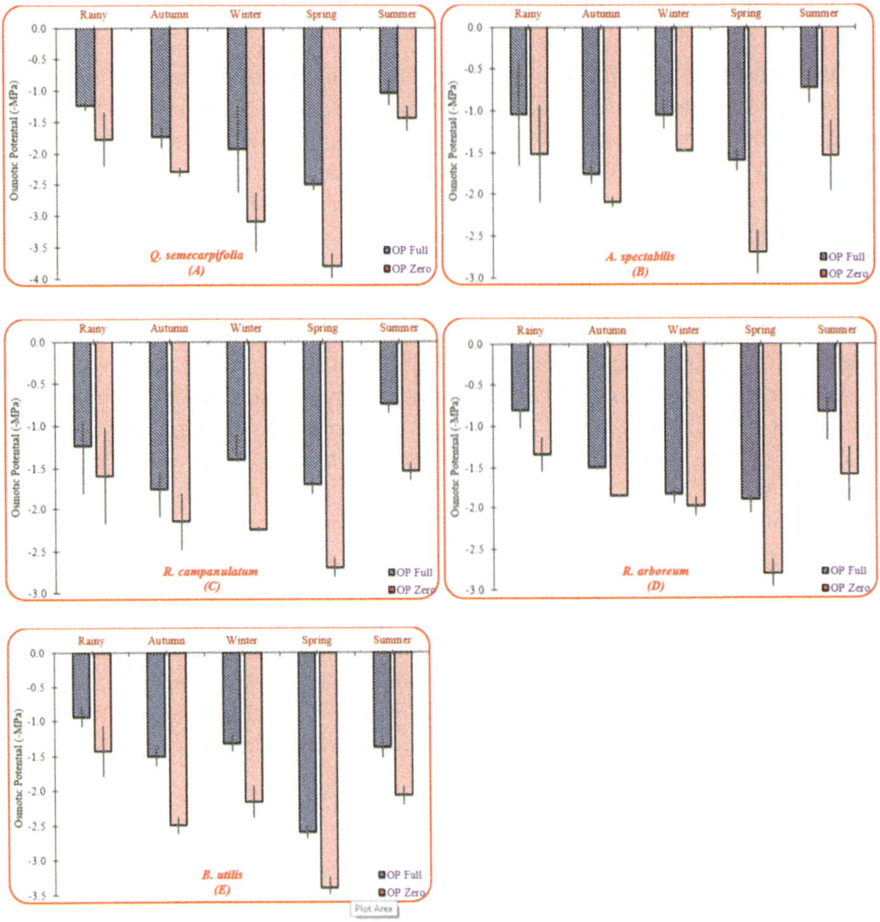

Fig. 16.3 (**a**–**e**) Seasonal changes in osmotic potential at full turgor and at zero turgor in treeline species. The data are average of 2 years. Figure (**a**) *Q. semecarpifolia*, (**b**) *A. spectabilis*, (**c**) *R. campanulatum*, (**d**) *R. arboreum* and (**e**) *B. utilis*

and *Q. semecarpifolia* (0.7 MPa) at zero turgor (Fig. 16.3). *R. campanulatum*, however, showed minimal capacity to adjust both at zero and full turgor. The RWC% across all studied seasons was generally high and ranged between 54.63 ± 1.46% during the summer season in *Q. semecarpifolia* and 93.26 ± 0.74% in *R. arboreum* during the autumn season. It was only in *Q. semecarpifolia* where RWC at zero turgor declined to 54.63 ± 1.46% during summer and 58.89 ± 5.99% during the autumn season.

16.4.5 Leaf Conductance

As the deciduous *B. utilis* was leafless during autumn and winter seasons leaf conductance in these two seasons were not estimated. Leaf conductance was measured over four seasons: summer, rainy, autumn and winter season for the other species. Morning conductance varied significantly across seasons ($P < 0.01$) but there was no significant variation across species. The afternoon conductance varied significantly across species as well as seasons ($P < 0.01$). In *Q. semecarpifolia*, *R. arboreum* and *B. utilis* the morning conductance was maximum during the rainy season and ranged between 204.2 ± 14.3 and 319.6 ± 15.46 m mol/m^2/s; however, in *R. campanulatum* it was in the summer season (175.9 ± 3.9 m mol/m^2/s) (Table 16.3).

In *Q. semecarpifolia* the maximum afternoon conductance was in autumn season (268.6 ± 15.1 m mol/m^2/s), in *R. arboreum* during rainy season (268.0 ± 24.7 m mol/m^2/s) and in *R. campanulatum* (243.5 ± 6.4 m mol/m^2/s) and *B. utilis* (314.8 ± 15.2 m mol/m^2/s) during summer season (Table 16.3).

At Chitkul treeline the morning leaf conductance values during the summer season in *R. campanulatum* and *B. utilis* trees were 165.0 ± 5.4 and 215.0 ± 23.1 m mol/m^2/s and afternoon conductance was 96.0 ± 4.2 and 132.3 ± 10.4 m mol/m^2/s which were similar to Tungnath site (Table 16.4). The morning leaf conductance of trees for autumn season was very high at Chitkul site. During autumn season the morning conductance of *R. campanulatum* was 783.8 ± 105.4 m mol/m^2/s and of *B. utilis* was 667.1 ± 123.1 m mol/m^2/s. The seedling leaf conductance for these species was exceptionally high 1147.7 ± 59.0

Table. 16.3 Seasonal patterns of morning and afternoon leaf conductance (m mol/m^2/s) in the treeline species at Tungnath treeline

Species	Seasons	Leaf conductance (m mol/m^2/S)	
		Morning	Afternoon
Q. semecarpifolia	Summer	43.9 ± 3.31	98.5 ± 4.23
	Rainy	204.2 ± 14.3	265.6 ± 10.1
	Autumn	152.6 ± 17.1	268.6 ± 15.1
	Winter	101.6 ± 4.01	49.5 ± 2.81
R. campanulatum	Summer	175.9 ± 3.9	243.5 ± 6.4
	Rainy	172.0 ± 10.1	151.05 ± 14.9
	Autumn	48.01 ± 4.26	115.7 ± 11.4
	Winter	28.9 ± 1.27	16.2 ± 2.34
R. arboreum	Summer	70.7 ± 4.9	129.5 ± 12.7
	Rainy	247.5 ± 42.3	268.0 ± 24.7
	Autumn	143.2 ± 34.6	230.9 ± 28.4
	Winter	149.7 ± 3.70	108.7 ± 5.5
B. utilis	Summer	210.9 ± 19.5	314.8 ± 15.2
	Rainy	319.6 ± 15.46	240.05 ± 12.8

Table. 16.4 Morning and afternoon leaf conductance (m mol/m^2/s) of *R. campanulatum* and *B. utilis* trees and seedlings during summer and autumn season at Chitkul treeline

		Leaf conductance (m mol/m^2/s)			
		Tree		Seedling	
Species	Seasons	Morning	Afternoon	Morning	Afternoon
R. campanulatum	Summer	165.0 ± 5.4	94.0 ± 6.6	96.0 ± 4.2	55.2 ± 7.7
	Autumn	783.8 ± 105.4	170.8 ± 66.7	1147.7 ± 59.0	245.2 ± 27.7
B. utilis	Summer	215.0 ± 23.1	103.0 ± 6.5	132.3 ± 10.4	93.7 ± 4.3
	Autumn	667.1 ± 123.1	210.1 ± 36.5	1323.3 ± 100.4	347.5 ± 14.03

and 1323.3 ± 100.4 m mol/m^2/s (Table 16.4). However, by afternoon leaf conductance values had declined by about 70% compared to morning conductance.

16.5 Discussion

The severity of drought stress can be judged by minimum level of water potential and the best indicator could be the mean water potential values of a specific species, as data of a single year or season might distort the water potential values as it is highly influenced by the environmental drivers (Hinckley et al. 1978). The data presented in the paper are the mean of 3 years and summer, rainy, autumn, winter and spring season for all parameters except water potential components (2 years).

The soil was generally moist across all seasons as is evident from soil water potential values ranging between −0.53 ± 0.02 MPa and −3.40 ± 0.15 MPa across all seasons and depths. The top layer of the soil had the most negative water potential that ranged between −1.29 ± 0.05 MPa and 3.40 ± 0.15 MPa across all seasons except the rainy season when soil water potential was relatively high. As the upper layer of the soil (0–10 cm) dries earlier because of evapo-transpiration the soil water potential values were always more negative at 10 cm soil depth than 60 cm. The most negative soil water potential −3.40 ± 0.15 MPa was recorded during the winter season in the top soil layer at 3400 m followed by the summer season −3.05 ± 0.25 MPa. The values of soil water potential during the summer season at the treeline sites are much higher than that of the low elevation sites where soil water potential can decline to relatively low level particularly in banj oak and pine forest. Tewari (1998) also reported very low soil water potential in banj oak forest (−5.6 MPa).

The mean Ψ_{PD} at the Himalayan treeline never declined below −1.2 MPa and it can be concluded from the present data that Himalayan treelines are not highly water stressed. Even during the winter months, the Ψ_{PD} values remained quite high. Compared to this, much lower winter water potential values have been reported for *Juniperus communis* and *Larix decidua* (less than −3.0 MPa) even though the latter is a deciduous species (Mayr et al. 2006a; b). In our deciduous species, *B. utilis* predawn water potential was Ψ_{PD} −0.34 ± 0.02 MPa. Richard and Bliss (1986) reported very low water potential −5.3 MPa during winters for *Larix lyallii;* however, *Pinus cembra* and *P. mugo* showed moderate water potential during winters.

All the investigated treeline species maintained a high daily change ($\Delta\Psi = \Psi_{MD} - \Psi_{PD}$) particularly during the rainy and winter seasons when the values ranged between 0.70 MPa and 0.82 MPa. In treeline species during rainy season both predawn water potential and daily change in it was the highest. While in lower forest species predawn water potential values were high but daily change in water potential is the lowest (Singh et al. 2006). This is in contrast to the species of lower forests in which daily change is minimal during the rainy season due to overcast skies and high atmospheric humidity (Tewari 1998). The treeline *Q. semecarpifolia* and *R. campanulatum* evergreen broad-leaved species even during the winter season had relatively high daily change varying between 0.70 to 0.80 MPa followed by the autumn. The daily change was minimum in all species during the summer season. It appears that all treeline species make maximum use of the growing season for dry matter accumulation.

Diurnal pattern of water potential during winter season showed pronounced decline between 8.30 and 10.30 a.m. across all tree species and this decline from predawn water potential ranged between −0.71 ± 0.02 MPa and − 2.25 ± 0.07 MPa. Replenishment of water potential began after 10.30 a.m. The possible explanation of this phenomenon can be that as sunlight strikes the leaves transpiration commences and exceeds the absorption rate. Absorption of water is impeded by frozen soil around roots. As air temperature rises and the frozen soil water gets thawed the water potentials begin to rise. The pattern of diurnal change (the water potential measured at every 2 h interval throught the day from pre-dawn to late in the afternoon) in water potential in the treeline areas is entirely different from the pattern in the tree species located between 1800 and 2200 m where most negative water potential are generally attained around midday.

All the treeline species lowered their osmotic potential during the spring season which coincides with the initiation of phenological activities in almost all species. However, they differed in the magnitude of osmotic potential lowering. *Q. semecarpifolia* showed a continuous decline in osmotic potential at both full and zero turgor adjusting by −1.26 MPa at full turgor and −2.02 MPa at zero turgor. In north American oak, *Q. Alba,* the values reported for maximum adjustment in full turgor and zero turgor are −1.30 MPa and −1.5 MPa (Parker et al. 1982). *A spectabilis* showed a much smaller adjustment between autumn and winter and larger decline −1.22 MPa at zero turgor between winter and spring. In *B. utilis* the magnitude of adjustment was amongst the highest in all treeline species being − 1.66 MPa at full turgor and −2.0 MPa at zero turgor. There being a continuous decline in osmotic potential from rainy to spring season. In *R. arboreum* and *R. campanulatum* winter to spring time change in osmotic potential both at zero and full turgor rise was minimal.

All the broad-leaved species carried out leaf conductance throughout the year but differed in the degree of fall of leaf conductance in response to adverse environmental factors. Keeping stomata open throughout the year signified that these species carry out photosynthesis throughout the year and accumulate dry matter for eventual use and growth. In *Q. semecarpifolia*, *R. arboreum* and *B. utilis* maximum conductance was during rainy season whereas in *R. campanulatum* during summer season. The deciduous *B. utilis* maintained high morning leaf conductance during the late spring season and appears to make maximum use of snow melt water. Its leaves are fully expanded by that time. Poudyal et al. (2004) had reported morning leaf conductance from 19 to 328 m mol/m^2/s in *Schima wallichii* and *Q. semecarpifolia.* The morning leaf conductance values of this study 43.9 ± 3.31 and 319.6 ± 15.4 m mol/m^2/s fall within the range to values reported by Poudyal et al. (2004). In a study carried out in the Nepal Himalayan region Poudyal et al. (2004) reported the leaf conductance value of *R. arboreum* between 28 and 219 m mol/m^2/

s the values of this study 70.7 ± 4.9 and 268.0 ± 24.7 m mol/m^2/s are similar to those values.

At the drier Chitkul treeline site very high morning leaf conductance was observed in seedlings and trees of *B. utilis* (1323.3 ± 100.4 m mol/m^2/s) and *R. campanulatum* (1147.7 ± 59.0 m mol/m^2/s) in mid-September. In trees the conductance was 667.06 ± 123.1 m mol/m^2/s in *B. utilis* and 783.86 ± 105.4 m mol/m^2/s in *R. campanuatum*. Comparable values of leaf conductance in trees ranging from 420 m mol/m^2/s to 450 m mol/m^2/s were reported in deciduous trees, Fraxinus *excelsior* and *Ulmus laevis* in well-watered and drought-treated sites by Eller et al. (2016). However, the exceptionally high conductance values as we observed in seedling have not been reported elsewhere. The lower atmospheric pressure at approximately 3400 m elevation of this treeline could also be contributing to these high conductance rates. Compared to this, timberlines in Europe and other areas of high latitudes are located at much lower elevation. The afternoon decline in leaf conductance in both trees and seedlings is more than 70%, indicating that the high conductance level was maintained only briefly.

16.6 Conclusion

The Himalayan treeline species are not severely water stressed, instead are subjected only to moderate stress which occurs during the summer and winter seasons. Some of our findings are difficult to explain and might need more observation. For example, exceptionally high leaf conductance at a dry site (Chitkul) and high daily change in tree water potential during monsoon period need more investigations for better explanation. During this period, trees in lower forests of the Himalayas hardly show any change in water potential. Studies in the inner Himalaya and low rainfall treeline areas across several sites are warranted to further explain the impact of water stress in the Himalayan region where stress may reach critical level and cause seedling mortality resulting in depression of the Himalayan timberlines. Long-term monitoring of water relation parameters is required in the treeline areas to understand the impacts of climate change and to explain the unanswered questions that have been raised by the this study. Reduced snow duration, snow cover and depth and enhanced evapo-transpiration rates may significantly impact the younger small-sized individuals that grow in the most severe microenvironment, an area where limited data could be generated. More frequent data collection (fortnightly) is required to relate the impact of tree water on the timing of phenological events of species.

Acknowledgement The authors of this chapter are thankful to Prof. S.P. Singh, Coordinator, NMHS-IHTRP for his constant guidance and valuable inputs in improving the content of this manuscript and Head, Department of Forestry & Environmental Science, D.S.B. Campus, Kumaun University, Nainital for providing laboratory facilities. The Financial support received from National Mission on Himalayan Studies (NMHS), Ministry of Environment, Forest and Climate Change, Government of India, is gratefully acknowledged.

References

Adams HD, Macalady AK, Breshears DD, Allen CD, Stephenson NL, Saleska SR, Huxman TE, McDowell NG (2010) Climate-induced tree mortality: earthsystem consequences. EOS Trans Am Geophys Union 91:153–154

Anderegg WRL, Callaway E (2012) Infestation and hydraulic consequences of induced carbon starvation. Plant Physiol 159:1866–1874

Anfodillo T, Rento S, Carraro V, Furlanetto L, Urbinati C, Carrer M (1998) Tree water relations and climatic variations at the alpine treeline: seasonal changes of sap flux and xylem water potential in *Larix deciduas* miller, *Picea abies* (L.) karst. And *Pinus cembra* L. Ann For Sci 55:159–172

Eller F, Jensen K, Reisdorff C (2016) Nighttime stomatal conductance differs with nutrient availability in two temperate floodplain tree species. Tree Physiol 37:428–440. https://doi.org/10.1093/treephys/tpw113

Hinckley TM, Lassoie JP, Running SW (1978) Temporal and spatial variations in the water status of forest trees. For Sci Monograph 20:56–72

Holbrook NM, Whitebeck JL, Mooney HA (1995) Drought response of neotropical dry forest trees. In: Bullock SH, Mooney HA, Medina E (eds) Seasonally dry tropical forest. University Press, Cambridge, pp 243–276

Houle G (1994) Spatiotemporal pattern in the components of regeneration of four sympatic tree species, *Acer rubrum, A. saccharum, Betula alleghaniennsis* and *Fagus grandifolia*. J Ecol 82:39–53

Kapnick S, Hall A (2010) Observed climate–snowpack relationships in California and their implications for the future. J Clim 23(13):3446–3456. https://doi.org/10.1175/2010JCLI2903.1

Korner C, Cochrane P (1985) Stomatal responses and water relations of *Eucalyptus pauciflora* in summer along an elevational gradient. Oecologia 66:443–455

Kramer PJ, Boyer JS (1995) Water relations of plants and soils. Academic Press, San Diego

Mayr S, Hacke U, Schmid P, Schwienbacher F, Gruber A (2006a) Frost drought in conifers at the alpine treeline: xylem dysfunction and adaptations. Ecology 87:3175–3185

Mayr S, Wieser G, Bauer H (2006b) Xylem temperatures during winter in conifers at the alpine timberline. Agric For Meteorol 137:81–88

Nautiyal MC, Nautiyal BP, Prakash V (2001) Phenology and growth form distribution in an alpine pasture at Tungnath, Garhwal, Himalaya. Mt Res Dev 21:168–174

Pallardy SD, Pereira JS, Parker WC (1991) Measuring the state of water in tree systems. In: Lassoie JP, Hinckley TM (eds) Techniques and approaches in Forest tree ecophysiology. CRC Press, Boca Raton, FL, pp 27–76

Parker WC, Pallardy SG, Hinckley TM, Teskey RO (1982) Seasonal changes in tissue water relations of three woody species of the Quercus-Carya forest type. Ecology 63:1259–1267

Poudyal K, Jha PK, Zobel DB, Thapa CB (2004) Patterns of leaf conductance and water potential of five Himalayan tree species. Tree Physiol 24:689–699

Rai ID, Adhikari BS, Rawat GS (2012) Floral diversity along sub-alpine ecosystems in Tungnath area of Kedarnath wildlife sanctuary, Uttarakhand. Indian For 138:927–940

Richard JH, Bliss LC (1986) Winter water relations of a deciduous treeline conifer, Larixlyallii Parl. Oecologia 69:16–24

Singh SP, Zobel DB, Garkoti SC, Tewari A, Negi CMS (2006) Pattern in water relations of central Himalayan trees. Trop Ecol 47:159–182

Singh U, Phulara M, David B, Ranhotra PS, Shekhr M, Bhattacharyya A, Dhyani R, Joshi R, Pal AK (2018) Static tree line of Himalayan silver fir since last several decades at Tungnath, western Himalaya. Trop Ecol 59(2):351–363

Tewari A (1998) Timing of drought effects on water relation of certain major forest types of lower and middle central Himalaya. Ph.D. Thesis, Kumaun University, Nainital

Tewari A (1999) Tree water relation study in Sal (*Shorea robusta* Gaertn) forests in Kumaun central Himalaya. J Environ Biol 20(4):351–355

Tewari A, Shah S, Singh N, Mittal A (2018) Treeline species in Western Himalaya are not water stressed: a comparison with low elevation species. Trop Ecol 59(2):313–325
Tranquillini W (1976) Water relations and alpine timberline. In: Lange OL, Kappen L, Schulze E-D (eds) Water and plant life, Ecological studies, vol 19. Springer, Berlin
Walck JL, Hidayati SN, Dixon KW, Thompson K, Poschlod P (2011) Climate change and plant regeneration from seed. Glob Chang Biol 17:2145–2161
Woodward FI (1987) Climate and plant distribution. Cambridge University Press, Cambridge
Yao Y, Zhang B (2012) The mass elevation effect of the Tibetan plateau and its implications for alpine treelines. Int J Climatol 35:1833–1846
Zobel DB, Garkoti SC, Singh SP, Tewari A, Negi CMS (2001) Patterns of water potential amongst forest types of the central Himalaya. Curr Sci 80:774–779

Herbaceous Vegetation Structure and Phenology at Treeline Ecotone in Relation to Natural Snowmelt

17

Bhupendra S. Adhikari and Rahul Kumar

Abstract

This study aims to find out the impact of early natural snowmelt on diversity, composition and phenology of herbaceous vegetation of treeline ecotone (3200–3300 m asl) at Tungnath, Uttarakhand Himalaya. Based on orography, four sites were selected across the treeline ecotone and within each site, two microsites based on snowmelt timing were identified. Across the sites, 86 species were recorded at and around treeline ecotone, 90% were perennial forbs, 84% were hemicryptophytes and 70% were native. Both herb species richness and species diversity were higher in exposed microsites. The total plant density ranged between 82 and 626 individuals m^{-2} in exposed and 69–288 individuals m^{-2} in unexposed microsites showing that a warming climate might favour an increase in richness and diversity. The early phenophase was noticed in the majority of species in exposed than at unexposed microsites. The timing of different phenophases varied among the sites. Comparing this study with previous studies either in the same or similar study areas based on 13 common species indicates that timing and duration of vegetative and flowering phenophase have advanced and lengthened, while fruiting and seed maturation have shortened presumably because of the increase in mean minimum temperature. This indicates that species might adopt different adaptations to a warming climate.

Keywords

Life form · Phenophase · Species diversity · Snowmelt · Treeline ecotone

B. S. Adhikari (✉) · R. Kumar
Department of Habitat Ecology, Wildlife Institute of India, Dehradun, Uttarakhand, India
e-mail: adhikaribs@wii.gov.in

S. P. Singh et al. (eds.), *Ecology of Himalayan Treeline Ecotone*,
https://doi.org/10.1007/978-981-19-4476-5_17

17.1 Introduction

The high mountain ecosystems are under tremendous pressure at present because of habitat destruction, landscape modification and loss in biodiversity (Korner 2003; Bruun et al. 2006) and are under the threat of climate change (Beniston 2003; Pauli et al. 2012). Some of the major consequences of climate warming are early snowmelt, increase in soluble N deposition in high mountain habitats, limitation of plant growth and diversity in terrestrial ecosystems (Hättenschwiler and Körner 1997; Tørseth and Semb 1997) may restrict plant growth in alpine species (Hiltbrunner et al. 2005; Gong et al. 2015). 'Alpine' is commonly used in a broad sense for the treeless areas above a low-temperature determined treeline in the high reaches of mountains, characterized by low temperature, precipitation in form of scanty rainfall, blizzard and snowstorms, high wind velocity and high intensity of ultraviolet (UV) radiation (Nautiyal et al. 2004). A treeline ecotone is a transition zone between a closed canopy upright forest (timberline) and low growing primarily herbaceous alpine. The treeline zone, elevational limit of a tree, is often controlled by temperature. This limit is directly affected by drastic drop in allocation of energy gained via photosynthesis (Körner 1998) and is modulated by several factors such as seasonal snowpack duration, winds, precipitation, soil and high intensity of UV radiation (Nautiyal et al. 2004). Due to the extreme climate, plants adapt to these conditions and in general are dwarfed, stunted, woolly or spiny, and develop a mosaic patch of different life forms (Walker et al. 1994). The community structure and composition show that seasonal fluctuations are strongly influenced by the extent to which periodic phenomena in the individuals are adjusted to each other (Kershaw 1973). The resulting hypothesis is that as the climate warms, trees will move up the slope towards higher reaches. Several studies have shown such geographic responses (Lloyd and Fastie 2002).

Snow is an important environmental factor controlling micro-climate and plant growth (Wipf 2010) and acts as an insulator during the cold season, protecting the plants and soil underneath (Sturm et al. 1997). The increase in soil moisture at the initial stage of the vegetative period due to early snowmelt changes the community composition, species richness and the occurrence patterns of individual species (Korner 2003; le Roux et al. 2013). Snow duration and extent determine the beginning and length of the growing period, and meltwater provides water and nutrients that are crucial for plant growth (Jones et al. 2001). To upland ecosystems, snowmelt timing, depth and duration of the snow cover are of special importance as they define the start and duration of the potential growing season (Inouye and Wielgolaski 2003), the temperatures under the snow (Pomeroy and Brun 2001) and characteristics of alpine plant communities (Korner 2003).

In alpine and arctic systems, plant communities' occurrence is closely associated with spatial snow distribution patterns (Odland and Munkejord 2008), showing the importance of snow depth and snowmelt with season length in structuring communities as changes in the snow cover can have a wide range of species-specific effects on alpine tundra plants (Wipf et al. 2009). Several workers (Campbell et al. 2005; Sturm et al. 2005) suggested that the snowmelt controls the biochemical, microbial and plant processes during winters, which persisted well into the growing season (Weih and Karlsson 2002). An early snowmelt results in the longer growing season and higher growing degree days, but it also increases frost damage (subzero temperature), earlier dehardening resulting in higher frost sensitivity (Inouye 2000; Gorsuch and Oberbauer 2002).

The changing climate has been affecting high altitude plant communities (Wipf and Rixen 2010) and plant phenology (Smith et al. 2012), which can potentially impact ecosystem services they provide (Winkler et al. 2018) and the range-restricted species (Parmesan 2006), particularly in polar and higher elevation (alpine) regions. The other drivers of change, such as extreme events, drought and forest fires, and anthropogenic disturbance have further intensified impact on the structural and functional attributes of alpine vegetation communities that harbours large number of high-value medicinal, aromatic and threatened plants (Xu et al. 2009; Singh et al. 2011). Therefore, alpine plants are sensitive to climate change and are considered as good indicators for understanding its impacts (Pauli et al. 2007) as compared to other ecosystems (Malik and Nautiyal 2016).

In the Himalayan region, the studies related to assessing the ecological impacts of winter precipitation (snowmelt) on alpine plant communities are very limited (Adhikari et al. 2018). Therefore, considering the gap, in this study we analyse the impact of snowmelt water and snow cover on phenology and structural and functional attributes of herbaceous plant communities at treeline ecotone.

17.2 Study Area

This study was conducted in Tungnath which lies in the upper catchment of Alaknanda (one of the two tributaries of river Ganga) and Mandakini rivers of Uttarakhand, India (Fig. 17.1). Alpine meadow communities consisting of forbs, grasses and sedges occur in Tungnath region along with *Rhododendron campanulatum* krummholz. A year is divisible into four seasons, viz. short summer (May to June), monsoon (July to mid–September), autumn (mid-September to October) and long winter (November–April). The period with snow cover is of about 4–5 months (December–April) and snowmelts during April–May between 3200 and 3300 m asl. The snowfall is staggered in the study region and area to become inaccessible.

17.3 Methods

17.3.1 Study Design

Treeline ecotone at Tungnath region (Photographs 17.1 and 17.2) is almost 2 km long lying between 3200 and 3300 m asl predominated by *Abies spectabilis* and *Quercus semecarpifolia* forests along with *Rhododendron* krummholz. Four summits namely Ravanshila, Chandrashila, Patankudi and Jhabra were selected based on the orographic lift for the natural snowmelt experiment (Table 17.1).

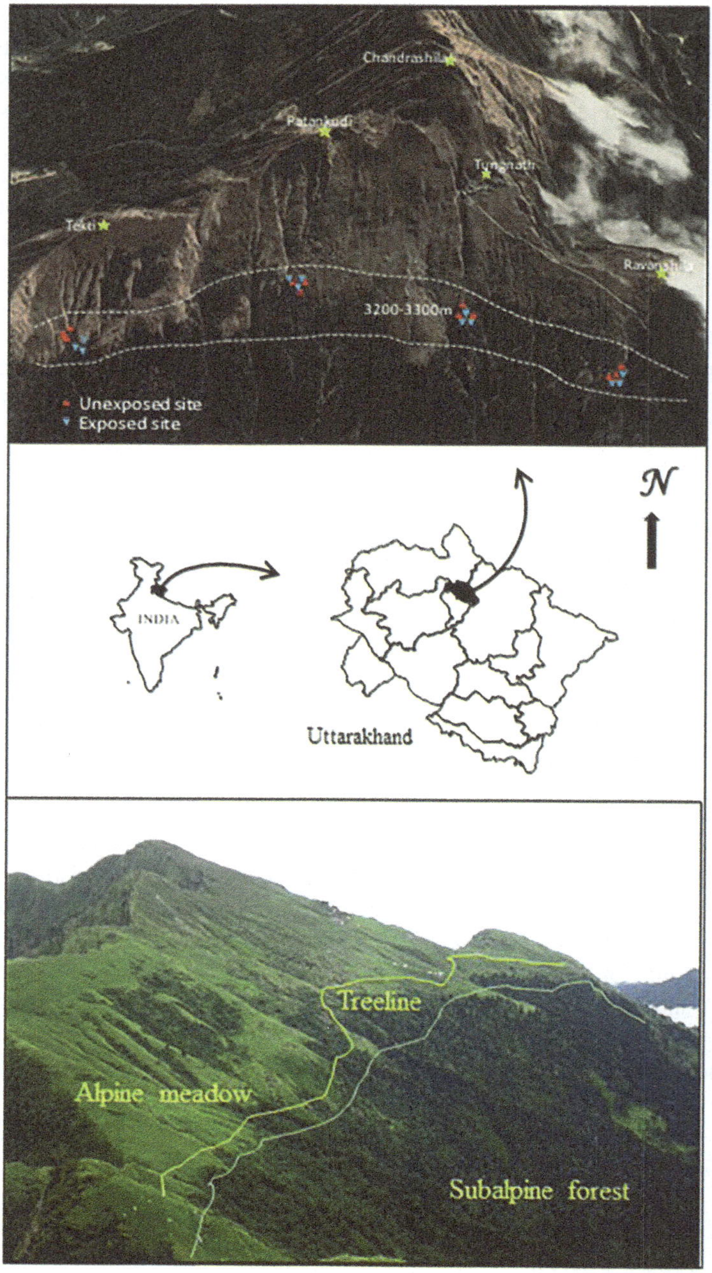

Fig. 17.1 Location of plots in exposed area (blue) and in depressed area (red) and study area in Tungnath region, Garhwal Himalaya

Photograph 17.1 Data collection from treeline ecotone region, Western Himalaya

Photograph 17.2 Treeline Ecotone region in Tungnath, Western Himalaya

Table 17.1 Characteristics of selected sites in treeline ecotone (3200–3300 m) in Tungnath region. CC: Canopy cover

Parameters	Ravanshilla	Chandrashilla	Patankuri	Jhabra
Site characters	Wide open edge with few scattered trees	Open slope with trees	Depressed slope with trees	Depressed slope with trees
Dominant species (canopy cover)	*Abies spectabilis* (7%) and *Rhododendroncampanulatum* (20%)	*Abies spectabilis* (15%) and *R. campanulatum*(35%)	*Quercus semecarpifolia* and *Abies spectabilis* (20%) and *R. campanulatum* (60%)	*R. arboreum* and *Quercus semecarpifolia* (15%) and *R. Campanulatum* (40%)
Slope (°)	35–40	30–35	40–55	35–45
Aspect	North	North-west	North-west	North-west
Soil moisture (0–20 cm; July)	48% ± 2.4%	45% ± 2.1%	45% ± 3.1%	46.5% ± 2.5%
Rock cover (%)	25	35	55	30
Wind velocity	Very high	High	Low	Low
Snow depth (m)	0.6–0.9	0.8–0.9	1.0–1.2	0.8–1.0
Snow melt	Very fast	Fast	Slow	Medium
Snow duration (days)	Exposed: 47.3 ± 1.4 Unexposed: 58.0 ± 1.5	Exposed: 58.7 ± 3.2 unexposed: 67.7 ± 1.4	Exposed: 74.7 ± 2.9 Unexposed: 79.0 ± 3.0	Exposed: 57.7 ± 1.4 Unexposed: 74.7 ± 2.9

These summits were further divided into two major snow cover regions (high snow cover: hereafter referred to as HSC and snow depth 1.0–1.5 m during January comprised of Patankudi and Jhabra summits and low snow cover: hereafter referred to as LSC and snow depth 0.6–1.0 m during January comprised of Ravanshila and Chandrashila summits). At each site, a plot (50x50m) was selected at the base of summits between 3200 and 3300 m asl. Within the selected plot two different microsites (namely, Exposed microsite: hereafter referred to as EX with canopy cover <20% and Unexposed microsite: hereafter referred to as UX with >20% canopy cover) were identified based on snowmelt timing. The UX microsites had higher snow deposition due to snowdrift and longer snow stay period due to *krummholz* cover resulting in late snowmelt (delayed 4–14 days). Within each microsite, 3 random quadrats (1x1m each) were permanently laid with a distance of 10–12 m to record plant abundance, richness and diversity.

17.3.2 Lifespan

The recorded plants were categorized based on their lifespan into the following categories: annuals, plants that live for one growing season, perennials; plants that grow and bloom over spring/summer season and die every autumn/winter and grow again from their rootstock; and biennials, plants which complete life cycle in two growing season, germinate, grow and survive through winters and next year grow more, bloom and die.

17.3.3 Growth Form Categories

The direction and extent of growth and branching on the main-shoot axis or axes determine the growth form of a plant. The plants were classified following Pérez-Harguindeguy et al. (2013) under the following growth forms, viz. climber, dwarf shrub, semi-basal, short-basal, erect leafy and tussock to compare with other studies conducted in western Himalaya.

17.3.4 Life Form Categories

Following Raunkier (1934) the plants were assigned to various life-form classes, viz. phanerophyte (surviving buds or shoot apices are borne on shoots into the air), chamaephyte (surviving buds/shoot apices are borne on shoots very close to the ground), hemicryptophyte (surviving buds/shoot apices are situated on the soil surface), geophyte (surviving buds/shoot apices are below soil surface) and therophyte (complete their life cycle from seed to seed and die within a season).

17.3.5 Nativity

The place of origin/first record of the species where it occurs naturally indicates the nativity of the species (Anonymous 1883–1970), while the introduction of the species in an area happened through anthropogenic activities and has not previously occurred in that area are known as 'Non-Native'. We used Index Kewensis Plantarum Phanerogamarum (Anonymous 1883–1970) to know the nativity of a species.

17.3.6 Species Richness, Diversity and Evenness

Species richness was taken as a count of the total number of species in a particular area. The diversity was calculated after Shannon and Wiener (1949). If Pi is the proportion of individuals (from the sample total) of species i, then diversity (H′) is

$$H' = -\sum_{i=1}^{s} (Pi)(\text{lnPi})$$

where P is the proportion (n/N) of individuals of one particular species found (n) to the total number of individuals found (N), ln is natural log and s is the number of species.

Evenness index was calculated using Pielou (1975).

$$Ep = \frac{H'}{H'\max}$$

where H' is Shannon & Wiener diversity index, $H'\max = \ln S$, is the natural log of S.

Sorenson Similarity Index (Is) between different sites was calculated following Sorensen (1948).

$$\text{Is} = \frac{2C}{A + B} \times 100$$

where Is the Sorenson Index of Similarity; C, the common species to both comparable sites; A, the total number of species in site A; and B, the total number of species in site B.

17.3.7 Simper

The contribution of each species to the dissimilarity (%) between two groups was calculated from the Bray–Curtis dissimilarity matrix through SIMPER analysis in PAST.

17.3.8 Phenological Observations

The phenological changes were observed visually to monitor how the species respond to climate variations on a monthly interval. Therefore, records of various phenophases such as growth initiation, vegetative phase, flowering, fruiting, seed formation and senescence of each species were taken from May to October in 2017 following the Biologische Bundesanstalt Bundessortenamt und Chemische Industrie (BBCH) scale (Hess et al. 1997). The complete existence of a particular phenophase was considered if 5% of the individuals showed that phenophase within the plot. Based on growth initiation time, the plants were divided into two categories: early growth species (up to mid-May) and late growth species (mid-May to June). The phenology of each species was observed within quadrats under each category in adjacent areas with similar micro-environment within selected plots to minimize the effect of herbivory.

17.4 Results

Plant growth period (the period from the beginning of growth to the time when about 80% of plant shoots senesced) was 6 months from mid-April to mid-October. Soil moisture values (45–48%) observed during the peak growing period (July) showed that it is not a limiting factor for plant growth. The number of snowfall days during winter months (December to April) in 2016–2017 was 39 days.

17.4.1 Lifespan Category, Growth Form, Growth Cycle and Life Form

Across the sites and within EX and UX areas, a total of 86 plant species were recorded, of which 90% were forbs and remaining 10% were climber (1%), shrub (3%) and grasses and sedges (6%), while 90.6% were perennials followed by annuals (8.2%) and biennials (1.2%; Table 17.2). Of the total species, 53% species were early growing (germination before July) while 47% were late growing species

Table 17.2 Per cent contribution of different growth cycle/growth forms (86 species)

Growth cycle	Total species (%)	Growth form (%)					
		Tall forbs	Med. forbs	Short forbs	Climber	Shrub	Grass & sedge
Short	5	7	–	4	100	–	–
Intermediate	31	52	15	25	–	67	–
Long	64	41	85	71	–	33	100
Total species (%)	100	34	23	33	1	3	6
Early growing (before May)	53	17	78	76	100	33	60
Late growing (after May)	47	83	22	24		67	40

(germination after June). Growth forms of the species were highest for semi basal/ short basal followed by erect leafy, dwarf shrub and climber/shrub/tussock. The maximum species had a long growth cycle (LGC) followed by an intermediate growth cycle (IGC) and a short growth cycle (SGC). Among various growth forms, tall and short forbs contributed the maximum followed by medium forbs and others (climbers, shrub/undershrub, grasses/sedges; Table 17.2). The majority of the species recorded in this study were native to the Himalayan region (70%) indicating a high conservation value of treeline in the Tungnath region.

17.4.2 Species Richness and Diversity Patterns

The species richness was almost equal in EX and (74) and UX (72) microsites. In general, it was noticed that the richness of the herbaceous species increased across the months from May to September (44–80 species, respectively) and then declined in October (56 species).

The species richness attained a peak between July and September (77–80 species) and 60% of species were common to both EX and UX microsites. Generally, species number in a month was markedly higher in EX microsite than in UX microsite, particularly in Jhabra (EX: 38, UX: 27) in July. Two-way ANOSIM showed significant variations in species richness between sites ($r = 0.93$) and across months ($r = 0.91$).

At EX microsites the species richness was highest in July for Chandrashila, Patankudi and Jhabra, while in August for Ravanshila. In UX microsites, species richness was highest in July for Ravanshila and Jhabra and in August for Chandrashila and Patankudi (Fig. 17.2). This could be because these microsites are dominated by krummholz, protecting species against environmental factors prevailing in other microsites. In general, species richness was higher in EX than UX microsites, probably due to the absence of canopy cover, ample sunlight and less competition for resources.

The diversity values from 2.10 to 3.11 in UX microsites and 2.41 to 3.21 in EX microsites and peaked during July for both EX and UX microsites (Fig. 17.3). The change in diversity after July was due to the senescence of early growth cycle species.

The evenness values ranged from 0.46 to 0.77 in UX microsites and 0.40 to 0.68 in EX microsites. The evenness values for Ravanshila, Patankudi and Jhabra sites showed a similar distribution pattern in both EX and UX microsites, while in Chandrashila species were much more evenly distributed in UX microsites. This could be due to the favourable condition for growth and dominance of *Ranunculus, Fragaria, Oxygraphis, Trachydium* and *Prunella* over other species.

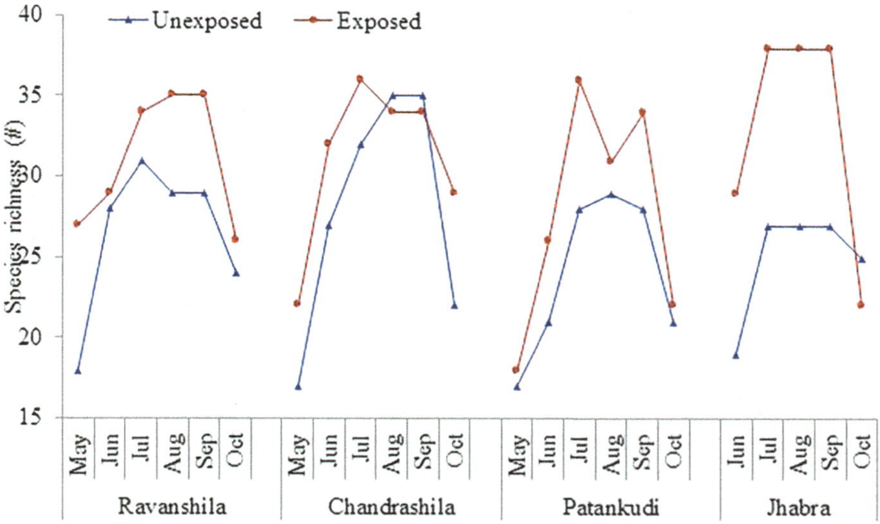

Fig. 17.2 Patterns of species richness (#) across spatio-temporal scale at treeline in Tungnath region

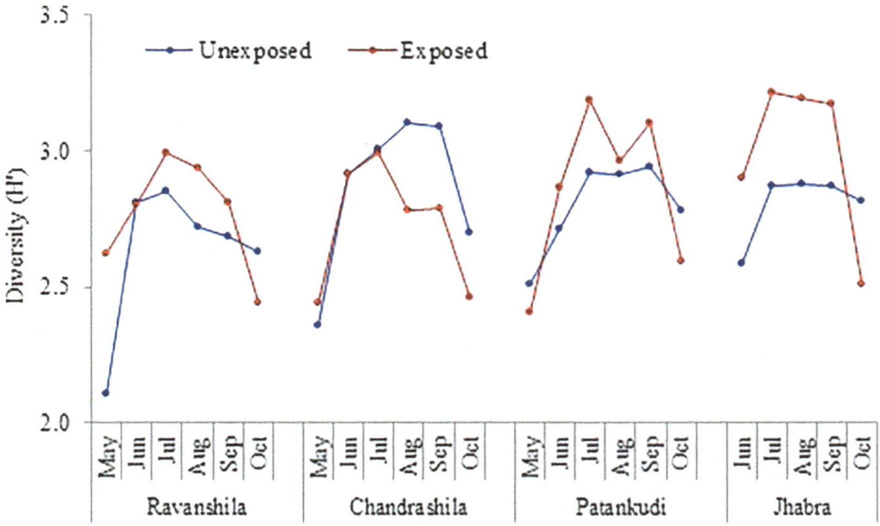

Fig. 17.3 Patterns of species diversity (H') across months in Tungnath region

17.4.3 Site and Microsite Similarity

The similarity among sites was highest between Ravanshila and Patankudi (66%) followed by Ravanshila and Chandrashila (65%), Chandrashila and Patankudi (63%), Ravanshila and Jhabra (62%) and Chandrashila and Jhabra (50%). The

Table 17.3 Similarity between months in exposed areas in upper inverted triangle and in unexposed areas in lower triangle

Months		Exposed microsites					
		May	June	July	August	September	October
Unexposed microsites	May	**100.0**	83.1	80.9	74.2	74.2	65.2
	June	81.0	**100.0**	92.6	81.5	85.2	66.7
	July	73.4	86.0	**100.0**	90.9	96.4	74.5
	August	70.9	86.0	95.4	**100.0**	91.7	71.6
	September	65.8	82.0	91.7	97.3	**100.0**	84.8
	October	63.3	62.0	66.1	68.5	81.7	**100.0**

Bold values represent the similarity for the same month

reason could be the similar number of species present in Ravanshila (45), Patankudi (46) and Jhabra (42), which is much less than the species present in Chandrashila (66). The micro-climate of Ravanshila and Patankudi is different than that of Chandrashila, which lies in between and have an admixture of micro-climate of both sites, which may lead the species to move in, hence the richness is high. Although the similarity between EX and UXmicrosites was 82% as a whole, the similarity between EX and UXmicrosites within each site was much less (Chandrashila 31%, Patankudi 34%, Ravanshila 36% and Jhabra 38%). Across months, the similarity was highest between microsites during peak growing season (July–September; Table 17.3).

SIMPER (Similarity Percentage) analysis suggested the primarily responsible taxa (species) for observed differences between groups of samples, that is, snow cover region and microsites. The overall dissimilarity between HSC and LSC region was 56.8% and the maximum contribution was generated by following species in decreasing order: *Ranunculus* > *Trachydium* > *Oxygraphis* > *Carex* > *Fragaria* > *Potentilla polyphylla* > *Prunella*. Comparing average dissimilarity between EX and UX microsites of HSC and LSC region, the average dissimilarity was 63% contributed in EX microsites by 5 species (*Ranunculus* > *Trachydium* > *Oxygraphis* > *Carex* > *Prunella*); however, in UX microsite it was 59.2% generated by 11 species (*Fragaria* > *Potentilla polyphylla* > *Ranunculus* > *Trachydium* > *Poa* > *Galium* > *Anemone* > *Polygonum filicaule* > *Polygonum vccinifolium* > *Carex* > *Lysimachia*).

Across sites the overall average dissimilarity in UX microsites was 64% between months and 72% between sites; however, in EX microsites it was 59% and 67%, respectively. Within sites, the maximum contribution to dissimilarity (up to 50%) as follows:

EX microsites: *Ranunculus* > *Trachydium* > *Carex* > *Oxygraphis* > *Fragaria* > *Anaphalis* > *Prunella* > *Koenigia* > *Anemone.*

UX microsites: *Fragaria* > *Ranunculus* > *Polygonum* > *Carex* > *Bistorta* > *Trachydium* > *Anemone* > *Poa* > *Polygonum* > *Circea* > *Potentilla.*

17.4.4 Habitat Preference of Plant Species

The species like *Bistorta amplexicaulis, Fragaria nubicola, Geranium wallichianum, Potentilla atrosanguinea* and *Viola biflora* were present in all microsites. The species specifically present in all UX microsites were *Myriactis wallichi* and *Veronica cana*, while *Anaphalis nepalensis, Kobresia, Oxygraphis polypetala, Polygonum filicaule, Potentilla lineata, Potentilla polyphylla* and *Selinum vaginatum* occurred in EX microsites. The distribution of certain species was restricted to few sites, such as *Bupleurum longicule, Rubus nepalensis, Swertia speciosa, Synotis alata, Arenaria, Euphorbia* sp. and *Taraxacum officinale* in Ravanshila; *Aster methodrus, Viburnum glanduliflorum, Swertia auriculata, Arisaema propinquim, Ligularia amplexicaulis, Halenia elliptica, Senecio graciliflorus, Trillium govanianum, Smilacina purpurea* and *Aster albescens* in Chandrashila; *Thalictrum foliolosum, Primula edgeworthii* and *Ligularia sibirica* in Patankudi and *Rumex nepalensis, Cynoglossum glochidiatum, Primula redii, Morina longifolia, Cerastium cerastoides* and *Persicaria nepalensis* in Jhabra. *Aster methodrus, Viburnum glanduliflorum, Arisaema propinquim, Ligularia amplexicaulis, Thalictrum foliolosum, Hemiphragma heterophyllum, Halenia elliptica, Trillium govanianum, Smilacina purpurea, Clematis, Parochetus communis, Galium asperifolium* and *Caltha palustris* were found in UX microsites, and *Swertia auriculata, Rumex nepalensis, Aster albescens, Cynoglossum glochidiatum, Primula redii, Ligularia sibirica, Goodyera repens, Persicaria nepalensis, Corydalis cornuta, Caltha palustris, Trachydium roylei, Gaultheria trichophylla, Gentiana argentea, Gentiana tubiflora, Gerbera gossypina* and *Morina polyphylla* in EX microsites.

17.4.5 Community Structure

In all microsites, the herbaceous communities in the initial phase were different in UX microsites. In Ravanshila and Jhabra sites, *Fragaria* and *Ranunculus* dominated the community in UX microsites, respectively and showed changes in August onwards in Ravanshila, while *Ranunculus* dominated the community throughout the growing period in Jhabra. It was interesting to note that in Chandrashila, the community changed after August, while in Patankudi the species dominance changed throughout the growing period in UX microsites. The community in Chandrashila and Jhabra was dominated by *Ranunculus* throughout the growing period in EX microsites, while *Anaphalis* dominated in Patankudi throughout the growing period, except May and August. *Ranunculus, Fragaria, Oxygraphis* and *Trachydium* dominated the vegetation in EX microsites of Ravanshila (Table 17.4).

Plant density among the sites (Chandrashila, Ravanshila and Patankudi) was more on EX microsites than on UX microsites, except Jhabra where the density values were not affected by exposure (Fig. 17.4). In moist microsites, water was not a limiting factor. In Chandrashila and Ravanshila microsites, plant density at EX

Table 17.4 Temporal changes in herbaceous community at different sites within unexposed and exposed areas. Species sequence arranged with decreasing density values

Site	Microsite	Months					
		May	June	July	August	September	October
Ravanshila	EX	*Ranunculus-Fragaria*	*Trachydium-Oxgraphis*	*Oxgraphis-Ranunculus*	*Ranunculus-Trachydium*		
	UX	*Fragaria-Ranunculus*	*Fragaria-Polygonum*		*Fragaria-Carex*		
Chandrashila	EX	*Ranunculus-Carex*			*Ranunculus-Trachydium*		
	UX	*Bistorta-Fragaria*				*Trachydium-Circaea*	
Patankudi	EX	*Nepeta-Caltha*	*Anaphalis-Polygonum*		*Carex-Anaphalis*	*Anaphalis-Carex*	
	UX	*Polygonum-Caltha*	*Bistorta-Caltha*	*Polygonum-Bistorta*		*Lysimachia-Bistorta*	*Bistorta-Potentilla*
Jhabra	EX	–	*Ranunculus-Fragaria*				*Ranunculus*
	UX	–	*Ranunculus-Poa*				*Ranunculus*

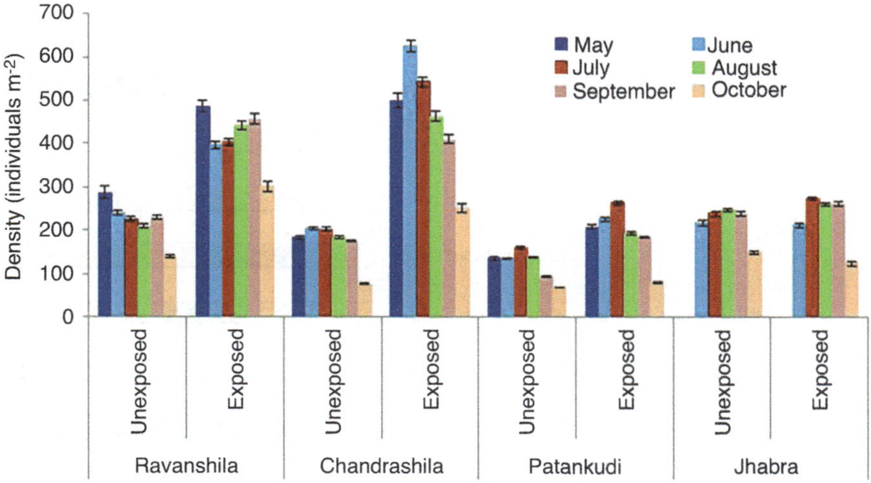

Fig. 17.4 Density (individuals m^{-2}) at different sites across the months at treeline in Tungnath region

microsites was more than twice than that on UX microsites. The snow effect was minimized because of the shadow of higher summit in Patankudi and Jhabra which resulted in faster snowmelt in Ravanshilla and Chandrashilla (45–65 days) as compared to Patankudi and Jhabra (55–80 days) for EX microsites. Plant densities generally were high initially in June–July, then declining from there with some deviations.

17.4.6 Phenological Trends across Microsites

The micro-climatic conditions, which include environmental conditions (aspect, slope, moisture, soil texture, ambient and soil temperature, rainfall etc.) and topographical features (gentle, steep, boundary/rocky) and anthropogenic factors (grazing, trampling, collection of NTFP) influenced the phenophases. The relationship between growth initiation and snowmelt was evident in several micro-sites. Growth in species was initiated immediately after the snowmelt in May. For example, immediately after snowmelt, *Anemone rivularis, Fragaria nubicola, Gentiana argentia, Lysimachia prolifera, Oxygraphis polypetala, Plantago major, Rannunculus hirtallus, Selinum vaginatum, Trachydium roylei* and *Gaultheria trichophylla* were observed in flowering, while *Arenaria, Caltha palustris, Carex setosa, Nepeta govaniana, Polygonum filicaule, Potentilla polyphylla, Primula denticulata, Trachydium roylei* and *Plantago himalaica* were in bud development phase. The timing of different phenophases varied among microsites and it was quite similar between Ravanshila and Chandrashila and between Patankudi and Jhabra, while more variability in phenophases was observed in extreme sites, that is, Ravanshila and Jhabra.

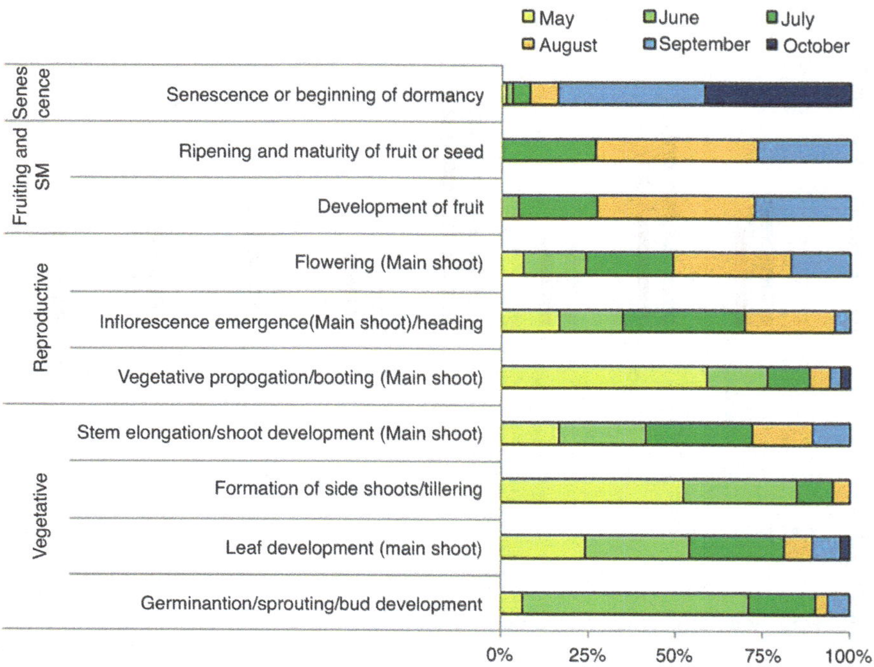

Fig. 17.5 Phenophases of species at temporal scale as per BBCH scale

In the early growing season (May–June) only vegetative and reproductive phenophases were observed in all the microsites, all phenophases were recorded from the mid-growing season (July–September; Fig. 17.5). *Goodyera ripens* germinated in EX microsites only during August and *Epilobiumroylei* showed vegetative reproduction in UXmicrosites in September only. Fruiting started in June (*Trachydium roylei* and *Plantago himalaica*) in EX microsites and in UX microsites it started in July and last till September.

Oxygraphis polypetala and *Ranunculus hirtelus* showed dual reproductive phenophase in months of May–June and August–September in both EX and UX microsites, while *Fragaria nubicola* showed dual reproductive phase in the month of June and August in EX microsite only. Of the total 86 species, only 23 species (26.7%) showed fruit and seed maturation phases. The absence of fruiting and seeding phenophase in most of the species (73.3%) at both EX and UX microsites and the abovementioned species in UX microsites may be due to an increase in mean minimum temperature in Tungnath region, which favours longer vegetative and reproductive phenophase.

17.5 Discussion

The precursor to initial plant growth is snowmelt water, which was received by the microsites in the form of snowfall (39 days; 4 days in December 2016 and 35 days during January to April in 2017), before the onset of monsoon. Due to heavy rainfall in recent years (241.1 ± 43.2 cm) (maximum in September) during 2008–2010 (Adhikari et al. 2012) and 370.7 cm (maximum in July) for 2017 (GBPIHE) in the region helped species to delay their senescence. The other reason could be much variation in temperature (minima and maxima) in the past (7.7 °C in October to 12.2 °C in June in 1984 (Sundriyal et al. 1987) as compared to 2017 and 2.4 °C in July to 9.9 °C in October (GBPIHED). Similarly, the mean monthly temperature was too high (13–23 °C) during the growth period (June to September) in the past (1988–1998) by Nautiyal et al. (2001) as compared 9.6 to 12.6 °C in 2008–2010 (Adhikari et al. 2012).

Comparing this study with that Nautiyal et al. (2001), we found a decrease in Long growth cycle (LGC) species and an increase in short growth cycle (SGC) and intermediate growth cycle (IGC) species for tall forbs. The percentage of LGC species increased and IGC species decreased for medium forbs, while SGC was absent in this study. For short forbs, per cent LGC increased, while IGC and SGC decreased in this study as compared to Nautiyal et al. (2001). We also noticed increase in LGC and IGC species in grasses/sedges, a climber was observed in LGC in this study.

Hemicryptophyte (83.5%) was the predominant life-form which is in conformity with other studies conducted at similar elevations in the alpine region of Garhwal Himalaya (Nautiyal 1996, Nautiyal et al. 2001; Sundriyal et al. 1987; Rawat 2007; Dad and Khan 2010; Vashistha et al. 2011, Kumar et al. 2016; Table 17.5). In general, the life-form spectrum of an area is affected by the grazing intensity as well as anthropogenic factors. The proportion of hemicryptophytes increased in the study area over the last three decades as compared to previous studies (Sundriyal et al. 1987), showing better adaptability of hemicryptophytes to grazing, trampling and other environmental factors, viz. wind, heavy rainfall/erosion, heavy frost than that of other life forms. The proportion of chamaephytes was reported to be high in Rudranath region (31%) by Ram and Arya (1991), in Central Himalaya (47%) by Pangtey et al. (1990) and in Yusmarg (46%) by Dhar and Kachroo (1983), could be due to less rainfall than that of this study site.

The nativity, first record/origin of the species and endemism of any biogeographic province denotes the high conservation value of the area (Anonymous 1883–1970). In Western Himalaya, very few studies (Samant et al. 1998, 2001; Samant and Joshi 2005) on nativity and endemism of the species are available. It indicates that the dominance of native species is quite prevalent in the region, while the introduction of non-natives due to livestock grazing, tourism and fuelwood collection in the region does not have any impact. The alpine meadows struggle with heavy tourism influx, high grazing intensity and exploitation of natural resources, such as high-value medicinal plants, imbalance the natural ecosystems. Samant et al. (1998) opined that the areas with a high percentage of native species

Table 17.5 Comparison of Raunkaier's life forms at various locations in Greater Himalaya. Numbers in parenthesis are the total number of species

Study area	Life form[a]						Reference
	Ph	Ch	He	Hy	Ge	Th	
Panwalikantha	8.3	6.8	44.7	–	35.0	5.3	Nautiyal (1996)
Central Himalaya (snowline)	–	46.7	30.0	–	18.3	5.0	Pangtey et al. (1990)
Yusmarg	10.4	46.4	25.4	–	10.2	6.2	Gupta and Kachroo (1981)
Tungnath	–	27.0	43.0	–	18.0	12.0	Sundriyal et al. (1987)
Tungnath (#171)	6.4	10.5	48.0	–	29.2	12	Nautiyal et al. (2001)
Tungnath (Grazed area; #68) Tungnath (Ungrazed area; #65)	4.4 6.2	16.2 15.4	50.0 49.2	- -	26.5 26.2	2.9 3.1	Vashistha et al. (2011) (26.47%), 26.2
Rudranath (#142)	2.8	31.0	24.6	–	28.9	12.7	Ram and Arya (1991)
Valley of flowers (#525)	7.5	6.1	68.4	0.4	11.8	5.0	Rawat (2007)
Tons valley (#761)	29.1	22.2	2.1	8.6	2.6	17.8	Rana et al. (2002)
Kashmir	5.7	13.8	48.3	–	13.8	18.4	Dad and Khan (2010)
Upper Dhauli Valley (#495)	11.9	5.1	61.4	0.2	10.5	11.1	Kumar et al. (2016)
Normal	46.6	9.0	26.0	2.0	4.0	13.0	Raunkaier (1934)
Tungnath (Treeline; # 86)	2.4	3.5	83.5	–	2.4	8.2	Present study

[a]Life form: *Ph*: Phanerophyte, *Ch*: Chaemophyte, *He*: Hemicryptophyte, *Hy*: Hydrophyte, *G*: Geophyte, *Th*: Therophyte

deserve more conservation, while due to overexploitation and habitat degradation allow the introduction of non-natives, which leads to less conservation value of the area. The loss of a native and endemic species can be due to direct competition between native and non-native plants, the introduction of non-natives in the area which alters ecosystem properties drastically (Vitousek 1986) and further leads to the community dominated by exotic annuals and through succession process (Stylinski and Allen 1999). The species richness across various sites in western Himalaya is reported to range from 27 to 56 species per site (average 30.5 ± 8.3 species; Rawat 2007), thus species richness of this study site (49.8 ± 5.5 species) is well within this range. The species richness values in this study per square meter (UX and EX microsites between 12.5 and 18.6 and 16.8 and 21.0 species m^{-2}, respectively) are similar to that (15–21 species m^{-2}) reported for Zemu valley in Sikkim by Tambe (2007). It shows that in alpine herbaceous communities the species richness does not differ between Tungnath, Uttarakhand and in Sikkim site, though in lower elevations Sikkim is far more species-rich than Uttarakhand. Across all microsites and months plant density ranged between 69 and

626 individuals m^{-2} in this study, which are higher than the values reported for the entire western Himalaya (152 ± 90 individuals m^{-2}) by Rawat (2007) and herb density reported for treeline zone (123–239 individuals m^{-2}) by Rai et al. (2012).

The average diversity ranged from 2.77 to 2.83 for this study, which are less than that reported for herb layer of treeline zone (3.06–3.25) by Rai et al. (2012), for south or south-east facing slopes of the Greater Himalaya (3.01–3.30) by Rawat (2007) and treeline gap (3.23) by Kala et al. (1998). The values of this study are higher than the values reported by Rawat (2007) for the entire greater Himalaya (2.1–2.4), Kala et al. (1998) for Valley of Flowers NP (2.47), Singh (1999) in Upper Tirthan Valley (2.39) between 3200 and 3300 m, Adhikari (2004) for Nanda Devi NP (0.587–2.088) and Tambe (2007) for the alpine landscape of Khangchendzonga NP (1.44–2.48), could be due to high grazing pressure and more moisture in the study area. Similar observations were also observed for high arctic tundra by Park et al. (2018).

It is interesting to note that while only 7 species were responsible for more than 50% dissimilarity between HSC and LSC regions, just 5 species contributed for dissimilarity of EX microsites between HSC and LSC against 11 species for UX microsites. Furthermore, all 5 species of EX microsites (*Ranunculus, Trachydium, Oxygraphis, Carex* and *Prunella*) contributed to the overall dissimilarity between HSC and LSC, while only 4 species (*Fragaria, Potentilla polyphylla, Ranunculus* and *Trachydium,* later 2 species common with EX microsites) out of 11 species contributed to overall dissimilarity. These 5 species search for places with ample sunlight and moisture, so that they can grow gregariously (> 50 individuals m^{-2}, mean average), as they are LGC species. Compared to the above 5 species, *Fragaria* and *Potentilla polyphylla* love to grow in moist and shady places with moderate abundance (30–45 individuals m^{-2}, mean average).

It is interesting to note that in EX microsites of Ravanshila, Chandrashila and Jhabra sites, *Ranunculus* contributed more for dissimilarity, while *Anaphalis* for Patankudi. This could be due to the fast melting of snow in the first three sites and late melting in Patankudi (Table 17.1). Among sites, *Fragaria* for Ravanshila and Chandrashila, *Bistorta* for Patankudi and *Ranunculus* for Jhabra contributed the most. It is evident that the contribution of these species is directly related to fast snowmelt and less grazing in the treeline ecotone of Ravanshila and Chandrashila, and high grazing pressure and late thawing of snow in Patankudi and Jhabra.

In general, phenophases were in advanced stages in EX than in UX microsites, which is similar to the findings of other experiments conducted in similar alpine environments (Wipf et al. 2009; Wipf 2010). Since different species respond differently to snowmelt effect, it indicates that apart from snowmelt, other factors also play major role (Gehrmann et al. 2018), as some contradictions were seen in different phenophases among different life forms. The absence of vegetative phenophase in some species may be contributed by the chilling effect due to frost and early snowmelt, where newly germinated buds either die or move to the next phenophase for stability.

Nautiyal et al. (2001) reported that immediately after the snowmelt 16 species were in the flowering phase, among them *Gentiana argentea* and *Oxygraphis polypetala* emerged early and most dominant in the Tungnath region (mainly alpine). Comparing this study results with Nautiyal et al. (2001) a general shift was observed in the vegetative phase by a month (i.e. June). Over a period of time (from 1984 to 2017) the duration of vegetative phenophase increased from 2 months (Sundriyal et al. 1987), to 4 months (Nautiyal et al. 2001) and 6 months (present study). It was also observed that the phenophase stability and duration was prominent in the studies conducted by Sundriyal et al. (1987) and Nautiyal et al. (2001), but in this study, phenophases overlapped resulting in no clear-cut dominance (stability) of phenophase. As such there was no change in the flowering period, but fruiting and seed maturation ended in September in this study as compared to October during 1988–1998 (Nautiyal et al. 2001). Senescence in this study started in May, which was 2 months prior to that reported by Nautiyal et al. (2001). The longer vegetative phenophase for this study could be due to higher precipitation in recent years (Adhikari et al. 2012; Adhikari et al. 2018) and less difference in minimum and maximum temperature in recent years (Adhikari et al. 2018) as compared to Sundriyal et al. (1987).

Most of the species showed senescence early in UX microsites than that of EX microsites, which may be due to an early drop in temperature in UX microsites than EX microsites. The above discrepancy may be due to a number of reasons like interspecific (early vs late flowering; Petraglia et al. 2014) or intraspecific (collective life form response to snowmelt timing, Khorsand et al. 2015; Livensperger et al. 2016) interactions or maybe due to micro-environmental changes (temperature). Considering early snowmelt (by March) in past few years, it is possible that certain species (*Primula edgeworthii, Anemone rivularis, Gentiana* spp., *Picrorhiza kurroa, Kobresia*) have started their growth in early April high water-absorbing ability at low soil temperature, which might be related to a high level of soluble carbohydrate (Mooney and Billings 1960) in early senescence. The species per cent in senescence increased sharply from August to September in this study is in conformity with the study conducted by Nautiyal et al. (2001).

The phenophases of 10 most common species namely *Anaphalis royleana, Anemone obtusiloba, Bupleurum longicaule, Danthonia cachemyriana, Geranium walliichianum, Oxygraphis polypetala, Pedicularis pectinata, Selinum vaginatum, Taraxacum officinale* and *Trachydium roylei* in this study were compared with the studies conducted by Sundriyal et al. (1987) for the year 1984, Nautiyal et al. (2001) for the year 1988–1998 (11 years; Fig. 17.6) in Tungnath region and Bijalwan et al. (2013) for the year 2008 in Dyara Bugyal. *Anaphalis royleana* had a much longer vegetative phase and consequently delayed flowering in 2017 than in 1984. The flowering period lagged by half month compared to 1984. The fruiting period duration was the same, but the timing of fruiting and senescence lagged behind by a month in 2017 as compared to 1984. *Anemone obtusiloba* had a much longer vegetative phase, and budding and flowering advanced by 1 and 3 weeks, respectively between 2008 and 2017. Fruiting was absent and senescence lagged by a month in 2017 than in 2008 (Bijalwan et al. 2013). In *Bupleurum longicaule*

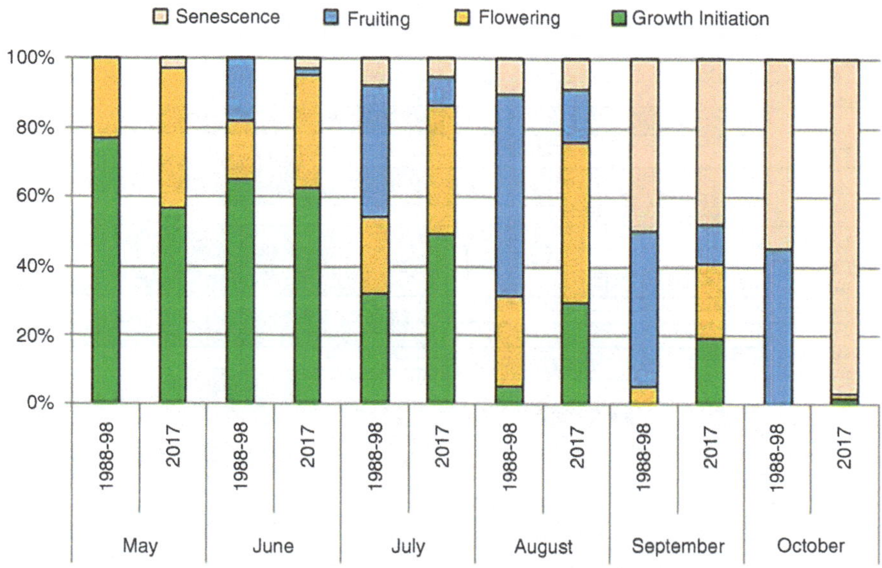

Fig. 17.6 Comparison of phenophases for the year 1988–1998 (Nautiyal et al. 2001) and 2017 (present study) in Tungnath region

flowering and fruiting phases advanced by more than a month and 3.5 weeks, respectively for the year 2017 as compared to 2008 (Bijalwan et al. 2013).

Danthonia cachemyriana had a much longer vegetative phase and its reproductive phase delayed by 4 weeks in 2017; however, fruiting was restricted to the last week of August in 2017 as compared to mid-August to mid-September in 1984 (Sundriyal et al. 1987). The senescence advanced by 2 weeks in 2017 as compared to 1984 (Sundriyal et al. 1987). The duration of vegetative, budding and flowering phenophases of *Geranium wallichianum* increased and advanced significantly in 2017 as compared to 2008. In this fruiting was delayed by 2 weeks and senescence advanced by a month in 2017 as compared to 2008 (Bijalwan et al. 2013). *Oxygraphis polypetala* had a much longer vegetative phase up to August in 2017 as compared to 1984, restricted to the first week of May. Budding, flowering and fruiting phases occurred in two different time periods between May–June and late July–September in 2017 as compared to 1984 (Sundriyal et al. 1987) budding and flowering phases were present in May and fruiting in June to early August. Senescence was delayed by one and a half months in 2017 than in 1984 (Sundriyal et al. 1987). The vegetative phase duration in *Pedicularispectinata* increased exponentially while the advancement of budding, flowering, fruiting and senescence phases was observed in 2017 than in 1984. The duration of the vegetative phase for *Selinum vaginatum* increased to early September and the budding and flowering delayed, fruiting phase was absent and there is no change in senescence between 2017 and 2001. The duration for the vegetative phase increased while reproductive and fruiting advanced for *Taraxacum officinale* during 2008–2017. For *Trachydium*

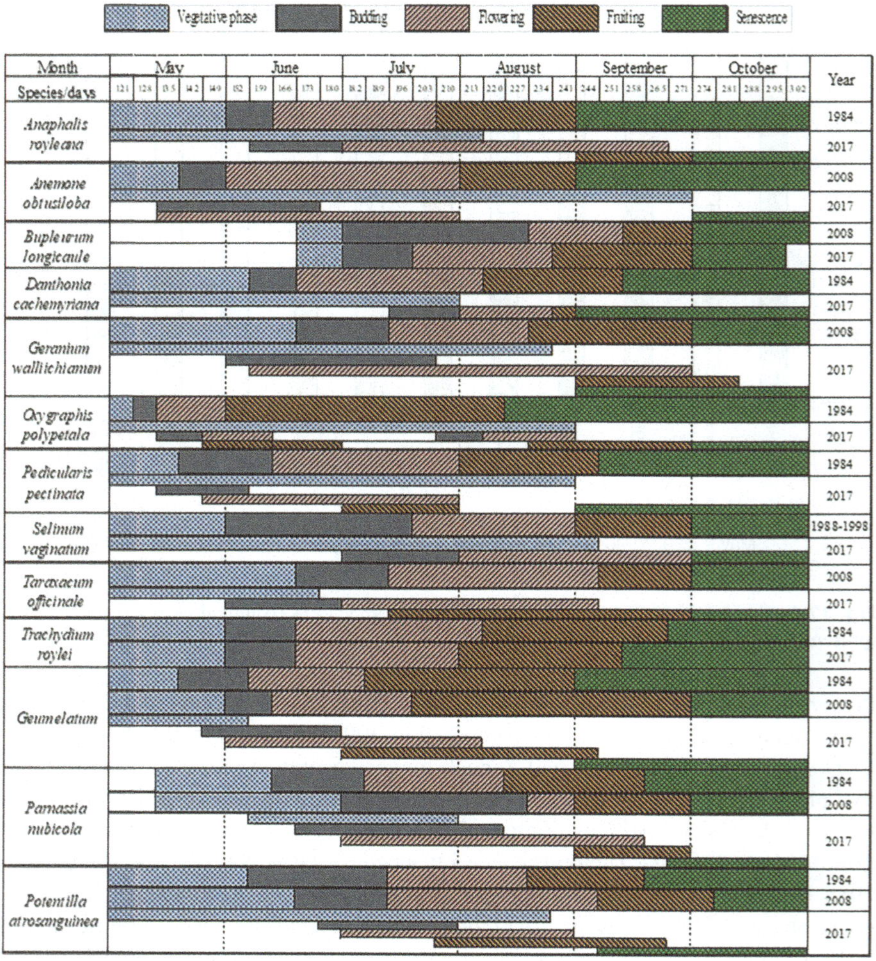

Fig. 17.7 Comparison of phenophases of common species studied during 1984 (Sundriyal et al. 1987), 2008 (Bijalwan et al. 2013) and 2017 (present study)

roylei, the flowering period shortened, and fruiting and senescence phases advanced between 1984 (Sundriyal et al. 1987).

The phenophases for *Geum elatum, Parnassia nubicola* and *Potentilla atrosanguinea* were compared at different sites, viz. Dyara in 2008 (Bijalwan et al. 2013; Sundriyal et al. 1987) with this study (Fig. 17.7). In Tungnath, the vegetative phase of *Geum elatum* extended by 3 weeks, budding lag behind a week, flowering and fruiting advanced by a week and no change in senescence was observed as compared to 1984. However, all the phenophases in Dyara, except the vegetative phase, lag behind 1 to 4 weeks in 2008 (Bijalwan et al. 2013). All phases of *Parnesia nubicola* were delayed by 1–4 weeks in 2017 as compared to 1984 (Sundriyal et al. 1987), while the reproductive phase was delayed as compared to

2008 (Bijalwan et al. 2013). *Potentilla atrosanguinea* had an extended vegetative phase, budding and flowering delayed and fruiting and senescence advanced in 2017 as compared to 1984 (Sundriyal et al. 1987).

The less variation in temperature (minima and maxima) during the peak growth period, that is, July–August helped species to remain in different phenophases for a longer period. A shift of phenophase timing and period was observed in general for the species common to past studies in the study site (Nautiyal et al. 2001; Sundriyal et al. 1987). Early onset of growth initiation and extended vegetative phenophase was observed, could be due to early snowmelt/soil water thawing leading to early soil moisture availability and ambient temperature which favoured early growth as suggested by several workers (Holway and Ward 1965; Kudo 1991; Ram et al. 1988; Kudo and Suzuki 1999; Kudo and Hirao 2006). Flowering and fruiting phenophases have also seen a shift (both advance and lag) in phenophase initiation and duration of the period in general and have extended compared to other studies (Sundriyal et al. 1987; Nautiyal et al. 2001). Variation in the phenophase period at spatio-temporal scale due to orography and micro-environmental variables were also observed (Nautiyal et al. 2001). The snowmelt timing with growth initiation and flowering has been observed (Fareed and Caldwell 1975; Bock 1976; Owen 1976), while ambient temperature during snowmelt was strongly favoured the growth initiation in alpine plants (Holway and Ward 1965; Kudo 1991; Kudo and Suzuki 1999; Kudo and Hirao 2006). The patterns of phenological events vary at spatio-temporal scale from species to species due to micro-environmental variables as well as orography of the region, which were also observed by Nautiyal et al. (2001) and the growth initiation depends on soil water availability due to snowmelt and rise in temperature (Ram et al. 1988).

17.6 Conclusion

The treeline ecotone at the Tungnath region is dominated by native hemicryptophytes. The plant density (325.8 ± 67.3 vs 183.9 ± 23.6), species richness (30.7 ± 1.22 vs 25.9 ± 1.10) and diversity (2.83 ± 0.05 vs 2.77 ± 0.05) were generally higher in EX microsites as compared to UX microsites. The herbaceous community changes across the months in all the microsites, except UX of Ravanshila and EX microsites of Chandrashila and Jhabra. Though the early snowmelt promotes species diversity and phonological events as observed in this study, there is a need for further experimentation on their relationships. The results of this study suggest that the herbaceous species in treeline ecotone are undergoing changes in phenological patterns due to the increase in temperature (minima) and precipitation period in the study area (1984 to 2017) in general. Some of the species entered into two distinct pheno-periods for reproductive and fruiting/seed maturation phenophases within the same growing period (*Oxygraphis, Fragaria* and *Ranunculus*), while some prefer vegetative propagation for continuity rather than progressing through fruiting and seeding instead. The early snowmelt and ambient temperature strongly favour early growth initiation, vegetative and flowering phases in a few of the species (of the total species (86) 16% in EX and 12% in UX microsites) in April. The availability of snowmelt water followed by monsoon promotes longer vegetative phenophase in most of the species (30% in EX and 31% in UX microsites of the total species). Duration of phenophase in general from the past indicates that they have become longer, while senescence has become truncated. This study provides an insight to the effect of changing climate on alpine vegetation, which requires further investigations to understand the changing dynamics of alpine communities in the light of changing temperature and precipitation regimes. The treeline research in the climate context is increasing and diversifying but important geographical gaps in the context to Himalaya remain. There is a need to shift from traditional monitoring and development and implementation of new tools and techniques to monitor alpine treeline biodiversity. Considering the current state of the literature on higher Himalaya, there is a need for establishing multi-dimensional and multi-institutional long term monitoring projects to understand the impact of changing climate and anthropogenic pressures on peri-glacial vegetation, with a strong emphasis on extreme climatic events in high altitudes ecosystems, viz. timberline, treeline and alpine.

References

Adhikari BS (2004) Ecological attributes of vegetation in and around Nanda Devi National Park. In: Biodiversity Monitoring Expedition Nanda Devi 2003. A report to the Ministry of Environment and Forests, Govt. of India Uttaranchal State Forest Department, Dehradun, pp 15–38

Adhikari BS, Rawat GS, Rai ID, Bhattacharyya S, Bharti RR (2012) Ecological assessment of timberline ecotone in Western Himalaya with special reference to climate change and anthropogenic pressures. Final Report, Wildlife Institute of India, Dehradun

Adhikari BS, Kumar R, Singh SP (2018) Early snowmelt impact on herb species composition, diversity and phenology in a western Himalayan treeline ecotone. Trop Ecol 59(2):365–382

Anonymous. Index Kewensis Plantarum Phanerogamarum. Vol 1–2:1883–1885 & 15 Suppl. 1886–1970. Clarendon Press, Oxford; pp 1883–1970

Beniston M (2003) Climatic change in mountain regions: a review of possible impacts. Clim Chang 59:5–31

Bijalwan R, Vats M, Joshi SP (2013) Plant phenological response to microclimatic variations in an alpine zone of Garhwal Himalaya. J Appl Nat Sci 5(1):47–52

Bock JH (1976) The effects of increased snowpack on the phenology and seed germinability of selected alpine species. In: Steinhoff HW, Ives JD (eds) Ecological impacts of snowpack augmentation in the San Juan Mountains of Colorado. Final report to the Division of Atmospheric Water Resources Management, Bureau of Reclamation, Denver, Colorado, USA

Bruun HH, Moen J, Virtanen R, Grytnes J-A, Oksanen L, Angerbjörn A (2006) Effects of altitude and topography on species richness of vascular plants, bryophytes and lichens in alpine communities. J Veg Sci 17:37–46

Campbell JL, Mitchell MJ, Groffman PM (2005) Winter in northeastern North America: an often overlooked but critical period for ecological processes. Front Ecol Environ 3:314–322

Dad M, Khan B (2010) Floristic composition of alpine grasslands in Bandipora, Kashmir. Grassl Sci 56:87–94

Dhar U, Kachroo P (1983) Alpine flora of Kashmir Himalaya. Scientific Publishers, Jodhpur

Fareed M, Caldwell MM (1975) Phenological patterns of two alpine tundra plant populations on Niwot Ridge, Colorado. Northw Sci 49:17–23

Gehrmann F, Hänninen H, Liu C, Saarinen T (2018) Phenological responses to small-scale spatial variation in snowmelt timing reveal compensatory and conservative strategies in subarctic-alpine plants. Plant Ecol Divers 10(5–6):453–468

Gong S, Guo R, Zhang T, Guo J (2015) Warming and nitrogen addition increase litter decomposition in a temperate meadow ecosystem. PLoS One 10:e0116013

Gorsuch DM, Oberbauer SF (2002) Effects of mid-season frost and elevated growing season temperature on stomatal conductance and specific xylem conductivity of the arctic shrub, Salix pulchra. Tree Physiol 22:1027–1034

Gupta VC, Kachroo P (1981) Relation between photosynthetic structure and standing biomass of meadowland communities of Yusmarg in Kashmir Himalayas

Hättenschwiler S, Körner C (1997) Annual CO_2 budget of spruce model ecosystems in the third year of exposure to elevated CO_2. Acta Oecol 18:319–325

Hess M, Barraljs G, Bleiholder H, Buhr L, Eggers T, Hack H, Stauss R (1997) Use of the extended BBCH scale – general for the descriptions of the growth stages of mono- and dicotyledonous weed species. Weed Res 37:433–441

Hiltbrunner E, Schwikowski M, Körner C (2005) Inorganic nitrogen storage in alpine snow pack in the Central Alps, Switzerland. Atmos Environ 39:2249–2259

Holway JG, Ward RT (1965) Phenology of alpine plants in northern Colorado. Ecology 461(2): 73–83

Inouye DW (2000) The ecological and evolutionary significance of frost in the context of climate change. Ecol Lett 3:457–463

Inouye DW, Wielgolaski FE (2003) High altitude climates. In: Schwartz MD (ed) Phenology: an integrative environmental science. Kluwer, Dordrecht, pp 195–214

Jones HG, Pomeroy JW, Walker DA, Hoham RW (2001) Snow ecology, 1st edn. Cambridge University Press, Cambridge

Kala CP, Rawat GS, Uniyal VK (1998) Ecology and conservation of the valley of flowers National Park, Garhwal Himalaya. Wildlife Institute of India, Dehra Dun, p 99

Kershaw RA (1973) Quantitative and dynamic plant ecology. Edward Arnold Ltd, London

Khorsand RR, Oberbauer SF, Starr G, La P, Puma I, Pop E, Ahlquist L, Baldwin T (2015) Plant phenological responses to a long-term experimental extension of growing season and soil warming in the tussock tundra of Alaska. Glob Chang Biol 21:4520–4532

Körner C (1998) A re-assessment of high elevation treeline positions and their explanation. Oecologia 115(4):445–459

Korner C (2003) Alpine plant life, 2nd edn. Springer Verlag, Berlin

Kudo G (1991) Effects of snow-free period on the phenology of alpine plants inhabiting snow patches. Arct Antarct Alp Res 23:436–443

Kudo G, Hirao AS (2006) Habitat-specific responses in the flowering phenology and seed set of alpine plants to climate variation: implications for global-change impacts. Popul Ecol 48:49–58

Kudo G, Suzuki S (1999) Flowering phenology of alpine plant communities along a gradient of snowmelt timing. Polar Biosci 12:100–113

Kumar A, Mitra M, Adhikari BS, Rawat GS (2016) Flora of Niti Valley: a cold arid region of Nanda Devi biosphere reserve, Western Himalaya, India. Check List 12(1):1824

Le Roux PC, Aalto J, Luoto M (2013) Soil moisture's underestimated role in climate change impact modelling in low-energy systems. Glob Chang Biol 19:2965–2975

Livensperger C, Steltzer H, Darrouzet-Nardi A, Sullivan PF, Wallenstein MD, Weintraub MN (2016) Earlier snowmelt and warming lead to earlier but not necessarily more plant growth. AOB Plants 8:plw021

Lloyd AH, Fastie CL (2002) Spatial and temporal variability in the growth and climate response of treeline trees in Alaska. Clim Chang 52(4):481–509

Malik ZA, Nautiyal MC (2016) Species richness and diversity along the altitudinal gradient inTungnath, the Himalayan benchmark site of HIMADRI. Trop Plant Res 3(2):396–407

Mooney HA, Billings WD (1960) The annual carbohydrate cycle of alpine plants as related to growth. Am J Bot 47(7):594–598

Nautiyal BP (1996) Studies on structure and function in an alpine meadow of Garhwal, central Himalaya. PhD Thesis, HNB Garhwal University, Srinagar, India
Nautiyal MC, Nautiyal BP, Prakash V (2001) Phenology and growth form distribution in an alpine pasture at Tungnath, Garhwal, Himalaya. Mt Res Dev 21(2):168–174
Nautiyal MC, Nautiyal BP, Prakash V (2004) Effect of grazing and climatic changes on alpine vegetation of Tungnath, Garhwal Himalaya, India. Environmentalist 24:125–134
Odland A, Munkejord HK (2008) Plants as indicators of snow layer duration in southern Norwegian mountains. Ecol Indic 81:57–68
Owen HE (1976) Phenological development of herbaceous plants in relation to snowmelt date. In: Steonhoff HW, Ives JD (eds) Ecological impacts of snowpack augmentation in the San Juan Mountains of Colorado. U.S. Department of Interior, Division of Atmospheric Water Resources Management, Bureau of Reclamation, Denver, Colorado, USA, pp 323–341
Pangtey YPS, Rawal RS, Bankoti NS, Samant SS (1990) Phenology of high-altitude plants of Kumaun in central Himalaya, India. Int J Biometeorol 342:122–127
Park JS, Son D, Lee YK, Yun JH, Lee EJ (2018) Multivariate relationship between snow melt and plant distribution in high arctic tundra. J Plant Biol 61(1):32–39
Parmesan C (2006) Ecological and evolutionary responses to recent climate change. Annu Rev Ecol Evol Syst 37:637–669
Pauli H, Gottfried M, Reiter K, Klettner C, Grabherr G (2007) Signals of range expansions and contractions of vascular plants in the high Alps: observations (1994-2004) at the GLORIA master site Schrankogel, Tyrol, Austria. Glob Chang Biol 13:147–156
Pauli H, Gottfried M, Dullinger S, Abdaladze O, Akhalkatsi M, Alonso JLB, Coldea G, Dick J, Erschbamer B, Calzado RF, Ghosn D, Holten JI, Kanka R, Kazakis G, Kollar J, Larsson P, Moiseev P, Moiseev D, Molau U, Mesa JM, Nagy L, Pelino G, Puscas M, Rossi G, Stanisci A, Syverhuset AO, Theurillat J-P, Tomaselli M, Unterluggauer P, Villar L, Vittoz P, Grabherr G (2012) Recent plant diversity changes on Europe's mountain summits. Science 336:353–355
Pérez-Harguindeguy N, Díaz S, Garnier E, Lavorel S, Poorter H, Jaureguiberry P, Bret-Harte MS, Cornwell WK, Craine JM, Gurvich DE, Urcelay C, Veneklaas EJ, Reich PB, Poorter L, Wright IJ, Ray P, Enrico L, Pausas JG, de Vos AC, Buchmann N, Funes G, Quétier F, Hodgson JG, Thompson K, Morgan HD, terSteege H, van der Heijden MGA, Sack L, Blonder B, Poschlod P, Vaieretti MV, Conti G, Staver AC, Aquino S, JHC C (2013) New handbook for standardised measurement of plant functional traits worldwide. Aust J Bot 61:167–234
Petraglia A, Tomaselli M, Petit Bon M, Delnevo N, Chiari G, Carbognani M (2014) Responses of flowering phenology of snowbed plants to an experimentally imposed extreme advanced snowmelt. Plant Ecol 215:759–768
Pielou EC (1975) Ecological diversity. Wiley, New York
Pomeroy JW, Brun E (2001) Physical properties of snow. In: Jones HG, Pomeroy JW, Walker DA et al (eds) Snow ecology. Cambridge University Press, Cambridge, pp 45–126
Rai ID, Adhikari BS, Rawat GS, Bargali K (2012) Community structure along timberline ecotone in relation to micro-topography and disturbances in Western Himalaya. Not Sci Biol 4(2):41–52
Ram J, Arya P (1991) Plant forms and vegetation analysis of an alpine meadow of central Himalaya, India. Proc Indian Natl Sci Acad 57(5):311–318
Ram J, Singh SP, Singh JS (1988) Community level phenology of grassland above treeline in central Himalaya, India. Arct Alp Res 20(3):325–332
Rana TS, Datt B, Rao RR (2002) Life forms and biological spectrum of the flora of Tons Valley, Garhwal Himalaya (Uttaranchal), India. Taiwania 47(2):164–169
Raunkier R (1934) The life form of plants and statistical geography. The Clarendon Press, Oxford
Rawat GS (2007) Alpine vegetation of the western Himalaya: species diversity, community structure, dynamics and aspects of conservation. D.Sc. Thesis, Kumaun University, Nainital
Samant SS, Joshi HC (2005) Plant diversity and conservation status of Nanda Devi National Park and comparison with highland National Parks of the Indian Himalayan region. Int J Biodivers Sci Manag 1:65–73

Samant SS, Dhar U, Palni LMS (1998) Medicinal plants of Indian Himalaya: diversity distribution potential values. GyanodayaPrakashan, Nainital
Samant SS, Joshi HC, Pant S, Arya SC (2001) Diversity, nativity and endemism of vascular plants of Valley of Flowers National Park. Himal Biosci Res Bia Bull 3(1–2):1–17
Shannon CE, Wiener W (1949) The mathematical theory of communication. University of Illinois Press, Urbana, IL
Singh S (1999) Assessment of floral and habitat diversity and collection of base line data to monitor vegetation of GHNP conservation area. Indian Institute of Remote Sensing Dehradun, India
Singh SP, Bassignana-Khadka I, Karky BS, Sharma E (2011) Climate change in the Hindu Kush Himalayas. ICIMOD, Kathmandu
Smith JG, Sconiers W, Spasojevic MJ, Ashton IW, Suding KN (2012) Phenological changes in alpine plants in response to increased snowpack, temperature, and nitrogen. Arct Antarct Alp Res 44(1):135–142
Sorensen T (1948) A method of establishing groups of equal amplitude in plant sociology based on similarity of species content and its application to analyses of the vegetation on Danish commons. Videnski Selskab Biol Skr 5:1–34
Sturm M, Schimel J, Michaelson G (2005) Winter biological processes could help convert arctic tundra to shrubland. Bioscience 55:17–26
Sturm W, Willmes K, Orgass B, Hartje W (1997) Do specific attention deficits need specific training? Neuropsychol Rehabil 7:81–103
Stylinski CD, Allen EB (1999) Lack of native species recovery following severe exotic disturbance in southern Californian shrublands. J Appl Ecol 36(4):544–554
Sundriyal RC, Joshi AP, Dhasmana R (1987) Phenology of high altitude plants at Tungnath in the Garhwal Himalaya. Trop Ecol 28:289–299
Tambe S (2007) Ecology and management of alpine landscape in Khangchendzonga National Park, Sikkim Himalaya. Ph.D. Thesis, FRI University, Dehradun, India
Tørseth K, Semb A (1997) Atmospheric deposition of nitrogen, Sulphur and chloride in two watersheds located in southern Norway. Ambio 26:258–265
Vashistha RK, Rawat N, Chaturvedi AK, Nautiyal BP, Prasad P, Nautiyal MC (2011) Characteristics of life-form and growth-form of plant species in an alpine ecosystem of north-west Himalaya. J For Res 22:501
Vitousek PM (1986) Biological invasions and ecosystem properties. Can species make a difference? In: Mooney HA, Drake JA (eds) Ecology of bilogical invasions of North America and Hawaii, Ecological Studies, vol 58. Springer-Verlag, New York, pp 163–178
Walker MD, Webber PJ, Arnold EH, Ebert-May D (1994) Effects of interannual climate variation on aboveground phytomass in alpine vegetation. Ecology 75(2):393–408
Weih M, Karlsson PS (2002) Low winter soil temperature affects summertime nutrient uptake capacity and growth rate of mountain birch seedlings in the subarctic, Swedish Lapland. Arct Antarct Alp Res 34:434–439
Winkler DE, Butz RJ, Germino MJ, Reinhardt K, Kueppers LM (2018) Snowmelt timing regulates community composition, phenology, and physiological performance of alpine plants. Fron Plant Sci 9:1140
Wipf S (2010) Phenology, growth, and fecundity of eight subarctic tundra species in response to snowmelt manipulations. Plant Ecol 207(1):53–66
Wipf S, Rixen C (2010) A review of snow manipulation experiments in Arctic and alpine tundra ecosystems. Polar Res 291:95–109
Wipf S, Stoeckli V, Bebi P (2009) Winter climate change in alpine tundra: plant responses to changes in snow depth and snowmelt timing. Clim Chang 94:105–121
Xu J, Grumbine ER, Shrestha A, Eriksson M, Yang X, Wang Y, Wilkes A (2009) The melting Himalayas: cascading effects of climate change on water, biodiversity, and livelihoods. Conserv Biol 23(3):520–530

Responses of Herbaceous Species of Alpine Treeline to Elevated CO_2 18

Sudeep Chandra, Vaishali Chandola, Ankit Singh, C. P. Singh, and M. C. Nautiyal

Abstract

Adaptation of plant species to cold environment and short growing season in the Himalayan high altitudes could render them highly sensitive to future climate change. In this study, we analyzed the effect of elevated CO_2 concentration on the growth, productivity, physiology, and various biochemical parameters of four alpine treeline herbaceous species, viz. *Acomastylis elata*, *Anaphalis nepalensis*, *Bistorta macrophylla,* and *Trillium govanianum*. We planted seedlings of the selected plant species in open top chambers and CO_2 concentration was raised from ambient (400 ppm) to elevated (650 ppm) levels. Elevated CO_2 stimulated net assimilation rate (34–38%), growth and productivity of *A. elata* and *A. nepalensis,* whereas *B. macrophylla* and *T. govanianum* showed decrease (18%) in photosynthesis. The sugar content in all the species increased (36–78%); however, foliar N decreased (17–37%), possibly due to dilution effect of high carbohydrate content. Reduced tissue N can probably affect the activity of key photosynthetic enzyme Rubisco and therefore, decreased carbon assimilation. From this study, it can be deduced that long-term effects of elevated CO_2 can be species specific and might be affected by availability of nutrients. More such studies in Himalayan regions involving different plant communities are needed to develop a better understanding of plant responses to climate change.

S. Chandra · V. Chandola · A. Singh · M. C. Nautiyal (✉)
High Altitude Plant Physiology Research Center, Hemwati Nandan Bahuguna Garhwal University, Srinagar Garhwal, Uttarakhand, India

C. P. Singh
EPSA, Space Applications Centre (SAC), Indian Space Research Organisation, Ahmedabad, Gujarat, India

S. P. Singh et al. (eds.), *Ecology of Himalayan Treeline Ecotone*,
https://doi.org/10.1007/978-981-19-4476-5_18

Keywords

Alpine treeline · Biomass · Elevated CO_2 · Herbaceous species · Photosynthesis

18.1 Introduction

The carbon dioxide (CO_2) concentration in the earth's atmosphere has varied over the geological time scale. CO_2 concentrations as low as 180–200 ppm had been known to exist during the last two ice ages around 13,000–30,000 and 140,000–160,000 years ago. Direct atmospheric sampling data have shown that CO_2 increased from 315 ppm in the late 1950s to 355 ppm in the early 1990s (Keeling et al. 1989) and has been rising continuously since then. The rise in atmospheric CO_2 has become more profound since the beginning of the industrial revolution and is expected to continue in the coming years with drastic effects on the earth's climate. The rise in atmospheric carbon dioxide will not only affect climatic events but is likely to have significant implications on the global carbon cycle and plants life too. Plants of the alpine regions are specifically more sensitive toward changes in the climatic factors chiefly temperature.

The upper limit of a closed forest is called as timberline, whereas a treeline is the upper limit of the "tree" life form showing an upright growth. The transitional zone between the timberline and treeline is called as the alpine treeline ecotone (Körner 2012). The trees that generally form treeline are *Abies spectabilis*, *Sorbus foliolosa*, *Betula utilis*, etc. Many herbaceous species are also found growing below canopies, along, and above treelines. Herbs are one of the major flowering plant life-forms in these regions. It has been estimated that approximately 2.6% land area besides Antarctica, covering the alpine belt, inhabits about 4% of all the flowering plant species on Earth. In other words, the alpine regions are quite rich in plant diversity than the lower elevation ecosystems (Körner 2018).

A number of herbaceous species can be restricted to the treeline or below it due to a more hospitable habitat than the open alpine meadows which are directly exposed to abiotic extremities, viz heavy snowfall, frost, and solar radiation. This higher distribution of herbaceous species is mainly based on their morphological adaptations, like small stature and dense structure, which help them restrict aerodynamic exchange with the atmosphere. This causes heat accumulation from solar radiation providing warm temperatures to operate unlike trees, which are upright and more ventilated. Alpine herbaceous plants produce short lived leaves and the sunlight heated soil provides a thermally buffered environment to their meristems which are positioned close to the ground in comparison to trees which form longer living and slowly maturing leaves along with meristems positioned above from the ground where they are fully exposed to cold temperatures (Körner 2018).

Increasing CO_2 concentration can cause positive effect on plant growth with an increase in leaf area (Allen Jr 1991). Plants that are exposed to higher levels of CO_2 usually have amplified growth, water use efficiency, and photosynthesis rate (Amthor 1995; Wittwer 1995). Being a very important raw material for the process of photosynthesis, an increase in atmospheric CO_2 concentration can have direct effect on plant photosynthesis. However, plants growing in alpine region and glacier fore-fields in the Swiss Alps were found to be carbon saturated at ambient CO_2 concentrations of 390 ppm (Körner 2018). It was further reported that doubling the concentration of atmospheric CO_2 for over four successive seasons had no effect on net primary productivity of alpine plants with exception of some species. Hence, in long run, a possibility of slow but certain shifts in species composition is expected with some species getting suppressed and some gaining ground (Körner 2018).

At higher altitudes, the impact of elevated CO_2 on plants can vary due to low ambient partial pressure of CO_2, short growing season, and a number of abiotic stresses mainly extreme low-temperature regimes. CO_2 exposure duration can also cause varying responses in plants as initial exposure can enhance the net assimilation rate and biomass (Ainsworth et al. 2008; Chaturvedi et al. 2009; Chaturvedi et al. 2013). With increasing duration of CO_2 exposure plant can show photosynthetic acclimation, lowering of leaf chlorophyll and nitrogen followed by declined (Zelikova et al. 2014) or no significant variation in biomass (Schäppi and Körner 1997; Ward et al. 1999).

Initial studies on the effect of warming have shown that the Himalayas have a higher warming, more prominent at higher altitudes, as compared to the global average rate (Shrestha et al. 1999; Liu and Chen 2000) with higher increases in the winter and autumn temperatures than the summer. There have been only a few studies regarding the impact of CO_2 enrichment on the Himalayan alpine species (Joshi et al. 2007; Chaturvedi et al. 2009; Chaturvedi et al. 2013) and one long-term study on alpine grassland biomass and community structure (Zhu et al. 2020). However, a considerate amount of work and long-term studies on warming and CO_2 enrichment effect on the European Alps treeline, community structure, and species have been carried out (Schäppi and Körner 1996; Schäppi and Körner 1997; Inauen et al. 2012; Dawes et al. 2013). The treeline in the Himalayan highlands ranges from 3200 m to 4900 mamsl (Singh et al. 2019) bearing the highest peaks of the world, whereas the treeline in the Alps ranges from 1750 m to slightly above 2350 mamsl from north to south (Paulsen and Körner 2001). Therefore, the responses of treeline species and community structure to climate change particularly relating to species migration and distribution patterns at the Himalayas might differentiate from the other alpine regions.

Since the alpine regions are dominated by herbaceous species which are also a key component of biodiversity, this study aims to estimate the effect of elevated CO_2 on growth-productivity, physiology, and biochemistry of alpine–treeline herbs. Due to low temperature and partial pressure of gases in the alpine regions, the effect of CO_2 on plant speciesof these regions can vary from species of lower elevations, however, the photosynthesis rate, pigment content, and biomass of the herbs were expected to increase as a result of CO_2 fertilization effect as found in some earlier studies. Since in the earlier studies these responses have been found varying in different species; therefore, four species were subjected to CO_2 treatment in this study.

In order to properly understand the treeline dynamics with reference to anthropogenic warming of the planet, it is essential to carry out more reliable assessments of forest treelines across the globe to preserve the ecological state of alpine ecotone biodiversity and to forge forest conservation policies (Holtmeier and Broll 2010; Mishra and Mainali 2017.

18.2 Materials and Methods

18.2.1 Study Site

This study was conducted at alpine field station of High Altitude Plant Physiology Research Center at Tungnath, (3400 m amsl, 30°14′N and 79°13E) in the Western-Central Himalaya, India. The study site is a part of Kedarnath Wild Life Sanctuary and exhibits a number of endemic species. The area usually remains snow covered from December to May and the growing period for herbaceous species is between May to October. The study area generally receives an annual rainfall ranging from of 300 to 500 cm with the month of July and August being the wettest. Daily mean temperatures from June to September fluctuate between 8 and 18 °C.

18.2.2 Growth Conditions

Two open top chambers (OTCs) were established at Tungnath. The OTCs were round in structure having 4 × 4 m (height × diameter) dimension. Due to high wind velocity at the study site, OTCs were preferred over free air CO_2 enrichment (FACE) experiment. The frame of the OTCs was made of galvanized iron pipes. The OTCs were covered with polycarbonate sheet which provided 90% transmittance of light. The OTCs were equipped with programmable logic controller coupled with supervisory control and data acquisition (SCADA) system which automatically recorded air temperature, relative humidity, and CO_2 concentration. The CO_2 concentration of one OTC was raised and maintained to 650 ppm during the growing period (June to mid-October). This performed automatically via a solenoid valve controlled by SCADA system.

18.2.3 Plant Materials

Morphologically similar seedlings of *Acomastylis elata* (Wall. ex G.Don) F.Bolle (Syn: *Geum elatum*, family: Rosaceae), *Anaphalis nepalensis* (Spreng.) Hand.-Mazz (Asteraceae), *Bistrota macrophylla* (D. Don) Soják (Polygonaceae), and *Trillium govanianum* Wall. ex. D. Don., (Melanthiaceae) were collected from surrounding areas and planted in triplicates (each having 10 seedlings) in each OTC. These species are naturally distributed from treeline to upper alpine regions, while *T. govanianum* grows in subalpine regions. Seedlings of all the 4 species were transplanted to OTCs and left for a month to acclimatize, after which the treatment was given. The OTCs were established at native site of these species, the soil inside the OTCs was not modified, and no fertilizers were added. Transplantation was done in the month of July in the first year (2019) and therefore, the treatment in the first season was given only for 30 ± 2 days. In the second year (2020), the treatment lasted from June to September. The seedlings growing under elevated CO_2 chamber were compared with seedlings growing at ambient CO_2 concentration in an OTC.

18.2.4 Gas Exchange Measurement

Leaf gas exchange parameters were recorded on leaves of nine randomly selected individuals (3 from each replicate) of each species from each chamber using Licor, Li-6400, a portable infrared gas analyzer (IRGA). The instrument was equipped with red blue LED light source and CO_2 mixer (for maintaining desired CO_2 concentration). The photosynthetic rate (Pn, μmol $CO_2/m^2/s$), stomatal conductance (gs, mol $H_2O/m^2/s$), transpiration rate (E, mmol $H_2O/m^2/s$), and intracellular CO_2 (Ci, μmol CO_2/mol air) were recorded between 9:00 and 11:00 h. The study was conducted under saturated light conditions (photosynthetic photon flux density—1250 μmol/ m^2-s). The water use efficiency (WUE = Pn/E) and intrinsic water use efficiency (iWUE = Pn/gs) were calculated.

18.2.5 Pigment, Carbohydrate, and Nitrogen Estimation

Leaf samples of all the fourspecies were collected (5 samples for each species) from both ambient (aCO_2) and elevated (eCO_2) chambers. Samples were further analyzed through UV-VIS spectrophotometer for chlorophyll a, b and total carotenoids following Holm et al. (1965) and total soluble sugars following (McCready et al., 1950). Total N was estimated from oven-dried leaves using Kjeldhal Nitrogen analyzer (Pelican Instruments).

18.2.6 Growth and Biomass

Plant height, leaf area, aboveground, and belowground biomass were recorded on five randomly selected individuals from each species grown under aCO_2 and eCO_2 chambers.

18.2.7 Data Analysis

All the growth parameters of plants from aCO_2 and eCO_2 were recorded after 120 days of CO_2 treatment. One way analysis of variance was performed to assess the effect of elevated CO_2 and further significant difference between ambient and elevated CO_2 was tested using paired sample t-test for each parameter.

18.3 Results

The daytime air temperature under eCO_2 increased by 0.63 °C and the temperature difference between aCO_2 and eCO_2 was greater in May and June. The total precipitation (Measured by tipping bucket rain gauge) during our study was 2130 mm and 60 to 80% of total precipitation took place between July and August (Fig. 18.1).

CO_2 enrichment experiment showed significant changes in physiology, growth, and biochemistry of herbaceous species; however, the effects are species specific.

18.3.1 Effect on Gaseous Exchange

CO_2 exposure strongly affected the physiological process of photosynthesis (Pn). Pn under eCO_2 significantly ($p = 0.05$) increased by 38 and 34% in *A. nepalensis* and *A. elata*, respectively; whereas it decreased (18%, $p < 0.05$) in *T. govanianum* and remained unaffected in *B. macrophylla*. The stomatal conductance (gs) under eCO_2 increased significantly in *A. elata* (+59%, $p < 0.01$); other species showed non-significant reduction in gs. CO_2 is considered as an ideal antitranspirant and in our observation we found reduced transpiration rate (E) under eCO_2 for

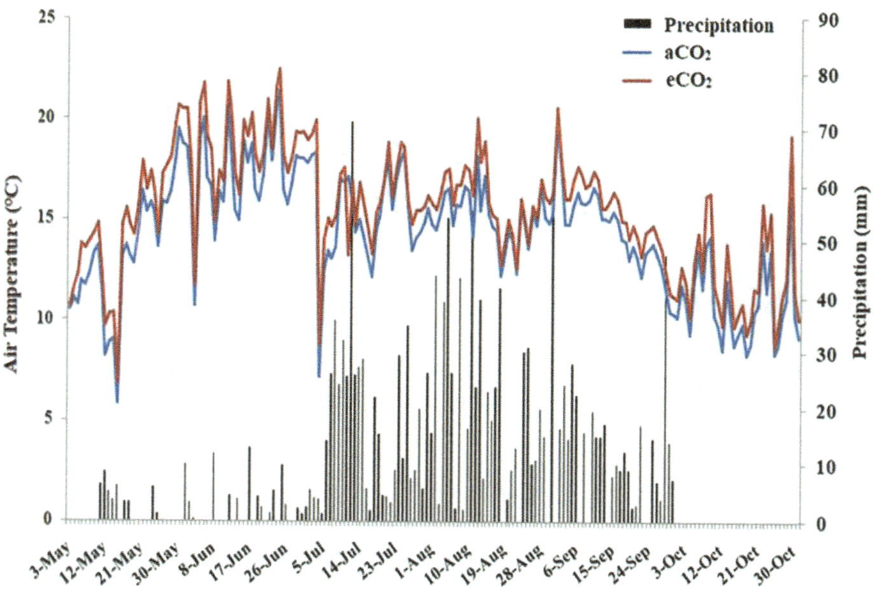

Fig. 18.1 Daily daytime air temperature in ambient and elevated open top chambers and the precipitation received by study area

Table 18.1 Gas exchange parameters of four herbaceous species grown under ambient and elevated CO_2

		Acomastylis elata	*Anaphalis nepalensis*	*Bistrota macrophylla*	*Trillium govanianum*
Photosynthesis rate	aCO_2	43.55 ± 7.45	15.10 ± 1.75	43.83 ± 10.99	18.4 ± 1.64
	eCO_2	58.36 ± 5.73	20.90 ± 3.65	40.23 ± 24.40	15.6 ± 1.04
p		0.046*	0.045*	0.83	0.037*
Stomatal conductance	aCO_2	1.42 ± 0.28	0.84 ± 0.17	1.64 ± 0.33	0.17 ± 0.02
	eCO_2	2.27 ± 0.29	0.68 ± 0.11	1.19 ± 0.35	0.13 ± 0.01
p		0.008**	0.11	0.18	0.06
Transpiration rate	aCO_2	5.47 ± 0.85	5.91 ± 1.31	9.18 ± 1.09	2.73 ± 0.21
	eCO_2	7.45 ± 1.09	3.82 ± 1.00	7.68 ± 0.37	1.71 ± 0.14
p		0.002**	0.04*	0.087	0.001**
WUE	aCO_2	9.95 ± 3.28	2.55 ± 0.65	4.86 ± 1.62	6.74 ± 0.62
	eCO_2	7.75 ± 1.22	5.47 ± 1.25	5.24 ± 2.86	9.13 ± 0.83
p		0.338	0.008**	0.86	0.016*
iWUE	aCO_2	29.54 ± 5.26	17.97 ± 3.79	28.45 ± 8.51	111.79 ± 19.37
	eCO_2	25.86 ± 3.63	35.89 ± 7.54	32.15 ± 10.76	124.13 ± 12.34
p		0.382	0.02*	0.74	0.41

The significant difference is shown by asterisk, $*p < 0.05$, $**p < 0.01$
WUE: Water use efficiency; *iWUE*: intrinsic Water use efficiency

A. nepalensis (−54%, $p < 0.05$), *B. macrophylla* (−54%, 0.08), and *T. govanianum* (−59%, <0.01); however, *A. elata* showed an increased E (+36%, <0.01). Elevated CO_2 improved the water use efficiency (WUE) in *A. nepalensis* (+2-folds) and *T. govanianum* (+35%), whereas WUE of *A. elata* and *B. macrophylla* was unaffected. The intrinsic WUE (iWUE) significantly improved by twofolds only in *A. nepalensis* (Table 18.1).

18.3.2 Effect on Pigments, Carbohydrates, and Nitrogen Content

CO_2 enrichment affected the pigment contents and the effect varied from species to species. The chl a and b (chlorophyll a and b) significantly ($p < 0.05$) increased in *A. elata* (+19%) and *A. nepalensis* (+21%) under eCO_2. Other species, i.e., *T. govanianum* and *B. macrophylla,* showed reduced chl a content under eCO_2; however, the chl b was unaffected. The total carotenoids showed most prominent response toward eCO_2. All species growing at eCO_2 chamber showed increased carotenoids (+23 to 46%). The TSS under eCO_2 increased significantly ($p < 0.05$) for *A. elata* (+46%), *A. nepalensis* (+36%), *T. govanianum* (+78%), and non-significantly for *B. macrophylla* (15%). The plant N in eCO_2 significantly

Table 18.2 Pigment contents, total soluble sugars and leaf nitrogen in four herbaceous species from ambient (aCO_2) and elevated (eCO_2) CO_2

		Acomastylis elata	*Anaphalis nepalensis*	*Bistrota macrophylla*	*Trillium govanianum*
Chl a	aCO_2	2.96 ± 0.17	1.33 ± 0.06	2.38 ± 0.37	4.37 ± 0.78
	eCO_2	3.54 ± 0.15	1.62 ± 0.14	2.09 ± 0.20	3.01 ± 0.08
p		0.011*	0.031*	0.259	0.04*
Chl b	aCO_2	2.17 ± 0.12	0.73 ± 0.08	1.47 ± 0.31	1.87 ± 0.12
	eCO_2	2.54 ± 0.08	0.93 ± 0.06	1.42 ± 0.18	1.93 ± 0.16
p		0.01*	0.025*	0.827	0.628
Carotenoids	aCO_2	1.11 ± 0.12	0.63 ± 0.05	0.98 ± 0.07	0.99 ± 0.2
	eCO_2	1.45 ± 0.12	0.92 ± 0.03	1.29 ± 0.06	1.22 ± 0.18
p		0.026*	0.001**	0.004**	0.22
TSS	aCO_2	26.1 ± 2.26	31.61 ± 3.48	22.57 ± 3.04	19.7 ± 4.43
	eCO_2	38.17 ± 4	43.1 ± 2.82	26.02 ± 1.81	35.22 ± 5.46
p		0.01*	0.011*	0.167	0.019*
N content	aCO_2	1.08 ± 0.05	1.11 ± 0.09	0.99 ± 0.12	1.10 ± 0.11
	eCO_2	0.92 ± 0.06	0.9 ± 0.08	0.94 ± 0.06	0.80 ± 0.08
p		0.028*	0.038*	0.536	0.019*

Significant difference is shown by asterisk, $*p < 0.05$, $**p < 0.01$

reduced for *A. elata* (−17%), *A. nepalensis* (−23%), and *T. govanianum* (−37%), although the effect of eCO2 on N reduction of *B. macrophylla* was non-significant (Table 18.2).

18.3.3 Effect on Growth and Biomass

Under eCO_2, the plant height and leaf area of *A. nepalensis* and *A. elata* significantly increased; however, the species *B. macrophylla* and *T. govanianum* did not show any significant change in morphometrics. Height of *A. nepalensis* and *A. elata* increased by 55 and 24%, respectively under eCO_2 (Fig. 18.2). Similarly, leaf area in these two species also increased by 97 and 32%, respectively (Fig. 18.3). The accumulation of dry matter in shoot and root system was measured and in comparison to aCO_2, the aboveground biomass in eCO_2 increased significantly by 50 and 116% in *A. elata* and *A. nepalensis*, respectively. The below ground biomass in these two species also increased, although the increment was significant only in *A. nepalensis* (+55%), while the other two species did not show any significant change in above and belowground biomass (Fig. 18.4).

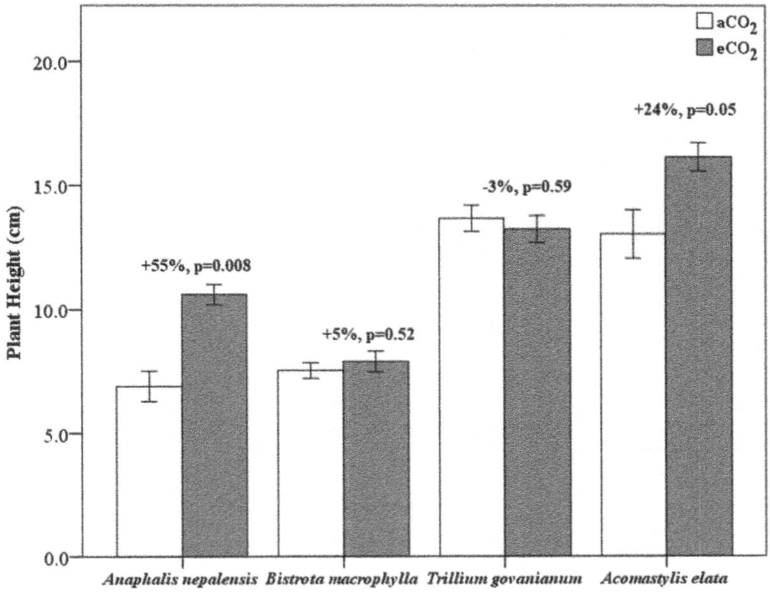

Fig. 18.2 Plant height of four herbaceous species under ambient (aCO_2) and elevated (eCO_2) CO_2. The significant difference was tested by paired sample t-test

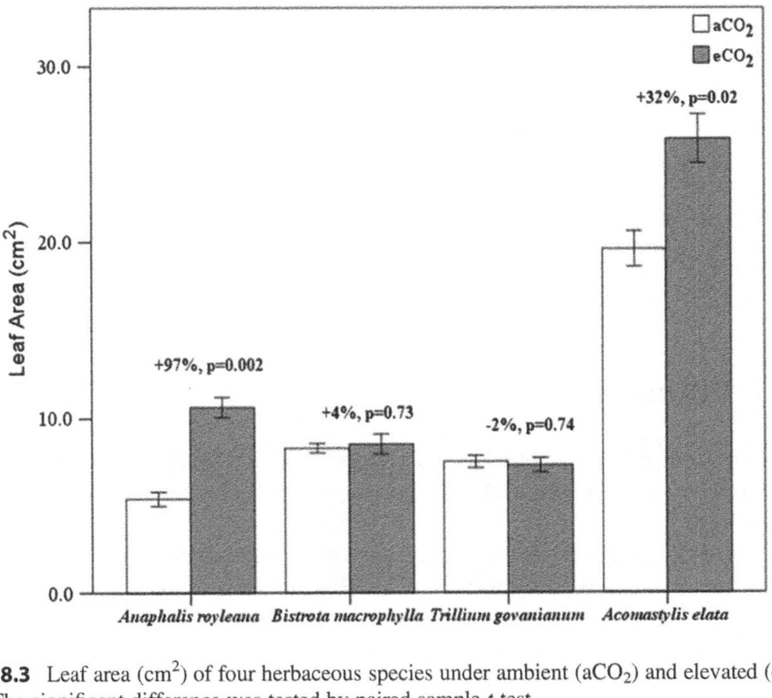

Fig. 18.3 Leaf area (cm^2) of four herbaceous species under ambient (aCO_2) and elevated (eCO_2) CO_2.The significant difference was tested by paired sample t-test

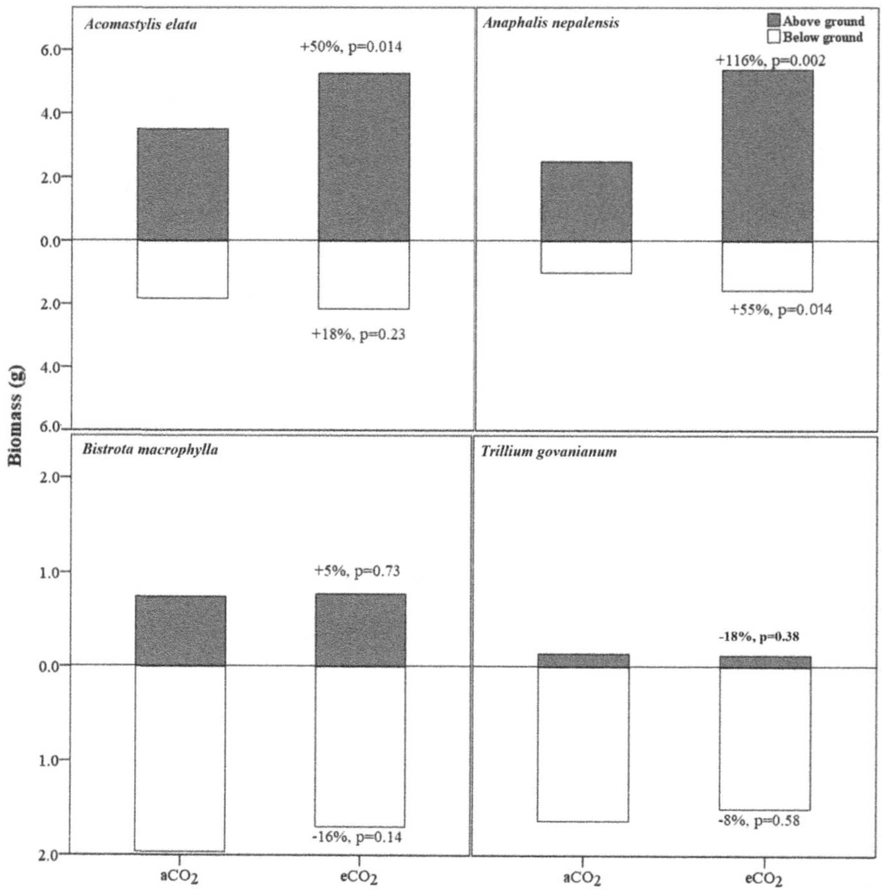

Fig. 18.4 Above and belowground biomass of four herbaceous species in ambient and elevated CO_2 chambers. The percentage increased or decreased was calculated and tested using paired sample *t*-test

18.4 Discussion

This study brings out the short-term exposure effects of CO_2 enrichment on four alpine treeline ecotone herbaceous species having different leaf structures and morphology.

Elevated CO_2 led to an increase in the daytime air temperature inside the chamber as compared to ambient CO_2 levels. This can be attributed to the heat absorbing and radiating nature of CO_2 and its activity as a greenhouse gas. Elevated CO_2 led to an increase in the photosynthesis rate in two species namely *A. elata* and *A. nepalensis*. Similar results were obtained in a number of other studies (Ainsworth and Long

2005; van der Kooi et al. 2016) where elevated CO_2 caused a rise in the photosynthetic rate with subsequent rise in growth and biomass. A higher CO_2 concentration has a fertilizing effect especially on C3 plants because Rubisco in C3 plants is not CO_2 saturated at its current atmospheric levels (Yu et al. 2012; Singh and Reddy 2016). Elevated CO_2 also reduces the instance of photorespiration, thereby having positive effects on the carboxylase activity of Rubisco (Zheng et al. 2019). However, the other two species, viz. *T. govanianum* and *B. macrophylla* either had negative or no significant effect on their photosynthetic activity as a result of elevated CO_2. This could be possibly because of differences in their carbon saturation threshold and the ability to adapt to a carbon richer environment. The down-regulation photosynthetic activity could also be due to lower concentration and activity of Rubisco. Kanemoto et al. (2009) also reported a decline in the photosynthetic activity of soybean plants under elevated CO_2. In this study, a decline in the plant nitrogen (N) content was also seen. N is an important element of tissue protein and amino acid, and under elevated CO_2 a decline in tissue N may affect the concentration of Rubisco.

The stomatal conductance in three out of the four studied species was found to be declined. Increased levels of atmospheric CO_2 cause smaller stomatal apertures, thereby decreasing leaf conductance for water vapor (Morison 1987). A decrease in stomatal conductance of these species also led to a decline in the transpiration rate. Stomatal conductance of *A. elata* increased along with an increase in its transpiration rate. Changes in stomatal conductance and transpiration rate are also greatly influenced by the species type (Ward et al. 2013; Haworth et al. 2013). Stomatal conductance and transpiration rates are also influenced by changes in stomatal density, which have been found to be either decreased or increased in different species under elevated CO_2 (Xu et al. 2016). The WUE was significantly increased in *T. govanianum* and *A. nepalensis* as a result of decline in the transpiration rate. In the other two species, viz. *A. elata* and *B. macrophylla* WUE remained unaffected. A significant increase in the pigment content, i.e., chlorophyll in *A. nepalensis* and *A. elata,* can be correlated with significantly higher Pn in both the species. Pigments trap light energy which is utilized to convert trapped carbon to carbohydrates by the process of photosynthesis. An increase in the carotenoid content of all the species has been found under elevated CO_2. Carotenoids moderate the effect of increased temperature by protecting the plants from photo-oxidative stress (Strzalka et al. 2003). Carotenoid content has shown variable responses under elevated CO_2 in a number of studies. Some studies have reported increase, whereas some had reported a decline in the carotenoid content with respect to elevated CO_2 (Loladze et al. 2019).

TSS increased under elevated CO_2 in all the studied species. Higher substrate availability can be the most probable reason for increase in TSS in comparison to ambient conditions. Higher Pn and TSS content leads to accumulation of carbohydrates stored as starch causing an increase in the biomass of the plants. The biomass accumulation was very significant in *A. nepalensis* and *A. elata* and so were the Pn rates and TSS content indicating the competence of these two species to utilize increased carbon more efficiently than the other two species. However, initial responses showed enhanced photosynthetic capacity and increase in biomass, but this might tend to saturate after a certain time period; since, with the continuously increasing or higher CO_2 levels, some or most plants initially grow rigorously, and later soil microbes enhance the immobilization of limiting nutrients thus hampering further growth (Shaw et al. 2002). A decline in the tissue nitrogen was observed in all the species under elevated CO_2. Previous studies have found that long-term elevated CO_2 leads to down-regulation of photosynthetic activity, termed as photosynthetic acclimation. This acclimation causes less uptake of nitrogen and thus a reduction in tissue nitrogen (Temperton et al. 2003; Leakey et al. 2006; Zheng et al. 2019), which then limits photosynthetic capacity (Ewa Jach and Ceulemans 1999).

An increase in the leaf area under CO_2 enrichment has been reported in a number of studies (Centritto et al. 1999; Pritchard et al. 1999; Masle 2000; Usuda 2006). Leaves are the point of interaction for carbon transfer and capture. Elevated CO_2 can bring changes in the internal structure of leaves (Pritchard et al. 1999). Stomatal density has been found to either decrease or increase with elevated CO_2 depending on the species. An additional variation is alteration in epicuticular wax on elevated CO_2 grown leaves (Thomas and Harvey 1983; Prior et al. 1997). Plant height increased in *A. nepalensis* and *A. elata*, whereas there was no significant change in the height of *T. govanianum* and *B. macrophylla*. The stems of *A. nepalensis* and *A. elata* were slightly branched having internodes, whereas *T. govanianum* and *B. macrophylla* had erect unbranched stems. The leaves of *T. govanianum* and *B. macrophylla* were also thinner in comparison to the thick and hairy leaves of *A. nepalensis* and *A. elata*. Elevated CO_2 can cause the stimulation of cell division at the shoot apical meristem (Pritchard et al. 1999) by reducing the time period between consecutive cell divisions as demonstrated by Masle (2000). A number of studies have shown an increase in leaf area and stem length or branch elongation in plants exposed to CO_2 enrichment without any change in the number of nodes (Downton et al. 1990; Pritchard et al. 1999). But these results have also varied among species as can be seen in this study.

18.5 Conclusion

From the results of the present and previous studies, a general assumption that can be drawn is that the response of any plant species of any ecosystem or form depends on a number of factors, viz its morphology, habitat and community structure, microbial and other associations, nutrient and water availability, carbon saturation threshold, limit to withstand environmental extremes, etc. Under elevated CO_2 the photosynthetic yield increases due to availability of carbon substrate but later due to impaired phloem loading capacity, source—sink imbalance can be observed in the long term. Changes in microbial activity under elevated CO_2 might also affect soil nutrients consequently affecting their availability and uptake. Overall the effect of CO_2 enrichment in high altitude are majorly species specific and an increased growth and productivity might be beneficial for establishment of an individual species but can adversely affect community structure and distribution of other species. Therefore, studies especially in the Himalayan alpine regions on effect of CO_2 enrichment and warming on community structure, species response, and distribution are required on a large scale to make precise predictions.

Acknowledgements We gratefully acknowledge the financial support from G.B. Pant National Institute of Himalayan Environment (GBPNIHE) under the National Mission on Himalayan Studies (NMHS) program and Space Applications Centre, Indian Space Research Organization, Ahmedabad, under SHRESTI - HIMADRI program. We would also like to thank Director HAPPRC, HNB Garhwal University for his support during the study and Prof. B. P. Nautiyal, VCSG Uttarakhand University of Horticulture and Forestry for reviewing the manuscript and providing valuable suggestions.

References

Ainsworth EA, Long SP (2005) What have we learned from 15 years of free-air CO_2 enrichment (FACE)? A meta-analytic review of the responses of photosynthesis, canopy properties and plant production to rising CO_2. New Phytol 165:351–372. https://doi.org/10.1111/j.1469-8137.2004.01224.x

Ainsworth EA, Leakey ADB, Ort DR, Long SP (2008) FACE-ing the facts: inconsistencies and interdependence among field, chamber and modeling studies of elevated [CO_2] impacts on crop yield and food supply. New Phytol 179(1):5–9. https://doi.org/10.1111/j.1469-8137.2008.02500.x

Allen Jr LH (1991) Effects of increasing carbon dioxide levels and climate change on plant growth, evapotranspiration, and water resources. In: Managing water resources in the west under conditions of climate uncertainty: a proceedings. The National Academies Press, Washington, DC; pp 101–147

Amthor JS (1995) Terrestrial higher-plant response to increasing atmospheric [CO_2] in relation to the global carbon cycle. Glob Chang Biol 1:243–274

Centritto M, Lee H, Jarvis PG (1999) Increased growth in elevated [CO_2]: an early, short-term response? Glob Chang Biol 5(6):623–633. https://doi.org/10.1046/j.1365-2486.1999.00263.x

Chaturvedi AK, Vashistha RK, Rawat N, Prasad P, Nautiyal MC (2009) Effect of CO2 enrichment on photosynthetic behavior of *Podophyllum hexandrum*an endangered medicinal herb. J Am Sci 5(5):113–118

Chaturvedi AK, Prasad P, Nautiyal MC (2013) Impact of elevated CO_2 on growth, morphology and dry matter partitioning in alpine growth forms of north western Himalayas. Indian J Plant Physiol 18(2):118–124

Dawes MA, Hagedorn F, Handa IT, Streit K, Ekblad A, Rixen C, Körner C, Hättenschwiler S (2013) An alpine treeline in a carbondioxide-rich world; synthesis of a nine-year free-air carbon dioxide enrichment study. Oecologia 171(3):623–637. https://doi.org/10.1007/s00442-012-2576-5

Downton WJS, Grant WJR, Chacko EK (1990) Effects of elevated carbon dioxide on the photosynthesis and early growth of mangosteen (*Garcinia mangostana* L.). Sci Hortic 44:215–225. https://doi.org/10.1016/0304-4238(90)90121-T

Ewa Jach M, Ceulemans R (1999) Effects of elevated atmospheric CO_2 on phenology, growth and crown structure of scots pine (*Pinus sylvestris*) seedlings after two years of exposure in the field. Tree Physiol 19(4–5):289–300. https://doi.org/10.1093/treephys/19.4-5.289

Haworth M, Elliott-Kingston C, McElwain JC (2013) Co-ordination of physiological and morphological responses of stomata to elevated [CO_2] in vascular plants. Oecologia 171:71–82. https://doi.org/10.1007/s00442-012-2406-9

Holm-Hansen O, Lorenzen CJ, Holmes RW, Strickland JD (1965) Fluorometric determination of chlorophyll. ICES J Mar Sci 30(1):3–15

Holtmeier KF, Broll G (2010) Altitudinal and polar treelines in the northern hemisphere causes and response to climate change (Obere und polareBaumgrenze auf der nördlichenHemisphäreUrsachen und Antwort auf den Klimawandel). Polarforschung 79:139–153

Inauen N, Körner C, Hiltbrunner E (2012) No growth stimulation by CO2 enrichment in alpine glacier forefield plants. Glob Chang Biol 18(3):985–999. https://doi.org/10.1111/j.1365-2486.2011.02584.x

Joshi SC, Chandra S, Palni LMS (2007) Differences in photosynthetic characteristics and accumulation of osmoprotectants in saplings of evergreen plants grown inside and outside a glasshouse during the winter season. Photosynthetica 45:594–600. https://doi.org/10.1007/s11099-007-0102-5

Kanemoto K, Yamashita Y, Ozawa T et al (2009) Photosynthetic acclimation to elevated CO_2 is dependent on N partitioning and transpiration in soybean. Plant Sci 177:398–403

Keeling CD, Bacastow RB, Carter AF, Piper SC, Whorf TP, Heimann M, Mook WG, Roeloffzen H (1989) A three-dimensional model of atmospheric CO_2 transport based on observed winds: 1. Analysis of observational data, in aspects of climate variability in the Pacific and the Western Americas, geophysical monograph 55, AGU, Washington, DC; pp 165–236

Körner C (2012) Alpine treelines functional ecology of the global high elevation tree limits. Springer, Basel

Körner C (2018) Alpine ecosystems and the high-elevation treeline. In: Sven Erik J, Brian DF (eds) Encyclopedia of ecology. Academic Press, Oxford, pp 407–413

Leakey ADB, Uribelarrea M, Ainsworth EA, Naidu SL, Rogers A, Ort DR, Long SP (2006) Photosynthesis, productivity and yield of maize are not affected by open-air elevation of CO2 concentration in the absence of drought. Plant Physiol 140(2):779–790. https://doi.org/10.1104/pp.105.073957

Liu XD, Chen BD (2000) Climatic warming in the Tibetan plateau during recent decades. Int J Climatol 20(14):1729–1742

Loladze I, Nolan JM, Ziska LH, Knobbe AR (2019) Rising atmospheric CO_2 lowers concentrations of plant carotenoids essential to human health: a meta-analysis. Mol Nutr Food Res 63:1801047

Masle J (2000) The effects of elevated CO_2 concentrations on cell division rates, growth patterns, and blade anatomy in young wheat plants are modulated by factors related to leaf position, vernalization, and genotype. Plant Physiol 122(4):1399–1416. https://doi.org/10.1104/pp.122.4.1399

Mccready RM, Guggolz J, Silviera V, Owen HS (1950) Determination of starch and amylase in vegetables. Anal Chem 22:1156–1158

Mishra NB, Mainali KP (2017) Greening and browning of the Himalaya: spatial patterns and the role of climatic change and human drivers. Sci Total Environ 587:326–339. https://doi.org/10.1016/j.scitotenv.2017.02.156

Morison JIL (1987) Intercellular CO_2 concentration and stomatal response to CO_2. In: Zeiger E, Farquhar GD, Cowan IR (eds) Stomatal function. Stanford University Press, Stanford, pp 229–252

Paulsen J, Körner C (2001) GIS-analysis of tree-line elevation in the Swiss Alps suggests no exposure effect. J Veg Sci 12:817–824. https://doi.org/10.2307/3236869

Prior SA, Pritchard SG, Runion GB, Rogers HH, Mitchell RE (1997) Influence of atmospheric CO2 enrichment, soil N, and water stress on needle *surface wax for*mation in *Pinus palustris* (Pinaceae). Am J Bot 84(8):1070–1077

Pritchard SG, Rogers HH, Prior SA, Peterson CM (1999) Elevated CO_2 and plant structure: a review. Glob Chang Biol 5(7):807–837. https://doi.org/10.1046/j.1365-2486.1999.00268.x

Schäppi B, Körner C (1996) Growth responses of an alpine grassland to elevated CO_2. Oecologia 105(1):43–52. https://doi.org/10.1007/BF00328790

Schäppi B, Körner C (1997) In situ effects of elevated CO_2 on the carbon and nitrogen status of alpine plants. Funct Ecol 11(3):290–299. https://doi.org/10.1046/j.1365-2435.1997.00084.x

Shaw MR et al (2002) Grassland responses to global environmental changes suppressed by elevated CO_2. Science 298:1987–1990

Shrestha AB, Wake CP, Mayewski PA, Dibb JE (1999) Maximum temperature trends in the Himalaya and its vicinity: an analysis based on temperature records from Nepal for the period 1971–94. J Clim 12(9):2775–2786

Singh SK, Reddy VR (2016) Methods of mesophyll conductance estimation: its impact on key biochemical parameters and photosynthetic limitations in phosphorus stressed soybean across CO_2. Physiol Plant 157:234–254

Singh SP, Sharma S, Dhyani PP (2019) Himalayan arc and treeline: distribution, climate change responses and ecosystem properties. Biodivers Conserv 28:1997–2016. https://doi.org/10.1007/s10531-019-01777-w

Strzalka K, Gugala AK, Latowski D (2003) Carotenoids and environmental stress in plants: significance of carotenoid-mediated modulation of membrane physical properties. Russ J Plant Physiol 50(2):168–173

Temperton VM, Grayston SJ, Jackson G, Barton CVM, Millard P, Jarvis PG (2003) Effects of elevated carbon dioxide concentration on growth and nitrogenfixation in *Alnusglutinosa*in a long-term field experiment. Tree Physiol 23:1051–1059. https://doi.org/10.1093/treephys/23.15.1051

Thomas JF, Harvey CN (1983) Leaf anatomy of four species grown under continuous CO_2 enrichment. Bot Gaz 144(3):303–309

Usuda H (2006) Effects of elevated CO_2 on the capacity for photosynthesis of a single leaf and a whole plant, and on growth in a radish. Plant Cell Physiol 47(2):262–269. https://doi.org/10.1093/pcp/pci244

van der Kooi CJ, Reich M, Löw M, De Kok LJ, Tausz M (2016) Growth and yield stimulation under elevated CO_2 and drought: a meta-analysis on crops. Environ Exp Bot 122:150–157. https://doi.org/10.1016/j.envexpbot.2015.10.004

Ward JK, Tissue DT, Thomas RB, Strain BR (1999) Comparative responses of model C3 and C4 plants to drought in low and elevated CO2. Glob Chang Biol 5(8):857–867

Ward EJ, Oren R, Bell DM et al (2013) The effects of elevated CO2 and nitrogen fertilization on stomatal conductance estimated from 11 years of scaled sap flux measurements at Duke FACE. Tree Physiol 33:135–151. https://doi.org/10.1093/treephys/tps118

Wittwer SH (1995) Food, climate and carbon dioxide – the global environment and world food production. CRC Press, Boca Raton, FL

Xu Z, Jiang Y, Jia B, Zhou G (2016) Elevated-CO2 response of stomata and its dependence on environmental factors. Front Plant Sci 7:657. https://doi.org/10.3389/fpls.2016.00657

Yu J, Chen L, Xu M (2012) Effects of elevated CO_2 on physiological responses of tall fescue to elevated temperature, drought stress, and the combined stress. Crop Science 52:1848–1858

Zelikova TJ, Blumenthal DM, Williams DG, Souza L, LeCain DR, Morgan J, Pendall E (2014) Long-term exposure to elevated CO2 enhances plant community stability by suppressing dominant plant species in a mixed-grass prairie. Proc Natl Acad Sci U S A 111(43):15456–15461

Zheng Y, Li F, Hao L et al (2019) Elevated CO_2 concentration induces photosynthetic down-regulation with changes in leaf structure, non-structural carbohydrates and nitrogen content of soybean. BMC Plant Biol 19:255. https://doi.org/10.1186/s12870-019-1788-9

Zhu J, Zhang Y, Yang X, Chen N, Jiang L (2020) Synergistic effects of nitrogen and CO_2 enrichment on alpine grassland biomass and community structure. New Phytol 228:1283–1294. https://doi.org/10.1111/nph.16767

19 Age-Girth Stand Structure of Himalayan Fir and Growth-NDVI Relationship in the Treeline Transects of Western Himalaya: An Ecological Perspective

Bency D. Chinthala, Ayushi Singh, Mayank Shekhar, Nidhi Tomar, Mohit Phulara, Arti Yadav, Pushpendra Pandey, Parminder S. Ranhotra, A. Bhattacharyya, Rajesh Joshi, and C. P. Singh

Abstract

Subalpine forests are the important indicators of climate change and the future biomass stock under forest densification and treeline advancement. Analysing the stand structure and growth behaviour of trees is essential for the assessment of the functioning and sustainable management of forest resources. We provided the girth and age stand structure of *Abies* spp. (silver fir) from the treeline ecotone in the moist transects of Western Himalaya, and the relationship between fir growth and normalized difference vegetation index (NDVI) for assessing the forest health. The age and girth class analysis of fir trees from the four different treeline transects revealed the presence of mixed age and girth classes within the treeline

B. D. Chinthala
Birbal Sahni Institute of Palaeosciences, Lucknow, Uttar Pradesh, India

Institute of Geography, University of Erlangen-Nürnberg, Erlangen, Germany

A. Singh · M. Shekhar · N. Tomar · P. S. Ranhotra (✉) · A. Bhattacharyya
Birbal Sahni Institute of Palaeosciences, Lucknow, Uttar Pradesh, India

M. Phulara
G. B. Pant National Institute of Himalayan Environment, Almora, Uttarakhand, India

A. Yadav · P. Pandey
Department of Geography, Lucknow University, Lucknow, Uttar Pradesh, India

R. Joshi
G. B. Pant National Institute of Himalayan Environment, Sikkim Regional Centre, Pangthang, Gangtok, Sikkim, India

C. P. Singh
Space Applications Centre (SAC), Indian Space Research Organisation (ISRO), Ahmedabad, Gujarat, India

S. P. Singh et al. (eds.), *Ecology of Himalayan Treeline Ecotone*,
https://doi.org/10.1007/978-981-19-4476-5_19

ecotone. Fir trees were established around 200 years ago in the transects with subsequent densification, but the fir treeline showed static behaviour during the later part of the twentieth century. We also found variations in the girth size increment with age amongst the fir trees, which is statistically significant for the fir trees growing near treeline ecotone. The relationships between the tree-ring width chronologies and NDVI suggested the role of temperature in controlling forest health. Such studies could help in extending the existing vegetation cover records to more past for better evaluation of changes in forest health and interaction with climatic conditions.

Keywords

Abies · Biomass · Himachal Pradesh · Kashmir · Tree-ring width · Uttarakhand · Vegetation cover

19.1 Introduction

Forest ecosystems have significant control on climate by managing the atmospheric carbon serving as a natural carbon sink and biomass storage (Pokhrel and Sherpa 2020). Globally, forests contribute around 80% aboveground and 40% belowground carbon storage of the total terrestrial carbon storage (Kirschbaum 1996; Pan et al. 2011). Studies have been carried out to understand the potential of different forest types for their carbon sequestration capacities by assessing the species diversity, forest health and aboveground biomass (Brown et al. 1999; Clark et al. 2001; Ketterings et al. 2001; Baishya et al. 2009; Ali et al. 2016; Ali 2019; Jang et al. 2019). The Himalayan region being sensitive to climatic changes is highly susceptible to variations in the biomass stock and rate of carbon sequestration. Forests of the Himalaya play a significant role in the regional and global carbon cycle (Brown et al. 1996; Upadhaya 2015) and also are the major source of livelihood for the Himalayan people (Bhutia et al. 2019). The Himalayan region, having a wide geographical area, is characterized by the subtropical to alpine forest types and glacial systems under different precipitation regimes. The region, being more vulnerable to climate change, has been well explored to understand the impact of climate change on glacial and biodiversity dynamics in the present and past using the dendrochronological approach (Fritts 1976). Also, studies are available on the quantification of biomass and carbon stocks from different forest types in the Himalaya (Sharma et al. 2010; Gairola et al. 2012; Sharma et al. 2018; Singh and Verma 2018), but the studies on a geographical scale are required.

In a forest ecosystem, the aboveground biomass is strongly associated with the stand structure of the trees, such as size (girth, height, canopy) and age stand (Lei et al. 2009; Wang et al. 2011; Zhang and Chen 2015; Ali et al. 2016; Dănescu et al. 2016; Ali 2019). The majority of the aboveground biomass assessments have considered the size of trees as the main parameter, whereas the stand age though a critical driver in aboveground carbon storage and productivity has been given less importance (Zhang and Chen 2015; Ali et al. 2016; KoÈhl et al. 2017). Studies have suggested that the mature and old forests with large trees contain high aboveground biomass (Brown and Lugo 1992; Clark and Clark 1996; Bunker et al. 2005; Chaturvedi et al. 2011) but have less potential for carbon sequestration as the older trees cease to grow and terminate the carbon storage process (Odum 1969; Terakunpisut and Gajaseni 2007; Baishya et al. 2009; Cao et al. 2019). A multi-site study by Luyssaert et al. (2008) indicated that both temperate and boreal forests showed declining net primary productivity (NPP) with forest age due to reduced leaf area and photosynthetic capacity (Zhou et al. 2015). Based on the tree-ring data of trees from natural tropical forests of Surinam, KoÈhl et al. (2017), on other hand, reported that the old-growth trees not only contribute to carbon stocks but maintain high carbon accumulation rates at later stages of their lifetime. Cao et al. (2019) pointed the relationship between forest total carbon storage and stand age to be site and species-specific. Studies revealed that the conifer tree species have high efficiency in carbon sequestration and serve as an important reservoir of carbon stocks (Gucinski et al. 1995; Laclau 2003; Negi et al. 2003; Singh and Verma 2018). In the Himalayan region, various species of conifers form dense forests from subtropical to subalpine altitudes (Champion and Seth 1968) and provide vital biomass storage and carbon sinks. Few conifer species form the upper conifer limit as pure stand at the upper temperate and subalpine altitudes in the glaciated and non-glaciated transects of the Himalayan region (Champion and Seth 1968; Ranhotra and Bhattacharyya 2013; Roy et al. 2021). However, the age structure of such conifer stands has not been attempted yet for their future carbon sequestration and storage potential. In this chapter, we provide the girth (diameter at breast height (DBH)) class and age stand of *Abies spectabilis* (Himalayan fir) growing at the temperate to subalpine altitudes in the transects under Indian summer monsoon (ISM) and winter westerly (WD) dominant regions of Western Himalaya. We established the age-girth (DBH) correlations to assess the growth variability amongst the fir trees and age-girth class distribution along the altitudinal gradients within and across the transects.

The unprecedented warming of the Earth's atmosphere (IPCC 2019) is producing a concern on ecological changes in the Himalayan region which is also under anthropogenic pressure. A rise in extreme climatic events could have a significant impact on forest health conditions. Vegetation is a sensitive indicator of climate change (Pauli et al. 2002; Nagy 2006; Jiapaer et al. 2015); the assessment of forest health is, therefore, crucial for understanding the interactions between the vegetation and climate. The normalized difference vegetation index (NDVI) is a parameter of remote-sensing taken as an effective measure of terrestrial vegetation health (greenness) and vegetation cover. NDVI approach, that also reflects the ecological changes in time (Gamon et al. 2015; Liu et al. 2021), is gaining importance in addressing different aspects of vegetation dynamics such as phenology, seasonal to annual variations in the vegetation coverage, spatial and qualitative changes in the distribution of vegetation and delineating treelines over the local and regional scales (Singh et al. 2012; Gamon et al. 2015; Bhavsar et al. 2017; Singh et al. 2018a; Anderson et al. 2020; Kumari et al. 2021). Within a forest ecosystem, the growth of species could be sensitive to various factors such as hydrothermal, seasonality and among species competition. Hence, a relationship possibly exists between the growth of individual species and forest health. Studies on the seasonal and centennial-scale variability assessment of vegetation cover based on the established relationships between tree-ring chronologies and NDVI datasets are available from different regions (Malmström et al. 1997; D'Arrigo et al. 2000; Kaufmann et al. 2004; He and Shao 2006; Lopatin et al. 2006; Leavitt et al. 2008; Liang et al. 2009; Wang et al. 2010; Beck et al. 2013; Wang et al. 2014; Shang et al. 2016; Zhang et al. 2018). Analysis of such relationships has not been attempted so far for the Himalayan region. Here we provided the trend analysis of NDVI for the past four decades from two transects, one each under the ISM and WD domain and discussed the role of climatic factors in controlling forest health. We also attempted the relationships between the developed tree-ring width chronologies of fir from the transects and the available NDVI datasets to test: (1) the robustness of such relationships in the Himalayan region and (2) the feasibility of this treeline conifer species to reconstruct the changes in forest health during past, i.e. beyond the measured records.

19.1.1 Study Area

Four treeline transects representing temperate to subalpine altitudes, namely Tungnath (TUN), Magguchatti (MAG), Chanshal (CHN) and Daksum-Sinthan (DAK-SNT), in Western Himalaya (Fig. 19.1), were investigated for the age-girth stand structure and the relationships between radial growth of fir and forest health (NDVI). Tungnath (TUN) and Magguchatti (MAG) transects are part of Kedarnath Wildlife Sanctuary in the Garhwal region of Uttarakhand, Western Himalaya. TUN site (Fig. 19.1c) forms the upper catchment (2700–3800 m asl) of river Alaknanda, a principal tributary of river Ganga, and is approachable by motor road till Chopta, a small settlement at ~2800 m asl altitude (Singh et al. 2018b). Treeline ecotone in the TUN transect (30° 29′ – 30° 30′ N and 79° 12′ – 79° 13′ E) is approachable by the mule track of ~3 km.

Site MAG (30° 37′ – 30° 38′ N and 78° 55′ – 78° 58′ E) is approachable by 9 km mule trek towards ENE direction from Triyuginarayan township at ~2300 m asl till the altitude of ~3400 m asl (Fig. 19.1c). MAG is with the highest elevation of ~3700 m asl and the transect opens to the west via Magguchatti pass.

Chanshal (CHN) area (31° 11′ – 31° 13′ N and 77° 58′ – 78° 00′ E) is located in the Shimla district of Himachal Pradesh and is approached by around 180 km of motorable road from the Shimla town via Rohru-Dodra-kwar-Larot localities (Fig. 19.1c). The highest elevation at CHN is ~4200 m asl and the Chanshal pass opens to the east of the transect.

Daksum-Sinthan (DAK-SNT) area (33° 34′ – 33° 35′ N and 75° 23′ – 75° 24′ E) is in the Anantnag district of south Kashmir, Jammu and Kashmir (Fig. 19.1b). Daksum (DAK) site (33°36′43"N 75°26′6″ E) is located at around 2400 m asl, 40 km east of Anantnag town, and about 85 km east from Srinagar city. Sinthan (SNT) Top (33°34' N and 75°30′ E) is the highest elevation of transect at ~3800 m

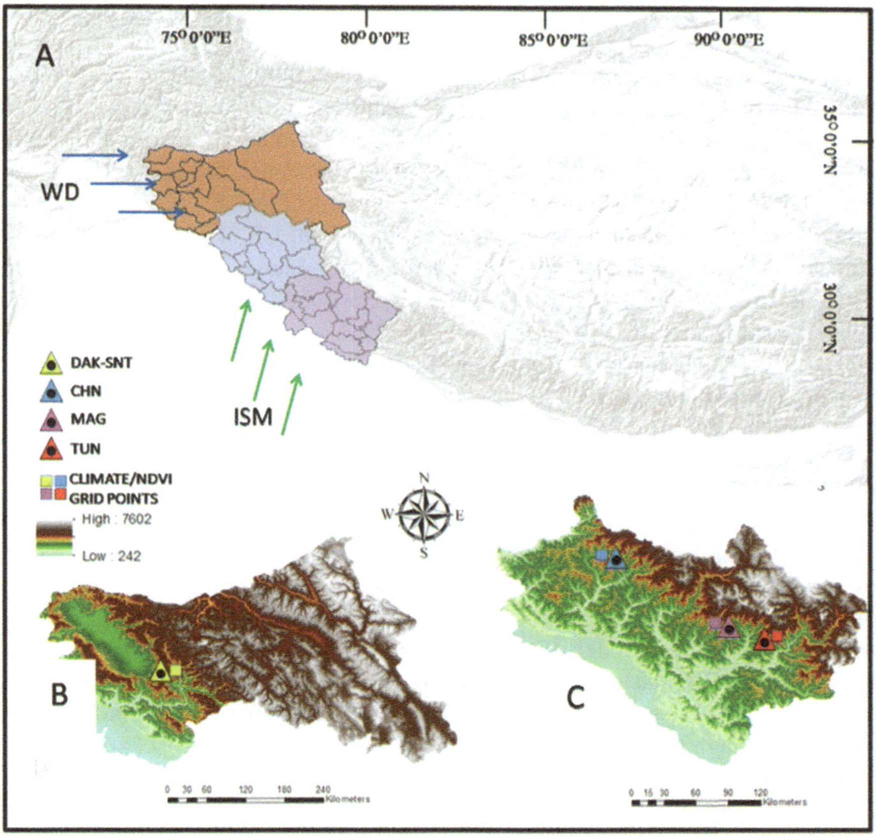

Fig. 19.1 Digital elevation map (**a**) state boundaries Kashmir, Himachal Pradesh and Uttarakhand, (**b**) study site Daksum-Sinthan (DAK-SNT) in Kashmir. (**c**) study sites Chanshal (CHN) in Himachal Pradesh, Tungnath (TUN) and Magguchatti (MAG) in Uttarakhand. Solid boxes denote the closest grid data points of climate (CRU-TS3.22 for TUN, MAG and CHN; GPCC for DAK-SNT) and NDVI

asl and around 30 km upstream connecting the Kashmir transect with Kishtwar in Jammu district (Nanda et al. 2018). The DAK-SNT transect is bisected by the snowmelt and groundwater stream that joins river Jhelum originating from the springs at Verinag downstream.

19.1.2 Climate and NDVI Data

The study sites have diverse climatic conditions, mainly in terms of precipitation. Sites TUN, MAG and CHN, are under the Indian summer monsoon (ISM) regime-receiving 70–80% of annual precipitation in the form of rainfall during the summer monsoon months. DAK-SNT area in J & K is influenced by the mid-latitudinal western disturbances (WD) and receives substantial winter precipitation (Fig. 19.1a). The long-term meteorological records are absent from the vicinity of all the four study sites. The meteorological station nearest to the study sites TUN and MAG in Uttarakhand is Mukteshwar that has a centennial long (since around 1898 CE) observed meteorological record. For the site CHN in Himachal Pradesh, the nearest meteorological station is at Shimla having the temperature and precipitation data since 1901 CE. The nearest meteorological station to DAK-SNT sites (J & K) is Srinagar providing the meteorological data since 1890 CE. To know the regional climate scenario, we also retrieved gridded climate dataset available through the Climatic Research Unit (CRU TS v. 3.23; 0.5° × 0.5° grid) University of East Anglia; (Harris et al. 2014) and GPCC (Udo et al. 2011) for the period 1901–2019 CE. The grid points are marked in Fig. 19.1b, c). Both gridded and meteorological station datasets of the study area show an increasing trend of MAT in the twentieth century.

Moreover, for the Tungnath area the weather record of the years 2008 to 2010 CE. (Adhikari et al. 2011) is also available through automated weather stations (AWS), which showed mean annual temperature (MAT) ~10.7 °C and 5.6 °C at ~2500 masl, and ~ 3300 masl altitudes, respectively. The warmest month remains July with a mean temperature of 12.56 ± 1.23 °C and January is the coldest month with a mean of ~3 °C at ~3300 masl. Annual precipitation ranges between 2400 and 3000 mm, of which 89.5% comes through ISM during June to September months. Snow cover, largely due to WD, lasts for 85 ± 22.7 days/year during winter months (Adhikari et al. 2011). The temperature and precipitation variabilities for the past century observed through CRU datasets for all the sites are shown in Fig. 19.2a–d.

For developing the relationship between radial growth of fir trees and NDVI, around 37 years (1982–2018 CE) gridded dataset of Normalized difference vegetation index in global inventory modelling and mapping studies (GIMMS-NDVI; 0.5° × 0.5° grid, source: http://apdrc.soest.hawaii.edu/data/data.php) were retrieved for the two sites (TUN in Uttarakhand and DAK-SNT in Kashmir) falling in their respective grids (Fig. 19.1). Figure 19.3 shows the mean monthly NDVI for the TUN and DAK-SNT sites. The mean annual NDVI data of four-decades have been plotted (Fig. 19.4) for the trend analysis. We found missing values in the NDVI datasets in the monsoon months (July–August) for TUN transect and in the winter months (January–February) for the DAK-SNT transect, which might be due to cloud cover in the monsoon months (JA) at TUN area and both cloud and snow cover during winter months (JF) due to the strong effect of WD over Kashmir region.

19.1.3 Vegetation Distribution

The forest vegetation of study sites resembled those of typical temperate and subalpine forests of the Western Himalaya. The common arboreal taxa growing in the studied transects in the treeline ecotone are *Abies spectabilis* (D.Don) Mirb, *Betula utilis* and *Rhododendron campanulatum. Abies spectabilis* (D.Don) Mirb, also called Himalayan silver fir, is an evergreen conifer forming the upper conifer

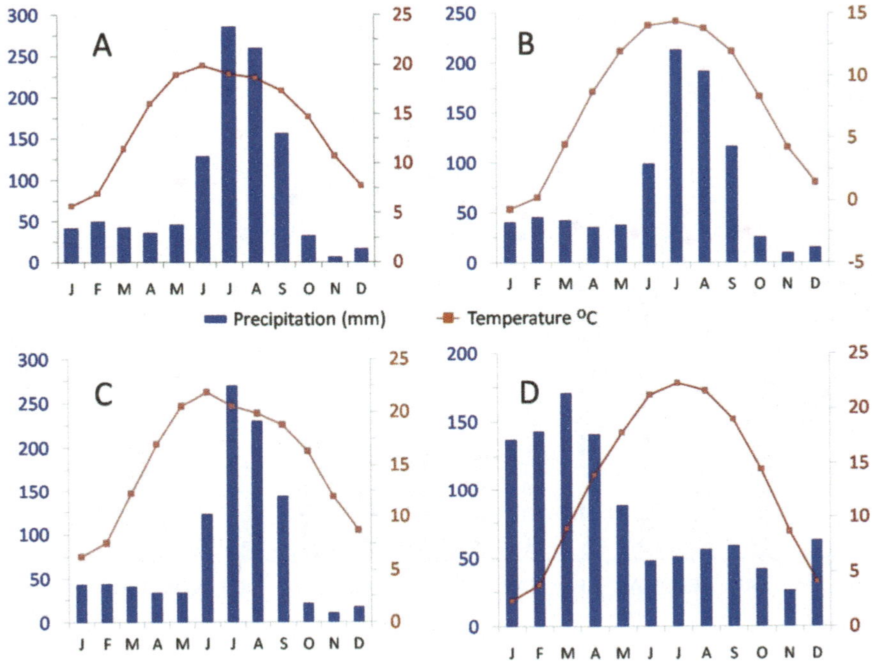

Fig. 19.2 Monthly mean temperature and precipitation data from CRU-TS3.22 for (**a**) Tungnath (TUN) Uttarakhand; (**b**) Magguchatti (MAG) Uttarakhand; (**c**) Chanshal (CHN) Himachal Pradesh. For (**d**) Daksum-Sinthan (DAK-SNT) Kashmir precipitation data source is GPCC

limit in most of the moist transects of the Himalayan region (Champion and Seth 1968; Ghimire et al. 2008; Gaire et al. 2011). Table 19.1 represents the distribution of major tree taxa along all the four studied transects. Presently, in the Tungnath (TUN) and Magguchatti (MAG) transects *A. spectabilis* is growing in association with broadleaved taxa, especially *Quercus semicarpifolia* (oak), *Rhododendron arboreum, R. campanulatum* and *Betula utilis* along with other species, viz *Acer caesium, Prunus cornuta* and *Sorbus foliolosa*e. The Chanshal (CHN) transect starting from the altitude of ~2800 m asl is well characterized by anthropogenic interference like the cultivation of various commercial fruit crops, especially *Malus* (apple) and *Juglans*. Above this altitude transect inhabits mixed conifers and broadleaved tree taxa namely *Quercus semicarpifolia, Abies spectabilis, Betula utilis* up to the elevation of ~3600 m asl. For many decades, the area is under human influence for firewood, fodder and also direct grazing of domestic animals (cattle) on grass meadows. In the Daksum-Sinthan (DAK-SNT) transect, there is the predominance of *Abies* from ~2200 m to 3500 m asl altitude. At the lower altitude, the trees of *Abies pindrow* form canopy with the other broadleaved taxa such as *Juglans and Rhododendron arboreum* until the altitude ~3000 m asl beyond which the *Abies pindrow* is replaced by the *Abies spectabilis*. These sites are also under the influence of anthropogenic activities with settlements and grazing etc.

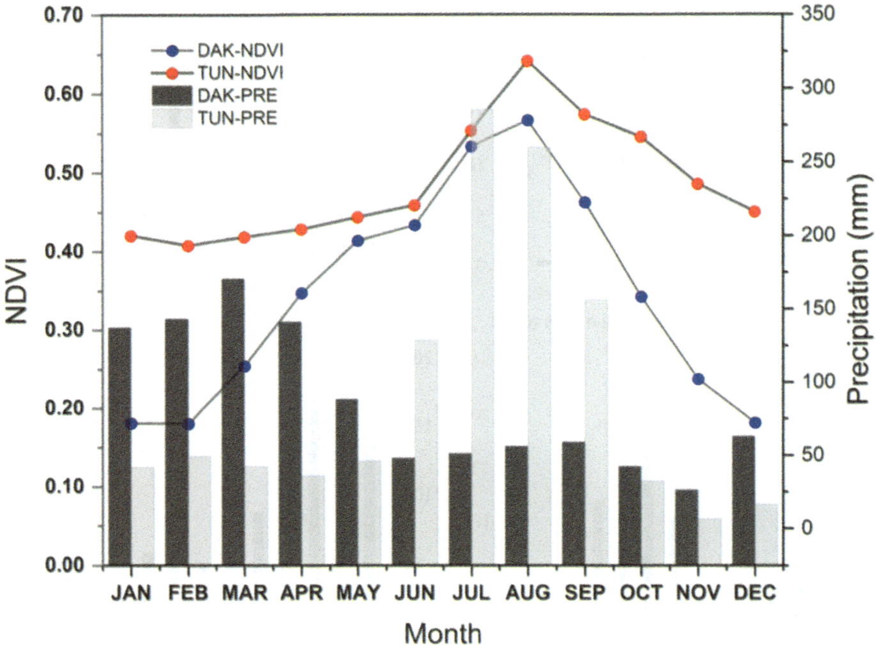

Fig. 19.3 Monthly NDVI and precipitation (PRE) data for the Kashmir (DAK-SNT) and Uttarakhand (TUN) study sites

19.2 Materials and Methods

19.2.1 Sample Collection, Processing and Data Generation

A good number of fir trees growing along the altitudinal gradient were sampled from the transects during different seasons between the years 2016 and 2019. At Tungnath (TUN) site, tree-ring cores were collected in May 2016 and June 2019 from *Abies spectabilis* growing in the treeline ecotone. From the Kashmir region, the fir trees growing around 2400 masl at Daksum (DAK) area were cored in May 2017 and from those growing at the Sinthan (SNT) area were collected in September 2017. The area Chanshal (CHN) in Himachal Pradesh and Magguchatti (MAG) in Uttarakhand was explored for the tree-ring samples of fir respectively in October and November months in the year 2019. At least two cores per tree were collected at the breast height (1.3 m above the ground) and the geographical locations and girth at breast height (GBH) were recorded for each tree. To get the longer chronological records, the trees with relatively thick girth were selected and cored. In addition, a number of fir trees at the sites TUN and CHN were randomly surveyed and measured for GBH to assess the girth class distribution. The extracted tree-ring cores were air-dried and mounted in the wooden frames. To make the boundaries of the adjacent

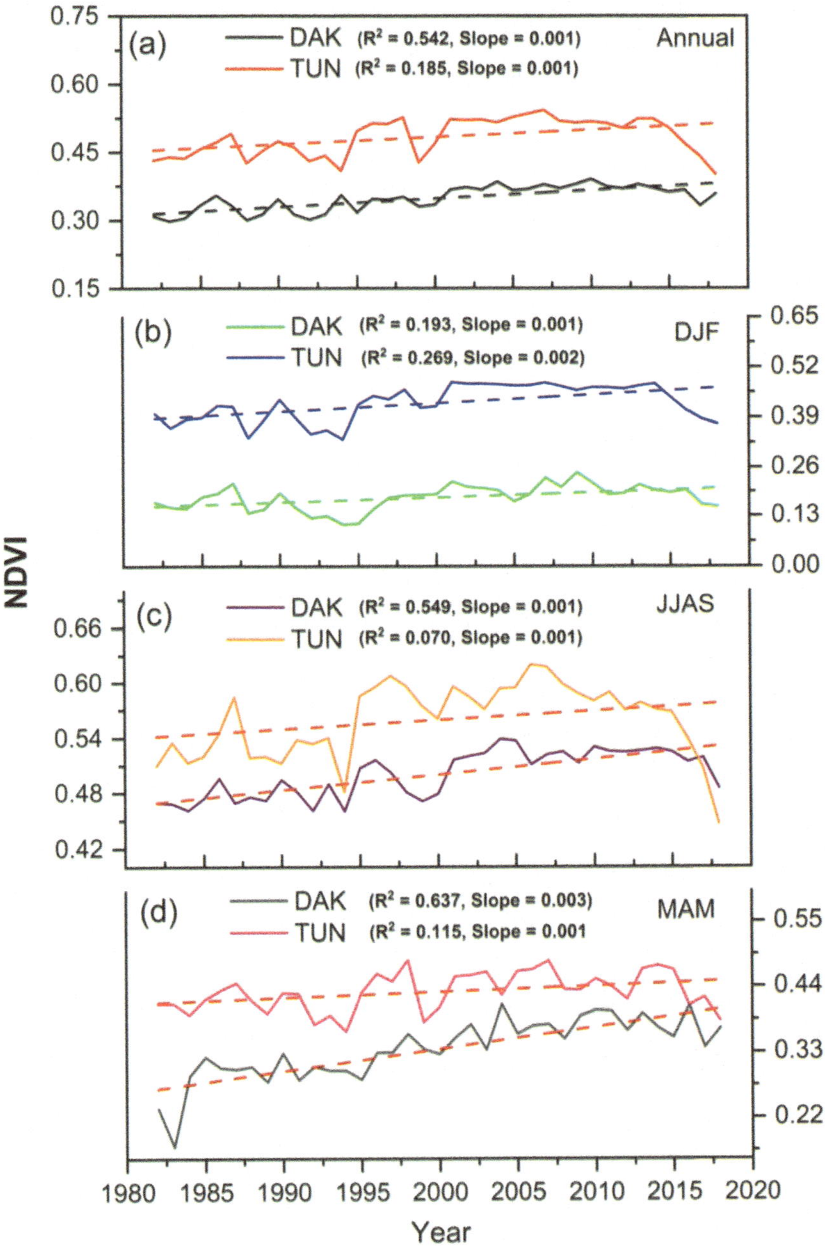

Fig. 19.4 NDVI trend for the period 1982 to 2018 for the Kashmir (DAK) and Uttarakhand (TUN) regions. (**a**) Mean annual NDVI. (**b**) Mean NDVI of winter months (December-January-February). (**c**) Mean NDVI of summer months (June-July-August). 9**d**) Mean NDVI of spring months (March-April-May)

Table 19.1 Distribution of major tree/shrub taxa along the transects at the four study sites

Altitude (m asl)	Uttarakhand		Himachal Pradesh	Kashmir (J & K)
	Tungnath	Magguchatti (Triyuginarayan)	Chanshal	Daksum & Sinthan
>3600	Grasses, rocky exposures and summit at ~3700 masl	Grasses, rocky exposures and summit at ~3700 masl	*Rhododendron campanulatum, R. anthopogon, Cassiope fastigiate, Juniperus communis,* summit at ~3900 m asl	*Grass meadow* with scattered growth of *R. campanulatum, R. anthopogon, Cassiope fastigiated, Juniperus communis,* Summit at ~4000 m asl
3400–3600	*Rhododendron campanulatum*	*Rhododendron campanulatum*	*Abies spectabilis, Rhododendron. Campanulatum, Betula utilis, Juniperus*	*Abies spectabilis, Pinus wallichiana, Betula utilis, Rhododendron campanulatum, Juniperus*
3000–3400	*Abies spectabilis, Quercus semicarpifolia, Rhododendron campanulatum, Betula utilis, Taxus bacata, Sorbus foliolosa*e	*Abies spectabilis, Quercus semicarpifolia, Rhododendron campanulatum, Betula utilis, Sorbus foliolosa*	*Quercus semicarpifolia, Abies spectabilis, Taxus wallichiana, Picea smithiana, Alnus nepalensis, acer, Juglans regia, Prunus sp.*	*Abies spectabilis, Pinus wallichiana, Betula utilis, acer, Rhododendron campanulatum, Juniperus*
2800–3000	*Abies spectabilis, Rhododendron arboreum, Quercus semicarpifolia, Acer caesium, Prunus cornuta*	*Abies spectabilis, Quercus, Rhododendron arboreum, Sorbus foliolosa*e	*Abies spectabilis, Pinus walichiana, Picea smithiana, Quercus* sp., *Alnus nepalensis, acer, Asculus, Juglans regia, Prunus cornuta*	*Abies spectabilis, Pinus wallichiana, Abies pindrow, Juglans, acer, Alnus nepalensis, Rhododendron arboreum*
2400–2800	*Quercus, Betula alnoides, Asculus, Juglans, Alnus*	*Quercus, Betula alnoides, Asculus, Juglans, Alnus,*	*Quercus, Pinus roxburgii, Asculus,* commercial fruit crops *Malus* (apple) and *Juglans*	*Abies pindrow, Pinus wallichiana, Juglans, Alnus nepalensis, Rhododendron arboreum, Juglans, Malus* etc.

annual growth-ring distinctly visible under the stereo-zoom binocular microscope, the upper surface of the core was cut by a sharp edge razor blade and then polished with coarse and fine grade sandpapers to enhance the surface resolution of cells of annual growth rings. Rings of each core were counted under the stereo-zoom microscope (Lieca) and each ring was assigned a calendar year following the standard method of cross-dating with necessary corrections (Camarero and Gutierrez 2004; Speer 2010; Gaire et al. 2014).

19.2.2 Development of Age-Girth-Altitude Relationships

Complete cores of the tree stems having the length from bark (outer) to pith (center) were taken to develop the age-girth relationship models. To correct the age for the core extraction height, a decade was added to the counted age of each tree (Gaire et al. 2011). To assess the growth dynamics and age-girth stand structure of the fir trees in the study sites, their girth class and age distribution were analysed by establishing the relationships between tree stem diameter at breast height (DBH) and age (Fig. 19.5a–d); and DBH and altitude (Fig. 19.6a–d). DBH of each tree was calculated by dividing the measured GBH (girth at breast height (circumference)) by 3.14 (π). DBH and age relationship model has also been tested separately for the trees growing at different elevations such as above and below the 3100 masl altitude respectively for the sites TUN (Fig. 19.5Aa, Ab) and DAK-SNT (Fig. 19.5Da, Db). This allowed assessing the stand structure and growth variability of fir at different

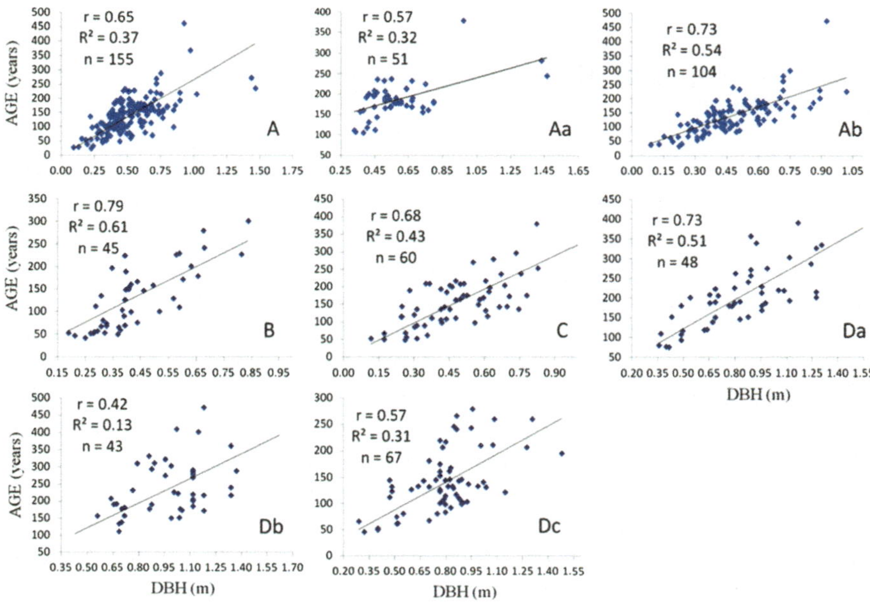

Fig. 19.5 Relationship between diameter at breast height (DBH) and Age of *Abies spectabilis* (fir) growing at (**a**) Tungnath (TUN); (**b**) Magguchatti (MAG); (**c**) Chanshal (CHN) and (**d**) Daksum-Sinthan (DAK-SNT)

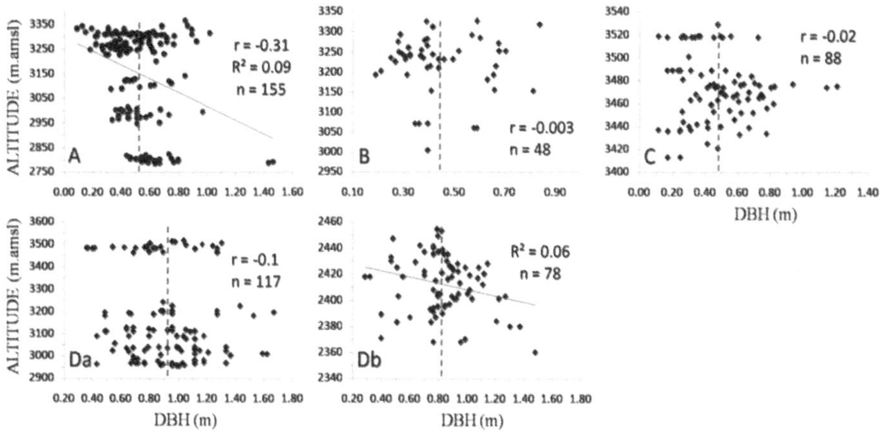

Fig. 19.6 Relationship between diameter at breast height (DBH) and Altitude of *Abies spectabilis* (fir) at (**a**) Tungnath (TUN); (**b**) Magguchatti (MAG); (**c**) Chanshal (CHN) and (**d**) Daksum-Sinthan (DAK-SNT). The vertical red dotted lines show the mean of DBH

elevational levels and the underlying factors contributing to elevation patterns. Few trees were found rotten inside and not suitable for the extraction of cores till pith length were neglected.

19.2.3 Development of Tree-Ring Chronology and Relationship with NDVI

Tree-ring chronologies of fir trees for the sites TUN and DAK-SNT were developed by measuring the ring widths to the nearest 0.001 mm precision using the LINTAB-6 measuring system coupled with the computer software TSAP-Win scientific version (Rinn 2003). Cross-dating and quality checks were done using the COFECHA (Holmes 1983) computer program. The computer program ARSTAN (Cook 1985) was used to get the final chronologies (Fig. 19.7a and b) by correcting and standardizing the ring-width datasets. ARSTAN removes the low–frequency variations occurring due to age, size and stand dynamics and enhances the common climatic signal in tree ring datasets (Cook et al. 1990). We used the cubic smoothing spline curve option to detrend the tree ring series with a 50% frequency response cut-off equal to 30% of the series length. The cubic spline fitting removes the low-frequency variance and the effect of localized disturbance events (Fritts 1976; Cook et al. 1990). Out of the three chronologies (Standard, Residual and Arstan) produced by ARSTAN program, we used the residual chronology which contains the high-frequency variations. Further, to determine the reliability and quality of the chronologies, we used various statistical parameters, viz. mean sensitivity (MS), standard deviation (SD), the average correlation between all series (Rbar), expressed population signal (EPS) and first-order autocorrelation (AC). The Rbar and EPS

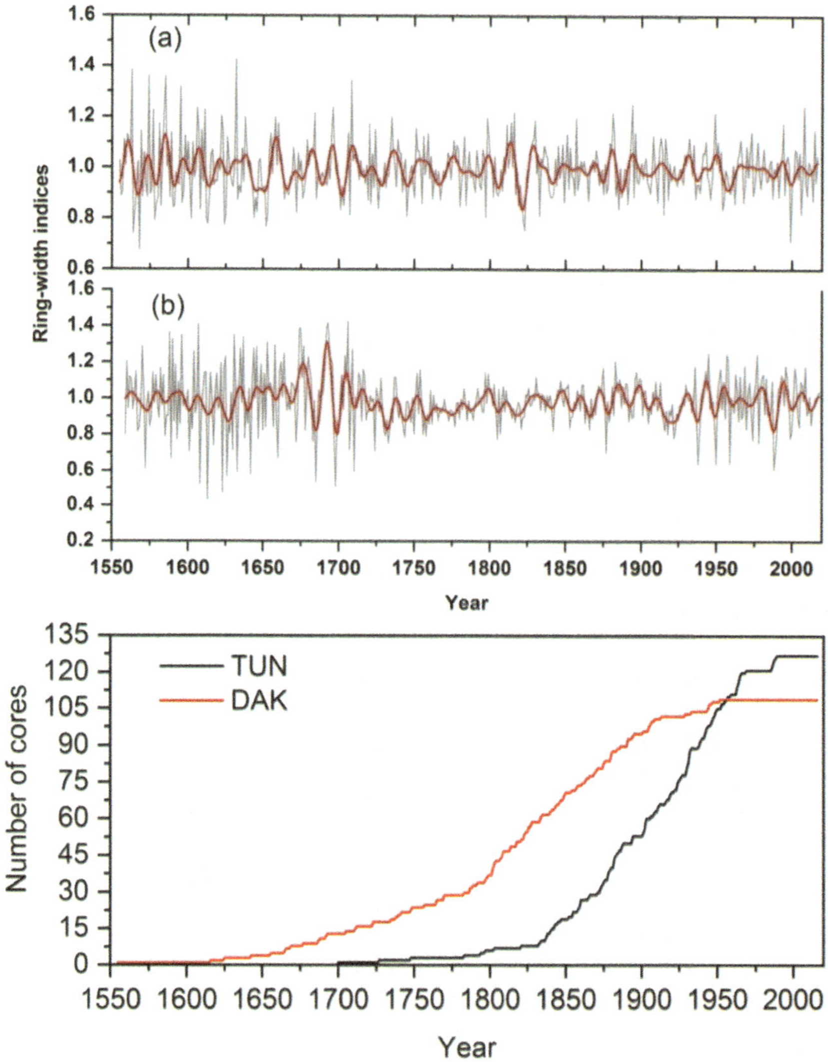

Fig. 19.7 Tree-ring Chronologies of *Abies spectabilis* from (**a**) Daksum-Sinthan (DAK-SNT) Kashmir and (**b**) Tungnath (TUN) Uttarakhand

values were computed using a 30-year moving window with a 15-year overlap. An EPS value of >0.85 is considered as a threshold for the reliability of time series (Wigley et al. 1984). The statistical details are provided in Table 19.2.

Table 19.2 Statistics of residual chronology of *Abies spectabilis* from Kashmir and Uttarakhand

Residual chronology statistics	Daksum-Sinthan (DAK-SNT) kashmir	Tungnath (TUN) Uttarakhand
Series intercorrelation	0.414	0.442
Core samples	109	96
Mean sensitivity	0.183	0.255
Standard deviation	0.230	0.687
Time span	1555–2017 C.E.	1559–2018 C.E.
Common period analysis	1881–2016 C.E	1888–2016 C.E.
All series Rbar	0.199	0.204
Within tree Rbar	0.229	0.194
Between tree Rbar	0.170	0.841
Signal to noise ratio	21.882	21.21
Expressed population signal	0.956	0.955
Year with EPS >0.85	1740 C.E.	1875 C.E.

19.3 Results

19.3.1 Upper Limit and Age-Girth Stand Structure of Abies

19.3.1.1 Tungnath Transect (Uttarakhand)

Along the Chopta-Tungnath (TUN) altitudinal transect, trees of *Abies spectabilis* (fir) formed the upper limit at 3335 m asl. Various correlation analyses were performed amongst age, DBH and altitude to analyse the age-girth stand structure of fir trees. Correlation between DBH and age of 155 fir trees along the Chopta-Tungnath transect was found positively significant ($r = 0.65$, $p < 0.01$) with 37% variability explained by DBH (Fig. 19.5a). Regression model based on cores of 51 fir trees growing at the lower part of the transect (below 3100 masl) showed a positive correlation ($r = 0.57$, $p < 0.01$) explaining 32% variability (Fig. 19.5Aa). For 104 fir trees growing above 3100 masl along the altitudinal transect, the regression model showed a significant positive correlation ($r = 0.73$, $p < 0.01$) explaining 53% variability (Fig. 19.5Ab). The DBH-age linear relationship equation model $Y = 287.72x$ (Eq –A) could be used to approximate the age of rotten and uncored fir trees growing above the 3100 m asl altitude.

All along the TUN altitudinal gradient, DBH ranged from 0.09 to 1.46 m (mean of 0.52 m), and DBH was negatively correlated with altitude ($r = -0.31$, $p < 0.01$), but explained only 9% variability in DBH by altitude (Fig. 19.6a). Though the trees with the highest DBH (1.43 and 1.46 m) were found growing at the altitudes 2786 and 2791 masl respectively, several high girth class trees were also found towards the upper ecotone limit. The oldest fir tree (472 years age; DBH - 0.97 m) was found growing at ~3284 masl, much above the altitude of maximum girth trees (DBH = 1.43 and 1.46 m) and calculated for age ~ 308 and 314 years respectively at the altitudes ~2790 masl. Another tree of 369 years of age was recorded at ~3000 masl altitude.

19.3.1.2 Magguchatti (Triyuginarayan), Uttarakhand

In the Magguchatti (MAG) transect the trees of *A. spectabilis* (fir) formed the upper limit at 3330 masl altitude. Correlation between DBH and age of 45 fir trees was found positively significant ($r = 0.79, p < 0.01$) with 61% variability explained by DBH (Fig. 19.5b). DBH ranged from 0.19 to 0.84 m (mean of 0.44 m) amongst the trees measured for girth. Correlation between DBH and altitude was found insignificant ($r = -0.003$) as most of the high girth class fir trees were found growing near the treeline limit (Fig. 19.6b). The tree of maximum age (~292 years) was reported at an altitude ~3318 masl.

19.3.1.3 Chanshal Transect (Uttarakhand)

In the Chanshal (CHN) transect *A. spectabilis* (fir) was dominantly growing above the altitude of 3400 masl and formed the upper limit at 3530 masl. Correlation analysis performed between DBH and age of 60 fir trees was positively significant ($r = 0.68, p < 0.01$) with 43% variability explained by DBH (Fig. 19.5c). DBH of fir trees along the studied gradient ranged from 0.115 to 1.21 m (mean of 0.48 m). The insignificant correlation ($r = -0.02$) between DBH and altitude indicates mixed girth size along the altitudinal gradient (Fig. 19.6c). The tree of the highest DBH (1.21 m) was found growing at 3475 masl. The tree with a maximum age of ~380 years was growing at 3460 masl. The age of rotten and uncored fir trees growing in the area could be approximated by using the linear equation model $y = 320.97x$ (Eq –B) for the DBH-age of fir trees.

19.3.1.4 Daksum-Sinthan Transect (Kashmir)

Here *A. spectabilis* occurs from 2900 masl to >3515 masl. Correlation between age and DBH increased with elevation as it was higher above 3100 m ($r = 0.73$, $p < 0.01$) with 51% variability explained by DBH based on 48 trees (Fig. 19.5Da), than between 2900 and 3100 masl ($r = 0.42, p < 0.01$) but with only 12% variability explained based on 43 trees (Fig. 19.5Db). DBH of fir trees along the studied gradient ranged from 0.37 to 1.31 m for the upper part of the transect and from 0.43 to 1.67 m for the lower part of the transect, with overall mean DBH of 0.92 m along the whole transect. The correlation between DBH and altitude of 117 fir trees growing along the Sinthan (SNT) transect was found insignificant ($r = -0.1$), which otherwise indicated mixed girth size along the altitudinal gradient (Fig. 19.7d); however, the fir trees with comparatively lower girth were found in the upper part of transact. The tree with the highest DBH (1.67 m) was found growing at altitude ~3196 masl and the tree of maximum age (~473 years) was growing at altitude ~2980 masl in the transect.

For the fir trees growing around Daksum (DAK) site which occurred between 2300 and 2500 masl, the correlation between DBH and age of 67 fir trees was found positively significant ($r = 0.57, p < 0.01$) with 31% variability explained by DBH (Fig. 19.5e). DBH of fir trees along the studied gradient ranged from 0.29 to 1.48 m (mean of 0.85 m) for the area. The fir trees growing at the DAK area were found younger than that growing at a higher altitude at the SNT area. The oldest tree found growing at DAK was ~280 years old at 2368 masl altitude.

19.3.2 Trend of NDVI and Correlation with Ring-Width Indices of Fir

The observed variations in the average monthly NDVI values for the study sites are presented in Fig. 19.3. At the ISM dominant TUN site, the NDVI was high during the monsoon and post-monsoon months (July to December), with a peak in August. The NDVI values during winter and pre-monsoon months (January to June) were relatively low with February being the lowest. Similar to this, at the WD dominant DAK-SNT, Kashmir area, the NDVI during summer and monsoon months (June to October) was high with August showing the highest value. However, the NDVI during post-monsoon, winter and pre-monsoon months (previous year November to current year May) remained very low as compared to that of TUN site, with January and February months having the least values. An overall increase in the values of mean annual as well as seasonal NDVI for the last 36 years (1982 to 2018 CE) has been observed at both the sites but was distinctly higher for the Kashmir site ($R^2 = 0.542$, Fig. 19.4). However, there was a noticeable decline in the mean annual NDVI during the past few years that is since 2007 at the TUN area and since 2010 CE at Kashmir. On a seasonal basis, the NDVI of monsoon months (JJAS) showed a significant decline ($R^2 = 0.763$) from 2006 CE till recent (2018 CE).

We found variation in the developed correlation between the fir tree-ring width indices and NDVI dataset for the common period (1982 to 2018 CE) from two sites (Fig. 19.8). For the DAK-SNT site in Kashmir, the correlation was negative during the winter months, January to March (JFM) and positive during the summer months May to July (MJJ) with March, June and July being significant (Fig. 19.8a). For the TUN site in Uttarakhand, the correlation was positive in most of the months but was significant for only the pre-monsoon months, February to April (Fig. 19.8b).

19.4 Discussion

19.4.1 Age-Girth Stand Structure of Fir Trees and Biomass Potential

The upper limits of *A. spectabilis* ecotone along the altitudinal transects at Chopta-Tungnath (TUN) and Magguchatti (MAG) in Uttarakhand are lower than at Chanshal (CHN) in Himachal Pradesh and Sinthan (SNT) in Kashmir (J & K), as well as in other transects of the central and Western Himalaya (Bhattacharyya et al. 2011; Gaire et al. 2011, 2014; Shrestha et al. 2014; Roy et al. 2021). Several factors such as summertime grazing, cutting of young trees by local people and also low summit height could account for the low silver fir treeline in TUN and MAG transects. The effects of summit syndrome could substantially lower the elevation of climatically governed timberlines (Odland 2015). The elevational distribution of

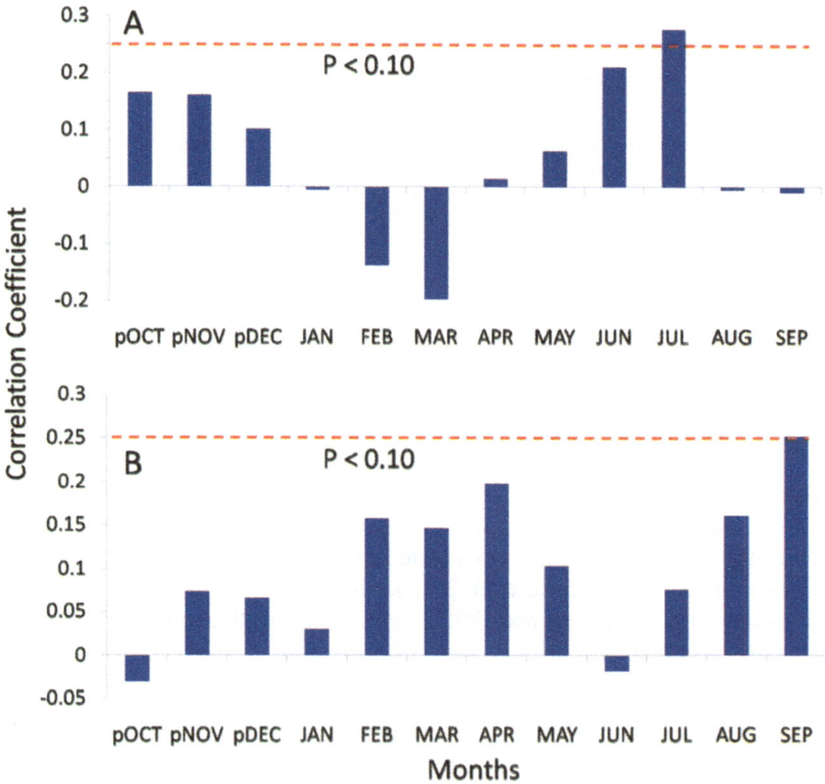

Fig. 19.8 Correlation of fir chronology with NDVI (GIMMS-NDVI) for the common period 1982 to 2016 CE. (**a**) Daksum-Sinthan (DAK-SNT) Kashmir, (**b**) Tungnath (TUN) Uttarakhand. Horizontal red dashed line represents the 90% confidence level

girth (DBH) for all the transects studied here (Fig. 19.6) showed that in the treeline ecotone the fir trees having the girth classes higher than the mean (TUN 0.52 m; MAG 0.44 m; CHN 0.48 m and SNT 0.92 m) are in good number, as also reported by Rai et al. (2012) for Tungnath transect. Moreover, the presence of a good number of fir trees having girth class less than mean value and mostly aged less than 150 years in the forest stand suggested the densification of fir forest within the ecotone limit during the twentieth century CE. For the TUN and DAK-SNT transects, the low (high) correlation value of DBH–age for the fir trees growing below (above) 3100 masl altitude (Fig. 19.5Aa and Db) explains non-uniformity in the growth of fir trees growing at lower altitudes. This variation might be due to the competition for space and nutrients from other dominantly growing broadleaved species at lower elevations. Furthermore, the trees of relatively younger girth and age reported growing below 3000 masl at TUN and at 2300–2400 masl at DAK might be due to clearing of high girth class and old aged trees for timber. Anthropogenic activities within the fir community (Rai et al. 2012) and grazing by herbivores at such approachable sites are also important factors determining the tree growth and population structure as well as treeline position at many areas (Cairns et al. 2007; Speed et al. 2011; Shrestha et al. 2014).

Variations in the stand structural attributes (tree size) of individual trees are crucial for the species sustenance, forest functioning (Clark 2010; Zhang and Chen 2015; Yuan et al. 2018) and enhancement of stand productivity or aboveground biomass in the forests (Dănescu et al. 2016; Ali and Mattsson 2017). Additionally, other factors such as environment, biodiversity matrix, site quality (productivity, disturbances) and stand age also influence the functioning of forests (Poorter et al. 2015; Zhang and Chen 2015; Ali et al. 2016; Fotis et al. 2017; Yuan et al. 2018). We found a mixed girth (DBH) and age class in stands of fir trees growing in the four studied transects. From subtropical forests in China, Ali et al. (2016) found that amongst the various drivers (stand age, stand structural diversity and species diversity) of aboveground carbon storage, the DBH and stand age showed significant positive influence on biomass. Pokhrel and Sherpa (2020) found no substantial relationship between tree species diversity and aboveground biomass from Chitwan-Annapurna landscape in central Nepal, which is similar to the observed weak relationships between carbon stock and species diversity from some other parts of Nepal (Karna 2012; Mandal et al. 2013). The correlations between biomass and stand parameters (tree height and DBH) were found positive, where the trees with larger DBH contributed more to biomass (Kirby and Potvin 2007; Baishya et al. 2009; Djuikouo et al. 2010; Pokhrel and Sherpa 2020). In our study, the stand structure of the fir population with uneven DBH class and age distribution proves the significant biomass storage as well as the potential for carbon sequestration. Standage being an important factor influencing the biomass with time (Ali et al. 2016; KoÈhl et al. 2017), our study sites could hold potential for future carbon sequestration due to the presence of young aged fir trees within the fir ecotone. However, several factors such as climate (temperature, rainfall pattern), transect topography and anthropogenic influence could be critical in influencing the process of carbon sequestration and biomass storage by altering the growth dynamics of trees. Despite climate warming, a recent study (Singh et al. 2018b) showed the static behaviour of fir treeline in the Tungnath transect, Uttarakhand, as also reported by Shrestha et al. (2014) and Gaire et al. (2011) from central Nepal. From Sygera Mountains, south-east Tibet, Liang et al. (2011) reported insignificant upslope movement of Smith fir stands since mid-twentieth century but with a considerable increase in the stand density. A similar insignificant change in the treeline position, but increase in stand density was reported from Rolwaling transect, Nepal by Schickhoff et al. (2015). A remote sensing-based study (Bharti et al. 2012) has shown an increase in biomass in a subalpine forest's canopy during 1980–2010 CE at Nanda Devi Biosphere Reserve, Uttarakhand. Additional to the climatic factors, the summit syndrome, anthropogenic activities and grazing were also taken as probable reasons behind the insignificant changes in the treeline, but the reported increase in the stand density of fir forest at the four sites during the last century remained favourable to biomass increase.

19.4.2 NDVI-Based Forest Health Assessment and Growth Response of Fir

The two sites, Tungnath (TUN) in Uttarakhand and Daksum-Sinthan (DAK-SNT) in Kashmir, studied for their NDVI analysis, differ in precipitation seasonality zones. Precipitation at TUN is mostly during the summer months (JJAS) through ISM, while the Kashmir area receives more precipitation during winter months in the form of snowfall through westerly disturbances (WD) (Fig. 19.2). Factors such as temperature, moisture and sunlight, affect the vegetation by influencing the rate of photosynthesis, respiration and transpiration (Zandalinas et al. 2018; Dusenge et al. 2019). In our study, the significant positive correlation between the temperature and NDVI for the common period (1982 to 2018 CE, provided in Table 19.3) points that the temperature has a pronounced effect on the seasonal vegetation cover and greenness by regulating the photosynthesis process. This correlation is more significant for the Tungnath site (Table 19.3). The temperature dependency of vegetation greenness has been discussed from other regions also. From east-coast, China, He et al. (2020) recorded maximum temperature and precipitation as important factors. Hao et al. (2012) from the upper catchments of the Yellow river basin found minimum temperature and precipitation as a significant factor with regard to NDVI. From Iraq, the NDVI variability had significant negative relation with temperature and positive with precipitation (Naif et al. 2020). However, from some arid regions such as the northern Loess Plateau of China (Ning et al. 2015), northern China (Wang et al. 2021) and northern Ethiopia (Siyum et al. 2018), the NDVI variations found significant relation with precipitation where moisture is the growth limiting factor. In south Asian region the temperature being an important controlling factor to the growth of vegetation has been extensively discussed based on tree-ring studies from different regions of Himalaya (Bhattacharyya and Yadav 1992; Yadav et al. 1997, 1999, 2011; Yadav and Singh 2002; Singh and Yadav 2014; Thapa et al. 2015; Shah et al. 2019). The lack of moisture stress at study sites could be a reason for insignificant correlations between precipitation and NDVI data (Table 19.3). Moreover, the Himalayan region being influenced by the local orographic rains, the relationships between precipitation and tree-rings remain variable (Singh and Yadav 2005). In our study, both the regions observed an increasing trend in the NDVI data since 1982 CE (Fig. 19.4) suggesting the expansion of vegetation cover.

Table 19.3 The correlation between the NDVI and climate (temperature and precipitation) for the Kashmir and Uttarakhand sites

		DAK-SNT site Kashmir		TUN site Uttarakhand	
	Correlation between NDVI and climate	Common period 1982 to 2018	Common period 2006 to 2018	Common period 1982 to 2018	Common period 2006 to 2018
1	NDVI and mean annual temperature	0.49	−0.24	0.61	−0.23
2	NDVI and annual precipitation	−0.07	−0.002	−0.42	−0.60

A similar trend has also been observed at many Himalayan sites such as the Hindukush region (Anderson et al. 2020), upper Khoh river basin, Uttarakhand (Kumari et al. 2021), Gori Ganga, Uttarakhand, (Parihar 2021), Gangotri and other regions of Uttarakhand (Singh et al. 2012) and Sikkim (Singh et al. 2018a) and also from North China (Liang et al. 2009). However, there is a noticeable decline in the NDVI values since 2006 CE at both the study sites and corresponds well with the rise in temperature (Fig. 19.4 and Table 19.3). Singh et al. (2018b), based on tree-ring studies of silver fir, discussed the possible role of warming temperatures resulting in pre-monsoon moisture stress conditions due to high evapotranspiration and decline of vegetation growth. Also in central Nepal, Panthi et al. (2017) observed the intensification of spring season drought conditions, a critical reason controlling the growth of fir trees at treeline ecotone.

The differential response between tree-ring width indices of fir and NDVI data for the two sites (Fig. 19.8) shows the effect of temperature controlled moisture variability on vegetation growth. Climatic factors, such as temperature and hydrologic conditions, greatly influence the vegetation density and also the tree-ring growth (Du et al. 2018; Liu et al. 2019). In our study, the tree-ring chronologies of fir reflected the NDVI changes for the summer months (MJJ) at Kashmir and for the pre-monsoon (FMAM) and monsoon (JAS) months at TUN, Uttarakhand. At the ISM dominant TUN area, the temperature rise during FMA helps in the melting of winter snow thus providing moisture for the photosynthesis process. In Kashmir, the negative response during FM months might be due to prolonged snow cover delaying the process of photosynthesis in vegetation by limiting the moisture availability under frosting conditions. The relationships, therefore, signify the role of temperature in controlling photosynthesis by regulating the moisture availability to plants. This is also evident by the positive correlations between inter-annual temperature and NDVI datasets (Table 19.3). The relationships between tree-ring width indices and NDVI though less attempted from different geographical regions correspond with our results. From north-east China (Liu et al. 2021) and Alatau mountains central Asia (Zhang et al. 2018), the respective tree-ring chronologies of *Pinus* and *Picea* spp. showed a significant positive correlation with the NDVI of July to October months. Various other studies also revealed that NDVI exhibits strong relationships with forest growth but diverse relationships may exist across different regions and forest types (Vicente-Serrano et al. 2016; Siyum et al. 2018) depending on the temperature and precipitation factors. Wang et al. (2004) reported a strong relationship between the tree-ring width of Oak and NDVI from the central Great Plains region of North America. Similarly, studies from Siberia (Kirdyanov et al. 2007; Bunn et al. 2013), arid and semi-arid regions in China (He and Shao 2006; Shi et al. 2015), Cyprus, Eastern Mediterranean (Coulthard et al. 2017), and the semi-arid woodlands in southern Ethiopia (Mokria et al. 2017) also recorded positive relationship between NDVI and tree growth but with varying strengths. A study from the high latitude temperature-limited regions such as boreal forests of North America, however, found no significant relationship between NDVI and tree-ring width (Beck et al. 2013). In this study, we found a relationship between NDVI and tree-ring width of *A. spectabilis,* but the variance explained is low for the reconstruction of past vegetation cover dynamics. Nevertheless, the study has shown the potential of developing such relationships from the Himalayan region with the enhancement of datasets for assessment of past changes in the vegetation cover using the tree-ring approach.

19.5 Conclusions

In this chapter, we attempted to characterize the girth and age stand structure of *A. spectabilis* population at the treeline ecotone zones from four transects in the Western Himalaya. We also assessed the fir growth and vegetation cover relationship using the NDVI approach. Despite having the dominance of different precipitation sources that is the ISM at Uttarakhand (TUN and MAG) and Himachal Pradesh (CHN), and the WD at Kashmir (DAK-SNT), the four study sites represented similarity in stand structure of fir showing the uneven (mixed) DBH class and age stand. The reported densification of fir forests during the twentieth century showed their suitability for future biomass increase. But also the increasing spring season drought conditions could affect the tree growth process. Inter-annual changes in the growth process of plants are reflected in the vegetation coverage and forest health in time which is captured by the NDVI, a remote sensing application. Both temperature and precipitation have an influence on plant growth and vegetation health. Based on the inter-annual climate and NDVI correlation results in this study, the warming temperature trend is found to have a significant role in the overall increase in vegetation coverage during the past four decades in the moist treeline transects of Western Himalaya. Also, the tree-ring width and NDVI relationships projected that the temperature-driven photosynthesis process during the pre-monsoon and summer months defines the tree growth, but with variability at the two different sites. At the WD dominated DAK-SNT area the summer months (JJ) are influential, whereas at the summer ISM dominated TUN area the melting of snow during pre-monsoon months (FMA) results in photosynthesis and initiation of earlywood formation. In addition to climate, the factors of transect topography and anthropogenic influence also appeared to be critical in influencing treeline and vegetation health. Such studies from different regions of Himalaya will provide regional understanding on the stand age structure of individual tree species and also on vegetation cover dynamics and forest functioning for the management of forest resources.

Acknowledgements The authors are grateful to the Director, Birbal Sahni Institute of Palaeosciences, Lucknow for the necessary support to carry out this work and permission to publish this work (BSIP/RDCC/Publication No. 33/2021-22). The authors extend sincere acknowledgement to MoEF & CC, New Delhi for financial support (1886/XII-86/2016) under National Mission on Himalayan Studies, and the Space Applications Centre (SAC) - ISRO, Ahmedabad, India, under the Studies on Harnessing Remote Sensing for Environment and Climate (SHRESTI) programme (SAC/EPSA/BPSG/ALPINE/SHRESTI/09/2019). The DAAD fellowship by FAU, Germany is duly acknowledged by the author BDC. The support provided by the Forest Departments of Uttarakhand, Himachal Pradesh and Kashmir state for sample collection is deeply accredited. We are highly grateful to Dr. R. S. Rawal (Ex-Director GBPNIHESD, Almora) and Prof. S. P. Singh, Coordinator of the project for their kind support and valuable inputs in this work.

References

Adhikari BS, Rawat GS, Rai ID, Bharti RR, Bhattacharyya S (2011) Ecological assessment of timberline ecotone in western Himalaya with special reference to climate change and anthropogenic pressures. IV Annual Report, Wildlife Institute of India, Dehradun, India

Ali A (2019) Forest stand structure and functioning: current knowledge and future challenges. Ecol Indic 98:665–677

Ali A, Mattsson E (2017) Disentangling the effects of species diversity, and intraspecific and interspecific tree size variation on aboveground biomass in dry zone homegarden agroforestry systems. Sci Total Environ 598:38–48

Ali A, Yan ER, Chen HY, Chang SX, Zhao YT, Yang XD, Xu MS (2016) Stand structural diversity rather than species diversity enhances aboveground carbon storage in secondary subtropical forests in eastern China. Biogeosciences 13(16):4627–4635

Anderson K, Fawcett D, Cugulliere A, Benford S, Jones D, Leng R (2020) Vegetation expansion in the subnival Hindu Kush Himalaya. Glob Chang Biol 26(3):1608–1625

Baishya R, Barik SK, Upadhaya K (2009) Distribution pattern of aboveground biomass in natural and plantation forests of humid tropics in Northeast India. Trop Ecol 50(2):295–304

Beck PS, Andreu-Hayles L, D'Arrigo R, Anchukaitis KJ, Tucker CJ, Pinzón JE, Goetz SJ (2013) A large-scale coherent signal of canopy status in maximum latewood density of tree rings at arctic treeline in North America. Glob Planet Chang 100:109–118

Bharti RR, Adhikari BS, Rawat GS (2012) Assessing vegetation changes in timberline ecotone of Nanda Devi National Park, Uttarakhand. Int J Appl Earth Obs Geoinf 18:472–479

Bhattacharyya A, Yadav RR (1992) Tree growth and recent climatic changes in the western Himalaya. Geophytology 22:255–260

Bhattacharyya A, Ranhotra PS, Gergan JT (2011) Vegetation vis-a-vis climate and glacier history during 12,400 to 5,400 yr BP from Dokriani transect, Garhwal Himalaya, India. J Geol Soc India 77:401–408

Bhavsar D, Kumar A, Roy A (2017) Applicability of NDVI temporal database for western Himalaya forest mapping using fuzzy-based PCM classifier. Eur Remote Sens 50(1):614–625

Bhutia Y, Gudasalamani R, Ganesan R, Saha S (2019) Assessing forest structure and composition along the altitudinal gradient in the state of Sikkim, eastern Himalayas, India. Forests 10(8):633

Brown S, Lugo AE (1992) Aboveground biomass estimates for tropical moist forests of the Brazilian Amazon. Interciencia 17(1):8–18

Brown S, Sathaye J, Cannell M, Kauppi PE (1996) Mitigation of carbon emissions to the atmosphere by forest management. Commonw For Rev 75:80–91

Brown SL, Schroeder P, Kern JS (1999) Spatial distribution of biomass in forests of the eastern USA. For Ecol Manag 123(1):81–90. https://doi.org/10.1016/S0378-1127(99)00017-1

Bunker DE, De Clerck F, Bradford JC, Colwell RK, Perfecto I, Phillips OL, Naeem SM (2005) Species loss and aboveground carbon storage in a tropical forest. Science 310:1029–1031

Bunn AG, Hughes MK, Kirdyanov AV, Losleben M, Shishov VV, Berner LT, Oltchev A, Vaganov EA (2013) Comparing forest measurements from tree rings and a space-based index of vegetation activity in Siberia. Environ Res Lett 8(3):035034

Cairns DM, Lafon C, Moen J, Young A (2007) Influences of animal activity on treeline position and pattern: implications for treeline response to climate change. Phys Geogr 28:419–433

Camarero JJ, Gutierrez E (2004) Pace and pattern of recent tree line dynamics: response of ecotones to climatic variability in the Spanish Pyrenees. Climate Change 63:181–200

Cao J, Gong Y, Adamowski JF, Deo RC, Zhu G, Dong X, Zhang X, Liu H, Xin C (2019) Effects of stand age on carbon storage in dragon spruce forest ecosystems in the upper reaches of the Bailongjiang River basin, China. Sci Rep 9:3005. https://doi.org/10.1038/s41598-019-39626-z

Champion HG, Seth SK (1968) A revised survey of forest types of India. Manager of Publications, Government of India, Delhi, p 404

Chaturvedi RK, Raghubanshi AS, Singh JS (2011) Carbon density and accumulation in woody species of tropical dry forest in India. For Ecol Manag 262:1576–1588

Clark DA, Brown S, Kicklighter DW, Chambers JQ, Thomlinson JR, Ni J, Holland EA (2001) Net primary production in tropical forests: an evaluation and synthesis of existing field data. Ecol Appl 11(2):371–384
Clark DB, Clark DA (1996) Abundance, growth and mortality of very large trees in neotropical lowland rain forest. For Ecol Manag 80(1–3):235–244
Clark J (2010) Individuals and the variation needed for high species diversity in forest trees. Science 327:1129–1132
Cook ER (1985) A time series analysis approach to tree ring standardization. PhD dissertation, University of Arizona, Tucson, AZ
Cook ER, Briffa KR, Shiyatov S, Mazepa V (1990) Tree-ring standardization and growth-trend estimation. In: Methods of dendrochronology: applications in the environmental sciences. Springer, Cham, pp 104–123
Coulthard BL, Touchan R, Anchukaitis KJ, Meko DM, Sivrikaya F (2017) Tree growth and vegetation activity at the ecosystem-scale in the eastern Mediterranean. Environ Res Lett 12(8):084008
D'Arrigo RD, Malmstrom CM, Jacoby GC, Los SO, Bunker DE (2000) Correlation between maximum latewood density of annual tree rings and NDVI based estimates of forest productivity. Int J Remote Sens 21:2329–2336. https://doi.org/10.1080/01431160050029611
Dănescu A, Albrecht AT, Bauhus J (2016) Structural diversity promotes productivity of mixed, uneven-aged forests in southwestern Germany. Oecologia 182(2):319–333
Djuikouo MNK, Doucet JL, Nguembou CK, Lewis SL, Sonk'e B (2010) Diversity and aboveground biomass in three tropical forest types in the dja biosphere reserve, Cameroon. Afr J Ecol 48(4):1053–1063
Du Q, Zhang M, Wang S, Che C, Ma R, Ma Z (2018) Changes in air temperature of China in response to global warming hiatus. Acta Geograph Sin 73:1748–1764
Dusenge ME, Duarte AG, Way DA (2019) Plant carbon metabolism and climate change: elevated CO 2 and temperature impacts on photosynthesis, photorespiration and respiration. New Phytol 221(1):32–49
Fotis AT, Murphy SJ, Ricart RD, Krishnadas M, Whitacre J, Wenzel JW (2017) Above-ground biomass is driven by mass-ratio effects and stand structural attributes in a temperate deciduous forest. J Ecol 106:561–570
Fritts HC (1976) Tree rings and climate. Academic Press, New York, p 567
Gaire NP, Dhakal YR, Lekhak HC, Bhuju DR, Shah SK (2011) Dynamics of Abies spectabilis in relation to climate change at the Treeline ecotone in Langtang National Park, Nepal. J Sci Tech 12:220–229
Gaire NP, Koirala M, Bhuju DR, Borgaonkar HP (2014) Treeline dynamics with climate change at the Central Nepal Himalaya. Clim Past 10:1277–1290. https://doi.org/10.5194/cp-10-1277-2014
Gairola SC, Sharma M, Ghildiyal SK, Suyal S (2012) Regeneration dynamics of dominant tree species along an altitudinal gradient in moist temperate transect slopes of the Garhwal Himalaya. J For Res 23:53–63
Gamon JA, Kovalchuck O, Wong CYS, Harris A, Garrity SR (2015) Monitoring seasonal and diurnal changes in photosynthetic pigments with automated PRI and NDVI sensors. Biogeosciences 12(13):4149–4159
Ghimire SK, Sapkota IB, Oli BR, Prajuli RR (2008) Non timber Forest products of Nepal Himalaya. WWF Nepal, Kathmandu, Nepal
Gucinski H, Vance E, Reiners WA (1995) Potential effects of global climate change. In: Ecophysiology of coniferous forests. Academic Press, New York, pp 309–331
Hao F, Zhang X, Ouyang W, Skidmore AK, Toxopeus AG (2012) Vegetation NDVI linked to temperature and precipitation in the upper catchments of Yellow River. Environ Model Assess 17(4):389–398
Harris IP, Jones D, Osborn TJ, Lister DH (2014) Updated high-resolution grids of monthly climatic observations—the CRU TS 3.10. Int J Climatol 34:623–642

He G, Li Z, Zhen X, Shen HJ, Zhang L, Zhang R (2020) Influence of variations of hydrothermal conditions on normalized difference vegetation index in typical temperature zones along the east coast of China. Front Earth Sci 394. https://doi.org/10.3389/feart.2020.574101

He J, Shao X (2006) Relationships between tree-ring width index and NDVI of grassland in Delingha. Chin Sci Bull 51(9):1106–1114

Holmes RL (1983) Computer–assisted quality control in tree–ring dating and measurement. Tree–Ring Bull 43:69–78

Intergovernmental Panel on Climate Change (IPCC) (2019) Climate change and land: An IPCC special report on climate change, desertification, land degradation, sustainable land management, food security, and greenhouse gas fluxes in terrestrial ecosystems. https://www.ipcc.ch/srccl/download/

Jang RH, Lee SY, Lee EP, Lee SI, Kim EJ, Lee SH, You YH (2019) Organic carbon distribution and budget of dominant woody plant community in the subalpine zone at volcanic Jeju Island, Korea. J Ecol Environ 43(1):1–10. https://doi.org/10.1186/s41610-019-0141-4

Jiapaer G, Liang S, Yi Q, Liu J (2015) Vegetation dynamics and responses to recent climate change in Xinjiang using leaf area index as an indicator. Ecol Indic 58:64–76. https://doi.org/10.1016/j.ecolind.2015.05.036

Karna YK (2012) Mapping above ground carbon using world view satellite image and lidar data in relationship with tree diversity and forests," M.S. thesis, University of Twente, Enschede, Netherlands

Kaufmann RK, D'Arrigo RD, Laskowski C, Myneni RB, Zhou L, Davi NK (2004) The effect of growing season and summer greenness on northern forests. Geophys Res Lett 31:L09205. https://doi.org/10.1029/2004GL019608

Ketterings QM, Coe R, Van Noordwijk M, Ambagau Y, Palm CA (2001) Reducing uncertainty in the use of allometric biomass equations for predicting above-ground tree biomass in mixed secondary forests. For Ecol Manag 146(1–3):199–209

Kirby KR, Potvin C (2007) Variation in carbon storage among tree species: implications for the management of a small-scale carbon sink project. For Ecol Manag 246(2):208–221

Kirdyanov AV, Vaganov EA, Hughes MK (2007) Separating the climatic signal from tree-ring width and maximum latewood density records. Trees 21(1):37–44

Kirschbaum MUF (1996) The carbon sequestration potential of tree plantations in Australia. Environmental management: the role of eucalypts and other fast growing species. CSIRO Forestry and Forest Products, Canberra 77–89

KoÈhl M, Neupane PR, Lotfiomran N (2017) The impact of tree age on biomass growth and carbon accumulation capacity: a retrospective analysis using tree ring data of three tropical tree species grown in natural forests of Suriname. PLoS One 12(8):e0181187. https://doi.org/10.1371/journal.pone.0181187

Kumari N, Srivastava A, Dumka UC (2021) A long-term spatiotemporal analysis of vegetation greenness over the himalayan region using google earth engine. Climate 9(7):109

Laclau P (2003) Biomass and carbon sequestration of ponderosa pine plantations and native cypress forests in Northwest Patagonia. For Ecol Manag 180:317–333

Leavitt SW, Chase TN, Rajagopalan B, Lee E, Lawrence PJ (2008) Southwestern U.S. tree-ring carbon isotope indices as a possible proxy for reconstruction of greenness of vegetation. Geophys Res Lett 35:L12704. https://doi.org/10.1029/2008GL033894

Lei X, Wang W, Peng C (2009) Relationships between stand growth and structural diversity in spruce-dominated forests in New Brunswick, Canada. Can J For Res 39:1835–1847

Liang E, Wang Y, Eckstein D, Luo T (2011) Little change in the fir tree-line position on the southeastern Tibetan plateau after 200 years of warming. New Phytol 190:760–769

Liang EY, Eckstein D, Liu HY (2009) Assessing the recent grassland greening trend in a long–term context based on tree–ring analysis: a case study in North China. Ecol Indic 9:1280–1283

Liu R, Song Y, Liu Y, Li X, Song H, Sun C, Wang L (2021) Changes in the tree-ring width-derived cumulative normalized difference vegetation index over Northeast China during 1825 to 2013 CE. Forests 12(2):241

Liu RS, Liu Y, Li Q, Song HM, Li XX, Sun CF, Cai QF, Song Y (2019) Seasonal palmer drought severity index reconstruction using tree–ring widths from multiple sites over the central–western Da Hinggan Mountains, China since 1825 AD. Clim Dyn 53:3661–3674

Lopatin E, Kolstrom T, Spiecker H (2006) Determination of forest growth trends in Komi Republic (northwestern Russia): combination of tree-ring analysis and remote sensing data. Boreal Environ Res 11(5):341

Luyssaert S, Schulze ED, BoÈrner A, Knohl A, Hessenmo Èller D, Law BE, Ciais P, Grace J (2008) Old-growth forests as global carbon sinks. Nature 455:213–215. https://doi.org/10.1038/nature07276. PMID: 18784722

Malmström CM, Thompson MV, Juday GP, Los SO, Randerson JT, Field CB (1997) Interannual variation in global-scale net primary production: testing model estimates. Global Biogeochem Cycles 11(3):367–392

Mandal RA, Dutta IC, Jha PK, Karmacharya S (2013) Relationship between carbon stock and plant biodiversity in collaborative forests in Terai, Nepal. Int Sch Res Notices 2013:625767

Mokria M, Tolera M, Sterck FJ, Gebrekirstos A, Bongers F, Decuyper M, Sass-Klaassen U (2017) The frankincense tree Boswellia neglecta reveals high potential for restoration of woodlands in the horn of Africa. For Ecol Manag 385:16–24

Nagy L (2006) European high mountain (alpine) vegetation and its suitability for indicating climate change impacts. In: Biology and environment: Proceedings of the royal Irish academy; pp 335–341

Naif SS, Mahmood DA, Al-Jiboori MH (2020) Seasonal normalized difference vegetation index responses to air temperature and precipitation in Baghdad. Open Agric 5(1):631–637

Nanda SA, Reshi ZA, Ul-haq M, Lone AB, Mir SA (2018) Taxonomic and functional plant diversity patterns along an elevational gradient through treeline ecotone in Kashmir. Trop Ecol 59(2):211–224

Negi JDS, Manhas RK, Chauhan PS (2003) Carbon allocation in different components of some tree species of India: a new approach for carbon estimation. Curr Sci 85:1528–1531

Ning T, Liu W, Lin W, Song X (2015) NDVI variation and its responses to climate change on the northern loess plateau of China from 1998 to 2012. Adv Meteorol 2015:1

Odland A (2015) Effect of latitude and mountain height on the timberline (*Betula pubescens ssp. czerpanovii*) elevation along the central Scandinavian mountain range. Fennia 193(2):260–270

Odum EP (1969) The strategy of ecosystem development. Science 164:262–270

Pan Y, Birdsey RA, Fang J, Houghton R, Kauppi PE, Kurz WA, Hayes D (2011) A large and persistent carbon sink in the world's forests. Science 333(6045):988–993

Panthi S, Bräuning A, Zhou ZK, Fan ZX (2017) Tree rings reveal recent intensified spring drought in the central Himalaya, Nepal. Glob Planet Chang 157:26–34

Parihar DS (2021) Timber line delineations using NDVI techniques in the Gori Ganga watershed of Kumaun Himalaya, Uttarakhand. J Res Environ Earth Sci 7(4):40–45

Pauli H, Gottfried M, Reiter K, Grabherr G (2002) High Mountain summits as sensitive indicators of climate change effects on vegetation patterns: the "multi summit-approach" of GLORIA (global observation research initiative in alpine environments). Adv Glob Chang Res 9:45–51. https://doi.org/10.1007/0-306-48051-4_6

Pokhrel S, Sherpa C (2020) Analyzing the relationship, distribution of tree species diversity, and above-ground biomass on the Chitwan-Annapurna landscape in Nepal. Int J For Res:2789753. https://doi.org/10.1155/2020/2789753

Poorter L, van der Sande MT, Thompson J, Arets EJMM, Alarcon A, Alvarez-Sanchez J, Ascarrunz N, Balvanera P, Barajas-Guzman G, Boit A, Bongers F, Carvalho FA, Casanoves F, Cornejo-Tenorio G, Costa FRC, de Castilho CV, Duivenvoorden JF, Dutrieux LP, Enquist BJ, Fernandez-Mendez F, Finegan B, Gormley LHL, Healey JR, Hoosbeek MR, Ibarra-Manriquez G, Junqueira AB, Levis C, Licona JC, Lisboa LS, Magnusson WE, Martinez-Ramos M, Martinez-Yrizar A, Martorano LG, Maskell LC, Mazzei L, Meave JA, Mora F, Munoz R, Nytch C, Pansonato MP, Parr TW, Paz H, Perez-Garcia EA, Renteria LY, Rodriguez-Velazquez J, Rozendaal DMA, Ruschel AR, Sakschewski B, Salgado-Negret B, Schietti J,

Simoes M, Sinclair FL, Souza PF, Souza FC, Stropp J, terSteege H, Swenson NG, Thonicke K, Toledo M, Uriarte M, van der Hout P, Walker P, Zamora N, Pena-Claros M (2015) Diversity enhances carbon storage in tropical forests. Glob Ecol Biogeogr 24:1314–1328

Rai ID, Adhikari BS, Rawat GS, Bargali K (2012) Community structure along timberline ecotone in relation to micro-topography and disturbances in Western Himalaya. Not Sci Biol 4(2):41–52

Ranhotra PS, Bhattacharyya A (2013) Modern Vegetational distribution and pollen dispersal study within Gangotri glacier transect, Garhwal Himalaya. J Geol Soc India 82:133–142

Rinn F (2003) TSAP-win: time series analysis and presentation for dendrochronology and related applications. Frank Rinn, Heidelberg

Roy I, Ranhotra PS, Shekhar M, Bhattacharyya A, Ghosh R, Sharma YK (2021) Modern pollen-vegetation relationships along the vegetation gradient in the Bhagirathi transect, Western Himalaya, India. J Geol Soc India 97:571–578. https://doi.org/10.1007/s12594-021-1732-0

Schickhoff U, Bobrowski M, Böhner J, Bürzle B, Chaudhary RP, Gerlitz L, Heyken H, Lange J, Müller M, Scholten T, Schwab N, Wedegärtner R (2015) Do Himalayan treelines respond to recent climate change? An evaluation of sensitivity indicators. Earth Syst Dyn 6:245–265

Shah SK, Pandey U, Mehrotra N, Wiles GC, Chandra R (2019) A winter temperature reconstruction for the Lidder transect, Kashmir, northwest Himalaya based on tree-rings of *Pinus wallichiana*. Clim Dyn 53(7–8):4059–4075

Shang HM, Wei WS, Yuan YJ, Yu SL, Zhang RB, Hong JC, Chen F, Zhang TW, Fan ZA (2016) Normalized vegetation variation index reconstruction based on the tree-ring width in Central Tibet. J Lanzhou Univ Nat Sci 52(1):18–24. (in Chinese, with English abstract)

Sharma CM, Baduni NP, Gairola S, Ghildiyal SK, Suyal S (2010) The effect of slope aspects on the forest composition, community structure and soil nutrient status of some major natural temperate forest types of Garhwal Himalayan. J For Res 21:331–337

Sharma CM, Tiwari OP, Rana YS, Krishan R, Mishra AK (2018) Elevational behaviour on dominance–diversity, regeneration, biomass and carbon storage in ridge forests of Garhwal Himalaya, India. For Ecol Manag 424:105–120

Shi Z, Xu L, Dong L, Gao J, Yang X, Lü S, Feng C, Shang J, Song A, Guo H, Zhang X (2015) Growth–climate response and drought reconstruction from tree-ring of Mongolian pine in Hulunbuir, Northeast China. J Plant Ecol 9(1):51–60

Shrestha KB, Hofgaard A, Vandvik V (2014) Recent treeline dynamics are similar between dry and Mesic areas of Nepal, central Himalaya. J Plant Ecol 1–12:347. https://doi.org/10.1093/jpe/rtu035

Singh CP, Mohapatra J, Pandya HA, Gajmer B, Sharma N, Shrestha DG (2018a) Evaluating changes in treeline position and land surface phenology in Sikkim Himalaya. Geocarto Int 35: 453. https://doi.org/10.1080/10106049.2018.1524513

Singh CP, Panigrahy S, Thapliyal A, Kimothi MM, Soni P, Parihar JS (2012) Monitoring the alpine treeline shift in parts of the Indian Himalayas using remote sensing. Curr Sci 102(4):559–562

Singh J, Yadav RR (2005) Spring precipitation variations over the western Himalaya, India, since AD 1731 as deduced from tree rings. J Geophys Res Atmos 110(D1). https://doi.org/10.1029/2004JD004855

Singh J, Yadav RR (2014) Chir pine ring-width thermometry in western Himalaya, India. Curr Sci:735–738. https://doi.org/10.18520/cs%2Fv106%2Fi5%2F735-738

Singh S, Verma AK (2018) Biomass and carbon stocks in different forest types of Western Himalaya. Trop Ecol 59(4):647–658

Singh U, Phulara M, David B, Ranhotra PS, Shekhar M, Bhattacharyya A, Dhyani R, Joshi R, Pal AK (2018b) Static tree line of Himalayan silver fir since last several decades at Tungnath, Western Himalaya. Trop Ecol 59(2):351–363

Siyum ZG, Ayoade JO, Onilude MA, Feyissa MT (2018) Relationship between space-based vegetation productivity index and radial growth of Main tree species in the dry Afromontane Forest remnants of northern Ethiopia. J Appl Sci Environ Manag 22(11):1781–1790

Speed JDM, Austrheim G, Hester AJ, Mysterud A (2011) Growth limitation of mountain birch caused by sheep browsing at the altitudinal treeline. For Ecol Manag 261:1344–1352

Speer JH (2010) Fundamentals of tree ring research. The University of Arizona Press, Tucson

Terakunpisut JN, Gajaseni RN (2007) Carbon sequestration potential in aboveground biomass of thong Phaphun national forest, Thailand. Appl Ecol Environ Res 5:93–102

Thapa UK, Shah SK, Gaire NP, Bhuju DR (2015) Spring temperatures in the far-western Nepal Himalaya since AD 1640 reconstructed from *Picea smithiana* tree-ring widths. Clim Dyn 45(7–8):2069–2081

Udo S, Andreas B, Peter F, Anja CM, Bruno R, Markus Z (2011) GPCC full data reanalysis version 6.0 at 0.5°: monthly land-surface precipitation from rain-gauges built on GTS-based and historic data. https://doi.org/10.5676/DWD_GPCC/FD_M_V7_050

Upadhaya K (2015) Structure and floristic composition of subtropical broad-leaved humid forest of Cherapunjee in Meghalaya, Northeast India. J Biodivers Manag For 4:2

Vicente-Serrano SM, Camarero JJ, Olano JM, MartínHernández N, Pea-Gallardo M, TomásBurguera M, Gazol A, Azorin-Molina C, Bhuyan U, Kenawy AE (2016) Diverse relationships between forest growth and the normalized difference vegetation index at a global scale. Remote Sens Environ 187:14–29

Wang J, Rich PM, Price KP, Kettle WD (2004) Relations between NDVI and tree productivity in the central Great Plains. Int J Remote Sens 25(16):3127–3138

Wang W, Lei X, Ma Z, Kneeshaw DD, Peng C (2011) Positive relationship between aboveground carbon stocks and structural diversity in spruce-dominated forest stands in New Brunswick, Canada. For Sci 57:506–515

Wang WZ, Liu XH, Chen T, An WL, Xu GB (2010) Reconstruction of regional NDVI using tree-ring width chronologies in the Qilian Mountains, northwestern China. Chin J Plant Ecol 34: 1033–1044

Wang X, Li Y, Wang X, Li Y, Lian J, Gong X (2021) Temporal and spatial variations in NDVI and analysis of the driving factors in the Desertified areas of northern China from 1998 to 2015. Front Environ Sci 9:633020

Wang YJ, Lu RJ, Ma YZ, Meng HW, Gao SY (2014) Response to climate change of different tree species and NDVI variation since 1923 in the middle arid region of Ningxia, China. Sci Cold Arid Reg 6(1):30–36. https://doi.org/10.3724/SP.J.1226.2014.00030

Wigley TM, Briffa KR, Jones PD (1984) On the average value of correlated time series, with applications in dendroclimatology and hydrometeorology. J Appl Meteorol Climatol 23(2): 201–213

Yadav RR, Braeuning A, Singh J (2011) Tree ring inferred summer temperature variations over the last millennium in western Himalaya, India. Clim Dyn 36:1545–1554

Yadav RR, Park WK, Bhattacharyya A (1997) Dendroclimatic reconstruction of April-May temperature fluctuations in the western Himalaya of India since A.D. 1698. Quat Res 48:187–191

Yadav RR, Park WK, Bhattacharyya A (1999) Spring-temperature variations in western Himalaya, India, as reconstructed from tree-rings: AD 1390-1987. The Holocene 9:85–90

Yadav RR, Singh J (2002) Tree-ring-based spring temperature patterns over the past four centuries in western Himalaya. Quat Res 57(3):299–305

Yuan Z, Wang S, Ali A, Gazol A, Ruiz-Benito P, Wang X, Loreau M (2018) Aboveground carbon storage is driven by functional trait composition and stand structural attributes rather than biodiversity in temperate mixed forests recovering from disturbances. Ann For Sci 75(3):1–13

Zandalinas SI, Mittler R, Balfagón D, Arbona V, Gómez-Cadenas A (2018) Plant adaptations to the combination of drought and high temperatures. Physiol Plant 162(1):2–12

Zhang T, Zhang R, Lu B, Mambetov BT, Kelgenbayev N, Dosmanbetov D, Huang L (2018) *Picea schrenkiana* tree-ring chronologies development and vegetation index reconstruction for the Alatau Mountains, Central Asia. Geochronometria 45(1):107–118

Zhang Y, Chen HYH (2015) Individual size inequality links forest diversity and above-ground biomass. J Ecol 103:1245–1252

Zhou T, Shi P, Jia G, Dai Y, Zhao X, Shangguan W, Du L, Wu H, Luo Y (2015) Age-dependent forest carbon sink: estimation via inverse modeling. J Geophys Res Biogeosci 120:2473–2492. https://doi.org/10.1002/2015JG002943

Response of Radial Growth in *Abies pindrow* (Royle ex D.Don) Royle to Climate at Treeline Ecotone in the northwestern Himalaya

20

Rayees A. Malik and Raman Sukumar

Abstract

The growth response of Himalayan pindrow fir (*Abies pindrow*) was investigated using tree rings at the treeline ecotone in Hirpora Wildlife Sanctuary in the northwestern Himalaya. The ring-width chronology was built based on 73 increment cores extracted from 40 healthy trees. The measurements were standardized to remove non-climatic growth trends, especially the age-related growth trends. The persistence, because of the effect of the previous year climate on current year growth, in standardized site chronology was removed by auto-regressive modelling. For dendroclimatic analysis, static and moving correlations were computed between site chronology and monthly temperature and precipitation data from Srinagar meteorological station. The climate data was divided into two parts pre-and post-1950 because of varying degrees of increase in the annual climate data during these two time periods. The growth rings formed post-1950 were bigger than those formed prior to 1950. Also, the growth-climate relationships varied between the time periods. The trees responded strongly to monthly temperature in the second part of the twentieth century, while the positive response to growing season precipitation weakened in the recent decades. The moving correlation analyses showed periods of strong and weak climatic responses consistent with the results from static correlations in the two studied time periods. This study will help in understanding the growth responses of Himalayan conifers to climate and hence would also aid in better predicting their future growth in response to predicted climate warming.

R. A. Malik (✉) · R. Sukumar
Centre for Ecological Sciences, Indian Institute of Science (IISc), Bangalore, India
e-mail: rayeesmalik@iisc.ac.in; rsuku@iisc.ac.in

S. P. Singh et al. (eds.), *Ecology of Himalayan Treeline Ecotone*,
https://doi.org/10.1007/978-981-19-4476-5_20

Keywords

Abies pindrow · Climate change · Dendroclimatology · Growth-climate relationships · Tree-ring width

20.1 Introduction

Forests not only play an important role in maintaining the mountain ecosystem functioning and also provide life-supporting ecosystem services that are valuable at local, regional and global scales. Worldwide, the ongoing climate change is leading to significant impacts on the forest ecosystems (Foster 2001; Field et al. 2007; Kirilenko and Sedjo 2007; Grimm et al. 2013). Climate change is expected to alter various eco-physiological processes resulting in shifting vegetation growth rates and distribution of species during the twenty-first century (Walther et al. 2002). Climate change is a great concern as it affects the human population, directly and indirectly, such as warmer winters, the spread of diseases, phenological shifts, seas level rise and frequent natural disasters (Kahn 2005; Webster 2005; Shepherd and Wingham 2007; IPCC 2014). High mountains around the World are among the most vulnerable ecosystems to climate change (Viviroli et al. 2011; Huggel et al. 2012; Rangwala and Miller 2012). The Himalaya is experiencing higher warming than average global warming (Bhutiyani et al. 2007; Shrestha et al. 2012) which can significantly impact the biodiversity, vegetation distribution and ecosystem structure in the region (Aryal et al. 2014; Rashid et al. 2015). The mean annual temperature in Himalaya is showing a more increasing trend than the global average (Pandey et al. 2018) up to 0.06–0.1 °C/y since the mid-1970s (Shrestha et al. 2012; Qin et al. 2013). More specifically, the mean annual temperature has increased by 1.6 °C during the last century in northwestern Himalaya with the maximum temperature showing a more rapid increase. Winter temperature has shown a more noticeable increase in the last century. The total annual precipitation has shown an overall decreasing trend in the northwestern Himalaya (Bhutiyani et al. 2007, 2009). The Himalaya plays an important role in climate regulation of the Indian subcontinent and supports the livelihood of millions of the people dependent on the ecosystem services provided by it (Gansser 1980; Singh and Singh 1987; Carrico et al. 2003; Anthwal et al. 2010; Viviroli et al. 2011; Ganjoo and Ota 2012). Despite being a global biodiversity hotspot, there are limited studies that have investigated the response of tree growth to changing climate in the Himalayan region (Schickhoff et al. 2015; Tiwari et al. 2017; Pandey et al. 2018). The plants growing in the Himalaya must quickly adapt to survive future climatic conditions. The variations in climate affect tree growth which significantly influence cambial phenology and wood formation (Camarero et al. 2010; Rossi et al. 2014).

Tree growth is largely influenced by endogenous factors like phytohormones and enzymes, and exogenous factors which include climatic variables, soil, slope and aspect. Climate is the most important factor for the regulation of growth particularly near the ecological boundaries of a species (Fritts 1972). Generally, the response of tree growth to climate is dependent on the habitat, species, age etc. Tree growth is sensitive to changes in regional environmental conditions (Korner 2007; McDowell et al. 2008). Given the predicted climate change, it is crucial to understand the tree growth and its response to climatic variability.

Himalayan conifers are known to have a huge dendroclimatic potential because of their age and climatic sensitivity (Ramesh et al. 1985, 1986a; Bhattacharyya et al. 1988). Several researchers have studied the role of climate in regulating the growth in various Himalayan conifers using annual growth rings (Borgaonkar et al. 1996, 2001; Yadav et al. 1997; Ahmed et al. 2010; Sohar et al. 2017) and wood micro-core samples for understanding the intra-annual dynamics of wood formation (Malik et al. 2020a, b). Trees in the treeline ecotone are strongly limited by temperature (Li et al. 2013, 2016, 2017), with an average growing season temperature around 6–7 °C (Körner 2003; Körner and Paulsen 2004). A detailed study on vascular cambium activity by Rossi et al., (2007a, b) at the alpine timberline showing that the cambium was active at a daily mean air temperature of 5.6–8.5 °C and a mean stem temperature of 7.2–9.0 °C. An increase in the growing season temperature would thus result in a longer time for cambial activity and hence more radial growth (Deslauriers et al. 2008a).

Tree rings have been commonly used as a proxy to investigate biological responses to regional/local climate as well as to study the temporal stability of these responses (Wang et al. 2005; Jump et al. 2007; Fan et al. 2009; Dang et al. 2013; He et al. 2013; Sohar et al. 2017). This approach provides a long time series at high spatiotemporal resolution and helps in better understanding of tree responses to ongoing climate change (Sohar et al. 2017). Many studies have shown that tree growth response to climate varies along the altitudinal range of a species (Kahle et al. 2002; Savva et al. 2006; Spiecker 2010; Dang et al. 2013; Kharal et al. 2015). Generally, trees respond positively to summer temperature in high altitudes to growing season precipitation in lower altitudinal limits. A better understanding of the climate-growth relationships in northwestern Himalaya will help in the assessment of the growth response of Himalayan conifers to changing climate and will help simulate the carbon sequestration potential of these forests in response to predicted climate change.

In the northwestern Himalayan region, tree rings have been widely used mostly for climate reconstruction (Ramesh et al. 1986a; Hughes 2001; Yadav 2009; Singh et al. 2009; Ram and Borgaonkar 2013, 2016; Yadav and Bhutiyani 2013; Shah et al. 2018). One of the most dominant conifer in the northwestern Himalayan coniferous forests is *Abies pindrow* (Royle ex D. Don) Royle. The climate sensitivity of *A. pindrow* has been utilized by dendroclimatologists to reconstruct regional temperature (Hughes 2001; Malik and Sukumar 2021) and precipitation (Borgaonkar et al. 1994a; Shah et al. 2018). Similarly, reconstruction of pre-monsoon relative humidity was carried out by Dhyani et al. (2021) using tree-ring data of *Pinus roxburghii* from Western Himalaya.

Generally, tree growth and climate relationships are assumed to be stable over time (Fritts et al. 1965). However, recent tree ring studies reported problems with this stable climate-growth relationship assumption. Some studies have found decrease or loss of temperature sensitivity (Jacoby and D'Arrigo 1995; Briffa et al. 1998; Smith et al. 1999; Jacoby et al. 2000; Solberg et al. 2002), whereas others have reported increased temperature sensitivity (Knapp et al. 2001; Wilmking 2005) and some studies have reported a shift from negative to positive temperature sensitivity or vice versa especially after the 1950s (Zhang et al. 2008; Wilmking and Myers-Smith 2008). These shifts in growth-climate relationships over time are probably because of some mechanisms, such as temperature-induced drought stress (Jacoby et al. 1996; Barber et al. 2000; Lloyd and Fastie 2002), delayed snowmelt and related change in seasonality (Vaganov et al. 1999), non-linear thresholds or time-dependent response to recent temperature warming (D'Arrigo et al. 2004; Wilmking et al. 2004; Rossi et al. 2007a), air pollution (Wilson and Elling 2004; Yonenobu and Eckstein 2006) and differential growth-climate relationships to mean, maximum and minimum temperatures (Wilson and Luckman 2002, 2003). There are other researchers who attribute these changed patterns in growth-climate relationships to methods of chronology development, biases in instrumental climate data and modelling (Cook and Peters 1997; Melvin et al. 2007; D'Arrigo et al. 2008).

In this study, we attempt to assess the response of radial growth of *Abies pindrow* (Royle ex D.Don) to monthly temperature and precipitation using standard dendroclimatological techniques. In addition, we investigated the temporal stability of growth-climate relationships using moving correlation analyses.

20.2 Materials and Methods

20.2.1 Study Site and Climate

This study was carried out in Hirpora Wildlife Sanctuary (HWS) in district Shopian, Jammu and Kashmir, India (33°40′N, 74°42′E). The HWS lies in the Pir-Panjal range of northwestern Himalaya approximately 70 km south-west of Srinagar city covering an area of approximately 341.25 km^2 (Ahmad et al. 2015) (Fig. 20.1). The Kashmir valley (Vale of Kashmir) is in the Northwest Himalaya lying between 33° 20′–34°54′ N latitudes and 73°55′–75°35′ E longitudes covering an area of nearly 15, 948 km^2 (Dar and Khuroo 2013) and is surrounded by the Karakoram range in the north, Pir-Panjal range in the south and west, Zanskar range in the east and Siwalik hills in the south (Pandey et al. 2016).

The vegetation of HWS is represented by moist temperate coniferous forest with sub-alpine and alpine grasslands at higher altitudes of the sanctuary. *A. pindrow* (silver fir/Himalayan pindrow fir) is the dominant conifer species, in association with *Pinus wallichiana* A. B. Jacks and *Picea smithiana* (Wall.) Boiss., growing on northern slopes from 2300 to 3200 m a.s.l. (Sharma and Baduni 2001). The forest type in HWS is a mixed coniferous forest with sub-alpine and alpine pastures at the higher altitudes (Ahmad et al. 2015).

The climate of the Kashmir valley is influenced by westerly disturbance (also known as 'westerlies') (Dimri et al. 2015). During winter, these westerlies originate from the Mediterranean and reach the Kashmir valley, bringing precipitation in the

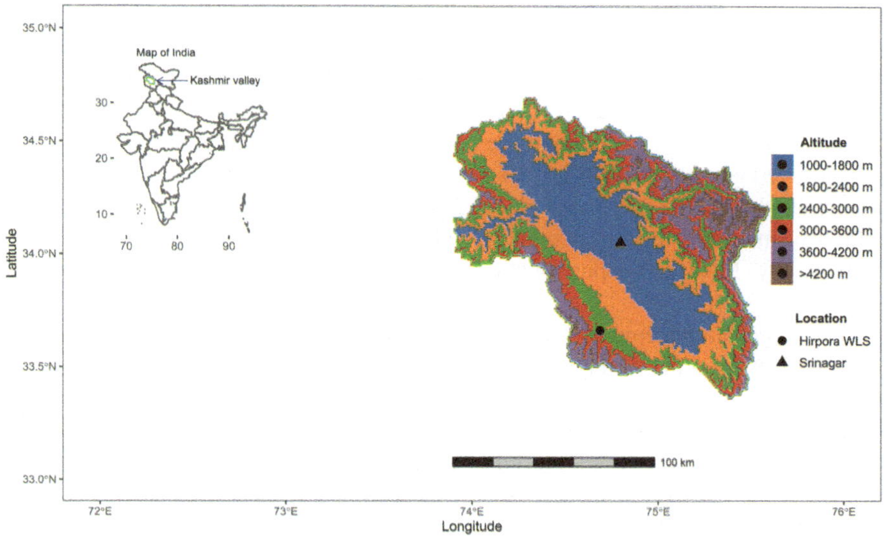

Fig. 20.1 Map showing the study site in the northwestern Himalaya

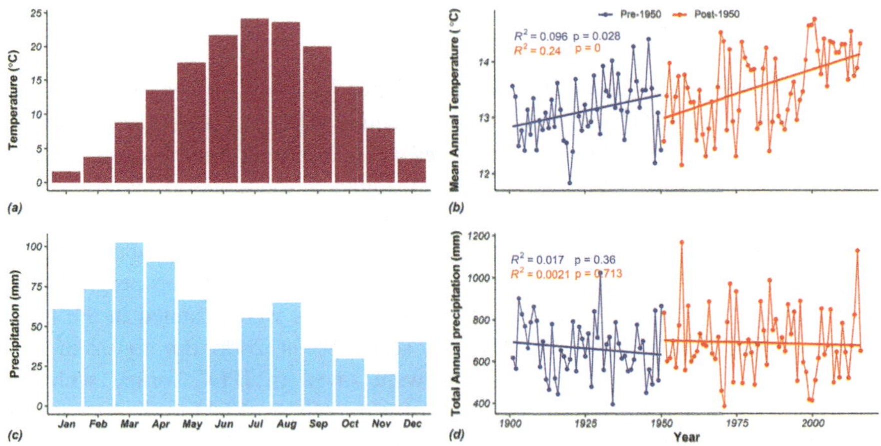

Fig. 20.2 Climate data of Srinagar meteorological station from 1901–2016. (**a**) Mean monthly temperature, (**b**) total monthly precipitation, (**c**) Mean annual temperature from 1901 to 2016 and (**d**) Total annual precipitation from 1901 to 2016

form of snow. Hirpora Wildlife Sanctuary has a sub-humid temperate climate. There is no weather station in the HWS, so climate data of the Srinagar meteorological station was used for this study (Fig. 20.2). The Srinagar station is the only station with more than 100 years of climate data in the Kashmir valley. The climate data from this station has earlier been used in many scientific studies (Ramesh et al. 1986b; Hughes 2001; Bhutiyani et al. 2007; Jeelani et al. 2012). Based on the monthly climatic data from Srinagar meteorological station, located at approximately 70 km from the study site, for the period 1901–2016, the mean annual temperature of the region is 13.4 °C, with July being the hottest month (mean maximum temperature = 24.2 °C) and January the coldest month (mean minimum temperature = 1.6 °C). Temperature goes to sub-zero in the winter months. The annual average precipitation is 680 mm, with March and April being the wettest months with an average rainfall of 102.8 and 90.8 mm respectively, while October and November are the driest months with total precipitation of 30 and 20.2 mm respectively. Most of the precipitation falls in the form of snow during the winter months.

The mean annual temperature has shown a significant increase in the last century as is evident from the Srinagar meteorological station data (Fig. 20.2). The magnitude of this increasing trend varies pre-and post-1950s with a greater increase post-1950s. There is no significant variation in the total annual precipitation in the two time periods. Because of the varying trends in the climate data of Srinagar, the growth-climate relationship analyses were done separately with these two time period climate data to understand the influence of recent climate warming on these relationships.

20.2.2 Data Collection

Near the highest distribution limits of the species in HWS, 40 healthy trees were selected for sampling. The diameter at breast height (DBH) of all the sampled trees was measured. A total of 73 increment cores were collected from the sampled trees. Usually, two cores per tree were collected parallel to the slope to avoid reaction wood. The samples were collected during 2014–2016. The cores were packed in straws and transported to the laboratory for further analysis. In the laboratory, the cores were glued to wooden holders and sequentially sanded using progressively finer grit sandpapers (80–3000 grit). The sanded cores were scanned in the high-resolution Espon Perfection V700 Photo scanner® at 2400 dpi to make ring boundaries visible. The scanned images were saved in JPEG format which is compatible with CooRecorder/CDendro software (Cybis Elektronik 2010). The ring-width measurements were done in CooRecorder/CDendro software package. Measurements were first visually cross-dated and then cross-checked with computer programme COFECHA (Holmes 1983) and dplR (Bunn 2010a, b) to ensure that each ring is assigned to its correct year of formation. COFECHA is a statistical tool in dendrochronology used for cross-dating purpose (Grissino-Mayer 2001; Speer 2010). The raw chronology was developed by taking the robust mean of all the series (Fig. 20.3). Various descriptive statistics such as mean ring-width, series-intercorrelation, the average mean sensitivity and first-order auto-correlation were computed for a general understanding of the site chronology. The ring width describes the radial growth in a site, mean sensitivity indicates the year-to-year variability between the consecutive tree-rings and the first-order auto-correlations describe the effect of the previous year growing conditions on the growth of the current year (Fritts 1972).

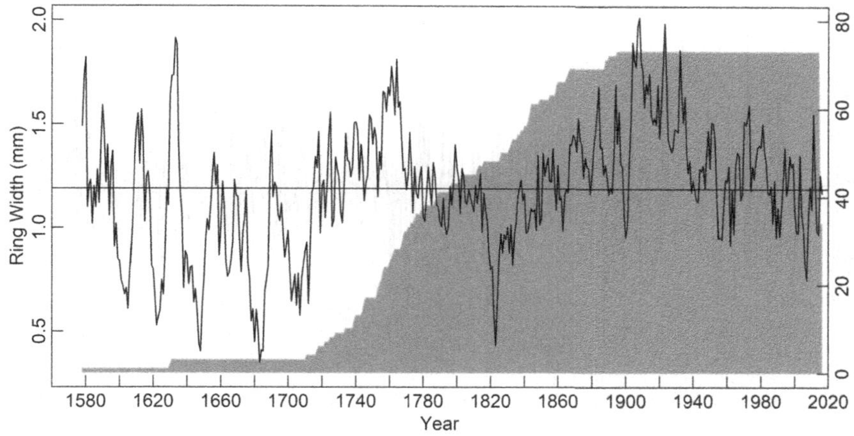

Fig. 20.3 Raw ring-width mean chronology. The bold horizontal line represents the mean ring-width, and the grey colour represents the sample depth

The age of the trees was calculated by counting the number of annual growth rings. The mean ring-width of five innermost existing growth rings was used to estimate the number of years to pith in samples, wherein the pith was missing using CooRecorder software (Cybis Elektronik 2010).

20.2.3 Chronology Development

The raw ring-width series were standardized to convert them into dimensionless indices using dplR package (Bunn 2008) in the R version 4.0.5 statistical programme (R Core Team 2020). The raw ring-width series were detrended using cubic spline curves of 32 years to remove the non-climatic trends due to tree age, size and forest stand dynamics and the ratios were calculated (Cook et al. 1990). The main criteria for detrending were to minimize the non-climatic low-frequency trend and maximize the high-frequency trend. The detrended series were averaged by computing a bi-weight robust mean to make the standardized site chronology (Cook 1985). The standardized site chronology showed high first-order auto-correlation values because of the impact of the previous year growing season on the current year growth. This persistence in the standardized chronology was removed by autoregressive modelling and the resulting residual series were averaged by computing bi-weight robust mean to make residual site chronology, henceforth referred to as Site chronology (Fig. 20.4) (Cook 1985).

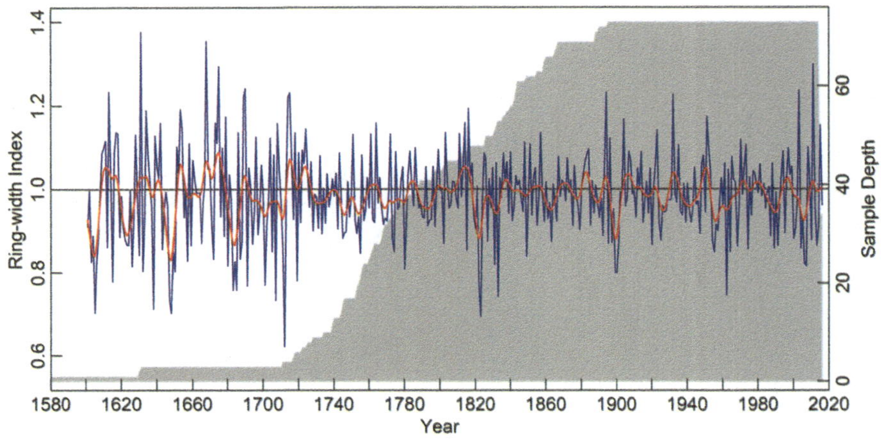

Fig. 20.4 Total ring-width site chronology (Residual). The horizontal black line indicates the long-term mean, grey colour indicates the sample depth and the red wavy line indicates 10-year moving average

20.2.4 Dendroclimatic Analysis

The relationships between site chronology and monthly meteorological records were analysed by using Pearson's correlation coefficients. The statistical significance of correlation coefficients was estimated by calculating bootstrapped confidence intervals. Monthly temperature and precipitation data from January to October, as well as mean of May–August temperature (mean-MJJA) and the sum of May–August precipitation (total-MJJA), was used for correlation analysis. The monthly correlation analysis was done for two time periods of 1901–1950 and 1951–2016 in addition to the full time period, henceforth referred to as pre-and post-1950s, to understand the impact of climate change on growth-climate relationships.

Further, the temporal stability of growth-climate relationships was investigated by running correlation analyses using 30-year moving intervals (climate data) with a 3-year offset to understand the variability of growth-climate relationships in smaller time periods of 30 years. The dynamic moving correlation analyses were run with the *treeclim* package (Zang and Biondi 2015) in R version 4.0.5 (R Core Team 2020).

20.3 Results

There is a stronger increasing trend in mean annual temperature after 1950s ($R^2 = 0.24$; $p < 0.001$) than before ($R^2 = 0.096$; $p = 0.028$), while there is no significant trend in total annual precipitation in both the analysed time periods ($p > 0.05$). This trend in mean annual temperature showed in growth-climate relationships in the two time periods of 1901–1950 and 1951–2016 respectively.

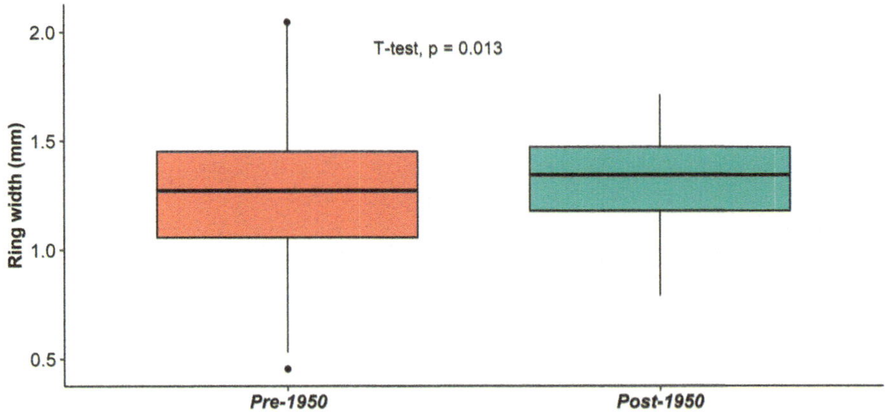

Fig. 20.5 Ring-width measurements pre- and post-1950 CE

The annual ring widths showed a significant variation between the two time periods ($p = 0.01$) (Fig. 20.5). The overall mean ring-width was 1.43 mm. The ring-width measured prior to 1950 were significantly smaller ($p = 0.013$) than the post-1950 time period with a mean ring-width of 1.25 and 1.33 mm respectively.

The average tree age was 229 years with a maximum age of 439 years (Table 20.1). There was a significant linear relationship between DBH and age of trees ($p < 0.01$). The age showed a linear increase with the increase in the diameter of individual trees (Fig. 20.6).

The chronology showed a good sensitivity to climatic conditions as depicted by a good mean sensitivity value (0.18). The first-order autocorrelation was high in the raw ring-width series (0.83) indicating the larger impact of previous year growth on current radial growth. The individual series showed lower auto-correlation after detrending and successive fitting of auto-regressive models (0.45). The expressed population signal (EPS) was more than 0.85 since 1765, which indicates that the chronology well represents the overall population, and can be reliably used for dendroclimatic studies. In addition, the strong series inter-correlation indicates the overall growth similarity between individual trees.

The response of radial growth to individual monthly climate changed during the two periods (Fig. 20.7). The positive response to January, February, March and April temperature in the pre-1950 changed to less positive or shifted to negative in the post-1950 time period, while May to October showing negative response in the pre-1950 period changed to less negative or responded positively in the later time period. The radial growth responded negatively to the mean of May–July temperature in the former time period, while the response was positive in the latter.

In pre-1950 period, the ring-width chronology showed strong negative response to October temperature ($r = -0.37, p < 0.01$), while the response was not significant to any other month temperature irrespective of the time period. The highest positive response was shown to January in 1901–2016 ($r = 0.18, p = 0.05$) and 1901–1950 ($r = 0.24, p = 0.09$) time periods, while July temperature showed maximum positive

Table 20.1 Summary statistics of tree ring width chronology of *Abies pindrow*

No. trees	No. cores	Time span	Mean age (yrs)	Mean ring-width (mm)	Mean sensitivity	Series inter-correlation	First-order auto-correlation	EPS > 0.85
40	73	1578–2016	229	1.43	0.18	0.45	0.83	1765–2016

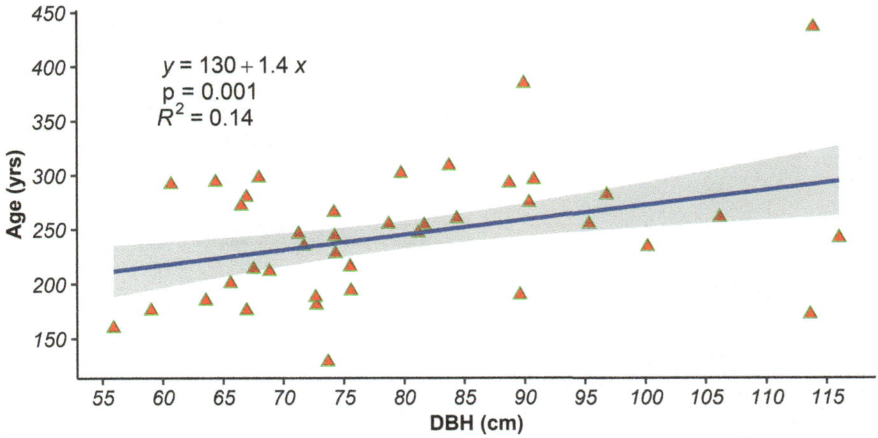

Fig. 20.6 Relationship between age and diameter at breast height (DBH) of sampled trees. The shaded area around the linear trend line indicates standard error (SE)

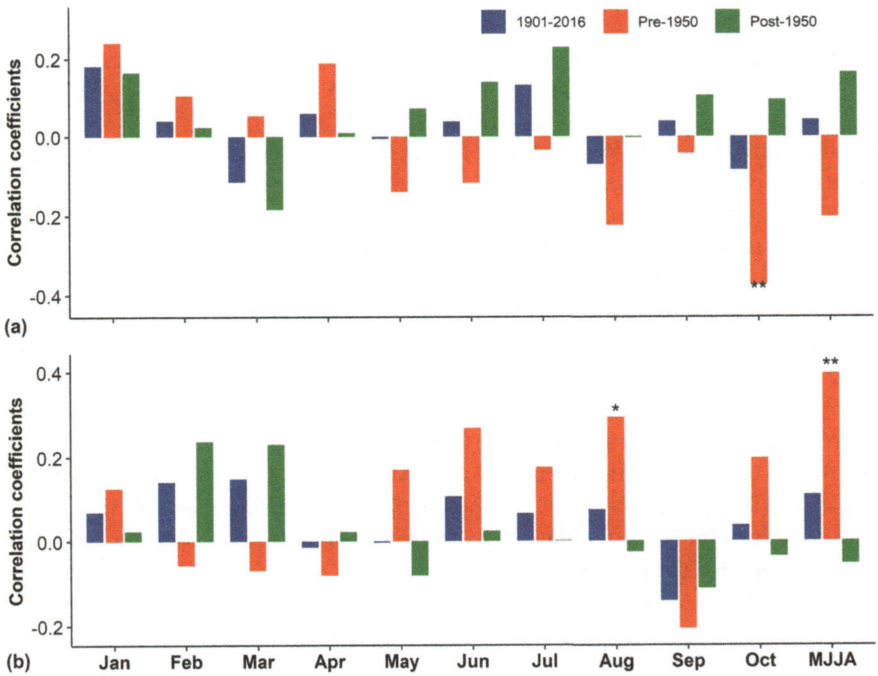

Fig. 20.7 Correlation coefficients between site chronology and monthly temperature (**a**) and monthly precipitation (**b**) for 1901–2016, 1901–1950 and 1951–2016 time periods. The asterisk marks * and ** above/below bars show statistical significance at $p < 0.05$ and $p < 0.01$ respectively

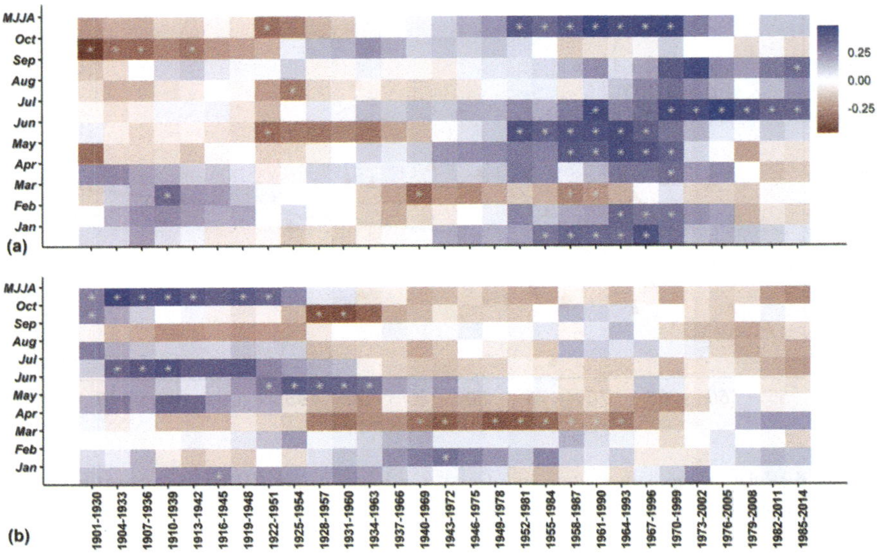

Fig. 20.8 Moving correlation coefficients between site chronology and monthly temperature (**a**) and monthly precipitation (**b**). The asterisk marks (*) represents 30-year window with statistically significant correlation coefficients

response during post-1950s time period ($r = 0.23$, $p = 0.06$). Likewise, the maximum negative response was shown to March temperature in 1901–2016 ($r = -0.11$, $p = 0.22$) and 1951–2016 ($r = -0.18$, $p = 0.14$) time periods, while the response was significantly negative to October temperature during 1901–1950 time period ($r = -0.37$, $p < 0.01$).

Similarly, the response to monthly precipitation varied in the two time periods as well. The overall response to growing season precipitation was weaker in the post-1950 time period. The response shifted from negative to either less negative or slightly positive, while there was a shift from positive to less positive or negative between the two studied time periods. Only August ($r = 0.29$, $p < 0.05$) and total-MJJA ($r = 0.40$, $p < 0.01$) precipitation showed significant positive response in the pre-1950s time period. None of the months in 1901–2016, as well as 1951–2016 time periods, responded significantly to radial growth at treeline site. The maximum positive response during 1901–2016 was shown to March precipitation ($r = 0.15$, $p = 0.11$), while August and total-MJJA precipitation responded positively during the pre-1950s time period. During 1951–2016 time period, the maximum positive response was shown to February precipitation ($r = 0.24$, $p = 0.05$). Similarly, the highest negative response was shown to September precipitation in all the studied time periods with correlation coefficients of −0.14, −0.21 and −0.11 in 1901–2016, pre-1950s and post-1950s time periods respectively.

The 30-year moving window analysis showed periods of low and high response to monthly climate data (Fig. 20.8). The general trend of responding to two time periods was seen in the moving correlation results as well. The strong negative response was shown to October month temperature during the 30-year window of 1901–1930 ($r = -0.45$, $p < 0.01$), while the strong positive response to shown to July temperature during 1976–2005 ($r = 0.50$, $p < 0.01$). The response to mean-MJJA temperature changed from being strongly negative ($r = -0.34$, $p < 0.05$) during 1922–1951 window to strongly positive ($r = 0.50$, $p < 0.01$) during the 30-year window of 1961–1990.

The moving correlation analysis with monthly precipitation data showed a strong negative response to October precipitation ($r = -0.41$, $p < 0.05$) during 1931–1960, while the response was strongly positive to July precipitation ($r = 0.44$, $p < 0.05$) during 1910–1939 time period. Likewise, the response to total growing season precipitation changed from strong negative ($r = -0.27$, $p > 0.05$) during 1985–2014 to strong positive ($r = 0.49$, $p < 0.01$) during the 1904–1933 time period.

20.4 Discussion

We present a dendroclimatology of *A. pindrow* that contributes to a better understanding of its response to recent climate change in the northwestern Himalaya. Due to its longevity and strong climate sensitivity, *A. pindrow* has been the subject of many dendroclimatological studies (Ramesh et al. 1985, 1986c; Borgaonkar et al. 1994b).

We found a linear relationship between tree age and diameter at breast-height (DBH). This relationship makes it easier to calculate the age of a forest stand at the treeline ecotone. Also, distinct differences were found between annual growth increments between two time periods. The relatively larger increase in average annual temperatures might have resulted in longer growing season duration and hence wider growth rings (Deslauriers et al. 2008b; Rossi et al. 2011; Malik et al. 2020a).

Radial growth in trees is affected by several biotic as well as abiotic factors (Briffa et al. 2004; Treydte et al. 2006; Yadav et al. 2011). Climate is one of the most important environmental factors affecting radial growth in trees. Temperate trees produce distinct growth rings which are easier to correlate with the regional climatic conditions. Trees annually record growth data in the form of the width of annual growth rings in their trunks, with wide rings during years of optimal climatic conditions and narrow rings during poor climatic conditions (Vaganov et al. 1999). Dendroclimatologists have developed time series of tree-ring widths from thousands of sites worldwide because of the well-established relationship between ring-width and climate (archived by the National Climate Data Center (NCDC) in the International Tree-Ring Data Bank (ITRDB; available online at http://www.ncdc.noaa.gov/paleo/treering.html). Several researchers have used statistical relationships between annual tree rings and climate to infer forest responses to future climate change.

Overall, the radial growth in pindrow fir responded positively to monthly temperature in recent decades. The positive response of tree growth to the temperature at the treeline site is more of a general phenomenon (Wilson and Hopfmüller 2001; Takahashi et al. 2003). Trees at treeline did not respond to growing season temperature in the first half of the twentieth century and responded positively in the second half. This indicates that radial growth trees at the tree line ecotone might be benefitted from climate warming. The higher temperatures could increase the rate of photosynthesis and thus enhancing radial growth (White 1987; Zhang and Hebda 2004). A significantly lower number of tracheids are produced at the higher altitudinal limits compared to lower limits in the northwestern Himalaya (Malik et al. 2020a) which could be because of favourable temperatures for a relatively short duration of time. Increasing temperatures might potentially lead to a lengthening of growing season duration and hence results in more wood formation (Deslauriers et al. 2008a, b; Rossi et al. 2008, 2016; Moser et al. 2010; Li et al. 2016).

The impact of precipitation was weaker at the tree line site. The weaker impact of growing season precipitation on pindrow fir growth at its highest distributional limit is likely caused by good water availability due to snowmelt and rainfall during the growing season (Kim 2000; Körner 2007). The decreasing importance of precipitation at higher altitudes was reported in many other studies as well (Kahle et al. 2002; Jump et al. 2007). High temperatures during summer promote radial growth at higher altitudes. Generally, precipitation and soil moisture increase with increasing altitude, whereas air temperature decreases with increasing altitude, suggesting that radial growth in *A. pindrow* at upper limits was limited more by low temperature than by precipitation (Bradley and Fritts 1978). These results were in agreement with other studies showing that radial growth responds positively to growing season precipitation at lower altitudes, while radial growth responds positively to growing season temperature at higher altitudes (Wilson and Hopfmüller 2001; Takahashi et al. 2003; Dang et al. 2013).

Also, there are significant impacts of previous year climatic conditions on current year growth as is evident from higher first-order auto-correlations. As found by other researchers working with subalpine conifers, the previous season climatic conditions influence bud formation, increased foliage and production of photosynthetic assimilates, which in turn impacts the wood production in the subsequent growing season (Bradley and Fritts 1978; Savva et al. 2006).

The impact of climate on pindrow fir growth was not stable over time. The climatic conditions especially temperature varied between the two selected time periods, pre- and post-1950. The variation in climatic conditions is depicted in the climate-growth relationships as well. The response to monthly climate shows a variation between the two selected time periods. Before 1950, the radial growth responds negatively to growing season monthly temperature and positively to precipitation. These effects of temperature and precipitation on *A. pindrow* growth, however, changed in recent decades. Growth responded positively to monthly temperatures, while the response to monthly precipitation diminished in the latter part of the twentieth century. Because of temporal variability in climate, response to climatic conditions has varied during the twentieth century generally referred to as the 'divergence effect' (Briffa et al. 1998). The variation in growth-climate relationships in time is more of a general phenomenon (Carrer et al. 2010). The shift in growth-climate relationships is a reliable indicator of varying tree growth response to climate (Di Filippo et al. 2007a, b). The temporal instability of growth-climate relationships might be because of possible changes in climate and other environmental conditions (Carrer and Urbinati 2006; Di Filippo et al. 2007a, b; Leonelli et al. 2009; Dang et al. 2013). Many recent studies also showed the instability of growth-climate relationships mostly in the northern latitudes (Jacoby et al. 1996, 2000; Briffa et al. 1998; Smith et al. 1999; Knapp et al. 2001; Solberg et al. 2002; Wilmking et al. 2004; Wilmking 2005). The variations in the growth-climate relationships in pindrow fir might be attributed to regional climate and other local environmental changes, which disrupted the growth-climate relationships and subsequently resulted in changed climate sensitivity and growth trends (Carrer and Urbinati 2006; Leonelli et al. 2009; Dang et al. 2013; Sohar et al. 2017). The appearance or loss of significance of climate sensitivity of certain climatic variables might be because of threshold-controlled mechanisms, as were detected in other climate-growth relationship studies (D'Arrigo et al. 2004; Wilmking et al. 2004; Carrer and Urbinati 2006; Rossi et al. 2007b; Dang et al. 2013).

The results from moving correlation analyses are in good agreement with the results from static correlation analyses for two time periods, pre-and post-1950. The response to mean-MJJA temperature changed from being negative in the first half of the twentieth century to being positive in the latter part of the century as shown in both the static and moving correlation analyses. The shifting response indicates the positive impact of ongoing warming summer temperatures on the radial growth and hence overall carbon sequestration of the forest stands towards higher altitudinal limits. The response to total-MJJA precipitation shifted from being strongly positive in the first part of the century to not responding in the later part of the climate data. The weaker response to growing season precipitation towards the higher altitudinal limits of the species could be possibly the result of enough availability of soil moisture, due to delayed snowmelt as well as good rainfall during this period.

The investigation of growth-climate relationships in different time periods gives a better indication of varying responses of radial growth to climate. Such studies are important for a better understanding of the growth responses of Himalayan tree species to climate change. Better clarity regarding growth responses to local climate will have management implications for afforestation programmes keeping in view the future carbon sequestration potential of Himalayan forests.

20.5 Conclusion

This study provides insights into the response of radial growth in *A. pindrow* to monthly temperature and precipitation at its highest distribution limit in the north-western Himalaya. The climate data used in this study showed an overall increase in annual temperature and more or less stable annual precipitation. The growth responded positively to the growing season monthly temperature in the recent decades. The positive response of monthly precipitation on radial growth decreased in recent decades. Specifically, the study provides evidence of varying response of pindrow fir growth to recent climate change. The radial growth is strongly assisted by recent warming, while the positive impact of precipitation diminished in recent decades. In general, pindrow fir in the northwestern Himalaya is responsive to growing season temperature, and therefore the increase in temperature in combination with enriched soil moisture will possibly lead to growth enhancement in this species near its higher distribution limits. We highlight the importance of studying the temporal stability of growth-climate relationships to provide evidence about the impacts of climate change on forest ecosystems.

Acknowledgements We thank Divecha Centre for Climate Change, IISc Bangalore for providing funding support for this project. R. Malik received the Merit Scholarship from Fonds de Recherche du Quebec – Nature et Technologies (FRQNT), Quebec, Canada. R. Malik also received Junior Research Fellowship from University Grants Commission (UGC), New Delhi. RS was a JC Bose Fellow during the tenure of this study. Also, thanks to Jammu and Kashmir Wildlife Department for giving permission to carry out this work in Hirpora Wildlife Sanctuary, J & K.

References

Ahmad F, ul Hassan Z, Farooq S, Shah JA (2015) Assessment of land use/land cover change in Hirpora wildlife sanctuary, Kashmir. Asian J Earth Sci 8:64–73. https://doi.org/10.3923/ajes.2015.64.73

Ahmed M, Khan N, Wahab M (2010) Climate response function analysis of Abies Pindrow (Royle) Spach. Preliminary results. Pak J Bot 42:165–171

Anthwal A, Gupta N, Sharma A et al (2010) Conserving biodiversity through traditional beliefs in sacred groves in Uttarakhand Himalaya, India. Resour Conserv Recycl 54:962–971. https://doi.org/10.1016/j.resconrec.2010.02.003

Aryal A, Brunton D, Raubenheimer D (2014) Impact of climate change on human-wildlife-ecosystem interactions in the trans-Himalaya region of Nepal. Theor Appl Climatol 115:517–529. https://doi.org/10.1007/s00704-013-0902-4

Barber VA, Juday GP, Finney BP (2000) Reduced growth of Alaskan white spruce in the twentieth century from temperature-induced drought stress. Nature 405:668–673. https://doi.org/10.1038/35015049

Bhattacharyya A, Lamarche V, Telewski F (1988) Dendrochronological reconnaissance of the conifers of Northwest India. Tree Ring Bull 48:21–30

Bhutiyani MR, Kale VS, Pawar NJ (2007) Long-term trends in maximum, minimum and mean annual air temperatures across the northwestern Himalaya during the twentieth century. Clim Chang 85:159–177. https://doi.org/10.1007/s10584-006-9196-1

Bhutiyani MR, Kale VS, Pawar NJ (2009) Climate change and the precipitation variations in the northwestern Himalaya: 1866–2006. Int J Climatol. https://doi.org/10.1002/joc.1920

Borgaonkar HP, Pant GB, Rupa Kumar K (1994a) Dendroclimatic reconstruction of summer precipitation at Srinagar, Kashmir, India, since the late-eighteenth century. The Holocene 4: 299–306

Borgaonkar HP, Pant GB, Rupa Kumar K (1994b) Dendroclimatic reconstruction of summer precipitation at Srinagar, Kashmir, India, since the late-eighteenth century. The Holocene 4: 299–306. https://doi.org/10.1177/095968369400400309

Borgaonkar HP, Pant GB, Rupa Kumar K (1996) Ring-width variations in Cedrus deodara and its climatic response over the western Himalaya. Int J Climatol 16:1409–1422

Borgaonkar HP, Kumar KR, Pant GB et al (2001) Climatic implications of tree-ring density variations in Himalayan conifers. Palaeobotanist 50:27–34

Bradley RS, Fritts HC (1978) Tree rings and climate. Arct Alp Res 10:144. https://doi.org/10.2307/1550667
Briffa KR, Schweingruber FH, Jones PD et al (1998) Reduced sensitivity of recent tree-growth to temperature at high northern latitudes. Nature 391:678–682. https://doi.org/10.1038/35596
Briffa KR, Osborn TJ, Schweingruber FH (2004) Large-scale temperature inferences from tree rings: a review. Glob Planet Chang 40:11–26. https://doi.org/10.1016/S0921-8181(03)00095-X
Bunn AG (2008) A dendrochronology program library in R (dplR). Dendrochronologia 26:115–124. https://doi.org/10.1016/j.dendro.2008.01.002
Bunn AG (2010a) Statistical and visual crossdating in R using the dplR library. Dendrochronologia 28:251–258. https://doi.org/10.1016/j.dendro.2009.12.001
Bunn AG (2010b) Statistical and visual crossdating in R using the dplR library. Dendrochronologia 28:251–258
Camarero JJ, Olano JM, Parras A (2010) Plastic bimodal xylogenesis in conifers from continental Mediterranean climates. New Phytol 185:471–480. https://doi.org/10.1111/j.1469-8137.2009.03073.x
Carrer M, Urbinati C (2006) Long-term change in the sensitivity of tree-ring growth to climate forcing in Larix decidua. New Phytol 170:861–872. https://doi.org/10.1111/j.1469-8137.2006.01703.x
Carrer M, Nola P, Motta R, et al (2010) Contrasting tree-ring growth to climate responses of Abies alba toward the southern limit of its distribution area published by : Wiley on behalf of Nordic society Oikos stable URL: https://www.jstor.org/stable/20779075 linked references are available on. 119:1515–1525. https://doi.org/10.1111/j.1
Carrico CM, Bergin MH, Shrestha AB et al (2003) The importance of carbon and mineral dust to seasonal aerosol properties in the Nepal Himalaya. Atmos Environ 37:2811–2824. https://doi.org/10.1016/S1352-2310(03)00197-3
Cook ER (1985) A time series analysis approach to tree ring standardization
Cook ER, Peters K (1997) Calculating unbiased tree-ring indices for the study of climatic and environmental change. The Holocene 7:361–370. https://doi.org/10.1177/095968369700700314
Cook ER, Briffa K, Shiyatov S, Mazepa V (1990) Tree-ring standardization and growth-trend estimation
Cybis Elektronik (2010) CDendro and CooRecorder., http://www.cybis.se/forfun/dendro/index.htm
D'Arrigo RD, Kaufmann RK, Davi N et al (2004) Thresholds for warming-induced growth decline at elevational tree line in the Yukon Territory, Canada, Global Biogeochem Cycles. 18. https://doi.org/10.1029/2004GB002249
D'Arrigo R, Wilson R, Liepert B, Cherubini P (2008) On the 'divergence problem' in northern forests: a review of the tree-ring evidence and possible causes. Glob Planet Chang 60:289–305. https://doi.org/10.1016/j.gloplacha.2007.03.004
Dang H, Zhang Y, Zhang K et al (2013) Climate-growth relationships of subalpine fir (Abies fargesii) across the altitudinal range in the Shennongjia Mountains, Central China. Clim Chang 117:903–917. https://doi.org/10.1007/s10584-012-0611-5
Dar GH, Khuroo AA (2013) Floristic diversity in the Kashmir Himalaya: Progress, problems and prospects. Sains Malays 42:1377–1386
Deslauriers A, Rossi S, Anfodillo T, Saracino A (2008a) Cambial phenology, wood formation and temperature thresholds in two contrasting years at high altitude in southern Italy. Tree Physiol 28:863–871. https://doi.org/10.1093/treephys/28.6.863
Deslauriers A, Rossi S, Anfodillo T, Saracino A (2008b) Cambial phenology, wood formation and temperature thresholds in two contrasting years at high altitude in southern Italy. Tree Physiol 28:863–871
Dhyani R, Shekhar M, Joshi R et al (2021) Reconstruction of pre-monsoon relative humidity since 1800 C.E. based on tree-ring data of Pinus roxburghii Sarg. (chir–pine) from Pithoragarh, Western Himalaya. Quat Int. https://doi.org/10.1016/j.quaint.2021.04.026

Di Filippo A, Biondi F, Čufar K et al (2007a) Bioclimatology of beech (Fagus sylvatica L.) in the eastern Alps: spatial and altitudinal climatic signals identified through a tree-ring network. J Biogeogr 34:1873–1892. https://doi.org/10.1111/j.1365-2699.2007.01747.x

Di Filippo A, Biondi F, Čufar K et al (2007b) Bioclimatology of beech (Fagus sylvatica L.) in the eastern Alps: spatial and altitudinal climatic signals identified through a tree-ring network. J Biogeogr 34:1873–1892. https://doi.org/10.1111/J.1365-2699.2007.01747.X

Dimri AP, Niyogi D, Barros AP et al (2015) Western disturbances: a review. Rev Geophys 53:225–246. https://doi.org/10.1002/2014RG000460

Fan Z-X, Bräuning A, Cao K-F, Zhu S-D (2009) Growth–climate responses of high-elevation conifers in the central Hengduan Mountains, southwestern China. For Ecol Manag 258:306–313. https://doi.org/10.1016/j.foreco.2009.04.017

Field CB, Lobell DB, Peters HA, Chiariello NR (2007) Feedbacks of terrestrial ecosystems to climate change. Annu Rev Environ Resour 32:1–29

Foster P (2001) The potential negative impacts of global climate change on tropical montane cloud forests. Earth-Sci Rev 55:73–106

Fritts HC (1972) Tree rings and climate. Sci Am 226:92–100. https://doi.org/10.1038/scientificamerican0572-92

Fritts HC, Smith DG, Cardis JW, Budelsky CA (1965) Tree-ring characteristics along a vegetation gradient in northern Arizona. Ecology 46:393–401. https://doi.org/10.2307/1934872

Ganjoo RK, Ota SB (2012) Mountain environment and early human adaptation in NW Himalaya, India: a case study of Siwalik Hill range and Leh valley. Quat Int 269:31–37. https://doi.org/10.1016/j.quaint.2011.11.033

Gansser A (1980) The significance of the Himalayan suture zone. Tectonophysics 62:37–52. https://doi.org/10.1016/0040-1951(80)90134-1

Grimm NB, Chapin FS, Bierwagen B et al (2013) The impacts of climate change on ecosystem structure and function. Front Ecol Environ 11:474–482

Grissino-Mayer HD (2001) Evaluating crossdating accuracy: a manual and tutorial for the computer program COFECHA. Tree-Ring Res 57:205–221. https://doi.org/S

He M, Yang B, Bräuning A (2013) Tree growth–climate relationships of Juniperus tibetica along an altitudinal gradient on the southern Tibetan plateau. Trees 27:429–439. https://doi.org/10.1007/s00468-012-0813-5

Holmes RL (1983) Computer-assisted quality control in tree-ring dating and measurement. Tree-Ring Bull. https://doi.org/10.1016/j.ecoleng.2008.01.004

Huggel C, Clague JJ, Korup O (2012) Is climate change responsible for changing landslide activity in high mountains? Earth Surf Process Landforms 37:77–91. https://doi.org/10.1002/esp.2223

Hughes MK (2001) An improved reconstruction of summer temperature at Srinagar, Kashmir since 1660 AD, based on tree-ring width and maximum latewood density of Abies pindrow [Royle] Spach. Palaeobotanist 50:13–19

IPCC (2014) Climate change 2013: the physical science basis: working group I contribution to the IPCC fifth assessment report. Annex 1 1–85. https://doi.org/10.1017/CBO9781107415324.Summary

Jacoby GC, D'Arrigo RD (1995) Tree ring width and density evidence of climatic and potential forest change in Alaska. Global Biogeochem Cycles 9:227–234. https://doi.org/10.1029/95GB00321

Jacoby GC, D'Arrigo RD, Davaajamts T (1996) Mongolian tree rings and 20th-century warming. Science (80-) 273:771–773. https://doi.org/10.1126/science.273.5276.771

Jacoby GC, Lovelius NV, Shumilov OI et al (2000) Long-term temperature trends and tree growth in the Taymir region of northern Siberia. Quat Res 53:312–318. https://doi.org/10.1006/qres.2000.2130

Jeelani G, Feddema JJ, van der Veen CJ, Stearns L (2012) Role of snow and glacier melt in controlling river hydrology in Liddar watershed (western Himalaya) under current and future climate. Water Resour Res 48. https://doi.org/10.1029/2011WR011590

Jump A, Hunt J, Penuelas J (2007) Climate relationships of growth and establishment across the altitudinal range of Fagus sylvatica in the Montseny Mountains, northeaast Spain. Ecoscience 14:507–518

Kahle H, Neumann U, Mäkinen H et al (2002) Radial growth variation of Norway spruce (Picea abies (L.) karst.) across latitudinal and altitudinal gradients in central and northern Europe. For Ecol Manag 171:243–259. https://doi.org/10.1016/S0378-1127(01)00786-1

Kahn ME (2005) The death toll from natural disasters: the role of income, geography, and institutions. Rev Econ Stat 87:271–284. https://doi.org/10.1162/0034653053970339

Kharal DK, Meilby H, Rayamajhi S et al (2015) Tree ring variability and climate response of Abies spectabilis along an elevation gradient in mustang, Nepal. Banko Janakari 24:3. https://doi.org/10.3126/banko.v24i1.13473

Kim E-SY-S (2000) Possibility of climate change and simulation of soil moisture content on Mt. Hallasan National Park, Chejudo Island, Korea. Korean J Ecol 23:117–123

Kirilenko AP, Sedjo RA (2007) Climate change impacts on forestry. Proc Natl Acad Sci 104: 19697–19702

Knapp PA, Soulé PT, Grissino-Mayer HD (2001) Detecting potential regional effects of increased atmospheric CO 2 on growth rates of western juniper. Glob Chang Biol 7:903–917. https://doi.org/10.1046/j.1365-2486.2001.00452.x

Körner C (2003) Plant ecology at high elevations. In: Alpine Plant Life. Springer, Berlin, pp 1–7

Korner C (2007) Significance of temperature in plant life. In: Plant growth and climate change. Blackwell Publishing Ltd, Oxford, pp 48–69

Körner C (2007) The use of 'altitude' in ecological research. Trends Ecol Evol 22:569–574. https://doi.org/10.1016/J.TREE.2007.09.006

Körner C, Paulsen J (2004) A world-wide study of high altitude treeline temperatures. J Biogeogr 31:713–732. https://doi.org/10.1111/j.1365-2699.2003.01043.x

Leonelli G, Pelfini M, Battipaglia G, Cherubini P (2009) Site-aspect influence on climate sensitivity over time of a high-altitude Pinus cembra tree-ring network. Clim Chang 96:185–201. https://doi.org/10.1007/s10584-009-9574-6

Li X, Liang E, Gricar J et al (2013) Age dependence of xylogenesis and its climatic sensitivity in Smith fir on the south-eastern Tibetan plateau. Tree Physiol 33:48–56

Li X, Rossi S, Liang E, Julio Camarero J (2016) Temperature thresholds for the onset of xylogenesis in alpine shrubs on the Tibetan plateau. Trees 30:2091–2099. https://doi.org/10.1007/s00468-016-1436-z

Li X, Liang E, Gričar J et al (2017) Critical minimum temperature limits xylogenesis and maintains treelines on the southeastern Tibetan plateau. Sci Bull 62:804–812

Lloyd AH, Fastie CL (2002) Spatial and temporal variability in the growth and climate response of treeline trees in Alaska. Clim Chang 52:481–509. https://doi.org/10.1023/a:1014278819094

Malik R, Sukumar R (2021) June–July temperature reconstruction of Kashmir Valley from tree rings of Himalayan Pindrow Fir. Atmosphere (Basel) 12:410. https://doi.org/10.3390/atmos12030410

Malik R, Rossi S, Sukumar R (2020a) Cambial phenology in Abies pindrow (Pinaceae) along an altitudinal gradient in northwestern Himalaya. IAWA J 41:186–201. https://doi.org/10.1163/22941932-bja10007

Malik R, Rossi S, Sukumar R (2020b) Variations in the timing of different phenological stages of cambial activity in Abies pindrow (Royle) along an elevation gradient in the north-western Himalaya. Dendrochronologia 59:125660. https://doi.org/10.1016/j.dendro.2019.125660

McDowell N, Pockman WT, Allen CD et al (2008) Mechanisms of plant survival and mortality during drought: why do some plants survive while others succumb to drought? New Phytol 178: 719–739

Melvin TM, Briffa KR, Nicolussi K, Grabner M (2007) Time-varying-response smoothing. Dendrochronologia 25:65–69. https://doi.org/10.1016/j.dendro.2007.01.004

Moser L, Fonti P, Buntgen U et al (2010) Timing and duration of European larch growing season along altitudinal gradients in the Swiss Alps. Tree Physiol 30:225–233. https://doi.org/10.1093/treephys/tpp108

Pandey U, Shah SK, Mehrotra N (2016) Tree-ring studies from Kashmir Valley: present status and future perspectives tree-ring studies from Kashmir Valley: present status. Geophytology 46: 207–220

Pandey S, Carrer M, Castagneri D, Petit G (2018) Xylem anatomical responses to climate variability in Himalayan birch trees at one of the world's highest forest limit. Perspect Plant Ecol Evol Syst 33:34–41

Qin C, Yang B, Melvin TM et al (2013) Radial growth of qilian juniper on the northeast tibetan plateau and potential climate associations. PLoS One. https://doi.org/10.1371/journal.pone.0079362

R Core Team (2020) R: a language and environment for statistical computing. R Found. Stat. Comput, Vienna, Austria. https//www.R-project.org/

Ram S, Borgaonkar HP (2013) Growth response of conifer trees from high-altitude region of western himalaya. Curr Sci 105(2):225–231

Ram S, Borgaonkar HP (2016) Reconstruction of heat index based on tree-ring width records of western Himalaya in India. Dendrochronologia 40:64–71. https://doi.org/10.1016/j.dendro.2016.06.003

Ramesh R, Bhattacharya SK, Gopalan K (1985) Dendroclimatological implications of isotope coherence in trees from Kashmir Valley, India. Nature 317:802–804. https://doi.org/10.1038/317802a0

Ramesh R, Bhattacharya SK, Gopalan K (1986a) Climatic correlations in the stable isotope records of silver fir (Abies pindrow) trees from Kashmir, India. Earth Planet Sci Lett 79:66–74

Ramesh R, Bhattacharya SK, Gopalan K (1986b) Climatic correlations in the stable isotope records of silver fir (Abies pindrow) trees from Kashmir, India. Earth Planet Sci Lett 79:66–74. https://doi.org/10.1016/0012-821X(86)90041-5

Ramesh R, Bhattacharya SKK, Gopalan K (1986c) Climatic correlations in the stable isotope records of silver fir (Abies pindrow) trees from Kashmir, India. Earth Planet Sci Lett 79:66–74. https://doi.org/10.1016/0012-821X(86)90041-5

Rangwala I, Miller JR (2012) Climate change in mountains: a review of elevation-dependent warming and its possible causes. Clim Chang 114:527–547. https://doi.org/10.1007/s10584-012-0419-3

Rashid I, Romshoo SA, Chaturvedi RK et al (2015) Projected climate change impacts on vegetation distribution over Kashmir Himalayas. Clim Chang 132:601–613. https://doi.org/10.1007/s10584-015-1456-5

Rossi S, Deslauriers A, Anfodillo T, Carraro V (2007a) Evidence of threshold temperatures for xylogenesis in conifers at high altitudes. Oecologia 152:1–12. https://doi.org/10.1007/s00442-006-0625-7

Rossi S, Deslauriers A, Anfodillo T, Carraro V (2007b) Evidence of threshold temperatures for xylogenesis in conifers at high altitudes. Oecologia 152:1–12

Rossi S, Deslauriers A, Griçar J et al (2008) Critical temperatures for xylogenesis in conifers of cold climates. Glob Ecol Biogeogr 17:696–707. https://doi.org/10.1111/j.1466-8238.2008.00417.x

Rossi S, Morin H, Deslauriers A, Plourde P-Y (2011) Predicting xylem phenology in black spruce under climate warming. Glob Chang Biol 17:614–625

Rossi S, Girard MJ, Morin H (2014) Lengthening of the duration of xylogenesis engenders disproportionate increases in xylem production. Glob Chang Biol 20:2261–2271

Rossi S, Anfodillo T, Čufar K et al (2016) Pattern of xylem phenology in conifers of cold ecosystems at the northern hemisphere. Glob Chang Biol 22:3804–3813. https://doi.org/10.1111/gcb.13317

Savva Y, Oleksyn J, Reich PB et al (2006) Interannual growth response of Norway spruce to climate along an altitudinal gradient in the Tatra Mountains, Poland. Trees 20:735–746. https://doi.org/10.1007/s00468-006-0088-9

Schickhoff U, Bobrowski M, Böhner J et al (2015) Do Himalayan treelines respond to recent climate change? An evaluation of sensitivity indicators. Earth Syst Dynam 6:245–265

Shah SK, Pandey U, Mehrotra N (2018) Precipitation reconstruction for the Lidder Valley, Kashmir Himalaya using tree-rings of Cedrus deodara. Int J Climatol 38:e758–e773. https://doi.org/10.1002/joc.5405

Sharma CM, Baduni NP (2001) Effect of aspect on the structure of some natural stands of Abies pindrow in Himalayan moist temperate forest. Environmentalist 20:309–317

Shepherd A, Wingham D (2007) Recent Sea-Level Contributions of the Antarctic and Greenland Ice Sheets. Science (80-) 315:1529–1532. https://doi.org/10.1126/science.1136776

Shrestha UB, Gautam S, Bawa KS (2012) Widespread climate change in the Himalayas and associated changes in local ecosystems. PLoS One 7:e36741. https://doi.org/10.1371/journal.pone.0036741

Singh JS, Singh SP (1987) Forest vegetation of the Himalaya. Bot Rev 53:80–192. https://doi.org/10.1007/BF02858183

Singh J, Yadav RR, Wilmking M (2009) A 694-year tree-ring based rainfall reconstruction from Himachal Pradesh, India. Clim Dyn 33:1149–1158. https://doi.org/10.1007/s00382-009-0528-5

Smith KT, Cufar K, Levanic T (1999) Temporal stability and dendroclimatology in silver fir and red spruce. In: Phyton - Annales Rei Botanicae, pp 117–122

Sohar K, Altman J, Lehečková E, Doležal J (2017) Growth-climate relationships of Himalayan conifers along elevational and latitudinal gradients. Int J Climatol 37:2593–2605. https://doi.org/10.1002/joc.4867

Solberg BO, Hofgaard A, Hytteborn H (2002) Shifts in radial growth responses of coastal Picea abies induced by climatic change during the 20th century, Central Norway. Ecoscience 9:79–88

Speer JHJH (2010) Fundamentals of tree-ring research. University of Arizona Press, Tucson, p 333. https://doi.org/10.1002/gea.20357

Spiecker H (2010) Growth responses of Abies alba Mill and Picea abies (L.) Karst at different altitudes in south-western Germany, vol. 8, pp 21–25

Takahashi K, Azuma H, Yasue K (2003) Effects of climate on the radial growth of tree species in the upper and lower distribution limits of an altitudinal ecotone on mount Norikura, Central Japan. Ecol Res 18:549–558. https://doi.org/10.1046/j.1440-1703.2003.00577.x

Tiwari A, Fan Z-X, Jump AS, Zhou Z-K (2017) Warming induced growth decline of Himalayan birch at its lower range edge in a semi-arid region of trans-Himalaya, Central Nepal. Plant Ecol 218:621–633. https://doi.org/10.1007/s11258-017-0716-z

Treydte KS, Schleser GH, Helle G et al (2006) The twentieth century was the wettest period in northern Pakistan over the past millennium. Nature 440:1179–1182. https://doi.org/10.1038/nature04743

Vaganov EA, Hughes MK, Kirdyanov AV et al (1999) Influence of snowfall and melt timing on tree growth in subarctic Eurasia. Nature 400:149–151. https://doi.org/10.1038/22087

Viviroli D, Archer DR, Buytaert W et al (2011) Climate change and mountain water resources: overview and recommendations for research, management and policy. Hydrol Earth Syst Sci 15: 471–504. https://doi.org/10.5194/hess-15-471-2011

Walther G-R, Post E, Convey P et al (2002) Ecological responses to recent climate change. Nature 416:389–395

Wang T, Ren H, Ma K (2005) Climatic signals in tree ring of Picea schrenkiana along an altitudinal gradient in the central Tianshan Mountains, northwestern China. Trees 19:736–742. https://doi.org/10.1007/s00468-005-0003-9

Webster PJ (2005) Changes in Tropical Cyclone Number, Duration, and Intensity in a Warming Environment. Science (80-) 309:1844–1846. https://doi.org/10.1126/science.1116448

White TL (1987) Drought tolerance of southwestern Oregon Douglas-fir. For Sci 33:283–293. https://doi.org/10.1093/forestscience/33.2.283

Wilmking M (2005) Increased temperature sensitivity and divergent growth trends in circumpolar boreal forests. Geophys Res Lett 32:L15715. https://doi.org/10.1029/2005GL023331

Wilmking M, Myers-Smith I (2008) Changing climate sensitivity of black spruce (Picea mariana mill.) in a peatland–forest landscape in interior Alaska. Dendrochronologia 25:167–175. https://doi.org/10.1016/j.dendro.2007.04.003

Wilmking M, Juday GP, Barber VA, Zald HSJ (2004) Recent climate warming forces contrasting growth responses of white spruce at treeline in Alaska through temperature thresholds. Glob Chang Biol 10:1724–1736. https://doi.org/10.1111/j.1365-2486.2004.00826.x

Wilson R, Elling W (2004) Temporal instability in tree-growth/climate response in the lower Bavarian Forest region: implications for dendroclimatic reconstruction. Trees - Struct Funct 18:19–28. https://doi.org/10.1007/s00468-003-0273-z

Wilson RJS, Hopfmüller M (2001) Dendrochronological investigations of Norway spruce along an elevational transect in the Bavarian Forest, Germany. Dendrochronologia 19:67–79

Wilson RJS, Luckman BH (2002) Tree-ring reconstruction of maximum and minimum temperatures and the diurnal temperature range in British Columbia, Canada. Dendrochronologia 20:257–268. https://doi.org/10.1078/1125-7865-00023

Wilson RJS, Luckman BH (2003) Dendroclimatic reconstruction of maximum summer temperatures from upper treeline sites in interior British Columbia, Canada. The Holocene 13: 851–861. https://doi.org/10.1191/0959683603hl663rp

Yadav RR (2009) Tree ring imprints of long-term changes in climate in western Himalaya, India. J Biosci 34:699–707. https://doi.org/10.1007/s12038-009-0058-7

Yadav RR, Bhutiyani MR (2013) Tree-ring-based snowfall record for cold arid western Himalaya, India since a.D. 1460. J Geophys Res Atmos 118:7516–7522. https://doi.org/10.1002/jgrd.50583

Yadav RR, Park W-K, Bhattacharyya A (1997) Dendroclimatic reconstruction of April–May temperature fluctuations in the Western Himalaya of India since A.D. 1698. Quat Res 48: 187–191. https://doi.org/10.1006/qres.1997.1919

Yadav RR, Braeuning A, Singh J (2011) Tree ring inferred summer temperature variations over the last millennium in western Himalaya, India. Clim Dyn 36:1545–1554. https://doi.org/10.1007/s00382-009-0719-0

Yonenobu H, Eckstein D (2006) Reconstruction of early spring temperature for central Japan from the tree-ring widths of Hinoki cypress and its verification by other proxy records. Geophys Res Lett 33. https://doi.org/10.1029/2006GL026170

Zang C, Biondi F (2015) Treeclim: an R package for the numerical calibration of proxy-climate relationships. Ecography (Cop) 38:431–436. https://doi.org/10.1111/ecog.01335

Zhang Q-B, Hebda RJ (2004) Variation in radial growth patterns of Pseudotsuga menziesii on the central coast of British Columbia, Canada. Can J For Res 34:1946–1954. https://doi.org/10.1139/x04-078

Zhang Y, Wilmking M, Gou X (2008) Changing relationships between tree growth and climate in Northwest China. In: Forest ecology. Springer, Dordrecht, pp 39–50

Pastoralism in Timberline Forests of Western Himalaya

21

Gajendra Singh, Sanjay Kumar Dwivedi, Sonam Bahuguna, and Naveen Chandra

Abstract

A landscape-level assessment on status of livestock population and impact of pastoral practices on timberline vegetation of the western Himalaya was conducted. We used a combination of field survey and remote sensing data (Landsat, Sentinal and Cartosat & LISS-IV merged) to quantify the cumulative impacts of livestock grazing, fuelwood extraction and other anthropogenic pressures on forest cover around tourist impacted and camping site areas. There are over 26,950 bovines, horses and mules of local and migratory communities graze across the timberline region >2700 masl in the western Himalaya. Grazing, fuelwood and fodder collection jointly have created about 4503.5 ha of artificial grasslands (*kharak*) in the timberline zone. The artificial grasslands are more in the enclosing or island timberline areas. The change analysis of artificial grasslands around the temporary huts where there is frequent grazing and tourist activities indicates a steady increase in the area of grazing land at the cost of forest cover around camping site. There was loss of forests about 30.3 ha during 1976–1990 (14 years), 53.67 ha during 1990–2005 (15 years) and 30.9 ha during 2005–2018 (13 years). It is also observed that many of the isolated camping sites are now connected due to loss of vegetation. Phytosociological analysis indicates that the conversion of regenerating individuals to mature tree around the cattle camping site is very low, leading to expanse in the gap areas at an alarming rate in the timberline ecotone.

Keywords

Fuelwood and fodder · Grazing · Livestock population · Migratory community

G. Singh (✉) · S. K. Dwivedi · S. Bahuguna · N. Chandra
Forestry and Climate Change Division, Uttarakhand Space Application Centre, Dehradun, Uttarakhand, India

S. P. Singh et al. (eds.), *Ecology of Himalayan Treeline Ecotone*,
https://doi.org/10.1007/978-981-19-4476-5_21

21.1 Introduction

A treeline represents the most striking vegetation boundary on the earth's surface; however, in reality, the transition from uppermost closed montane forests to the treeless alpine vegetation is commonly not a line, but a steep gradient or ecotone of increasing stand fragmentation and stuntedness of trees (Körner 1998; Körner and Paulsen 2004). The treeline ecotones are of considerable ecological interest as they serve as indicators of response of plant species to environmental gradients, as loci for species diversity and as between ecosystem transition reflected by spatial flux of species, matter and energy (Malanson 1997). There is a common thinking that altitudinal timberlines are currently advancing towards higher altitudes, and along latitudinal gradient they are perceived to be moving further north in the northern hemisphere (Arno 1984; Parmesan 2006; Holtmeier and Broll 2007). However, this trend has many anomalies due to its dependency upon local and regional conditions such as edaphic factors, orographic influences, herbivory, fire, disease and anthropogenic influences (Cairns and Moen 2004; Weiser and Tuasz 2007).

The Himalaya is the home to many unique and diverse human groups, living in the river valleys and mountain slopes that differ from each other in terms of language, culture, tradition, religion and patterns of resource use. They have been subsisting on the Himalayan natural resources for millennia. These forests also play an important role in soil formation and mitigation of the adverse climate, cooling and purifying the atmosphere. In recent decades, demand for socioeconomic development and dependency on natural resources have increased immensely. A steady increase in human population, tourism, over exploitation of resources, widespread logging, overgrazing and removal of understory species (e.g. bamboos) (Gorrie 1937; Chaturvedi 1985) have increased the anthropogenic pressure on these forests. The local inhabitants also burn these forests from time to time in order to encourage growth of grasses at the expense of natural forests. All the gentle and accessible areas in the temperate and sub-alpine regions have undergone extensive habitat degradation, with over 70% of the natural vegetation have been reported lost (Singh 1991).

Pastoralism has been a major component of the economy of local people in the Himalaya, and its role has been far more important compared to other related activities, such as collection of medicinal plants and others. Overgrazing is a common problem that is now occurring on more than half of the world's pastureland, in both developing and developed countries (Pimentel and Burgess 2013). Grazing affects vegetation by defoliation through consumption by animals and by causing physical damage through trampling, dung pat deposits and urine (Mayer et al. 2006). The most important aspect of grazing that affects vegetation composition is selection of some plants and avoidance of others by herbivores, together with intensity and frequency of grazing. Livestock grazing in natural ecosystems results in changes in vegetation structure and composition (Stern et al. 2002).

Although several authors have worked on the ecology of high-altitude forests along the elevation gradients touching the treeline zone (Singh and Singh 1992; Garkoti and Singh 1994; Adhikari et al. 1995; Rawal and Dhar 1997), in-depth studies on the ecology, spatial extent of grazing land on Himalayan timberline are very few (Schickhoff 2006). Recently, Bharti et al. (2012), Rai et al. (2012), Rai (2012) studied ecology, extent and changes along the timberline ecotone in the western Himalaya with rigorous field data and observations. These studies indicated either increase or decrease of vegetation cover along timberline zone (3000–4000 m), but none of them reported a clear shift in the upper limit of forests. Regeneration of tree species towards higher alpine slopes adjacent to treeline has been reported from several localities in the western Himalaya (Dubey et al. 2003; Rai et al. 2012) mainly in the protected areas.

Studies are scarce on the impact of pastoral practices, tourism and other recent changes in the forest cover at a landscape level in the Himalayan region. This study was conducted with the following broad objectives: (i) to assess the status of livestock grazing in the timberline ecotone of the western Himalaya, (ii) to quantify the cumulative impacts of livestock grazing, fuelwood extraction and other anthropogenic pressures on timberline and adjoining forests and grasslands.

21.2 Material and Methods

21.2.1 Study Area

Timberline ecotone of the western Himalaya was surveyed intensively in the state of Uttarakhand above 2700 m elevation. The camping sites across the timberline zone situated in six districts of the state namely Pithoragarh, Bageshwar, Chamoli, Rudraprayag, Tehri and Uttarkashi were surveyed to assess the livestock population and impact on vegetation. To document the formation and changes within the forests due to landslides, and livestock grazing, an extensive survey was conducted in the Dugalbitta-Chopta transect (16 km^2 area), in the fringes of Kedarnath Wildlife Sanctuary. The area has Tungnath temple (3450 m), which is one of the important shrines of Hindus. Five tree species dominated the forest along timberline ecotone, viz. *Abies spectabilis*, *Betula utilis*, *Pinus wallichiana*, *Rhododendron arboreum* and *Quercus semecarpifolia*. *B. utilis* and *A. spectabilis* are restricted to the moist north-east to north-west facing slopes, whereas dry south-west to south-east facing slopes are dominated by *Q. semecarpifolia* and *R. arboreum* forests in the timberline. These forests contribute in maintaining hydrological balance and life support system for the local inhabitants.

21.2.2 Data Collection and Analysis

Geometrically corrected Cartosat-1 & LISS-IV MX fused 2.5 m high-resolution satellite data of the year 2015-18 provided by National Remote Sensing Centre (NRSC), Hyderabad, were used for mapping of grazing-induced grasslands (*khark*) in the timberline zone (>2700 m) in the state of Uttarakhand. Landsat data for the year 1976 (MSS), 1990 (TM), 2005 (ETM+) and Sentinel-2 data available at http://glcf.umiacs.umd.edu/data/landsat/ in UTM projection and WGS 84 datum were downloaded and used for assessing the changing pattern of artificial grasslands (gap analysis) in the Dugalbitta-Chopta area in the fringes of Kedarnath WLS. The ancillary database on settlements was created using topo-maps on 1:50,000 scale in the GIS domain (ARC GIS). The satellite data were processed using Arc GIS computer package.

Since the timberline ecotone is characterized by complex terrain, there was a profound effect of elevation, aspect and slope on the vegetation. The similar features may show different reflectance, or different features may have similar reflectance. Therefore, 'headup' approach of onscreen visual interpretation method was followed. The interpretation of Landsat MSS, TM, ETM+, Cartosat-1 & LISS-IV MX was done on 1:50,000 scale to map the camping sites. At places, Google earth software was also used for more accuracy. Timberline *kharks* of the state were mapped and validated based on the ground information. Ground truth points for mapping were collected from >150 localities based on various field visits.

The livestock population was calculated by headcount during various field surveys. The previous literatures available on certain valleys, viz. Niti valley, were taken to assess for overall status of livestock in the timberline zone of the state. Vegetation sampling was done following Misra (1968) and Kershaw (1973) in the disturbed and less disturbed timberline forests. We used 100 m^2 area (10 × 10 m^2 plots) for sampling the tree layer which has been found adequate for sampling forest vegetation in the Himalayan region (Singh and Singh 1992; Adhikari et al. 1995; Rai 2012). At each site, ten random plots were laid out covering the treeline ecotone at a distance of 10 m between subsequent plots. Saplings (<20 cm circumference at breast height, cbh) and seedlings of tree species were recorded within 10 × 10 m quadrats to observe the regeneration status. The structural aspects of vegetation such as density, frequency, abundance and dominance of constituent species were determined following Misra (1968).

21.3 Result and Discussion

21.3.1 Livestock Grazing Practices in the Timberline of Western Himalaya

Timberline of the western Himalaya is used for grazing by five pastoral communities: (a) nomadic, (b) semi-nomadic, (c) nuclear transhumance, (d) trans-migratory and (e) sedentary (Rawat 2007). Of all the five, three forms, viz nuclear transhumance, trans-migratory and sedentary (resident) of pastoral practices, are prevalent in the western Himalaya. Mainly the Changpa herders in the Changthang plateau of Ladakh practice true nomadic pastoralism. The *Gujjars* (the buffalo herders in Greater Himalaya and the Shivaliks) have over the years shifted from nomadic to semi-nomadic lifestyle. They lead a pastoral life style traversing almost the entire elevation ranges from less than 1000–3500 masl in the western Himalaya to graze their animals (buffaloes, cows and mules). The *Gaddis* and *Bakarwals* of Himachal Pradesh and Jammu & Kashmir follow semi-nomadic lifestyle (only few members of a family move long distance with their livestock). Many agro-pastoral communities in Uttarakhand and Himachal Pradesh practice nuclear transhumance where a part of the family moves to higher elevations (2500–4000 m) closer to treeline along with surplus cattle as their predominant occupation. Transmigration (seasonal altitudinal movement by the entire family along with the livestock) is practiced mainly by the Bhotiya communities in various parts of Uttarakhand, such as Byans, Darma, Johar and the upper basins of the Alaknanda and Bhagirathi basin. Currently 30–40% households and only few members in each family along with most of the livestock, except milk cattle, migrate to the distant alpine valleys, making available several alpine pastures for other communities belonging to the lower parts of the state and the *Gaddi* herders from Himachal Pradesh. In Chamoli, Tehri, Uttarkashi districts, several families drive their scrub cattle to sub-alpine and alpine areas for free grazing during the snow free period (June–October).

These grazers stay in the timberline zone for almost 6 months from May–June to September–October. Livestock mainly feed on small regenerating plants and in this process they gradually eliminate the understory vegetation of the forest. People, who carry livestock with them, meet their needs for fuelwood and timber for making huts and poles for fencing from the surrounding forests. As a result of regular lopping and logging, the canopy density continues to decrease, as younger plants are not available to occupy the open canopies.

Many families residing in the cool temperate region of the western Himalaya (viz Dharchula, Munsyari, Kapkot) practice nuclear transhumance where only 20–30% of the original households and only few members in each family migrate to these valleys. In addition, several non-migrant communities drive their scrub cattle to sub-alpine and alpine areas for free grazing during summer monsoon. This practice leads to faster degradation of rangelands that includes loss of soil and moisture (Rawat 2005). In the timberline region of the western Himalaya, the communities raise mainly sheep, goats, buffaloes, mules and horses and occasionally *jhokpa* and *jhuma* (cross breed of cow and yak). These animals are used for breeding, transportation of goods and production of milk, meat, hides and wool. Buffaloes produce milk, butter and cheese (*khoya*) that are occasionally used for local consumption and mostly sold by the Gujjars in local markets. Sheep is used as pack animals; meat is sold and fur is used to make local carpets. The herding communities in the western Himalaya move northward to timberline regions during April–May and stay there until September. In September–October, as the temperature falls and fodder for animals decreases, they travel back to their villages in Bhabar-Shivalik belt (in case of Gujjars and Gaddies) and stay there throughout the winter season.

21.3.2 Livestock Density in the Timberline of Western Himalaya

It is estimated that about 26,000–27,000 bovines (buffaloes, cows), horses and mules graze in timberline ecotone of Uttarakhand. The densities of horses and mules were observed more in the surrounding and in route to pilgrimage. Buffaloes and scrub cattle were found more in the island (Table 21.1) type of timberline areas. The

Table 21.1 Livestock density (bovines, horses and mules) in various timberline forests of Uttarakhand, western Himalaya

Major valleys, from where livestock go to timberline areas	Districts	Bovines/horses/mules
Byans	Pithoragarh	3000
Darma		600
Panchachuli basin		500
Ralam		400
Johar		1500
Pindari-Sunderdhunga	Bageshwar	50
Niti-Gamsali[a]	Chamoli	350
Badrinath		350
Kedarnath WLS[b]	Chamoli-Rudraprayag	10,000
Pawalikantha	Rudraprayag	550
Khatling-Chouki	Tehri	800
Nilang valley[c]	Uttarkashi	50
Kyarkoti-Kandara-Gidara		900
Kush Kalyan		650
Dayara		4500
Kanasar-Hanuman ganga		750
Harkidun-Kedarkantha		2000
Total		26,950

[a]Mitra et al. (2013), [b]Singh and Rawat (2012), [c]Rawat (2005)

number of cattle belonging to semi-nomadic community has increased from 150 to 600 in part of Kedarnath sanctuary between 1984–85 and 2007 (Thakur et al. 2011). A large number of bovines/horses/mules were recorded in the Dayara (4500), Byans (3000), Kedarnath WLS (1500), Johar (1500) areas. Livestock density in the timberline of the western Himalaya (Uttarakhand) is much higher (33.92/km^2) than the other regions of the Indian Himalaya (Rawat 2007). The highest densities of sheep and goats were observed in the fringes of Govind WLS (50 sheep and goats, and 10 bovids and equids per sq. km). Except core areas of Nanda Devi NP, Valley of Flowers NP and part of Gangotri NP, most of the timberline areas had high grazing pressure (mean density of sheep and goats 28.23 $\pm$ 17.90/km^2, and bovid and equids 2.15 $\pm$ 2.87/km^2).

21.3.3 Impact of Pastoralism on Timberline

The excessive grazing specially by large animals in the timberline ecotone causes soil erosion, degradation and clearance of forests which further converts into artificial grasslands, locally known as *Kharak.*

The forest gap *(kharak)* map of timberline vegetation of Uttarakhand state was prepared for the year 2018 using cartosata & LISS-IV merged product (Fig. 21.1). In

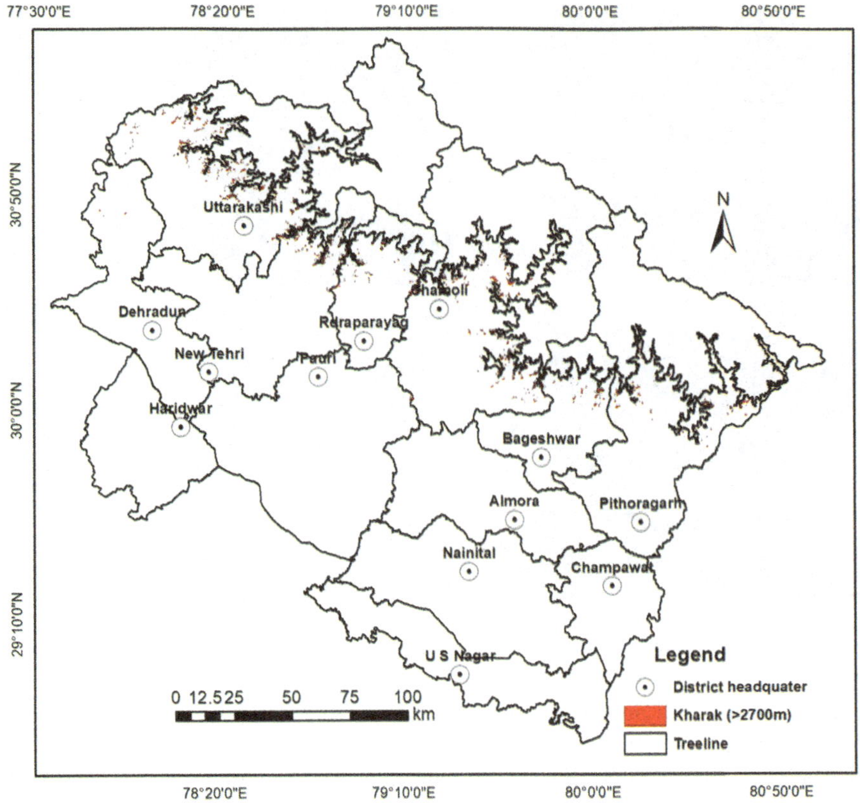

Fig. 21.1 Distribution of artificial grasslands (*Kharak*) in the timberline zone of the western Himalaya

the timberline ecotone (>2700 masl) of the western Himalaya (Uttarakhand), about 4503.5 ha (45.03 km^2) area of artificial grassland/*kharak* (gaps) has been created owing to continuous livestock grazing and camping by migratory communities. There are two types of timberline reported in the state: continuous timberline and enclosing/island timberline (Shah and Sharma 2018) in the western Himalaya. Among these, island type of timberline those are isolated from the continuous timberlines, and situated in the outer fringes of the continuous timberline are having more kharak (gaps) which indicates the higher degradation..

21.3.3.1 Forest Dynamics in the Timberline

In recent times, to perceive the changes in timberline resource, the gap area (*kharak*) change analysis becomes a central component. The time series gap analysis for the years 1976–2018 was conducted in the timberline region where there is frequent summer grazing by Gujjar and local communities and tourist activities also persist during May–October. The rapid analysis was attempted in such a manner that a direct focus on the grazing pressure and fuelwood extraction over forested area could

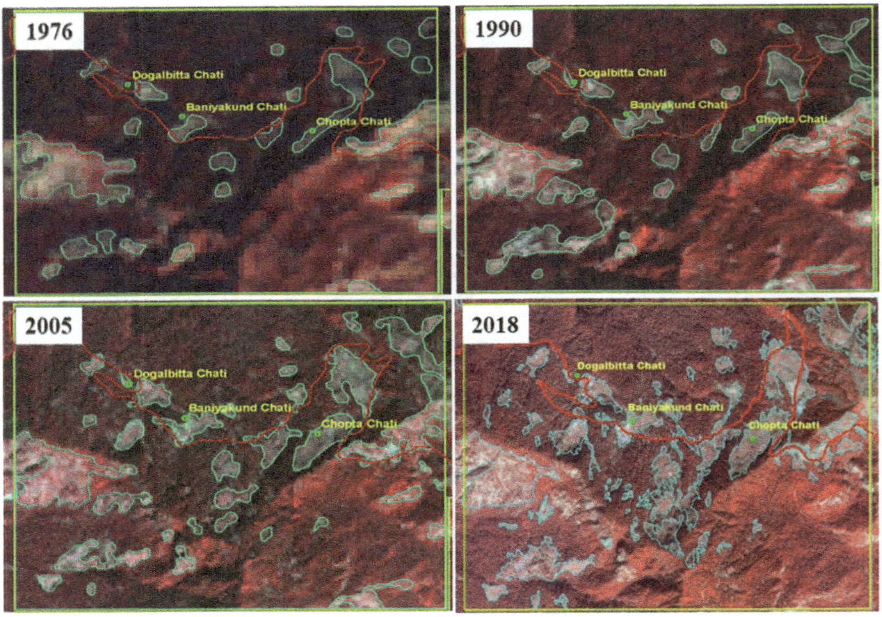

Fig. 21.2 Gap area analysis across the years in the fringes of Kedarnath WS

be revealed. The GIS analysis of the area showed that initially the area of grazing land was 135.6 ha in 1976, and it gradually increased to 168.8 ha (1990), 222.5 ha (2005) and 253.4 ha (2018) (Fig. 21.2). This indicates that around the camping sites temporary huts where there is frequent grazing and tourist activities, gap areas have increased considerably. The forest area converted into the grassland was 30.3 ha during 1976–1990 (14 years), 53.67 ha during 1990–2005 (15 years) and 30.9 ha during 2005–2018 (13 years) (Fig. 21.3). Owing to tremendous pressure, it was also observed that the widening of forest gaps is still rampant in the areas. That will hamper the regenerating capacity of underneath vegetation.

The frequent cutting and lopping of trees by pastoral communities for fuelwood, fodder and fencing the temporary huts and grazing of large body animals attributed to the expansion of open area called *kharak*. The next step is the loss of tree cover between the adjacent camping sites, and conversion of the landscape, with isolated tress. These sunny open areas are preferred sites during wet monsoon period not only for grazing animals but also for trekking. Along the timberline of pilgrimage routes, where seasonal huts/shops (roadside refreshment establishments) locally known as 'dhabas', although fuelwood consumption is restricted to 6–8 months (April–October), the use of fuelwood is appreciably high (90–120 kg/dhaba/day), where the tourist inflow continued throughout the season over the common villagers (20–22 kg/household/day) of the region (Singh et al. 2010). These activities create cumulative impacts on the forested areas and the *kharaks* keep increasing in size. *Rhododndron arboreum*, *R. campanulatum*, *Quercus semecarpifolia* and *Betula utilis* constitute the main fuelwood resource. Herders and local people use fuelwood

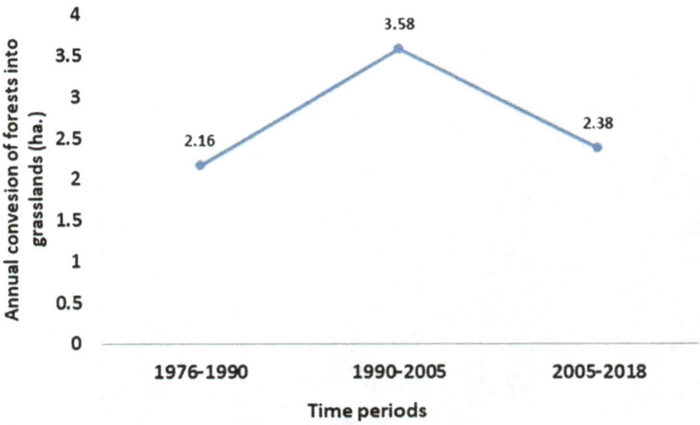

Fig. 21.3 Annual average loss of forest cover in three time periods

for space heating, water boiling, lightings and cooking as well. This is apparently the main reason for the high firewood demand at timberline altitude (Fig. 21.4).

It is estimated that the fuelwood requirement of timberline habitation (temporary shops and migratory communities) is although restricted only for 6–8 months, but due to high tourist activities and cold environment each of the dhaba owners and migratory Gujjars consumes around 100–120 kg/day (Singh et al. 2010). In the tourist-affected areas, fuelwood consumption of 10–14 per dhaba per year is more than a village having 55–60 households.

21.3.3.2 Impact of Livestock Grazing and Fuelwood Collection on Timberline Vegetation

Livestock graze the forests situated around the temporary huts of local and migratory Gujjar communities within the 2–3 km area for 6–8 months. Apart from that, they collect fodder, fuelwood and timber. The forests situated away from these camping sites and which are inaccessible to large animals like buffaloes showed higher tree density and regenerating individuals (Table 21.2). *Q. floribunda*, *Q. semecarpifolia*, *R. arboreum*, *Betula utilis* and *R. campanulatum* fulfil the requirements of fodder and fuelwood around the timberlines. The densities of trees, shrubs and saplings were found to be significantly higher in less disturbed forests, whereas the seedlings density was found to be higher in highly disturbed forests with low canopy cover.

Thick weed and low canopy cover around camp sites indicate the high collection (lopping/logging) of species for fuelwood, fodder and livestock grazing. The ground cover was higher in highly disturbed forests where about 37% trees were lopped mainly for fuelwood requirement in the region, which promote weedy species to take over the ground. Champion (1939) and Upreti et al. (1985) pointed out that the proportion of *R. arboreum* and *Abies* spp. in the disturbed forests was higher when compared to the forests at undisturbed sites. These species are unpalatable, and the fuelwood obtained is also of inferior quality in case of *R. arboreum* and *Abies* is

Fig. 21.4 (**a**) Forest gaps (*kharak*) created due to excessive grazing in the Dugalbitta area, (**b**) collection of *R. arboreum* for fencing, (**c**) Livestock grazing in forest gaps and (**d**) Absence of regenerating individuals and young trees in the kharaks

Table 21.2 Status of timberline vegetation at different anthropogenic pressure zones in WH

Parameters	Less disturbed Timber line forest (>2–3 km away from camping sites)	Highly disturbed Timberline forests (within 2–3 km of camping sites)
Tree density/ha	470 ± 30.4	219 ± 11.2
Shrub density/ha	4123 ± 99.2	3466 ± 128.8
Herb density/m^2	226 ± 7.97	1029 ± 84.62
Seedling density/ha	263 ± 27.3	686 ± 37.5
Sapling density/ha	438 ± 17.4	219 ± 13.9
No. of lopped tree/ha	5	80
Canopy cover (%)	41.6	25
Herbaceous ground cover (%)	27	64.6
Weed cover (%)	15.5[a]	35[b]

[a]Strobilantheus atropupureus dominated site, [b]Trachediumroylei dominated site

banned for commercial exploitation. They escape the onslaught when compared to *Q. semecarpifolia* and *B. utilis*, which is highly preferred for fuelwood and fodder. The high anthropogenic pressure is reported more on the island/enclosing timberline owing to closer to the village settlement and easy excess to these regions. Similar degradation of timberline has also been reported in other parts of the western Himalaya, viz. the Great Himalayan National Park (GHNP) (Singh 1999).

Each livestock has one or more unique feature of its own, sheep preferred sedges to grasses, goat avoided the forb community and buffaloes, cows, mules and horses graze for a long period without moving far. Due to difference in the grazing characteristic, any vegetation community or species will be over exploited within the carrying capacity of the area. Thus, continued grazing might have widened the range of camp sites or kharak. At places, landslide and deposition of debris also lead to formation of open areas in the treeline ecotone. In the timberline region, high dominance of *Trachydium roylei* indicates the high canopy opening and grazing. Due to excessive grazing and trampled by livestock, most of the herbaceous species, for example, *Geum elatum*, *Geranium wallichianum*, *Potentilla atronsanguinea* and *Dactylorhiza hatagirea* cannot grow luxuriantly, but species such as *Trachydium roylei*, *Anemone rivularis*, *Primula denticulata*, *Taraxacum officinale*, *Rumex nepalensis*, *Polygonum amplexicaule*, *P. polystachyum* and *Impatiens sulcata* showed a tendency to increase under heavy grazing in the region. *Trachydium roylei* has deep root network which when established would not easily allow trees to establish again (Rawat 2007).

The pressure on the timberline region in terms of livestock grazing reaches its peak during May–September when in addition to local communities, the right-holder (Gujjar) and mules in the tourist areas from outside areas graze their livestock in the region. There are various environmental issues linked with excessive grazing, fodder and firewood collection in the timberline forests. Moreover, there is an increasing danger of human–wildlife conflict due to habitat fragmentation and destruction by humans, and resource competition, as the forests and alpine meadows provide habitat to many wild mountain ungulates and carnivores including bharal (*Pseudois nayaur*), Himalayan tahr (*Hemitragus jemlahicus*) and leopard (*Panthera pardus*) (Mitra et al. 2013).

21.4 Conclusion

The timberline forests not only play an important role in providing ecosystem services and support regional biodiversity but also serve as an important habitat for a number of high-altitude mammals and birds during winter season. Based on above finding, it is concluded that overall magnitude of anthropogenic pressure was significant in the area. The kharaks (forest gaps) (migratory summer camp) are scattered across the timberline zone. Various anthropogenic disturbances widening the forest gaps in the timberline region of western Himalaya. Increased degree of disturbance beyond a certain limit caused loss in plant diversity and brought about change in vegetation characteristics. Such types of disturbance should be controlled for good vegetation health of the area, taking rotational grazing into consideration. Some areas designated as stopovers (camping sites) for pastoralist groups migrating to treeline become nutrient-rich areas and restrict the growth of tree seedlings. Management authorities need to rationalize livestock grazing within high altitude areas in order to pursue conservation objectives and check further degradation. The study reveals that current levels of pastoral practices and consumption of fuelwood by the herders in the treeline ecotone are clearly unsustainable. Therefore, appropriate management interventions and amicable solutions for better livestock management need to be initiated without further delay. For better management of fuelwood requirement, the Forest Department can provide the fallen and dead trees to the dhaba owners, Gujjars and villagers at low cost. Alternative option could be to provide gas stoves and cylinders at subsidized rates to reduce their dependency on treeline species. Chronic disturbances at treeline region do not provide time to ecosystem recovery and arrest the regeneration.

The documentation of migratory routes on maps and details of transhumant pastoralists' activities can help in identifying degrading areas to understand the current status and future impact of transhumant pastoralism in this region. Quantitative information on the extent of degradation would be necessary in order to monitor the condition to formulate recovery plans. Currently, separate management plan for treeline ecotone is lacking.

Acknowledgments The authors are thankful to the Director, Uttarakhand Space Application Centre (USAC), Dehradun, and Uttarakhand Forest Department for the necessary permission and support. We duly acknowledge GB Pant National Institute of Himalayan Environment (GBPNIHE), Ministry of Environment, Forest and Climate Change (MoEF & CC) for financial assistance under National Mission for Himalayan Studies (NMHS) program (NMHS-2019-20/MG/399). The senior author would like to thank National Remote Sensing Centre (NRSC), Hyderabad, for providing Cartosat-1 & LISS-IV merged satellite data under SISDP update project which was used for digitizing artificial grasslands.

References

Adhikari BS, Rawat YS, Singh SP (1995) Structure and function of high altitude forests of central Himalaya II. Nutrient dynamics. Ann Bot 75:249–258

Arno (1984) Timberline. Mountain and Arctic Forest Frontiers, Seattle

Bharti RR, Adhikari BS, Rawat GS (2012) Assessing vegetation changes in timberline ecotone of Nanda Devi National Park, Uttarakhand. Int J Appl Earth Obs Geoinf 18:472–479

Cairns D, Moen J (2004) Herbivory influence tree lines. J Ecol 92:1019–1024

Champion HG (1939) The relative stability of the Indian vegetation types. J Ind Bot Soc 8(1):1–12

Chaturvedi AN (1985) Fuel and fodder trees for man-made forests in hills. In: Environmental regeneration in Himalaya: concepts and strategies. Central Himalayan Environment Association and Gyanodaya Prakashan, Nainital

Dubey B, Yadav RR, Singh J, Chaturvedi R (2003) Upward shift of Himalayan pine in Western Himalaya, India. Curr Sci 85:1135–1136

Garkoti SC, Singh SP (1994) Nutrient cycling in the three central Himalayan forests ranging from close canopied to open canopied treeline forests, India. Arct Alp Res 26:339–348

Gorrie RM (1937) Tree lopping on a permanent basis. The. Indian Forester LXIII:29–31

Holtmeier FK, Broll G (2007) Treeline advance driving processes and adverse factors. Landsc Online 1:1–33. https://doi.org/10.3097/L0.200701

Kershaw KA (1973) Quantitative and dynamic plant ecology. Edward Arnold Ltd., London, p 308

Körner C (1998) A re-assessment of high elevation treeline position and their explanation. Oecologia 115:445–459

Körner C, Paulsen J (2004) A world-wide study of high altitude treeline temperatures. J Biogeogr 31:713–732

Malanson GP (1997) Effects of feedbacks and seed rain on ecotone patterns. Landsc Ecol 12:27–38

Mayer AC, Stockli V, Konold W, Kreuzer M (2006) Influence of cattle stocking rate on browsing of Norway spruce in subalpine wood pastures. Agrofor Syst 66:143–149

Misra R (1968) Ecology work book. Oxford and IBH Publ, New Delhi

Mitra M, Kumar A, Adhikari BS, Rawat GS (2013) A note on transhumant pastoralism in Niti valley, Western Himalaya, India. Pastoralism Res Policy Pract 3:29

Parmesan C (2006) Ecological and evolutionary responses to recent climate change. Ann Rev Ecol Evol System 37:637–669

Pimentel D, Burgess M (2013) Soil erosion threatens food production. Agriculture 3:443–463

Rai ID (2012) Ecological attributes of timberline vegetation with special reference to climatic variability in Kedarnath Wildlife Sanctuary, Western Himalaya, Ph.D Thesis, Kumaun University, Nainital

Rai ID, Adhikari BS, Rawat GS, Bargali K (2012) Community structure along timberline ecotone in relation to micro-topography and disturbances in Western Himalaya. Not Sci Biol 4(2):41–52

Rawal RS, Dhar U (1997) Sensitivity of timberline flora in Kumaon Himalaya, India: conservation implications. Arct Alp Res 29:112–121

Rawat GS (2005) Alpine meadows of Uttaranchal. Bishen Singh Mahendra Pal Singh, Dehradun

Rawat GS (2007) Pastoral practices, wild mammals and conservation status of alpine meadows in Western Himalaya. J Bombay Nat Hist Soc 104(1):251–257

Schickhoff U (2006) The upper timberline in the Himalayas, Hindu Kush and Karakoram: a review of geographical and ecological aspects. In: Broll G, Keplin B (eds) Mountain ecosystems: studies in treeline ecology. Springer-Verlag, Berlin, pp 275–354

Shah P, Sharma S (2018) Topographical characterization of high altitude timberline in the Indian central Himalayan region. Trop Ecol 59(2):187–196

Singh SP (1991) Structure and function of low and high altitude grazingland ecosystems and the impact of the livestock component in the Central Himalaya. Final technical report, Dept. of Envi. Govt. of India, New Delhi

Singh SK (1999) A study on the plant community composition and species diversity in Great Himalayan National Park, western Himalaya. Ph.D. Thesis, Kumaun University Nainital

Singh G, Rawat GS (2012) Depletion of oak (*Quercus* spp.) forests in the Western Himalaya: grazing, fuelwood and fodder collection. In: Okia CA (ed) Global perspectives on sustainable Forest management. Intech Publisher, Crotia

Singh JS, Singh SP (1992) Forests of Himalaya: structure, functioning and impact of man. Gyanodaya Prakashan, Nainital

Singh G, Rawat GS, Verma D (2010) Comparative study of fuelwood consumption by villagers and seasonal "Dhaba owners" in the tourist affected regions of Garhwal Himalaya, India. Energy Policy 38(4):1895–1899

Stern M, Quesada M, Stoner KE (2002) Changes in composition and structure of a tropical dry forest following intermittent cattle grazing. Int J Trop Biol 50:1021–1034

Thakur AK, Singh G, Singh S, Rawat GS (2011) Impact of pastorl practices on forest cover and regeneration in the outer finges of Kedarnath wildlife sanctuary, Western Himalaya. J Indian Soc Remote Sens 39(1):127–134

Upreti N, Tewari JC, Singh SP (1985) The oak forests of the Kumaun Himalaya (India): composition, diversity and regeneration. Mt Res Dev 5(2):163–174

Weiser J, Tuasz M (2007) Current concepts for treelife limitation at the upper timberline. In: Wieser G, Tauz M (eds) Tree at their upper limit. Springer, Chan, pp 1–18

Treeline Dynamics in Nepal Himalaya in a Response to Complexity of Factors 22

Narayan Prasad Gaire, Ze-Xin Fan, Parveen K. Chhetri, Santosh K. Shah, Dinesh Raj Bhuju, Jinniu Wang, Bimal Sharma, Peili Shi, and Yub Raj Dhakal

N. P. Gaire (✉)
CAS Key Lab of Tropical Forest Ecology, Xishuangbanna Tropical Botanical Garden, Chinese Academy of Sciences, Mengla, Yunnan, People's Republic of China

Department of Environmental Science, Patan Multiple Campus, Tribhuvan University, Lalitpur, Nepal
e-mail: narayan.gaire@pmc.tu.edu.np

Z.-X. Fan (✉) · B. Sharma
CAS Key Lab of Tropical Forest Ecology, Xishuangbanna Tropical Botanical Garden, Chinese Academy of Sciences, Mengla, Yunnan, People's Republic of China
e-mail: fanzexin@xtbg.org.cn; bimal@xtbg.ac.cn

P. K. Chhetri
Department of Earth Science and Geography, California State University, Carson, CA, USA
e-mail: pchhetri@csudh.edu

S. K. Shah
Birbal Sahni Institute of Paleosciences, Lucknow, Uttar Pradesh, India

D. R. Bhuju
Nepal Academy of Science and Technology, Lalitpur, Nepal
e-mail: drbhuju@cdes.edu.np

J. Wang
Chengdu Institute of Biology, Chinese Academy of Sciences, Chengdu, People's Republic of China
e-mail: wangjn@cib.ac.cn

P. Shi
Institute of Geographic Sciences and Natural Resources Research, Chinese Academy of Sciences, Beijing, People's Republic of China
e-mail: shipl@igsnrr.ac.cn

Y. R. Dhakal
Siddhartha Environmental Services, Kathmandu, Nepal

S. P. Singh et al. (eds.), *Ecology of Himalayan Treeline Ecotone*,
https://doi.org/10.1007/978-981-19-4476-5_22

Abstract

The Nepal Himalaya is experiencing a higher temperature rise with erratic rainfall and more drought events in recent decades. Alpine treelines are proven bio-indicator and bio-monitor to understand the environmental impacts at high elevation. This chapter provides an overview of studies conducted at treelines of the Nepal Himalaya. Treeline elevation ranges from ~3400 to 4500 masl, regardless of the treeline type. The elevation decreases from east to west and is composed of multiple species, usually three to six species with some exception of mono-species. The studies found that site- and species-specific regeneration has maintained the treeline dynamics. The average rate of upward treeline shift is 0.46 m/year with the site- and species-specific differences in the rate ranging from 0 to 2.6 m/year. Various biotic and abiotic factors including winter and summer temperature and spring month's precipitation are responsible for the treeline dynamics. The study indicated that unidirectional upward shifting of the treeline is not the sole response of the treeline in the Nepal Himalaya to climate change. For further understanding, integrative research considering multiple species, multiple sites, and the assessment of multiple factors is recommended.

Keywords

Climate · Geology · Nepal Himalaya · Soil characteristics · Treeline dynamics · Treeline shift · Tree ring

22.1 Introduction

The Himalayan region is experiencing rapid climate change compared to the other areas with similar altitudes or latitudes (IPCC 2013; Shrestha et al. 2012; Ren et al. 2017). In the Nepal Himalaya, the temperature is rising, and hydrological extremes are increasing. The available meteorological records have shown an increase in average annual maximum and mean temperatures in Nepal by a rate of 0.04–0.06 °C/year (DHM 2017; Karki et al. 2019; Shrestha et al. 2012; Thakuri et al. 2019). Generally, higher elevation areas are getting warmer at a higher rate than low elevation regions (DHM 2017; Shrestha et al. 2012; Thakuri et al. 2019). Precipitation data analyses reveal no specific nation-wise trend, but overall extreme events have increased (DHM 2017; Gaire et al. 2017b; Qi et al. 2013; Ren et al. 2017; Talchabhadel et al. 2018; Zhan et al. 2017). Similarly, there is an increasing trend in drought events in the central Himalaya (Bhandari et al. 2019; Gaire et al. 2019; Panthi et al. 2017). These climatic changes have severely affected the physical, biological, and social environment systems (Anderson et al. 2020; Aryal et al. 2017; Baniya et al. 2018; Gaire et al. 2017a; 2014; Salerno et al. 2015; Shrestha et al. 2012; Sigdel et al. 2020a, 2018). But compared to the impacts of climate change on physical environments in Nepal, few studies have been carried out on biological systems (Anderson et al. 2020; Lamsal et al. 2017; Shrestha et al. 2012) such as treeline.

The alpine treeline represents the upper elevational limit of forest and tree growth on high mountains due to environmental constraints such as cold temperatures, drought, low nutrient availability, waterlogging, and stormy winds (Holtmeier 2009; Holtmeier and Broll 2009; Körner 2012). Globally occurring treelines can be defined and categorized into four primary forms (Harsch and Bader 2011): (1) Diffused, characterized by a gradual decrease in the height of single-stemmed trees as well as tree density along the treeline ecotone; (2) Abrupt, characterized by a continuous forest >3 m tall directly bordering low alpine vegetation. Tree height, as well as density thus, changes rapidly; (3) Island, characterized by clumped patches or linear strips (fingers) of krummholz or trees above the continuous forest limit, and (4) Krummholz, characterized by severely stunted or deformed multi-stemmed trees (Harsch and Bader 2011). The sensitivity and response of these different treelines to climate change vary depending on the prevailing environmental constraints (Harsch and Bader 2011).

In more than 200 years of research on treeline worldwide (Holtmeier 2009; Holtmeier and Broll 2009; Körner 2012), several hypotheses have been proposed to explain responsible factors for the upper elevational limit of tree life (Körner 2012, 1998; Li et al. 2008; Müller et al. 2016a; Weiser and Tausz 2007). Among them, four environmental and two biological hypotheses have been widely used to interpret the alpine treeline formation (Körner 2012, 1998; Li et al. 2008). These include (1) Climatic stress as a result of repeated damage by freezing, frost desiccation, and phototoxic effects, (2) Disturbance due to mechanical damage by wind, ice blasting, snow break, and avalanches, (3) Insufficient tissue maturation because of the short growing season, (4) Temperature-limited reproduction, (5) Direct growth limitation, and (6) Carbon limitation (Körner 2012, 1998; Li et al. 2008). Treeline controlling factors could be of different spatial scales like global or continental climatic zone (temperature), to regional (precipitation, seasonal, and interannual climatic variability) to landscape (wind flow pattern, growing season) to site or site mosaic (soil, exposure) (Holtmeier 2009).

Many authors have described the worldwide positions of treelines (e.g., Arno 1984; Baumgartner 1980; Körner 2012, 1998; Miehe et al. 2007; Troll 1973; Wardle 1974). Harsch et al. (2009) did a meta-analysis of literature that reported treeline dynamics after 1900 CE. In recent years, treeline research in the mountainous area of Asia has increased rapidly (Bhuju et al. 2010; Camarero et al. 2021; Du et al. 2017; Liang et al. 2016, 2011; Schickhoff 2005; Schickhoff et al. 2015; Shen et al. 2018, 2014; Singh et al. 2018; Tiwari and Jha 2018; Wang et al. 2019a, 2019b, 2016; Yadava et al. 2017; Zhou et al. 2018). In both hemispheres, the upper treeline rises from high latitudes toward the subtropics and gradually declines in the tropical equatorial zone (Körner 2012, 1998; Miehe et al. 2007; Weiser and Tausz 2007). The altitudes of treelines depend not only on latitudes but also on continentality, especially in temperate interior zones of North America and Asia, where treelines are nearly as high as those of the tropics (Schickhoff 2005; Weiser and Tausz 2007). Tree species growing at treeline varies across the region, and major tree genera of treeline ecotone include summer green species such as *Larix*, *Betula*, and *Nothofagus*, and evergreen species such as *Picea*, *Pinus*, *Abies*, *Juniperus*, *Araucaria*, *Podocarpus*, and *Eucalyptus* (Holtmeier 2009; Körner 2012; Miehe et al. 2007).

As most of the high elevation climatic treelines are shaped due to low-temperature limitations (Körner 2012; Körner and Paulsen 2004), natural treeline ecotone is taken as sensitive biomonitors and bioindicators of past and recent climate change (Camarero and Gutiérrez 2004; Kullman 1998). Treeline is also assumed to represent an ideal early warning feature that responds structurally, positionally, and physiognomically to climate change, particularly to temperature increase (Camarero et al. 2021; Chhetri and Cairns 2015; Gaire et al. 2017a, 2014; Holtmeier 2009; Holtmeier and Broll 2005; Körner 2012; Kullman 2001, 1998; Sigdel et al. 2018; Tiwari and Jha 2018). As the treeline is a space- and time-related phenomenon (Holtmeier and Broll 2005), the response of the treeline to climate change is not unidirectional; it not only varies spatiotemporally but also concerning treeline form and structure (Harsch et al. 2009; Harsch and Bader 2011; Holtmeier 2009).

Several dendroecological studies have documented that trees at the treeline often respond to temperature warming with changes in the growth and recruitment phase including upward advances in the position (Aakala et al. 2014; Camarero et al. 2021; Chhetri and Cairns 2015; Du et al. 2017; Gaire et al. 2017a; Hofgaard et al. 2013; Kirdyanov et al. 2012; Kullman and Öberg 2009; Liang et al. 2011; Öberg and Kullman 2012; Sigdel et al. 2018; Wang et al. 2019a; Yadava et al. 2017). A meta-analysis of a global data-set based on 166 sites since 1900 CE showed that the treeline either advanced (52% of sites) or remained unchanged (47%), while only a few treelines (1%) declined under heavy anthropogenic disturbance (Harsch et al. 2009). Treelines that experienced strong winter warming and treelines with a diffuse form are more likely to advance (Harsch et al. 2009; Harsch and Bader 2011).

The Himalayan mountain system provides opportunities to study the impacts of environmental change in the biological communities considering the treeline as a reference point. Earlier, Schickhoff (2005) has given an overview of the treeline research in the Hindu Kush Karakorum Himalaya (HKKH) region. Some gaps identified by the study included relationships of the Himalayan timberlines to other ecological conditions and processes such as carbon balance, freezing and frost drought, soil temperatures, wind, snow cover, and soils, regeneration. This study also emphasized the necessity of more systematic, interdisciplinary research to get in-depth insights into the many abiotic, biotic, and anthropogenic factors. It indicated the importance of knowing the complex ecological and socioeconomic processes at various spatial and temporal scales expressed in present spatial and physiognomic timberline structures. IPCC's AR4 also identified research gaps in the Himalayan region (IPCC 2007), motivating many treeline researchers. Therefore, this study aims to provide an overview of the studies on the treeline in the Nepal Himalaya.

22.2 Materials and Methods

This study reviewed published and unpublished literature about the treeline/timberline research in the Nepal Himalaya. Literature was searched in Google Scholar, Web of Knowledge Science, and Research Gate. We used search term Treeline*Nepal, Timberline*Nepal, Forest Line*Nepal, Treeline*Himalaya, Soil*treeline*Himalaya, Soil*treeline*Nepal, Timberline*Himalaya, vegetation*Himalaya, sub-alpine vegetation*Nepal, Paleoclimatology*Nepal, dendrochronology*Nepal, etc. Similarly, the reference lists of the published literature were also checked to get information about additional studies. Those pieces of literature which were written in English and published were mostly focused; however, some literature written in non-English language but cited in a previous review paper on Himalayan treeline (Schickhoff 2005) was also checked as far as possible. Then the available literature was reviewed and categorized based on the scope of the studies. As studies have used timberline, treeline, forest limit almost in an interchangeable way, we used both (treeline and timberline) terms in this paper interchangeably. Our review covered the studies dealing with the treeline ecotone region (closed canopy forest to tree species limit). The review emphasized the literature related to treeline patterns, processes, and dynamics, though other aspects have also been briefly explained. The result of this review is presented under different subheadings.

22.3 Results and Discussion

22.3.1 Treeline Studies in Nepal: Spatial Coverage

Studies in the Nepal Himalaya have used timberline, treeline, forest limit almost in an interchangeable way. Researchers who focused only on treeline studies have defined treeline considering tree height and canopy coverage percentage criteria. However, earlier vegetation surveys have simply mentioned timberline, treeline, forest limit without a proper definition. The dendroecological studies have generally used a 2 m height criterion to define a tree and demarcate a treeline (Chhetri and Cairns 2015; Gaire et al. 2017a, 2014, 2011; Shrestha et al. 2015a; Sigdel et al. 2018; Suwal et al. 2016; Tiwari et al. 2017a).

Treeline studies have covered the treeline ecotone from the far-east to far-west of the Nepal Himalaya (Fig. 22.1). The majority of earlier studies were part of the forest and vegetation surveys carried out through a series of expeditions. However, integrative studies covering diverse aspects at treeline are very limited in number. The majority of treeline studies are from central Nepal, which might be related to accessibility (Fig. 22.1; Table 22.1). However, at least some aspect of treeline is studied in all high altitude protected areas in the Nepal Himalaya (Fig. 22.1; Table 22.1) namely the Kanchenjunga Conservation Area (KCA) (Bhuju et al. 2019, 2016; Sigdel et al. 2020b, 2018), Makalu Barun National Park (MBNP) (Chhetri and Cairns 2016, 2015; Chhetri et al. 2018, 2016), Sagarmatha National

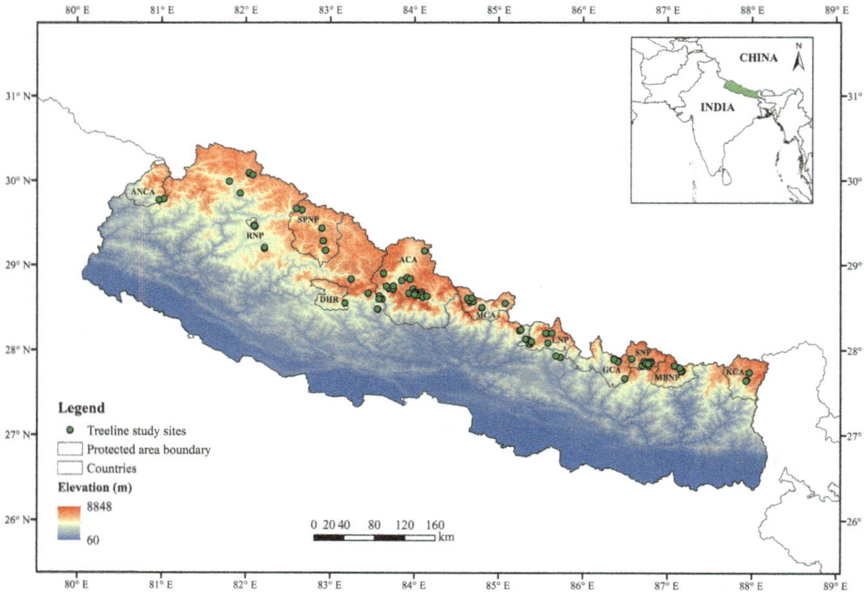

Fig. 22.1 Map showing locations of the sites in the Nepal Himalaya in which some aspect of treeline is studied or mentioned in the published literature. The abbreviated letters in the map indicate name of protected areas and refer text for full form

Park (SNP) (Bhuju et al. 2010; Gaire et al. 2017a, 2015; Liang et al. 2014; Pandey et al. 2020, 2018; Sigdel et al. 2020b, 2018; Tenca and Carrer 2010), the Gaurishankar Conservation Area (GCA) (Bürzle et al. 2018, 2017; Müller et al. 2017, 2016a, b; Panthi et al. 2021, 2020; Schickhoff et al. 2016a, b, 2015; Schwab et al. 2020, 2018, 2017, 2016, 2016), Langtang National Park (LNP) (Dawadi et al. 2013; Gaire et al. 2011, 2010; Liang et al. 2014; Rayback et al. 2017; Schickhoff et al. 2015; Shrestha et al. 2017, 2015a, b; Sigdel et al. 2018; Suwal et al. 2016), the Manaslu Conservation Area (MCA) (Gaire et al. 2014, 2020, 2014; Liang et al. 2014; Rana et al. 2017, 2016; Suwal et al. 2016), the Annapurna Conservation Area (ACA) (Dhakal et al. 2016; Kharal et al. 2017, 2015; Tiwari et al. 2017a, 2017b; Shrestha et al. 2007; Sigdel et al. 2020b, 2018; Udas 2009), Dhorpatan Hunting Reserve (DHR) (Chhetri and Cairns 2018), Shey Phoksundo National Park (SPNP) (Dhakal et al. 2016), Rara National Park (RNP) (Bhuju et al. 2016; Panthi et al. 2020), and the Api-Namapa Conservation Area (ANCA) (Bhuju et al. 2016).

22.3.2 Temporal Dimensions of Treeline Studies in Nepal Himalaya

Compared to the studies in other mountain systems in the world (some having over 200 years of research history) (Holtmeier 2009; Körner 2012), treeline research in the Nepal Himalaya has a recent history (Fig. 22.2). The studies that focused only on

Table 22.1 Summary of findings of selected studies from treeline ecotones from the Nepal Himalaya with major focus on dynamics

Reference	PAs	Site	Species	Key Findings
Schickhoff (2005)	Himalaya	Himalaya	Review	Reviewed existing literature from HKH treeline and highlighted research gaps
Shrestha et al. (2007)	ACA	Manang	*BU*	Regeneration of treeline birch (*B. utilis*) was good. The treeline was abrupt and the anthropogenic type without the Krummholz zone. Treeline not shifting in recent years
Ghimire and Lekhak (2007)	ACA	Manang	*AS, BU, PW*	The size class distribution of *A. spectabilis* resembled the reverse J-shaped curve indicating regeneration
Gaire et al. (2010)	LNP	Lauribina	*AS, BU, RC, SM, SS*	Reverse-J to bell shaped size class distribution of treeline-forming species
Gaire et al. (2011)	LNP	Lauribina	*AS*	Regeneration of *A. spectabilis* is high with domination of young in stand age at treeline with potential for shifting upward
Sujakhu et al. (2013)	MCA	Prok, Gorkha	TreeVegetation (*BU, AS, RC,* etc.)	Density of *B. utilis* increased with an increase in elevation, whereas the density of other tree species decreased with an increase in elevation. Deviation from inverse J-shaped curve. Species-specific regeneration pattern
Gaire et al. (2014)	MCA	Kalchuman, Gorkha	*AS, BU*	Regeneration of *A. spectabilis* is better than *B. utilis*. Upward shifting of *A. spectabilis* at the rate of 2.61 m/year since 1850 CE while upper distribution limit of *B. utilis* was stagnant in the past few decades
Gaire et al. (2015)	SNP	Dole and Phortse, Khumbu	*AS, BU, RC, SM, SS*	Species-specific regeneration pattern at treeline with high regeneration of *Abies* and *rhododendron*

(continued)

Table 22.1 (continued)

Reference	PAs	Site	Species	Key Findings
Chhetri and Cairns (2015)	MBNP	Barun valley	*AS*	Young recruitment history at treeline. Treeline position is not shifting up substantially
Shrestha et al. (2015a, b)	LNP, ACA	Lauribina-Langtang and Manang	*AS, PW*	Similar treeline dynamics over the past six decades in both xeric and mesic treeline site. Stationary treelines over the past six decades; however a potential for treeline advance depending on future land use and climate change
Schickhoff et al. (2015)	GCA	Rolwaling valley	Mix	Intense recruitment of treeline trees suggests a great potential for future treeline advance. Competitive abilities of seedlings within Krummholz thickets and dwarf scrub heaths will be a major source of variation in treeline dynamics
Dhakal et al. (2016)	ACA, SPNP	Manang, Mustang, Dolpa	*AS, BU, PW*	Site- and species- specific regeneration dynamics. Stable to shifting treeline position
Chhetri et al. (2016)	MBNP	Barun valley	*AS*	The *A. spectabilis* population showed the reverse-J-shaped size and age distribution curves. Positive relationships between recruitment and high temperatures during winter months (January, February and December) and August month.
Suwal et al. (2016)	MCA, GCA	Samagaun-Gorkha, Rolwaling valley	*AS*	Upward shift of both the upper species limit and the tree limit. Land-use change under a warming climate facilitated upslope expansion of *A. spectabilis*
Bhuju et al. (2016)	ANCA, RNP, KCA	Ghunsa, Chhuchemara, Api-Himal	*AS, BU*	Site- and species- specific regeneration dynamics. Stable to shifting treeline position

(continued)

Table 22.1 (continued)

Reference	PAs	Site	Species	Key Findings
Rana et al. (2016)	MCA	Prok, Gorkha	Tree vegetation	Regeneration of *R. campanulatum* is good compared to other treeline-forming species
Müller et al. (2016a, b)	GCA	Rolwaling valley		Growing season mean soil temperature at treeline in Rolwaling (7.6 ± 0.6 °C) is 1.2 °C higher than global average (6.4 ± 0.7 °C). Decoupling treeline shift and global warming could be related to nutrient availability (N, K, Mg) in soils and foliage
Bobrowaski et al. (2017)	Himalaya	Himalaya	*BU*	Prediction of future distribution of *B. utilis* suggest that *B. utilis* does not reach the upper elevational limit in vast areas of its potential distribution range due to anthropogenically caused treeline depressions
Gaire et al. (2017a, b)	SNP	Dole, Khumbu	*AS, BU*	Spatiotemporally heterogeneous regeneration with a higher regeneration of *A. spectabilis* compared to *B. utilis*. During the past 200 years, *A. spectabilis* shifted upward by about 0.93 m/year and *B. utilis* by 0.42 m/year, with stabilization during the second half of the twentieth century
Chhetri et al. (2017)	MBNP	Barun valley and Manang	*AS, BU, PW*	Topography and human disturbances are major controlling factors in treeline pattern at Barun and Manang.
Tiwari et al. (2017)	ACA	Chimang, Mustang	*AS*	Abundant seedling recruitment, higher regenerative inertia and colonization with a consistent upward range shift of the *A. spectabilis* treeline

(continued)

Table 22.1 (continued)

Reference	PAs	Site	Species	Key Findings
Drolinger et al. (2017)	GCA	Rolwaling valley		Decreasing nutrient concentrations in soils and trees with increasing elevation across a treeline ecotone in Rolwaling Himal, Nepal
Rana et al. (2017)	MCA	Prok, Gorkha	*RC*	Regeneration of *R. campanulatum* is high along with upward migration at a rate of 24.7 m per decade
Müller et al. (2017)	GCA	Rolwaling valley		Soil N and P availability and soil C:P and N:P ratios decrease significantly as elevation increases, especially at the transition from Krummholz (dominated by *R. campanulatum*) to the alpine tundra (dwarf scrub heath)
Schwab et al. (2017)	GCA	Rolwaling valley	Tree vegetation	Krummholz belt will primarily prevent the upward migration of other tree species and constrain the future response of Himalayan Krummholz treelines to climate warming
Basnet and Gaire (2017)	ACA	Thulokharka, Myagdi	*AS*	The per-decade upward range shifting rate of *A. spectabilis*was 13.6 m in the SE aspect and 5.6 m in the NW aspect
Chhetri et al. (2018)	MBNP	Barun valley	*AS, BU, PW*	Under projected climate conditions, regional increase in suitable habitat for treeline species *A. spectabilis, B. utilis,* and *P. wallichiana* predicted potential for northward and upslope advance
Chhetri and Cairns (2018)	DHR	Dhorpatan, Baglung	*BU*	Treeline ecotone formed by a young stand but treeline is stable and low regeneration at the treeline ecotone. Establishment above the forest line is limited by a lack of moisture, absence of suitable microsites, and presence of herbivores

(continued)

Table 22.1 (continued)

Reference	PAs	Site	Species	Key Findings
Sigdel et al. (2018)	KCA, SNP, LNP, ACA, Jumla, Humla	20 plot in six sites	*AS, BU*	Treelines shifted upward in response to recent climate warming with varying shift rates primarily mediated by spring precipitation. The rate of the upward shift was higher in the wettest eastern Himalayas compared to dry sites
Schwab et al. (2020)	GCA	Rolwaling valley	Mix	Differential competitive abilities of treeline-forming species. Treeline has responded to climate warming in terms of stand densities, seed-based regeneration and growth patterns of trees; the treeline position is relatively stable
Sigdel et al. (2020a, b)	KCA, SNP, LNP, ACA, Jumla, Humla	Six sites	*AS, BU*	Tree aggregation weakens treeline responsiveness to climate warming, and thus warming-induced drought stress tends to lower treeline shift rates by enhancing clustering
Panthi et al. (2020)	GCA, RNP	Rolwaling and Rara	*AS*	The iWUE (intrinsic water use efficiency) increased over time. At high elevations in wet region positive growth trends, suggesting that warming has favored tree growth in regions where temperature most strongly limits growth
Mainali et al. (2020)	LNP	Langtang	*AS, RC*	Elevational limits of *R. campanulatum* have shifted upward (due to young plants above treeline benefited from facilitation of recruitment by surrounding vegetation, allowing upward expansion of recruitment, and temperature amelioration to mature plants increased adult survival)

(continued)

Table 22.1 (continued)

Reference	PAs	Site	Species	Key Findings
Sharma et al. (2020)	ACA	Mustang	*RC*	Regeneration of *R. campanulatum* was fairly good in above and below *Betula* treeline

PAs: Protected areas; *AS*: *Abies spectabilis*, *BU*: *Betula utilis*, *PW*: *Pinus wallichiana*, *RC*: *Rhododendron campanulatum*, *SM*: *Sorbus microphylla*, *SS*: *Salix* sp., *ACA*: Annapurna Conservation Area, *ANCA*: Api-Namapa Conservation Area, *DHR*: Dhorpatan Hunting Reserve, *GCA*: Gaurishankar Conservation Area, *KCA*: Kanchenjunga Conservation Area, *LNP*: Langtang National Park, *MBNP*: Makalu-Barun National Park, *MCA*: Manaslu Conservation Area, *RNP*: Rara National Park, *SNP*: Sagarmatha National Park, *SPNP:* Shey Phoksundo National Park

treeline ecotones have been published mainly after 2000 CE (Fig. 22.2), although the earlier vegetation surveys or expeditions reported about the treeline/timberline vegetation or position (Schickhoff 2005), their focus was not on the treeline patterns, processes, drivers, and dynamics. The first systematic treeline study in perspective of changing climate began in 2007, when a team of Nepal Academy of Science and Technology, the University of Padova, and Resources Himalaya Foundation established permanent plots at the treeline at Pangboche (4050 masl), SNP to conduct long-term monitoring on the impact of climate change in the high altitude forests (Bhuju et al. 2010). When climate change issues came to the forefront in recent decades (IPCC 2014, 2013, 2007), the treeline research in the Nepal Himalaya also increased, resulting in some important publications (Table 22.1).

22.3.3 Methodological Aspect or Scope of Studies

Studies in the treeline ecotone in the Nepal Himalaya are presented broadly categorizing into the following approaches: dendrochronology (dendroecological, dendroclimatological), vegetation survey, ecological (vegetation and edaphic), paleontological, remote sensing/modeling, and other (general overview, and review or theoretical studies) (Figs. 22.3 and 22.4). In those studies conducted since the year 2000 CE, the dendrochronological approach was the most widely used (45.7%), followed by ecological (vegetation survey or vegetation plus soil aspect) (27.2%), remote sensing (11.1%), and other elements (13.6%) (Fig. 22.4). Those studies that have used tree-ring to estimate the population age structure at the treeline are considered dendroecological (Speer 2010). Those studies that use tree-ring data to check the response of trees to climate variables and/or also to reconstruct past climate are categorized into dendroclimatological. In the dendroecological studies, researchers have used belt transect of 20–30 m width and variable-length (100–300 m) covering the entire treeline ecotone for field survey (Chhetri and Cairns 2015; Gaire et al. 2017a, 2014; Sigdel et al. 2018; Suwal et al. 2016; Tiwari et al. 2017a). Similarly, some studies have used quadrat (20 × 20 m or 10 × 10 m) sampling (Basnet and Gaire 2017; Gaire et al. 2011; Panthi et al. 2020; Rana et al. 2017; Schickhoff et al. 2016a, 2016b, 2015; Schwab et al. 2017, 2016). Remote

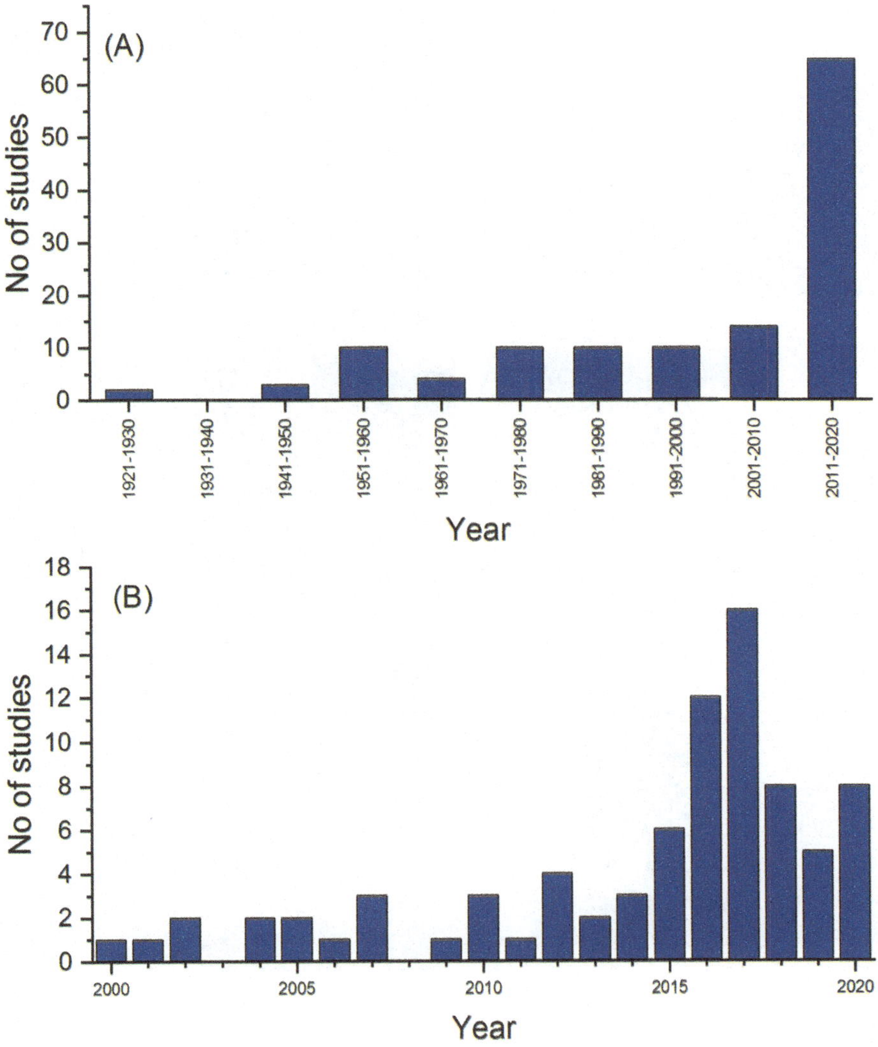

Fig. 22.2 Temporal dimension of the treeline research at the central Himalaya for total period (**a**) and since 2000 CE (**b**)

sensing and GIS approach is also used to assess vegetation dynamics, phenological change, or species distribution modeling to measure and predict change in the distribution of treeline-forming species or overall vegetation over time (Bobrowski et al. 2017; Chhetri et al. 2018, 2017; Hamid et al. 2018; Schickhoff et al. 2015; Shrestha et al. 2012). A study successfully used unmanned aerial system imagery for species-level vegetation mapping in a Himalayan treeline ecotone (Mishra et al. 2018). However, assessing the treeline change at a small plot scale by comparing the repeated photography of historical images is not found. Vegetation and edaphic factors along elevation gradient including treeline ecotone have also been explored

Fig. 22.3 Dendroecological survey at treeline ecotone. Layout of transect plot (**a**), recording of seedlings (**b**), measurement of height of saplings (**c**), collection of tree cores (**d–f**)

(Bürzle et al. 2017; Ghimire et al. 2010; Ghimire and Lekhak 2007; Müller et al. 2017, 2016a, b; Schwab et al. 2017; Shrestha et al. 2007). Studies have also explained palaeovegetation around the treeline ecotone (Miehe et al. 2009, 2002; Schlütz and Zech 2004). Gerlitz et al. (2016) analyzed temperature lapse rates and precipitation gradients in a Himalayan treeline environment, taking the reference of Rolwaling valley's treeline. They indicated that the local-scale climatic conditions are influenced by both large-scale atmospheric parameters and topographic characteristics. Dendrochronological studies are carried out in the treeline ecotone throughout Nepal Himalaya (Fig. 22.5; Table 22.2), and the majority of the cases have focused on either *Abies spectabilis* or *Betula utilis* (Fig. 22.5; Table 22.2).

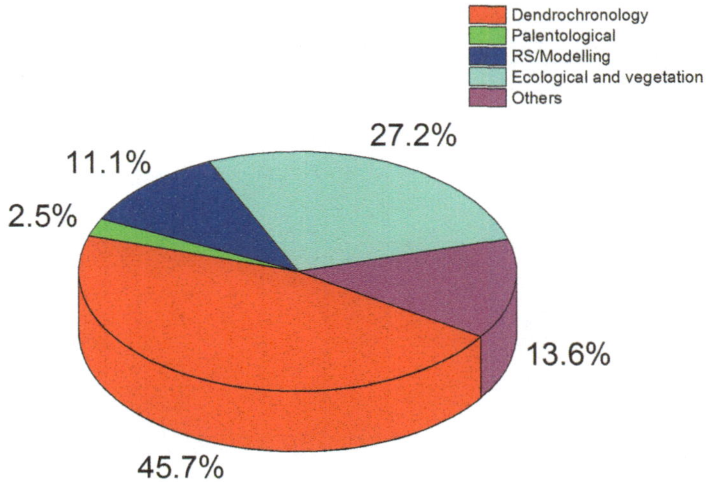

Fig. 22.4 Research focus of the studies carried out in treeline ecotone in the Nepal Himalaya since 2000 CE

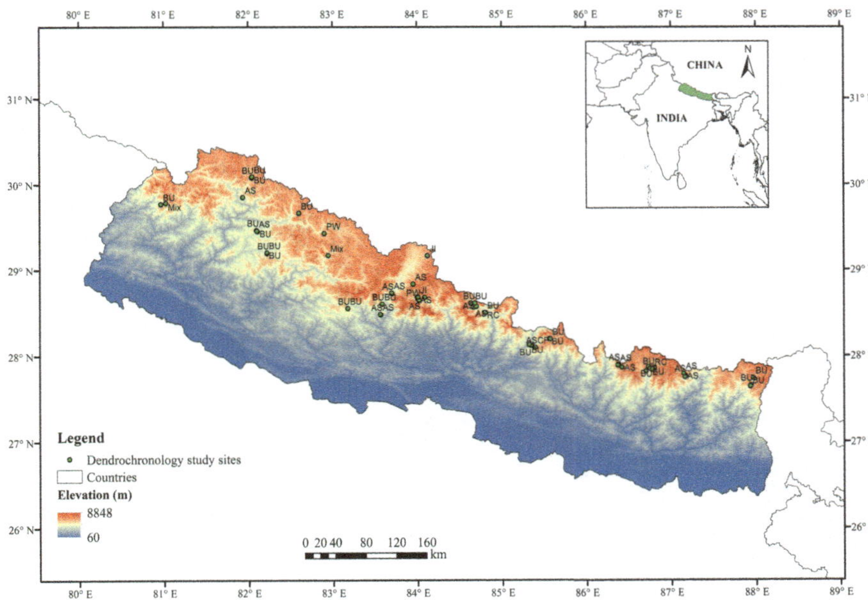

Fig. 22.5 Dendrochronological study sites at the timberline region of the Nepal Himalaya. The species include *Abies spectabilis* (AS), *Betula utilis* (BU), *Rhododendron campanulatum* (RC), *Juniperus indica* (JI), and Mix (more than one species per site)

Table 22.2 An overview of growth–climate relationship of treeline-forming species from treeline or timberline regions in the Nepal Himalaya

Reference	Site name/ Location	Species	Parameters	Response with mean temperature												
				J	F	M	A	M	J	J	A	S	O	MAM	JJAS	Others
Sano et al. (2005)	Humla	AS	RW											–		
		AS	MLD											+	JJ+	AS+
		AS	MXD													AS+
		AS	MND											+	JJ+	AS+
Tenca and Carrer (2010)	Khumbu, SNP	AS	RW								+					
		BU	RW							+		+				
Gaire et al. (2011)	Lauribina, LNP	AS	RW			–	–							–		
Sano et al. (2012)	Humla	AS	$\delta^{18}O$						+						–	
Dawadi et al. (2013)	Langtang, LNP	BU	RW					–								
Gaire et al. (2014)	Prok, MCA	AS	RW					–	–	–						
Liang et al. (2014)	SKB1, SNP	BU	RW													
	SKB2, SNP	BU	RW				–							–		
	SKB3,SNP	BU	RW			–	–							–		
	SKB4, SNP	BU	RW													
	LT1, Langtang	BU	RW													
	LT4, Langtang	BU	RW					–						–		
	MSB1, Manaslu	BU	RW													
	MSB2, Manaslu	BU	RW													
	MPB1, Manaslu	BU	RW				–		+					–		
Shrestha et al. (2015a, b, 2017)	Lauribina_TL	AS	RW							+						
	Manang_TL	PW	RW													
	Manang_TL	PW	RW	–												
Chherti et al. (2016)	Barun valley	AS	RW		+											DJF+
Basnet and Gaire (2017)	ACA, Myagdi	AS	RW								+					

Gaire et al. (2017a, b)	Khumbu, SNP	AS	RW								+				JJA +	
	Khumbu, SNP	BU	RW			–		–					–	–		
Kharal et al. (2017)	Manang	AS	RW				–	–						–		
Rayback et al. (2017)	Langtang	AS	RW		+		+									
Tiwari et al. (2017a)	ATE-Mustang	AS	RW				–							–		
	CTF-Mustang	AS	RW				–							–		
Tiwari et al. (2017b)	MT1-Mustang	BU	RW			–							–	–		
	MT2-Mustang	BU	RW			–							–	–		
	DT-Mustang	BU	RW										–			
Pandey et al. (2018)	Khumbu, SNP	BU	RW	+						+	+	+			+	
Schawb et al. (2018)	Rolwaling	AS	RW				–	–						–		DJF-
		AS	BI	–	–										–	DJF-
Rai et al. (2019)	Manang	AS	RW						+		+					
Panthi et al. (2020)	Rara	AS	RW													
	Gaurishankar	AS	RW													
Gaire et al. (2020)	Prok, MCA	AS	RW	+							+					DJF+
Brunello et al. (2019)	Mustang	JI	$\delta^{18}O$													
Pandey et al. (2020)	Khumbu, SNP	AS	RW													
		BU	RW													
		AS	$\delta^{13}C$	+	+	+	+		+		+	+	+	+	+	+
		BU	$\delta^{13}C$	+	+	+	+	+	+		+	+	+	+	+	+
		AS	$\delta^{18}O$	+	+	+	+		+		+		+	+	+	+
		BU	$\delta^{18}O$	+	+	+	+	+	+		+		+	+	+	+

(continued)

Table 22.2 (continued)

Reference	Site name/ Location	Species	Parameters	Response with precipitation J	F	M	A	M	J	J	A	S	O	MAM	JJAS	Others
Sano et al. (2005)	Humla	AS	RW											+		
		AS	MLD											–		
		AS	MXD													
		AS	MND												JJ-	
Tenca and Carrer (2010)	Khumbu, SNP	AS	RW													
		BU	RW		+						+					
Gaire et al. (2011)	Lauribina, LNP	AS	RW													
Sano et al. (2012)	Humla	AS	$\delta^{18}O$						–		–	–			–	
Dawadi et al. (2013)	Langtang, LNP	BU	RW			+ +					–			+		
Gaire et al. (2014)	Prok, MCA	AS	RW		–											
Liang et al. (2014)	SKB1, SNP	BU	RW			+		+						+		
	SKB2, SNP	BU	RW			+	+	+			–			+		
	SKB3,SNP	BU	RW			+								+		
	SKB4, SNP	BU	RW			+		+				–		+		
	LT1, Langtang	BU	RW			+		+						+		
	LT4, Langtang	BU	RW				+	+			–			+		
	MSB1, Manaslu	BU	RW					+						+		
	MSB2, Manaslu	BU	RW			+		+						+		
	MPB1, Manaslu	BU	RW											+		

Shrestha et al. (2015a, b, 2017)	Lauribina_TL	AS	RW					–			–					
	Manang_TL	PW	RW		–											
	Manang_TL	PW	RW	–			+									
Chherti et al. (2016)	Barun valley	AS	RW													
Basnet and Gaire (2017)	ACA, Myagdi	AS	RW								–		+			
Gaire et al. (2017a, b)	Khumbu, SNP	AS	RW													
	Khumbu, SNP	BU	RW			+		+						+		
Kharal et al. (2017)	Manang	AS	RW						–							
Rayback et al. (2017)	Langtang	AS	RW													
Tiwari et al. (2017a)	ATE-Mustang	AS	RW													
	CTF-Mustang	AS	RW					+						+		
Tiwari et al. (2017b)	MT1-Mustang	BU	RW	–		+								+		
	MT2-Mustang	BU	RW			+								+		
	DT-Mustang	BU	RW					+						+		
Pandey et al. (2018)	Khumbu, SNP	BU	RW											+		
Schawb et al. (2018)	Rolwaling	AS	RW	+			+			–						
		AS	BI	+				–						–	–	
Rai et al. (2019)	Manang	AS	RW					+	+	+	+			+	+	
Panthi et al. (2020)	Rara	AS	RW				+	+								
	Gaurishankar	AS	RW								–					
Gaire et al. (2020)	Prok, MCA	AS	RW	–			+				–					
Brunello et al. (2019)	Mustang	JI	$\delta^{18}O$						–	–						
Pandey et al. (2020)	Khumbu, SNP	AS	RW													
		BU	RW				+			–						DJF+
		AS	$\delta^{13}C$					+			–					
		BU	$\delta^{13}C$				+	–						+		
		AS	$\delta^{18}O$						–		–			+		
		BU	$\delta^{18}O$					+	–							

Only significant response (95% or more confidence limit) is presented. (– = negative; + = positive), *AS*: *Abies spectabilis*, *BU*: *Betula utilis*, *PW*: *Pinus wallichiana*, *JI*: *Juniperus indica*; *RW*: ring width, *MLD*: mean density, *MXD*: maximum density, *MND*: minimum density, $\delta^{18}O$: Oxygen stable isotope, $\delta^{13}C$: carbon stable isotope; *MAM*: March–May or spring season, *JJAS*: June–September or summer season, *JJ*: June–July; *AS*: August–September, *DJF*: December–February or winter

22.3.4 Treeline Vegetation and Sensitivity of Treeline Types/Forms

Several physiognomy and floristic composition related studies have been conducted in the treeline ecotone of Nepal Himalaya. Schickhoff (2005) reviewed the existing literature and explained the floristic composition and physiognomy of the treelines in the HKKH region. Available studies have indicated that generally three to six species form treelines at a site in the Nepal Himalaya (Table 22.1). The primary species are *Abies spectabilis*, *Betula utilis*, *Juniperus indica*, *Juniperus recurva*, *Juniperus wallichiana*, *Larix griffithiana*, *Pinus wallichiana*, *Sorbus microphylla*, *Acer* spp., *Salix* spp., *Quercus semecarpifolia*, *Rhododendron campanulatum*, and *Rhododendron lanatum* (Bhuju et al. 2016; Dhakal et al. 2016; Numata 1967; Schickhoff 2005; Yoda 1967). Generally, treelines are formed by multiple tree species; however, a pure stand treeline, generally of *Abies*, *Betula*, *Juniperus*, or *Pinus* species are also observed. The tree species composition varies from one aspect to another. North-facing slopes along the NW-SE gradient show a floristic change from deciduous Betula to evergreen *Rhododendron*-dominated upper timberlines, while the *Juniperus* spp. (*Juniperus recurva*, *J. wallichiana*, *J. indica*) are the main treeline-forming tree species on south-facing slopes throughout the Himalaya (Schickhoff 2005). Physiognomically, high coniferous forests give way to medium-sized broad leaved trees stands and finally to a krummholz belt (Schickhoff 2005); however, in some areas (e.g., Chuchemara in RNP and Majphal in Dolpa) closed forests directly give way to treeless alpine grass vegetation (Bhuju et al. 2016; Gaire et al. 2015).

There is a slight difference in species association in different treeline sites across the Nepal Himalaya (Table 22.1). For example, the treeline in the Ghunsa valley in Kanchenjunga is formed by *A. spectabilis*, *B. utilis*, *J. indica*, *R. campanulatum*, *S. microphylla*, and occasionally by *L. griffithiana*. In Dole, Phortse, Debuche, Pangboche of SNP (Mt. Everest), *A. spectabilis*, *B. utilis*, *J. recurva*, *R. campanulatum*, *S. microphylla*, and *Salix* spp. are the treeline-forming species, while some *Acer campbellii* and *R. campylocarpum* are also present at the timberline at Debuche (Bhuju et al. 2010; Gaire et al. 2015; Miehe 1991, 1989; Schickhoff 2005). In Rolwaling valley at the GCA, *A. spectabilis*, *B. utilis*, *R. campanulatum*, *S. microphylla* are the predominant treeline-forming species in NE and NW aspects, while *Acercaudatum* and *Prunusrufa* occasionally occur (Schickhoff et al. 2015; Schwab et al. 2017). At the Kalchuman Lake area in the MCA, the treeline ecotone is primarily formed by *A. spectabilis* and *B. utilis* with patches of *R. campanulatum*, *S. microphylla* growing at higher elevations than *Abies* and *Betula* in northern slope (Gaire et al. 2014; Rana et al. 2016). In the treeline near the Phoksundo Lake area *J. indica* and *P. wallichiana* are the major species of the southern aspect, while *B. utilis*, *P. wallichiana*, and *R. campanulatum* are the treeline species in the northern aspect (Dhakal et al. 2016). For the Humla–Jumla area, *B. utilis* and *R. campanulatum* grow at northern slopes, and *J. indica* in southern slopes (Dobremez and Shrestha 1980; Williams 1953 cited in Schickhoff 2005; Sigdel et al. 2018). At the treeline ecotone of the Api-Namapa region, *A. spectabilis*, *B. utilis*, *R. campanulatum*, and *S. microphylla* are principal treeline-forming species in NE and NW aspects (Bhuju et al. 2016). The *Q. semecarpifolia* also grows up to the timberline region in some sites in western Nepal including ANCA (Tabata et al. 1988).

All major forms of the treeline (Harsch et al. 2009; Harsch and Bader 2011) exist in the Himalayan region (Fig. 22.6). The treelines in the Himalaya can be categorized into climatic, orographic/edaphic, and anthropogenic according to the causative factor of the treeline formation (Schickhoff et al. 2015). Based on extensive field surveys along the entire Himalayan mountain arc, Schickhoff et al. (2015) estimated 85–90% of treelines to be anthropogenic largely influenced by pastoralism and only 10–15% to be orographic/edaphic and climatically determined. The sensitivity of climatic treelines to climatic change varies from several years to a few decades (Harsch et al. 2009; Schickhoff et al. 2015). Orographic treelines result from debris slides, rock falls, and snow avalanches that prevent the establishment of trees; therefore, these treelines will be less responsive to climate change (Holtmeier 2009; Schickhoff et al. 2015). According to Schickhoff et al. (2015) edaphic treelines in the Himalayas will become more sensitive to a warming climate once pedogenetic processes accelerate and favor the establishment of tree seedlings and tree invasion. Rangelands above the treeline in the Himalaya have been used for pastoral activities during the summer season since historical time. Because of the economic development and social transformation in recent years, there is a decline in pastoral activities in many high elevation areas (Aryal et al. 2018, 2017). Therefore, anthropogenic treelines have shown considerable dynamics in recent decades with high regeneration, increased tree establishment within the treeline ecotone, and/or invasion into treeless areas above the anthropogenic forest limit (Schickhoff et al. 2015; Shrestha et al. 2015a, 2007; Suwal et al. 2016). However, to isolate the influence of the changing anthropogenic activities and the climate change in those treelines further experiments are required.

22.3.5 Treeline Elevations

An extensive overview of the treeline position in the HKKH region is given by Schickhoff (2005). This study indicated that upper timberline elevations in the Himalaya increase along two gradients: a NW-SE gradient (corresponds to higher temperature sums at same elevations along the mountain arc) and peripheral-central direction from the Himalayan south slope to the Great Himalayan range (combined effects of continentality and mass-elevation, both leading to higher temperature sums). Thus, potentially natural timberline elevations are higher at south-facing slopes than north-facing (Schickhoff 2005). Figure 22.7 presents a variation of the treeline position along the longitudinal gradient, irrespective of aspects and treeline form in the Nepal Himalaya, which shows a general east-west decreasing treeline position. In peripheral region, the *Betula* treeline at Kanchenjunga was at 4132 masl, while it is at 3845 masl in the Api-Nampa region (Bhuju et al. 2016). A similar east-west general decreasing pattern was found for *A. spectabilis* treeline (Bhuju et al. 2016). However, in the central direction/region of a large mountain system, this

Fig. 22.6 Different form and structure of treeline in the Nepal Himalaya. Orographic treeline at Prok, MCA (**a**); Krummholz treeline at Dole in SNP (**b**); Abrupt/Anthropogenic treeline at Chuchemara, RNP (**c**); Diffuse treeline at Ghunsa valley, KCA (**d**), Gosaikunda valley, LNP (**e**) and Barun valley, MBNP (**f**)

general east-west decreasing pattern does not follow. For example, in the Annapurna range, the treeline positions vary rapidly within the site and reach up to 4400 masl (Schickhoff 2005). Similarly, *B. utilis* at treeline in the northern slope of Mt. Kanjirowa in Dolpa was above 4270 masl (Dhakal et al. 2016), it was at 4243 masl in Humla, but it descends to 4217 masl at the Mt. Kanchenjunga region (Sigdel et al. 2018). At the treeline in the northern aspect, generally, *R. campanulatum* was found to be growing at the highest elevations, followed by *S. microphylla*, *B. utilis*, and *A. spectabilis*. However, in some sites (e.g., Pangboche in SNP), *A. spectabilis* grew at a higher elevation than the *B. utilis* (Gaire et al. 2017a, 2015). Therefore,

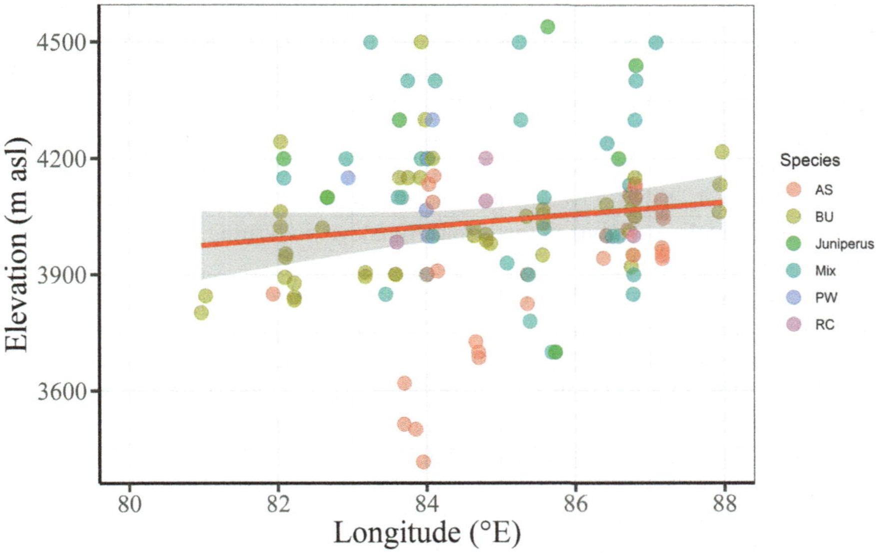

Fig. 22.7 Variation of the treeline position in the west-east direction (longitude)of the Nepal Himalaya, regardless of aspects and treeline form. The treeline species include *Abies spectabilis* (AS), *Betula utilis* (BU), *Rhododendron campanulatum* (RC), *Juniperus* sps., *Pinus wallichiana* (PW), and Mix (combination of two or more than two treeline-forming species)

there is a broader pattern and site-level modifications or specificities in treeline positions in the Nepal Himalaya.

22.3.6 Stand Structure and Regeneration at Treeline Ecotone

Studies have indicated species-specific and site-dependent variations in plant structural parameters (densities and size class). Along the elevation gradient of the timberline ecotone of the Kalchuman Lake area in the MCA, *R. campanulatum* is the most dominant species, followed by *S. microphylla* (4000 masl), while *B. utilis* is dominant at lower reaches (3600–3900 masl) (Rana et al. 2016). There was a lack of seedlings of *Betula,* which might be due to unfavorable microhabitat (dense Krummholz of *Rhododendron* tree and shrubs) (Gaire et al. 2014; Rana et al. 2016). The regeneration of the *Betula* at the Samagaun area in the MCA was not sustainable as sapling density was higher than seedling density and tree density higher than sapling and seedling (Sujakhu et al. 2013). Similar poor regeneration of *A. spectabilis* was observed at the treeline of the Chimang area in Mustang with a lower number of seedlings and saplings compared to tree individuals (Tiwari et al. 2017a). However, the regeneration of *A. spectabilis* at the treeline ecotone of Luribina, LNP was high compared to the other treeline-forming species (e.g., *B. utilis*), though plot-level differences existed (Gaire et al. 2010). The higher seedling densities indicate stand densification and possible range expansion of the *Abies* at the treeline (Gaire et al.

2011, 2010). At Rolwaling valley in the GCA, there was prolific regeneration across the treeline ecotone on NW and NE slopes with the seedling establishment of *Betula*, *Abies*, and *Sorbus* to some extent far above the upper limit of the adult tree (Schickhoff et al. 2015). However, the density of the seedlings and saplings sharply decreased toward the alpine tundra, in particular with *Abies* and *Betula*, and the average maximum seedling/sapling density was distinctly higher on the NE slope (Schickhoff et al. 2015). Overall, there is a considerable variation in population density at the treeline ecotone in the Nepal Himalaya, indicating site-level regeneration dynamics.

Ecological studies from the treeline ecotone have also used DBH, height class distribution, and density diameter curves to assess regeneration at the treeline ecotone (Gaire et al. 2015, 2010; Ghimire et al. 2010; Ghimire and Lekhak 2007; Schickhoff et al. 2015; Schwab et al. 2016; Shrestha et al. 2007). On the other hand, dendroecological studies are based on the age-class distribution, in addition to DBH and height class, to assess regeneration (Chhetri and Cairns 2018, 2016, 2015; Gaire et al. 2017a, 2014, 2011; Shrestha et al. 2015a; Sigdel et al. 2018; Suwal et al. 2016; Tiwari et al. 2017a). Studies have observed reverse-J to bell-shaped distribution (example of DBH distribution in Fig. 22.8) of these parameters indicating continuous regeneration to stable (mature) stage of tree species at the studied treeline sites

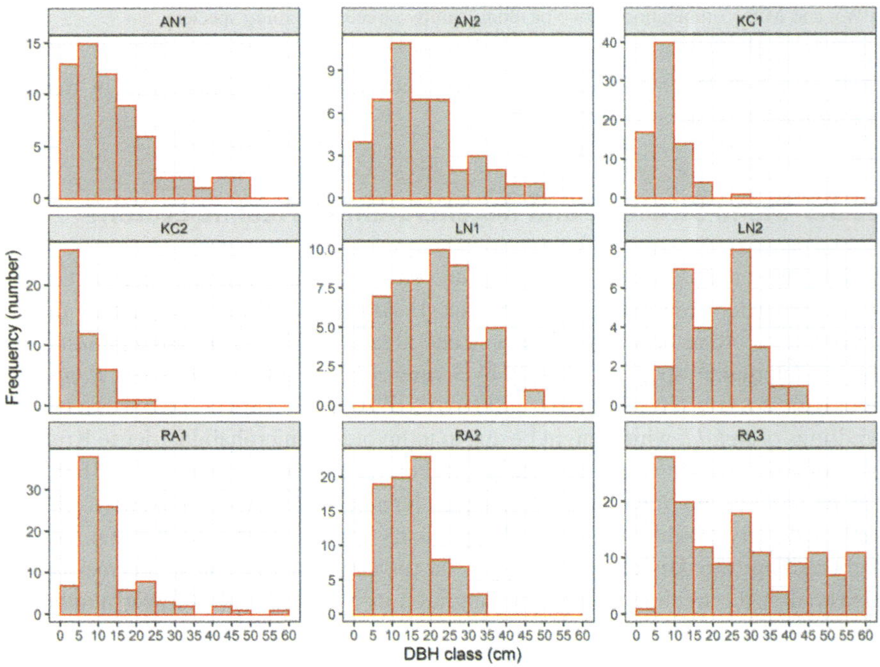

Fig. 22.8 Total DBH distribution (5 cm class) in different treeline regions (Api-Nampa = AN1, AN2; Kanchenjungha = KC1, KC2; Langtang = LN1, LN2, and Rara = RA1, RA2, RA3) in the Nepal Himalaya

(Bhuju et al. 2016; Gaire et al. 2015, 2010; Schickhoff et al. 2015; Schwab et al. 2016; Shrestha et al. 2007). However, gaps in some size class distribution have also been observed, indicating possible episodic regeneration or some disturbance at treeline during the past. Considering species-wise regeneration, in most of the sites, *Abies* and *R. campanulatum* have better regeneration compared to other treeline-forming *Betula, Juniperus,* and *Pinus* (Bhuju et al. 2016; Gaire et al. 2015, 2010; Rana et al. 2016; Schickhoff et al. 2015; Schwab et al. 2016; Shrestha et al. 2007; Sigdel et al. 2018). The Schwab et al. (2016) found overall sustainable regeneration in all elevation zones of the treeline ecotone in Rolwaling valley.

The maximum age of the trees at timberline ranges from less than 100 to over 300 years (Basnet and Gaire 2017; Chhetri and Cairns 2018, 2015; Gaire et al. 2017a, 2014, 2011; Rana et al. 2017; Shrestha et al. 2015a; Sigdel et al. 2018; Suwal et al. 2016; Tiwari et al. 2017a). The oldest age of *A. spectabilis* at the different treeline ranges from about 50 years to more than 200 years, indicating recent recruitment history at the treeline. Over 600 years old *A. spectabilis* trees growing at the lower elevations (~3700 masl) have also been observed. Over 450-year-old *B. utilis* trees have been reported from the timberline (approximately 4050 masl) in the Nepal Himalaya (Dawadi et al. 2013; Liang et al. 2014). At the mixed-species treeline ecotone, generally, *Betula* trees are older than the other species (Gaire et al. 2017a, 2014; Rana et al. 2017; Sigdel et al. 2018); however, in some treeline sites (e.g., Pangboche, SNP) it was found that *Abies* at a higher elevation was older than *Betula* (Gaire et al. 2017a). *R. campanulatum* and *S. microphylla* were at the higher elevations than *Abies* and *Betula* in most of the studied treelines from northern slopes (Gaire et al. 2017a, 2014, 2011; Rana et al. 2017, 2016; Schickhoff et al. 2015). The age structure of *R. campanulatum* is studied in Manaslu, Langtang, and Sagarmatha, and a study found more than 164 years old trees at the treeline in Manaslu (Rana et al. 2017). The exact age of *S. microphylla* trees in the Nepal Himalaya at their maturity has not been explored yet; however, over 100 years old *Sorbus* trees at the treeline ecotone of the SNP have been observed (authors field observation). Likewise, we have limited information on the age structure of the *Juniperus* species at the treeline ecotone of the Nepal Himalaya, though it is a long-living tree species that can survive thousands of years (Shao et al. 2009; Yadav et al. 2006). Studies have shown continuously to sporadic regeneration at the treeline as revealed from reverse-J shaped to uni- to multi-modal bell-shaped age-class distribution (e.g., Fig. 22.9), depending on the site and species (Chhetri and Cairns 2018, 2015; Gaire et al. 2017a, 2014, 2011; Rana et al. 2017; Shrestha et al. 2015a; Sigdel et al. 2018; Suwal et al. 2016; Tiwari et al. 2017a). The overall observation of age structure from different treeline of the Nepal Himalaya shows a young history of the treeline community, indicating a possible dynamic state of the central Himalayan treeline with rapid warming trends.

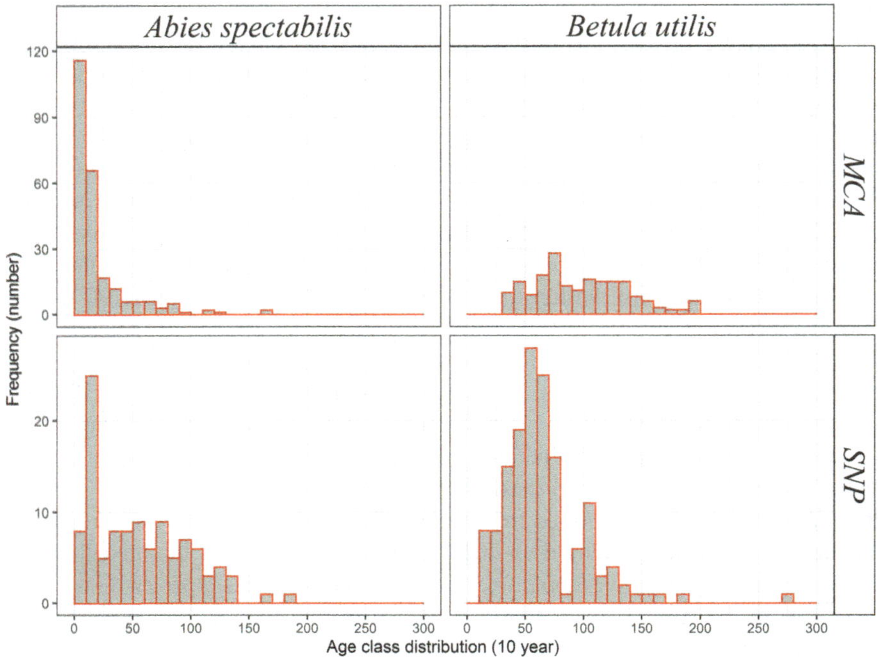

Fig. 22.9 Age-class distribution (10-year class) of the treeline-forming species (*Abies spectabilis and Betula utilis*) at some (MCA and SNP) of the treeline ecotone of the Nepal Himalaya

22.3.7 Geology and Soil Characteristics at Treeline

Not much is known about the geological condition/division focusing in the treeline ecotone in the Nepal Himalaya. Information on the soil properties and soil nutrients is essential from a plant growth perspective and for climate change mitigation because soil stores a significant fraction of the global carbon stock. Limited studies have analyzed soil characteristics of treeline ecotone (e.g., Bäumler and Zech 1994; Müller et al. 2016a, 2016b). The soil properties can affect alpine treelines and their dynamics (Müller et al. 2016a). At the treeline ecotone in Rolwaling valley, GCA, the soil texture at 10 cm depth was sand or loamy sand and a significantly decreasing nutrient availability in soils and foliage with elevation was observed (Müller et al. 2017, 2016a). Similarly, Drollinger et al. (2017) found a significantly decreasing trend in soil nutrients and foliar macronutrient concentrations (nitrogen, magnesium, potassium, and foliar phosphorus) along elevation gradient across the treeline ecotone in Rolwaling Himal, Nepal. Shortage of some of these nutrients at the treeline ecotone can hinder the upward shifting of the treeline or treeline-forming species (Müller et al. 2017).

Müller et al. (2016b) presented a 2½-year (2013–2015) soil temperature monitoring data from the treeline ecotone in Rolwaling Himal. They found 7.6 ± 0.6 °C growing season mean soil temperature at 10 cm depth, which is 1.2 °C higher compared to the suggested 6.4 ± 0.7 °C for alpine treelines (Körner 2012; Körner and Paulsen 2004). While in a year-long (1999–2000) monitoring of the soil temperature at a *Betula* treeline near Kiyangjing Gompa of Langtang, Körner (2012) found mean growing season soil temperature of 5.8 °C (Minimum = −2.5 °C, maximum = 9.2 °C, median = 6.8 °C) with a 142-day vegetative growing period, which is lower than the global average value (Körner 2012). In the treeline ecotone of the GCA, there was a spatiotemporal variation in soil temperature, significantly decreasing with increasing elevations in the NW aspect, as well as no significant decrease at the NE aspect (Müller et al. 2016b). Similarly, there were some differences in the seasonal variation in soil temperature along the elevation gradient in different aspects of the mountain slope. The mean soil temperature in spring was commonly higher in the NE aspect than the NW, while an opposite spatial trend for autumn and winter was found (Müller et al. 2016b). The soil temperature in the treeline is influenced by the tree physiognomy and forest stand structure (Müller et al. 2016b). Similarly, there is a decreasing water availability with increasing elevation in the treeline ecotone at site-level study. The moisture availability is influenced by the soil texture (low water-holding capacity in shallow and coarse-grained soils) and tree physiognomy (Müller et al. 2016b). Shrestha et al. (2007) indicated that moisture plays a vital role in the spatial (horizontal) distribution of treeline-forming *B. utilis* in the trans-Himalayan region in Nepal. From these studies, it can be inferred that variation in micro-topographic conditions including soil properties can affect regeneration and tree growth at the treeline.

22.3.8 Tree-Ring Chronologies and Growth of the Trees at Treelines

Dendrochronological studies have extensively carried out from the timberline region of the Nepal Himalaya using multi-proxy tree-ring parameters (Fig. 22.5; Table 22.2). More than 40 tree-ring width chronologies of *A. spectabilis* (22), *B. utilis* (15), *P. wallichiana* (4), *R. campanulatum* (1) and *J. indica* (1) (Fig. 22.5) have been published. One xylem wood density (Sano et al. 2005), two oxygen stable isotope ($\delta^{18}O$) (Pandey et al. 2020; Sano et al. 2012), and one blue light intensity (Schwab et al. 2018) chronologies of *A. spectabilis*, and one each chronology of oxygen stable isotope for *J. indica* (Brunello et al. 2019) and *B. utilis* (Pandey et al. 2020) have also been developed from the timberline region of Nepal. Moreover, four carbon stable isotope ($\delta^{13}C$) chronologies have also been prepared using *A. spectabilis* (3) and *B. utilis* (1) treelines from Rara, Gaurishankar, and Sagarmatha (Pandey et al. 2020; Panthi et al. 2020). The oxygen stable isotope chronology of *J. indica* tree is from a site at 4400 masl in the trans-Himalayan region of the upper Mustang (Brunello et al. 2019), which captured long-term changes in the strength of the Indian summer monsoon. Schwab et al. (2018) presented the first blue intensity chronology of *A. spectabilis* from the timberline region (Rolwaling) in Nepal, spanning 1819–2012 CE.

Looking at the individual chronologies, some tree-ring chronologies of *A. spectabilis* from the timberline regions indicated growth enhancement during recent years, while others showed no trend or slight growth decline (Camarero et al. 2021; Panthi et al. 2020; Thapa et al. 2017). Growth of the fir at the Rolwaling in the GCA increased in the late twentieth century, but growth is decreasing in the recent few years (Schwab et al. 2018). There was no persistent and significant growth trend in the *B. utilis* chronologies in most of the twentieth century in most of the studied timberline sites. However, the growth is declining in the recent few decades/years in most sites (Dawadi et al. 2013; Gaire et al. 2017a; b; Liang et al. 2014; Tenca and Carrer 2010; Tiwari et al. 2017a; b). A *P. wallichiana* chronology from the timberline of the Dolpa has a positive growth trend in most of the nineteenth and twentieth centuries with the growth decline in the recent few decades (Gaire et al. 2019). A similar recent growth decline was observed in the chronology of the same pine species from the Manang valley (Shrestha et al. 2015b). As different studies have used different detrending options to produce site chronologies (standard, residual, or signal-free), some variations in the trend could be due to different deternding methods, in addition to different climatic regimes and site conditions. If all the reported chronologies can be analyzed using similar detrending options (or develop BAI chronologies), then the result will be more robust for the comparative purpose. Only limited studies (14 chronologies) have reported annual radial growth (raw measurement) at treeline ecotone, which shows average radial growth of *Abies* ranged from 0.94 to 4.6 mm/year while *Betula* growth from 0.86 to 1.05 mm/year. Temporal tree-height increment at the treeline ecotone in the Nepal Himalaya is also reported in some sites (e.g., Shrestha et al. 2015a; Tiwari et al. 2017a; Sigdel et al. 2018). The comparisons of this reported vertical increment indicate site and species-specific growth rate of the treeline-forming species.

22.3.9 Climatic Control of Tree Growth at the Timberline Region

Growth climate responses of timberline forming *A. spectabilis*, *B. utilis*, *J. indica*, *P. wallichiana*, and *R. campanulatum* have been thoroughly analyzed from the different regions of the Nepal Himalaya using various (ring-width, wood density, stable isotopes, wood-anatomy) tree-ring parameters as ring-width, wood density, stable isotopes, wood-anatomy (Table 22.2). There is no chronology published for other timberline forming species like *Sorbus* and *Acer*. Species-specific growth climate response is also observed from the sites in which the response of the multiple species is analyzed, indicating differential sensitivity and vulnerability of the species to climate change.

A. spectabilis growing in the treeline has generally responded positively to the spring season precipitation and negatively to the spring season temperature (Gaire et al. 2011; Panthi et al. 2020; Sano et al. 2005; Schwab et al. 2018; Tiwari et al. 2017a) and positively to the winter and/or summer season temperature (Chhetri and Cairns 2016; Gaire et al. 2020, 2017a; Panthi et al. 2020). However, there also exists between site differences in the species responses. It is observed that the juveniles are more sensitive to temperature, whereas the adults are more sensitive to moisture availability (Tiwari 2020). The growth of the *B. utilis* in most of the studied site is generally limited by the moisture stress during the spring season (March–June), revealing a positive correlation with the precipitation and negative with the temperature of the spring season (Dawadi et al. 2013; Gaire et al. 2017a; Liang et al. 2014; Tiwari et al. 2017b). Xylem wood anatomical responses of the *Betula* trees to climate variability at SNP indicated that mean ring width, mean vessel area, and ring-specific hydraulic conductivity were positively correlated with summer temperatures of the previous and current year. In contrast, the mean vessel area was significantly associated with March precipitations (Pandey et al. 2018). The climatic response of *P. wallichiana* from the timberline region of the Nepal Himalaya indicated that moisture availability during and preceding growth season is crucial for the species (Bräuning 2004; Gaire et al. 2019; Shrestha et al. 2015b). Similarly, there is no study on the growth-climate response of the juniper tree-ring width chronology from the timberline region of the Nepal Himalaya. However, two *J. recurva* sites from eastern Nepal were used in the tree-ring network-based temperature reconstruction (Cook et al. 2003).

Stable isotopes in tree-rings provide valuable information about past climate, and studies have also analyzed the climatic sensitivity by using stable isotopes of carbon and oxygen (Brunello et al. 2019; Pandey et al. 2020; Panthi et al. 2020; Sano et al. 2005). Climate response analysis revealed that the tree-ring $\delta^{18}O$ of *A. spectabilis* was primarily controlled by the amount of precipitation and relative humidity during the monsoon season (June–September) and the temperature of the summer season (Sano et al. 2012). The annual $\delta^{18}O$ of cellulose of *J. indica* tree-ring in Mustang shows summer monsoon signal (Brunello et al. 2019). The cellulose stable isotopes ($\delta^{13}C$ and $\delta^{18}O$) and intrinsic water use efficiency (iWUE) analysis of *A. spectabilis* and *B. utilis* at the treeline in SNP indicated an increase in photosynthetic efficiency (Pandey et al. 2020). Climate change progressively reduces physiological limitations due to low temperatures and low spring precipitations, thus potentially facilitating a further altitudinal forest advance (Pandey et al. 2020). From these studies, it can be inferred that stable isotopes can be a useful variable to study the climatic sensitivity and dynamics of the treeline trees.

The tree-ring wood density-based climatic response study of *A. spectabilis* from treeline ecotone in the Humla region indicated that the pre-monsoon temperature and precipitation primarily control the ring width and minimum density, while the latewood density by the late monsoon temperature (Sano et al. 2005). During recent years, the blue intensity (BI) parameter, measured as the value of the blue light absorbed by wood surface recorded directly from the RGB (red, green, and blue light components) digital image (Rydval et al. 2014), has increasingly been applied as a reliable surrogate for maximum latewood density (Björklund et al. 2015, 2014; Kaczka et al. 2018; McCarroll et al. 2002; Schwab et al. 2018). However, this is a new parameter for application in the tree-ring field in Nepal. The climatic response of BI chronology of *A. spectabilis* from Rolwaling valley is not stable temporally, and response of the static analysis is different from that obtained in 30-year moving correlation (Schwab et al. 2018). This study opened a new avenue of application of BI to study the climatic response of timberline-forming species in Nepal.

With climate change, divergence in the growth or growth climate response is observed in many cold regions of the world. In the Nepal Himalaya, studies have used moving correlations between the chronology and climatic parameters to check the temporal stability of the response and divergence issue (Gaire et al. 2020; Schwab et al. 2018; Shrestha et al. 2015b). The climatic response of the tree growth at the treeline is temporally stable to unstable depending on the parameter considered and the site conditions. Shrestha et al. (2015b) studied the temporal response of two treeline-forming species (*A. spectabilis* and *P. wallichiana*) in Langtang and Manang at two different elevations (treeline and forest line) and found that growth is strongly affected by the growing-season and non-growing-season factors. However, there are inconsistencies in terms of the climate–growth relationship across space and over time. Similarly, Schwab et al. (2018) found a temporally unstable dendroclimatic signal of *A. spectabilis* at the treeline of the Rolwaling valley. Likewise, moving correlation analysis revealed a persistent response of fir tree growth at the treeline of the MCA to May and August temperatures and January and June precipitation; however, growth response to the climate of the other months is temporally unstable (Gaire et al. 2020).

Climate change may affect the treeline communities in different ways. If drought conditions increase in the future, trees from dry treeline regions will be adversely affected, and tree mortality can occur. However, in the humid treeline sites, the growth of the trees may further enhance, and a range of the treeline could expand to higher elevations. Though a limited number of studies draw an inference about divergent issues, the spatially and temporally unstable response indicates the necessity of designing holistic forest management and conservation policies and programs to save the treeline communities from the adverse impacts of climate change.

22.3.10 Timberline Region as a Source of Palaeoclimate or Long-Term Climate Information

Trees or shrubs species growing at the timberline region can provide valuable proxy climatic sources in the form of different proxies such as tree-rings, stable isotopes, or pollen-spores (Fritts 1976; Speer 2010). Several dendrochronological and some paleontological studies have been carried from the timberline region of the Nepal Himalaya. Some have reconstructed the past climate or other environmental changes (Gaire et al. 2019; Liang et al. 2019; Miehe et al. 2009, 2002; Panthi et al. 2021; Sano et al. 2012, 2005; Schlütz and Zech 2004; Yonebayashi and Minaki 1997). For example, a temperature reconstruction of the spring-summer season (March–September) of western Nepal for the past 249 years (1752–2000 CE) by using ring width and wood density of *A. spectabilis* from the Humla region in western Nepal shows a warming trend from the 1750s until approximately 1790 CE followed by cooling until 1810 CE then by a gradual warming trend extending to 1950 CE and a notable cold period continuing up to 2000 (Sano et al. 2005). However, the study did not find a consistent warming trend that was found in higher latitude over the reconstruction period (Sano et al. 2005). In another study, Gaire et al. (2019) reconstructed 319 years (1697–2015 CE) spring-summer season (February–August) scPDSI using the timberline *P. wallichiana* from the trans-Himalayan region (Dolpa). They found that the area was under slightly dry conditions for most of the reconstructed period, with below-normal scPDSI values. A regional spring season precipitation reconstruction back to 1650 CE by using the tree-ring chronology of the timberline *B. utilis* from SNP, LNP, and the MCA region shows fluctuations in the precipitation over time with several dry and wet episodes (Liang et al. 2019). In addition to persistent pre-monsoon droughts in the central Himalayas related to prolonged La Niña events, the study also indicates that large-scale volcanic eruptions in the tropical regions can cause drought events in the Himalayan region in the year following the eruptions (Liang et al. 2019). In addition to tree-ring proxy-based climate studies, paleontological studies have analyzed palaeovegetation, palaeoclimate as well as influence of human activities in the high elevation region in the Himalaya (Schlütz and Zech 2004; Yonebayashi and Minaki 1997; Miehe and Miehe 2000; Miehe et al. 2002, 2009).

22.3.11 Treeline Dynamics in the Central Himalaya

Treeline dynamics include changes in the regeneration and stand density at the current treeline ecotone as well as upward/downward shifting of treeline position in response to global warming and other environmental changes. Alpine climatic treelines are responding to temperature increase either by advancing to higher elevation or latitude or remaining stable but with an increase in regeneration and stand density (Harsch et al. 2009; Holtmeier 2009; Holtmeier and Broll 2005;

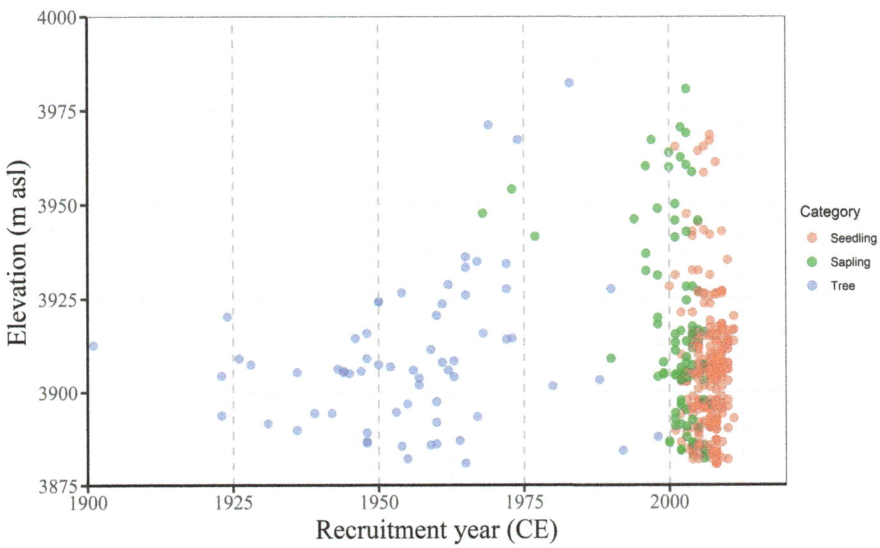

Fig. 22.10 Example of temporal treeline dynamics in response to environmental change. The *Abies spectabilis* treeline dynamics in the Langtang National Park region in the Nepal Himalaya

Körner 2012). We have observed similar processes in the treeline ecotone of the Nepal Himalaya with the environmental changes. Figure 22.10 presents *Abies* treeline dynamics in the LNP region. Though several studies have assessed the treeline dynamics in the Nepal Himalaya, only a few studies have explicitly mentioned treeline shifting and its rate (Basnet and Gaire 2017; Chhetri and Cairns 2015; Gaire et al. 2017a, 2014; Rana et al. 2017; Sigdel et al. 2018; Suwal et al. 2016; Tiwari et al. 2017a). Most of the treeline dynamics studies have observed stand densification at the treeline, even if they are not advancing at present (Chhetri and Cairns 2015; Gaire et al. 2017a, 2014; Schickhoff et al. 2015; Shrestha et al. 2015a). This also indicates the establishment of species and in-fillings process at treeline ecotone (Körner 2012). The average upward shifting rate at 47 sites/transects for which shifting reported was 0.46 m/year ranged from zero (static) to 2.6 m (Fig. 22.11). The upward shifting is higher in *A. spectabilis* and *R. campanulatum*

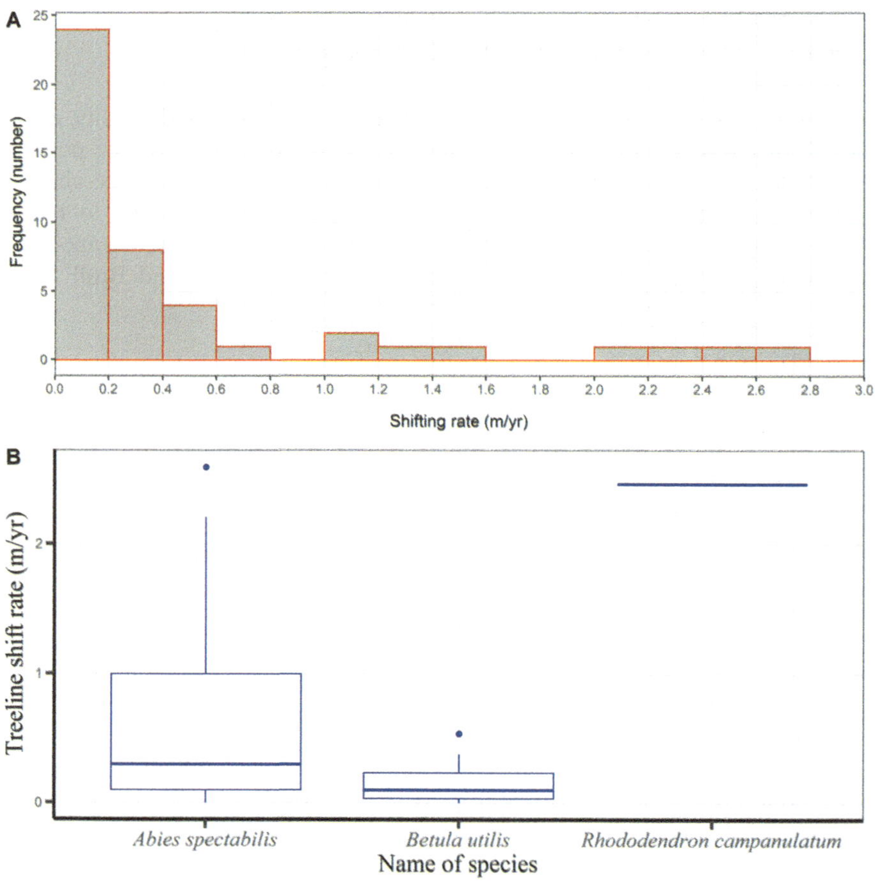

Fig. 22.11 (**a**) Frequency distribution of treeline (or tree species limit) shifting (advancement) rate at the Nepal Himalaya. (**b**) Species-wise shifting rate

compared to *B. utilis* treeline (Basnet and Gaire 2017; Chhetri and Cairns 2016; Gaire et al. 2017a, 2014; Rana et al. 2017; Sigdel et al. 2018; Suwal et al. 2016; Tiwari et al. 2017a). Among the studied species, *A. spectabilis* is in much dynamic condition (regeneration and shifting) followed by *R. campanulatum*, *P. wallichiana*, and *B. utilis*. In some locations, the upper limit advancement of *A. spectabilis* is up to 3.4 m per year (Suwal et al. 2016). Those studies that have not calculated treeline shifting rate have observed the presence of seedlings and saplings at higher elevations than the current treeline (e.g., Gaire et al. 2011, 2010; Schickhoff et al. 2015; Shrestha et al. 2015a), indicating the potential for shifting in future.

The rate of treeline shifting in Nepal is comparable to other parts of the Himalayas (Bharti et al. 2012; Dubey et al. 2003; Panigrahy et al. 2010; Singh et al. 2012; Yadava et al. 2017; Singh et al. 2018), some regions in China (Du et al. 2017), and global pattern (Harsch et al. 2009). A global meta-analysis of literature about treeline dynamics in different regions of the world also showed the nearly 50/50% scenario on the treeline shifting and static (Harsch et al. 2009). Similarly, the rate of shifting and response of the treeline to climate change also governs by the treeline forms (Harsch et al. 2009; Harsch and Bader 2011). Past studies indicated that micro-topography and associated microclimate are very important for the early stage seedling survival in the treeline ecotone (Holtmeier 2009; Körner 2012; Liang et al. 2016; Schickhoff et al. 2015). Additionally, non-climatic drivers can interact with the influence of climate change, resulting instability and shifting of the treeline ecotone (Liang et al. 2016; Wang et al. 2019a, 2019b, 2016). Concurrent to the scientific observations, local peoples also perceived changes in the treeline vegetation in response to climate change in the Nepal Himalaya (Aryal et al. 2017).

22.3.12 Predicting Future Treeline Position

A few modeling exercises have been done to detect past changes and to predict future changes in the treeline ecotone or distribution of the treeline-forming species in Nepal or the entire Himalaya (Bobrowski et al. 2017; Chhetri et al. 2018; Hamid et al. 2018; Schickhoff et al. 2015). Chhetri et al. (2018) used the Maxent species distribution modeling to predict the likelihood of treeline advance in the Nepal Himalaya of three dominant treeline-forming species *A. spectabilis*, *B. utilis*, and *P. wallichiana*. They found a regional increase in suitable habitat for all the three treeline species, predicting a potential for northward and upslope advance (few meters to few hundred meters) under the projected climatic conditions (Fig. 22.12). Ecological niche modeling suggested that the bioclimatic conditions for range expansion of *B. utilis* treeline will be created in the Himalaya during the coming decades (Schickhoff et al. 2015). The species can advance upward and northward in response to climate change (Schickhoff et al. 2015). Another modeling study on the potential distribution of *B. utilis* in the Himalayas (Bobrowski et al. 2017) indicated that the projected suitable area in the Himalayan Mountains varies considerably, with the most extensive distribution in the western and central Himalayan region. There is a difference between potential and actual distribution in the

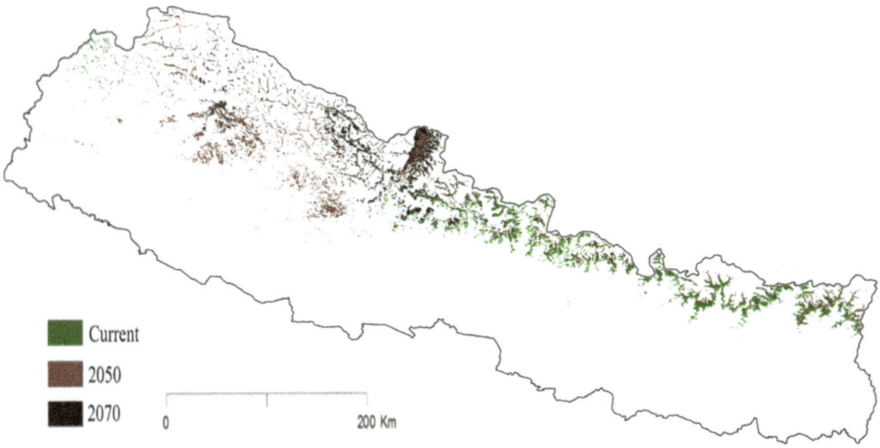

Fig. 22.12 Application of Maxent species distribution modeling approach to predict the likelihood of treeline advance in the Nepalese Himalayas. Figure showing current and future suitable habitat of one of the dominant treeline-forming species (*Abies spectabilis*) in Representative Concentration Pathway (RCP 4.5) scenarios

eastern Himalaya, indicating decreasing competitiveness of *B. utilis* under more oceanic conditions on the east part of the mountain system (Bobrowski et al. 2017). Also, anthropogenic activities hinder the *Betula* from reaching the upper elevational limit in vast areas of its potential distribution range (Bobrowski et al. 2017).

22.3.13 Drivers of Treeline Dynamics

Regeneration and successful establishment of the seedlings at the treeline ecotone is the fundamental aspect of the treeline dynamics in response to the environmental change. Several biotic and abiotic factors including climatic factors are responsible for the regeneration and successful establishment of the trees at the treeline ecotone (Batllori et al. 2009; Holtmeier 2009; Körner 2012; Müller et al. 2016a). From a global perspective, temperature (Körner 2012) is the main climatic driver for the treeline formation and treeline dynamics in the Nepal Himalaya; however, at a site level, precipitation is also an important factor controlling the regeneration and treeline dynamics. In addition to climate, several other biophysical factors are also responsible for the treeline pattern and dynamics in the Himalaya.

Many studies have found a positive relationship between regeneration of treeline trees and temperature (Chhetri and Cairns 2016; Gaire et al. 2017a; Lv and Zhang 2012; Schickhoff et al. 2015). Gaire et al. (2014) found that high temperature throughout the year with moist summer is beneficial for the regeneration of *A. spectabilis* at the alpine treeline ecotone of the Manaslu region. Chhetri et al. (2016) found a positive relationship between the recruitment numbers of *Abies* from treeline in Barun valley and mean temperature during winter months (January,

February, December) and season, and summer months (August). In SNP, there was a species- and site-specific response of regeneration at treeline to climatic variables (Gaire et al. 2017a). The regeneration of *A. spectabilis* was favored by high temperatures throughout the year with sufficient moisture, while the climatic response of the regeneration of *B. utilis* was spatiotemporally different, though both temperature and precipitation played an important role in the regeneration (Gaire et al. 2017a). Precipitation is another climatic factor controlling tree growth and the regeneration of the trees in the studied treeline ecotone (Gaire et al. 2017a; Liang et al. 2014; Schickhoff et al. 2015). In a network of 20 treeline sites (the GCA, SNP, LNP, the ACA, Jumla, and Humla), Sigdel et al. (2018) found a slower upward shift of treelines (mainly *Betula*) in response to recent climate warming, and its pace primarily controlled by spring precipitation in association with changes in mean annual maximum temperature. The study also noted that the treeline shifts in the Himalayas are better explained by interactions between precipitation and temperature than either factor alone.

At local and regional scales, factors like land-use change, forest fire, moisture or drought, disturbance events, species interactions, frost damage, light intensity, and geomorphic processes can control the dynamics of treeline ecotone (Elliott and Cowell 2015; Körner 2012; Liang et al. 2016; Lyu et al. 2016; Munier et al. 2012; Schickhoff et al. 2015; Shen et al. 2018; Sigdel et al. 2018; Wang et al. 2019a, 2019b, 2016). A study covering different treeline sites across Nepal found that tree-to-tree interaction is also playing a role in the slowdown of treeline shifting in the Himalaya (Sigdel et al. 2020b). In the treeline ecotone of the GCA, there was a negative correlation between the abundance and density of *R. campanulatum* and the recruitment of other tree species (Schickhoff et al. 2015). Therefore, the dense *Rhododendron* Krummholz belt can act as an effective barrier for upslope migration of other tree species at the treeline ecotone (Chhetri and Cairns 2015; Schickhoff et al. 2015). However, Mainali et al. (2020) argued that *A. spectabilis* can out-compete the *R. campanulatum* at the treeline ecotone and can move upward in the mix treeline of the *Abies* and *Rhododendron* species in the Langtang. Chhetri et al. (2017) indicated that the topography and human disturbances are major controlling factors in treeline patterns at the Barun and Manang areas in the Nepal Himalaya. Other studies also indicate the importance of land use change in the recent treeline dynamics in some sites (Chhetri et al. 2017; Shrestha et al. 2015a; Suwal et al. 2016). We have observed fire events in some of the treeline sites in the Nepal Himalaya; however, the impact of fire events on the treeline dynamics in the Himalaya has not been studied yet. Microsites, microhabitat, and habitat preferences also play some role in the regeneration which ultimately can influence treeline dynamics (Bürzle et al. 2018; Körner 2012; Mong and Vetaas 2006; Shrestha et al. 2007).

22.3.14 Impacts of the Treeline Shifting in the Himalaya

Changes in the stand density and upward advancement of the treeline ecotone or treeline-forming species can have several positive and negative implications. The positive impacts of the treeline advance are that trees can grow in the higher and wider geographical areas and sequester more carbon in their biomass, helping to mitigate climate change. Local communities can use the increasing biomass of these trees and other ecosystem services to improve their livelihood. On the other hand, an upward shifting of the treeline can adversely affect the distribution and abundance of endemic and medicinal plant communities which are found at or above treeline (Tiwari et al. 2020). Similarly, upward movement of the treeline encroaches the alpine area in the Himalaya, which might affect the region's plant diversity and ultimately to the livelihood of the rangeland-dependent transhumant or pastoralist communities (Aryal et al. 2018, 2017). Furthermore, treeline shifting can impact the habitat of endangered snow leopard because Forrest et al. (2012) indicated that about 30% of snow leopard habitat in the Himalaya might be lost due to treeline shifting and consequent shrinking of the alpine zone. The musk deer prefers dense forest and sparse ground/crown cover (Aryal et al. 2010); therefore, any change in treeline structure can expand or reduce the habitat and distribution of the musk deer.

22.4 Synthesis, Research Gaps, and Way Forward

There is rapid climate change in the Nepal Himalaya with an impact on diverse sectors, including forest and biodiversity. In the central Himalaya in Nepal, considerable progress was found in the treeline research during recent two to three decades. The overall conclusion of the treeline research reveals that treeline in the Nepal Himalaya is at dynamic state and vary with site and species. Both climatic and non-climatic factors are responsible for the recent dynamics of the treeline ecotone. Depending on the study site and species, temperature and precipitation are the primary climatic drivers for the growth and regeneration at the treeline. Several studies in treeline physiognomy, treeline position, regeneration, and stand structure has been conducted in Nepal Himalaya. Similarly, geo-referenced treeline elevation or tree species limit is now archived for over 50 transects for future monitoring of the exact shift in treeline or species limit position. Some initiative is there to unravel the treeline dynamics and regulatory factors; however, integrative researches are still limited. Though there are several studies on the tree-growth climate relationship, majority are from *Abies* and *Betula* trees. Calculations of the climatic response of multiple species present within a treeline site are still lacking, and hence the climatic response of the other treeline-forming species is either very less studied or has not been tested yet that has limited our understanding of growth-climate relationships and vulnerabilities of treeline plant communities with the ongoing environmental changes including climate change. On-site recording and monitoring of the climatic variables (temperature, precipitation, humidity, sunshine hour, wind speed, cloud cover, etc.) are crucial in treeline research, but is still significantly lacking. Only a few studies focused on the edaphic aspect of the treeline; therefore, more studies are recommended. Likewise, limited studies have analyzed the influence of tree-to-tree interaction on the regeneration, distribution pattern (clustering), but comprehensive study verified from experimental manipulation and verification is still lacking.

Future research can initiate new or strengthen the existed studies. Studies are still insufficient in the following aspects: permanent plot establishment and long-term monitoring, edaphic element, impacts of inter or intraspecific competition, the impact of fire and post-fire regeneration, autecology of treeline-forming species, plant phenology and their changes over time, cambium dynamics and growth monitoring using dendrometer data, wood anatomical response, state of the resilience of treeline trees to the extreme climate events, sap flow studies, CO_2 enrichment experiment, drought experiment, long-term growth studies, tree ontogenic effect on growth and their climatic response, genetic basis of treeline dynamics, grazing exclusion-inclusion experiment, seed production and seed dispersal at treeline, seedlings transplantation experiment, frost events during the growing season, permafrost, effect of wind, effect of human impacts, etc. For the vegetation or species distribution modeling of treeline-forming species in response to the environmental change, many input variables are necessary for the robustness of the model. Integrative and holistic studies are essential to conclude the formation of the Himalayan treeline, their pattern, and their dynamics in response to environmental change, including climate change. Similarly, impacts of treeline shifting in the ecological and socio-ecological sectors need to be explored more and appropriately documented.

Acknowledgments The authors are deeply indebted to the various workers who dedicated a lifetime to treeline studies in Nepal Himalaya. NPG is grateful to the Chinese Academy of Sciences for CAS-PIFI Postdoc Fellowship (2019PC0102). This study was also jointly supported by the National Natural Science Foundation of China (Grant No. 41661144045). SKS is thankful to the Director, Birbal Sahni Institute of Palaeosciences for the kind support given. We thank editors and reviewers for their constructive comments in the earlier version of the manuscript.

References

Aakala T, Hari P, Dengel S, Newberry SL, Mizunuma T, Grace J (2014) A prominent stepwise advance of the tree line in Northeast Finland. J Ecol 102:1582–1591

Anderson K, Fawcett D, Cugulliere A, Benford S, Jones D, Leng R (2020) Vegetation expansion in the subnival Hindu Kush Himalaya. Glob Chang Biol 26:1608–1625

Arno SF (1984) Timberline. Mountain and Arctic Forest Frontiers, Seattle

Aryal A, Raubenheimer D, Subedi S, Kattel B (2010) Spatial habitat overlap & habitat preference of Himalayan musk deer 'Moschuschrysogaster' in Sagarmatha (Mt. Everest) National Park, Nepal. Curr Res J Biol Sci 2:217–225

Aryal S, Ghimire SK, Dhakal YR, Gaire NP, Bhandari S (2017) Perceptions of agro-pastoralists towards the change in temperature and precipitation in the trans-Himalayan regions of Nepal. Banko Janakari 27:21–30

Aryal S, Panthi J, Dhakal YR, Gaire NP, Karki K, Joshi NR (2018) Historically evolved practices of the Himalayan transhumant pastoralists and their implications for climate change adaptation. Int J Glob Warm 14:356–371

Baniya B, Tang Q, Huang Z, Sun S, Techato KA (2018) Spatial and temporal variation of NDVI in response to climate change and the implication for carbondynamics in Nepal. Forest 9:329

Basnet P, Gaire NP(2017) Growth and dynamics of *Abies spectabilis* with climate change at tree-line ecotone in the western Nepal Himalaya. In: Proceedings from the International Conference: Wild harvests, governance, and livelihoods in Asia" held in Kathmandu; pp. 71–84

Batllori E, Camarero JJ, Ninot JM, Gutiérrez E (2009) Seedling recruitment, survival and facilitation in alpine *Pinus uncinata*tree line ecotones. Implications and potential responses to climate warming. Glob Ecol Biogeogr 18:460–472

Baumgartner A (1980) Mountain climates from a perspective of forest growth. In: Benecke U, Davis MR (eds) Mountain environments and subalpine tree growth. New Zealand Forest Service, Wellington, pp 27–40

Bäumler R, Zech W (1994) Soils of the high mountain region of eastern Nepal: classification, distribution and soil-forming processes. Catena 22:85–103

Bhandari S, Gaire NP, Shah SK, Speer JH, Bhuju DR, Thapa UK (2019) A 307-year tree-ring SPEI reconstruction indicates modern drought in western Nepal Himalayas. Tree-Ring Res 75:73–85

Bharti RR, Adhikari BS, Rawat GS (2012) Assessing vegetation changes in timberline ecotone of Nanda Devi national park, Uttarakhand. Int J Appl Earth Obs Geoinf 18:472–479

Bhuju DR, Carrer M, Gaire NP, Soraruf L, Riondato R, Salerno F, Maharjan SR (2010) Dendroecological study of high altitude forest at Sagarmatha National Park, Nepal. In: Jha PK, Khanal IP (eds) Contemporary research in Sagarmatha (Mt. Everest) region, Nepal. Nepal Academy of Science and Technology, Lalitpur, pp 119–130

Bhuju DR, Shah SK, Gaire NP (2016) Environmental reconstruction and impact of climate change on vegetation at tree-lines of Nepal Himalaya. Annual Rep Pro Nat Foundation Jpn 24:170–181

Bhuju DR, Adhikari K, Bashila SR, Shrestha L, Upadhyay S (2019) Livelihood strategy under climate change stress: assessing climate vulnerability and preparing an adaptation framework with communities in Kanchenjungha conservation area. Resources Himalaya Foundation and International Centre for Integrated Mountain Development, Kathmandu, Nepal

Björklund JA, Gunnarson BE, Seftigen K, Esper J, Linderholm HW (2014) Blue intensity and density from northern Fennoscandian tree rings, exploring the potential to improve summer temperature reconstructions with earlywood information. Clim Past 10:877–885

Björklund J, Gunnarson BE, Seftigen K, Zhang P, Linderholm HW (2015) Using adjusted blue intensity datato attain high-quality summer temperature information: a case study from Central Scandinavia. Holocene 25:547–556

Bobrowski M, Gerlitz L, Schickhoff U (2017) Modelling the potential distribution of *Betula utilis* in theHimalaya. Glob Ecol Conserv 11:69–83

Bräuning A (2004) Tree-ring studies in the Dolpo-Himalya (western Nepal). TRACE – tree rings in archaeology. Climatol Ecol 2:8–12

Brunello CF, Andermann C, Helle G, Comiti F, TononG TA, Hovius N (2019) Hydroclimatic seasonality recorded by tree ring $\delta^{18}O$ signature across a Himalayan altitudinal transect. Earth Planet Sci Lett 518:148–159

Bürzle B, Schickhoff U, Schwab N, Oldeland J, Müller M, Böhner J, Chaudhary RP, Scholten T, Dickoré WB (2017) Phytosociology and ecology of treeline ecotone vegetation in Rolwaling Himal, Nepal. Phytocoenologia 47:197–220

Bürzle B, Schickhoff U, Schwab N, Wernicke LM, Müller YK, Böhner J, Chaudhary RP, Scholten T, Oldeland J (2018) Seedling recruitment and facilitation dependence on safe site characteristics in a Himalayan treeline ecotone. Plant Ecol 219:115–132

Camarero JJ, Gutiérrez E (2004) Pace and pattern of recent treeline dynamics: response of ecotones to climatic variability in the Spanish Pyrenees. Clim Chang 63:181–200

Camarero JJ, Gazol A, Sánchez-Salguero R, Fajardo A, McIntire EJB, Gutiérrez E, Batllori E, Boudreau S, Carrer M, Diez J, Dufour-Tremblay G, Gaire NP, Hofgaard A, Jomelli V,

Kirdyanov A, Lévesque E, Liang E, Linares JC, Mathisen IE, Moiseev PA, Sangüesa-Barreda G, Shrestha KB, Toivonen JM, Tutubalina OV, Wilmking M (2021) Global fading of the temperature–growth coupling at alpine and polar treelines. Glob Chang Biol 27(9): 1879–1889

Chhetri PK, Cairns DM (2015) Contemporary and historic population structure of *Abies spectabilis* at treeline in Barun valley, eastern Nepal Himalaya. J Mt Sci 12:558–570

Chhetri PK, Cairns DM (2016) Dendroclimatic response of *Abies spectabilis* at treeline ecotone of Barun Valley, eastern Nepal Himalaya. J For Res 27:1163–1170

Chhetri PK, Cairns DM (2018) Low recruitment above treeline indicates treeline stability under changing climate in Dhorpatan hunting reserve, Western Nepal. Phys Geogr 39:329–342

Chhetri PK, Bista R, Cairns DM (2016) Population structure and dynamics of *Abies spectabilis* at treeline ecotone of Barun Valley, Makalu Barun National Park, Nepal. Acta Ecol Sin 36:269–274

Chhetri PK, Shrestha KB, Cairns DM (2017) Topography and human disturbances are major controlling factors in treeline pattern at Barun and Manang area in the Nepal Himalaya. J. Mt. Sci. 14:119–127

Chhetri PK, Gaddis KD, Cairns DM (2018) Predicting the suitable habitat of treeline species in theNepalese Himalayas under climate change. Mt Res Dev 38:153–163

Cook ER, Krusic PJ, Jones PD (2003) Dendroclimatic signals in long tree-ringchronologies from the Himalayas of Nepal. Int J Climatol 23:707–732

Dawadi B, Liang E, Tian L, Devkota LP, Yao T (2013) Premonsoon precipitation signal in tree rings of timberline *Betula utilis* in the Central Himalayas. Quat Int 283:72–77

Dhakal YR, Gaire NP, Aryal S, Shah SK, Bhandari S, Kunwar U, Rayamajhi S (2016) Treeline shift in Central Nepal Himalaya and climate reconstruction of past millennia. In: Bhuju DR, McLaughlin K, Sijapati J, Devkota BD, Shrestha N, Ghimire GP, Neupane PK (eds) Building knowledge for climate resilience in Nepal. Nepal Academy of Science and Technology, Khumaltar, Lalitpur

DHM (2017) Observed climate trend analysis in the districts and physiographic regions of Nepal (1971–2014). Department of Hydrology and Meteorology, Kathmandu

Dobremez JF, Shrestha TB(1980) Carte écologique du Népal. Région Jumla-Saipal 1:250 000. Cahiers Népalais 9. Paris

Drollinger S, Müller M, Kobl T, Schwab N, Böhner J, Schickhoff U, Scholten T (2017) Decreasing nutrient concentrations in soils and trees with increasing elevation across a treeline ecotone in Rolwaling Himal, Nepal. J Mt Sci 14:843–858

Du H, Liu J, Li MH, Büntgen U, Yang Y, Wang L, Wu Z, He HS (2017) Warming-induced upward migration of the alpine treeline inthe Changbai Mountains, Northeast China. Glob Chang Biol 24:1256–1266

Dubey B, Yadav RR, Singh J, Chaturvedi R (2003) Upward shift of Himalayan pine in Western Himalaya, India. Curr Sci 85:1135–1136

Elliott GP, Cowell CM (2015) Slope aspect mediates fine-scale tree establishment patterns at upper treeline during wet and dry periods of the 20th century. Arct Antarct Alp Res 47:681–692

Forrest JL, Wikramanayake E, Shrestha R, Areendran G, Gyeltshen K, Maheshwari A, Mazumdar S, Naidoo R, Thapa GJ, Thapa K (2012) Conservation and climate change: assessing the vulnerability of snow leopard habitat to treeline shift in the Himalaya. Biol Conserv 150: 129–135

Fritts HC (1976) Tree rings and climate. University Press Cambridge, Cambridge, p 567

Gaire NP, Dhakal YR, Lekhak H, Bhuju DR, Shah SK (2010) Vegetation dynamics in treeline ecotoneof Langtang National Park, Central Nepal. Nepal J Sci Technol 11:107–114

Gaire NP, Dhakal YR, Lekhak HC, Bhuju DR, Shah SK (2011) Dynamics of *Abies spectabilis* in relationto climate change at the treeline ecotone in Langtang National Park. Nepal J Sci Technol 12:220–229

Gaire NP, Koirala M, Bhuju DR, Borgaonkar HP (2014) Treeline dynamics with climate change at thecentral Nepal Himalaya. Clim Past 10:1277–1290

Gaire NP, Rana P, Koirala M, Bhuju DR, Carrer M (2015) Study of treelineecotone to assess long-term impact of environmental change in Mt. Everestregion, Nepal Himalaya. FUUAST J. Biol. 5:1–11

Gaire NP, Koirala M, Bhuju DR, Carrer M (2017a) Site- and species-specific treeline responses toclimatic variability in eastern Nepal Himalaya. Dendrochronologia 41:44–56

Gaire NP, Bhuju DR, Koirala M, Shah SK, Carrer M, Timilsena R (2017b) Tree-ring based spring precipitationreconstruction in western Nepal Himalaya sinceAD 1840. Dendrochronologia 42: 21–30

Gaire NP, Dhakal YR, Shah SK, Fan ZX, Bräuning A, Thapa UK, Bhandari S, Aryal S, Bhuju DR (2019) Drought (scPDSI) reconstruction of trans-Himalayan region of central Himalaya using *Pinus wallichiana* tree-rings.Palaeogeography. Palaeoclimatol Palaeoecol 514:251–264

Gaire NP, Fan ZX, Bräuning A, Panthi S, Rana P, Shrestha A, Bhuju DR (2020) *Abies spectabilis* shows stable growth relations to temperature, but changing response to moisture conditions along an elevation gradient in the central Himalaya. Dendrochronologia 60:125675

Gerlitz L, Bechtel B, Böhner J, Bobrowski M, BürzleB MM, Scholten T, Schickhoff U, Schwab N, Weidinger J (2016) Analytic comparison of temperaturelapse rates and precipitation gradients in a Himalayantreeline environment: implications for statistical downscaling. In: Singh RB, Schickhoff U, Mal S (eds) Climate change, glacier response and vegetation dynamicsin the Himalaya. Springer, Cham, pp 49–64

Ghimire B, Lekhak HD (2007) Regeneration of *Abies spectabilis* (D. Don) Mirb. in subalpine forest of UpperManang, North-central Nepal. In: Chaudhary RP, Aase TH, Vetaas OR, Subedi BP (eds) Local effectsof global changes in the Himalayas: Manang, Nepal. Tribhuvan University, Nepal and University of Bergen, Norway, pp 139–149

Ghimire B, Mainali KP, Lekhak HD, Chaudhary RP, Ghimeray AK (2010) Regeneration of *Pinuswallichiana* AB Jackson in a trans-Himalayan dry valley of north-Central Nepal. Himal J Sci 6:19–26

Hamid M, Khuroo AA, Charles B, Ahmad R, Singh CP, Aravind NA (2018) Impact of climate change on the distribution rangeand niche dynamics of Himalayan birch, a typical treelinespecies in Himalayas. Biodivers Conserv 28:2345–2370

Harsch MA, Bader MY (2011) Treeline form—a potential key to understanding treeline dynamics. Glob Ecol Biogeogr 20:582–596

Harsch MA, Hulme PE, McGlone MS, Duncan RP (2009) Are treelinesadvancing? A global meta-analysis of treeline response to climate warming. Ecol Lett 12:1040–1049

Hofgaard A, Tømmervik H, Rees G, Hanssen F (2013) Latitudinal forest advancein northernmost Norway since the early 20th century. J Biogeogr 40:938–949

Holtmeier FK (2009) Mountain timberlines.Ecology, patchiness, and dynamics. Springer, Dordrecht

Holtmeier FK, Broll G (2005) Sensitivity and response of northern hemisphere altitudinal and polar treelines to environmental change at landscape and local scales. Glob Ecol Biogeogr 14:395–410

Holtmeier FK, Broll G (2009) Altitudinal and polar Treelines in the northern hemisphere –causes and response to climate change. Polarforschung 79:139–153

IPCC (2007) Climate change 2007: impacts, adaptation and vulnerability. In: Parry ML, Canziani OF, Palutikof JP, van der Linden PJ, Hanson CE (eds) Contribution of working group II to the fourthassessment report of the IPCC. Cambridge University Press, Cambridge, p 976

IPCC (2013) Climate change 2013: The Physical Science Basis. Contribution of Working Group I to the Fifth Assessment Reportof the Intergovernmental Panel on Climate Change, Cambridge University Press, Cambridge, United Kingdom and New York, NY T.F

IPCC (2014) Climate change 2014: impacts, adaptation, and vulnerability. Part A: global and sectoral aspects. Contribution of working group II to the fifth assessment report of the intergovernmentalpanel on climate change. Cambridge University Press, Cambridge, UK; pp 1132

Kaczka RJ, Spyt B, Janecka K, Beil I, Büntgen U, Scharnweber T, Nievergelt D, Wilmking M (2018) Different maximum latewood density and blue intensity measurements techniques reveal similar results. Dendrochronologia 49:94–101

Karki R, ul Hasson S, Gerlitz L, Talchabhadel R, Schickhoff U, Scholten T, Böhner J (2019) Rising mean and extreme near-surface air temperature across Nepal. Int J Climatol 40:2445–2463

Kharal DK, Bhuju DR, Gaire NP, Rayamajhi S, Meilby H, Chaudhary A (2015) Population structure and distribution of *Abies spectabilis* (D. Don) in Central Nepal Himalaya: a comparison with the total woody vegetation of the forests at the three different elevation ranges in Manang District. BankoJanakari 25:3–14

Kharal DK, Thapa UK, St. George S, Meilby H, Rayamajhi S, Bhuju DR (2017) Tree-climate relations along an elevational transect in Manang Valley, Central Nepal. Dendrochronologia 41: 57–64

Kirdyanov AV, Hagedorn F, Knorre AA, Fedotova EV, Vaganov A, Naurzbaev MM, Moiseev PA, Rigling A (2012) 20th century treelineadvance and vegetation changes along an altitudinal transect in the PutoranaMountains, Northern Siberia. Boreas 41:56–67

Körner C (1998) A re-assessment of high elevation treeline position and their explanation. Oecologia 115:445–459

Körner C (2012) Alpine treelines: functional ecology of the global high elevation tree limits. Springer, Cham

Körner C, Paulsen J (2004) A worldwide study of high altitude treelinetemperatures. J Biogeogr 31: 713–732

Kullman L (1998) Tree-limits and montane forests in the Swedish Scandes: sensitive biomonitors of climate change and variability. Ambio 27:312–321

Kullman L (2001) 20th century climate warming trend and tree-limit rise in the southern Scandes of Sweden. Ambio 30:72–80

Kullman L, Öberg L (2009) Post-little ice age tree line rise and climate warming inthe Swedish Scandes: a landscape ecological perspective. J Ecol 97:415–429

Lamsal P, Kumar L, Shabani F, Atreya K (2017) Thegreening of the Himalayas and Tibetan plateau under climatechange. Glob Planet Chang 159:77–92

Li MH, Xiao WF, Wang SG, Cheng GW, Cherubini P, Cai XH, Liu XL, Wang XD, Zhu WZ (2008) Mobile carbohydrates in Himalayan treeline trees: evidence for carbon gain limitation but not for growth limitation. Tree Physiol 28:1287–1296

Liang EY, Wang Y, Eckstein D, Luo T (2011) Little change in the fir tree-lineposition on the southeastern Tibetan plateau after 200 years of warming. New Phytol 190:760–769

Liang EY, Dawadi B, Pederson N, Eckstein D (2014) Is the growth of birch at theupper timberline in the Himalayas limited by moisture or by temperature? Ecology 95:2453–2465

Liang E, Wang Y, Piao S, Lu X, Camarero JJ, Zhu H, Ellison AM, Ciais P, Peñuelas J (2016) Species interactions slow warming-induced upward shifts of treelines on the Tibetan plateau. Proc Natl Acad Sci U S A 113:4380–4385

Liang EY, Dawadi B, Pederson N, Piao S, Zhu H, Sigdel SR, Chen D (2019) Strong link between large tropical volcanic eruptions and severe droughts prior to monsoon in the Central Himalayas revealed by tree-ring records. Sci Bull 64:1018–1023

Lv LX, Zhang QB (2012) Asynchronous recruitment history of *Abies spectabilis*along an altitudinal gradient in the Mt. Everest region. J Plant Ecol 5:147–156

Lyu L, Zhang QB, Deng X, Mäkinen H (2016) Fine-scale distribution of treeline trees and the nurse plant facilitation on the eastern Tibetan plateau. Ecol Indic 66:251–258

Mainali K, Shrestha BB, Sharma RK, Adhikari A, Gurarie E, Singer M, Parmesan C (2020) Contrasting responses to climate change at Himalayan treelines revealed by population demographics of two dominant species. Ecol Evol 10:1209–1222

McCarroll D, Pettigrew E, Luckman A, Guibal F, Edouard JL (2002) Blue reflectance provides a surrogate forlatewood density of high-latitude pine tree rings. Arct Antarct Alp Res 34:450–453

Miehe G (1989) Vegetation patterns of Mt. Everest as influenced by monsoon and föhn. Vegetatio 79:21–32

Miehe G (1991) DieVegetationskarte des KhumbuHimal (Mt. Everest-Südabdachung) 1:50.000. Gefügemuster der Vegetation und Probleme der Kartierung. Erdkunde 45:81–94
Miehe G, Miehe S (2000) Comparative high mountainresearch on the treeline ecotone under human impact. Erdkunde 54:34–50
Miehe G, Miehe S, Schlütz F (2002) Vegetationskundliche und palynologischeBefundeausdemMuktinath-Tal (Tibetischer Himalaya, Nepal). Erdkunde 56: 268–285
Miehe G, Miehe S, Vogel J, Co S, Duo L (2007) Highest treeline in the northern hemisphere found in southern Tibet. Mt Res Dev 27:169–173
Miehe G, Miehe S, Schlütz F (2009) Early human impact in the forest ecotone of southern high Asia (Hindu Kush, Himalaya). Quat Res 71:255–265
Mishra NB, Mainali KP, Shrestha BB, Radenz J, Karki D (2018) Species-level vegetation mapping in a Himalayan treeline ecotone using unmanned aerial system (UAS) imagery. ISPRS. Int J Geo-Inf 7:445
Mong CE, Vetaas OR (2006) Establishment of *Pinus wallichiana*on a Himalayan glacier foreland: stochastic distribution or safe sites? Arct Antarct Alp Res 38:584–592
Müller M, Schickhoff U, Scholten T, Drollinger S, Böhner J, Chaudhary RP (2016a) How do soil properties affect alpine treelines? General principles in a global perspective and novel findings from Rolwaling Himal, Nepal. Progr Phys Geogr Earth Environ 40:135–160
Müller M, Schwab N, Schickhoff U, Böhner J, Scholten T (2016b) Soil temperature and soil moisture patterns in a Himalayan alpine treeline ecotone. Arct Antarct Alp Res 48:501–521
Müller M, Oelmann Y, Schickhoff U, Böhner J, Scholten T (2017) Himalayan treeline soil and foliar C:N:P stoichiometry indicate nutrient shortage with elevation. Geoderma 291:21–32
Munier A, Hermanutz L, Jacobs JD, Lewis K (2012) The interacting effects oftemperature, ground disturbance, and herbivory on seedling establishment:implications for treeline advance with climate warming. Plant Ecol 210:19–30
Numata M (1967) Notes on botanical trip in eastern Nepal, I. Chiba Univ. J Coll Arts Sci 5:57–74
Öberg L, Kullman L (2012) Contrasting short-term performance of mountainbirch (*Betula pubescens* ssp. czerepanovii) treeline along a latitudinalcontinentality-maritimity gradient in the southern Swedish Scandes. Fennia 190:19–40
Pandey S, Carrer M, Castagneri D, Petit G (2018) Xylem anatomical responses to climate variability in Himalayan birch trees at one of the world's highest forest limit. Perspect Plant Ecol Evol Syst 33:34–41
Pandey S, Cherubini P, Saurer M, Carrer M, Petit G (2020) Effects of climate change on treeline trees in Sagarmatha (Mt. Everest, central Himalaya). J Veg Sci 31(6):1144–1153
Panigrahy S, Anitha D, Kimothi MM, Singh SP (2010) Timberline changedetection using topographic map and satellite imagery. Trop Ecol 51:87–91
Panthi S, Bräuning A, Zhou ZK, Fan ZX (2017) Tree rings reveal recent intensified spring drought in the central Himalaya. Nepal Glob Planet Chang 157:26–34
Panthi S, Fan ZX, van der Sleen P, Zuidema PA (2020) Long-term physiological and growth responses of Himalayan fir to environmental change are mediated by mean climate. Glob Chang Biol 26:1778–1794
Panthi S, Fan ZX, Bräuning A (2021) Ring widths of *rhododendron* shrubs reveal a persistent winter warming in the central Himalaya. Dendrochronologia 65:125799
Qi W, Zhang Y, Gao J, Yang X, Liu L, Khanal NR (2013) Climate change on thesouthern slope of Mt. Qomolangma (Everest) region in Nepal since 1971. J Geogr Sci 23:595–611
Rai S, Dawadi B, Wang Y, Lu X, Ru H, Sigdel SR (2019) Growth response of *Abiesspectabilis* to climate along an elevation gradient of the Manang valley in the Central Himalayas. J For Res 31: 2245–2254
Rana P, Koirala M, Bhuju DR, Boonchird C (2016) Population structure of *Rhododendron campanulatum* D. Don and associated tree species along the elevational gradient of Manaslu conservation area, Nepal. J Inst Sci Technol 21:95–102

Rana P, Bhuju DR, Koirala M, Boonchird C (2017) Dendroecological studies of *rhododendron campanulatum*D.Don along the elevational gradient of Manaslu conservation area, Nepal Himalaya. Pak J Bot 49:1749–1755

Rayback SA, Shrestha KB, Hofgaard A (2017) Growth variable-specific moisture and temperature limitations in co-occurring alpine tree and shrub species, Central Himalayas, Nepal. Dendrochronologia 44:193–202

Ren YY, Ren GY, Sun XB et al (2017) Observed changes in surface air temperature and precipitation in the Hindu Kush Himalayan region during 1901-2014. Adv Clim Chang Res 8:148–156

Rydval M, Larsson L-Å, McGlynn L, Gunnarson BE, Loader NJ, Young GHF, Wilson R (2014) Blueintensity for dendroclimatology: should we have the blues? Experiments from Scotland. Dendrochronologia 32:191–204

Salerno F, Guyennon N, Thakuri S, Viviano G, Romano E, Vuillermoz E, Cristofanelli P, Stocchi P, Agrillo G, Ma Y, Tartari G (2015) Weakprecipitation, warm winters and springs impact glaciers of south slopes of Mt.Everest (central Himalaya) in the last two decades (1994–2013). Cryosphere 9:1229–1247

Sano M, Furuta F, Kobayashi O, Sweda T (2005) Temperature variations since themid-18th century for Western Nepal, as reconstructed from tree-ring width anddensity of *Abies spectabilis*. Dendrochronologia 23:83–92

Sano M, Ramesh R, Sheshshayee MS, Sukumar R (2012) Increasing aridity over the past 223 years in Nepal Himalayainferred from a tree ring $\delta^{18}O$ chronology. The Holocene 22:809–817

Schickhoff U (2005) The upper timberline in the Himalayas, Hindu Kush and Karakorum: a review of geographicaland ecological aspects. In: Broll G, Keplin B (eds) Mountain ecosystems. Studies in treeline ecology. Springer, Berlin, pp 275–354

Schickhoff U, Bobrowski M, Böhner J, Bürzle B, Chaudhary RP, Gerlitz L, Heyken H, Lange J, Müller M, Scholten T, Schwab N, Wedegärtner R (2015) Do Himalayan treelines respond to recent climate change? An evaluation of sensitivity indicators. Earth Syst Dynam 6:245–265

Schickhoff U, Bobrowski M, Böhner J, Bürzle B, Chaudhary RP, Gerlitz L, Lange J, Müller M, Scholten T, Schwab N (2016a) Climate change and treeline dynamics in the Himalaya. In: Singh RB, Schickhoff U, Mal S (eds) Climate change, glacier response, and vegetation dynamics in the Himalaya. Springer, Cham, pp 271–306

Schickhoff U, Singh RB, Mal S (2016b) Climate change and dynamics of glaciers and vegetation in theHimalaya: an overview. In: Singh RB, Schickhoff U, Mal S (eds) Climate change, glacier response, and vegetation dynamics in the Himalaya. Springer, Cham, pp 1–26

Schlütz F, Zech W (2004) Palynological investigations on vegetation and climate change in theLate quaternary of Lake Rukche area, GorkhaHimal, Central Nepal. Veg Hist Archaeobot 13:81–90

Schwab N, Schickhoff U, Müller M, Gerlitz L, Bürzle B, Böhner J, Chaudhary RP, Scholten T (2016) Treeline responsiveness to climate warming: insights from a krummholz treeline in Rolwaling Himal, Nepal. In: Singh RB, Schickhoff U, Mal S (eds) Climate change, glacier response, and vegetation dynamics in the Himalaya. Springer, Cham, pp 307–345

Schwab N, Schickhoff U, Bürzle B, Müller M, Böhner J, Chaudhary RP, Scholten T, Oldeland J (2017) Implications of tree species – environment relationships for the responsiveness of Himalayan krummholz treelines to climate change. J Mt Sci 14:453–473

Schwab N, Kaczka RJ, Janecka K, Böhner J, Chaudhary RP, Scholten T, Schickhoff U (2018) Climate change-induced shift of tree growth sensitivity at a central Himalayan treeline ecotone. Forests 9:267

Schwab N, Janecka K, Kaczka RJ, Böhner J, Chaudhary RP, Scholten T, Schickhoff U (2020) Ecological relationships at a near-natural treeline, Rolwaling valley, Nepal Himalaya: implications for the sensitivity to climate change. Erdkunde 74:15–44

Shao X, Wang S, Zhu H, Xu Y, Liang E, Yin ZY, Xu X, Xiao Y (2009) A 3585-year ring-width dating chronology of Qilian juniper from the northeastern Qinghai-Tibetan plateau. IAWA J 30: 379–394

Sharma KP, Tiwari A, Shrestha BB (2020) Changes in regeneration and leaf traits of Rhododendron campanulatum along a treeline ecotone in central Nepal. J. Mt. Sci. 17:602–613

Shen W, Zhang L, Liu X, Luo T (2014) Seed-based treeline seedlings arevulnerable to freezing events in the early growing season under a warmerclimate: evidence from a reciprocal transplant experiment in the SergyemlaMountains, Southeast Tibet. Agric For Meteorol 187:83–92

Shen W, Zhang L, Guo Y, Luo T (2018) Causes for treeline stability under climate warming: evidence from seed and seedling transplant experiments in Southeast Tibet. For Ecol Manag 408:45–53

Shrestha BB, Ghimire B, Lekhak HD, Jha PK (2007) Regeneration of treeline birch (*Betula utilis* D. Don) forest in a trans-Himalayan dry valley in Central Nepal. Mt Res Dev 27:259–267

Shrestha UB, Gautam S, Bawa KS (2012) Widespread climate change in the Himalayas and associated changes in local ecosystems. PLoS One 7:e36741

Shrestha KB, Hofgaard A, Vandvik V (2015a) Recent treeline dynamics are similar between dry and Mesic areas of Nepal, central Himalaya. J Plant Ecol 8:347–358

Shrestha KB, Hofgaard A, Vandvik V (2015b) Tree growth response to climatic variability in two climatically contrasting treeline ecotone areas, central Himalaya, Nepal. Can J For Res 45: 1643–1653

Shrestha KB, Chhetri PK, Bista R (2017) Growth responses of *Abies spectabilis* to climate variations along an elevational gradient in Langtang National Park in the central Himalaya, Nepal. J For Res 22:274–281

Sigdel SR, Wang Y, Camarero JJ, Zhu H, Liang E, Peñuelas J (2018) Moisture-mediated responsivenessof treeline shifts to global warming in the Himalayas. Glob Chang Biol 24: 5549–5559

Sigdel SR, Zhang H, Zhu H, Muhammad S, Liang E (2020a) Retreating glacier and advancing forest over the past 200 years in the Central Himalayas. Geophys Res J Geophys Res Biogeosci 125:e2020JG005751

Sigdel SR, Liang E, Wang Y, Dawadi B, Camarero JJ (2020b) Tree-to-tree interactions slow down Himalayan treeline shifts as inferred from tree spatial patterns. J Biogeogr 47:1816–1826

Singh CP, Panigrahy S, Thapliyal A, Kimothi MM, Soni P, Parihar JS (2012) Monitoring the alpine treeline shift in parts of the Indian Himalayas usingremote sensing. Curr Sci 102:559–562

Singh U, Phulara M, David B, Ranhotra PS, Shekhar M, Bhattacharyya A, Dhyani R, Joshi R, Pal AK (2018) Static tree line of Himalayan silver fir since last several decades at Tungnath, Western Himalaya. Trop Ecol 59:351–363

Speer JH (2010) Fundamentals of tree ring research. The University of Arizona Press, Tucson

Sujakhu H, Gosai KR, Karmacharya SB (2013) Forest structure and regeneration pattern of *Betula utilis*D. Don in Manaslu conservation area, Nepal. Ecoprint 20:107–113

Suwal MK, Shrestha KB, Guragain L, Shakya R, Shrestha K, Bhuju DR, Vetaas OR (2016) Land-use change under a warming climate facilitatedupslope expansion of Himalayan silver fir (*Abies spectabilis*(D. Don) Spach). Plant Ecol 217:993–1002

Tabata H, Tsuchiya K, Shimzu Y, Fuzita N, Matsui K, Koike F, Yumoto T (1988) Vegetation and climatic changes in Nepal Himalayas I. Vegetation and climatic changes in Nepal Himalayas as the basis of palaeoecological studies. Proc Indian Natl Sci Acad 54A(4):530–537

Talchabhadel R, Karki R, Thapa BR, Maharjan M, Parajuli B (2018) Spatio-temporal variability of extreme precipitation in Nepal. Int J Climatol 38:4296–4313

Tenca A, Carrer M (2010) Growth climate response at high elevation: comparing Alps and Himalayas. Tree Rings Archaeol Climatol Ecol 8:89–97

Thakuri S, Dahal S, Shrestha D, Guyennon N, Romano E, Colombo N, Salerno F (2019) Elevation-dependent warming of maximum air temperature in Nepal during 1976–2015. Atmos Res 228: 261–269

Thapa UK, St. George S, Kharal DK, Gaire NP (2017) Tree growth across the Nepal Himalaya during the last four centuries. Prog Phys Geogr 41:478–495

Tiwari A (2020) Age-dependent growth responses to climate from trees in Himalayan treeline. Nepal J Zool 4:16–22

Tiwari A, Jha PK (2018) An overview of treeline response to environmental changes in Nepal Himalaya. Trop Ecol 59:273–285

Tiwari A, Fan ZX, Jump AS, Li SF, Zhou ZK (2017a) Gradual expansion of moisture sensitive*Abies spectabilis* forest in the trans-Himalayan zone ofcentral Nepal associated with climate change. Dendrochronologia 41:34–43

Tiwari A, Fan ZX, Jump AS, Zhou ZK (2017b) Warming induced growth decline of Himalayan birch at its lower range edge in a semi-arid region of trans- Himalaya, Central Nepal. Plant Ecol 218:621–633

Tiwari A, Uprety Y, Rana SK (2020) Plant endemism in the Nepal Himalayas and phytogeographical implications. Plant Divers 4:174–182

Troll C (1973) The upper timberlines in different climatic zones. Arct Alp Res 5(3):A 3-A18

Udas E(2009) Influence of climate variability on the growth performance of *Abies spectabilis* at tree line of west-Central Nepal. Master Thesis, University of Greifswald, Germany

Wang Y, Pederson N, Ellison AM, Buckley HL, Case BS, Liang E, Camarero JJ (2016) Increased stem density and competition may diminish the positive effects of warming at alpine treeline. Ecology 97:1668–1679

Wang Y, Sylvester SP, Lu X, Dawadi B, Sigdel SR, Liang E, Camarero JJ (2019a) The stability of spruce treelines on the eastern Tibetan plateau over the last century is explained by pastoral disturbance. For Ecol Manag 442:34–45

Wang Y, Case B, Lu X, Ellison AM, Peñuelas J, Zhu H, Liang E, Camarero JJ (2019b) Fire facilitates warming-induced upward shifts of alpine treelines by alteringinterspecific interactions. Trees 33:1051–1061

Wardle P (1974) Alpine timberlines. In: Ives JD, Barry RG (eds) Arctic and alpine environments. Routledge, London, pp 371–402

Weiser G, Tausz M (eds) (2007) Trees at their upper limit: tree life limitations at the alpine timberline. Springer, Cham, p 232

Williams LHJ (1953) The 1952 expedition to western Nepal. J R Hort Soc 78:323–337

Yadav RR, Singh J, Dubey B, Mishra KG (2006) A 1584-year ring-width chronology of juniper from Lahul, HimachalPradesh: prospects of developing millennia long climaterecords. Curr Sci 90:1122–1126

Yadava AK, Sharma YK, Dubey B, Singh J, Singh V, Bhutiyani MR, Yadav RR, Misra KG (2017) Altitudinal treeline dynamics of Himalayan pine in western Himalaya, India. Quat Int 444:44–52

Yoda K (1967) A preliminary survey of the forest vegetation of eastern Nepal. II. General description, structure and floristic composition of the sample plots chosen from different vegetation zones. J Coll Arts Sci Chiba Univ 5:99–140

Yonebayashi C, Minaki M (1997) Late quaternary vegetation and climatic history of eastern Nepal. J Biogeogr 24:837–843

Zhan YJ, Ren GY, Shrestha AB et al (2017) Change in extreme precipitation events over the Hindu Kush Himalayan region during 1961-2012. Adv Clim Chang Res. 8:166–175

Zhou TY, Gaire NP, Liao LB, Zheng LL, Wang JN, Sun J, Wei YQ, Xie Y, Wu Y (2018) Spatio-temporal dynamics of two alpine treeline ecotones and ecological characteristics of their dominate species at the eastern margin of Qinghai-Xizang plateau. Chin J Plant Ecol 42: 1082–1093

Epilogue: Way Forward and Policy Recommendations for Treeline Ecotone Research in the Himalaya

This book brings out that treeline ecotones are distinct and diverse systems that are structured by complex interactions among various factors, such as vegetation, climate, soils, snow, and topography. In spite of remoteness, they have been under anthropogenic influence since times immemorial. Pastoralism in the high elevation zones of Himalaya and reindeer hunting in Boreal treelines are some examples of human influence in these systems. While their sensitivity to climate change is well recognized, responses of treeline ecotone to changes in climate are complex and poorly understood. In many areas treelines are stationary in spite of decades of global warming. It is well established that treeline dynamics are constrained by several non-climatic factors, such as nutrient availability, soil conditions, landscape fragmentation, or species-specific traits, including dispersal capacity, competition with ground vegetation, because of not having mycorrhizal fungal symbionts, and new pests and pathogens. There is a need to undertake studies that improve the scientific understanding of the structural organization and functioning of the treeline ecotones in the Himalaya to explain ecotone processes, such as seedling establishment and recruitment, demography and population ecology, forest growth responses in relation to the physical and biotic factors. In addition, it is important to study the processes related to the formation of treelines at three hierarchical levels, namely (1) first-level processes including growth limitation, dieback, and mortality, (2) second-level processes leading to gradients in the first-level processes mentioned above, including reduced cell formation due to a lack of suitable temperature or nutrients, loss of foliage because of frost, wind, damage due to snow mould and fire, dispersal limitations, etc., and (3) third-level processes of species interactions leading to modifications in the second-level processes. It would help in analysing treeline ecotone patterns across Himalaya and linking them to underlying processes. This would also allow comparisons with the global studies and help in better management of this sensitive ecosystem particularly under changing environmental conditions.

This book summarizes the state-of-the-art knowledge on various facets of the Himalayan treeline ecotone and sheds light on a way forward based on knowledge gaps and concerns of conservation. This treatise makes a strong case for treating the Himalayan treeline ecotone as a distinct ecological entity that requires long-term studies across the Himalayan arc for a more detailed understanding of the structural

S. P. Singh et al. (eds.), *Ecology of Himalayan Treeline Ecotone*,
https://doi.org/10.1007/978-981-19-4476-5

complexity of treeline ecotone across the Himalaya and the factors that explain this complexity. A critical perusal of the studies so far carried out on various facets of treeline ecotone and those included in the chapters of this book reveal that multi-site and comprehensive studies in different parts of the Himalaya are lacking, that not much is known about the type of treeline forms from west to east Himalayan Arc, and which types of treeline forms are more sensitive or resistant to climate change. Therefore, it is imperative to undertake well-replicated and detailed studies on the treeline ecotones across the east-to-west Himalayan Arc covering various aspects, such as treeline form and position, species richness and diversity with an emphasis on hitherto unexplored taxa, such as lower groups of angiosperms, all groups of animals and microbes (particularly mutualists), phenological and physiological responses of key ecotonal species to changing climate, soil physical, chemical, and colloidal properties, nutrient dynamics and establishment of ecological sites in the treeline ecotone for monitoring over a long period are required to discern the interactions between patterns and processes both in space and time for better management of this sensitive ecotone in the face of changing climate and other anthropogenic influences.

Treeline Research: Way Forward

Given that the studies so far carried out on treeline ecotone in the Indian Himalayan region are scattered and limited, it is imperative to focus attention on this distinct ecological entity that is treated as the part of a continuum consisting of forests, timberline, treeline and alpine tundra, ice-covered and barren areas with biotic and abiotic components interconnected and interacting in multiple ways. Following are the research questions that should represent a way forward for urgently required studies on treeline ecotone.

1. How is climate change affecting Elevation-Dependent Warming (EDW) and temperature lapse rate (TLR)? It calls for long-term data from several representative sites. How seasonal changes in TLR are likely to affect biotic components, and how TLR is influenced by widespread pollution in the big cities in adjacent plains need consideration. The inconsistencies in lapse rate, both seasonally and annually, across the mountain regions are largely due to insufficient climatic stations and limited temporal and spatial coverage; long-term climate observations are essential to study the change in TLRs *vis-a-vis* change in climate and its impacts on treeline environments in the Himalayan region. Further, there is a need to examine whether the shallow TLRs in the Himalaya are due to EDW under the influence of global climate change. A comprehensive study of TLR is also required to increase our understanding of Himalayan climate, its controlling mechanisms and impacts on critical ecosystems through improved and long-term observations.
2. Do taxonomic, functional, and phylogenetic diversity patterns vary in the treeline ecotone with elevation and across Himalayan arc? Such studies should not be

restricted only to vascular plants, but extended to less-studied taxa including pteridophytes, bryophytes, and lichens. Even the studies on vascular plants have very rarely covered the Himalayan Arc. For example, the eastern Himalaya is much less investigated than the western Himalaya. Thus, it is high time that detailed and holistic landscape-level studies are undertaken on species diversity with a focus on beta diversity as well as the contributing process of species turnover and nestedness. Particularly interesting would be the studies explicating the role of mutualists (mycorrhizas and dark septate hyphae etc.) and other belowground organisms in the growth of certain treeline species and their interaction under various climate change scenarios. Equally valuable would be to discern the relative role of inorganic and organic forms of nutrients, particularly nitrogen, phosphorus, iron, etc., in the growth and performance of species in the treeline ecotone.

3. How do eco-physiological characteristics of plant species in the treeline ecotone relate to plant gas exchange, soil-plant relationships (seed set and desiccation of seed and seedling mortality due to freezing of soil) and biotic interactions (pollinators, microbial mutualism, refugia) need special attention given the paucity of information about the eco-physiology of plant species in this ecotone. Study of factors that influence the establishment and survival of new seedlings and upward movement of the treeline species, particularly in the context of climate change in the Himalayan treeline ecotone would be of interest and importance in treeline ecotone conservation and management. Particularly, the studies aimed at understanding the impact of water stress on seedling mortality resulting in depression of the timberlines in the inner Himalaya and low rainfall treeline areas are necessary and need to be taken up on priority. Long term monitoring of water relations of key treeline species is required to understand the impacts of climate change on these species. Reduced snow duration, snow cover and depth, and enhanced evapotranspiration rates may significantly impact the younger small-sized individuals that grow in the most severe microenvironments but limited data is presently available on this aspect. More frequent data collection (fortnightly) is required to relate soil tree water status with the timing of phenological events of species. The studies examining the response of treeline trees to pre-monsoon temperature and humidity conditions need to be urgently undertaken. How does climate change affect air and soil temperature, soil moisture, and nutrient availability in the Himalayan treeline ecotones and how these changes influence the structural organization and functional integrity of treeline ecotones also merits consideration.
4. How does pastoralism affect the treeline ecotone and which areas are more degraded and require urgent restoration efforts? In this regard documentation of migratory routes and details of transhumant pastoralists' activities can help in identifying degraded and degrading areas to formulate recovery plans. Currently, a separate management plan for the treeline ecotone is lacking.

In view of the above, well-planned long-term studies are likely to open new vistas in our understanding of the seasonal, annual, and more long-term changes in the treeline ecotone and its response to changing environment which is of pivotal importance in formulating effective strategies for conservation of such fragile ecosystems in the Himalaya for larger and longer benefit of the dependent and marginalized human population. Efforts would be required to enhance interdependence and alignment in research purposes and approaches. Networking of scientists working on Himalayan ecology in different countries with common research objectives and methodologies is necessary to obtain data on a large scale which is necessary to draw reliable conclusions and inferences about the patterns and processes that shape the Himalayan biodiversity. Additionally, it is also suggested that remotely sensed data could be used for regular monitoring of the Himalayan treeline for documenting any changes that are taking place in the treeline position and other allied features. The Dynamic Global Vegetation Model (DGVM) needs to be developed and used to understand how climate and vegetation interact to define the past, current, and future distribution of vegetation, particularly in the Himalayan context.

Policy Recommendations

- ***Development of a policy research framework for treeline ecotone research***
  A policy framework needs to be developed through a wider consultative process and brainstorming so as to include all the elements required to better understand the patterns in the treeline ecotone across the entire Himalayan landscape and identify the underlying processes that control them. In this regard, the research infrastructure created under the IHTR project can be helpful. The term 'research infrastructure' is used here in its broadest context to include data collected on treeline position, temperature lapse rates (which can be continued following the methods already in place with some adjustments, instruments placed in the field, and a research scholar familiar with the sites and research methods), patterns in species richness, water relations and phenological studies on treeline species, and reconstruction of past climate using the dendrochronological approach. This already existing database can be augmented with relatively small budgetary support to generate reliable long-term data. The study sites could serve as long-term ecological research (LTER) sites that are necessary for studying ecological processes over relevant spatial and temporal scales. The services of the project coordinator could also be availed.
- ***Strengthen transboundary cooperation and expanded team science approach***
  Transboundary cooperation with scientists working in different countries on various aspects of the Himalayan treeline ecotone needs to be strengthened. A few research scholars and field staff could be trained to collect comparable data in different countries.
- ***Data collection, storage, and use***
  An arrangement for storage of uniformly collected data and sharing should be in place. This part is generally weak in Himalayan countries. The task could be

accomplished by some national/international organizations in collaboration with universities/institutes.

- ***Central instrumentation, remote sensing products, and software facility***
  To ensure optimal use of resources, a central facility needs to be created where all the necessary equipment, remote sensing products, and softwares are available for use by team members. The said facility could be used as a training centre for researcher in treeline ecotone research.
- ***Funding for long-term research in the Himalaya***
  Long-term monitoring of patterns and processes in the Himalaya is necessary for better prediction of likely impacts of anthropogenic stresses and their management. As stated earlier, the minimum budgetary allocation of Rs. 50–60 lakhs per year (INR) initially for a period of about 10–15 years would be needed for such studies.
- ***Research-society interface***
  The research teams need to have frequent interactions with local habitants (who are dependent on the Himalayan ecosystem goods and services) in order to incorporate their perceptions and perspectives into research programmes and policy formulations.
- ***Education and Awareness***
  All the stakeholders, including policy planners, local population, students, tourists, nomads, pastoralists, and members of local institutions such as panchayats, etc., need to be educated about the importance of ecosystems to their social and economic well-being, threats faced by these ecosystems, and measures that need to be taken to manage them through a participatory approach.

The manufacturer's authorised representative in the EU is Springer Nature Customer Service Centre GmbH, Europaplatz 3, 69115 Heidelberg, Germany. If you have any concerns regarding our products, please contact ProductSafety@springernature.com

Printed and bound by CPI Group (UK) Ltd, Croydon, CR0 4YY
30/09/2026
02225244-0019